The Properties of Gases and Liquids

The Properties of Gases and Liquids

ROBERT C. REID
Professor of Chemical Engineering
Massachusetts Institute of Technology

JOHN M. PRAUSNITZ
Professor of Chemical Engineering
University of California at Berkeley

THOMAS K. SHERWOOD
Late Professor of Chemical Engineering
Massachusetts Institute of Technology
and
Visiting Professor of Chemical Engineering
University of California at Berkeley

Third Edition

McGRAW-HILL BOOK COMPANY

New York St. Louis San Francisco Auckland Bogotá Düsseldorf
Johannesburg London Madrid Mexico Montreal
New Delhi Panama Paris São Paulo
Singapore Sydney Tokyo Toronto

Library of Congress Cataloging in Publication Data

Reid, Robert C
  The properties of gases and liquids, their
estimation and correlation.

  Includes bibliographies.
  1. Gases.   2. Liquids.   I. Prausnitz, J. M.,
joint author.   II. Sherwood, Thomas Kilgore,
1903-1976 joint author.   III. Title.
TP242.R4   1977   660'.04'2   76-42204
ISBN 0-07-051790-8

1234567890   KPKP   786543210987

The editors for this book were Jeremy Robinson and Joan Zseleczky,
the designer was Naomi Auerbach, and the production supervisor
was Frank P. Bellantoni.

Printed and bound by The Kingsport Press.

Dedication

The manuscript for this book was completed late in 1975. Thomas K. Sherwood died on January 14, 1976, at the age of 72, after a brief illness.

Because of his pioneering books and research publications, Tom Sherwood was a world-famous chemical engineer. He received many honors from professional societies; he was a member of the National Academy of Sciences and a founding member of the National Academy of Engineering. Officially retiring after 40 years of teaching and research at MIT, Tom came to the University of California at Berkeley, where he participated in the educational and research activities of the Chemical Engineering Department as a visiting professor and consultant.

Because of his wide experience and his penetrating intellect, Tom Sherwood had an unusually broad perspective on chemical engineering. There was no aspect of chemical engineering which was strange to him; no matter what the topic of discussion might be, he was able to make a useful contribution and he did so gladly, often with disarming charm, usually with refreshing humor, and always with utter honesty.

There are few men who possess both remarkable technical and personal qualifications. Tom was one of these few. In his exemplary life and work he beneficially influenced the lives and work of countless former students, friends, and colleagues. In dedicating this book to Tom Sherwood, we honor not only his memory but also the values which he taught us: science and scholarship for meeting human needs.

ROBERT C. REID
JOHN M. PRAUSNITZ

v

Contents

Preface

Reliable values of the properties of materials are necessary in the design and operation of industrial processes. An enormous amount of data has been collected and correlated over the years, but the rapid advance of technology into new fields seems always to maintain a significant gap between demand and availability. The engineer is still required to rely primarily on common sense, on experience, and on a variety of methods for estimating physical properties.

This book presents a critical review of various estimation procedures for a limited number of properties of gases and liquids—critical and other pure-component properties, P-V-T and thermodynamic properties of pure components and mixtures, vapor pressures and phase-change enthalpies, standard enthalpies of formation, standard Gibbs energies of formation, heat capacities, surface tensions, viscosities, thermal conductivities, diffusion coefficients, and phase equilibria. Comparisons of experimental and estimated values are normally shown in tables to indicate the degree of reliability. Most methods are illustrated by examples. The procedures described are necessarily limited to those which appear to the authors to have the greatest practical use and validity. Wherever possible, we have included recommendations delineating the best methods of estimating each property and the most reliable technique for extrapolating or interpolating available data.

Although the book is intended to serve the practicing engineer, especially the process or chemical engineer, all engineers and scientists dealing with gases and liquids may find the book of value.

The first edition of this book was published in 1958 and the second in 1966. Each revision is essentially a *new* book, as many estimation methods are proposed each year and, over an 8- to 10-year span, most earlier methods are modified or displaced by more accurate or more general techniques. Most new methods are still empirical in nature although there are often theoretical bases for the correlation; whenever possible, the theory is outlined to provide the user with the raison d'être of the proposed estimation method.

The third edition introduces two major changes: Chap. 8 gives a concise discussion of methods for correlating and estimating vapor-liquid and liquid-liquid equilibrium ratios, and Appendix A is a data bank which shows the more important physical constants for some 468 common organic and inorganic chemicals. This tabulation was begun by Nancy Reid, who searched the literature for over a year. The first draft was reviewed by several industrial corporations, who cooperated by supplying missing values and noting discrepancies from their own extensive data banks. With sincere appreciation we acknowledge the help of D. R. Vredeveld and T. J. Farrell, of Union Carbide Corp., C. F. Spencer and S. B. Adler, of M. W. Kellogg-Pullman Co.; R. H. Johnson and L. Domash, of Exxon Research and Engineering Co.; A. H. Larsen, of Monsanto Co.; P. L. Chueh, of Shell Development Co.; and C. F. Chueh, of Halcon, Inc.

Although the final choice of recommended methods rested solely with ourselves, we were aided by a number of people who read draft chapters and made valuable comments and criticisms. In other cases, we received copies of unpublished work that, in several instances, proved to be better than published correlations. Particular thanks are extended to T. E. Daubert and R. P. Danner, of Pennsylvania State University, in relating A.P.I. work to our studies; to J. Erbar, of Oklahoma State University, for providing his Soave mixture parameters and for showing us his new developments in liquid viscosity; to A. Vetere, of Snam Progetti, for unpublished correlations of several properties, and to P. L. Chueh (Shell Development) and S. W. Benson (Stanford Research Institute) for new group contributions for ideal-gas properties. D. Ambrose (National Physical Laboratory, Teddington) provided valuable criticism and data on the vapor pressures of many compounds, while D. T. Jamieson (National Engineering Laboratory, Glasgow) was extremely helpful in reviewing the presentation dealing with liquid thermal conductivities. D. R. Vredeveld, T. J. Farrell, T. S. Krolikowski, and E. Buck (Union Carbide) were exceptionally cooperative in critically reviewing our work and in providing new ideas and techniques based on their extensive experience. Others to whom special thanks are due include R. F. Fedora (Jet Propulsion Laboratory), B.-I. Lee (Mobil Oil Corp.), L.-C. Yen (The Lummus Co.), E. A. Harlacher (Continental Oil Co.), L. I. Stiel (Allied Chemical Corp.), F. A. L. Dullien (University of Waterloo), E. A. Mason (Brown University), D. van Velzen (Euratom), P. E. Liley (Purdue University), and D. Reichenberg (National Physical Laboratory, Teddington).

We are grateful to Maria Tseng for her patience, dedication, and skill in typing most of the manuscript, and to Juan San José for his many helpful comments and suggestions.

To Nancy Reid we owe an especial expression of appreciation. Besides her role in preparing the data bank, she demonstrated countless times her uncanny ability to locate references and has offered valuable criticism and encouragement.

ROBERT C. REID
JOHN M. PRAUSNITZ
THOMAS K. SHERWOOD

Chapter One

The Estimation of Physical Properties

1-1 Introduction

The structural engineer cannot design a bridge without knowing the properties of steel and concrete. Scientists and other engineers are more often concerned with the properties of gases and liquids, and it is with these that this book deals. The chemical or process engineer, in particular, finds knowledge of physical properties of fluids essential to the design of many kinds of industrial equipment. Even the theoretical physicist must touch base occasionally by comparing theory with measured properties.

The physical properties of every pure substance depend directly on the nature of the molecules of which it consists. The ultimate generalization of physical properties of fluids will require a complete understanding of molecular behavior, which we do not yet have. Though its origins are ancient, the molecular theory was not generally accepted until about the beginning of the nineteenth century. Since then, many pieces of the puzzle of molecular behavior have fallen into place, and a useful, though incomplete, generalization has been developed.

The laws of Charles and Gay-Lussac were combined with Avogadro's hypothesis to form the gas law, $PV = NRT$, which was perhaps the first important correlation of properties. Deviations from the ideal-gas law, though often small, were tied to the fundamental nature of the molecules. The equation of van der Waals, the virial equation, and other equations of state express these quantitatively and have greatly facilitated progress in the development of a basic molecular theory.

The original "hard-sphere" kinetic theory of gases was perhaps the greatest single contribution to progress in understanding the statistical behavior of molecules. Physical, thermodynamic, and even transport properties were related quantitatively to molecular properties. Deviations from the hard-sphere kinetic theory led inevitably to studies of the interaction of molecules, based on the realization that molecules attract each other when separated and repel each other when they come very close. The semiempirical potential functions of Lennard-Jones and others describe the attraction and repulsion in an approximate quantitative fashion. More recently potential functions which allow for the shape of molecules and the special nature of polar molecules have been developed.

Although allowance for the forces of attraction and repulsion between molecules is a development of the last 60 years, the concept is not new. About 1750 Boscovich suggested that molecules (which he referred to as atoms) are "endowed with potential force, that any two atoms attract or repel each other with a force depending on their distance apart. At large distances the attraction varies as the inverse square of the distance. The ultimate force is a repulsion which increases without limit as the distance decreases without limit, so that the two atoms can never coincide" [4].

From one viewpoint the development of a comprehensive molecular theory would appear to be complete. J. C. Slater observes[†] that "[in nuclear physics] we are still seeking the laws; whereas in the physics of atoms, molecules, and solids, we have found the laws and are exploring the deductions from them." The suggestion that, in principle, everything is known about molecules is of little comfort to the engineer who needs to know the high-pressure behavior of a new chemical in order to design a commercial plant.

Paralleling the continuing refinement of the molecular theory has been the development of thermodynamics and its application to properties. The two are intimately related and interdependent; Carnot was an engineer interested in steam engines, but the (second) law was shown by Clausius, Kelvin, Maxwell, and Gibbs to have broad applications in all branches of science. The Clausius-Clapeyron

†J. C. Slater, "Modern Physics," McGraw-Hill, New York, 1955.

equation, for example, provides an extremely useful method of obtaining enthalpies of vaporization from the more easily measured vapor pressures.

The second law led to the concept of chemical potential, which is basic to an understanding of chemical and phase equilibria, and Maxwell's equations provide ways to obtain many important thermodynamic properties of a substance from P-V-T relations.   Since derivatives are required for this latter purpose, the P-V-T function must be known quite accurately.   This partly explains the great interest in deviations from the ideal-gas law.

In spite of the tremendously impressive developments of the molecular theory, the engineer frequently finds need for values of physical properties which have not been measured and which cannot be calculated from existing theory.   The "International Critical Tables," "Landolt-Börnstein," and many handbooks are convenient sources, and the CAChE Committee [2] has published a valuable compendium which lists the major reference works and indicates the properties covered in each.   But it is inconceivable that experimental data will ever be available on the many thousands of compounds of interest in science and industry.   The rate of accumulation of new data appears to be decreasing, while the need for accurate design data is increasing.   Data on mixtures are particularly scarce.   The process engineer is frequently called upon to design a plant to produce a new chemical; because it is new, little is yet known about its properties.

1-2 Estimation of Properties

In the all too frequent situation where no experimental value of the needed property is to be found, it must be estimated or predicted. "Estimation" and "prediction" are used as if they were synonymous, although the former carries the frank implication that the result may be only approximately correct.   Estimates may be based on theory, on correlations of experimental values, or on a combination of the two.

A theoretical relation which is not generally valid may serve adequately in many cases.   The engineer is quite justified in using $PV = NRT$ to relate mass and volumetric flow rates of air through an air-conditioning unit and in using Dalton's law to calculate the mass fraction of water in saturated air from the vapor pressure.   However, he must be able to judge the pressure above which this simple calculation leads to unacceptable error.

Completely empirical correlations are tempting to use outside the narrow range of properties on which they are based but are to be avoided.   In general, the less the empiricism the more valid the correlation.

Most of the better estimation methods involve equations based on the *form* of a theory, with empirical correlations of the constants which the incomplete theory does not provide. The introduction of empiricism into minor parts of a theoretical relation is a powerful method of developing an excellent correlation. For example, the van der Waals equation of state is a modification of the simple $PV = NRT$:

$$\left(P + \frac{a}{V^2}\right)(V - b) = RT \tag{1}$$

where V is the volume per mole. This is based on the reasoning that the pressure on a container wall, exerted by the impinging molecules, is lessened because of the attraction by the mass of molecules in the bulk gas, which will be greater if the gas density is large. Furthermore, the space in which the molecules move is less than the total volume by the covolume b of the molecules themselves. The "constants" a and b have some theoretical basis but are perhaps better thought of as empirical. The correlation of a and b in terms of other properties of a substance is an example of the use of an empirically modified theoretical form.

There are many examples of this approach to the development of a correlation useful for estimation purposes. Several methods of estimating diffusion coefficients in low-pressure binary gas systems are empirical modifications of the equation given by the simple kinetic theory. Almost all the better estimation procedures are based on correlations developed in this way.

1-3 Types of Estimation Methods

An *ideal* system for the estimation of a physical property would (1) provide reliable physical and thermodynamic data, for pure substances and for mixtures, at any temperature and pressure; (2) indicate the state (solid, liquid, or gas); (3) require a minimum of input data; (4) choose the least-error route; (5) indicate the probable error; and (6) minimize computation time. Few of the available methods approach this ideal, but many of them serve remarkably well.

In many practical cases, the most generally accurate method may not be the best for the purpose. Many engineering judgments properly require only approximate estimates, and a simple estimation method, requiring little or no input data, is preferred over a complex but more accurate correlation. The simple gas law is used at low to modest pressures by engineers, although more accurate correlations are available. It is not easy to provide guidance on when to reject the simpler in favor of the more complex but more accurate method.

The Law of Corresponding States   This expresses the generalization that those properties dependent on intermolecular forces are related to the critical properties in the same way for all compounds.   It is the single most important basis for the development of correlations and estimation methods.   Van der Waals showed it to be theoretically valid for all substances whose P-V-T properties could be expressed by a two-constant equation of state, such as Eq. (1).   It is similarly valid if the intermolecular potential function requires only two characteristic parameters.   It holds well, not only for the simplest molecules, but in many other instances where molecular orientation is not important, as it is for polar and hydrogen-bonded molecules.

The relation of pressure to volume at constant temperature is different for different substances, but if pressure, volume, and temperature are related to the corresponding critical properties, the function connecting the *reduced properties* becomes the same for each substance.

The reduced property is commonly expressed as a *fraction* of the critical property: $P_r \equiv P/P_c$, $V_r \equiv V/V_c$, $T_r \equiv T/T_c$.   Figure 1-1 shows how well this works to relate P-V-T data for methane and nitrogen.   In effect, the critical point is taken as the origin.   It is seen that the data for saturated liquid and saturated vapor coincide well for the two substances.   The isotherms (constant T_r), of which only one is shown,

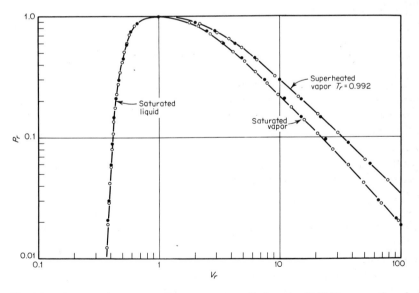

Fig. 1-1   The law of corresponding states as applied to the P-V-T properties of methane and nitrogen.   Experimental values [3]: ∘ methane, • nitrogen.

agree equally well.   It is fortunate that the expression of the reduced property as a simple *fraction* of the critical value works so well.

An important application of the law of corresponding states is in the correlation of P-V-T data using the compressibility factor $Z \equiv PV/RT$.   The law suggests a correlation of Z/Z_c as a function of P_r and T_r.   But since Z_c for many nonpolar substances is almost constant near 0.27, the correlation is simplified to Z as a function of P_r and T_r for a large group of compounds.   Correlations of gas-law deviations on this basis are presented in Chap. 3.

The successful application of the law of corresponding states in the correlation of P-V-T data has encouraged many similar correlations of other   properties   which   depend   primarily   on   intermolecular forces.   Many   of   these   have   proved   invaluable   to   the   practicing engineer.   Modifications of the law to improve accuracy or ease of use are common.   Good correlations of high-pressure gas viscosity have been obtained by expressing η/η_c as a function of P_r and T_r.   But since η_c is seldom known and not easily estimated, this quantity has been replaced in other correlations by η_c°, η_T°, or the group $M^{1/2}P_c^{2/3}/T_c^{1/6}$. η_c° is the viscosity at T_c but at low pressure, η_T° is the viscosity at the temperature of interest, again at low pressure; and the group involving M, P_c, and T_c is suggested by dimensional analysis.   Many other alternatives to the use of η_c might be proposed, each being modeled on the law of corresponding states but empirical as applied to transport properties.

Statistical mechanics provides a similarly valid framework for the development of estimation methods.   Equations describing various properties, including transport properties, are derived, starting with an expression for the potential-energy function for molecular inter-actions.   This last is largely empirical, but the resulting equations for properties are surprisingly insensitive to the potential function from which they stem, and two-constant potential functions serve remarkably well for some systems.   As in the case of the law of corresponding states, deviations from the resulting equations may be handled empirically.

Nonpolar and Polar Molecules   Spherically symmetric molecules, for example, CH_4, are well fitted by a two-constant law of corresponding states.   Nonspherical and weakly polar molecules do not fit poorly, but the deviations are often great enough to encourage the development of correlations involving a third parameter.   One possible third parameter for nonpolar molecules allows for nonsphericity.   Most employ Z_c, the acentric factor, or the Riedel factor.   The last two, which are related, depend on the deviation of the vapor–pressure–temperature function for the compound from that which might be expected for a substance consisting of   spherically   symmetric   molecules. Typical   correlations express the dimensionless property as a function of P_r, T_r, and the chosen third parameter.

The properties of strongly polar molecules do not fit well the two- or three-constant correlations which do so well for nonpolar molecules. An additional parameter involving the dipole moment is suggested, since polarity is measured by dipole moments. This works only moderately well, since polar molecules are more individualistic than the nonpolar compounds and not easily characterized. Parameters have also been developed in attempts to characterize quantum effects which may be important at low temperatures.

In summary, estimation of various properties which depend on intermolecular forces are based on correlations of the nondimensional property with two to four parameters, two of which are usually P_r and T_r.

Structure   All properties are related to molecular structure; it is this which determines the magnitude and predominant type of the intermolecular forces. Structure determines the energy-storage capacity of a molecule and thus its heat capacity.

The relevant characteristic of structure is variously related to the atoms, atomic groups, bond type, etc., to which weighting factors are assigned and the property determined, usually by an algebraic operation. The weighted characteristics are often added to obtain the property directly or to obtain a correction to some approximate theory or simple rule. Lydersen's method for estimating T_c, for example, starts with the loose rule that the ratio of the normal boiling temperature to the critical temperature is about $2:3$. Additive structural increments based on bond types are then used to obtain a parameter to correct this ratio empirically.

Some of the better correlations of ideal-gas heat capacities employ theoretical values of C_p° (which are intimately related to structure) to obtain a polynomial expressing C_p° as a function of temperature, with the constants determined by the constituent atoms, atomic groups, and types of bonds.

1-4   Organization of Book

Reliable experimental data are always to be preferred over values obtained by even the best estimation methods.

The various estimation methods to be described involve correlations of experimental data. The best are based on theory, with empirical corrections for the theory's defects. Others, including those stemming from the law of corresponding states, are based on generalizations which are partly empirical but which have application to a remarkably wide range of properties.   Purely empirical correlations are useful only when applied to situations very similar to those used to establish the correlation.

The text includes many numerical examples of the use of the estimation methods described, especially those methods which are recommended. Almost all of these are designed to explain the calculation procedure for a single property.   Many engineering design problems require the estimation of a number of properties, and the error in each contributes to the overall result, some being more important than others.   Very often the result is found adequate for engineering purposes, in spite of the large measure of empiricism incorporated in so many of the estimation procedures.

As an example, consider the case of a chemist who has synthesized a new compound, which has the chemical formula CCl_2F_2 and boils at $-20.5°C$ at atmospheric pressure.   Using only this information, is it possible to obtain a useful prediction of whether or not the substance has the thermodynamic properties which might make it a practical refrigerant?

Figure 1-2 shows portions of a Mollier diagram developed by the prediction methods described in later chapters.   The dashed curves and points are developed from estimates of liquid and vapor heat capacities, critical properties, vapor pressure, enthalpy of vaporization, and pressure corrections to ideal-gas enthalpies and entropies.   The substance is, of course, a well-known refrigerant, and its known properties are shown by the solid curves.

For a standard refrigeration cycle operating between $48.9°C$ ($120°F$) and $-6.7°C$ ($20°F$), the evaporator and condenser pressures are estimated to be 2.4 and 12.2 atm, vs. the known values 2.4 and 11.7 atm.   The estimate of the heat absorption in the evaporator checks closely, and the estimated volumetric vapor rate to the compressor also shows good agreement: 296 vs. 304 ft^3/h ton of refrigeration.   (This number is indicative of the physical size of the compressor.)   Constant-entropy lines are not shown on Fig. 1-2, but it is found that the constant-entropy line through the point for the low-pressure vapor essentially coincides with the saturated-vapor curve.   The estimated coefficient of performance (ratio of refrigeration rate to isentropic compression power) is estimated to be 3.8; the value obtained from the published data is 3.5.   This last is not a particularly good check, but is perhaps remarkable in view of the fact that the only data used for the estimate were the normal boiling point and the chemical formula.

Most estimation methods require parameters which are characteristic of single pure components or of constituents of a mixture of interest.   The more important of these are considered in Chap. 2, and tables of values for common substances are provided in Appendix

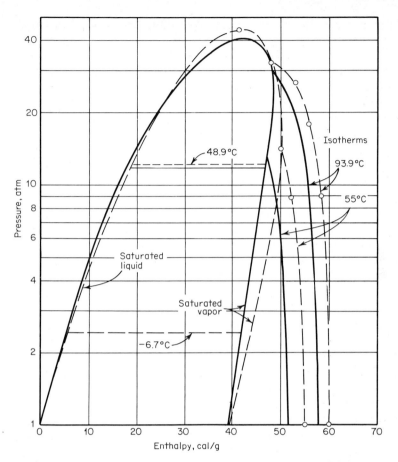

Fig. 1-2 Mollier diagram for dichlorodifluoromethane. The solid lines represent published data. Dashed lines and points represent values obtained by estimation methods when only the chemical formula and the normal boiling temperature are known.

A.   Thermodynamic properties (such as enthalpy and heat capacity) are treated in Chaps. 3 to 5.   The more accurate equations of state are employed, but the basic thermodynamic relationships are developed, so that other equations of state can be introduced if thought to be more applicable for a particular purpose.

Chapters 6 and 7 deal with vapor pressures and enthalpies of vaporization of pure substances, heat capacity, enthalpy of formation, and entropy.   Chapter 8 is a discussion of techniques for the estimation and correlation of phase equilibria.   Chapters 9 to 11 describe estimation methods for viscosity, thermal conductivity, and diffusion coefficients.   Surface tension is treated briefly in Chap. 12.

The literature searched was voluminous, and the lists of references following each chapter represent but a fraction of the material examined. Of the many estimation methods available, only a few were selected for detailed discussion. These were selected on the basis of their generality, accuracy, and availability of the required input data. Our tests of all methods were more extensive than suggested by the abbreviated tables comparing experimental with estimated values. However, no comparison is adequate to indicate *expected* errors for new compounds. The arithmetic *average* errors given in the comparison tables represent but a crude overall evaluation; the inapplicability of a method for a few compounds may so increase the average error as to distort judgment of the method's merit.

Many estimation methods are of such complexity that a computer is required to use them. This is less of a handicap than it once was, since computers have become widely available, as are programs for machine-computation estimation. Electronic desk computers, which have become so popular in recent years, have made the more complex correlations practical. Accuracy, of course, is not necessarily enhanced by greater complexity.

The scope of the book is necessarily limited. The properties discussed were selected arbitrarily because they were believed to be fundamental in nature and of wide interest, especially to chemical engineers. Electrical properties are not included, nor are chemical properties other than certain thermodynamically derived properties such as enthalpy and free energy of formation. The difficult area of polymers and crystals is treated by Bondi [1] and van Krevelen [5].

The book is intended to provide useful estimates of a limited number of physical properties of fluids. It is to be hoped that the need for such estimates, and for a book of this kind, may diminish as more experimental values become available and as the continually developing molecular theory approaches perfection. In the meantime, estimation methods must be used in process-design calculations and for many other purposes in engineering and applied science.

REFERENCES

1. Bondi, A.: "Physical Properties of Molecular Crystals, Liquids, and Glasses," Wiley, New York, 1968.
2. "CAChE Physical Properties Data Book," National Academy of Engineering, Washington, 1972.
3. Din, F. (ed.): "Thermodynamic Functions of Gases," vol. 3, Butterworth, London, 1961.
4. Quoted from James Clerk Maxwell, Atoms, "Encyclopaedia Britannica," 9th ed., 1875.
5. van Krevelen, D. W.: "Properties of Polymers: Correlations with Chemical Structure," Elsevier, Amsterdam, 1972.

Chapter Two

Pure-Component Constants

2-1 Scope

Chemical engineers normally deal with mixtures rather than pure materials. However, the chemical compositions of most mixtures of interest (except, perhaps, in the petroleum industry) are known. Thus, for both pure components and mixtures, one is concerned with specific chemical entities. Also, few mixture–property-estimation correlations to date have incorporated true *mixture parameters*. Instead, the techniques employ parameters which relate only to pure components, which may exist alone or may constitute the mixture. These *pure-component constants* are then used with the state variables such as temperature, pressure, and composition to generate property-estimation methods.

In this chapter, we introduce the more common and useful pure-component constants and show how they can be estimated if no experimental data are available. Those which are discussed would be likely candidates for inclusion into a *data base* for computer-based property-estimation systems.

Some useful pure-component constants, however, are not covered in this chapter, as it is more convenient and appropriate to discuss them in subsequent chapters. For example, the liquid specific volume (or density) at some reference state (for example, 20°C, 1 atm) is a useful pure-component constant, yet it is more easily introduced in Chap. 3, where volumetric properties are covered.

2-2 Critical Properties

Critical temperature, pressure, and volume represent three widely used pure-component constants. Yet recent, new experimental measurements are almost nonexistent. In Appendix A, we have tabulated the critical properties of many materials. In most instances the values given were those measured experimentally. When estimated, the method of Lydersen was normally employed.

Excellent, comprehensive reviews of critical properties are available. Kudchadker, Alani, and Zwolinski [12] cover organic compounds, and Mathews [16] reviews inorganic substances.

Lydersen's Method [15]   This estimation method employs structural contributions to estimate T_c, P_c, and V_c. The relations are

$$T_c = T_b[0.567 + \Sigma\Delta_T - (\Sigma\Delta_T)^2]^{-1} \qquad (2\text{-}2.1)$$

$$P_c = M(0.34 + \Sigma\Delta_p)^{-2} \qquad (2\text{-}2.2)$$

$$V_c = 40 + \Sigma\Delta_v \qquad (2\text{-}2.3)$$

The units employed are kelvins, atmospheres, and cubic centimeters per gram mole. The Δ quantities are evaluated by summing contributions for various atoms or groups of atoms as shown in Table 2-1. To employ this method, only the normal boiling point T_b and the molecular weight M are needed. Errors associated with these relations vary as summarized in Table 2-2.

Spencer and Daubert [31] made an extensive evaluation of available methods to estimate the critical properties of *hydrocarbons*. For critical temperature, they found Lydersen's method the most accurate; however, by modifying the constants in a technique proposed by Nokay [20] they were able to reduce the average error slightly. The Nokay relation is

$$\log T_c = A + B \log SG + C \log T_b \qquad (2\text{-}2.4)$$

where T_c = critical temperature, K, T_b = normal boiling-point temperature, K, SG = specific gravity of liquid hydrocarbon at 60°F relative to water at same temperature.

The constants A, B, and C were determined by Spencer and Daubert from a multiple least-squares program employing experimental

TABLE 2-1   Lydersen's Critical-Property Increments [15]†

	Δ_T	Δ_p	Δ_v
Nonring Increments			
—CH₃	0.020	0.227	55
—CH₂	0.020	0.227	55
—CH	0.012	0.210	51
—C—	0.00	0.210	41
=CH₂	0.018	0.198	45
=CH	0.018	0.198	45
=C—	0.0	0.198	36
=C=	0.0	0.198	36
≡CH	0.005	0.153	(36)
≡C—	0.005	0.153	(36)
Ring Increments			
—CH₂—	0.013	0.184	44.5
—CH	0.012	0.192	46
—C—	(−0.007)	(0.154)	(31)
=CH	0.011	0.154	37
=C—	0.011	0.154	36
=C=	0.011	0.154	36

TABLE 2-1 Lydersen's Critical-Property Increments [15]† *(Continued)*

	Δ_T	Δ_p	Δ_v
Halogen Increments			
—F	0.018	0.224	18
—Cl	0.017	0.320	49
—Br	0.010	(0.50)	(70)
—I	0.012	(0.83)	(95)
Oxygen Increments			
—OH (alcohols)	0.082	0.06	(18)
—OH (phenols)	0.031	(−0.02)	(3)
—O— (nonring)	0.021	0.16	20
—O— (ring)	(0.014)	(0.12)	(8)
$-\overset{\mid}{C}=O$ (nonring)	0.040	0.29	60
$-\overset{\mid}{C}=O$ (ring)	(0.033)	(0.2)	(50)
$H\overset{\mid}{C}=O$ (aldehyde)	0.048	0.33	73
—COOH (acid)	0.085	(0.4)	80
—COO— (ester)	0.047	0.47	80
=O (except for combinations above)	(0.02)	(0.12)	(11)
Nitrogen Increments			
—NH₂	0.031	0.095	28
$-\overset{\mid}{N}H$ (nonring)	0.031	0.135	(37)
$-\overset{\mid}{N}H$ (ring)	(0.024)	(0.09)	(27)
$-\overset{\mid}{N}-$ (nonring)	0.014	0.17	(42)

Nitrogen Increments (*Continued*)

$-\overset{\|}{N}-$ (ring)	(0.007)	(0.13)	(32)
—CN	(0.060)	(0.36)	(80)
—NO₂	(0.055)	(0.42)	(78)

Sulfur Increments

—SH	0.015	0.27	55
—S— (nonring)	0.015	0.27	55
—S— (ring)	(0.008)	(0.24)	(45)
=S	(0.003)	(0.24)	(47)

Miscellaneous

$-\overset{\|}{\underset{\|}{Si}}-$	0.03	(0.54)
$-\overset{\|}{\underset{\|}{B}}-$	(0.03)	

†There are no increments for hydrogen.   All bonds shown as free are connected with atoms other than hydrogen.   Values in parentheses are based upon too few experimental values to be reliable.   From vapor-pressure measurements and a calculational technique similar to Fishtine [6], it has been suggested that the $\overset{\diagdown\!\diagup}{\underset{\|}{C}}$—H ring increment common to two condensed saturated rings be given the value of $\Delta_T = 0.064$.

TABLE 2-2   Estimated Errors Associated with Lydersen's Critical-Property Estimation Method

Property	Equation	Typical error
T_c	(2-2.1)	Usually less than 2%; up to 5% for higher molecular-weight (>100) nonpolar materials; errors uncertain for molecules with multifunctional polar groups, e.g., glycols
P_c	(2-2.2)	See error for T_c, but double numerical values shown
V_c	(2-2.3)	Similar to P_c; perhaps somewhat greater; fewer data upon which to base correlation

TABLE 2-3   Spencer and Daubert Constants for Use in Nokay's Eq. (2-2.4) for Predicting Critical Temperature

Family of compounds	A	B	C
Alkanes (paraffins)	1.359397	0.436843	0.562244
Cycloalkanes (naphthenes)	0.658122	−0.071646	0.811961
Alkenes (olefins)	1.095340	0.277495	0.655628
Alkynes (acetylenes)	0.746733	0.303809	0.799872
Alkadienes (diolefins)	0.147578	−0.396178	0.994809
Aromatics	1.057019	0.227320	0.669286

data.   These constants are given in Table 2-3.   At the present time, there is no way to generalize the Nokay correlation should the hydrocarbon contain characteristics of more than a single family, e.g., an aromatic with olefinic side chains.

Many other critical-temperature estimation methods have been proposed; several are discussed in an earlier edition of this book [27].   Also, more recently, Rao et al. [25] correlated the critical temperature with the molar refraction and the parachor (see Chap. 12) for several homologous series, while Mathur et al. [17] employed the molecular weight as the correlating factor.   Gold and Ogle [7] have made an extensive comparison between experimental critical temperatures and those estimated by several methods.   They conclude that Lydersen's method is the most accurate.   The Nokay method was included in their testing but not with the modified constants given in Table 2-3.

In critical-pressure estimations, though many techniques have been suggested, the Lydersen method has been found to be easy to use and the most accurate both for hydrocarbons [31] and for organic compounds in general [7].

To estimate critical volumes, the Lydersen method was found to be the most reliable by Gold and Ogle [7] after an extensive comparison between experimental values and those calculated from various estimation schemes.   Spencer and Daubert found, however, that *for hydrocarbons* the method of Riedel [24, 28] is somewhat more accurate.   In this method

$$V_c = \frac{RT_c}{P_c}[3.72 + 0.26(\alpha_c - 7.0)]^{-1} \qquad (2\text{-}2.5)$$

$$\alpha_c = 0.9076\left[1.0 + \frac{(T_b/T_c)\ln P_c}{1.0 - T_b/T_c}\right] \qquad (2\text{-}2.6)$$

with P_c in atmospheres. α_c is the Riedel factor (see Chap. 6) and the group in the square brackets of Eq. (2-2.5) is the inverse of the critical compressibility factor (see below).   To obtain better predictions of V_c with Riedel's form, Spencer and Daubert used experimental values of T_c

and P_c in Eq. (2-2.5); if estimated values had been employed, the predictive accuracy would have been less.

Another group-contribution method for critical volumes has been suggested by Vetere [34]. It is quite similar to the Lydersen form [Eq. (2-2.3)], i.e.,

$$V_c = 33.04 + \left[\sum_i (\Delta V_i M_i) \right]^{1.029} \qquad (2\text{-}2.7)$$

where ΔV_i is given in Table 2-4 for many groups and M_i is the molecular weight of the group. In most cases, Eq. (2-2.7), with Table 2-4, yields more accurate estimations of the critical volume than any other method. Fedors [5] has also published a group-contribution method for V_c which is quite reliable.

At the critical point the critical compressibility factor is

$$Z_c = \frac{P_c V_c}{R T_c} \qquad (2\text{-}2.8)$$

Obviously, if one knows P_c, V_c, and T_c, it is readily determined. Alternatively, it can be estimated by special techniques [27]. These

TABLE 2-4   Vetere Group Contributions to Estimate Critical Volumes

Group	ΔV_i	Group	ΔV_i
Nonring:		$-\overset{\|}{C}{=}O$ (nonring)	1.765
In linear chain:			
CH_3, CH_2, CH, C	3.360	$-\overset{\|}{C}{=}O$ (ring)	1.500
In side chain			
CH_3, CH_2, CH, C	2.888	$H\overset{\|}{C}{=}O$ (aldehyde)	2.333
${=}CH_2, {=}\overset{\|}{C}H, {=}\overset{\|}{C}{-}$	2.940	$-COOH$	1.652
${=}C{=}$	2.908	$-COO{-}$	1.607
${\equiv}CH, {\equiv}C{-}$	2.648		
Ring:		$-NH_2$	2.184
CH_2, CH, C	2.813		
${=}\overset{\|}{C}H, {=}\overset{\|}{C}{-}$	2.538	$-\overset{\|}{N}H$ (nonring)	2.333
		$-\overset{\|}{N}H$ (ring)	1.736
F	0.770		
Cl	1.237	$-\overset{\|}{N}{-}$ (nonring)	1.793
Br	0.899		
I	0.702	$-\overset{\|}{N}{-}$ (ring)	1.883
		$-CN$	2.784
$-OH$ (alcohols)	0.704	$-NO_2$	1.559
$-OH$ (phenols)	1.553		
$-O{-}$ (nonring)	1.075	$-SH$	1.537
$-O{-}$ (ring)	0.790	$-S{-}$ (nonring)	0.591
$-O{-}$ (epoxy)	-0.252	$-S{-}$ (ring)	0.911

methods are not discussed here since Eq. (2-2.8) is usually sufficiently accurate to obtain Z_c even if some or all of the other critical properties must be estimated.   Values of Z_c are rarely needed except as a parameter in estimation techniques for other properties, and these are normally rather insensitive to the exact values of Z_c.   Values of Z_c for a number of common substances are tabulated in Appendix A.   In all instances, they were obtained from Eq. (2-2.8).

Another rarely used method to estimate critical properties is to employ correlations for other physical properties in which critical properties are embedded as nondimensionalizing parameters.   With low-temperature data only, it is then possible to extract approximate values of the critical properties.   As an example of this approach, consider the Gunn-Yamada estimation method for saturated liquid volumes, as discussed in Sec. 3-15 [9].   The saturated liquid molal volume V is given in Eq. (3-15.1) as

$$\frac{V}{V_{sc}} = V_r^{(0)}(1 - \omega\Gamma)$$ (2-2.9)

where $V_r^{(0)}$ and Γ are shown as functions of reduced temperature T/T_c in Eqs. (3-15.4) to (3-15.6). ω is the acentric factor described in Sec. 2-3, and V_{sc} is a scaling parameter characteristic of the substance.   From vapor-pressure data around the normal boiling point and a first estimate of P_c and T_c, a value of ω is determined from Eq. (2-3.1).   Then with this ω and values of the liquid molal volume V for at least two temperatures, T_c can be found from Eq. (2-2.9) in a trial-and-error iterative calculation.   A value of V_{sc} can then be found.   Next, the critical pressure and volume are determined by the relations

$$P_c = \frac{(0.2920 - 0.0967\omega)RT_c}{V_{sc}}$$ (2-2.10)

$$V_c = \frac{(0.2918 - 0.0928\omega)RT_c}{P_c}$$ (2-2.11)

From these calculated values of T_c and P_c, a new estimate of ω is made and the procedure repeated until there is no change in ω in the loop.   Good liquid volumetric data (over a reasonable temperature range) and the services of a computer are necessary.   However, as shown by Gunn and Yamada, the level of accuracy to estimate the critical properties is equal to or often better than that obtained from the Lydersen method.   A similar scheme has been suggested by Pitzer and Brewer [14] and Gunn and Mahajan [8] using two vapor pressures and a single liquid density.

Example 2-1   Estimate the critical properties of pentafluorotoluene using the Lydersen method.   The normal boiling point is 390.65 K, and $M = 182.1$.

solution From Table 2-1

$$\Sigma\Delta_T = 6(=\overset{|}{C}-)_{ring} + (CH_3) + 5(F)$$
$$= (6)(0.011) + 0.020 + (5)(0.018) = 0.176$$
$$\Sigma\Delta_p = (6)(0.154) + 0.227 + (5)(0.224) = 2.271$$
$$\Sigma\Delta_v = (6)(36) + 55 + (5)(18) = 361$$

From Eqs. (2-2.1) to (2-2.3)

$$T_c = \frac{390.65}{0.567 + 0.176 - (0.176)^2} = 549 \text{ K}$$

$$P_c = \frac{182.1}{(0.34 + 2.271)^2} = 26.7 \text{ atm}$$

$$V_c = 40 + 361 = 401 \text{ cm}^3/\text{g-mol}$$

Experimental values for T_c and P_c reported by Ambrose and Sprake [1] are 566 K and 30.8 atm. Errors are 3.0 and 13.3 percent, respectively. Pentafluorotoluene was not used by Lydersen in developing the method. No experimental critical volume is available for comparison.

Example 2-2 Using Vetere's method, estimate the critical volume of isobutanol.
  solution From Table 2-4

Group	ΔV_i	M_i	$\Delta V_i M_i$
—CH$_3$	3.360	15.03	50.50
—CH$_3$	3.360	15.03	50.50
—$\overset{|}{\underset{|}{CH}}$	3.360	13.02	43.75
—CH$_2$—	3.360	14.03	47.14
—OH	0.704	17.01	11.98
			203.87

$V_c = 33.04 + (203.87)^{1.029} = 270.9 \text{ cm}^3/\text{g-mol}$

The experimental value for V_c is 273 cm^3/g-mol.

2-3 Acentric Factor

One of the more common pure-component constants is the acentric factor [22, 23], which is defined as

$$\omega = -\log P_{vp_r} \text{ (at } T_r = 0.7) - 1.000 \qquad (2\text{-}3.1)$$

To obtain values of ω, the vapor pressure at $T_r = T/T_c = 0.7$ is required as well as the critical pressure.

As originally proposed, ω was to represent only the acentricity or nonsphericity of a molecule. For monatomic gases, ω is, therefore, essentially zero. For methane, it is still very small. However, for higher-molecular-weight hydrocarbons, ω increases and often rises with polarity.   At present, it is widely used as a parameter which in some manner measures the complexity of a molecule with respect to both the geometry and the polarity.   Application of correlations employing the acentric factor should be limited to normal fluids; in no case should such correlations be used for H_2, He, Ne, or for strongly polar and/or hydrogen-bonded fluids.

We show in Appendix A the acentric factor for many materials. These data were obtained, in most cases, from the best experimental or estimated values of T_c and P_c as well as from vapor-pressure data at $T_r = T/T_c = 0.7$.   Passut and Danner [21] tabulate ω for 192 hydrocarbons.

If acentric factors are needed for a material not listed in Appendix A, several estimation techniques are available.   The simplest, and the one recommended here, is to determine first the critical temperature and pressure.   Then, locate at least one other boiling point, for example, T_b at $P = 1$ atm, and use a vapor-pressure correlation from Chap. 6 to determine P_{vp} at $T_r = 0.7$.   With this value, Eq. (2-3.1) can then be used to find ω.   As an example, if the vapor-pressure correlation chosen were

$$\log P_{vp} = A + \frac{B}{T} \tag{2-3.2}$$

with A and B found from the sets $(T_c, P_c; T_b, P = 1)$, then

$$\omega = \frac{3}{7}\frac{\theta}{1-\theta} \log P_c - 1 \tag{2-3.3}$$

where P_c is in atmospheres and $\theta \equiv T_b/T_c$.   This relation was first suggested by Edmister [4].

Similarly, if the Lee-Kesler vapor-pressure relation [Eqs. (6-2.6) to (6-2.8)] were used,

$$\omega = \frac{-\ln P_c - 5.92714 + 6.09648\theta^{-1} + 1.28862 \ln \theta - 0.169347\theta^6}{15.2518 - 15.6875\theta^{-1} - 13.4721 \ln \theta + 0.43577\theta^6} \tag{2-3.4}$$

Lee and Kesler [13] report that Eq. (2-3.4) yields values of ω very close to those selected by Passut and Danner [21] in their critical review of ω for hydrocarbons.

Example 2-3   Estimate the acentric factor for n-octane.   From Appendix A, $T_b = 398.8$ K, $T_c = 568.8$ K, and $P_c = 24.5$ atm.
  solution $\theta = T_b/T_c = 398.8/568.8 = 0.701$.   With Eq. (2-3.3),

$$\omega = \frac{3}{7}\frac{0.701}{1-0.701} \log 24.5 - 1 = 0.396$$

With Eq. (2-3.4),

$$\omega = \frac{-\ln 24.5 - 5.92714 + \dfrac{6.09648}{0.701} + 1.28862 \ln 0.701 - (0.169347)(0.701)^6}{15.2518 - \dfrac{15.6875}{0.701} - 13.4721 \ln 0.701 + (0.43577)(0.701)^6}$$

$$= 0.396$$

The value of ω selected for Appendix A and based upon experimental vapor pressures is 0.394.

In many instances in the literature, one finds ω related to Z_c by

$$Z_c = 0.291 - 0.080\omega \qquad (2\text{-}3.5)$$

This equation results from applying a P-V-T correlation using ω (see Chap. 3) to the critical point, where $Z = Z_c$. Equation (2-3.5) is only *very* approximate, as the interested reader can readily show from Appendix A. In fact, should Z_c be defined by Eq. (2-3.5), it would be best to consider this Z_c as a new parameter rather than as the true critical compressibility factor defined by Eq. (2-2.8).

2-4 Freezing and Boiling Points

Ordinarily, when one refers to a freezing or boiling point, there is an implied condition that the pressure is 1 atm. A more exact terminology for these temperatures might be the *normal freezing* and *normal boiling points.*

In Appendix A, values for T_f and T_b are given for many substances. Methods for estimating T_b are generally poor. They are summarized elsewhere [27]; most involve group-contribution techniques which are devised for homologous series with no more than one functional group attached to a hydrocarbon framework.

The estimation of the normal freezing point is complicated by the fact that $T_f = \Delta H_{\text{fus}}/\Delta S_{\text{fus}}$ and whereas ΔH_{fus} depends primarily upon intermolecular forces, ΔS_{fus} is a function of the molecular symmetry. As noted by Bondi [2], ΔS_{fus} is larger when the molecule can assume many orientations in the liquid phase relative to the solid. Thus ΔS_{fus} is smaller for spherical, rigid molecules, and T_f is higher than for molecules of the same size which are anisometric and flexible. An interpolative method was, however, suggested by Eaton [3] to correlate normal freezing points of a homologous series. For such a series, a plot is made of $(T_b - T_f)/T_f$ vs. molecular weight. Except perhaps for the first members of a series, this type of plot yields a straight line. Interpolation or a reasonable extrapolation allows one to estimate T_f for members of the family whose freezing points are not known; however, an accurate value of T_b is desirable.

2-5 Dipole Moments

Dipole moments of molecules are often required in property correlations for polar materials.   The best source of this constant is from the compilation by McClellan [18], which has, to a large degree, superseded prior summaries such as those given by Smith [29] and Smyth [30].   For those rare occasions when one may be forced to estimate a value, there are vector group-contribution methods, although they ordinarily require considerable effort.   Most such methods are summarized in the text by Minkin et al. [19].

Dipole moments for many materials are listed in Appendix A; no temperature effect is shown, as dipole moments are insensitive to this variable.   Also, we have not noted whether the dielectric constants were measured in the gas phase or in a solvent because differences between such measurements are ordinarily small.

Dipole moments are expressed in debye units, 1 debye being equivalent to 10^{-18} $(\text{dyn cm}^4)^{1/2}$.   Thus, the physical unit for this property is $[(\text{energy})(\text{volume})]^{1/2}$.

2-6 Stiel Polar Factor

Equation (6-2.6) relates the reduced vapor pressure to the acentric factor and reduced temperature.   If this estimated reduced vapor pressure is called $P_{\text{vp}_r,\text{normal}}$, the Stiel polarity factor X is given as

$$X \equiv \log \frac{P_{\text{vp}_r}}{P_{\text{vp}_r,\text{normal}}} \qquad \text{at } T_r = 0.6 \qquad (2\text{-}6.1)$$

If the material has a vapor pressure well correlated by Eq. (6-2.6), X will, of course, be zero.   Polar materials have been shown to deviate from this Pitzer vapor-pressure correlation at low values of T_r.   [Note that all must fit at $T_r = 0.7$ by the definition of the acentric factor in Eq. (2-3.1).]   The definition of X thus quantifies this deviation.   If the

TABLE 2-5   Stiel Polar Factors [32]

	X		X
Methanol	0.037	Water	0.023
Ethanol	0.0	Hydrogen chloride	0.008
n-Propanol	−0.057	Acetone	0.013
Isopropanol	−0.053	Methyl fluoride	0.012
n-Butanol	−0.07	Ethylene oxide	0.012
Dimethylether	0.002	Methyl acetate	0.005
Methyl chloride	0.007	Ethyl mercaptan	0.004
Ethyl chloride	0.005	Diethyl ether	−0.003
Ammonia	0.013		

Pitzer vapor-pressure correlation is extended to $T_r = 0.6$ and expressed analytically, it can be shown [10] that

$$X = \log P_{vp_r} \text{ (at } T_r = 0.6) + 1.70\omega + 1.552 \qquad (2\text{-}6.2)$$

Values of the Stiel polar factor are known for only a few materials; a convenient tabulation is given in Table 2-5.

2-7 Potential Force Constants

In most theory-based estimation techniques, there is a need to express the energy of interaction between molecules as a function of the separation distance. In analytical representations of such relations, various parameters appear which are characteristic of the molecules involved. The famous Lennard-Jones 12-6 potential

$$\phi(r) = 4\epsilon \left[\left(\frac{\sigma}{r}\right)^{12} - \left(\frac{\sigma}{r}\right)^{6} \right] \qquad (2\text{-}7.1)$$

is an example of such a relation between the interacting energy $\phi(r)$ and separation distance r. The scaling factors ϵ and σ have, of course, units of energy and length. Other more realistic potential functions would have different scaling parameters.

If there were a truly satisfactory intermolecular potential function, the scaling parameters would certainly be useful pure-component constants. Unfortunately, at the present time, we can obtain relations only slightly more realistic than Eq. (2-7.1) and this at the expense of adding more undetermined scaling factors, e.g., the three-parameter Kihara potential.

Values of ϵ and σ and similar parameters are normally found by employing a theoretical relation to calculate some property and regressing experimental data to obtain satisfactory values of ϵ and σ. When this is done, an interesting result is found. From any specific property, many *sets* of ϵ and σ are usually found which are satisfactory in the sense that when any set is used to compute property values, all yield about the same result. Hu, Chappelear, and Kobayashi [11] have shown clearly that ϵ-σ sets determined from second virial coefficients, viscosity, and diffusion coefficients all differ but the intersection of these sets will lead to a single pair of ϵ-σ values which is satisfactory for computing all these properties. Reichenberg [26] has shown that the form of the Lennard-Jones potential is such that when regressing experimental data to find "best" values of ϵ/k and σ, one cannot separate these potential parameters. That is, the ϵ/k and σ have essentially collapsed together into a single parameter for any particular property. For any reasonable choice of ϵ/k, there is then a corresponding σ, and this ϵ/k-σ pair is satisfactory for property estimation. Other ϵ/k-σ sets are applicable for other properties, and it was the intersection of these sets that was found by Hu et al., as noted above. Most estimation techniques to date have been based on ϵ/k-σ back-calculated from a single property [27] and, as such, are of limited use.

Should one find it necessary to estimate Lennard-Jones or Kihara potential parameters, the study by Tee, Gotoh, and Stewart [33] is very helpful. For nonpolar fluids, they obtained preferred values of the parameters ϵ and σ for the Lennard-Jones potential by using theoretical relations and regressing *both* experimental viscosity and second-virial-coefficient data to obtain a best fit. Correlations were then devised to relate these parameters to T_c, P_c, V_c, and ω. For example, in the Lennard-Jones form, the best correlation for both the viscosity and second virial coefficient is

$$\sigma\left(\frac{P_c}{T_c}\right)^{1/3} = 2.3551 - 0.087\omega \tag{2-7.2}$$

$$\frac{\epsilon}{kT_c} = 0.7915 + 0.1693\omega \tag{2-7.3}$$

where ω = acentric factor
σ = potential length constant, Å
P_c = critical pressure, atm
T_c = critical temperature, K
ϵ = potential-energy constant, ergs
k = Boltzmann's constant = 1.3805 ergs/K

Different equations are recommended if only viscosities (or second virial coefficients) are to be estimated. Theoretical equations for calculating viscosity are covered in Chap. 9; theoretical relations for the second virial coefficient are considered briefly in Chap. 3.

Values of ϵ/k and σ are listed for the 12-6 Lennard-Jones potential in Appendix C.

NOTATION

k = Boltzmann's constant
M = molecular weight; M_i, of group i
P = pressure, atm; P_c, critical pressure; P_r, reduced pressure, P/P_c; P_{vp}, vapor pressure; P_{vp_r}, P_{vp}/P_c
R = gas constant
SG = specific gravity of liquid at 60°F
T = temperature, K; T_c, critical temperature; T_r, reduced temperature, T/T_c; T_b, normal boiling point; T_f, freezing point
V = molal volume, cm³/g-mol; V_c, critical volume
X = Stiel polar factor, Eq. (2-6.2)
Z = compressibility factor; Z_c, at the critical point

Greek

α_c = Riedel factor, Eq. (2-2.6)

$\Delta_T, \Delta_p, \Delta_v$ = Lydersen et al. factors, Table 2-1
$\qquad \Delta V_i$ = Vetere contribution to V_c, Table 2-4
$\qquad \epsilon$ = potential-energy constant, Sec. 2-7
$\qquad \theta = T_b / T_c$
$\qquad \sigma$ = potential-length constant, Sec. 2-7
$\qquad \phi$ = potential energy of interaction
$\qquad \omega$ = Pitzer acentric factor

REFERENCES

1. Ambrose, D., and C. H. S. Sprake: *J. Chem. Soc.*, **1971A**: 1264.
2. Bondi, A.: "Physical Properties of Molecular Crystals, Liquids and Glasses," chap. 6, Wiley, New York, 1968.
3. Eaton, E. O.: *Chem. Technol.*, June 1971, p. 362.
4. Edmister, W. C.: *Pet. Refiner*, **37**(4): 173 (1958).
5. Fedors, R. F.: *Polymer Lett. Ed.*, **11**: 767 (1973).
6. Fishtine, S. H.: *Ind. Eng. Chem. Fundam.*, **2**: 149 (1963).
7. Gold, P. I., and G. J. Ogle: *Chem. Eng.*, **75**(21): 185 (1968).
8. Gunn, R. D., and V. J. Mahajan, Corresponding States Theory for High Boiling Compounds, paper presented at *Nat. Meet. AIChE, New Orleans, La., March 1974.*
9. Gunn, R. D., and T. Yamada: *AIChE J.*, **17**: 1341 (1971).
10. Halm, R. L., and L. I. Stiel: *AIChE J.*, **16**: 3 (1970), **17**: 259 (1971).
11. Hu, A. T., P. S. Chappelear, and R. Kobayashi, *AIChE J.*, **16**: 490 (1970).
12. Kudchadker, A. P., G. H. Alani, and B. J. Zwolinski, *Chem. Rev.*, **68**: 659 (1968).
13. Lee, B. I., and M. G. Kesler: *AIChE J.*, **21**: 510 (1975).
14. Lewis, G. N., and M. Randall: "Thermodynamics," 2d ed., rev. by K. S. Pitzer and L. Brewer, app. 1, McGraw-Hill, New York, 1961.
15. Lydersen, A. L.: Estimation of Critical Properties of Organic Compounds, *Univ. Wisconsin Coll. Eng., Eng. Exp. Stn. Rep.* 3, Madison, Wis., April 1955.
16. Mathews, J. F.: *Chem. Rev.*, **72**: 71 (1972).
17. Mathur, B. C., S. H. Ibrahim, and N. R. Kuloor, *Chem. Eng.*, **76**(6): 182 (1969).
18. McClellan, A. L.: "Tables of Experimental Dipole Moments," Freeman, San Francisco, 1963.
19. Minkin, V. I., O. A. Osipov, and Y. A. Zhdanov: "Dipole Moments in Organic Chemistry," trans. from the Russian by B. J. Hazard, Plenum, New York, 1970.
20. Nokay, R.: *Chem. Eng.*, **66**(4): 147 (1959).
21. Passut, C. A., and R. P. Danner: *Ind. Eng. Chem. Process. Des. Dev.*, **12**: 365 (1973).
22. Pitzer, K. S.: *J. Am. Chem. Soc.*, **77**: 3427 (1955).
23. Pitzer, K. S., D. Z. Lippmann, R. F. Curl, C. M. Huggins, and D. E. Peterson: *J. Am. Chem. Soc.*, **77**: 3433 (1955).
24. Plank, R., and L. Riedel: *Ing. Arch.*, **16**: 255 (1948).
25. Rao, M. B., D. S. Viswanath, and N. R. Kuloor: *J. Indian Inst. Sci.*, **51**(3): 233 (1969).
26. Reichenberg, D.: *AIChE J.*, **19**: 854 (1973).
27. Reid, R. C., and T. K. Sherwood: "The Properties of Gases and Liquids," 2d ed., chap. 2, McGraw-Hill, New York, 1966.
28. Riedel, L.: *Chem. Ing. Tech.*, **26**: 83 (1954).
29. Smith, J. W.: "Electric Dipole Moments," chap. 3, Butterworth, London, 1955.
30. Smyth, C. P.: "Dielectric Behavior and Structure," pp. 16–50, McGraw-Hill, New York, 1955.
31. Spencer, C. F., and T. E. Daubert: *AIChE J.*, **19**: 482 (1973).
32. Stiel, L. I.: private communication, April 1972.
33. Tee, L. S., S. Gotoh, and W. E. Stewart: *Ind. Eng. Chem. Fundam.*, **5**: 356, 363 (1966).
34. Vetere, A.: private communication, December 1973; February 1976.

Chapter Three

Pressure-Volume-Temperature Relationships of Pure Gases and Liquids

3-1 Scope

Methods are presented in this chapter for estimating the volumetric behavior of *pure* gases and liquids as functions of temperature and pressure. Extension to mixtures is given in Chap. 4. Emphasis is placed on equations of state which are most applicable to computer-based property-estimation systems.

The equations of state described in this chapter are employed in Chap. 5 to determine thermodynamic departure functions and partial molal properties.

3-2 Two-Parameter Correlations

The nonideality of a gas is conveniently expressed by the *compressibility factor Z*, where

$$Z = \frac{PV}{RT} \tag{3-2.1}$$

where V = molar volume

P = absolute pressure

T = absolute temperature

R = universal gas constant

If V is in cubic centimeters per gram mole, P in atmosphere, and T in kelvins, then $R = 82.04$.[†] For an ideal gas $Z = 1.0$. For real gases, Z is normally less than unity, except at very high pressures or temperatures. Equation (3-2.1) can also be used to define Z for a liquid; in this case it is normally much less than unity.

The compressibility factor is often correlated with the reduced temperature T_r and pressure P_r as

$$Z = f(T_r, P_r) \qquad (3\text{-}2.2)$$

where $T_r = T/T_c$ and $P_r = P/P_c$. The function $f(\ )$ has been obtained from experimental P-V-T data by Nelson and Obert [83], and the final curves are shown in Figs. 3-1 to 3-3. Except as noted below, the use of

[†]For English units, V is in ft³/lb-mol, P in psia, T in °R, and $R = 10.73$. For SI, V is in m³/kg-mol, P in N/m², T in K, and $R = 8314$; in this case the units of R are N m/kg-mol K or J/kg-mol K.

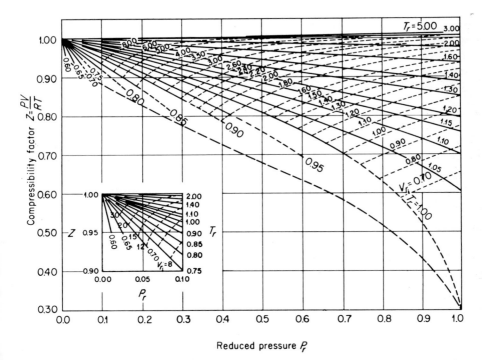

Fig. 3-1 Generalized compressibility chart. (*From Ref. 83.*)

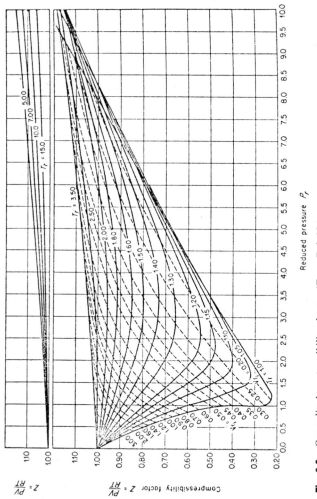

Fig. 3-2 *Generalized compressibility chart.* *(From Ref. 83.)*

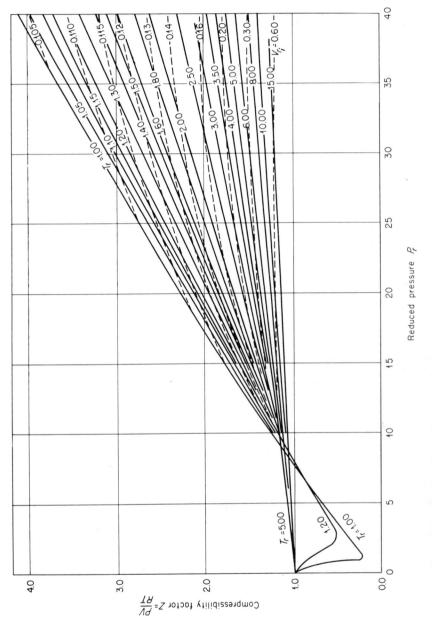

Fig. 3-3 Generalized compressibility chart. (*From Ref. 83.*)

29

these figures to obtain Z at a given T_r and P_r should lead to errors less than 4 to 6 percent except near the critical point, where Z is very sensitive to both T_r and P_r.

Figures 3-1 to 3-3 should not be used for strongly polar fluids, nor are they recommended for helium, hydrogen, or neon unless special, modified critical constants are used [77, 81, 84]. For very high pressures or very high temperatures, the reduced pressure-temperature-density charts of Breedveld and Prausnitz [16] are useful.

Many graphs similar to those in Figs. 3-1 to 3-3 have been published. All differ somewhat, as each reflects the author's choice of experimental data and how the data are smoothed. Those shown are as accurate as any two-parameter plots published, and they have the added advantage that volumes can be found directly. The reduced-volume curves are based on *ideal reduced* volumes, as the scaling factor is RT_c/P_c rather than V_c [120],

$$V_{r_i} = \frac{V}{RT_c/P_c} \qquad (3\text{-}2.3)$$

Equation (3-2.2) is an example of the law of corresponding states. This law, though not exact, suggests that reduced configurational properties† of all fluids are essentially the same if compared at equal reduced temperatures and pressures. For P-V-T properties, this law gives

$$V_r = \frac{V}{V_c} = \frac{(Z/Z_c)(T/T_c)}{P/P_c} = f_1(T_r, P_r)$$

or

$$Z = Z_c f_2(T_r, P_r) \qquad (3\text{-}2.4)$$

Except for highly polar fluids or fluids composed of large molecules, values of Z_c for most organic compounds range from 0.27 to 0.29. If it is assumed to be a constant, Eq. (3-2.4) reduces to Eq. (3-2.2). In Sec. 3-3, Z_c is introduced as a third correlating parameter (in addition to T_c and P_c) to estimate Z, but not in the form of Eq. (3-2.4).

In Eq. (3-2.2), T_c and P_c are *scaling factors* to nondimensionalize T and P. Other scaling factors have been proposed, but none has been widely accepted.‡ A convenient tabulation of T_c and P_c is presented in Appendix A, and methods to estimate them are described in Sec. 2-2.

---

†A reduced property is the ratio

$$\frac{\text{Property}}{\text{Property at the critical point}}$$

‡For example, using the Lennard-Jones constants described in Sec. 2-7, a reduced temperature can be defined as $T/(\epsilon/k)$ and a reduced pressure as $P/(\epsilon/\sigma^3)$.

3-3 Three-Parameter Correlations

Equation (3-2.2) is a two-parameter equation of state; the two parameters are T_c and P_c. That is, knowing T_c and P_c for a given fluid, it is possible to estimate the volumetric properties at various temperatures and pressures. The calculation may involve the use of Figs. 3-1 to 3-3, or one may employ an analytical function for $f(\ )$ in Eq. (3-2.2). Both methods are only approximate. Many suggestions have been offered which retain the general concept yet allow an increase in accuracy and applicability. In general, the more successful modifications have involved the inclusion of an additional *third parameter* into the function expressed by Eq. (3-2.2). Many have been suggested. Most are related to the reduced vapor pressure at some specified reduced temperature or to some volumetric property at or near the critical point, although one recent correlation employs the molar polarizability as the third parameter [95]. Two common and well-tested three-parameter correlations are described below.

Assume that there are different, but unique, functions $Z = f(T_r, P_r)$ for each class of pure components with the same Z_c. Then, for each Z_c we have a different set of Figs. 3-1 to 3-3. All fluids with the same Z_c values then follow the Z-T_r-P_r behavior shown on charts drawn for that particular Z_c. Such a structuring indeed leads to a significant increase in accuracy. This is exactly what was done in the development of the Lydersen-Greenkorn-Hougen Tables, which first appeared in 1955 [70] and were later modified [47]. Here, Z is tabulated as a function of T_r and P_r with separate tables for various values of Z_c. Edwards and Thodos [31] have also utilized Z_c in a correlation to estimate *saturated* vapor densities of nonpolar compounds.

An alternate third parameter is the Pitzer acentric factor [90, 92, 94], defined in Sec. 2-3. This factor is an indicator of the nonsphericity of a molecule's force field; e.g., a value of $\omega = 0$ denotes rare-gas spherical symmetry. Deviations from simple-fluid behavior are evident when $\omega > 0$. Within the context of the present discussion, it is assumed that all molecules with equal acentric factors have identical $Z = f(T_r, P_r)$ functions, as in Eq. (3-2.2). However, rather than prepare separate Z, T_r, P_r tables for different values of ω, it was suggested that a linear expansion could be employed

$$Z = Z^{(0)}(T_r, P_r) + \omega Z^{(1)}(T_r, P_r) \qquad (3\text{-}3.1)$$

Thus, the $Z^{(0)}$ function would apply to spherical molecules, and the $Z^{(1)}$ term is a *deviation function*.

Pitzer et al. tabulated $Z^{(0)}$ and $Z^{(1)}$ as functions of T_r and P_r [93], and Edmister has shown the same values graphically [29]. Several modifications as well as extensions to wider ranges of T_r and P_r have been

TABLE 3-1 Values of $Z^{(0)}$ [59]

T_r	P_r 0.010	0.050	0.100	0.200	0.400	0.600	0.800
0.30	0.0029	0.0145	0.0290	0.0579	0.1158	0.1737	0.2315
0.35	0.0026	0.0130	0.0261	0.0522	0.1043	0.1564	0.2084
0.40	0.0024	0.0119	0.0239	0.0477	0.0953	0.1429	0.1904
0.45	0.0022	0.0110	0.0221	0.0442	0.0882	0.1322	0.1762
0.50	0.0021	0.0103	0.0207	0.0413	0.0825	0.1236	0.1647
0.55	0.9804	0.0098	0.0195	0.0390	0.0778	0.1166	0.1553
0.60	0.9849	0.0093	0.0186	0.0371	0.0741	0.1109	0.1476
0.65	0.9881	0.9377	0.0178	0.0356	0.0710	0.1063	0.1415
0.70	0.9904	0.9504	0.8958	0.0344	0.0687	0.1027	0.1366
0.75	0.9922	0.9598	0.9165	0.0336	0.0670	0.1001	0.1330
0.80	0.9935	0.9669	0.9319	0.8539	0.0661	0.0985	0.1307
0.85	0.9946	0.9725	0.9436	0.8810	0.0661	0.0983	0.1301
0.90	0.9954	0.9768	0.9528	0.9015	0.7800	0.1006	0.1321
0.93	0.9959	0.9790	0.9573	0.9115	0.8059	0.6635	0.1359
0.95	0.9961	0.9803	0.9600	0.9174	0.8206	0.6967	0.1410
0.97	0.9963	0.9815	0.9625	0.9227	0.8338	0.7240	0.5580
0.98	0.9965	0.9821	0.9637	0.9253	0.8398	0.7360	0.5887
0.99	0.9966	0.9826	0.9648	0.9277	0.8455	0.7471	0.6138
1.00	0.9967	0.9832	0.9659	0.9300	0.8509	0.7574	0.6353
1.01	0.9968	0.9837	0.9669	0.9322	0.8561	0.7671	0.6542
1.02	0.9969	0.9842	0.9679	0.9343	0.8610	0.7761	0.6710
1.05	0.9971	0.9855	0.9707	0.9401	0.8743	0.8002	0.7130
1.10	0.9975	0.9874	0.9747	0.9485	0.8930	0.8323	0.7649
1.15	0.9978	0.9891	0.9780	0.9554	0.9081	0.8576	0.8032
1.20	0.9981	0.9904	0.9808	0.9611	0.9205	0.8779	0.8330
1.30	0.9985	0.9926	0.9852	0.9702	0.9396	0.9083	0.8764
1.40	0.9988	0.9942	0.9884	0.9768	0.9534	0.9298	0.9062
1.50	0.9991	0.9954	0.9909	0.9818	0.9636	0.9456	0.9278
1.60	0.9993	0.9964	0.9928	0.9856	0.9714	0.9575	0.9439
1.70	0.9994	0.9971	0.9943	0.9886	0.9775	0.9667	0.9563
1.80	0.9995	0.9977	0.9955	0.9910	0.9823	0.9739	0.9659
1.90	0.9996	0.9982	0.9964	0.9929	0.9861	0.9796	0.9735
2.00	0.9997	0.9986	0.9972	0.9944	0.9892	0.9842	0.9796
2.20	0.9998	0.9992	0.9983	0.9967	0.9937	0.9910	0.9886
2.40	0.9999	0.9996	0.9991	0.9983	0.9969	0.9957	0.9948
2.60	1.0000	0.9998	0.9997	0.9994	0.9991	0.9990	0.9990
2.80	1.0000	1.0000	1.0001	1.0002	1.0007	1.0013	1.0021
3.00	1.0000	1.0002	1.0004	1.0008	1.0018	1.0030	1.0043
3.50	1.0001	1.0004	1.0008	1.0017	1.0035	1.0055	1.0075
4.00	1.0001	1.0005	1.0010	1.0021	1.0043	1.0066	1.0090

1.000	1.200	1.500	2.000	3.000	5.000	7.000	10.000
0.2892	0.3470	0.4335	0.5775	0.8648	1.4366	2.0048	2.8507
0.2604	0.3123	0.3901	0.5195	0.7775	1.2902	1.7987	2.5539
0.2379	0.2853	0.3563	0.4744	0.7095	1.1758	1.6373	2.3211
0.2200	0.2638	0.3294	0.4384	0.6551	1.0841	1.5077	2.1338
0.2056	0.2465	0.3077	0.4092	0.6110	1.0094	1.4017	1.9801
0.1939	0.2323	0.2899	0.3853	0.5747	0.9475	1.3137	1.8520
0.1842	0.2207	0.2753	0.3657	0.5446	0.8959	1.2398	1.7440
0.1765	0.2113	0.2634	0.3495	0.5197	0.8526	1.1773	1.6519
0.1703	0.2038	0.2538	0.3364	0.4991	0.8161	1.1241	1.5729
0.1656	0.1981	0.2464	0.3260	0.4823	0.7854	1.0787	1.5047
0.1626	0.1942	0.2411	0.3182	0.4690	0.7598	1.0400	1.4456
0.1614	0.1924	0.2382	0.3132	0.4591	0.7388	1.0071	1.3943
0.1630	0.1935	0.2383	0.3114	0.4527	0.7220	0.9793	1.3496
0.1664	0.1963	0.2405	0.3122	0.4507	0.7138	0.9648	1.3257
0.1705	0.1998	0.2432	0.3138	0.4501	0.7092	0.9561	1.3108
0.1779	0.2055	0.2474	0.3164	0.4504	0.7052	0.9480	1.2968
0.1844	0.2097	0.2503	0.3182	0.4508	0.7035	0.9442	1.2901
0.1959	0.2154	0.2538	0.3204	0.4514	0.7018	0.9406	1.2835
0.2901	0.2237	0.2583	0.3229	0.4522	0.7004	0.9372	1.2772
0.4648	0.2370	0.2640	0.3260	0.4533	0.6991	0.9339	1.2710
0.5146	0.2629	0.2715	0.3297	0.4547	0.6980	0.9307	1.2650
0.6026	0.4437	0.3131	0.3452	0.4604	0.6956	0.9222	1.2481
0.6880	0.5984	0.4580	0.3953	0.4770	0.6950	0.9110	1.2232
0.7443	0.6803	0.5798	0.4760	0.5042	0.6987	0.9033	1.2021
0.7858	0.7363	0.6605	0.5605	0.5425	0.7069	0.8990	1.1844
0.8438	0.8111	0.7624	0.6908	0.6344	0.7358	0.8998	1.1580
0.8827	0.8595	0.8256	0.7753	0.7202	0.7761	0.9112	1.1419
0.9103	0.8933	0.8689	0.8328	0.7887	0.8200	0.9297	1.1339
0.9308	0.9180	0.9000	0.8738	0.8410	0.8617	0.9518	1.1320
0.9463	0.9367	0.9234	0.9043	0.8809	0.8984	0.9745	1.1343
0.9583	0.9511	0.9413	0.9275	0.9118	0.9297	0.9961	1.1391
0.9678	0.9624	0.9552	0.9456	0.9359	0.9557	1.0157	1.1452
0.9754	0.9715	0.9664	0.9599	0.9550	0.9772	1.0328	1.1516
0.9865	0.9847	0.9826	0.9806	0.9827	1.0094	1.0600	1.1635
0.9941	0.9936	0.9935	0.9945	1.0011	1.0313	1.0793	1.1728
0.9993	0.9998	1.0010	1.0040	1.0137	1.0463	1.0926	1.1792
1.0031	1.0042	1.0063	1.0106	1.0223	1.0565	1.1016	1.1830
1.0057	1.0074	1.0101	1.0153	1.0284	1.0635	1.1075	1.1848
1.0097	1.0120	1.0156	1.0221	1.0368	1.0723	1.1138	1.1834
1.0115	1.0140	1.0179	1.0249	1.0401	1.0747	1.1136	1.1773

TABLE 3-2 Values of $Z^{(1)}$ [59]

T_r	P_r						
	0.010	0.050	0.100	0.200	0.400	0.600	0.800
0.30	-0.0008	-0.0040	-0.0081	-0.0161	-0.0323	-0.0484	-0.0645
0.35	-0.0009	-0.0046	-0.0093	-0.0185	-0.0370	-0.0554	-0.0738
0.40	-0.0010	-0.0048	-0.0095	-0.0190	-0.0380	-0.0570	-0.0758
0.45	-0.0009	-0.0047	-0.0094	-0.0187	-0.0374	-0.0560	-0.0745
0.50	-0.0009	-0.0045	-0.0090	-0.0181	-0.0360	-0.0539	-0.0716
0.55	-0.0314	-0.0043	-0.0086	-0.0172	-0.0343	-0.0513	-0.0682
0.60	-0.0205	-0.0041	-0.0082	-0.0164	-0.0326	-0.0487	-0.0646
0.65	-0.0137	-0.0772	-0.0078	-0.0156	-0.0309	-0.0461	-0.0611
0.70	-0.0093	-0.0507	-0.1161	-0.0148	-0.0294	-0.0438	-0.0579
0.75	-0.0064	-0.0339	-0.0744	-0.0143	-0.0282	-0.0417	-0.0550
0.80	-0.0044	-0.0228	-0.0487	-0.1160	-0.0272	-0.0401	-0.0526
0.85	-0.0029	-0.0152	-0.0319	-0.0715	-0.0268	-0.0391	-0.0509
0.90	-0.0019	-0.0099	-0.0205	-0.0442	-0.1118	-0.0396	-0.0503
0.93	-0.0015	-0.0075	-0.0154	-0.0326	-0.0763	-0.1662	-0.0514
0.95	-0.0012	-0.0062	-0.0126	-0.0262	-0.0589	-0.1110	-0.0540
0.97	-0.0010	-0.0050	-0.0101	-0.0208	-0.0450	-0.0770	-0.1647
0.98	-0.0009	-0.0044	-0.0090	-0.0184	-0.0390	-0.0641	-0.1100
0.99	-0.0008	-0.0039	-0.0079	-0.0161	-0.0335	-0.0531	-0.0796
1.00	-0.0007	-0.0034	-0.0069	-0.0140	-0.0285	-0.0435	-0.0588
1.01	-0.0006	-0.0030	-0.0060	-0.0120	-0.0240	-0.0351	-0.0429
1.02	-0.0005	-0.0026	-0.0051	-0.0102	-0.0198	-0.0277	-0.0303
1.05	-0.0003	-0.0015	-0.0029	-0.0054	-0.0092	-0.0097	-0.0032
1.10	-0.0000	0.0000	0.0001	0.0007	0.0038	0.0106	0.0236
1.15	0.0002	0.0011	0.0023	0.0052	0.0127	0.0237	0.0396
1.20	0.0004	0.0019	0.0039	0.0084	0.0190	0.0326	0.0499
1.30	0.0006	0.0030	0.0061	0.0125	0.0267	0.0429	0.0612
1.40	0.0007	0.0036	0.0072	0.0147	0.0306	0.0477	0.0661
1.50	0.0008	0.0039	0.0078	0.0158	0.0323	0.0497	0.0677
1.60	0.0008	0.0040	0.0080	0.0162	0.0330	0.0501	0.0677
1.70	0.0008	0.0040	0.0081	0.0163	0.0329	0.0497	0.0667
1.80	0.0008	0.0040	0.0081	0.0162	0.0325	0.0488	0.0652
1.90	0.0008	0.0040	0.0079	0.0159	0.0318	0.0477	0.0635
2.00	0.0008	0.0039	0.0078	0.0155	0.0310	0.0464	0.0617
2.20	0.0007	0.0037	0.0074	0.0147	0.0293	0.0437	0.0579
2.40	0.0007	0.0035	0.0070	0.0139	0.0276	0.0411	0.0544
2.60	0.0007	0.0033	0.0066	0.0131	0.0260	0.0387	0.0512
2.80	0.0006	0.0031	0.0062	0.0124	0.0245	0.0365	0.0483
3.00	0.0006	0.0029	0.0059	0.0117	0.0232	0.0345	0.0456
3.50	0.0005	0.0026	0.0052	0.0103	0.0204	0.0303	0.0401
4.00	0.0005	0.0023	0.0046	0.0091	0.0182	0.0270	0.0357

			P_r				
1.000	1.200	1.500	2.000	3.000	5.000	7.000	10.000
-0.0806	-0.0966	-0.1207	-0.1608	-0.2407	-0.3996	-0.5572	-0.7915
-0.0921	-0.1105	-0.1379	-0.1834	-0.2738	-0.4523	-0.6279	-0.8863
-0.0946	-0.1134	-0.1414	-0.1879	-0.2799	-0.4603	-0.6365	-0.8936
-0.0929	-0.1113	-0.1387	-0.1840	-0.2734	-0.4475	-0.6162	-0.8606
-0.0893	-0.1069	-0.1330	-0.1762	-0.2611	-0.4253	-0.5831	-0.8099
-0.0849	-0.1015	-0.1263	-0.1669	-0.2465	-0.3991	-0.5446	-0.7521
-0.0803	-0.0960	-0.1192	-0.1572	-0.2312	-0.3718	-0.5047	-0.6928
-0.0759	-0.0906	-0.1122	-0.1476	-0.2160	-0.3447	-0.4653	-0.6346
-0.0718	-0.0855	-0.1057	-0.1385	-0.2013	-0.3184	-0.4270	-0.5785
-0.0681	-0.0808	-0.0996	-0.1298	-0.1872	-0.2929	-0.3901	-0.5250
-0.0648	-0.0767	-0.0940	-0.1217	-0.1736	-0.2682	-0.3545	-0.4740
-0.0622	-0.0731	-0.0888	-0.1138	-0.1602	-0.2439	-0.3201	-0.4254
-0.0604	-0.0701	-0.0840	-0.1059	-0.1463	-0.2195	-0.2862	-0.3788
-0.0602	-0.0687	-0.0810	-0.1007	-0.1374	-0.2045	-0.2661	-0.3516
-0.0607	-0.0678	-0.0788	-0.0967	-0.1310	-0.1943	-0.2526	-0.3339
-0.0623	-0.0669	-0.0759	-0.0921	-0.1240	-0.1837	-0.2391	-0.3163
-0.0641	-0.0661	-0.0740	-0.0893	-0.1202	-0.1783	-0.2322	-0.3075
-0.0680	-0.0646	-0.0715	-0.0861	-0.1162	-0.1728	-0.2254	-0.2989
-0.0879	-0.0609	-0.0678	-0.0824	-0.1118	-0.1672	-0.2185	-0.2902
-0.0223	-0.0473	-0.0621	-0.0778	-0.1072	-0.1615	-0.2116	-0.2816
-0.0062	0.0227	-0.0524	-0.0722	-0.1021	-0.1556	-0.2047	-0.2731
0.0220	0.1059	0.0451	-0.0432	-0.0838	-0.1370	-0.1835	-0.2476
0.0476	0.0897	0.1630	0.0698	-0.0373	-0.1021	-0.1469	-0.2056
0.0625	0.0943	0.1548	0.1667	0.0332	-0.0611	-0.1084	-0.1642
0.0719	0.0991	0.1477	0.1990	0.1095	-0.0141	-0.0678	-0.1231
0.0819	0.1048	0.1420	0.1991	0.2079	0.0875	0.0176	-0.0423
0.0857	0.1063	0.1383	0.1894	0.2397	0.1737	0.1008	0.0350
0.0864	0.1055	0.1345	0.1806	0.2433	0.2309	0.1717	0.1058
0.0855	0.1035	0.1303	0.1729	0.2381	0.2631	0.2255	0.1673
0.0838	0.1008	0.1259	0.1658	0.2305	0.2788	0.2628	0.2179
0.0816	0.0978	0.1216	0.1593	0.2224	0.2846	0.2871	0.2576
0.0792	0.0947	0.1173	0.1532	0.2144	0.2848	0.3017	0.2876
0.0767	0.0916	0.1133	0.1476	0.2069	0.2819	0.3097	0.3096
0.0719	0.0857	0.1057	0.1374	0.1932	0.2720	0.3135	0.3355
0.0675	0.0803	0.0989	0.1285	0.1812	0.2602	0.3089	0.3459
0.0634	0.0754	0.0929	0.1207	0.1706	0.2484	0.3009	0.3475
0.0598	0.0711	0.0876	0.1138	0.1613	0.2372	0.2915	0.3443
0.0565	0.0672	0.0828	0.1076	0.1529	0.2268	0.2817	0.3385
0.0497	0.0591	0.0728	0.0949	0.1356	0.2042	0.2584	0.3194
0.0443	0.0527	0.0651	0.0849	0.1219	0.1857	0.2378	0.2994

published [68, 107].   Tables 3-1 and 3-2 list those prepared by Lee and Kesler [59].   The method of calculation is described later in Sec. 3-9.   With Tables 3-1 and 3-2, Z can be determined for both gases and liquids.[†]   The $Z^{(0)}$ table agrees well with that presented originally by Pitzer et al. over the range of T_r and P_r common to both.   The deviation-function table of Lee and Kesler (Table 3-2) differs somewhat from that of Pitzer and Curl, but extensive testing [59, 124] indicates the new table is the more accurate.

Tables 3-1 and 3-2 were not intended to be applicable for strongly polar fluids, though they are often so used with surprising accuracy except at low temperatures near the saturated-vapor region.   Though none has been widely adopted, special techniques have been suggested to modify Eq. (3-3.1) to polar materials [33, 40, 65, 86, 119, 126].

Considerable emphasis has been placed on the Pitzer-Curl generalized relation.   It has proved to be accurate and general when applied to pure gases.   Only the acentric factor and critical temperature and pressure need be known.   It is probably the most successful and useful result of corresponding-states theory [60, 116, 117].

Example 3-1   Estimate the specific volume of dichlorodifluoromethane vapor at 20.4 atm and 366.5 K.

solution   From Appendix A, $T_c = 385.0$ K, $P_c = 40.7$ atm, and $\omega = 0.176$.

$$T_r = \frac{366.5}{385.0} = 0.952 \qquad P_r = \frac{20.4}{40.7} = 0.501$$

From Fig. 3-1, $Z = 0.77$ and

$$V = \frac{ZRT}{P} = \frac{(0.77)(82.04)(366.5)}{20.4}$$

$$= 1134 \text{ cm}^3/\text{g-mol}$$

The value reported in the literature is 1109 cm³/g-mol [4].

If the Pitzer-Curl method were to be used, from Tables 3-1 and 3-2, $Z^{(0)} = 0.759$ and $Z^{(1)} = -0.085$.   From Eq. (3-3.1),

$$Z = 0.759 + 0.176(-0.085) = 0.744$$

$$V = \frac{ZRT}{P} = 1097 \text{ cm}^3/\text{g-mole}$$

3-4 Analytical Equations of State

An analytical equation of state is an algebraic relation between pressure, temperature, and molar volume.   As indicated in later sections of this chapter, many different forms have been suggested.   All, however,

†For mixtures, see Eqs. (4-6.3) to (4-6.7).

should satisfy the thermodynamic stability criteria at the critical point,

$$\left(\frac{dP}{dV}\right)_{T_c} = 0 \tag{3-4.1}$$

$$\left(\frac{d^2P}{dV^2}\right)_{T_c} = 0 \tag{3-4.2}$$

In addition, dependable equations of state should predict certain other characteristics that are usually found in all fluids. These have been pointed out by Martin [72–76] and Abbott [1]. Finally, each should reduce to the ideal-gas law as the pressure approaches zero.

The most famous analytical equation of state is that suggested by van der Waals,

$$\left(P + \frac{a}{V^2}\right)(V - b) = RT \tag{3-4.3}$$

where a and b are specific constants. With Eqs. (3-4.1) and (3-4.2),

$$a = \frac{27}{64}\frac{R^2 T_c^2}{P_c} \tag{3-4.4}$$

$$b = \frac{RT_c}{8P_c} \tag{3-4.5}$$

Though we discuss several more accurate equations of state in the succeeding sections of this chapter, the van der Waals relation is still useful to provide an approximate, yet simple, analytical representation of the behavior of a real gas [97]. Vera and Prausnitz [131] have discussed the generalized van der Waals theory and have shown that, with certain assumptions, one can derive several of the more contemporary analytical equations of state, including the Redlich-Kwong form, which is discussed next.

3-5 Redlich-Kwong Equation of State

The most successful two-parameter equation of state was formulated in 1949 by Redlich and Kwong [101]. As originally proposed,

$$P = \frac{RT}{V - b} - \frac{a}{T^{0.5}V(V + b)} \tag{3-5.1}$$

No extensive tabulation of a and b exists for pure compounds, but if Eq. (3-5.1) is used with Eqs. (3-4.1) and (3-4.2), it is readily shown that

$$a = \frac{\Omega_a R^2 T_c^{2.5}}{P_c} \tag{3-5.2}$$

$$b = \frac{\Omega_b RT_c}{P_c} \tag{3-5.3}$$

where Ω_a and Ω_b are pure numbers, i.e.,

$$\Omega_a = [(9)(2^{1/3} - 1)]^{-1} = 0.4274802327 \cdots \tag{3-5.4}$$

$$\Omega_b = \frac{2^{1/3} - 1}{3} = 0.086640350 \cdots \tag{3-5.5}$$

Thus, with values of the critical temperature and pressure for any material, a and b are easily determined.

The Redlich-Kwong equation of state is a cubic equation in volume or in Z. It can be expressed in this fashion as

$$Z^3 - Z^2 + (A* - B*^2 - B*)Z - A*B* = 0 \tag{3-5.6}$$

where

$$A* = \frac{\Omega_a P_r}{T_r^{2.5}} = \frac{aP}{R^2 T^{2.5}} \tag{3-5.7}$$

$$B* = \frac{\Omega_b P_r}{T_r} = \frac{bP}{RT} \tag{3-5.8}$$

Equations (3-5.1) and (3-5.6) are identical in form. They are generally known as the *original* Redlich-Kwong equation to distinguish them from many modified forms proposed after 1949. Before noting these modifications, it is interesting to mention that in a recent examination of a large number of two-constant equations of state, the general form of Eq. (3-5.1) was chosen by Bjerre and Bak [14] as the most accurate.

The success of the original Redlich-Kwong form as a reliable two-constant equation has encouraged others to improve its accuracy and range [46]. Only a few are noted here to indicate the approaches employed.

The simplest modification is to employ Eqs. (3-5.6) to (3-5.8) but make Ω_a and Ω_b functions of the acentric factor [17, 21, 22, 32]. Another involves the incorporation of a deviation function, i.e.,

$$Z = Z_{RK} + \Delta Z \tag{3-5.9}$$

where Z_{RK} is the value of Z as found with the original Redlich-Kwong equation and ΔZ is a correction term. ΔZ can be expressed in many forms [2, 37, 99, 100, 102]; most are complex functions of T_r, P_r and, in some instances, ω. As an illustration, consider the Gray, Rent, Zudkevitch expression for ΔZ [37].

$$\Delta Z = D_1 T_r^2 P_r^2 \exp\{-[7000(1 - T_r)^2 + 770(1.02 - P_r)^2]\}$$
$$+ \omega(-0.464419 + 0.424568 T_r^2) \frac{P_r}{T_r^4 + P_r^4}$$
$$+ \omega(D_3 + D_4 T_r) \frac{P_r^2}{(1 + T_r)^4 + P_r^4}$$
$$+ [D_2 + \omega(D_5 + D_6 T_r)] \frac{P_r^3}{(1 + T_r)^4 + P_r^4} \tag{3-5.10}$$

where

$$D_1 = -0.04666626 \qquad D_2 = -0.11386032 \qquad D_3 = -41.76451266$$
$$D_4 = 40.47298767 \qquad D_5 = 12.55135462 \qquad D_6 = -12.5583112$$

The Gray et al. modification was developed for gases to cover a temperature range up to $T_r = 1.1$ when $P_r \leq 2$. Compressibility factors calculated from Eqs. (3-5.9) and (3-5.10) agree well with the original Pitzer-Curl tables. When used for *hydrocarbons*, Sokolov et al. [109] found another modification [2, 99] slightly more accurate.

The most successful modifications of the Redlich-Kwong equation result when the parameters a and b are assumed temperature-dependent [8, 17, 108, 133, 136, 137]. Chaudron et al. [17], for example, have expressed both a and b as polynomials in inverse reduced temperature and by regressing experimental data for pure components have obtained constants for 25 substances. It is claimed that when these constants are used to estimate volumetric properties in the liquid, vapor, or supercritical state, the average deviation is less than 1.5 percent. Another popular modification may be expressed as

$$Z = \frac{PV}{RT} = \frac{V}{V-b} - \frac{\Omega_a}{\Omega_b} \frac{b}{V+b} F \tag{3-5.11}$$

when Ω_a and Ω_b are given by Eqs. (3-5.4) and (3-5.5), b is obtained from Eq. (3-5.3), and F varies, depending on the suggested modification as shown below:

Original Redlich-Kwong: $\qquad F = T_r^{-1.5}$ $\qquad\qquad$ (3-5.12)

TABLE 3-3   Comparison of F Functions

	Reduced temperature				
	0.6	0.8	1.0	1.2	1.4
$\omega = 0$					
Original Redlich-Kwong	2.15	1.40	1.00	0.761	0.604
Wilson	2.05	1.39	1.00	0.738	0.551
Barnés-King	2.04	1.36	1.00	0.785	0.643
Soave	2.05	1.38	1.00	0.759	0.594
$\omega = 0.2$					
Original Redlich-Kwong	2.15	1.40	1.00	0.761	0.604
Wilson	2.26	1.47	1.00	0.684	0.459
Barnés-King	2.32	1.45	1.00	0.727	0.547
Soave	2.31	1.47	1.00	0.713	0.523

Wilson [136]:
$$F = 1 + (1.57 + 1.62\omega)(T_r^{-1} - 1) \qquad (3\text{-}5.13)$$

Barnés-King [8]:
$$F = 1 + (0.9 + 1.21\omega)(T_r^{-1.5} - 1) \qquad (3\text{-}5.14)$$

Soave [108]:
$$F = \frac{1}{T_r}[1 + (0.480 + 1.574\omega - 0.176\omega^2)(1 - T_r^{0.5})]^2 \qquad (3\text{-}5.15)$$

Except at reduced temperatures above about 1.0, the various F functions agree surprisingly well, as shown in Table 3-3.

West and Erbar [135] have evaluated the Soave relation for light hydrocarbon systems and report it to be very accurate when used to predict vapor-liquid equilibria and enthalpy departures. Barnés and King [8] have applied their form to both hydrocarbon and nonhydrocarbon mixtures, while Wilson [136] has shown that his form well predicts enthalpy departures of both polar (including ammonia) and nonpolar materials. These relations are discussed further in Chap. 5.

3-6   Barner-Adler Equation of State

Barner and Adler [6] modified the Joffe equation of state [49, 50] to obtain a P-V-T relation that is most accurate in the saturated or slightly superheated gas region. The equation and constants are shown in Table 3-4, and, as an example, the excellent fit between calculated and experimental values of Z for n-heptane is shown in Fig. 3-4. The range of the equation is $V_r > 0.6$, $T_r < 1.5$.

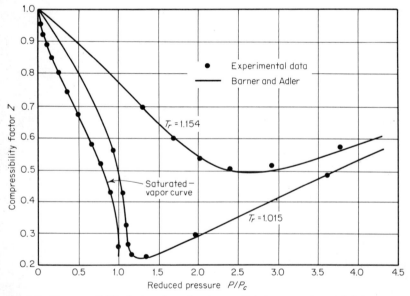

Fig. 3-4   Compressibility factors for n-heptane. Points = experimental data, —— = Ref. 6. (*From Ref. 6.*)

TABLE 3-4   Barner-Adler Modification of the Joffe Equation of State

$$P = \frac{RT}{V-b} - \frac{af_a}{V(V-b)} + \frac{cf_c}{V(V-b)^2} - \frac{df_d}{V(V-b)^3} + \frac{ef_e}{V(V-b)^4} \quad (3\text{-}6.1)$$

$$h = 1 - [(\tfrac{8}{5})(0.3361 + 0.0713\omega)]^{1/2} \quad (3\text{-}6.2)$$

$$a = \frac{R^2 T_c^2}{4P_c}(5h - 1) + (\tfrac{5}{2})(1 - h)^2 \quad (3\text{-}6.3)$$

$$b = \frac{RT_c}{4P_c}(5h - 1) \quad (3\text{-}6.4)$$

$$c = \frac{5R^3 T_c^3}{32P_c^2}(1 - h)^3 \quad (3\text{-}6.5)$$

$$d = \frac{5R^4 T_c^4}{256P_c^3}(1 - h)^4 \quad (3\text{-}6.6)$$

$$e = \frac{R^5 T_c^5}{1024P_c^4}(1 - h)^5 \quad (3\text{-}6.7)$$

$$f_a = 1 - A\left(1 - \frac{1}{T_r}\right) \quad (3\text{-}6.8)$$

$$f_c = 1 - C\left(1 - \frac{1}{T_r}\right) \quad (3\text{-}6.9)$$

$$f_d = D_1 + \frac{D_2}{T_r} - \frac{D_3}{T_r^2} \quad (3\text{-}6.10)$$

$$f_e = E_1 + \frac{E_2}{T_r^2} - \frac{E_3}{T_r^4} \quad (3\text{-}6.11)$$

$$A = \frac{0.904 + 3.716\omega}{(5h - 1) + (\tfrac{5}{2})(1 - h)^2} \quad (3\text{-}6.12)$$

$$C = \frac{(32)(0.043 + 0.17\omega)}{5(1 - h)^3} \quad (3\text{-}6.13)$$

$$D_1 = -(0.30 + 6.28\omega^{2/3}) \quad (3\text{-}6.14)$$

$$D_2 = 1.89 + 13.59\omega^{2/3} \quad (3\text{-}6.15)$$

$$D_3 = 0.59 + 7.31\omega^{2/3} \quad (3\text{-}6.16)$$

$$E_1 = 0.23 - 2.58\omega^{2/3} \quad (3\text{-}6.17)$$

$$E_2 = 1.25 + 8.99\omega^{2/3} \quad (3\text{-}6.18)$$

$$E_3 = 0.48 + 6.41\omega^{2/3} \quad (3\text{-}6.19)$$

The constants shown in Table 3-4 are related to composition in Sec. 4-4. Applications to the thermodynamic properties of gases are covered in Chap. 5. As written (see Table 3-4), the equation of state applies to 1 mol of gas. To obtain the extensive form, one simply replaces V by $\underline{V}/N$, where $\underline{V}$ is the total volume and N the total number of moles.

3-7 Sugie-Lu Equation of State

In 1971, Sugie and Lu [122] proposed an equation of state that is similar to that of Barner-Adler and is reliable over the same range, i.e., in the

TABLE 3-5 Sugie and Lu Equation of State

$$P = \frac{RT}{V - b + c} - \frac{aT^{-0.5}}{(V + c)(V + b + c)} + \sum_{j=1}^{10} \frac{d_j T + e_j T^{-0.5}}{V^{j+1}}$$

where

$$a = a^* \frac{R^2 T_c^{2.5}}{P_c} \qquad b = b^* \frac{RT_c}{P_c} \qquad c = c^* \frac{RT_c}{P_c} \qquad d_j = d_j^* \frac{R^{j+1} T_c^j}{P_c^j} \qquad e_j = e_j^* \frac{R^{j+1} T_c^{j+1.5}}{P_c^j}$$

$a^* = 0.42748$	$d_4^* = 2.8115 \times 10^{-3} - 9.8715 \times 10^{-3} \omega$
$b^* = 0.08664$	$d_5^* = -1.1178 \times 10^{-3} + 6.6578 \times 10^{-4} \omega$
$c^* = \dfrac{1 - 3Z_c}{3}$	$d_6^* = 2.3658 \times 10^{-5} + 4.6647 \times 10^{-5} \omega$
	$d_7^* = 1.6314 \times 10^{-5} - 2.6384 \times 10^{-5} \omega$
$d_1^* = 9.7806 \times 10^{-2} + 7.0750 \times 10^{-1} \omega$	$d_8^* = -2.6225 \times 10^{-7} + 4.4515 \times 10^{-7} \omega$
$d_2^* = -6.5927 \times 10^{-2} - 3.0890 \times 10^{-1} \omega$	$d_9^* = -1.1441 \times 10^{-7} + 1.8492 \times 10^{-8} \omega$
$d_3^* = 1.4085 \times 10^{-2} + 1.0353 \times 10^{-1} \omega$	$d_{10}^* = 2.6681 \times 10^{-9} + 1.3076 \times 10^{-8} \omega$

$$e_1^* = -\sum_{j=1}^{10} \frac{(j-2)(j-3)}{2} \frac{d_j^*}{Z_c^{j-1}} - \sum_{j=4}^{10} \frac{(j-2)(j-3)}{2} \frac{e_j^*}{Z_c^{j-1}}$$

$$e_2^* = \sum_{j=1}^{10} (j-1)(j-3) \frac{d_j^*}{Z_c^{j-2}} + \sum_{j=4}^{10} (j-1)(j-3) \frac{e_j^*}{Z_c^{j-2}}$$

$$e_3^* = -\sum_{j=1}^{10} \frac{(j-1)(j-2)}{2} \frac{d_j^*}{Z_c^{j-3}} - \sum_{j=4}^{10} \frac{(j-1)(j-2)}{2} \frac{e_j^*}{Z_c^{j-3}}$$

$$e_4^* = 2.1163 \times 10^{-3} + 5.8262 \times 10^{-3} \omega$$

$$e_5^* = 4.3405 \times 10^{-5} - 4.6678 \times 10^{-4} \omega$$

$$e_6^* = -1.9517 \times 10^{-5} + 8.8237 \times 10^{-5} \omega$$

$$e_7^* = -9.1644 \times 10^{-7} + 4.7942 \times 10^{-6} \omega$$

$$e_8^* = 2.1117 \times 10^{-8} - 4.7493 \times 10^{-8} \omega$$

$$e_9^* = -1.4070 \times 10^{-8} - 1.3246 \times 10^{-8} \omega$$

$$e_{10}^* = 3.1756 \times 10^{-9} - 8.3832 \times 10^{-9} \omega$$

saturated or slightly superheated gas region. The equation and constants are shown in Table 3-5. The accuracy of this relation is about the same as that found for the Barner-Adler method discussed in Sec. 3-6. It also works surprisingly well in many cases for liquid-saturated volumes.

The constants in Table 3-5 are related to composition in Sec. 4-5, and the equation is applied to gas-phase thermodynamic properties in Chap. 5.

3-8 Benedict-Webb-Rubin Equation of State

To many, the very name *equation of state* brings to mind the Benedict-Webb-Rubin equation [10]. It has been extremely valuable in correlating both liquid and vapor thermodynamic and volumetric data for light hydrocarbons and their mixtures. Expressed in terms of the molar density ρ, it is

$$P = RT\rho + \left(B_0 RT - A_0 - \frac{C_0}{T^2}\right)\rho^2 + (bRT - a)\rho^3$$

$$+ a\alpha\rho^6 + \frac{c\rho^3}{T^2}(1 + \gamma\rho^2)\exp(-\gamma\rho^2) \qquad (3\text{-}8.1)$$

The eight constants are normally determined from pure-component volumetric data though new techniques using multiproperty data have been suggested [25, 26, 63, 105].

Several tabulations of the Benedict-Webb-Rubin constants are now available [13, 24, 62, 88]. In Table 3-6, we show the values recommended by Cooper and Goldfrank [24] for 33 materials, together with the range of applicability of the constants. A few of the values of these tabulated Benedict-Webb-Rubin constants were not the "best-fit" values to P-V-T data but were adjusted slightly to improve a generalized correlation for constants in homologous series. They were employed by Johnson and Colver [52] in their computer program to estimate densities of both gases and liquids. Note that the units for each constant are those applicable to the cgs system, i.e., liters, kelvins, gram moles, and atmospheres. Also, as noted in Table 3-6, the temperature range is almost always such that $T_r > 0.6$.

Another reliable tabulation of Benedict-Webb-Rubin constants, presented by Orye [88] is shown in Table 3-7. The constant C_0 is now temperature-dependent. The numerical values differ greatly from those based on the cgs system in Table 3-6 since English units (pounds per square inch absolute, cubic feet, pound moles, and degrees Rankine) are employed in Table 3-7. In most cases, these constants agree with those originally selected by Benedict et al. [10].

One should never mix constants from different tabulations; e.g., for a specific fluid, values of some constants should not be obtained from

TABLE 3-6 Benedict-Webb-Rubin Coefficients [24]

Compound	Vapor Density, g-mol/l	Vapor Temp., °C	Liquid Pressure, atm	Liquid Temp., °C	a, (l/g-mol)³ atm	A_0, (l/g-mol)² atm	b, (l/g-mol)²	B_0, l/g-mol	c, (l/g-mol)³ K² atm	C_0, (l/g-mol)² K² atm	α, (l/g-mol)³	γ, (l/g-mol)²
Methane	2.0	−70	4.5	−140	4.94000E − 2	1.85500	3.38004E − 3	4.26000E − 2	2.54500E + 3	2.225700E + 4	1.24359E − 4	6.0000E − 3
	18.0	200	41.8	−85								
	0.75	0	b	b	4.35200E − 2	1.79894	2.52033E − 3	4.54625E − 2	3.58780E + 3	3.18382E + 4	3.30000E − 4	1.05000E − 2
	12.5	350										
Ethane	0.5	25	5.5	−50	3.45160E − 1	4.15556	1.11220E − 2	6.27724E − 2	3.27670E + 4	1.79592E + 5	2.43389E − 4	1.18000E − 2
	10.0	275	41.4	25								
Propane	1.0	96.8	2.0	−25	9.4700E − 1	6.87225	2.25000E − 2	9.73130E − 2	1.29000E + 5	5.08256E + 5	6.07175E − 4	2.20000E − 2
	9.0	275	28.0	75								
n-Butane	0.5	150	1.2	4.0	1.88231	1.00847E + 1	3.99983E − 2	1.24361E − 1	3.16400E + 5	9.92830E + 5	1.10132E − 3	3.40000E − 2
	7.0	300	22.5	121								
Isobutane	0.5	104.4	1.0	−12	1.93763	1.023264E + 1	4.24352E − 2	1.37544E − 1	2.86010E + 5	8.49943E + 5	1.07408E − 3	3.40000E − 2
	7.0	237.8	27.2	119								
n-Pentane	0.46	140	2.1	60	4.07480	1.21794E + 1	6.68120E − 2	1.56751E − 1	8.24170E + 5	2.12121E + 6	1.81000E − 3	4.75000E − 2
	4.8	280	25.5	180								
Isopentane	0.5	130	0.34	0	3.75620	1.27959E + 1	6.68120E − 2	1.60053E − 1	6.95000E + 5	1.74632E + 6	1.70000E − 3	4.63000E − 2
	5.0	280	21.5	160								
2,2-Dimethyl-propane	1.0	160	b	b	3.4905	1.29635E + 1	6.68120E − 2	1.70580E − 1	5.46E + 5	1.273E + 6	2.0E − 3	5.0E − 2
	6.0	275										
n-Hexane	2.5	225	1.0	70	7.11671	1.44373E + 1	1.09131E − 1	1.77813E − 1	1.51276E + 6	3.31935E + 6	2.81086E − 3	6.66849E − 2
	5.0	275	23.8	220								
Isohexane	1.5	250	d	d	7.4986	1.4930E + 1	1.215E − 1	1.729E − 1	1.400E + 6	2.8500E + 6	2.35E − 3	6.20E − 2
	5.5	275										

a Range of application

3-Methyl-pentane	3.2 / 6.0	250 / 275	d	d	5.9716	1.2203E + 1	1.1224E − 1	8.1505E − 2	9.5556E + 5	2.2125E + 6	2.25E − 3	6.2890E − 2
2,2-Dimethyl-butane	1.8 / 5.0	925 / 275	d	d	1.0108E + 1	1.1842E + 1	1.400E − 1	1.9214E − 1	1.7483E + 6	3.3595E + 6	2.189E − 3	5.6500E − 2
2,3-Dimethyl-butane	1.5 / 5.0	250 / 275	d	d	4.6956	1.6430E + 1	7.900E − 2	1.9000E − 1	1.1346E + 6	2.5534E + 6	3.5948E − 3	7.5E − 2
n-Heptane	1.0 / 4.0	275 / 350	1.0 / 13.6	100 / 221	1.036475E + 1	1.75206E + 1	1.51954E − 1	1.9900E − 1	2.47000E + 6	4.74574E + 6	4.35611E − 3	9.00000E − 2
3-Methyl-hexane	1.5 / 5.0	250 / 275	d	d	7.5854	1.4310E + 1	1.4321E − 1	9.1423E − 2	1.3252E + 6	3.1564E + 6	2.8155E − 3	7.446E − 2
2,2-Dimethyl-pentane	1.8 / 5.0	225 / 275	d / d	d / d	1.1786E + 1	1.2423E + 1	1.7721E − 1	2.0246E − 1	2.2586E + 6	5.1237E + 6	2.764E − 3	6.799E − 2
n-Nonane	c	c	0 / 6.5	204 / 238	5.51599E + 1	−1.31560E − 2	8.56466E − 1	−2.32095	7.81821E + 5	−3.20417E + 6	2.32899E − 3	0.0
n-Decane	c	c	0 / 5.3	38 / 238	1.25122E + 2	−3.58180E − 2	1.96701	−6.23189	4.42954E + 3	1.31900E + 5	2.14459E − 3	0.0
Ethylene	1.0 / 12.8	0 / 198	2.1 / 31.0	−90 / −10	2.59000E − 1	3.33958	8.6000E − 3	5.56833E − 2	2.1120E + 4	1.31140E + 5	1.78000E − 4	9.23000E − 3
Propylene	0.5 / 8.0	25 / 300	2.1 / 24.9	30 / 60	7.74056E − 1	6.11220	1.87059E − 2	8.50647E − 2	1.02611E + 5	4.39182E + 5	4.55696E − 4	1.89900E − 2
Isobutylene	1.0 / 7.0	150 / 275	1.0 / 27.2	−7 / 123	1.69270	8.95325	3.48156E − 2	1.16025E − 1	2.74920E + 5	9.27280E + 5	9.10889E − 4	2.95945E − 2
Propyne	0.3 / 11.4	50 / 200	c	c	6.970948E − 1	5.1079342	1.482999E − 2	6.94603E − 2	1.0984375E + 5	6.4062824E + 5	2.7363248E − 4	1.245167E − 2
Benzene	0.6 / 8.1	240 / 355	c	c	5.570	6.509772	7.663E − 2	5.03055E − 2	1.176418E + 6	3.42997E + 6	7.001E − 4	2.930E − 2
Ammonia	0 / 2.6	27 / 307	d	d	1.0354029E − 1	3.7892819	7.1958516E − 4	5.1646121E − 2	1.575329E + 2	1.7857089E + 5	4.6541779E − 5	1.9805156E − 2

45

TABLE 3-6 Benedict-Webb-Rubin Coefficients [24] (Continued)

Compound	Range of application[a] Vapor Den-sity, g-mol/l	Vapor Temp., °C	Liquid Pres-sure, atm	Liquid Temp., °C	a, (l/g-mol)³ atm	A_0, (l/g-mol)² atm	b, (l/g-mol)²	B_0, l/g-mol	c, (l/g-mol)³ K²-atm	C_0, (l/g-mol)² K² atm	α, (l/g-mol)³	γ, (l/g-mol)²
Ammonia (Continued)	0.02 / 29.8	c	20 / 800	37 / 127	1.6797763E − 5	2.0259528	2.8513822E − 3	3.3649627E − 3	1.7292401E + 2	6.0409764E + 4	1.6797770E − 5	6.6251015E − 4
Argon	— / —	−111 / 327	d	d	2.88358E − 2	8.23417E − 1	2.15289E − 3	2.2282597E − 2	7.982437E + 2	1.314125E + 4	3.558895E − 5	2.3382711E − 3
Carbon dioxide	0 / 14.5	up to 138	0 / 66	−23 / 31	1.36814E − 1	2.73742	4.1239E − 3	4.99101E − 2	1.49180E + 4	1.38567E + 5	8.47E − 5	5.394E − 3
Carbon monoxide	0.15 / 9.0	−140.2 / −25	3.13 / 34.53	−180 / −25	3.665E − 2	1.34122	2.63158E − 3	5.45425E − 2	1.040E + 3	8.56209E − 3	1.35E − 4	6.0E − 3
Helium	2.0 / 50.0	−270 / −253	0 / 100	−270 / −267	−5.7339E − 4	4.0969E − 2	−1.9727E − 7	2.3661E − 2	−5.521E − 3	1.6227E − 1	−7.2673E − 6	7.7942E − 4
Nitrogen	0.02 / 24.34	−170 / 93	d	d	2.5102E − 2	1.053642	2.3277E − 3	4.07426E − 2	7.2841E + 2	8.05900E + 3	1.272E − 4	5.300E − 3
Nitric oxide	0.04 / 1.6	5 / 105	c	c	−3.50821484E − 1	2.19573852	−7.53154391E − 3	6.04508814E − 2	−1.152372289E + 4	1.79557089E + 4	1.563969033E − 5	2.0E − 3
Nitrous oxide	0.0 / 25.0	−30 / 150	d	d	1.6774177E − 1	2.5441140	5.1001644E − 3	3.9458452E − 2	1.5946E + 4	1.4793968E + 5	6.5559433E − 5	4.8129414E − 3
Oxygen	0.0 / 2.4	27 / 727	c	d	1.62689940E − 1	9.50851963E − 1	3.58834736E − 3	3.5328505E − 8	1.28273741E + 4	3.26435918E + 4	−3.927058894	3.01E − 2
Sulfur dioxide	0.0 / 22.8	10 / 250	d	d	8.446E − 1	2.12044	1.4653E − 2	2.6182E − 2	1.1335E + 5	7.9384E + 5	7.1955E − 5	5.9236E − 3

[a] To conserve space, ranges are shown with the lower value above and the higher value below; i.e., the first vapor-density values for methane are to be read "2.0 to 18.0." [b] Not studied. [c] Not recommended for gas phase. [d] Not recommended for liquid region. [e] Not given.

Table 3-6 and others from Table 3-7 even though both are converted to the same units.   All constants for a given fluid should always be obtained from the same reference.

A number of authors have recommended that below the normal boiling point, C_0 be a function of temperature to enable the equation to reproduce vapor pressures more accurately [5, 7, 10, 54, 64, 110, 113, 115, 143].   The most quantitative analysis of this problem was given by Orye [88] and is to be used in conjunction with Table 3-7 (*not* Table 3-6).   The temperature dependence of C_0 is expressed as

$$C_0^{1/2}(T) = C_0^{1/2}(T_0) - \Delta C_0^{1/2}(T) \tag{3-8.2}$$

In Eq. (3-8.2), $C_0(T_0)$ is the value of C_0 in Table 3-7. ΔC_0 is given in Eq. (3-8.3) for all compounds *except* n-C_8, n-C_9, n-C_{10}, and H_2S.

$$\Delta C_0^{1/2}(T) = Q_1 \Theta^2 + Q_2 \Theta^3 + Q_3 \Theta^4 + Q_4 \Theta^5 \tag{3-8.3}$$

where $\Theta \equiv (T - T_0)/T_0$ and T, and T_0 are in degrees Rankine.   For n-C_8, n-C_9, n-C_{10}, and H_2S,

$$\Delta C_0^{1/2}(T) = Q_1 + Q_2 T + Q_3 T^2 + Q_4 T^3 \tag{3-8.4}$$

The Q and T_0 functions are given in Table 3-7 for the compounds studied by Orye.   Also shown in this table are values of T_1 (°R).   This is the lowest temperature where Eq. (3-8.3) or (3-8.4) applies; should one wish to extrapolate C_0 to even lower temperatures, the column DCDT may be used.   Here,

$$\text{DCDT} \equiv \frac{d(\Delta C_0^{1/2})}{dT} \qquad \text{at } T_1 \tag{3-8.5}$$

For those materials wherein DCDT is zero, the temperature T_1 is either zero or so low that no extrapolation to lower temperatures is warranted.   If the derivative of C_0 with respect to temperature is desired at any temperature between T_1 and T_0, it should be obtained by differentiating Eq. (3-8.3) or (3-8.4).

Hydrogen and nitrogen are special cases.   For nitrogen, no constant is temperature-dependent.   For hydrogen, C_0 is temperature-independent, but the constants b and γ vary as shown:

$$b = 0.08682 + \frac{197.194}{RT} \tag{3-8.6}$$

$$\gamma = \begin{cases} 0.8281 & T > 459.69°\text{R} \\ 0.8281 - 0.00051(459.69 - T) \\ 0.6751 & T < 160.69°\text{R} \end{cases} \tag{3-8.7}$$

In all calculations using Orye's equations and tables, English units are used, and $R = 10.73$ psi ft³/lb-mol °R.

Kaufman [54] has reported Benedict-Webb-Rubin constants for 1-

TABLE 3-7 Orye Tabulation of Benedict-Webb-Rubin Constants

English units are to be used, i.e., pounds per square inch, cubic feet, pound moles, and degrees Rankine [88]

	B_0	A_0	C_0	b	a	c	α	γ
C_1	0.6924010	6995.2500	0.2757630E+09	0.8673250	2984.120	0.4981060E+09	0.511172	1.539610
$C_2^=$	0.8919800	12593.6000	0.1602280E+10	2.2067800	15645.500	0.4133600E+10	0.731661	2.368440
C_2	1.0055400	15670.7000	0.2194270E+10	2.8539300	20850.200	0.6413140E+10	1.000440	3.027900
$C_3^=$	1.3626300	23049.2000	0.5365970E+10	4.7999700	46758.600	0.2008300E+11	1.873120	4.693250
C_3	1.5588400	25915.3999	0.6209930E+10	5.7735500	57248.000	0.2524870E+11	2.495770	5.645240
$i\text{-}C_4$	2.2032900	38587.3999	0.1038470E+11	10.8889999	117047.000	0.5597770E+11	4.414960	8.724470
$i\text{-}C_4^=$	1.8585800	33762.8999	0.1132960E+11	8.9937499	102251.000	0.5380720E+11	3.744170	7.594010
$n\text{-}C_4$	1.9921100	38029.5996	0.1213050E+11	10.2636000	113705.000	0.6192560E+11	4.526930	8.724470
$i\text{-}C_5$	2.5638600	48253.5996	0.2133670E+11	17.1441000	226902.000	0.1362050E+12	6.987770	11.880700
$n\text{-}C_5$	2.5109600	45928.7998	0.2591720E+11	17.1441000	246148.000	0.1613060E+12	7.439920	12.188600
$n\text{-}C_6$	2.8483500	54443.3999	0.4055620E+11	28.0031998	429901.000	0.2960770E+12	11.553600	17.111500
$n\text{-}C_7$	3.1878200	66070.5996	0.5798400E+11	38.9916997	626106.000	0.4834270E+12	17.905600	23.094200
$n\text{-}C_8$	2.4316500	55471.7996	0.8810000E+11	73.0545998	1259500.000	0.8823030E+12	17.942100	24.568000
$n\text{-}C_9$	2.6158700	60351.5000	0.1219000E+12	100.3569994	1825500.000	0.1398020E+13	23.910500	29.996700
$n\text{-}C_{10}$	2.7671700	64360.5000	0.1608000E+12	130.2536983	2497800.000	0.2040170E+13	30.294000	35.421700
C_6H_6	0.8057382	24548.4800	0.4190775E+11	19.6634109	336468.000	0.2302473E+12	2.877295	7.518409
H_2	0.3339370	585.1270	0.4110000E+07	0.0868200	98.599	0.1423170E+07	0.479116	0.828100
N_2	0.7336413	4496.9410	0.7195331E+08	0.5084650	900.070	0.1072668E+09	1.198385	1.924507
CO_2	0.7994960	10322.7999	0.1698000E+10	1.0582000	8264.460	0.2919710E+10	0.348000	1.384000
H_2S	0.5582140	11701.1000	0.2360000E+10	1.1354300	8758.270	0.3660180E+10	0.289043	1.168890

C₀ Temperature Dependence

	T_0	T_1	DCDT	Q_1	Q_2	Q_3	Q_4
C_1	289.69	159.69	0.00000	0.1165490E + 05	0.3367750E + 05	0.6926850E + 05	0.5279740E + 05
$C_2^=$	369.69	159.69	0.00000	0.3404110E + 05	0.1493090E + 06	0.3106840E + 06	0.2039330E + 06
C_2	424.69	159.69	0.00000	0.1599320E + 05	0.1018840E + 05	0.2180010E + 05	0.1962190E + 05
$C_3^=$	459.69	159.69	0.00000	0.4823550E + 05	0.1412580E + 06	0.2164240E + 06	0.1202830E + 06
C_3	509.69	159.69	0.00000	0.8239730E + 04	−0.4473340E + 05	−0.6114080E + 05	−0.2319690E + 05
$i\text{-}C_4$	484.69	269.69	−22.12256	0.1078770E + 06	0.7314030E + 06	0.2187170E + 07	0.2231160E + 07
$i\text{-}C_4^=$	529.69	269.69	−32.83256	0.1061340E + 06	0.5923800E + 06	0.1565130E + 07	0.1416790E + 07
$n\text{-}C_4$	529.69	269.69	−34.32522	0.9480760E + 05	0.5009210E + 06	0.1358720E + 07	0.1266080E + 07
$i\text{-}C_5$	514.69	309.69	−29.57480	0.1040800E + 06	0.4990670E + 06	0.1385770E + 07	0.1479550E + 07
$n\text{-}C_5$	629.69	309.69	−32.49820	0.4517410E + 05	0.1520220E + 06	0.4747210E + 06	0.4837050E + 06
$n\text{-}C_6$	629.69	309.69	−40.00000	0.9696390E + 05	0.1985950E + 06	0.2112580E + 06	0.9049590E + 05
$n\text{-}C_7$	669.69	309.69	−62.00000	0.3805060E + 05	−0.2426230E + 06	−0.5757700E + 06	−0.3495650E + 06
$n\text{-}C_8$	1009.69	459.69	−33.88000	0.4763729E + 05	−0.1977239E + 03	0.2315804	−0.8194300E − 04
$n\text{-}C_9$	1059.69	459.69	−40.76000	0.5239423E + 05	−0.2271831E + 03	0.2585502	−0.8627400E − 04
$n\text{-}C_{10}$	1059.69	484.69	−41.04000	0.7181043E + 05	−0.3177407E + 03	0.3794226	−0.1355970E − 03
C_6H_5	671.69	492.00	−20.44400	0.2507454E + 05	0.2649263E + 06	0.1263452E + 07	0.1843210E + 07
H_2	0.00	0.00	0.00000	0.0000000	0.0000000	0.0000000	0.0000000
N_2	0.00	0.00	0.00000	0.0000000	0.0000000	0.0000000	0.0000000
CO_2	449.69	333.34	−17.1300	0.5051527E + 04	0.2958724E + 05	0.5381413E + 06	0.1204973E + 07
H_2S	649.69	383.29	−25.00000	0.2543442E + 05	−0.1153709E + 03	0.1630762	−0.7045528E − 04

TABLE 3-8   Kaufman's Benedict-Webb-Rubin Constants

English units are used, i.e., pounds per square inch, cubic feet, pound moles, and degrees Rankine [54]

Substance	B_0	A_0	b	a	$c \times 10^{-9}$	α	γ
1-Butene	1.858581	34,146.5	8.933752	101,605	53.80707	3.74417	7.59454
cis-2-Butene	1.953850	37,041.5	9.865122	115,822	65.36291	4.34468	8.38628
1-Pentene	2.049167	41,683.1	10.85090	136,693	88.81093	5.011754	9.22429
1,3-Butadiene	1.529053	27,980.91	7.185653	84,056.1	47.96007	2.918084	6.03404
Methylacetylene	1.112799	19,262.64	3.80588	42,113.1	21.50043	1.124815	3.19593

Coefficients of C_0 polynomial
$$C_0 \times 10^{-10} = A_1 + A_2 T_r + A_3 T_r^2 + A_4 T_r^3 + A_5 T_r^4$$

Substance	A_1	A_2	A_3	A_4	A_5
cis-2-Butene	−0.290680	7.32702	−12.6342	9.74800	−2.84040
1-Pentene	−3.58381	29.4598	−62.1758	58.9453	−20.9321
Methylacetylene	0.298698	1.22342	−0.794138	−0.196283	−0.251860
1,3-Butadiene	0.308014	2.02193	−3.15673	1.20545	

butene, *cis*-2-butene, 1-pentene, methylacetylene, and 1,3-butadiene. These are shown in Table 3-8; again, English units are employed. Note that no C_0 value is shown. This constant is expressed as a polynomial in reduced temperature:

$$C_0(T) = \sum_{i=1}^{5} A_i T_r^{i-1} \tag{3-8.8}$$

where the coefficients A_i are shown in Table 3-8.

In summary, to obtain pure-component Benedict-Webb-Rubin constants at temperatures greater than about 0.6 times the critical, either Table 3-6 or Table 3-7 (or 3-8) may be used. If the temperature is less than $0.6T_c$, Table 3-7 (or 3-8) should be used and C_0 considered a function of temperature.

3-9   Generalized Benedict-Webb-Rubin Equations

The success of the original Benedict-Webb-Rubin equation to calculate volumetric and thermodynamic properties of pure gases and liquids has led to a number of studies wherein the equation or a modification of it has been generalized to apply to many types of compounds [30, 87, 121, 138]. Starling [112] and Starling and Han [114] have proposed an 11-constant generalized Benedict-Webb-Rubin equation in which the constants are functions of T_c, V_c, and the acentric factor. Light hydrocarbon mixtures typical of LNG and LPG were the focus of interest.

Lee and Kesler [59] developed a modified Benedict-Webb-Rubin equation within the context of Pitzer's three-parameter correlation. To employ the analytical form, care must be taken in the method of solution. The compressibility factor of a real fluid is related to properties of a *simple fluid* ($\omega = 0$) and those of *n*-octane as a reference fluid. Assume that Z is to be calculated for a fluid at some temperature and pressure. First, using the critical properties of this fluid, determine T_r and P_r. Then determine an *ideal-reduced* volume of a simple fluid with Eq. (3-9.1).

$$\frac{P_r V_r^{(0)}}{T_r} = 1 + \frac{B}{V_r^{(0)}} + \frac{C}{(V_r^{(0)})^2} + \frac{D}{(V_r^{(0)})^5}$$

$$+ \frac{c_4}{T_r^3 (V_r^{(0)})^2} \left[\beta + \frac{\gamma}{(V_r^{(0)})^2} \right] \exp \left[-\frac{\gamma}{(V_r^{(0)})^2} \right] \quad (3\text{-}9.1)$$

where
$$B = b_1 - \frac{b_2}{T_r} - \frac{b_3}{T_r^2} - \frac{b_4}{T_r^3}$$

$$C = c_1 - \frac{c_2}{T_r} + \frac{c_2}{T_r^3} \qquad D = d_1 + \frac{d_2}{T_r}$$

$V_r^{(0)} = P_c V^{(0)} / RT_c$, and the constants are given in Table 3-9 for a simple fluid. With $V_r^{(0)}$, the simple-fluid compressibility factor is calculated,

$$Z^{(0)} = \frac{P_r V_r^{(0)}}{T_r} \quad (3\text{-}9.2)$$

Next, *using the same reduced temperature and pressure as determined above*, Eq. (3-9.1) is again solved for $V_r^{(0)}$ but with the reference-fluid constants in Table 3-9; call this value $V_r^{(R)}$. Then

$$Z^{(R)} = \frac{P_r V_r^{(R)}}{T_r} \quad (3\text{-}9.3)$$

The compressibility factor Z for the fluid of interest is then calculated

$$Z = Z^{(0)} + \left(\frac{\omega}{\omega^R} \right) (Z^{(R)} - Z^{(0)}) \quad (3\text{-}9.4)$$

where $\omega^R = 0.3978$.

TABLE 3-9   Lee-Kesler Constants for Eq. (3-9.1)

Constant	Simple fluid	Reference fluid	Constant	Simple fluid	Reference fluid
b_1	0.1181193	0.2026579	c_3	0.0	0.016901
b_2	0.265728	0.331511	c_4	0.042724	0.041577
b_3	0.154790	0.027655	$d_1 \times 10^4$	0.155488	0.48736
b_4	0.030323	0.203488	$d_2 \times 10^4$	0.623689	0.0740336
c_1	0.0236744	0.0313385	β	0.65392	1.226
c_2	0.0186984	0.0503618	γ	0.060167	0.03754

Equation (3-9.4) was used to generate the $Z^{(0)}$ and $Z^{(1)}$ values in Tables 3-1 and 3-2.   While tested primarily on hydrocarbons, average errors normally were less than 2 percent for both the vapor and liquid phases.   The reduced-temperature range is 0.3 to 4, and the reduced-pressure range is 0 to 10.   The application of Lee-Kesler equations to mixtures is covered in Sec. 4-7 and to thermodynamic properties in Sec. 5-4.

3-10   Lee-Erbar-Edmister Equation of State

As in the Benedict-Webb-Rubin equation, the Lee-Erbar-Edmister form was devised primarily for hydrocarbon systems [58] and is a modification of an earlier form [57].   The equations are shown in Table 3-10, but discussion of the utility and accuracy of the equation is delayed until Sec. 5-4 because the primary object of this equation is not to estimate volumetric properties but to provide a reliable base for determining enthalpies and phase-equilibrium constants of hydrocarbon mixtures. To employ Eq. (3-10.1) for pure fluids, only values of T_c, P_c, and ω must be specified.   For mixtures, rules are given in Sec. 4-7 to relate the constants to composition.

TABLE 3-10   Lee-Erbar-Edmister Equation of State

$$P = \frac{RT}{V-b} - \frac{a}{V(V-b)} + \frac{bc}{V(V-b)(V+b)} \tag{3-10.1}$$

$$b = (0.086313 + 0.002\omega)\frac{RT_c}{P_c} \tag{3-10.2}$$

$$a = [(0.246105 + 0.02869\omega) - (0.037472 + 0.149687\omega)T_r$$
$$+ (0.16406 + 0.023727\omega)T_r^{-1} + (0.04937 + 0.132433\omega)T_r^{-2}]\frac{R^2T_c^2}{P_c} \tag{3-10.3}$$

$$c = [(0.451169 + 0.00948\omega)T_r^{-1/2} + (0.387082 + 0.078842\omega)T_r^{-2}]\frac{R^2T_c^2}{P_c} \tag{3-10.4}$$

For Hydrogen

$$b = \frac{0.0982RT_c}{P_c} \tag{3-10.5}$$

$$a = (0.277844 + 0.118992T_r^{-1} + 0.075088T_r^{-2})\frac{R^2T_c^2}{P_c} \tag{3-10.6}$$

$$c = (0.67254T_r^{-1/2} + 0.194258T_r^{-2})\frac{R^2T_c^2}{P_c} \tag{3-10.7}$$

3-11   Virial Equation

Most analytical equations of state discussed so far are explicit in pressure and of a form resembling a polynomial series in inverse volume.   This might have been expected if one examines the virial equation of state

$$P = \frac{RT}{V} + \frac{RTB}{V^2} + \frac{RTC}{V^3} + \cdots \qquad (3\text{-}11.1)$$

which can be derived from statistical mechanics.   The parameters $B, C, \ldots$ are called the second, third, ... virial coefficients and are functions only of temperature for a pure fluid.   Much has been written about this particular equation, and reviews are available [e.g., 78, 129].   One reason for its popularity is that the coefficients $B, C, \ldots$ can be related to parameters characterizing the intermolecular potential function.   For our purpose, the virial equation is most useful when truncated to retain only the second virial term and arranged to eliminate the molar volume.   That is, if Eq. (3-2.1) is used in Eq. (3-11.1) to eliminate V and the series truncated to retain terms only in the zeroth or first power in pressure,

$$Z = 1 + \frac{BP}{RT} \qquad (3\text{-}11.2)$$

Equation (3-11.2) should not be used if $\rho > \rho_c/2$.   Since B is a function only of temperature, Eq. (3-11.2) predicts that Z is a linear function of pressure along an isotherm.   Examination of Fig. 3-1 shows that this is not a bad assumption at low values of P_r.

A compilation of second-virial coefficients is given by Dymond and Smith [28].   To estimate values, a number of techniques are available.   Most are based on the integration of a theoretical expression relating intermolecular energy to the distance of separation between molecules.   With our present limited ability to determine such energies, however, it is more common to employ corresponding-states relations for estimating B.

For nonpolar molecules, Tsonopoulos [127] has modified a simple expansion proposed earlier by Pitzer and Curl [91]

$$\frac{BP_c}{RT_c} = f^{(0)} + \omega f^{(1)} \qquad (3\text{-}11.3)$$

$$f^{(0)} = 0.1445 - \frac{0.330}{T_r} - \frac{0.1385}{T_r^2} - \frac{0.0121}{T_r^3} - \frac{0.000607}{T_r^8} \qquad (3\text{-}11.4)$$

$$f^{(1)} = 0.0637 + \frac{0.331}{T_r^2} - \frac{0.423}{T_r^3} - \frac{0.008}{T_r^8} \qquad (3\text{-}11.5)$$

In Fig. 3-5, Eq. (3-11.3) is shown to correlate well experimental second-virial-coefficient data for benzene [127].   Except at high temperatures,

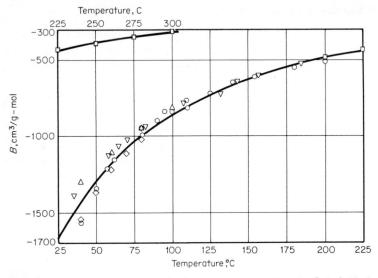

Fig. 3-5 Second virial coefficient for benzene. ——— Eq. (3.11-3), ○ Ref. 39, △ Ref. 55, ▽ Ref. 15, ◇ Ref. 142, □ Ref. 23, ◇ Ref. 9. (*From Ref. 127.*)

B is negative, and, by Eq. (3-11.2), the compressibility factor is less than unity.

Equation (3-11.3) should be considered applicable only for nonpolar or slightly polar materials.   Several modifications have been suggested for highly polar compounds [85, 86, 106, 123]; Polak and Lu [94] suggest that the Stockmayer potential function be used for polar gases, and they have regressed experimentally determined virial coefficients to determine the best intermolecular potential parameters.   Johnson and Eubank [53] evaluated a number of possible intermolecular potentials that might be used for polar gases.   A simple extension of Eq. (3-11.3) was also presented by Halm and Stiel [42] using the polar parameter X as a measure of polarity (see Sec. 2-6).   Tsonopoulos [127] recommends that Eq. (3-11.3) be modified by the addition of another term $f^{(2)}$, where

$$f^{(2)} = \frac{a}{T_r^6} - \frac{b}{T_r^8} \tag{3-11.6}$$

Neither a nor b can be estimated with much accuracy; b, however, is zero for materials exhibiting no hydrogen bonding; in this case, for ketones, aldehydes, nitrides, and ethers, approximately,

$$a = -2.140 \times 10^{-4} \mu_r - 4.308 \times 10^{-21} \mu_r^8 \tag{3-11.7}$$

where
$$\mu_r = \frac{10^5 \mu_p^2 P_c}{T_c^2} \tag{3-11.8}$$

where μ_p = dipole moment, debyes
P_c = critical pressure, atm
T_c = critical temperature, K

In hydrogen-bonded fluids (alcohols, water, etc.), a and b are constants specific for each material. For straight- and branched-chain alcohols, a is essentially constant and equal to 0.0878 (for phenol, however, it is -0.136). For alkyl halides, $a \approx 0$, irrespective of μ_r, except for mono-alkyl halides, where specific values of a are recommended [128]. b usually ranges from 0.04 to 0.06, and values for several alcohols are given by Tsonopoulos [127]. Fluids which hydrogen-bond tend to associate, and B exhibits a strong temperature dependence at low reduced temperatures (where the b/T_r^8 term predominates).

Example 3-2   Estimate the second virial coefficient of methyl isobutyl ketone at 120°C.
  solution   From Appendix A, $T_c = 571$ K, $P_c = 32.3$ atm, $\omega = 0.400$, and $\mu_p = 2.8$ debyes. With $T_r = (120 + 273)/571 = 0.689$, then with Eqs. (3-11.4) and (3-11.5), $f^{(0)} = -0.675$, $f^{(1)} = -0.690$, and with Eq. (3-11.8), $\mu_r = (10^5)(2.8)^2(32.3/571^2) = 77.7$. From Eqs. (3-11.7) and (3-11.6), $a = -0.0167$ and $f^{(2)} = -0.156$. Since methyl isobutyl ketone is non-hydrogen-bonded, $b = 0$. Then, using Eqs. (3-11.3) and (3-11.6),

$$\frac{(B)(32.3)}{(82.04)(571)} = -0.675 + (0.400)(-0.690) + (-0.156)$$

$$B = -1605 \text{ cm}^3/\text{g-mol}$$

The experimental value is $-1580 \text{ cm}^3/\text{g-mol}$ [44].

Vetere [132] recommends a slightly different way to estimate second virial coefficients for polar molecules, i.e.,

$$\frac{BP_c}{RT_c} = g^{(0)} + \omega g^{(1)} + \omega_p g^{(2)} \tag{3-11.9}$$

$$g^{(0)} = 0.1445 - \frac{0.330}{T_r} - \frac{0.1385}{T_r^2} - \frac{0.0121}{T_r^3} \tag{3-11.10}$$

$$g^{(1)} = 0.073 + \frac{0.46}{T_r} - \frac{0.50}{T_r^2} - \frac{0.097}{T_r^3} - \frac{0.0073}{T_r^8} \tag{3-11.11}$$

$$g^{(2)} = 0.1042 - \frac{0.2717}{T_r} + \frac{0.2388}{T_r^2} - \frac{0.0716}{T_r^3} + \frac{1.502 \times 10^{-4}}{T_r^8} \tag{3-11.12}$$

ω is the acentric factor and ω_p is defined as

$$\omega_p = \frac{T_b^{1.72}}{M} - 263 \tag{3-11.13}$$

If ω_p is calculated to be negative, it should be considered zero. T_b is the normal boiling temperature in kelvins, and M is the molecular weight.

In Vetere's method, the $g^{(0)}$ and $g^{(1)}$ functions are identical to those given by Pitzer and Curl [91]; the term $\omega_p g^{(2)}$ was added to include polar compounds, but water, CH_3OH, and C_2H_5OH cannot be treated.

Example 3-3   Repeat Example 3-2 using Vetere's correlation for B.

solution   For methyl isobutyl ketone at 120°C, $T_r = 0.689$, and with $\omega = 0.400$, $T_b = 389.6$ K, and $M = 100.161$,

$$\omega_p = \frac{(389.6)^{1.72}}{100.161} - 263 = 22.2$$

From Eqs. (3-11.10) to (3-11.12), we get $g^{(0)} = -0.663$, $g^{(1)} = -0.753$, and $g^{(2)} = -3.05 \times 10^{-3}$; thus, with Eq. (3-11.9),

$$\frac{(B)(32.3)}{(82.04)(571)} = -0.663 + (0.400)(-0.753) + (22.2)(-3.05 \times 10^{-3})$$

$$B = -1496 \text{ cm}^3/\text{g-mol}$$

3-12 Discussion of Relations to Estimate Pure-Gas *P-V-T* Properties

One corresponding-states correlation [Eq. (3-3.1)] and eight analytical equations of state were introduced. All were developed for broad usage, i.e., to estimate volumetric and thermodynamic properties of pure components and mixtures. At this point, however, we are only considering their applicability to estimate gas-phase properties of pure compounds; in later chapters, we reconsider the same relations when applied to mixtures and to estimating thermodynamic properties.

For *hydrocarbons*, the Soave modification of the Redlich-Kwong equation [Eqs. (3-5.11) and (3-5.15)], the Lee-Kesler modification of the Benedict-Webb-Rubin equation [Eqs. (3-9.1) and (3-9.4)], and the Lee-Erbar-Edmister equation (3-10.1) all are generalized equations that are accurate within 1 to 2 percent except perhaps in the region near the critical point. The Lee-Kesler form can also be used for estimating the volumetric properties of liquids. The Benedict-Webb-Rubin equation (3-8.1) with specific constants from either Table 3-6 or 3-7 is also accurate and well tested.

For *nonhydrocarbons*, the methods of Barner-Adler [Eq. (3-6.1)], Sugie-Lu [Table 3-5], or the Lee-Kesler modification of the Benedict-Webb-Rubin equation [Eqs. (3-9.1) and (3-9.4)] are recommended. Errors are probably within 2 to 3 percent except for highly polar fluids or when volumetric properties are estimated near the critical point.

The original Redlich-Kwong equation (3-5.1) and the truncated virial equation (3-11.2) are less accurate than the equations noted above. They are, however, less complex and are often useful in multiple, iterative calculations where computation time is important. Also, their accuracy is still good, except for polar materials; i.e., predictions of gas-phase volumetric properties are normally not in error by more than 5 percent. The truncated virial, however, is limited to densities less than half the critical.

Except for the Benedict-Webb-Rubin equation of state, all require the critical temperature, critical pressure, and usually acentric factor as input parameters.   The choice of which one to use probably does not rest upon the applicability of the equation to predicting gas-phase volumetric properties but more upon the accuracy obtained when enthalpies or phase-equilibrium constants (or vapor-phase component fugacities) are predicted.   These are discussed in Chap. 5.

3-13 *P-V-T* Properties of Liquids— General Considerations

Liquid specific volumes are relatively easy to measure.   For most common organic liquids, at least one experimental value is available.   There are a number of references in which these experimental volumes (or densities) are tabulated or in which constants are given to allow one to calculate them rapidly with an empirical equation [3, 27, 36, 61, 79, 80, 82, 125, 134].   Ritter, Lenoir, and Schweppe [104] have published convenient monographs to estimate saturated-liquid densities as a function of temperature for some 90 liquids covering, primarily, hydrocarbons and hydrocarbon derivatives.   In Appendix A, single-liquid densities are tabulated for many compounds at a given temperature.

3-14 Estimation of the Liquid Molal Volume at the Normal Boiling Point

A number of additive methods are discussed by Partington [89].   Each element and certain bond linkages are assigned numerical values so that the molal volume at the normal boiling point can be calculated by the addition of these values in a manner similar to that described in Chap. 2 for estimating the critical volume.

Additive Methods   Schroeder [89] has suggested a novel and simple additive method for estimating molal volumes at the normal boiling point.   His rule is to count the number of atoms of carbon, hydrogen, oxygen, and nitrogen, add 1 for each double bond, and multiply the sum by 7.   This gives the volume in cubic centimeters per gram mole.   This rule is surprisingly good, giving results within 3 to 4 percent except for highly associated liquids.   Table 3-11 indicates the value to be assigned each atom or functional group.   The accuracy of this method is shown in Table 3-12, where molal volumes at the normal boiling point are compared with experimental values for a wide range of materials.   The average error for the compounds tested is 3.0 percent.[†]

†Schroeder's original rule has been expanded to include halogens, sulfur, and triple bonds.

Additive volumes published by Le Bas [56] represent a refinement of Schroeder's rule. Volume increments from Le Bas are shown in Table 3-11, and calculated values of V_b are compared with experimental values in Table 3-12. The average error for the compounds tested is 4.0 percent. Although the average error in this case is greater than that found by Schroeder's increments, the method appears to be more general and as accurate as Schroeder's for most of the compounds tested; i.e., the average error is not particularly representative.

Other additive methods are discussed by Fedors [35].

TABLE 3-11   Additive-volume Increments for the Calculation of Molal Volumes V_b

	Increment, cm³/g-mol	
	Schroeder	Le Bas†
Carbon	7	14.8
Hydrogen	7	3.7
Oxygen (except as noted below)	7	7.4
In methyl esters and ethers	.....	9.1
In ethyl esters and ethers	.....	9.9
In higher esters and ethers	.....	11.0
In acids	.....	12.0
Joined to S, P, N	.....	8.3
Nitrogen	7	
Doubly bonded	.....	15.6
In primary amines	.....	10.5
In secondary amines	.....	12.0
Bromine	31.5	27
Chlorine	24.5	24.6
Fluorine	10.5	8.7
Iodine	38.5	37
Sulfur	21	25.6
Ring, three-membered	−7	−6.0
Four-membered	−7	−8.5
Five-membered	−7	−11.5
Six-membered	−7	−15.0
Naphthalene	−7	−30.0
Anthracene	−7	−47.5
Double bond between carbon atoms	7	
Triple bond between carbon atoms	14	

†The additive-volume procedure should not be used for simple molecules. The following approximate values are employed in estimating diffusion coefficients by the methods of Chap. 11: H_2, 14.3; O_2, 25.6; N_2, 31.2; air, 29.9; CO, 30.7; CO_2, 34.0; SO_2, 44.8; NO, 23.6; N_2O, 36.4; NH_3, 25.8; H_2O, 18.9; H_2S, 32.9; COS, 51.5; Cl_2, 48.4; Br_2, 53.2; I_2, 71.5.

TABLE 3-12 Comparison of Calculated and Experimental Liquid Molal Volumes at the Normal Boiling Point

Compound	Molal volume, cm³/g-mol		Percent error† when calculated by method of:		
	Exp. V_b	Ref.	Tyn and Calus	Schroeder	Le Bas
Methane	37.7	58	−6.7	−7.2	−21.5
Propane	74.5	114	0.2	3.3	−0.7
Heptane	162	58	1.8	−0.6	0.5
Cyclohexane	117	58	−1.2	1.7	1.0
Ethylene	49.4	58	−6.0	−0.8	−10
Benzene	96.5	58	−0.1	1.6	−0.5
Fluorobenzene	102	58	−0.9	−0.5	−1.0
Bromobenzene	120	58	1.6	2.1	−1.6
Chlorobenzene	115	58	0.0	0.0	1.7
Iodobenzene	130	58	1.9	−0.4	−0.5
Methanol	42.5	58	−0.5	−1.2	−13
n-Propyl alcohol	81.8	58	−1.4	2.7	−0.5
Dimethyl ether	63.8	58	2.0	−1.3	−4.5
Ethyl propyl ether	129	58		−2.3	−0.5
Acetone	77.5	58	−0.6	−0.6	−4.5
Acetic acid	64.1	58	−2.7	−1.7	6.7
Isobutyric acid	109	58	0.3	−3.7	3.5
Methyl formate	62.8	58	0.0	0.3	−0.3
Ethyl acetate	106	58	0.9	−0.9	2.5
Diethylamine	109	58	3.5	2.8	2.7
Acetonitrile	57.4	58	10	−2.4	
Methyl chloride	50.6	58	−0.8	3.7	−0.2
Carbon tetrachloride	102	58	1.0	2.8	11
Dichlorodifluoromethane	80.7	55	−0.8	−4.6	0.9
Ethyl mercaptan	75.5	58	0.9	2.0	2.5
Diethyl sulfide	118	58	1.3	0.9	3.2
Phosgene	69.5	58	0.2	0.7	2.7
Ammonia	25.0	58	1.5	12	
Chlorine	45.5	58	−2.1	7.7	8.1
Water	18.7	57	3.5	12	
Hydrochloric acid	30.6	58	−6.8	2.9	−7.5
Sulfur dioxide	43.8	113	0.0	−12	−3.7
Average error			1.9	3.1	3.9

†Error $= \dfrac{\text{calc.} - \text{exp.}}{\text{exp.}} \times 100.$

Tyn and Calus Method [130] V_b is related to the critical volume by

$$V_b = 0.285 V_c^{1.048} \qquad (3\text{-}14.1)$$

where V_b and V_c are both expressed in cubic centimeters per gram mole. This simple relation is generally accurate within 3 percent except for the low-boiling permanent gases (He, H_2, Ne, Ar, Kr) and some polar nitrogen and phosphorus compounds (HCN, PH_3). A similar relation was suggested earlier by Benson [11], but in that case the critical pressure was also employed in the correlation.

Recommendation   The Tyn and Calus method is recommended for estimating liquid molal volumes at the boiling point. The average error for 32 compounds is only 2 percent, as shown in Table 3-12. A reliable value of the critical volume must be available, however.

Example 3-4   Estimate the molal volume of liquid chlorobenzene at its normal boiling point. The critical volume is $308\,cm^3/g$-mol (Appendix A). The experimental value is $115\,cm^3/g$-mol.

solution   *Schroeder   Method.*   From   Table   3-11, $C = 7$, $H = 7$, $Cl = 24.5$,   the ring $= -7$, and each double bond $= 7$. Therefore, for C_6H_5Cl,

$$V_b = (6)(7) + (5)(7) + 24.5 - 7 + (3)(7) = 115\ cm^3/g\text{-mol}$$

$$\text{Error} = \frac{115 - 115}{115} \times 100 = 0\%$$

Le Bas Method.   From   Table   3-11, $C = 14.8$, $H = 3.7$, $Cl = 24.6$,   and   the   ring $= -15.0$. Therefore,

$$V_b = (6)(14.8) + (5)(3.7) + 24.6 - 15.0 = 117\ cm^3/g\text{-mol}$$

$$\text{Error} = \frac{117 - 115}{115} \times 100 = +1.7\%$$

Tyn and Calus Method.   With Eq. (3-14.1),

$$V_b = 0.285\,V_c^{1.048} = (0.285)(308^{1.048}) = 115\ cm^3/g\text{-mol}$$

$$\text{Error} = \frac{115 - 115}{115} \times 100 = 0\%$$

3-15   Estimation of Liquid Densities

Even if no data are available, there are a number of techniques available to estimate pure-liquid specific volumes or densities. Some of the more accurate are described below. All are based on some form of the law of corresponding states, and all are sufficiently complex algebraically to necessitate the aid of a computer for easy utilization if a large number of calculations is to be made.

Gunn and Yamada [38]   This method is limited to *saturated*-liquid volumes

$$\frac{V}{V_{sc}} = V_r^{(0)}(1 - \omega\Gamma) \tag{3-15.1}$$

where V is the liquid specific volume and V_{sc} is a scaling parameter which is defined in terms of the volume at $T_r = 0.6$

$$V_{sc} = \frac{V_{0.6}}{0.3862 - 0.0866\omega} \tag{3-15.2}$$

where $V_{0.6}$ is the saturated-liquid molar volume at a reduced temperature of 0.6. If $V_{0.6}$ is not available, then, approximately, V_{sc} can be estimated by

$$V_{sc} = \frac{RT_c}{P_c}(0.2920 - 0.0967\omega) \tag{3-15.3}$$

In most cases V_{sc} is close to V_c. However, if the saturated-liquid molar volume is available at *any* temperature, V_{sc} can be eliminated as shown later in Eq. (3-15.7).

In Eq. (3-15.1), $V_r^{(0)}$ and Γ are functions of reduced temperature, and ω is the acentric factor. For $0.2 \leq T_r \leq 0.8$

$$V_r^{(0)} = 0.33593 - 0.33953\,T_r + 1.51941\,T_r^2 - 2.02512\,T_r^3 + 1.11422\,T_r^4 \tag{3-15.4}$$

For $0.8 < T_r < 1.0$

$$\begin{aligned} V_r^{(0)} = 1.0 + 1.3(1 - T_r)^{1/2} \log(1 - T_r) \\ - 0.50879(1 - T_r) - 0.91534(1 - T_r)^2 \end{aligned} \tag{3-15.5}$$

For $0.2 \leq T_r < 1.0$

$$\Gamma = 0.29607 - 0.09045\,T_r - 0.04842\,T_r^2 \tag{3-15.6}$$

With a known value of V at some reference temperature, that is, V^R at T^R, Eq. (3-15.1) can be applied at any other T as well as at T^R to cancel V_{sc}. Thus

$$\frac{V}{V^R} = \frac{V_r^{(0)}(T_r)[1 - \omega\Gamma(T_r)]}{V_r^{(0)}(T_r^R)[1 - \omega\Gamma(T_r^R)]} \qquad \text{where } T_r^R = T^R/T_c \tag{3-15.7}$$

This method appears to be one of the most accurate available for saturated-liquid volumes. It should not, however, be used above $T_r = 0.99$; the $V_r^{(0)}$ function becomes undefined at $T_r = 1$. It is said to be applicable for nonpolar and slightly polar compounds although acetonitrile and alcohols were fitted well. Lu et al. [68] have proposed a similar analytical relation for estimating liquid molar volumes; they also employed the Gunn and Yamada scaling volume. The accuracy of this alternate method is similar to that of the original Gunn-Yamada relation described above.

Yen and Woods Lydersen, Greenkorn, and Hougen [70] developed a corresponding-states method to estimate pure-liquid densities at any pressure and at temperatures below $T_r = 1$ in terms of T_r, P_r, and the critical-compressibility factor Z_c. The correlation was originally given in tabular form, but Yen and Woods [141] modified the correlation, improved the accuracy, and expressed the results in analytical form.

With ρ_s, the saturated molar liquid density, ρ_c, the critical density, and T_r, the reduced temperature,

$$\rho_s/\rho_c = 1 + \sum_{j=1}^{4} K_j (1 - T_r)^{j/3} \qquad (3\text{-}15.8)$$

where $K_1 = 17.4425 - 214.578Z_c + 989.625Z_c^2 - 1522.06Z_c^3 \qquad (3\text{-}15.9)$

$K_2 = -3.28257 + 13.6377Z_c + 107.4844Z_c^2 - 384.211Z_c^3$
$$\text{if } Z_c \leqslant 0.26 \quad (3\text{-}15.10)$$

$K_2 = 60.2091 - 402.063Z_c + 501.0Z_c^2 + 641.0Z_c^3$
$$\text{if } Z_c > 0.26 \quad (3\text{-}15.11)$$

$K_3 = 0 \qquad (3\text{-}15.12)$

$K_4 = 0.93 - K_2 \qquad (3\text{-}15.13)$

To determine subcooled liquid densities at a pressure P (greater than the vapor pressure P_{vp}) the following equation was proposed:

$$\frac{\rho - \rho_s}{\rho_c} = \Delta\rho_r + \delta_{Z_c} \qquad (3\text{-}15.14)$$

The first term on the right corrects for the pressure difference $\Delta P_r \equiv (P - P_{vp})/P_c$ for compounds with $Z_c = 0.27$. The second term then accounts for variations in Z_c away from the 0.27 reference. If $Z_c = 0.27$, $\delta_{Z_c} = 0$.

$$\Delta\rho_r = E + F \ln \Delta P_r + G \exp (H \, \Delta P_r) \qquad (3\text{-}15.15)$$

The constants E, F, G, and H are given as functions of T_r in Table 3-13. If, however, $P/P_c < 0.2$, instead of using Eq. (3-15.15) one should determine $\Delta\rho_r$ at $P_r = 0.2$ and use Eq. (3-15.16),

$$\Delta\rho_r = \Delta\rho_r \, (\text{at } P_r = 0.2) \frac{\Delta P_r}{0.2} \qquad (3\text{-}15.16)$$

If Z_c differs from 0.27, δ_{Z_c} is nonzero and is given as

$$\delta_{Z_c} = I + J \ln \Delta P_r + K \exp (L \, \Delta P_r) \qquad (3\text{-}15.17)$$

The constants are also shown in Table 3-13. The lowercase constants used to determine I, J, K, and L are given only for discrete values of Z_c, that is, 0.23, 0.25, and 0.29 (they are zero for 0.27) in Table 3-14. In most computations, interpolation is necessary.

When this method was tested with almost 100 liquids, polar and nonpolar, saturated and subcooled, from temperatures around the freezing point to near the critical point and to reduced pressures of 30, errors were normally less than 3 to 6 percent. If critical densities are unavailable or in doubt, a different calculational technique may be

TABLE 3-13   Constants for Eqs. (3-15.15) and (3-15.17)

$$E = 0.714 - 1.626(1 - T_r)^{1/3} - 0.646(1 - T_r)^{2/3} + 3.699(1 - T_r) - 2.198(1 - T_r)^{4/3}$$

$$F = \frac{0.268 T_r^{2.0967}}{1.0 + 0.8 (-\ln T_r)^{0.441}}$$

$$G = 0.05 + 4.221(1.01 - T_r)^{0.75} \exp[-7.848(1.01 - T_r)]$$

$$H = -10.6 + 45.22(1 - T_r)^{1/3} - 103.79(1 - T_r)^{2/3} + 114.44(1 - T_r) - 47.38(1 - T_r)^{4/3}$$

$$I = a_1 + a_2(1 - T_r)^{1/3} + a_3(1 - T_r)^{2/3} + a_4(1 - T_r) + a_5(1 - T_r)^{4/3}$$

$$J = b_1 + b_2(1 - T_r)^{1/3} + b_3(1 - T_r)^{2/3} + b_4(1 - T_r) + b_5(1 - T_r)^{4/3}$$

$$K = c_1 + c_2 T_r + c_3 T_r^2 + c_4 T_r^3$$

$$L = d_1 + d_2(1 - T_r)^{1/3} + d_3(1 - T_r)^{2/3} + d_4(1 - T_r) + d_5(1 - T_r)^{4/3}$$

used.   Noting that Eqs. (3-15.8) and (3-15.15) contain ρ_c only on the left-hand side, one can use as a reference density a well-authenticated saturated-liquid density at a low temperature.   When these equations are written with this known reference density and divided into the working equations, ρ_c cancels from all forms.   Computation times are not appreciably increased by this modification.

TABLE 3-14   Constants in Eq. (3-15.17)

	$Z_c = 0.29$	$Z_c = 0.25$	$Z_c = 0.23$
a_1	−0.0817	0.0933	0.0890
a_2	0.3274	−0.3445	−0.4344
a_3	−0.5014	0.4042	0.7915
a_4	0.3870	−0.2083	−0.7654
a_5	−0.1342	0.05473	0.3367
b_1	−0.0230	0.0220	0.0674
b_2	−0.0124	−0.003363	−0.06109
b_3	0.1625	−0.07960	0.06261
b_4	−0.2135	0.08546	−0.2378
b_5	0.08643	−0.02170	0.1665
c_1	0.05626	0.01937	−0.01393
c_2	−0.3518	−0.03055	−0.003459
c_3	0.6194	0.06310	−0.1611
c_4	−0.3809	0	0
d_1	−21.0	−16.0	−6.550
d_2	55.174	30.699	7.8027
d_3	−33.637	19.645	15.344
d_4	−28.100	−81.305	−37.04
d_5	23.277	47.031	20.169

Chueh and Prausnitz [20], **Lyckman, Eckert, and Prausnitz** [69]   Starting from an earlier corresponding-states correlation [69], Chueh and Prausnitz correlated the liquid density with P, T, Z_c, P_c, T_c, and the acentric factor as

$$\rho = \rho_s\left[1 + \frac{9Z_cN(P - P_{vp})}{P_c}\right]^{1/9} \tag{3-15.18}$$

where P_{vp} = vapor pressure
P = system pressure
P_c = critical pressure
Z_c = critical-compressibility factor

N is a function of the acentric factor ω and the reduced temperature:

$$N = (1.0 - 0.89\omega)[\exp(6.9547 - 76.2853\,T_r$$
$$+ 191.3060\,T_r^2 - 203.5472\,T_r^3 + 82.7631\,T_r^4)] \tag{3-15.19}$$

The term ρ_s refers to the saturated-liquid molar density at T_r and P_{vp}. It is given as

$$\frac{\rho_c}{\rho_s} = V_r^{(0)} + \omega V_r^{(1)} + \omega^2 V_r^{(2)} \tag{3-15.20}$$

In Eq. (3-15.20), the V_r functions may be expressed as

$$V_r^{(j)} = a_j + b_j T_r + c_j T_r^2 + d_j T_r^3 + \frac{e_j}{T_r} + f_j \ln(1 - T_r) \tag{3-15.21}$$

The coefficients are

j	a_j	b_j	c_j	d_j	e_j	f_j
0	0.11917	0.009513	0.21091	-0.06922	0.07480	-0.084476
1	0.98465	-1.60378	1.82484	-0.61432	-0.34546	0.087037
2	-0.55314	-0.15793	-1.01601	0.34095	0.46795	-0.239938

The value of ρ_c in Eq. (3-15.20) can be eliminated if another reference density is used in the same general way as that described for both the Gunn-Yamada and Yen-Woods methods.

The accuracy of this method is comparable to that of Yen and Woods. The logarithmic function in Eq. (3-15.21), however, causes problems if T_r is close to unity.

Liquid Compressibility Factors   One may define the compressibility factor for a liquid by Eq. (3-2.1), but P must then be equal to or greater than the saturation vapor pressure at T. Except at high reduced temperatures, the liquid compressibility factor is almost proportional to pressure, at constant reduced temperature.   Since molar volume is proportional to Z/P, it is essentially invariant with pressure.

Most liquid-compressibility correlations are written in the form of Eq. (3-3.1) [18, 19, 48, 67] though in one, the Stiel polar factor is also employed [118].   In Tables 3-1 and 3-2, $Z^{(0)}$ and $Z^{(1)}$ are shown for both the vapor and the liquid phase.   If these tables are used to estimate liquid molar volumes, it is recommended that interpolation be made by calculating $Z^{(0)}/P_r$ and $Z^{(1)}/P_r$ at nearby grid points and then interpolating.   This technique is illustrated in Example 3-5.

Example 3-5   Estimate the saturated-liquid molar volume of ethyl mercaptan at 150°C.   The experimental value is 95.0 cm³/g-mol.

   solution   From Appendix A, $T_b = 308.2$ K, $T_c = 499$ K, $M = 62.134$, $P_c = 54.2$ atm, $V_c = 207$ cm³/g-mol, $Z_c = 0.274$, and $\omega = 0.190$.   The liquid density at 20°C is 0.839 g/cm³, where $V = 74.05$ cm³/g-mol.   When $T = 150°C$,

$$T_r = \frac{150 + 273}{499} = 0.848$$

   Gunn and Yamada Method.   The reference temperature T^R is 20°C, and $T_r^R = 0.587$.   With Eqs. (3-15.4) to (3-15.6),

$$V_r^{(0)} = 0.487 \qquad \text{at } T_r = 0.848$$
$$V_r^{(R)} = 0.383 \qquad \text{at } T_r^R = 0.587$$
$$\Gamma = \begin{cases} 0.185 & \text{at } T_r = 0.848 \\ 0.226 & \text{at } T_r^R = 0.587 \end{cases}$$

Then, with Eq. (3-15.7), where $V^R = 74.05$ cm³/g-mol,

$$V = 74.05 \frac{(0.487)[1 - (0.190)(0.185)]}{(0.383)[1 - (0.190)(0.226)]} = 94.9 \text{ cm}^3/\text{g-mol}$$

$$\text{Error} = \frac{94.9 - 95.0}{95.0} \times 100 = 0.1\%$$

   Yen and Woods.   First the K values in Eqs. (3-15.9) to (3-15.13) are found using $Z_c = 0.274$:

$$K_1 = 1.635 \qquad K_2 = 0.843 \qquad K_3 = 0 \qquad K_4 = 0.0872$$

Then

$$\frac{\rho_s(T_r = 0.848)}{\rho_c} = 1 + 1.635(1 - 0.848)^{1/3} + 0.843(1 - 0.848)^{2/3} + 0.0872(1 - 0.848)^{4/3}$$
$$= 2.12$$
$$V = \frac{V_c}{2.12} = \frac{207}{2.12} = 97.6 \text{ cm}^3/\text{g-mol}$$
$$\text{Error} = \frac{97.6 - 95.0}{95.0} \times 100 = 2.8\%$$

   Chueh and Prausnitz Method.   The $V_r^{(0)}$, $V_r^{(1)}$, and $V_r^{(2)}$ values for $T_r = 0.848$ and 0.587 are shown below as determined from Eq. (3-15.21):

T_r	$V_r^{(0)}$	$V_r^{(1)}$	$V_r^{(2)}$
0.848	0.485	−0.00907	−0.206
0.587	0.386	−0.118	0.0824

Using Eq. (3-15.20) with $\omega = 0.190$,

At $T_r = 0.848$:

$$\frac{\rho_c}{\rho_s} = 0.485 + (0.190)(-0.00907) + (0.190)^2(-0.206) = 0.476$$

At $T_r = 0.587$:

$$\frac{\rho_c}{\rho_s} = 0.386 + (0.190)(-0.118) + (0.190)^2(0.0824) = 0.367$$

From only the first of these two relations,

$$\frac{\rho_c}{\rho_s} = \frac{V_s}{V_c} = 0.476$$

$$V_s = (0.476)(207) = 96.0 \text{ cm}^3/\text{g-mol}$$

If, however, the expressions at $T_r = 0.848$ and 0.587 are divided into each other, ρ_c cancels. With V at $T_r = 0.587 = 74.05 \text{ cm}^3/\text{g-mol}$,

$$\frac{V_s}{V_s \text{ at } T_r = 0.587} = \frac{0.476}{0.367}$$

$$V_s = \frac{(74.05)(0.476)}{0.367} = 95.8 \text{ cm}^3/\text{g-mol}$$

The errors in these two cases are 1.1 and 0.8 percent, respectively.   Usually it is preferable to make use of a reference density other than that at the critical point.

   Compressibility-Factor Method.   Tables 3-1 and 3-2 are used to estimate the liquid compressibility factor.   The reduced temperature is 0.848.   To obtain the vapor pressure at 150°C, the wide-range Harlacher vapor-pressure equation may be used (see Sec. 6-6 and Appendix A).   The value found is 17.2 atm.   Thus $P_r = 17.2/54.2 = 0.317$.   Then Z at $T_r = 0.848$ and $P_r = 0.317$ is found from Tables 3-1 and 3-2.   Interpolation is not easily done.   Let us assume that $T_r = 0.848 \approx 0.85$.   Then, since $Z \approx P_r$, we can make the following table for $T_r = 0.85$:

P_r	0.4	0.6	0.8
$Z^{(0)}/P_r$	0.1653	0.1638	0.1626
$Z^{(1)}/P_r$	-0.0670	-0.0652	-0.0636

Since, with our approximate estimate of the vapor pressure, we know $P_r \approx 0.3$, extrapolation of the ratios $Z^{(0)}/P_r$ and $Z^{(1)}/P_r$ to this reduced pressure may be readily accomplished; that is, $Z^{(0)}/P_r \approx 0.166$, $Z^{(1)}/P_r \approx -0.069$.   Then

$$V = \frac{ZRT}{P} = \frac{Z}{P_r} \frac{RT}{P_c}$$

$$= \frac{RT}{P_c} \left(\frac{Z^{(0)}}{P_r} + \omega \frac{Z^{(1)}}{P_r} \right)$$

$$= \frac{(82.04)(150 + 273)}{54.2} [0.166 + (0.190)(-0.069)]$$

$$= 97.9 \text{ cm}^3/\text{g-mol}$$

$$\text{Error} = \frac{97.9 - 95.0}{95.0} \times 100 = 3.0\%$$

3-16   Discussion of Liquid-Density Estimation Methods

Emphasis has been placed on corresponding-states correlations which are primarily suitable for machine computation.   No note was taken of the fact that certain of the gas-phase equations of state are also applicable to determine liquid volumes, e.g., the Benedict-Webb-Rubin equation for hydrocarbons, as these are normally less accurate than the ones mentioned here.   All methods (except the liquid-compressibility technique) require at least one liquid density; often this is the critical density, though all may be arranged so as to employ, as a reference density, a value at any specified temperature and pressure.

For *saturated* liquids below $T_r = 0.99$, the Gunn-Yamada technique [Eq. (3-15.1)] appears preferable, as it is slightly more accurate. Spencer and Danner [111] surveyed all available predictive equations for saturated-liquid density, and, after extensive testing, they also found the Gunn-Yamada correlation to be the most accurate. However, they did report somewhat better results with the Rackett equation [96] if the critical-compressibility factor in this latter correlation is replaced by an empirical constant characteristic of the compound under study.   Such constants are tabulated for many materials.   Yamada and Gunn [139] also proposed the Rackett equation with a slight modification.   Their form may be written

$$V = V^R Z_{cr} \exp \phi(T_r, T_r^R) \tag{3-16.1}$$

where

$$Z_{cr} = 0.29056 - 0.08775\omega \tag{3-16.2}$$

and

$$\phi(T_r, T_r^R) = (1 - T_r)^{2/7} - (1 - T_r^R)^{2/7} \tag{3-16.3}$$

V^R is a liquid molar volume at a reference (reduced) temperature of T_r^R.   With this simple form, the accuracy is nearly equal to that of the Gunn-Yamada correlation; for most nonpolar saturated liquids, errors are less than 1 percent.

Example 3-6   Repeat Example 3-4 using the Gunn-and-Yamada modification of the Rackett equation.

   solution   From Example 3-4, $V^R = 74.05 \text{ cm}^3/\text{g-mol}$, $T_r = 0.848$, $T_r^R = 0.587$, and $\omega = 0.190$.   With Eqs. (3-16.1) to (3-16.3),

$$Z_{cr} = 0.29056 - (0.08775)(0.190) = 0.274$$
$$\phi(T_r, T_r^R) = (1 - 0.848)^{2/7} - (1 - 0.587)^{2/7} = -0.193$$
$$V = (74.05)(0.276)^{-0.193} = 94.9 \text{ cm}^3/\text{g-mol}$$
$$\text{Error} = \frac{94.9 - 95.0}{95.0} \times 100 = -0.1\%$$

For a more general correlation, applying to both saturated and compressed liquids, either the Yen-Woods [Eq. (3-15.8)] or the Chueh et al. correlation [Eq. (3-15.18)] may be used.   The latter is not recommended above $T_r \approx 0.99$. Both have been extensively tested with pure materials, both polar and nonpolar. Neither was developed specifically for polar materials, but both allow reasonable estimations to be made and only for accurate work would one have recourse to modified correlations which are specific for polar liquids [41, 51, 118]. In a rather complete study of correlation methods to estimate the effect of both temperature and pressure on pure hydrocarbons, Rea et al. [98] recommended the Chueh et al. correlation for pressure effects but suggest a modified Rackett equation [96] for determining the saturated-liquid volume. In addition, they also expressed Lu's graphical correlation for pressure effects on liquid volumes [66] in analytical form to make it usable in computer estimation techniques. Harmens [43] has also studied low-molecular-weight hydrocarbons and proposed an equation relating the orthobaric liquid density to reduced temperature.

For very highly compressed liquids, special correlation methods [34, 45] may be valuable.

Correlations that relate ρ_L to ρ_v and temperature at saturation are covered elsewhere [103].

NOTATION

In many equations in this chapter, special constants are defined and usually denoted $a, b, \ldots, A, B, \ldots$. They are not defined in this Notation, as they apply to the specific equation and do not occur elsewhere in the chapter.

B = second virial coefficient
F = Redlich-Kwong temperature-dependent function, see Eqs. (3-5.11) to (3-5.15)
M = molecular weight
P = pressure; P_{vp}, vapor pressure; P_c, critical pressure; P_r, reduced pressure, P/P_c
R = gas constant
T = absolute temperature; T_c, critical temperature; T_r, reduced temperature, T/T_c; T_b, normal boiling point
V = molar volume; V_c, critical volume; V_r, reduced volume, V/V_c; V_b, at normal boiling point; V_{sc}, scaling volume, see Eq. (3-15.3); V_{ri}, ideal reduced volume, $V/(RT_c/P_c)$
X = Stiel polar factor, see Sec. 2-6
Z = compressibility factor, PV/RT; Z_c, at the critical point; $Z^{(0)}$, simple fluid compressibility factor, Eq. (3-3.1) and Table 3-1; $Z^{(1)}$, deviation compressibility factor, Eq. (3-3.1) and Table 3-2

Greek

μ_p = dipole moment, debyes; μ_r, reduced dipole moment, see Eq. (3-11.8)
ρ = molar density; ρ_c, critical density; ρ_r, reduced density, ρ/ρ_c; ρ_b, at normal boiling point; ρ_s, saturated-liquid molar density
Ω_a, Ω_b = Redlich-Kwong parameters, see Eqs. (3-5.4) and (3-5.5)
ω = Pitzer acentric factor, see Sec. 2-3

Superscripts

R = reference fluid or reference state

REFERENCES

1. Abbott, M. M.: *AIChE J.*, **19**: 596 (1973).
2. Ackerman, F. J.: M.S. thesis, UCRL-10650, University of California, Berkeley, February 1963.
3. American Chemical Society: Physical Properties of Chemical Compounds, *Adv. Chem. Ser.*, vols. 15, 22, and 29.
4. "ASHRAE Thermodynamic Properties of Refrigerants," 1969, p. 45.
5. Barner, H. E., and S. B. Adler: *Hydrocarbon Process.*, **47**(10): 150 (1968).
6. Barner, H. E., and S. B. Adler: *Ind. Eng. Chem. Fundam.*, **9**: 521 (1970).
7. Barner, H. E., and W. C. Schreiner: *Hydrocarbon Process.*, **45**(6): 161 (1966).
8. Barnés, F. J.: Ph.D. thesis, Department of Chemical Engineering, University of California, Berkeley, 1973; C. J. King, personal communication, 1974.
9. Baxendale, J. H., and B. V. Enüstün: *Phil. Trans. R. Soc. Lond.*, **A243**: 176 (1951).
10. Benedict, M., G. B. Webb, and L. C. Rubin: *Chem. Eng. Prog.*, **47**: 419 (1951); *J. Chem. Phys.*, **8**, 334 (1940), **10**: 747 (1942).
11. Benson, W. S.: *J. Phys. Colloid Chem.*, **52**: 1060 (1948).
12. Beyer, H. H., and R. G. Griskey: *AIChE J.*, **10**: 764 (1964).
13. Bishnoi, P. R., R. D. Miranda, and D. B. Robinson: *Hydrocarbon Process.*, **53**(11), 197 (1974).
14. Bjerre, A., and T. A. Bak: *Acta Chem. Scand.*, **23**: 1733 (1969).
15. Bottemly, G. A., and T. H. Spurling: *Aust. J. Chem.*, **19**: 1331 (1966).
16. Breedveld, G. J. F., and J. M. Prausnitz: *AIChE J.*, **19**: 783 (1973).
17. Chaudron, J., L. Asselineau, and H. Renon: *Chem. Eng. Sci.*, **28**: 839 (1973).
18. Chen, T.-T., and G.-J. Su: personal communication, 1975.
19. Chen, T.-T., and G.-J. Su: *AIChE J.*, **21**: 397 (1975).
20. Chueh, P. L., and J. M. Prausnitz: *AIChE J.*, **13**: 1099 (1967), **15**: 471 (1969).
21. Chueh, P. L., and J. M. Prausnitz: *AIChE J.*, **13**: 1099, 1107 (1967).
22. Chueh, P. L., and J. M. Prausnitz: *Ind. Eng. Chem. Fundam.*, **6**: 492 (1967).
23. Connolly, J. F., and G. A. Kandalic: *Phys. Fluids*, **3**: 463 (1960).
24. Cooper, H. W., and J. C. Goldfrank: *Hydrocarbon Process.*, **46**(12): 141 (1967).
25. Cox, K. W.: S.M. thesis, University of Oklahoma, Norman, 1968.
26. Cox, K. W., J. L. Bono, Y. C. Kwok, and K. E. Starling: *Ind. Eng. Chem. Fundam.*, **10**: 245 (1971).
27. Doolittle, A. K.: *AIChE J.*, **6**: 150, 153, 157 (1960).
28. Dymond, J. H., and E. B. Smith: "The Virial Coefficient of Gases," Clarendon Press, Oxford, 1969.
29. Edmister, W. C.: *Petrol. Refiner*, **37**(4): 173 (1958).
30. Edmister, W. C., J. Vairogs, and A. J. Klekers: *AIChE J.*, **14**: 479 (1968).
31. Edwards, M. N. B., and G. Thodos: *J. Chem. Eng. Data*, **19**: 14 (1974).
32. Estes, J. M., and P. C. Tully: *AIChE J.*, **13**: 192 (1967).
33. Eubank, P. T., and J. M. Smith: *AIChE J.*, **8**: 117 (1962).
34. Ewbank, W. J., and D. G. Harden: *J. Chem. Eng. Data*, **12**: 363 (1967).
35. Fedors, R. F.: *Poly. Eng. Sci.*, **14**: 147, 153 (1974).
36. Francis, A. W.: *Chem. Eng. Sci.*, **10**: 37 (1959); *Ind. Eng. Chem.*, **49**: 1779 (1957).
37. Gray, R. D., Jr., N. H. Rent, and D. Zudkevitch: *AIChE J.*, **16**: 991 (1970).
38. Gunn, R. D., and T. Yamada: *AIChE J.*, **17**: 1341 (1971).
39. Hajjar, R. F., W. B. Kay, and G. F. Leverett: *J. Chem. Eng. Data*, **14**: 377 (1969).
40. Halm, R. L., and L. I. Stiel: *AIChE J.*, **13**: 351 (1967).
41. Halm, R. L., and L. I. Stiel: *AIChE J.*, **16**: 3 (1970).
42. Halm, R. L., and L. I. Stiel: *AIChE J.*, **17**: 259 (1971).
43. Harmens, A.: *Chem. Eng. Sci.*, **20**: 813 (1965), **21**: 725 (1966).
44. Hauthal, W. H., and H. Sackman: *Proc. 1st Int. Conf. Calorimetry Thermodyn.*, *Warsaw*, p. 625 (1969).

45. Hayward, A. T. J.: *Br. J. Appl. Phys.*, **18**: 965 (1967).
46. Horvath, A. L.: *Chem. Eng. Sci.*, **29**: 1334 (1974).
47. Hougen, O., K. M. Watson, and R. A. Ragatz: "Chemical Process Principles," pt. II, Wiley, New York, 1959.
48. Hsi, C. H., and B. C.-Y. Lu: *AIChE J.*, **20**: 616 (1974).
49. Joffe, J.: *Chem. Eng. Prog.*, **45**: 160 (1949).
50. Joffe, J.: *J. Am. Chem. Soc.*, **69**: 540 (1947).
51. Joffe, J., and D. Zudkevitch: *Chem. Eng. Prog. Symp. Ser.*, **70**(140): 22 (1974).
52. Johnson, D. W., and C. P. Colver: *Hydrocarbon Process.*, **47**(12): 79 (1968), **48**(1): 127 (1969).
53. Johnson, J. R., and P. T. Eubank: *Ind. Eng. Chem. Fundam.*, **12**: 156 (1973); Intermolecular Force Constants of Highly Polar Gases, paper presented at *161st Nat. Meet. ACS, Los Angeles, Calif., April 1971.*
54. Kaufman, T. G.: *Ind. Eng. Chem. Fundam.*, **7**: 115 (1968).
55. Knoebel, D. H., and W. C. Edmister: *J. Chem. Eng. Data*, **13**: 312 (1968).
56. Le Bas, G.: "The Molecular Volumes of Liquid Chemical Compounds," Longmans, Green, New York, 1915.
57. Lee, B. I., and W. C. Edmister: *AIChE J.*, **17**: 1412 (1971); *Ind. Eng. Chem. Fundam.*, **10**: 32 (1972).
58. Lee, B. I., J. H. Erbar, and W. C. Edmister: *AIChE J.*, **19**: 349 (1973); *Chem. Eng. Prog.*, **68**(9): 83 (1972).
59. Lee, B. I., and M. G. Kesler: *AIChE J.*, **21**: 510 (1975).
60. Leland, T. W., Jr., and P. S. Chappelear: *Ind. Eng. Chem.*, **60**(7): 15 (1968). See also: G. D. Fisher and T. W. Leland, Jr., *Ind. Eng. Chem. Fundam.*, **9**: 537 (1970).
61. Li, K., R. L. Arnett, M. B. Epstein, R. B. Ries, L. P. Butler, J. M. Lynch, and F. D. Rossini: *J. Phys. Chem.*, **60**: 1400 (1956).
62. Lin, C.-J., and S. W. Hopke: *Chem. Eng. Prog. Symp. Ser.*, **70**(140): 37 (1974).
63. Lin, C.-J., Y. C. Kwok, and K. E. Starling: *Can. J. Chem. Eng.*, **50**: 644 (1972).
64. Lin, M. S., and L. M. Naphtali: *AIChE J.*, **9**: 580 (1963).
65. Lo, H. Y., and L. I. Stiel: *Ind. Eng. Chem. Fundam.*, **8**: 713 (1969).
66. Lu, B. C.-Y.: *Chem. Eng.*, **66**(9): 137 (1959).
67. Lu, B. C.-Y., C. Hsi, S.-D. Chang, and A. Tsang: *AIChE J.*, **19**: 748 (1973).
68. Lu, B. C.-Y., J. A. Ruether, C. Hsi, and C.-H. Chiu: *J. Chem. Eng. Data*, **18**: 241 (1973).
69. Lyckman, E. W., C. A. Eckert, and J. M. Prausnitz: *Chem. Eng. Sci.*, **20**: 703 (1965).
70. Lydersen, A. L., R. A. Greenkorn, and O. A. Hougen: "Generalized Thermodynamic Properties of Pure Fluids," *Univ. Wisconsin, Coll. Eng., Eng. Exp. St. Rep. 4, Madison, Wis.,* October 1955.
71. Martin, J. J.: *AIChE J.*, **18**: 248 (1972).
72. Martin, J. J.: *Chem. Eng. Prog. Symp. Ser.*, **59**(44): 120 (1963).
73. Martin, J. J.: *Ind. Eng. Chem.*, **59**(12): 34 (1967).
74. Martin, J. J., and Y.-C. Hou: *AIChE J.*, **1**: 142 (1955).
75. Martin, J. J., R. M. Kapoor, and N. DeNevers: *AIChE J.*, **5**: 159 (1959).
76. Martin, J. J., and T. G. Stanford: *Chem. Eng. Prog. Symp. Ser.*, **70**(140): 1 (1974).
77. Maslan, F. D., and T. M. Littman: *Ind. Eng. Chem.*, **45**: 1566 (1953).
78. Mason, E. A., and T. H. Spurling: "The Virial Equation of State," Pergamon, New York, 1968.
79. "Matheson Gas Data Book," 5th ed., 1971.
80. Morecroft, D. W.: *J. Inst. Petrol.*, **44**: 433 (1958).
81. Morgan, R. A., and J. A. Childs: *Ind. Eng. Chem.*, **37**: 667 (1945).
82. Nakanishi, K., M. Kurata, and M. Tamura: *Chem. Eng. Data Ser.*, **5**: 210 (1960).
83. Nelson, L. C., and E. F. Obert: *Trans. ASME*, **76**: 1057 (1954).
84. Newton, R. H.: *Ind. Eng. Chem.*, **27**: 302 (1935).
85. O'Connell, J. P., and J. M. Prausnitz: "Advances in Thermophysical Properties at Extreme Temperatures and Pressures," p. 19, ASME, New York, 1965.

86. O'Connell, J. P., and J. M. Prausnitz: *Ind. Eng. Chem. Process Des. Dev.*, **6**: 245 (1967).
87. Opfell, J. B., B. H. Sage, and K. S. Pitzer: *Ind. Eng. Chem.*, **48**: 2069 (1956).
88. Orye, R. V.: *Ind. Eng. Chem. Process Des. Dev.*, **8**: 579 (1969).
89. Partington, J.: "An Advanced Treatise on Physical Chemistry," vol. I, "Fundamental Principles: The Properties of Gases," Longmans, Green, New York, 1949.
90. Pitzer, K. S., and R. F. Curl: *J. Am. Chem. Soc.*, **77**: 3427 (1955).
91. Pitzer, K. S., and R. F. Curl: *J. Am. Chem. Soc.*, **79**: 2369 (1957).
92. Pitzer, K. S., and R. F. Curl: "The Thermodynamic Properties of Fluids," Institution of Mechanical Engineers, London, 1957.
93. Pitzer, K. S., D. Z. Lippmann, R. F. Curl, C. M. Huggins, and D. E. Petersen: *J. Am. Chem. Soc.*, **77**: 3433 (1955).
94. Polak, J., and B. C.-Y. Lu: *Can. J. Chem. Eng.*, **50**: 553 (1972).
95. Prasad, D. H. L., and D. S. Viswanath, Dept. of Chemical Engineering, Indian Institute of Science, Bangalore, India: personal communication, 1974.
96. Rackett, H. G.: *J. Chem. Eng. Data*, **15**: 514 (1970).
97. Rao, A. S., and D. S. Viswanath: *Ind. J. Tech.*, **9**: 476 (1971).
98. Rea, H. E., C. F. Spencer, and R. P. Danner: *J. Chem. Eng. Data*, **18**: 227 (1973).
99. Redlich, O., F. J. Ackerman, R. D. Gunn, M. Jacobson, and S. Lau: *Ind. Eng. Chem. Fundam.*, **4**: 369 (1965).
100. Redlich, O., and A. K. Dunlop: *Chem. Eng. Prog. Symp. Ser.*, **59**: 95 (1963).
101. Redlich, O., and J. N. S. Kwong: *Chem. Rev.*, **44**: 233 (1949).
102. Redlich, O., and V. B. T. Ngo: *Ind. Eng. Chem. Fundam.*, **9**: 287 (1970).
103. Reid, R. C., and T. K. Sherwood: "Properties of Gases and Liquids," 2d ed., pp. 95–107, McGraw-Hill, New York, 1966.
104. Ritter, R. B., J. M. Lenoir, and J. L. Schweppe: *Petrol. Refiner*, **37**(11): 225 (1958).
105. Roberts, F. D.: M.S. thesis, University of Oklahoma, Norman, 1968.
106. Rowlinson, J. S.: *Trans. Faraday Soc.*, **45**: 975 (1949).
107. Satter, A., and J. M. Campbell: *Soc. Petrol. Eng. J.*, December **1963**: 333.
108. Soave, G.: *Chem. Eng. Sci.*, **27**: 1197 (1972).
109. Sokolov, B. I., A. A. Kharchenko, O. N. Man'kovskii, and A. N. Davydov: *Russ. J. Phys. Chem.*, **47**(2): 239 (1973).
110. Sood, S. K., and G. G. Haselden: *AIChE J.*, **16**: 891 (1970).
111. Spencer, C. F., and R. P. Danner: *J. Chem. Eng. Data*, **17**: 236 (1972).
112. Starling, K. E.: *Hydrocarbon Process.*, **50**(3): 101 (1971).
113. Starling, K. E.: paper presented at *NGPA Annual Conv., March 17–19, 1970, Denver, Colo.*
114. Starling, K. E., and M. S. Han: *Hydrocarbon Process.*, **51**(5): 129 (1972), **51**(6): 107 (1972).
115. Starling, K. E., and J. E. Powers: paper presented at *159th ACS Meet. Feb. 22–27, 1970, Houston, Tex.*
116. Stiel, L. I.: *Chem. Eng. Sci.*, **27**: 2109 (1972).
117. Stiel, L. I.: *Ind. Eng. Chem.*, **60**(5): 50 (1968).
118. Stipp, G. K., S. D. Bai, and L. I. Stiel: *AIChE J.*, **19**: 1227 (1973).
119. Storvick, T. S., and T. H. Spurling: *J. Phys. Chem.*, **72**: 1821 (1968); Suh, K. W., and T. S. Storvick, *J. Phys. Chem.*, **71**: 1450 (1967).
120. Su, G.-J., and C.-H. Chang: *Ind. Eng. Chem.*, **38**: 802, 803 (1969).
121. Su, G. S., and D. S. Viswanath: *AIChE J.*, **11**: 205 (1965).
122. Sugie, H., and B. C.-Y. Lu: *AIChE J.*, **17**: 1068 (1971).
123. Suh, K. W., and T. S. Storvick: *J. Phys. Chem.*, **72**: 1821 (1968).
124. Tarakad, R., and T. E. Daubert, Pennsylvania State University, University Park: private communication, Sept. 23, 1974.
125. "Technical Data Book, Petroleum Refining," American Petroleum Institute, 2d ed., Washington, D.C., 1971.
126. Tseng, J. K., and L. I. Stiel: *AIChE J.*, **17**: 1283 (1971).
127. Tsonopoulos, C.: *AIChE J.*, **20**: 263 (1974).

128. Tsonopoulos, C.: *AIChE J.*, **21**: 827 (1975).
129. Tsonopoulos, C., and J. M. Prausnitz: *Cryogenics*, October **1969**: 315.
130. Tyn, M. T., and W. F. Calus: *Processing*, **21**(4): 16 (1975).
131. Vera, J. H., and J. M. Prausnitz: *Chem. Eng. J.*, **3**: 1 (1972).
132. Vetere, A.: private communication, 1974.
133. Vogl, W. F., and K. R. Hall: *AIChE J.*, **16**: 1103 (1970).
134. Weast, R. C.: "Handbook of Chemistry and Physics," 49th ed., Chemical Rubber Co., Cleveland, 1968.
135. West, E. W., and J. H. Erbar: An Evaluation of Four Methods of Predicting the Thermodynamic Properties of Light Hydrocarbon Systems, paper presented at *52d Annu. Meet. NGPA, Dallas, Tex., March 26–28, 1973.*
136. Wilson, G. M.: *Adv. Cryog. Eng.*, **9**: 168 (1964), **11**: 392 (1966).
137. Wilson, G. M.: *65th Natl. Meet., AIChE, Cleveland, May 4–7, 1969,* pap. 15C.
138. Yamada, T.: *AIChE J.*, **19**: 286 (1973).
139. Yamada, T., and R. D. Gunn: *J. Chem. Eng. Data*, **18**: 234 (1973).
140. Yang, C.-L., and E. F. Yendall: *AIChE J.*, **17**: 596 (1971).
141. Yen, L. C., and S. S. Woods: *AIChE J.*, **12**: 95 (1966).
142. Zaalishvili, Sh. D., and Z. S. Belousova: *Russ. J. Phys. Chem.*, **38**: 269 (1964).
143. Zudkevitch, D., and T. G. Kaufman: *AIChE J.*, **12**: 577 (1966).

Chapter Four

Mixture Combination Rules

4-1 Scope

Chapter 3 reviewed methods for calculating the P-V-T properties of gases and liquids. To extend these methods to mixtures, they must be modified to include the additional variable of composition. In essentially all cases this is accomplished by making the scaling factors or equation-of-state constants a function of composition. Many algebraic relations have been suggested, although it can be shown (Sec. 4-9) that essentially all can be derived from a single general expression. One must keep in mind that the rules covered in this chapter are, with a single exception [Eq. (4-8.1)], essentially empirical and have resulted after many trials and comparisons of calculated mixture properties with experimental data. Our treatment is to present first the recommended mixing rules for all gas P-V-T estimation methods covered in Chap. 3 and then follow with a general discussion and comparison. Rules applicable to liquid mixtures are covered in Sec. 4-10.

In Chap. 5 we examine the thermodynamic properties of both pure components and mixtures; for the former, one may simply use the constants given with each method outlined in Chap. 3.   For the latter, the equation-of-state parameters must be determined separately for each mixture with a different composition.

4-2   Corresponding-States Methods for Mixtures

In Sec. 3-3, two corresponding-states methods are discussed.   The compressibility factor is related to the reduced temperature, reduced pressure, and, by Eq. (3-3.1), the acentric factor.   To apply these same methods to mixtures, rules must be devised to relate the pseudocritical mixture constants to composition.   Many such rules have been suggested [11, 15, 21].

For the pseudocritical temperature T_{c_m}, a simple mole-fraction average method is usually satisfactory.   This rule, often called *Kay's rule* [13], is

$$T_{c_m} = \sum_j y_j T_{c_j} \tag{4-2.1}$$

Comparison of T_{c_m} from Eq. (4-2.1) with values determined from other, more complicated rules shows that the differences are usually less than 2 percent if, for all components,

$$0.5 < \frac{T_{c_i}}{T_{c_j}} < 2 \qquad \text{and} \qquad 0.5 < \frac{P_{c_i}}{P_{c_j}} < 2^{\dagger}$$

For the pseudocritical pressure, a simple mole-fraction average of the pure-component critical pressures is normally not satisfactory unless all components have similar critical pressures or critical volumes.   The simplest rule which gives acceptable results is the modified Prausnitz and Gunn combination [18]

$$P_{c_m} = \frac{R \left(\sum_j y_j Z_{c_j} \right) T_{c_m}}{\sum_j y_j V_{c_j}} \tag{4-2.2}$$

In all cases, the mixture acentric factor is given by the approximation [12]:

$$\omega_m = \sum_j y_j \omega_j \tag{4-2.3}$$

No binary (or higher) interaction parameters are included in Eqs. (4-2.1) to (4-2.3); thus these mixing rules cannot truly reflect mixture properties.   Yet surprisingly good results are often obtained when these simple pseudomixture parameters are used in corresponding-states calculations to determine mixture properties.

Less satisfactory results are found for mixtures of dissimilar compo-

†Ref. 21.

nents, especially if one or more of the components is polar or shows any tendency to associate into dimers, etc.

However, if one has available some experimental data for any of the possible binaries in the mixture, it is frequently worthwhile to use these data to modify the pseudocritical rules. Though many options are open, one which has often proved successful is to change Eq. (4-2.1) from a linear to a quadratic form:

$$T_{c_m} = \sum_i \sum_j y_i y_j T_{c_{ij}} \tag{4-2.4}$$

with $\qquad T_{c_{ii}} = T_{c_i} \qquad$ and $\qquad T_{c_{ij}} = k_{ij}^* \frac{(T_{c_i} + T_{c_j})}{2} \qquad$ (4-2.5)[†]

From the available data, the best values of the binary constants k_{ij}^* are back-calculated by trial and error. If k_{ij}^* is assumed equal to unity, Eq. (4-2.4) reduces to Eq. (4-2.1). With k_{ij}^* values for all possible binary sets, multicomponent-mixture properties can then be estimated. In Table 4-3, k_{ij}^* values are tabulated for many binary pairs [1].

There is nothing rigorous in this approach. First, Eq. (4-2.5) may not be the best form for expressing $T_{c_{ij}}$. From molecular theory, the geometric mean of T_{c_i} and T_{c_j} could more easily be defended. Second, no ternary or higher interaction parameters are considered. Third, it is probably too much to expect that for a given binary, there is but a single value of k_{ij}^* that is not a function of composition, temperature, or pressure. Only by experience have we obtained some confidence in approaches of this sort.

We consider next how the analytical equations of state mentioned in Chap. 3 can be modified for mixtures.

4-3 Redlich-Kwong Equation of State

For the original Redlich-Kwong equation of state introduced in Sec. 3-5, the parameters A^* and B^*, defined in Eqs. (3-5.7) and (3-5.8), must be determined for mixtures. As suggested in the original paper,

$$A_m^{*1/2} = \sum_j y_j A_j^{*1/2} \tag{4-3.1}$$

$$B_m^* = \sum_j y_j B_j^* \tag{4-3.2}$$

Assuming Ω_a and Ω_b are constants [as in Eqs. (3-5.4) and (3-5.5)], Eqs. (4-3.1) and (4-3.2) imply that Eqs. (3-5.7) and (3-5.8) can be used directly

---

[†]In some articles, $T_{c_{ij}} = \sqrt{T_{c_i} T_{c_j}}(1 - k_{ij})$. It is clear that k_{ij}^* and k_{ij} are not the same, although they are easily related to one another.

to determine A_m^* and B_m^* if the pseudocritical temperature and pressure of the mixture are defined as

$$T_{c_m} = \left\{ \frac{\left[\sum_j y_j (T_{c_j}^{5/2}/P_{c_j})^{1/2} \right]^2}{\sum_j y_j (T_{c_j}/P_{c_j})} \right\}^{2/3} \tag{4-3.3}$$

$$P_{c_m} = \frac{T_{c_m}}{\sum_j y_j (T_{c_j}/P_{c_j})} \tag{4-3.4}$$

These same pseudocritical mixing rules are applicable to the Gray-Rent-Zudkevitch modification of the Redlich-Kwong equation and can

TABLE 4-1 F_m Functions for Eq. (4-3.6)

Equation	F_m
Original Redlich-Kwong	$\dfrac{\left[\sum_j y_j (T_{c_j}/T_{r_j}^{1.5}P_{c_j})^{1/2} \right]^2}{\sum_j y_j (T_{c_j}/P_{c_j})}$
Wilson	$\sum_j y_j F_j$ F_j is determined from Eq. (3-5.13)
Barnés-King	$\dfrac{\sum_i \sum_j y_i y_j [(T_{c_i}/P_{c_i} + T_{c_j}/P_{c_j}) F_{ij}/2]}{\sum_j y_j (T_{c_j}/P_{c_j})}$ $F_{ij} = 1 + (0.9 + 1.21\omega_{ij}) \left[\left(\dfrac{T}{T_{c_{ij}}} \right)^{-1.5} - 1 \right]$ $\omega_{ij} = \dfrac{\omega_i + \omega_j}{2}$ $T_{c_{ij}}$ = interaction critical temperature, obtained from experimental data by back-calculation or from a rule such as Eq. (4-2.5)
Soave	$\dfrac{\sum_i \sum_j y_i y_j (1 - \bar{k}_{ij})[(T_{c_i} T_{c_j}/P_{c_i} P_{c_j}) F_i F_j]^{1/2}}{\sum_j y_j T_{c_j}/P_{c_j}}$ F_i, F_j are determined from Eq. (3-5.15); the interaction parameter $\bar{k}_{ij}$ is approximately zero for hydrocarbon-hydrocarbon binary pairs; for a few other systems, $\bar{k}_{ij}$ is shown in Table 4-2

TABLE 4-2 $\bar{k}_{ij}$ **Values for the Soave Modification of the Redlich-Kwong Equation** [8]

	Carbon dioxide	Hydrogen sulfide	Nitrogen	Carbon monoxide
Methane	0.12	0.08	0.02	−0.02
Ethylene	0.15	0.07	0.04	
Ethane	0.15	0.07	0.06	
Propylene	0.08	0.07	0.06	
Propane	0.15	0.07	0.08	
Isobutane	0.15	0.06	0.08	
n-Butane	0.15	0.06	0.08	
Isopentane	0.15	0.06	0.08	
n-Pentane	0.15	0.06	0.08	
n-Hexane	0.15	0.05	0.08	
n-Heptane	0.15	0.04	0.08	
n-Octane	0.15	0.04	0.08	
n-Nonane	0.15	0.03	0.08	
n-Decane	0.15	0.03	0.08	
n-Undecane	0.15	0.03	0.08	
Carbon dioxide	. . .	0.12	. . .	−0.04
Cyclohexane	0.15	0.03	0.08	
Methyl cyclohexane	0.15	0.03	0.08	
Benzene	0.15	0.03	0.08	
Toluene	0.15	0.03	0.08	
o-Xylene	0.15	0.03	0.08	
m-Xylene	0.15	0.03	0.08	
p-Xylene	0.15	0.03	0.08	
Ethylbenzene	0.15	0.03	0.08	

be used to calculate T_{r_m} and P_{r_m} in Eq. (3-5.10) to determine ΔZ_m. In this relation,

$$\omega_m = \sum_j y_j \omega_j \qquad (4\text{-}3.5)$$

For the Redlich-Kwong modification given in Eq. (3-5.11), for mixtures

$$Z_m = \frac{V_m}{V_m - b_m} - \frac{\Omega_a}{\Omega_b} \frac{b_m}{V_m + b_m} F_m \qquad (4\text{-}3.6)$$

Ω_a and Ω_b are defined in Eqs. (3-5.4) and (3-5.5). All authors recommend

$$b_m = \sum_j y_j b_j \qquad (4\text{-}3.7)$$

where b_j is given in Eq. (3-5.3), i.e.,

$$b_j = \frac{\Omega_b R T_{c_j}}{P_{c_j}} \qquad (4\text{-}3.8)$$

The factor F_m, however, is expressed in different ways. Table 4-1 shows the recommended relations for the original Redlich-Kwong equation,† the Wilson [27], Barnés-King [2], and Soave [4, 22] modifications.

4-4 Barner-Adler Equation of State

This equation of state was described in Sec. 3-6. To employ it for mixtures, one must relate T_{c_m}, P_{c_m}, and ω_m to composition. The recommended equations are

$$T_{c_m} = \sum_i \sum_j y_i y_j T_{c_{ij}} \tag{4-4.1}$$

$$T_{c_{ij}} = k_{ij}^* \frac{(T_{c_i} + T_{c_j})}{2} \tag{4-4.2}$$

$$V_{c_m} = \sum_i \sum_j y_i y_j V_{c_{ij}} \tag{4-4.3}$$

$$V_{c_{ij}} = \left(\frac{V_{c_i}^{1/3} + V_{c_j}^{1/3}}{2} \right)^3 \tag{4-4.4}$$

$$P_{c_m} = \frac{R Z_{c_m} T_{c_m}}{V_{c_m}} \tag{4-4.5}$$

$$Z_{c_m} = 0.291 - 0.08\omega_m \tag{4-4.6}$$

$$\omega_m = \sum_j y_j \omega_j \tag{4-4.7}$$

The computation of T_{c_m} and P_{c_m} is straightforward if k_{ij}^* values for all possible binaries are known. These parameters are, however, a function of the binary ij and not of either i or j alone. Each is related to a similar interaction parameter first proposed by Chueh and Prausnitz [6]. Table 4-3 shows a number of k_{ij}^* parameters given by Barner and Quinlan [1]; these were back-calculated from the Chueh-Prausnitz parameters. Most k_{ij}^* values are near unity. However, even a small variation from unity can appreciably affect estimated properties.

In simple binaries, k_{ij}^* can be approximately correlated with the ratio of the critical volumes of pure i and j, as shown in Fig. 4-1. k_{ij}^* is assumed to be independent of temperature, pressure, or composition. If values are not given in Table 4-3 and Fig. 4-1 appears inappropriate, one must either assume a value of k_{ij}^* or obtain at least some experimental P-V-T data from which it can be estimated.

As $y_A \to 1$, P_{c_m} does not necessarily reduce to the critical pressure P_{c_A} of the pure component A unless $Z_{c_A} = 0.291 - 0.08\omega_A$.

---

†These reduce to Eqs. (4-3.1) and (4-3.2).

TABLE 4-3   Barner and Quinlan k_{ij}^* Values [1]

Component i	Component j	k_{ij}^*	Component i	Component j	k_{ij}^*
Methane	Ethylene	1.01	n- or Isobutane	Isobutane	1.00
	Ethane	1.03		n-Pentane	1.00
	Propylene	1.06		Isopentane	1.00
	Propane	1.07		n-Hexane	1.02
	n-Butane	1.11		n-Heptane	1.03
	Isobutane	1.11		Cyclohexane	1.01
	n-Pentane	1.15	n- or Isopentane	Isopentane	1.00
	Isopentane	1.15		n-Hexane	1.00
	n-Hexane	1.19		n-Heptane	1.01
	n-Heptane	1.22		n-Octane	1.02
	Cyclohexane	1.16		Cyclohexane	1.00
	Naphthalene	1.23	n-Hexane	n-Heptane	1.00
Ethylene	Ethane	1.00		n-Octane	1.01
	Propylene	1.02		Toluene	0.98
	Propane	1.02	Cyclohexane	n-Heptane	1.00
	n-Butane	1.05		n-Octane	1.00
	Isobutane	1.05		Toluene	0.99
	n-Pentane	1.08	n-Heptane	n-Octane	1.01
	Isopentane	1.08	Nitrogen	Methane	0.97
	n-Hexane	1.11		Ethylene	1.01
	Cyclohexane	1.09		Ethane	1.02
	n-Heptane	1.13		n-Butane	1.13
	Benzene	1.07		1-Pentene	1.13
	Naphthalene	1.15		1-Hexene	1.25
Ethane	Propylene	1.01		n-Hexane	1.26
	Propane	1.01		n-Heptane	1.31
	n-Butane	1.03		n-Octane	1.34
	Isobutane	1.03	Argon	Oxygen	0.99
	n-Pentane	1.05		Nitrogen	0.99
	Isopentane	1.05	Carbon dioxide	Ethylene	0.94
	n-Hexane	1.08		Ethane	0.92
	n-Heptane	1.10		Propylene	0.93
	Cyclohexane	1.06		Propane	0.93
	Benzene	1.04		n-Butane	0.93
	Naphthalene	1.11		Naphthalene	1.07
Propylene	Propane	1.00	Hydrogen sulfide	Methane	0.93
	n-Butane	1.01		Ethane	0.92
	Isobutane	1.01		Propane	0.92
	n-Pentane	1.02		n-Pentane	0.96
	Isopentane	1.03		Carbon dioxide	0.92
	Benzene	1.03	Acetylene	Ethylene	0.94
Propane	n-Butane	1.01		Ethane	0.92
	Isobutane	1.01		Propylene	0.95
	n-Pentane	1.01		Propane	0.94
	Isopentane	1.02	Hydrogen chloride	Propane	0.88
	Benzene	1.00			

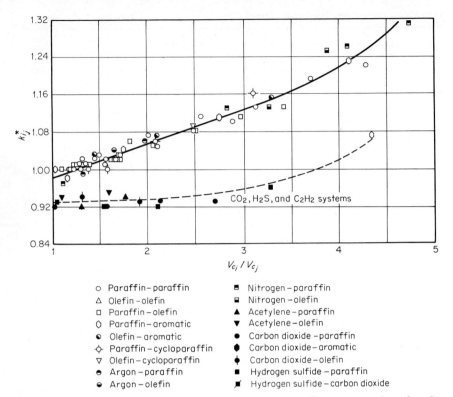

Fig. 4-1 Plot of pseudocritical temperature-interaction coefficients vs. ratios of molar critical volumes. (*From Ref. 1.*)

4-5 Sugie-Lu Equation of State

For the Sugie-Lu equation of state discussed in Sec. 3-7 and summarized in Table 3-5, the following rules have been suggested for obtaining T_{c_m}, ρ_{c_m}, and Z_{c_m}:

$$T_{c_m} = \left\{ \frac{\left[\sum_i \sum_j y_i y_j (1 - k'_{ij})(T_{c_i}^{2.5}/P_{c_i})(T_{c_j}^{2.5}/P_{c_j}) \right]^{1/2}}{\sum_j y_i (T_{c_j}/P_{c_j})} \right\}^{2/3} \tag{4-5.1}$$

$$P_{c_m} = \frac{T_{c_m}}{\sum_j (y_j T_{c_j}/P_{c_j})} \tag{4-5.2}$$

$$\omega_m = \sum_j y_j \omega_j \tag{4-5.3}$$

$$Z_{c_m} = 0.291 - 0.080 \omega_m \tag{4-5.4}$$

The binary interaction parameter in this case is k'_{ij} and is normally near zero. It is not simply related to the k^*_{ij} factor in the Barner-Adler equation.   Almost no values of k'_{ij} are available, and it must either be assumed zero or determined from experimental data.   If k'_{ij} is assumed zero, the rules for pseudocritical temperature and pressure collapse to those given for the original Redlich-Kwong equation, Eqs. (4-3.3) and (4-3.4).

4-6 Benedict-Webb-Rubin Equation of State

The Benedict-Webb-Rubin equation of state was introduced in Sec. 3-8.   As originally proposed, the eight constants are related to composition by an equation of the form

$$\chi_m = \left(\sum_j y_j \chi_j^{1/r} \right)^r \qquad (4\text{-}6.1)$$

Suggested values of r for each Benedict-Webb-Rubin constant are

Constant (χ)	r	Constant (χ)	r
B_0	1	b	3
A_0	2	c	3
C_0	2	α	3
a	3	γ	2

Although these simple mixing rules have been employed in numerous studies, variations have been proposed to increase accuracy and to broaden applicability.   Stotler and Benedict [25], Orye [16], Gugnoni et al. [10], and Bishnoi and Robinson [3] have recommended that the A_0 rule be modified to

$$A_{0_m} = \sum_i \sum_j y_i y_j L_{ij} A_{0_i}^{1/2} A_{0_j}^{1/2} \qquad (4\text{-}6.2)$$

If the interaction parameter L_{ij} is unity, Eq. (4-6.2) reduces to the original Benedict-Webb-Rubin mixing rule.   Variations of L_{ij} from unity tend to be more important for mixtures at low temperatures. L_{ij} has been determined for CO_2-ethane mixtures by Gugnoni et al. [10], and it was found to be a weak function of temperature ranging from 0.87 at 10°C to 0.82 at -57°C.   Normally, however, it is assumed independent of temperature, pressure, and composition.   Approximate values of L_{ij} can be determined by equating L_{ij} to $1 - k_{ij}$, where k_{ij} interaction parameters are reported by Chueh and Prausnitz [6].

Other authors [3, 24, 25] have suggested modifications to the original Benedict-Webb-Rubin mixing rules, but with the exception of Eq. (4-6.2) none has been widely adopted.

For the modified Benedict-Webb-Rubin equation proposed by Lee and Kesler (Sec. 3-9), the recommended mixing rules are

$$T_{c_m} = \frac{1}{8V_{c_m}} \sum_i \sum_j y_i y_j (V_{c_i}^{1/3} + V_{c_j}^{1/3})^3 (T_{c_i} T_{c_j})^{1/2} \qquad (4\text{-}6.3)$$

$$V_{c_m} = \frac{1}{8} \sum_i \sum_j y_i y_j (V_{c_i}^{1/3} + V_{c_j}^{1/3})^3 \qquad (4\text{-}6.4)$$

$$V_{c_i} = \frac{(0.2905 - 0.085\omega_i)RT_{c_i}}{P_{c_i}} \qquad (4\text{-}6.5)$$

$$\omega_m = \sum_i y_i \omega_i \qquad (4\text{-}6.6)$$

$$P_{c_m} = \frac{(0.2905 - 0.085\omega_m)RT_{c_m}}{V_{c_m}} \qquad (4\text{-}6.7)$$

These rules should also be used when Tables 3-1 and 3-2 are applied to mixtures.   Testing has been limited to hydrocarbon mixtures.

4-7   Lee-Erbar-Edmister Equation of State

The Lee-Erbar-Edmister equation of state introduced in Sec. 3-10 was designed primarily for hydrocarbon-vapor mixtures.   The constants for pure materials are shown in Eqs. (3-10.1) to (3-10.4).   For mixtures, the combination rules are shown in Table 4-4.

TABLE 4-4   Lee-Erbar-Edmister Mixing Rules

$b_m = \sum_j y_j b_j$	$(4\text{-}7.1)$
$a_m = \sum_i \sum_j y_i y_j \alpha_{ij} (a_i a_j)^{1/2}$	$(4\text{-}7.2)$
$c_m = \sum_i \sum_j y_i y_j \beta_{ij} (c_i c_j)^{1/2}$	$(4\text{-}7.3)$
$\beta_{ij} = \left[\dfrac{T_{c_i} + T_{c_j}}{2(T_{c_i} T_{c_j})^{1/2}} \right]^{m_1} \qquad \alpha_{ij} = \beta_{ij}^{m_2}$	$(4\text{-}7.4)$

$$\text{where } m_1 = \begin{cases} -0.3 & \text{if } i \text{ or } j \text{ is } H_2 \\ 2.0 & \text{if neither } i \text{ nor } j \text{ is } H_2 \text{ but either contains } CO_2, N_2, \text{ or } CH_4 \\ 0 & \text{if neither } i \text{ nor } j \text{ is } H_2, CO_2, N_2, \text{ or } CH_4 \end{cases}$$

$$m_2 = \begin{cases} -1.1 & \text{if } i \text{ or } j \text{ is } H_2 \\ -0.8 & \text{if neither } i \text{ nor } j \text{ is } H_2 \text{ but either is } CO_2 \\ 0 & \text{if neither } i \text{ nor } j \text{ is } H_2 \text{ or } CO_2 \end{cases}$$

4-8 Second Virial Coefficients for Mixtures

The truncated virial equation shown in Sec. 3-11 is the only equation for real gases for which an exact relation is known for mixture coefficients:

$$B_m = \sum_i \sum_j y_i y_j B_{ij} \tag{4-8.1}$$

For example, for a ternary mixture of 1, 2, and 3,

$$B_m = y_1^2 B_1 + y_2^2 B_2 + y_3^2 B_3 + 2y_1 y_2 B_{12} + 2y_1 y_3 B_{13} + 2y_2 y_3 B_{23} \tag{4-8.2}$$

The pure-component second virial coefficients, B_1, B_2, B_3, can often be found as described in Sec. 3-11 [e.g., Eqs. (3-11.3) to (3-11.5)] from the critical temperatures, critical pressures, and acentric factors of the pure materials. For the mixture interaction virials B_{ij}, if one still wished to employ Eq. (3-11.3), combination rules must be devised to obtain $T_{c_{ij}}$, $P_{c_{ij}}$, and ω_{ij}. This problem has been discussed from a theoretical point of view by Leland and Chappelear [15] and Ramaiah and Stiel [20]. For typical engineering calculations involving normal fluids, the following simple rules are useful [17, 26]:

$$T_{c_{ij}} = (T_{c_i} T_{c_j})^{1/2} (1 - k_{ij}) \tag{4-8.3}$$

$$V_{c_{ij}} = \left(\frac{V_{c_i}^{1/3} + V_{c_j}^{1/3}}{2} \right)^3 \tag{4-8.4}$$

$$Z_{c_{ij}} = \frac{Z_{c_i} + Z_{c_j}}{2} \tag{4-8.5}$$

$$\omega_{ij} = \frac{\omega_i + \omega_j}{2} \tag{4-8.6}$$

$$P_{c_{ij}} = \frac{Z_{c_{ij}} R T_{c_{ij}}}{V_{c_{ij}}} \tag{4-8.7}$$

These relations are probably reliable only for molecules which do not differ greatly in size or chemical structure. For a first approximation, the binary constant k_{ij} can be set equal to zero. Lists of these constants for a variety of binary systems have been published [6]. Also, if the molecule pairs contain a quantum gas (H_2, He, or Ne), modified critical constants are recommended [17].

4-9 Mixing Rules

We do not intend to consider any theory that might suggest the form of the mixing rules outlined in this chapter. This topic is discussed elsewhere [15, 21]. It is usually found, however, that if one can relate an equation-of-state interaction parameter, in a physical sense, to a term which depends on the volume of i and j, normally it is of the form

$$Y_{ij}^{1/3} = \frac{Y_i^{1/3} + Y_j^{1/3}}{2} \tag{4-9.1}$$

with Y proportional to V. This relation results from the assumption that the interaction molecular diameter σ is an arithmetic average of the molecular diameters of i and j:

$$\sigma_{ij} = \frac{\sigma_i + \sigma_j}{2} \tag{4-9.2}$$

If Y is proportional to σ^3, Eq. (4-9.1) results.

Similarly, if any constant represents or is proportional to an attractive energy of interaction, theory indicates that, very approximately,

$$W_{ij} = (W_i W_j)^{1/2} \tag{4-9.3}$$

with W proportional to energy. Since energies of interaction are often associated with critical temperature, Eq. (4-9.3) is normally found in mixture-combination rules for $T_{c_{ij}}$; e.g., see Eq. (4-8.3).

In addition to these rough rules, we should note that theory gives the mixture second virial B_m as a general quadratic in mole fraction [see Eq. (4-8.1)] and the mixture third virial as a general cubic in mole fraction, etc. Such expressions have prompted many investigators to formulate combination rules to match the virial expansion for a mixture.

For example, suppose one desired to express a mixture parameter Q_m in terms of composition as a general quadratic:

$$Q_m = \sum_i \sum_j y_i y_j Q_{ij} \tag{4-9.4}$$

Terms such as Q_{ii} and Q_{jj} involve only pure-component properties. The difficulty arises when interaction terms Q_{ij} are considered. If Q_{ij} is set equal to the arithmetic average, then

$$Q_m = \sum_j y_j Q_j \qquad Q_{ij} = \frac{Q_i + Q_j}{2} \tag{4-9.5}$$

If, instead, Q_{ij} is chosen as the geometric mean,

$$Q_m = \left(\sum_j y_j Q_j^{1/2} \right)^2 \qquad Q_{ij} = (Q_i Q_j)^{1/2} \tag{4-9.6}†$$

Equations (4-9.5) and (4-9.6) are special cases of Eq. (4-6.1). There are many examples of both Eqs. (4-9.5) and (4-9.6) given in specific methods presented earlier in the chapter. To introduce empirically

†This form is meaningful only if all $Q > 0$.

determined binary interaction parameters, one has many choices.   Two obvious ones are to define

$$Q_{ij} = L_{ij}^* \frac{Q_i + Q_j}{2} \tag{4-9.7}$$

and

$$Q_{ij} = L_{ij}^{**}(Q_iQ_j)^{1/2} \tag{4-9.8}$$

With Eqs. (4-9.4) and (4-9.7),

$$Q_m = \sum_i \sum_j y_iy_jL_{ij}^* \frac{Q_i + Q_j}{2} \tag{4-9.9}$$

For a binary of 1 and 2, Eq. (4-9.9) becomes

$$Q_m = y_1^2Q_1 + y_2^2Q_2 + L_{12}^*y_1y_2(Q_1 + Q_2) \tag{4-9.10}$$

and if $L_{12}^* = 1.0$, Eq. (4-9.10) reduces to Eq. (4-9.5).

With Eqs. (4-9.4) and (4-9.8),

$$Q_m = \sum_i \sum_j y_iy_jL_{ij}^{**}Q_i^{1/2}Q_j^{1/2} \tag{4-9.11}$$

For a binary of 1 and 2, Eq. (4-9.11) becomes

$$Q_m = y_1^2Q_1 + y_2^2Q_2 + 2L_{12}^{**}Q_1^{1/2}Q_2^{1/2} \tag{4-9.12}$$

Equation (4-9.11) is illustrated by the Benedict-Webb-Rubin quadratic rule, Eq. (4-6.2).   For $L_{12}^{**} = 1.0$, Eq. (4-9.6) is recovered.

The discussion thus far has dealt only with ways in which Eq. (4-9.4) can be expanded.   In some cases, however, a general cubic is chosen to represent the mixture property:

$$Q_m = \sum_i \sum_j \sum_k y_iy_jy_kQ_{ijk} \tag{4-9.13}$$

If, for all i, j, k,

$$Q_{ijk} = (Q_iQ_jQ_k)^{1/3} \tag{4-9.14}$$

Equation (4-6.1) is obtained with $r = 3$.   If an arithmetic mean is taken instead, Eq. (4-6.1) is still obtained but $r = 1$.   Only seldom is a ternary interaction parameter introduced (see, however, Ref. 14).

Generally, when one expresses Q_m as

$$Q_m = \sum_1 \cdots \sum_n y_1 \cdots y_nQ_{1 \cdots n} \tag{4-9.15}$$

if,

$$Q_{1 \cdots n} = \prod^n Q_i^{1/n}$$

then Eq. (4-6.1) results with $r = n$.   However, if

$$Q_{1 \cdots n} = \frac{\sum^n Q_i}{n}$$

Q_m can be determined by a simple mole-fraction average.

In summary, it is sufficient to note that all mixing rules suggested in this chapter are empirical and most are simplified forms of either Eq. (4-9.4) or (4-9.13).   Any particular rule must be justified by comparison of predicted with experimental properties.   Such comparisons are made later in Chap. 5.   In general, the introduction of binary interaction parameters into equation-of-state mixing rules is often not important when overall properties of the mixture are calculated, for example, V_m, H_m.   However, they may become very important when partial molal properties are calculated.   This importance is amplified when determining partial molal properties of a component whose mole fraction in the mixture is small.

4-10   Mixing Rules for Liquid Mixtures

For mixtures of liquids containing components which do not differ greatly, at modest pressures it is often a good rule to assume Amagat's law

$$V_m{}^L = \sum_j x_j V_j^L \tag{4-10.1}$$

where V_j^L refers to the molar volume of pure liquid j at the same temperature as that of the mixture.

To adapt the pure-liquid density correlations given in Sec. 3-15 to mixtures, pseudocritical mixture parameters must be chosen.   No rules have yet been formulated for the Gunn-Yamada method.   In the Yen-Woods correlation [Eqs. (3-15.8) to (3-15.17)] mole-fraction averages are suggested for T_{c_m}, V_{c_m}, and Z_{c_m} but for P_{c_m}, Eq. (4-2.2) should be used.

Although these simple mixture rules are often satisfactory at low temperatures, problems arise near the true critical point of the mixture.   The pseudocritical values given are *not* the true critical values, yet at the true critical, $V^L \rightarrow V_{c_m}$ (true); this is not reflected in Yen and Woods' method if one employs only pseudocritical parameters.

This critical-point problem was taken into account in the Chueh and Prausnitz–Lyckman, Eckert, and Prausnitz liquid-density correlation given in Eqs. (3-15.18) to (3-15.21).   A pseudocritical temperature is defined as

$$T_{c_m} = \sum_i \sum_j \phi_i \phi_j T_{c_{ij}} \tag{4-10.2}$$

where ϕ_i is a volume fraction defined by

$$\phi_i = \frac{x_i V_{c_i}}{\sum_j x_j V_{c_j}} \tag{4-10.3}$$

and $$T_{c_{ij}} = (1 - k_{ij}{}^L)(T_{c_i} T_{c_j})^{1/2} \tag{4-10.4}$$

$k_{ij}{}^L$ is a binary interaction parameter which must be found from experimental data although it can be correlated with pure-component critical volumes for paraffin-paraffin mixtures [7]. Normally, $k_{ij}{}^L$ is between 0 and 0.2.

Two regions are considered, one less than $T_{r_m} = T/T_{c_m} < 0.93$ and the other with $1.0 > T_{r_m} > 0.93$. For $T_{r_m} < 0.93$, Eqs. (3-15.18) to (3-15.21) are employed as for pure materials with T_{c_m} found from Eqs. (4-10.2) to (4-10.4) and

$$\frac{1}{\rho_{c_m}} = V_{c_m} = \sum_j x_j V_{c_j} \tag{4-10.5}$$

$$\omega_m = \sum_j \phi_j \omega_j \tag{4-10.6}$$

For $T_{r_m} > 0.93$, to assure that at $T_{r_m} = 1.0$, $V_{r_m} = 1.0$, *corrected* pseudo-critical temperatures and volumes are defined:

$$T'_{c_m} = T_{c_m} + (T_{c_T} - T_{c_m})D(T'_{r_m}) \tag{4-10.7}$$

$$V'_{c_m} = V_{c_m} + (V_{c_T} - V_{c_m})D(T'_{r_m}) \tag{4-10.8}$$

T_{c_m} and V_{c_m} are the pseudocritical values given above in Eqs. (4-10.2) and (4-10.5); T_{c_T} and V_{c_T} are the *true* mixture critical temperatures and volumes. $D(T'_{r_m})$ is a deviation function expressed in terms of T'_{r_m}:

$$D(T'_{r_m}) = \begin{cases} 0 & T'_{r_m} < 0.93 \\ 1.0 & T'_{r_m} = 1.0 \\ \exp\left[(T'_{r_m} - 1)\left(2901.01 - 5738.92 T'_{r_m}\right.\right. \\ \qquad \left.\left. + 2849.85 T'^2_{r_m} + \frac{1.74127}{1.01 - T'_{r_m}}\right)\right] & 0.93 \le T'_{r_m} < 1.0 \end{cases}$$

$$\tag{4-10.9}$$

Equation (4-10.7) must be solved by trial and error for T'_{c_m}. Techniques for estimating true mixture properties T_{c_T} and V_{c_T} are covered in Chap. 5. Estimates of liquid specific volumes near the critical point are sensitive to the values of T_{c_T}, as might be expected since liquid volumes change rapidly with temperature in this region.

Several investigators have suggested techniques whereby the Rackett liquid-density equation can be adapted to include *hydrocarbon* mixtures [19, 23]. Chiu et al. [5] propose a different correlation between reduced density and reduced temperature but employ identical mixing rules for the pseudocritical temperature. These relations are limited to hydrocarbon systems at low temperatures, i.e., far removed from the critical point.

Won and Prausnitz [29] indicate that the Cailletet and Mathias law of rectilinear diameters can be extended to mixtures by expressing the arithmetic average vapor and liquid mixture density as a linear function of pressure from low pressures to the mixture critical point.

Gubbins [9] has reviewed the theory of simple liquid mixtures with emphasis on perturbation methods.

Recommendations   To approximate the volume of liquid mixtures at low to moderate pressures, Amagat's law, Eq. (4-10.1) is normally satisfactory.   This assumes that when mixing two or more liquids at constant temperature and pressure, the volumes are additive.   If, however, the mixture is much above the boiling point, or if the components are polar, significant errors may result [28].

A more accurate liquid-density estimation method for mixtures that is applicable over wide temperature and pressure ranges (up to the critical point) is given by Eqs. (3-15.18) to (3-15.21) and modified for mixtures in Eqs. (4-10.2) to (4-10.9).   To employ this method, pure-component critical properties must be known.   Also, if it is applied at reduced pseudocritical temperatures in excess of 0.93, the true critical temperature of the mixture must be available.   Finally, for accurate work, the binary interaction parameter k_{ij}^{L} [Eq. (4-10.4)] should be available.   To date, this parameter can only be estimated a priori for aliphatic hydrocarbon mixtures [7].   In other liquid mixtures, even a limited amount of experimental data is extremely helpful in determining this parameter approximately.

NOTATION

$A^* =$ Redlich-Kwong parameter, see Eq. (3-5.7)
$b =$ Redlich-Kwong parameter, see Eq. (4-3.8)
$B =$ second virial coefficient
$B^* =$ Redlich-Kwong parameter, see Eq. (3-5.8)
$D(T'_{r_m}) =$ deviation function in Eqs. (4-10.7) to (4-10.9)
$F =$ Redlich-Kwong parameter, see Eq. (4-3.6)
$H =$ enthalpy
$L_{ij} =$ Benedict-Webb-Rubin interaction parameter for A_0
$P =$ pressure; P_c, critical pressure; P_r, reduced pressure, P/P_c; P_{cm}, pseudocritical mixture pressure
$Q =$ generalized property
$R =$ gas constant
$T =$ temperature; T_c, critical temperature; T_r, reduced temperature, T/T_c; T_{cT}, true critical temperature of a mixture; T_{cm}, pseudocritical mixture temperature
$V =$ volume; V_c, critical volume; V_r, reduced volume, V/V_c; V_{cT}, true critical volume for a mixture; V_{cm}, pseudocritical mixture volume
$W =$ generalized energy parameter
$y_j =$ mole fraction of component j
$Y =$ generalized volume parameter
$Z_c =$ critical compressibility factor

Greek

α_{ij} = Lee-Erbar-Edmister parameter
β_{ij} = Lee-Erbar-Edmister parameter
σ = characteristic molecular dimension
ϕ_j = volume fraction of j defined by Eq. (4-10.3)
χ = generalized Benedict-Webb-Rubin parameter
ω = acentric factor

Superscript

L = liquid

Subscripts

j = component j
ij = interaction between i and j
m = mixture

REFERENCES

1. Barner, H. E., and C. W. Quinlan: *Ind. Eng. Chem. Process Des. Dev.*, **9**: 407 (1969).
2. Barnés, F. J.: Ph.D. thesis, Department of Chemical Engineering, University of California, Berkeley, 1973; C. J. King, personal communication, 1974.
3. Bishnoi, P. R., and D. B. Robinson: *Can. J. Chem. Eng.*, **50**: 101, 506 (1972).
4. Chaudron, J., L. Asselineau, and H. Renon: *Chem. Eng. Sci.*, **28**: 1991 (1973).
5. Chiu, C.-H., C. Hsi, J. A. Ruether, and B. C.-Y. Lu: *Can. J. Chem. Eng.*, **51**: 751 (1973).
6. Chueh, P. L., and J. M. Prausnitz: *Ind. Eng. Chem. Fundam.*, **6**: 492 (1967).
7. Chueh, P. L., and J. M. Prausnitz: *AIChE J.*, **13**: 1099 (1967).
8. Erbar, J. H.: personal communication, 1973.
9. Gubbins, K. E.: *AIChE J.*, **19**: 684 (1973).
10. Gugnoni, R. J., J. W. Eldridge, V. C. Okay, and T. J. Lee: *AIChE J.*, **20**: 357 (1974).
11. Gunn, R. D.: *AIChE J.*, **18**: 183 (1972).
12. Joffe, J.: *Ind. Eng. Chem. Fundam.*, **10**: 532 (1971).
13. Kay, W. B.: *Ind. Eng. Chem.*, **28**: 1014 (1936).
14. Lee, B. I., J. H. Erbar, and W. C. Edmister: *AIChE J.*, **19**: 349 (1973); *Chem. Eng. Prog.*, **68**(9): 83 (1972).
15. Leland, T. W., Jr., and P. S. Chappelear: *Ind. Eng. Chem.*, **60**(7): 15 (1968).
16. Orye, R. V.: *Ind. Eng. Chem. Process Des. Dev.*, **8**: 579 (1969).
17. Prausnitz, J. M.: "Molecular Thermodynamics of Fluid-Phase Equilibria," pp. 128–129, Prentice-Hall, New York, 1969.
18. Prausnitz, J. M., and R. D. Gunn: *AIChE J.*, **4**: 430, 494 (1958).
19. Rackett, H. G.: *J. Chem. Eng. Data*, **16**: 308 (1971).
20. Ramaiah, V., and L. I. Stiel: *Ind. Eng. Chem. Process Des. Dev.*, **11**: 501 (1972), **12**: 305 (1973).
21. Reid, R. C., and T. W. Leland, Jr.: *AIChE J.*, **11**: 228 (1965), **12**: 1227 (1966).
22. Soave, G.: *Chem. Eng. Sci.*, **27**: 1197 (1972).
23. Spencer, C. F., and R. P. Danner: *J. Chem. Eng. Data*, **18**: 230 (1973).
24. Starling, K. E., and J. E. Powers: paper presented at *159th ACS Meet., Houston, Tex., Feb. 22–27, 1970.*
25. Stotler, H. H., and M. Benedict: *Chem. Eng. Prog. Symp. Ser.*, **49**(6): 25 (1953).
26. Vetere, A.: personal communication, 1973.
27. Wilson, G. M.: *Adv. Cryog. Eng.*, **9**: 168 (1964), **11**: 392 (1966).
28. Winnick, J., and J. Kong: *Ind. Eng. Chem. Fundam.*, **13**: 292 (1974).
29. Won, K. W., and J. M. Prausnitz: *AIChE J.*, **20**: 200 (1974).

Chapter Five

Thermodynamic Properties

5-1 Scope

In this chapter we first develop relations to calculate the Helmholtz and Gibbs energies, enthalpies, entropies, and fugacity coefficients. These relations are then used with equation-of-state correlations from Chap. 3 to develop estimation techniques for enthalpy and entropy departure functions and fugacity-pressure ratios for pure components and mixtures. In Sec. 5-5 derivative properties are introduced, and in Sec. 5-6 methods are presented for determining the heat capacities of real gases. The true critical properties of mixtures are discussed in Sec. 5-7, while heat capacities of liquids are treated in Sec. 5-8. Fugacity coefficients of components in gas mixtures are considered in Sec. 5-9.

5-2 Fundamental Thermodynamic Principles

Enthalpy, internal energy, entropy, fugacity, etc., are useful thermodynamic properties. In analyzing or designing process equipment, a variation in these properties can often be related to operating variables, e.g., the temperature rise of a fluid in a heat exchanger. It is therefore important to estimate such property variations as the temperature, pressure, and other independent variables of a system change.

The variation in any thermodynamic property between two states is independent of the path chosen to pass from one state to the other. For example, with a pure fluid or a mixture of a fixed composition, if the difference in enthalpy between states P_1, T_1 and P_2, T_2 is desired, there are an infinite number of possible calculational paths, all of which give the same numerical result. Two of the most obvious are illustrated in Eqs. (5-2.2) and (5-2.3):

$$H = f(P, T)$$

$$dH = \left(\frac{\partial H}{\partial P}\right)_T dP + \left(\frac{\partial H}{\partial T}\right)_P dT \tag{5-2.1}$$

$$H_2 - H_1 = \int_{P_1}^{P_2} \left(\frac{\partial H}{\partial P}\right)_{T_1} dP + \int_{T_1}^{T_2} \left(\frac{\partial H}{\partial T}\right)_{P_2} dT \tag{5-2.2}$$

$$H_2 - H_1 = \int_{P_1}^{P_2} \left(\frac{\partial H}{\partial P}\right)_{T_2} dP + \int_{T_1}^{T_2} \left(\frac{\partial H}{\partial T}\right)_{P_1} dT \tag{5-2.3}$$

In the first method, a stepwise process is visualized whereby the temperature is held constant at T_1 and the isothermal variation in H is determined from P_1 to P_2; this change is then added to the isobaric variation in H with T from T_1 to T_2 at pressure P_2. The second method is similar; but now the variation in H is first determined at P_1 from T_1 to T_2, and then the variation of H from P_1 to P_2 is determined at T_2. These paths are shown schematically in Fig. 5-1, where Eq. (5-2.2) is illustrated by path ADC, whereas Eq. (5-2.3) refers to path ABC. The net $\Delta H = H_2 - H_1$ represents the change AC. Obviously any other convenient path is possible, for example, $AEFGHC$, but to

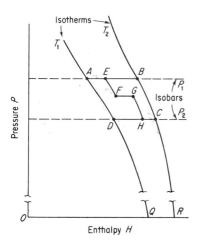

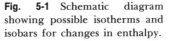

Fig. 5-1 Schematic diagram showing possible isotherms and isobars for changes in enthalpy.

calculate ΔH by this path, values of $(\partial H/\partial P)_T$ and $(\partial H/\partial T)_P$ must be available for the various isotherms and isobars.

The partial derivatives of enthalpy (or other thermodynamic properties) can be visualized as slopes of the isotherms or isobars in Fig. 5-1. To obtain numerical values of such derivatives, both a pressure and a temperature must be specified; i.e., values of these derivatives are also functions of two independent intensive variables in the same way that $H, U, \ldots$ are functions of two such variables.

In Chap. 7, the quantity $(\partial H/\partial T)_{P^\circ}$ is considered. This derivative is called the *constant-pressure heat capacity* C_p°, and the superscript indicates that the pressure level is zero absolute pressure; i.e., the fluid is an ideal gas.

Rarely are heat capacities available at high pressures. Therefore, the usual path for determining $H_2 - H_1$ in Fig. 5-1 with values of C_p° is $AQRC$, that is,

$$\Delta H = \int_{P_1}^{P^\circ} \left(\frac{\partial H}{\partial P}\right)_{T_1} dP + \int_{T_1}^{T_2} C_p^\circ \, dT + \int_{P^\circ}^{P_2} \left(\frac{\partial H}{\partial P}\right)_{T_2} dP \qquad (5\text{-}2.4)$$

or

$$\Delta H = (H^\circ - H_{P_1})_{T_1} + \int_{T_1}^{T_2} C_p^\circ \, dT - (H^\circ - H_{P_2})_{T_2} \qquad (5\text{-}2.5)$$

The terms $(H^\circ - H_{P_1})_{T_1}$ and $(H^\circ - H_{P_2})_{T_2}$ are called *departure functions*. They relate a thermodynamic property (enthalpy in this case) at some P, T to a reference state ($P = 0$, for enthalpy) *at the same temperature*. It is shown in Sec. 5-3 that departure functions can be calculated solely from P-V-T data or, equivalently, from an equation of state. On the other hand, the term $\int C_p^\circ \, dT$ is evaluated in the ideal-gas state and values of C_p° are estimated as described in Chap. 7.

5-3 Departure Functions

Let $\mathscr{L}$ be the value of some thermodynamic property of a pure component (or a mixture with a fixed composition) at some P, T. If $\mathscr{L}^\circ$ is defined to be the value of $\mathscr{L}$ at the same temperature (and at the same composition if a mixture) but in an *ideal-gas state* and *at a reference pressure* P°, then a *departure function* is defined as $\mathscr{L} - \mathscr{L}^\circ$ or $\mathscr{L}^\circ - \mathscr{L}$. In the reference state at T, P°, the molal volume V° would be given by

$$V^\circ = \frac{RT}{P^\circ} \qquad (5\text{-}3.1)$$

As shown below, departure functions can be expressed in terms of the P-V-T properties of a fluid. Two general approaches are used. The first is more convenient if the P-V-T properties of a fluid are characterized by an equation of state explicit in pressure. All the analytical equations of state described in Chap. 3 are of this form. The second is more useful when temperature and pressure are the independent variables. The corresponding-states correlations in Chap. 3, for example, are expressed as $Z = f(T_r, P_r)$.

For the first form, we develop the departure function for the Helmholtz energy A, and from this result all other departure functions can readily be obtained. At constant temperature and composition, the variation in the Helmholtz energy with molar volume V is

$$dA = -P\,dV \tag{5-3.2}$$

Integrating at constant temperature and composition from the reference volume $V°$ to the system volume V gives

$$A - A° = -\int_{V°}^{V} P\,dV \tag{5-3.3}$$

The evaluation of Eq. (5-3.3) is inconvenient since one limit of the integration refers to the real state but the other to the reference, ideal-gas state. Thus, we break the integral into two parts:

$$A - A° = -\int_{\infty}^{V} P\,dV - \int_{V°}^{\infty} P\,dV \tag{5-3.4}$$

The first integral requires real-gas properties, that is, $P = f(V)$ at constant temperature T, while the second is written for an ideal gas and can be integrated immediately. Before doing so, however, to avoid the difficulty introduced by the infinity limit, we add and subtract $\int_{\infty}^{V}(RT/V)\,dV$ from the right-hand side. Then

$$A - A° = -\int_{\infty}^{V}\left(P - \frac{RT}{V}\right)dV - RT\ln\frac{V}{V°} \tag{5-3.5}$$

The departure function for A [Eq. (5-3.5)] depends upon the choice of $V°$. Note that $A - A°$ does *not* vanish even for an ideal gas unless $V°$ is chosen to equal V. Other departure functions are readily obtained from Eq. (5-3.5)

$$\Delta S° < S - S° = \frac{-\partial}{\partial T}(A - A°)_v \tag{5-3.6}$$

$$S = \Delta S° + S° = \int_{\infty}^{V}\left[\left(\frac{\partial P}{\partial T}\right)_v - \frac{R}{V}\right]dV + R\ln\frac{V}{V°} \tag{5-3.7}$$

$$H - H° = (A - A°) + T(S - S°) + RT(Z - 1) \tag{5-3.8}$$

$$U - U° = (A - A°) + T(S - S°) \tag{5-3.9}$$

$$G - G° = (A - A°) + RT(Z - 1) \tag{5-3.10}$$

Also, although not strictly a departure function, the fugacity-pressure ratio can be expressed in a similar manner:

$$\ln \frac{f}{P} = \frac{A - A^\circ}{RT} + \ln \frac{V}{V^\circ} + (Z - 1) - \ln Z$$

$$= -\frac{1}{RT} \int_\infty^V \left(P - \frac{RT}{V} \right) dV + (Z - 1) - \ln Z \qquad (5\text{-}3.11)$$

where $Z = PV/RT$.

Therefore, from any pressure-explicit equation of state and a definition of the reference state (P° or V°), all departure functions can readily be found.

Example 5-1   Derive the departure functions for a pure material or for a mixture of constant composition using the Redlich-Kwong equation of state (3-5.1).
  solution   Using Eq. (3-5.1),

$$P = \frac{RT}{V - b} - \frac{a}{T^{0.5}V(V + b)}$$

Then, with Eq. (5-3.5),

$$A - A^\circ = -\int_\infty^V \left[\frac{RT}{V - b} - \frac{RT}{V} - \frac{a}{T^{0.5}V(V + b)} \right] dV - RT \ln \frac{V}{V^\circ}$$

$$= -RT \ln \frac{V - b}{V} - \frac{a}{T^{0.5}b} \ln \frac{V + b}{V} - RT \ln \frac{V}{V^\circ}$$

For entropy, enthalpy, and internal energy with Eqs. (5-3.7) to (5-3.9),

$$S - S^\circ = -\left[\frac{\partial (A - A^\circ)}{\partial T} \right]_V$$

$$= R \ln \frac{V - b}{V} - \frac{a}{2bT^{1.5}} \ln \frac{V + b}{V} + R \ln \frac{V}{V^\circ}$$

$$H - H^\circ = (A - A^\circ) + T(S - S^\circ) + RT(Z - 1)$$

$$= PV - RT - \frac{3a}{2bT^{0.5}} \ln \frac{V + b}{V}$$

$$= \frac{bRT}{V - b} - \frac{a}{T^{0.5}(V + b)} - \frac{3a}{2bT^{0.5}} \ln \frac{V + b}{V}$$

$$U - U^\circ = (A - A^\circ) + T(S - S^\circ) = -\frac{3a}{2bT^{0.5}} \ln \frac{V + b}{V}$$

For $\ln (f/P)$, with Eq. (5-3.11), where

$$Z = \frac{PV}{RT} = \frac{V}{V - b} - \frac{a}{RT^{1.5}(V + b)}$$

we have

$$\ln \frac{f}{P} = \frac{b}{V - b} - \frac{a}{RT^{1.5}(V + b)} - \ln \frac{V - b}{V} - \frac{a}{bRT^{1.5}} \ln \frac{V + b}{V} - \ln \left[\frac{V}{V - b} - \frac{a}{RT^{1.5}(V + b)} \right]$$

Finally, from Eq. (5-3.10)

$$G - G^\circ = \frac{bRT}{V - b} - \frac{a}{T^{0.5}(V + b)} - RT \ln \frac{V - b}{V} - \frac{a}{bT^{0.5}} \ln \frac{V + b}{V} - RT \ln \frac{V}{V^\circ}$$

From Eqs. (5-3.5) to (5-3.11), or with Example 5-1, one can see that the departure functions $H - H°$, $U - U°$, and $\ln (f/P)$ do not depend upon the value of the reference-state pressure $P°$ (or $V°$). In contrast, $A - A°$, $S - S°$, and $G - G°$ do depend upon $P°$ (or $V°$). Either of two common reference states is normally chosen. In the first, $P°$ is set equal to a unit pressure, for example, 1 atm if that is the pressure unit chosen. Then, from Eq. (5-3.1), $V° = RT$, but it is necessary to express R in the same units of pressure. In the second reference state, $P° = P$, the system pressure. Then $V/V° = Z$, the compressibility factor. Other reference states can be defined, for example, $V° = V$, but the two noted above are the more common.

An alternate calculational path to obtain departure functions is more convenient if the equation of state is explicit in volume or if pressure and temperature are the independent variables. In such cases, we again choose as a reference state an ideal gas at the same temperature and composition as that of the system under study. The reference pressure is $P°$, and Eq. (5-3.1) applies. We begin, however, with the Gibbs energy rather than the Helmholtz energy. The analog to Eq. (5-3.4) is

$$G - G° = \int_{P°}^{P} V \, dP = \int_{0}^{P} V \, dP + \int_{P°}^{0} V \, dP$$

$$= \int_{0}^{P} \left(V - \frac{RT}{P} \right) dP + RT \ln \frac{P}{P°} \qquad (5\text{-}3.12)$$

$$= RT \int_{0}^{P} (Z - 1) \, d \ln P + RT \ln \frac{P}{P°} \qquad (5\text{-}3.13)$$

For entropy,

$$S - S° = \frac{-\partial}{\partial T} (G - G°)_P$$

$$= R \int_{0}^{P} \left[1 - Z - T \left(\frac{\partial Z}{\partial T} \right)_P \right] d \ln P - R \ln \frac{P}{P°} \qquad (5\text{-}3.14)$$

and for the enthalpy and internal-energy departure functions,

$$H - H° = (G - G°) + T(S - S°) \qquad (5\text{-}3.15)$$

$$U - U° = (G - G°) + T(S - S°) - RT(Z - 1) \qquad (5\text{-}3.16)$$

The Helmholtz-energy departure function is

$$A - A° = (G - G°) - RT(Z - 1) \qquad (5\text{-}3.17)$$

while

$$\ln \frac{f}{P} = \left(\frac{G - G°}{RT} \right) - \ln \frac{P}{P°} \qquad (5\text{-}3.18)$$

Again, simple algebraic substitution shows that the departure functions $H - H°$, $U - U°$, and $\ln (f/P)$ do not depend upon the choice of $P°$ (or $V°$).

5-4   Evaluation of Departure Functions

The departure functions shown in Eqs. (5-3.5) to (5-3.11) or in Eqs. (5-3.12) to (5-3.18) can be evaluated with P-V-T data and, where necessary, a definition of the reference state. Generally, either an analytical equation of state or some form of the law of corresponding states is used to characterize P-V-T behavior although, if available, experimental P-V-T data for a pure substance or a given mixture can be employed.

Departure Functions from Equations of State   Several analytical equations of state were introduced in Chap. 3. All are pressure-explicit. In the same manner as that shown in Example 5-1, departure functions can be determined. $A - A°$ and $S - S°$ for seven equations of state are shown in Table 5-1. With $A - A°$ and $S - S°$, $H - H°$, $U - U°$, etc., are readily found from Eqs. (5-3.8) to (5-3.11). In each case reference is made to the appropriate equations or tables in Chaps. 3 and 4, where the characteristic parameters of the equation are defined.

The Lee-Kesler correlation [Eqs. (3-9.1) to (3-9.4)] is not shown in Table 5-1. To employ it to calculate thermodynamic properties, care must be exercised in following the procedure recommended by the authors. The method is illustrated with the enthalpy departure function, and analogous rules apply to the entropy departure function, fugacity-pressure ratio, and derivative functions (Sec. 5-5).

Given a pressure and temperature, the first step involves the calculation of the reduced temperature and pressure. If the fluid is pure, T_c and P_c values can be found in Appendix A or estimated by relations given in Chap. 2. For mixtures, the appropriate pseudocritical properties are determined from Eqs. (4-6.3) to (4-6.7). With T_r and P_r following the procedure given in Sec. 3-9, $V_r^{(0)}$, $V_r^{(R)}$, $Z^{(0)}$, and $Z^{(R)}$ are determined. From T_r, $V_r^{(0)}$, and $Z^{(0)}$ the simple fluid enthalpy departure function can then be found:

$$\left(\frac{H° - H}{RT_c}\right)^{(0)} = - T_r\left[Z^{(0)} - 1 - \frac{b_2 + 2b_3/T_r + 3b_4/T_r^2}{T_r(V_r^{(0)})}\right.$$
$$\left. - \frac{c_2 - 3c_3/T_r^2}{2T_r(V_r^{(0)})^2} + \frac{d_2}{5T_r(V_r^{(0)})^5} + 3E\right] \tag{5-4.1}$$

where $E = \dfrac{c_4}{2T_r^3\gamma}\left\{\beta + 1 - \left[\beta + 1 + \dfrac{\gamma}{(V_r^{(0)})^2}\right]\exp\left[-\dfrac{\gamma}{(V_r^{(0)})^2}\right]\right\}$ $\tag{5-4.2}$

Next, using the same T_r but $V_r^{(R)}$ and $Z^{(R)}$, recompute Eq. (5-4.1) using the reference-fluid constants in Table 3-9; call this departure function $[(H° - H)/RT_c]^{(R)}$. The departure function for the real fluid is then

$$\frac{H° - H}{RT_c} = \left(\frac{H° - H}{RT_c}\right)^{(0)} + \frac{\omega}{\omega^R}\left[\left(\frac{H° - H}{RT_c}\right)^{(R)} - \left(\frac{H° - H}{RT_c}\right)^{(0)}\right] \tag{5-4.3}$$

TABLE 5-1 Departure Functions from Analytical Equations of State

Only $A - A°$ and $S - S°$ departure functions are given; $H - H°$, $U - U°$, $G - G°$, and $\ln(f/P)$ are readily obtained from Eqs. (5-3.8) to (5-3.11)

Redlich-Kwong, Secs. 3-5 and 4-3

$$P = \frac{RT}{V - b} - \frac{a}{T^{1/2}V(V + b)}$$

$$Z = \frac{V}{V - b} - \frac{a}{RT^{3/2}(V + b)}$$

$$A - A° = -RT \ln \frac{V - b}{V} - \frac{a}{T^{1/2}b} \ln \frac{V + b}{V} - RT \ln \frac{V}{V°}$$

$$S - S° = R \ln \frac{V - b}{V} - \frac{a}{2bT^{3/2}} \ln \frac{V + b}{V} + R \ln \frac{V}{V°}$$

a, b are given in Eqs. (3-5.2) to (3-5.8) and (4-3.1) and (4-3.2)

Soave modification of Redlich-Kwong, Secs. 3-5 and 4-3

$$P = \frac{RT}{V - b} - \frac{\Omega_a}{\Omega_b} \frac{RTbF}{V(V + b)}$$

$$Z = \frac{V}{V - b} - \frac{\Omega_a}{\Omega_b} \frac{bF}{V + b}$$

$$A - A° = RT\left(-\ln \frac{V - b}{V} - \frac{\beta}{b} \ln \frac{V + b}{V} - \ln \frac{V}{V°}\right)$$

where

$$\beta = \sum_i \sum_j y_i y_j (1 - \bar{k}_{ij})\left(\frac{\Omega_{a_i}}{\Omega_{b_i}} \frac{\Omega_{a_j}}{\Omega_{b_j}} b_i b_j F_i F_j\right)^{1/2}$$

$$S - S° = R\left(\ln \frac{V - b}{V} - \frac{\gamma}{b} \ln \frac{V + b}{V} + \ln \frac{V}{V°}\right)$$

where

$$\gamma = -\frac{1}{2}\sum_i \sum_j y_i y_j (1 - \bar{k}_{ij})\left(\frac{\Omega_{a_i}}{\Omega_{b_i}} \frac{\Omega_{a_j}}{\Omega_{b_j}} b_i b_j F_i F_j\right)^{1/2}\left(\frac{f\omega_j}{F_j^{1/2}} + \frac{f\omega_i}{F_i^{1/2}}\right)$$

$$f\omega_j = 0.480 + 1.574\omega_j - 0.176\omega_j^2$$

F_i, F_j, b_i, b_j, and $\bar{k}_{ij}$ are given in Eqs. (3-5.3) and (3-5.15) and Tables 4-1 and 4-2

Barner-Adler, Secs. 3-6 and 4-4

$$P = \frac{RT}{V}\left[\frac{V - a*b}{V - b} + \frac{c*b^2}{(V - b)^2} - \frac{d*b^3}{(V - b)^3} + \frac{e*b^4}{(V - b)^4}\right]$$

$$Z = \frac{V - a*b}{V - b} + \frac{c*b^2}{(V - b)^2} - \frac{d*b^3}{(V - b)^3} + \frac{e*b^4}{(V - b)^4}$$

TABLE 5-1 Departure Functions from Analytical Equations of State (*Continued*)

Only $A - A°$ and $S - S°$ departure functions are given; $H - H°$, $U - U°$, $G - G°$, and $\ln (f/P)$ are readily obtained from Eqs. (5-3.8) to (5-3.11)

Barner-Adler, Secs. 3-6 and 4-4 (*Continued*)

$$A - A° = RT\left[(a* + c* + d* + e* - 1)\ln\frac{V-b}{V} + \frac{b}{V-b}(c* + d* + e*)\right.$$

$$\left. - \frac{b^2}{2(V-b)^2}(d* + e*) + \frac{b^3 c*}{3(V-b)^3} - \ln\frac{V}{V°}\right]$$

$$S - S° = \left(R - \frac{af_a'}{b} - \frac{cf_c'}{b^2} - \frac{df_d'}{b^3} - \frac{ef_e'}{b^4}\right)\ln\frac{V-b}{V}$$

$$- \frac{1}{V-b}\left(\frac{cf_c'}{b} + \frac{df_d'}{b^2} + \frac{cf_e'}{b^3}\right)$$

$$+ \frac{1}{(V-b)^2}\left(\frac{df_d'}{2b} + \frac{ef_e'}{2b^2}\right) - \frac{1}{(V-b)^3}\frac{cf_e'}{3b} + R\ln\frac{V}{V°}$$

The constants are given in Eqs. (3-6.1) to (3-6.16) and (4-4.1) to (4-4.7) and

$$a* = \frac{af_a}{bRT} \qquad d* = \frac{df_d}{b^3 RT}$$

$$c* = \frac{cf_c}{b^2 RT} \qquad e* = \frac{ef_e}{b^4 RT}$$

$$f_a' = -\frac{A}{TT_r} \qquad f_d' = -\frac{D_2}{TT_r} + \frac{2D_3}{TT_r^2}$$

$$f_c' = -\frac{C}{TT_r} \qquad f_e' = -\frac{2E_2}{TT_r} + \frac{4E_3}{TT_r^4}$$

Sugie-Lu, Secs. 3-7 and 4-5

$$P = \frac{RT}{V-b+c} - \frac{a}{T^{1/2}(V+c)(V+b+c)} + \sum_{j=1}^{10}\frac{d_j T + e_j T^{-0.5}}{V^{j+1}}$$

$$Z = \frac{V}{V-b+c} - \frac{aV}{RT^{3/2}(V+c)(V+b+c)} + \sum_{j=1}^{10}\frac{d_j + e_j T^{-1.5}}{RV^j}$$

$$A - A° = -RT\ln\frac{V-b+c}{V} - \frac{a}{bT^{1/2}}\ln\frac{V+b+c}{V+c}$$

$$+ \sum_{j=1}^{10}\frac{d_j T + e_j T^{-0.5}}{jV^j} - RT\ln\frac{V}{V°}$$

$$S - S° = R\ln\frac{V-b+c}{V} - \frac{a}{2bT^{3/2}}\ln\frac{V+b+c}{V+c}$$

$$- \sum_{j=1}^{10}\frac{d_j - e_j T^{-1.5}/2}{jV^j} + R\ln\frac{V}{V°}$$

The constants are shown in Table 3-5 and in Eqs. (4-5.1) to (4-5.4)

Benedict-Webb-Rubin, Secs. 3-8 and 4-6

$$P = \frac{RT}{V} + \frac{B_0 RT - A_0 - C_0 T^{-2}}{V^2} + \frac{bRT - a}{V^3}$$

$$+ \frac{a\alpha}{V^6} + \frac{c}{V^3 T^2}\left(1 + \frac{\gamma}{V^2}\right)\exp\left(-\frac{\gamma}{V^2}\right)$$

$$Z = 1 + \frac{B_0 RT - A_0 - C_0 T^{-2}}{RTV} + \frac{bRT - a}{RTV^2}$$

$$+ \frac{a\alpha}{RTV^5} + \frac{c}{RT^3 V^2}\left(1 + \frac{\gamma}{V^2}\right)\exp\left(-\frac{\gamma}{V^2}\right)$$

$$A - A^\circ = \frac{B_0 RT - A_0 - C_0 T^{-2}}{V} + \frac{bRT - a}{2V^2} + \frac{a\alpha}{5V^5}$$

$$+ \frac{c}{T^2 \gamma}\left[1 - \left(1 + \frac{\gamma}{2V^2}\right)\exp\left(-\frac{\gamma}{V^2}\right)\right] - RT \ln \frac{V}{V^\circ}$$

$$S - S^\circ = -\frac{B_0 R + 2C_0 T^{-3}}{V} - \frac{bR}{2V^2}$$

$$+ \frac{2c}{T^3 \gamma}\left[1 - \left(1 + \frac{\gamma}{2V^2}\right)\exp\left(-\frac{\gamma}{V^2}\right)\right] + R \ln \frac{V}{V^\circ}$$

Note: All constants assumed independent of temperature

The constants are given in Table 3-6 or 3-7 and in Eq. (4-6.1)

Lee-Erbar-Edmister, Secs. 3-10 and 4-7

$$P = \frac{RT}{V - b} - \frac{a}{V(V - b)} + \frac{bc}{V(V - b)(V + b)}$$

$$Z = \frac{V}{V - b} - \frac{a/RT}{V - b} + \frac{bc/RT}{(V - b)(V + b)}$$

$$A - A^\circ = \frac{a - bRT}{b} \ln \frac{V - b}{V} - \frac{c}{2b} \ln \frac{V^2 - b^2}{V^2} - RT \ln \frac{V}{V^\circ}$$

$$S - S^\circ = \frac{bR - a'}{b} \ln \frac{V - b}{V} + \frac{c'}{2b} \ln \frac{V^2 - b^2}{V^2} + R \ln \frac{V}{V^\circ}$$

$$a' = \frac{da}{dT}$$
$$c' = \frac{dc}{dT}$$

See Eqs. (3-10.3) and (3-10.4) for a, c

For b and for the mixture rules, see Eq. (3-10.2) and Table 4-4

Virial, Secs. 3-11 and 4-8

$$P = \frac{RT}{V} + \frac{BRT}{V^2}$$

$$Z = 1 + \frac{B}{V}$$

$$A - A^\circ = \frac{BRT}{V} - RT \ln \frac{V}{V^\circ}$$

$$S - S^\circ = -\frac{BR}{V} - \frac{RT}{V}\frac{dB}{dT} + R \ln \frac{V}{V^\circ}$$

Eq. (3-11.3) shows $B = f(T)$; mixing rules are given in Eqs. (4-8.1) to (4-8.7)

TABLE 5-2 Lee-Kesler Residual Enthalpy [51]

Simple fluid $\left(\dfrac{H^\circ - H}{RT_c}\right)^{(0)}$

T_r	P_r						
	0.010	0.050	0.100	0.200	0.400	0.600	0.800
0.30	6.045	6.043	6.040	6.034	6.022	6.011	5.999
0.35	5.906	5.904	5.901	5.895	5.882	5.870	5.858
0.40	5.763	5.761	5.757	5.751	5.738	5.726	5.713
0.45	5.615	5.612	5.609	5.603	5.590	5.577	5.564
0.50	5.465	5.463	5.459	5.453	5.440	5.427	5.414
0.55	0.032	5.312	5.309	5.303	5.290	5.278	5.265
0.60	0.027	5.162	5.159	5.153	5.141	5.129	5.116
0.65	0.023	0.118	5.008	5.002	4.991	4.980	4.968
0.70	0.020	0.101	0.213	4.848	4.838	4.828	4.818
0.75	0.017	0.088	0.183	4.687	4.679	4.672	4.664
0.80	0.015	0.078	0.160	0.345	4.507	4.504	4.499
0.85	0.014	0.069	0.141	0.300	4.309	4.313	4.316
0.90	0.012	0.062	0.126	0.264	0.596	4.074	4.094
0.93	0.011	0.058	0.118	0.246	0.545	0.960	3.920
0.95	0.011	0.056	0.113	0.235	0.516	0.885	3.763
0.97	0.011	0.054	0.109	0.225	0.490	0.824	1.356
0.98	0.010	0.053	0.107	0.221	0.478	0.797	1.273
0.99	0.010	0.052	0.105	0.216	0.466	0.773	1.206
1.00	0.010	0.051	0.103	0.212	0.455	0.750	1.151
1.01	0.010	0.050	0.101	0.208	0.445	0.728	1.102
1.02	0.010	0.049	0.099	0.203	0.434	0.708	1.060
1.05	0.009	0.046	0.094	0.192	0.407	0.654	0.955
1.10	0.008	0.042	0.086	0.175	0.367	0.581	0.827
1.15	0.008	0.039	0.079	0.160	0.334	0.523	0.732
1.20	0.007	0.036	0.073	0.148	0.305	0.474	0.657
1.30	0.006	0.031	0.063	0.127	0.259	0.399	0.545
1.40	0.005	0.027	0.055	0.110	0.224	0.341	0.463
1.50	0.005	0.024	0.048	0.097	0.196	0.297	0.400
1.60	0.004	0.021	0.043	0.086	0.173	0.261	0.350
1.70	0.004	0.019	0.038	0.076	0.153	0.231	0.309
1.80	0.003	0.017	0.034	0.068	0.137	0.206	0.275
1.90	0.003	0.015	0.031	0.062	0.123	0.185	0.246
2.00	0.003	0.014	0.028	0.056	0.111	0.167	0.222
2.20	0.002	0.012	0.023	0.046	0.092	0.137	0.182
2.40	0.002	0.010	0.019	0.038	0.076	0.114	0.150
2.60	0.002	0.008	0.016	0.032	0.064	0.095	0.125
2.80	0.001	0.007	0.014	0.027	0.054	0.080	0.105
3.00	0.001	0.006	0.011	0.023	0.045	0.067	0.088
3.50	0.001	0.004	0.007	0.015	0.029	0.043	0.056
4.00	0.000	0.002	0.005	0.009	0.017	0.026	0.033

1.000	1.200	1.500	2.000	3.000	5.000	7.000	10.000
5.987	5.975	5.957	5.927	5.868	5.748	5.628	5.446
5.845	5.833	5.814	5.783	5.721	5.595	5.469	5.278
5.700	5.687	5.668	5.636	5.572	5.442	5.311	5.113
5.551	5.538	5.519	5.486	5.421	5.288	5.154	4.950
5.401	5.388	5.369	5.336	5.270	5.135	4.999	4.791
5.252	5.239	5.220	5.187	5.121	4.986	4.849	4.638
5.104	5.091	5.073	5.041	4.976	4.842	4.704	4.492
4.956	4.945	4.927	4.896	4.833	4.702	4.565	4.353
4.808	4.797	4.781	4.752	4.693	4.566	4.432	4.221
4.655	4.646	4.632	4.607	4.554	4.434	4.303	4.095
4.494	4.488	4.478	4.459	4.413	4.303	4.178	3.974
4.316	4.316	4.312	4.302	4.269	4.173	4.056	3.857
4.108	4.118	4.127	4.132	4.119	4.043	3.935	3.744
3.953	3.976	4.000	4.020	4.024	3.963	3.863	3.678
3.825	3.865	3.904	3.940	3.958	3.910	3.815	3.634
3.658	3.732	3.796	3.853	3.890	3.856	3.767	3.591
3.544	3.652	3.736	3.806	3.854	3.829	3.743	3.569
3.376	3.558	3.670	3.758	3.818	3.801	3.719	3.548
2.584	3.441	3.598	3.706	3.782	3.774	3.695	3.526
1.796	3.283	3.516	3.652	3.744	3.746	3.671	3.505
1.627	3.039	3.422	3.595	3.705	3.718	3.647	3.484
1.359	2.034	3.030	3.398	3.583	3.632	3.575	3.420
1.120	1.487	2.203	2.965	3.353	3.484	3.453	3.315
0.968	1.239	1.719	2.479	3.091	3.329	3.329	3.211
0.857	1.076	1.443	2.079	2.807	3.166	3.202	3.107
0.698	0.860	1.116	1.560	2.274	2.825	2.942	2.899
0.588	0.716	0.915	1.253	1.857	2.486	2.679	2.692
0.505	0.611	0.774	1.046	1.549	2.175	2.421	2.486
0.440	0.531	0.667	0.894	1.318	1.904	2.177	2.285
0.387	0.466	0.583	0.777	1.139	1.672	1.953	2.091
0.344	0.413	0.515	0.683	0.996	1.476	1.751	1.908
0.307	0.368	0.458	0.606	0.880	1.309	1.571	1.736
0.276	0.330	0.411	0.541	0.782	1.167	1.411	1.577
0.226	0.269	0.334	0.437	0.629	0.937	1.143	1.295
0.187	0.222	0.275	0.359	0.513	0.761	0.929	1.058
0.155	0.185	0.228	0.297	0.422	0.621	0.756	0.858
0.130	0.154	0.190	0.246	0.348	0.508	0.614	0.689
0.109	0.129	0.159	0.205	0.288	0.415	0.495	0.545
0.069	0.081	0.099	0.127	0.174	0.239	0.270	0.264
0.041	0.048	0.058	0.072	0.095	0.116	0.110	0.061

TABLE 5-3 Lee-Kesler Residual Enthalpy [51]

Deviation function $\left(\dfrac{H^\circ - H}{RT_c}\right)^{(1)}$

T_r	P_r						
	0.010	0.050	0.100	0.200	0.400	0.600	0.800
0.30	11.098	11.096	11.095	11.091	11.083	11.076	11.069
0.35	10.656	10.655	10.654	10.653	10.650	10.646	10.643
0.40	10.121	10.121	10.121	10.120	10.121	10.121	10.121
0.45	9.515	9.515	9.516	9.517	9.519	9.521	9.523
0.50	8.868	8.869	8.870	8.872	8.876	8.880	8.884
0.55	0.080	8.211	8.212	8.215	8.221	8.226	8.232
0.60	0.059	7.568	7.570	7.573	7.579	7.585	7.591
0.65	0.045	0.247	6.949	6.952	6.959	6.966	6.973
0.70	0.034	0.185	0.415	6.360	6.367	6.373	6.381
0.75	0.027	0.142	0.306	5.796	5.802	5.809	5.816
0.80	0.021	0.110	0.234	0.542	5.266	5.271	5.278
0.85	0.017	0.087	0.182	0.401	4.753	4.754	4.758
0.90	0.014	0.070	0.144	0.308	0.751	4.254	4.248
0.93	0.012	0.061	0.126	0.265	0.612	1.236	3.942
0.95	0.011	0.056	0.115	0.241	0.542	0.994	3.737
0.97	0.010	0.052	0.105	0.219	0.483	0.837	1.616
0.98	0.010	0.050	0.101	0.209	0.457	0.776	1.324
0.99	0.009	0.048	0.097	0.200	0.433	0.722	1.154
1.00	0.009	0.046	0.093	0.191	0.410	0.675	1.034
1.01	0.009	0.044	0.089	0.183	0.389	0.632	0.940
1.02	0.008	0.042	0.085	0.175	0.370	0.594	0.863
1.05	0.007	0.037	0.075	0.153	0.318	0.498	0.691
1.10	0.006	0.030	0.061	0.123	0.251	0.381	0.507
1.15	0.005	0.025	0.050	0.099	0.199	0.296	0.385
1.20	0.004	0.020	0.040	0.080	0.158	0.232	0.297
1.30	0.003	0.013	0.026	0.052	0.100	0.142	0.177
1.40	0.002	0.008	0.016	0.032	0.060	0.083	0.100
1.50	0.001	0.005	0.009	0.018	0.032	0.042	0.048
1.60	0.000	0.002	0.004	0.007	0.012	0.013	0.011
1.70	0.000	0.000	0.000	-0.000	-0.003	-0.009	-0.017
1.80	-0.000	-0.001	-0.003	-0.006	-0.015	-0.025	-0.037
1.90	-0.001	-0.003	-0.005	-0.011	-0.023	-0.037	-0.053
2.00	-0.001	-0.003	-0.007	-0.015	-0.030	-0.047	-0.065
2.20	-0.001	-0.005	-0.010	-0.020	-0.040	-0.062	-0.083
2.40	-0.001	-0.006	-0.012	-0.023	-0.047	-0.071	-0.095
2.60	-0.001	-0.006	-0.013	-0.026	-0.052	-0.078	-0.104
2.80	-0.001	-0.007	-0.014	-0.028	-0.055	-0.082	-0.110
3.00	-0.001	-0.007	-0.014	-0.029	-0.058	-0.086	-0.114
3.50	-0.002	-0.008	-0.016	-0.031	-0.062	-0.092	-0.122
4.00	-0.002	-0.008	-0.016	-0.032	-0.064	-0.096	-0.127

			P_r				
1.000	1.200	1.500	2.000	3.000	5.000	7.000	10.000
11.062	11.055	11.044	11.027	10.992	10.935	10.872	10.781
10.640	10.637	10.632	10.624	10.609	10.581	10.554	10.529
10.121	10.121	10.121	10.122	10.123	10.128	10.135	10.150
9.525	9.527	9.531	9.537	9.549	9.576	9.611	9.663
8.888	8.892	8.899	8.909	8.932	8.978	9.030	9.111
8.238	8.243	8.252	8.267	8.298	8.360	8.425	8.531
7.596	7.603	7.614	7.632	7.669	7.745	7.824	7.950
6.980	6.987	6.997	7.017	7.059	7.147	7.239	7.381
6.388	6.395	6.407	6.429	6.475	6.574	6.677	6.837
5.824	5.832	5.845	5.868	5.918	6.027	6.142	6.318
5.285	5.293	5.306	5.330	5.385	5.506	5.632	5.824
4.763	4.771	4.784	4.810	4.872	5.008	5.149	5.358
4.249	4.255	4.268	4.298	4.371	4.530	4.688	4.916
3.934	3.937	3.951	3.987	4.073	4.251	4.422	4.662
3.712	3.713	3.730	3.773	3.873	4.068	4.248	4.497
3.470	3.467	3.492	3.551	3.670	3.885	4.077	4.336
3.332	3.327	3.363	3.434	3.568	3.795	3.992	4.257
3.164	3.164	3.223	3.313	3.464	3.705	3.909	4.178
2.471	2.952	3.065	3.186	3.358	3.615	3.825	4.100
1.375	2.595	2.880	3.051	3.251	3.525	3.742	4.023
1.180	1.723	2.650	2.906	3.142	3.435	3.661	3.947
0.877	0.878	1.496	2.381	2.800	3.167	3.418	3.722
0.617	0.673	0.617	1.261	2.167	2.720	3.023	3.362
0.459	0.503	0.487	0.604	1.497	2.275	2.641	3.019
0.349	0.381	0.381	0.361	0.934	1.840	2.273	2.692
0.203	0.218	0.218	0.178	0.300	1.066	1.592	2.086
0.111	0.115	0.108	0.070	0.044	0.504	1.012	1.547
0.049	0.046	0.032	-0.008	-0.078	0.142	0.556	1.080
0.005	-0.004	-0.023	-0.065	-0.151	-0.082	0.217	0.689
-0.027	-0.040	-0.063	-0.109	-0.202	-0.223	-0.028	0.369
-0.051	-0.067	-0.094	-0.143	-0.241	-0.317	-0.203	0.112
-0.070	-0.088	-0.117	-0.169	-0.271	-0.381	-0.330	-0.092
-0.085	-0.105	-0.136	-0.190	-0.295	-0.428	-0.424	-0.255
-0.106	-0.128	-0.163	-0.221	-0.331	-0.493	-0.551	-0.489
-0.120	-0.144	-0.181	-0.242	-0.356	-0.535	-0.631	-0.645
-0.130	-0.156	-0.194	-0.257	-0.376	-0.567	-0.687	-0.754
-0.137	-0.164	-0.204	-0.269	-0.391	-0.591	-0.729	-0.836
-0.142	-0.170	-0.211	-0.278	-0.403	-0.611	-0.763	-0.899
-0.152	-0.181	-0.224	-0.294	-0.425	-0.650	-0.827	-1.015
-0.158	-0.188	-0.233	-0.306	-0.442	-0.680	-0.874	-1.097

TABLE 5-4 Lee-Kesler Residual Entropy [51]

Simple fluid $\left[\dfrac{S^\circ - S}{R}\right]^{(0)}$

T_r	P_r						
	0.010	0.050	0.100	0.200	0.400	0.600	0.800
0.30	11.614	10.008	9.319	8.635	7.961	7.574	7.304
0.35	11.185	9.579	8.890	8.205	7.529	7.140	6.869
0.40	10.802	9.196	8.506	7.821	7.144	6.755	6.483
0.45	10.453	8.847	8.157	7.472	6.794	6.404	6.132
0.50	10.137	8.531	7.841	7.156	6.479	6.089	5.816
0.55	0.038	8.245	7.555	6.870	6.193	5.803	5.531
0.60	0.029	7.983	7.294	6.610	5.933	5.544	5.273
0.65	0.023	0.122	7.052	6.368	5.694	5.306	5.036
0.70	0.018	0.096	0.206	6.140	5.467	5.082	4.814
0.75	0.015	0.078	0.164	5.917	5.248	4.866	4.600
0.80	0.013	0.064	0.134	0.294	5.026	4.649	4.388
0.85	0.011	0.054	0.111	0.239	4.785	4.418	4.166
0.90	0.009	0.046	0.094	0.199	0.463	4.145	3.912
0.93	0.008	0.042	0.085	0.179	0.408	0.750	3.723
0.95	0.008	0.039	0.080	0.168	0.377	0.671	3.556
0.97	0.007	0.037	0.075	0.157	0.350	0.607	1.056
0.98	0.007	0.036	0.073	0.153	0.337	0.580	0.971
0.99	0.007	0.035	0.071	0.148	0.326	0.555	0.903
1.00	0.007	0.034	0.069	0.144	0.315	0.532	0.847
1.01	0.007	0.033	0.067	0.139	0.304	0.510	0.799
1.02	0.006	0.032	0.065	0.135	0.294	0.491	0.757
1.05	0.006	0.030	0.060	0.124	0.267	0.439	0.656
1.10	0.005	0.026	0.053	0.108	0.230	0.371	0.537
1.15	0.005	0.023	0.047	0.096	0.201	0.319	0.452
1.20	0.004	0.021	0.042	0.085	0.177	0.277	0.389
1.30	0.003	0.017	0.033	0.068	0.140	0.217	0.298
1.40	0.003	0.014	0.027	0.056	0.114	0.174	0.237
1.50	0.002	0.011	0.023	0.046	0.094	0.143	0.194
1.60	0.002	0.010	0.019	0.039	0.079	0.120	0.162
1.70	0.002	0.008	0.017	0.033	0.067	0.102	0.137
1.80	0.001	0.007	0.014	0.029	0.058	0.088	0.117
1.90	0.001	0.006	0.013	0.025	0.051	0.076	0.102
2.00	0.001	0.006	0.011	0.022	0.044	0.067	0.089
2.20	0.001	0.004	0.009	0.018	0.035	0.053	0.070
2.40	0.001	0.004	0.007	0.014	0.028	0.042	0.056
2.60	0.001	0.003	0.006	0.012	0.023	0.035	0.046
2.80	0.000	0.002	0.005	0.010	0.020	0.029	0.039
3.00	0.000	0.002	0.004	0.008	0.017	0.025	0.033
3.50	0.000	0.001	0.003	0.006	0.012	0.017	0.023
4.00	0.000	0.001	0.002	0.004	0.009	0.013	0.017

			P_r				
1.000	1.200	1.500	2.000	3.000	5.000	7.000	10.000
7.099	6.935	6.740	6.497	6.182	5.847	5.683	5.578
6.663	6.497	6.299	6.052	5.728	5.376	5.194	5.060
6.275	6.109	5.909	5.660	5.330	4.967	4.772	4.619
5.924	5.757	5.557	5.306	4.974	4.603	4.401	4.234
5.608	5.441	5.240	4.989	4.656	4.282	4.074	3.899
5.324	5.157	4.956	4.706	4.373	3.998	3.788	3.607
5.066	4.900	4.700	4.451	4.120	3.747	3.537	3.353
4.830	4.665	4.467	4.220	3.892	3.523	3.315	3.131
4.610	4.446	4.250	4.007	3.684	3.322	3.117	2.935
4.399	4.238	4.045	3.807	3.491	3.138	2.939	2.761
4.191	4.034	3.846	3.615	3.310	2.970	2.777	2.605
3.976	3.825	3.646	3.425	3.135	2.812	2.629	2.463
3.738	3.599	3.434	3.231	2.964	2.663	2.491	2.334
3.569	3.444	3.295	3.108	2.860	2.577	2.412	2.262
3.433	3.326	3.193	3.023	2.790	2.520	2.362	2.215
3.259	3.188	3.081	2.932	2.719	2.463	2.312	2.170
3.142	3.106	3.019	2.884	2.682	2.436	2.287	2.148
2.972	3.010	2.953	2.835	2.646	2.408	2.263	2.126
2.178	2.893	2.879	2.784	2.609	2.380	2.239	2.105
1.391	2.736	2.798	2.730	2.571	2.352	2.215	2.083
1.225	2.495	2.706	2.673	2.533	2.325	2.191	2.062
0.965	1.523	2.328	2.483	2.415	2.242	2.121	2.001
0.742	1.012	1.557	2.081	2.202	2.104	2.007	1.903
0.607	0.790	1.126	1.649	1.968	1.966	1.897	1.810
0.512	0.651	0.890	1.308	1.727	1.827	1.789	1.722
0.385	0.478	0.628	0.891	1.299	1.554	1.581	1.556
0.303	0.372	0.478	0.663	0.990	1.303	1.386	1.402
0.246	0.299	0.381	0.520	0.777	1.088	1.208	1.260
0.204	0.247	0.312	0.421	0.628	0.913	1.050	1.130
0.172	0.208	0.261	0.350	0.519	0.773	0.915	1.013
0.147	0.177	0.222	0.296	0.438	0.661	0.799	0.908
0.127	0.153	0.191	0.255	0.375	0.570	0.702	0.815
0.111	0.134	0.167	0.221	0.325	0.497	0.620	0.733
0.087	0.105	0.130	0.172	0.251	0.388	0.492	0.599
0.070	0.084	0.104	0.138	0.201	0.311	0.399	0.496
0.058	0.069	0.086	0.113	0.164	0.255	0.329	0.416
0.048	0.058	0.072	0.094	0.137	0.213	0.277	0.353
0.041	0.049	0.061	0.080	0.116	0.181	0.236	0.303
0.029	0.034	0.042	0.056	0.081	0.126	0.166	0.216
0.021	0.025	0.031	0.041	0.059	0.093	0.123	0.162

TABLE 5-5 Lee-Kesler Residual Entropy [51]

Deviation function $\left[\dfrac{S°-S}{R}\right]^{(1)}$

T_r	P_r						
	0.010	0.050	0.100	0.200	0.400	0.600	0.800
0.30	16.782	16.774	16.764	16.744	16.705	16.665	16.626
0.35	15.413	15.408	15.401	15.387	15.359	15.333	15.305
0.40	13.990	13.986	13.981	13.972	13.953	13.934	13.915
0.45	12.564	12.561	12.558	12.551	12.537	12.523	12.509
0.50	11.202	11.200	11.197	11.192	11.182	11.172	11.162
0.55	0.115	9.948	9.946	9.942	9.935	9.928	9.921
0.60	0.078	8.828	8.826	8.823	8.817	8.811	8.806
0.65	0.055	0.309	7.832	7.829	7.824	7.819	7.815
0.70	0.040	0.216	0.491	6.951	6.945	6.941	6.937
0.75	0.029	0.156	0.340	6.173	6.167	6.162	6.158
0.80	0.022	0.116	0.246	0.578	5.475	5.468	5.462
0.85	0.017	0.088	0.183	0.408	4.853	4.841	4.832
0.90	0.013	0.068	0.140	0.301	0.744	4.269	4.249
0.93	0.011	0.058	0.120	0.254	0.593	1.219	3.914
0.95	0.010	0.053	0.109	0.228	0.517	0.961	3.697
0.97	0.010	0.048	0.099	0.206	0.456	0.797	1.570
0.98	0.009	0.046	0.094	0.196	0.429	0.734	1.270
0.99	0.009	0.044	0.090	0.186	0.405	0.680	1.098
1.00	0.008	0.042	0.086	0.177	0.382	0.632	0.977
1.01	0.008	0.040	0.082	0.169	0.361	0.590	0.883
1.02	0.008	0.039	0.078	0.161	0.342	0.552	0.807
1.05	0.007	0.034	0.069	0.140	0.292	0.460	0.642
1.10	0.005	0.028	0.055	0.112	0.229	0.350	0.470
1.15	0.005	0.023	0.045	0.091	0.183	0.275	0.361
1.20	0.004	0.019	0.037	0.075	0.149	0.220	0.286
1.30	0.003	0.013	0.026	0.052	0.102	0.148	0.190
1.40	0.002	0.010	0.019	0.037	0.072	0.104	0.133
1.50	0.001	0.007	0.014	0.027	0.053	0.076	0.097
1.60	0.001	0.005	0.011	0.021	0.040	0.057	0.073
1.70	0.001	0.004	0.008	0.016	0.031	0.044	0.056
1.80	0.001	0.003	0.006	0.013	0.024	0.035	0.044
1.90	0.001	0.003	0.005	0.010	0.019	0.028	0.036
2.00	0.000	0.002	0.004	0.008	0.016	0.023	0.029
2.20	0.000	0.001	0.003	0.006	0.011	0.016	0.021
2.40	0.000	0.001	0.002	0.004	0.008	0.012	0.015
2.60	0.000	0.001	0.002	0.003	0.006	0.009	0.012
2.80	0.000	0.001	0.001	0.003	0.005	0.008	0.010
3.00	0.000	0.001	0.001	0.002	0.004	0.006	0.008
3.50	0.000	0.000	0.001	0.001	0.003	0.004	0.006
4.00	0.000	0.000	0.001	0.001	0.002	0.003	0.005

				P_r				
1.000	1.200	1.500	2.000	3.000	5.000	7.000	10.000
16.586	16.547	16.488	16.390	16.195	15.837	15.468	14.925
15.278	15.251	15.211	15.144	15.011	14.751	14.496	14.153
13.896	13.877	13.849	13.803	13.714	13.541	13.376	13.144
12.496	12.482	12.462	12.430	12.367	12.248	12.145	11.999
11.153	11.143	11.129	11.107	11.063	10.985	10.920	10.836
9.914	9.907	9.897	9.882	9.853	9.806	9.769	9.732
8.799	8.794	8.787	8.777	8.760	8.736	8.723	8.720
7.810	7.807	7.801	7.794	7.784	7.779	7.785	7.811
6.933	6.930	6.926	6.922	6.919	6.929	6.952	7.002
6.155	6.152	6.149	6.147	6.149	6.174	6.213	6.285
5.458	5.455	5.453	5.452	5.461	5.501	5.555	5.648
4.826	4.822	4.820	4.822	4.839	4.898	4.969	5.082
4.238	4.232	4.230	4.236	4.267	4.351	4.442	4.578
3.894	3.885	3.884	3.896	3.941	4.046	4.151	4.300
3.658	3.647	3.648	3.669	3.728	3.851	3.966	4.125
3.406	3.391	3.401	3.437	3.517	3.661	3.788	3.957
3.264	3.247	3.268	3.318	3.412	3.569	3.701	3.875
3.093	3.082	3.126	3.195	3.306	3.477	3.616	3.796
2.399	2.868	2.967	3.067	3.200	3.387	3.532	3.717
1.306	2.513	2.784	2.933	3.094	3.297	3.450	3.640
1.113	1.655	2.557	2.790	2.986	3.209	3.369	3.565
0.820	0.831	1.443	2.283	2.655	2.949	3.134	3.348
0.577	0.640	0.618	1.241	2.067	2.534	2.767	3.013
0.437	0.489	0.502	0.654	1.471	2.138	2.428	2.708
0.343	0.385	0.412	0.447	0.991	1.767	2.115	2.430
0.226	0.254	0.282	0.300	0.481	1.147	1.569	1.944
0.158	0.178	0.200	0.220	0.290	0.730	1.138	1.544
0.115	0.130	0.147	0.166	0.206	0.479	0.823	1.222
0.086	0.098	0.112	0.129	0.159	0.334	0.604	0.969
0.067	0.076	0.087	0.102	0.127	0.248	0.456	0.775
0.053	0.060	0.070	0.083	0.105	0.195	0.355	0.628
0.043	0.049	0.057	0.069	0.089	0.160	0.286	0.518
0.035	0.040	0.048	0.058	0.077	0.136	0.238	0.434
0.025	0.029	0.035	0.043	0.060	0.105	0.178	0.322
0.019	0.022	0.027	0.034	0.048	0.086	0.143	0.254
0.015	0.018	0.021	0.028	0.041	0.074	0.120	0.210
0.012	0.014	0.018	0.023	0.035	0.065	0.104	0.180
0.010	0.012	0.015	0.020	0.031	0.058	0.093	0.158
0.007	0.009	0.011	0.015	0.024	0.046	0.073	0.122
0.006	0.007	0.009	0.012	0.020	0.038	0.060	0.100

TABLE 5-6 Lee-Kesler Fugacity-Pressure Ratio [51]

Simple fluid $\left(\log \dfrac{f}{P}\right)^{(0)}$

T_r	P_r						
	0.010	0.050	0.100	0.200	0.400	0.600	0.800
0.30	-3.708	-4.402	-4.696	-4.985	-5.261	-5.412	-5.512
0.35	-2.471	-3.166	-3.461	-3.751	-4.029	-4.183	-4.285
0.40	-1.566	-2.261	-2.557	-2.848	-3.128	-3.283	-3.387
0.45	-0.879	-1.575	-1.871	-2.162	-2.444	-2.601	-2.707
0.50	-0.344	-1.040	-1.336	-1.628	-1.912	-2.070	-2.177
0.55	-0.008	-0.614	-0.911	-1.204	-1.488	-1.647	-1.755
0.60	-0.007	-0.269	-0.566	-0.859	-1.144	-1.304	-1.413
0.65	-0.005	-0.026	-0.283	-0.576	-0.862	-1.023	-1.132
0.70	-0.004	-0.021	-0.043	-0.341	-0.627	-0.789	-0.899
0.75	-0.003	-0.017	-0.035	-0.144	-0.430	-0.592	-0.703
0.80	-0.003	-0.014	-0.029	-0.059	-0.264	-0.426	-0.537
0.85	-0.002	-0.012	-0.024	-0.049	-0.123	-0.285	-0.396
0.90	-0.002	-0.010	-0.020	-0.041	-0.086	-0.166	-0.276
0.93	-0.002	-0.009	-0.018	-0.037	-0.077	-0.122	-0.214
0.95	-0.002	-0.008	-0.017	-0.035	-0.072	-0.113	-0.176
0.97	-0.002	-0.008	-0.016	-0.033	-0.067	-0.105	-0.148
0.98	-0.002	-0.008	-0.016	-0.032	-0.065	-0.101	-0.142
0.99	-0.001	-0.007	-0.015	-0.031	-0.063	-0.098	-0.137
1.00	-0.001	-0.007	-0.015	-0.030	-0.061	-0.095	-0.132
1.01	-0.001	-0.007	-0.014	-0.029	-0.059	-0.091	-0.127
1.02	-0.001	-0.007	-0.014	-0.028	-0.057	-0.088	-0.122
1.05	-0.001	-0.006	-0.013	-0.025	-0.052	-0.080	-0.110
1.10	-0.001	-0.005	-0.011	-0.022	-0.045	-0.069	-0.093
1.15	-0.001	-0.005	-0.009	-0.019	-0.039	-0.059	-0.080
1.20	-0.001	-0.004	-0.008	-0.017	-0.034	-0.051	-0.069
1.30	-0.001	-0.003	-0.006	-0.013	-0.026	-0.039	-0.052
1.40	-0.001	-0.003	-0.005	-0.010	-0.020	-0.030	-0.040
1.50	-0.000	-0.002	-0.004	-0.008	-0.016	-0.024	-0.032
1.60	-0.000	-0.002	-0.003	-0.006	-0.012	-0.019	-0.025
1.70	-0.000	-0.001	-0.002	-0.005	-0.010	-0.015	-0.020
1.80	-0.000	-0.001	-0.002	-0.004	-0.008	-0.012	-0.015
1.90	-0.000	-0.001	-0.002	-0.003	-0.006	-0.009	-0.012
2.00	-0.000	-0.001	-0.001	-0.002	-0.005	-0.007	-0.009
2.20	-0.000	-0.000	-0.001	-0.001	-0.003	-0.004	-0.005
2.40	-0.000	-0.000	-0.000	-0.001	-0.001	-0.002	-0.003
2.60	-0.000	-0.000	-0.000	-0.000	-0.000	-0.001	-0.001
2.80	0.000	0.000	0.000	0.000	0.000	0.000	0.001
3.00	0.000	0.000	0.000	0.000	0.001	0.001	0.002
3.50	0.000	0.000	0.000	0.001	0.001	0.002	0.003
4.00	0.000	0.000	0.000	0.001	0.002	0.003	0.004

			P_r				
1.000	1.200	1.500	2.000	3.000	5.000	7.000	10.000
-5.584	-5.638	-5.697	-5.759	-5.810	-5.782	-5.679	-5.461
-4.359	-4.416	-4.479	-4.547	-4.611	-4.608	-4.530	-4.352
-3.463	-3.522	-3.588	-3.661	-3.735	-3.752	-3.694	-3.545
-2.785	-2.845	-2.913	-2.990	-3.071	-3.104	-3.063	-2.938
-2.256	-2.317	-2.387	-2.468	-2.555	-2.601	-2.572	-2.468
-1.835	-1.897	-1.969	-2.052	-2.145	-2.201	-2.183	-2.096
-1.494	-1.557	-1.630	-1.715	-1.812	-1.878	-1.869	-1.795
-1.214	-1.278	-1.352	-1.439	-1.539	-1.612	-1.611	-1.549
-0.981	-1.045	-1.120	-1.208	-1.312	-1.391	-1.396	-1.344
-0.785	-0.850	-0.925	-1.015	-1.121	-1.204	-1.215	-1.172
-0.619	-0.685	-0.760	-0.851	-0.958	-1.046	-1.062	-1.026
-0.479	-0.544	-0.620	-0.711	-0.819	-0.911	-0.930	-0.901
-0.359	-0.424	-0.500	-0.591	-0.700	-0.794	-0.817	-0.793
-0.296	-0.361	-0.437	-0.527	-0.637	-0.732	-0.756	-0.735
-0.258	-0.322	-0.398	-0.488	-0.598	-0.693	-0.719	-0.699
-0.223	-0.287	-0.362	-0.452	-0.561	-0.657	-0.683	-0.665
-0.206	-0.270	-0.344	-0.434	-0.543	-0.639	-0.666	-0.649
-0.191	-0.254	-0.328	-0.417	-0.526	-0.622	-0.649	-0.633
-0.176	-0.238	-0.312	-0.401	-0.509	-0.605	-0.633	-0.617
-0.168	-0.224	-0.297	-0.385	-0.493	-0.589	-0.617	-0.602
-0.161	-0.210	-0.282	-0.370	-0.477	-0.573	-0.601	-0.588
-0.143	-0.180	-0.242	-0.327	-0.433	-0.529	-0.557	-0.546
-0.120	-0.148	-0.193	-0.267	-0.368	-0.462	-0.491	-0.482
-0.102	-0.125	-0.160	-0.220	-0.312	-0.403	-0.433	-0.426
-0.088	-0.106	-0.135	-0.184	-0.266	-0.352	-0.382	-0.377
-0.066	-0.080	-0.100	-0.134	-0.195	-0.269	-0.296	-0.293
-0.051	-0.061	-0.076	-0.101	-0.146	-0.205	-0.229	-0.226
-0.039	-0.047	-0.059	-0.077	-0.111	-0.157	-0.176	-0.173
-0.031	-0.037	-0.046	-0.060	-0.085	-0.120	-0.135	-0.129
-0.024	-0.029	-0.036	-0.046	-0.065	-0.092	-0.102	-0.094
-0.019	-0.023	-0.028	-0.036	-0.050	-0.069	-0.075	-0.066
-0.015	-0.018	-0.022	-0.028	-0.038	-0.052	-0.054	-0.043
-0.012	-0.014	-0.017	-0.021	-0.029	-0.037	-0.037	-0.024
-0.007	-0.008	-0.009	-0.012	-0.015	-0.017	-0.012	0.004
-0.003	-0.004	-0.004	-0.005	-0.006	-0.003	0.005	0.024
-0.001	-0.001	-0.001	-0.001	0.001	0.007	0.017	0.037
0.001	0.001	0.002	0.003	0.005	0.014	0.025	0.046
0.002	0.003	0.003	0.005	0.009	0.018	0.031	0.053
0.004	0.005	0.006	0.008	0.013	0.025	0.038	0.061
0.005	0.006	0.007	0.010	0.016	0.028	0.041	0.064

TABLE 5-7 Lee-Kesler Fugacity-Pressure Ratio [51]

Deviation function $\left(\log \dfrac{f}{P}\right)^{(1)}$

T_r	P_r						
	0.010	0.050	0.100	0.200	0.400	0.600	0.800
0.30	-8.778	-8.779	-8.781	-8.785	-8.790	-8.797	-8.804
0.35	-6.528	-6.530	-6.532	-6.536	-6.544	-6.551	-6.559
0.40	-4.912	-4.914	-4.916	-4.919	-4.929	-4.937	-4.945
0.45	-3.726	-3.728	-3.730	-3.734	-3.742	-3.750	-3.758
0.50	-2.838	-2.839	-2.841	-2.845	-2.853	-2.861	-2.869
0.55	-0.013	-2.163	-2.165	-2.169	-2.177	-2.184	-2.192
0.60	-0.009	-1.644	-1.646	-1.650	-1.657	-1.664	-1.671
0.65	-0.006	-0.031	-1.242	-1.245	-1.252	-1.258	-1.265
0.70	-0.004	-0.021	-0.044	-0.927	-0.934	-0.940	-0.946
0.75	-0.003	-0.014	-0.030	-0.675	-0.682	-0.688	-0.694
0.80	-0.002	-0.010	-0.020	-0.043	-0.481	-0.487	-0.493
0.85	-0.001	-0.006	-0.013	-0.028	-0.321	-0.327	-0.332
0.90	-0.001	-0.004	-0.009	-0.018	-0.039	-0.199	-0.204
0.93	-0.001	-0.003	-0.007	-0.013	-0.029	-0.048	-0.141
0.95	-0.001	-0.003	-0.005	-0.011	-0.023	-0.037	-0.103
0.97	-0.000	-0.002	-0.004	-0.009	-0.018	-0.029	-0.042
0.98	-0.000	-0.002	-0.004	-0.008	-0.016	-0.025	-0.035
0.99	-0.000	-0.002	-0.003	-0.007	-0.014	-0.021	-0.030
1.00	-0.000	-0.001	-0.003	-0.006	-0.012	-0.018	-0.025
1.01	-0.000	-0.001	-0.003	-0.005	-0.010	-0.016	-0.021
1.02	-0.000	-0.001	-0.002	-0.004	-0.009	-0.013	-0.017
1.05	-0.000	-0.001	-0.001	-0.002	-0.005	-0.006	-0.007
1.10	-0.000	-0.000	0.000	0.000	0.001	0.002	0.004
1.15	0.000	0.000	0.001	0.002	0.005	0.008	0.011
1.20	0.000	0.001	0.002	0.003	0.007	0.012	0.017
1.30	0.000	0.001	0.003	0.005	0.011	0.017	0.023
1.40	0.000	0.002	0.003	0.006	0.013	0.020	0.027
1.50	0.000	0.002	0.003	0.007	0.014	0.021	0.028
1.60	0.000	0.002	0.003	0.007	0.014	0.021	0.029
1.70	0.000	0.002	0.004	0.007	0.014	0.021	0.029
1.80	0.000	0.002	0.003	0.007	0.014	0.021	0.028
1.90	0.000	0.002	0.003	0.007	0.014	0.021	0.028
2.00	0.000	0.002	0.003	0.007	0.013	0.020	0.027
2.20	0.000	0.002	0.003	0.006	0.013	0.019	0.025
2.40	0.000	0.002	0.003	0.006	0.012	0.018	0.024
2.60	0.000	0.001	0.003	0.006	0.011	0.017	0.023
2.80	0.000	0.001	0.003	0.005	0.011	0.016	0.021
3.00	0.000	0.001	0.003	0.005	0.010	0.015	0.020
3.50	0.000	0.001	0.002	0.004	0.009	0.013	0.018
4.00	0.000	0.001	0.002	0.004	0.008	0.012	0.016

				P_r			
1.000	1.200	1.500	2.000	3.000	5.000	7.000	10.000
-8.811	-8.818	-8.828	-8.845	-8.880	-8.953	-9.022	-9.126
-6.567	-6.575	-6.587	-6.606	-6.645	-6.723	-6.800	-6.919
-4.954	-4.962	-4.974	-4.995	-5.035	-5.115	-5.195	-5.312
-3.766	-3.774	-3.786	-3.806	-3.845	-3.923	-4.001	-4.114
-2.877	-2.884	-2.896	-2.915	-2.953	-3.027	-3.101	-3.208
-2.199	-2.207	-2.218	-2.236	-2.273	-2.342	-2.410	-2.510
-1.677	-1.684	-1.695	-1.712	-1.747	-1.812	-1.875	-1.967
-1.271	-1.278	-1.287	-1.304	-1.336	-1.397	-1.456	-1.539
-0.952	-0.958	-0.967	-0.983	-1.013	-1.070	-1.124	-1.201
-0.700	-0.705	-0.714	-0.728	-0.756	-0.809	-0.858	-0.929
-0.499	-0.504	-0.512	-0.526	-0.551	-0.600	-0.645	-0.709
-0.338	-0.343	-0.351	-0.364	-0.388	-0.432	-0.473	-0.530
-0.210	-0.215	-0.222	-0.234	-0.256	-0.296	-0.333	-0.384
-0.146	-0.151	-0.158	-0.170	-0.190	-0.228	-0.262	-0.310
-0.108	-0.114	-0.121	-0.132	-0.151	-0.187	-0.220	-0.265
-0.075	-0.080	-0.087	-0.097	-0.116	-0.149	-0.180	-0.223
-0.059	-0.064	-0.071	-0.081	-0.099	-0.132	-0.162	-0.203
-0.044	-0.050	-0.056	-0.066	-0.084	-0.115	-0.144	-0.184
-0.031	-0.036	-0.042	-0.052	-0.069	-0.099	-0.127	-0.166
-0.024	-0.024	-0.030	-0.038	-0.054	-0.084	-0.111	-0.149
-0.019	-0.015	-0.018	-0.026	-0.041	-0.069	-0.095	-0.132
-0.007	-0.002	0.008	0.007	-0.005	-0.029	-0.052	-0.085
0.007	0.012	0.025	0.041	0.042	0.026	0.008	-0.019
0.016	0.022	0.034	0.056	0.074	0.069	0.057	0.036
0.023	0.029	0.041	0.064	0.093	0.102	0.096	0.081
0.030	0.038	0.049	0.071	0.109	0.142	0.150	0.148
0.034	0.041	0.053	0.074	0.112	0.161	0.181	0.191
0.036	0.043	0.055	0.074	0.112	0.167	0.197	0.218
0.036	0.043	0.055	0.074	0.110	0.167	0.204	0.234
0.036	0.043	0.054	0.072	0.107	0.165	0.205	0.242
0.035	0.042	0.053	0.070	0.104	0.161	0.203	0.246
0.034	0.041	0.052	0.068	0.101	0.157	0.200	0.246
0.034	0.040	0.050	0.066	0.097	0.152	0.196	0.244
0.032	0.038	0.047	0.062	0.091	0.143	0.186	0.236
0.030	0.036	0.044	0.058	0.086	0.134	0.176	0.227
0.028	0.034	0.042	0.055	0.080	0.127	0.167	0.217
0.027	0.032	0.039	0.052	0.076	0.120	0.158	0.208
0.025	0.030	0.037	0.049	0.072	0.114	0.151	0.199
0.022	0.026	0.033	0.043	0.063	0.101	0.134	0.179
0.020	0.023	0.029	0.038	0.057	0.090	0.121	0.163

Equation (5-4.3) may be rewritten as

$$\frac{H°-H}{RT_c} = \left(\frac{H°-H}{RT_c}\right)^{(0)} + \omega\left(\frac{H°-H}{RT_c}\right)^{(1)} \tag{5-4.4}$$

if the deviation function for enthalpy is defined as

$$\left(\frac{H°-H}{RT_c}\right)^{(1)} = \frac{1}{\omega^R}\left[\left(\frac{H°-H}{RT_c}\right)^{(R)} - \left(\frac{H°-H}{RT_c}\right)^{(0)}\right] \tag{5-4.5}$$

With Eqs. (5-4.1) and (5-4.5), Tables 5-2 and 5-3 were developed [51].   In these calculations ω^R was set equal to 0.3978.

Entropy   Analogous expressions for the entropy departure functions are

$$\left(\frac{S°-S}{R}\right)^{(0)} = -\ln\frac{P°}{P} - \ln Z^{(0)} + \frac{b_1 + b_3/T_r^2 + 2b_4/T_r^3}{V_r^{(0)}}$$

$$+ \frac{c_1 - 2c_3/T_r^3}{2(V_r^{(0)})^2} + \frac{d_1}{5(V_r^{(0)})^5} - 2E \tag{5-4.6}$$

$$\frac{S°-S}{R} = \left(\frac{S°-S}{R}\right)^{(0)} + \frac{\omega}{\omega^R}\left[\left(\frac{S°-S}{R}\right)^{(R)} - \left(\frac{S°-S}{R}\right)^{(0)}\right] \tag{5-4.7}$$

If Eq. (5-4.6) is written as

$$\frac{S°-S}{R} = -\ln\frac{P°}{P} + \left(\frac{S°-S}{R}\right)^{(0)} + \omega\left(\frac{S°-S}{R}\right)^{(1)} \tag{5-4.8}$$

the simple-fluid entropy departure $(\ )^{(0)}$ and the deviation function $(\ )^{(1)}$ can be found in Tables 5-4 and 5-5.

Fugacity-Pressure Ratio   As with the enthalpy departure calculation,

$$\left(\ln\frac{f}{P}\right)^{(0)} = Z^{(0)} - 1 - \ln Z^{(0)} + \frac{B}{V_r^{(0)}} + \frac{C}{2(V_r^{(0)})^2} + \frac{D}{5(V_r^{(0)})^5} + E \tag{5-4.9}$$

$$\ln\frac{f}{P} = \left(\ln\frac{f}{P}\right)^{(0)} + \frac{\omega}{\omega^R}\left[\left(\ln\frac{f}{P}\right)^{(R)} - \left(\ln\frac{f}{P}\right)^{(0)}\right] \tag{5-4.10}$$

and when Eq. (5-4.10) is written as

$$\ln\frac{f}{P} = \left(\ln\frac{f}{P}\right)^{(0)} + \omega\left(\ln\frac{f}{P}\right)^{(1)} \tag{5-4.11}$$

the simple-fluid and deviation functions can be found from Tables 5-6 and 5-7.   (Note that $\log\frac{f}{P}$, not $\ln\frac{f}{P}$, is obtained from these tables.)

In addition to the equations shown in this section for estimating enthalpy departures, Yen and Alexander [103] have used the Lydersen, Greenkorn, and Hougen compressibility-factor tables [54] and Eqs. (5-3.13) to (5-3.15) to obtain a generalized relation for $(H°-H)/T_c =$

TABLE 5-8 Yen and Alexander Enthalpy Departure Equations [103]

$\dfrac{H° - H}{T_c}$ in Btu/lb mol °R or cal/g-mol K

Superheated vapor, $P_r = 0.01$–30

$$\frac{H° - H}{T_c} = \frac{mP_r\left(1 - \dfrac{P_r}{X_0}\right)}{\exp\left[-C_1(P_r)^2\right]\{\ \} + C_4 + C_5 P_r + C_6 P_r^2}$$

where

$$\{\ \} = 1 - C_2 - C_4 - C_5 P_r + C_2\left[\frac{1}{\pi}\arctan\,(C_3 - C_3 P_r) + 0.50\right]^2$$

$Z_c = 0.29$: Reduced-temperature range: $0.8 < T_r < 8.0$:

$C_1 = 166\exp\,(-5.16 T_r) + 0.017$

$C_2 = 0.62\exp\,[-18.4(T_r - 1.0)] + 0.05$

$C_3 = \exp\left(\dfrac{38.8}{T_r} - 34.2\right)$

$C_4 = 0.989(T_r - 0.75)^{1.63}\exp\,[-0.3175(T_r - 0.75)]$

$C_5 = 0.215\exp\,[-0.1045(T_r^{4.935})]$

$C_6 = 0.0564(T_r - 0.5)^{4.67}\exp\,[-2.869(T_r - 0.5)]$

$X_0 = 1.15 - 0.314(T_r - 8)^3$

$m = \dfrac{1 - 0.0001499 T_r^{9.17}\exp\,(-1.297 T_r)}{0.1879 + 1.0826(T_r - 0.65)^{1.1726}}$

$Z_c = 0.27$: Reduced-temperature range: $0.9 < T_r < 4.0$:

$C_1 = \dfrac{40\exp\,(-5.7 T_r)}{T_r - 0.81} + 0.01$

$C_2 = \dfrac{0.35\exp\,[-26.2(T_r - 1.0)]}{T_r} + 0.31$

$C_3 = 17.25\exp\,[-16.7(T_r - 1.0)] + 2.75$

$C_4 = 0.444(T_r - 2.0) - 0.0215(T_r - 2.0)^2 + 0.0610(T_r - 2.0)^3 - 0.0404(T_r - 2.0)^4 + 0.6564$

$C_5 = \begin{cases} -0.0697(T_r - 2.0) + 0.0734(T_r - 2.0)^2 - 0.0533(T_r - 2.0)^5 + 0.0125 & \text{if } T_r \leqslant 2.25 \\ 0 & \text{if } T_r > 2.25 \end{cases}$

$C_6 = 0.00301(T_r - 0.8)\exp\,[-0.87(T_r - 0.8)^2]$

$X_0 = 8.5 + 10.5(T_r - 4.0)^2$

$m = \dfrac{1.0}{0.1052 + 1.2044(T_r - 0.4429)^{2.135}}$

$Z_c = 0.25$: Reduced-temperature range: $0.75 < T_r < 2.0$:

$C_1 = 0.87\exp\,[9.845(1.0 - T_r)] + 0.023955\exp\,(1.2405 T_r)$

$C_2 = 0.47148\exp\,[39.35(1.0 - T_r)] + 0.142$

$C_3 = 3918\exp\,[122(1 - T_r)] + 2.0$

$C_4 = 1.1938 T_r - 0.20726 T_r^2 - 0.90649$

TABLE 5-8   Yen and Alexander Enthalpy Departure Equations   (Continued)

$\dfrac{H° - H}{T_c}$ in Btu/lb mol °R or cal/g-mol K

Superheated vapor, $P_r = 0.01$–30   (Continued)

$C_5 = 3.0 \exp(-2.52 T_r)$

$C_6 = 0.004277(T_r - 0.895) \exp[-0.946(T_r - 0.895)^2]$

$X_0 = -60 T_r + 170$

$m = \dfrac{1}{0.0853 + 1.5078(T_r - 0.5952)^{1.7336}}$

$Z_c = 0.23$: Reduced-temperature range: $0.8 < T_r < 1.7$:

$C_1 = \dfrac{T_r}{0.003147 \exp(6.0759 T_r) - 61.12(T_r - 1.0) \exp[-19.933(T_r - 1.0)]}$
$$+ 1.9936 \exp\{-0.911[10(T_r - 0.88)]^{10}\}$$

$C_2 = \exp(17.671 - 17.63 T_r) + 0.17 - 1.1738 \exp[-17.193(|T_r - 1.05|)]$

$C_3 = \exp(52.268 - 49.188 T_r) + 24.249 \exp[-920.5(T_r - 1.1)^2]$

$C_4 = 0.0032456 \exp(3.2917 T_r) - \exp(17.3 T_r - 30.6)$

$C_5 = 1.9772 \exp(-2.1839 T_r)$

$C_6 = 0.000336 \exp(1.61 T_r) - 0.0000054 T_r^{11.5}$

$X_0 = 515 - 250 T_r$

$m = \dfrac{1}{-0.023 + 1.0743(T_r - 0.4146)^{1.8103}}$

Saturated vapor

$Z_c = 0.29$:	$\dfrac{H° - H}{T_c} = \dfrac{5.4 P_r^{0.6747}}{1 + 1.227(-\ln P_r)^{0.503}}$

$Z_c = 0.27$:	$\dfrac{H° - H}{T_c} = \dfrac{5.8 P_r^{0.63163}}{1 + 1.229(-\ln P_r)^{0.55456}}$

$Z_c = 0.25$:	$\dfrac{H° - H}{T_c} = \dfrac{6.5 P_r^{0.62252}}{1 + 0.76218(-\ln P_r)^{0.53042}}$

$Z_c = 0.23$:	$\dfrac{H° - H}{T_c} = \dfrac{7.0 P_r^{0.65135}}{1 + 0.75727(-\ln P_r)^{0.46108}}$

Saturated liquid

$Z_c = 0.29$:	$\dfrac{H° - H}{T_c} = \dfrac{5.4 + 3.6485(-\ln P_r)^{0.33464}}{1.0 - 0.0056942(\ln P_r)}$

$Z_c = 0.27$:	$\dfrac{H° - H}{T_c} = \dfrac{5.8 + 5.19(-\ln P_r)^{0.4963}}{1.0 - 0.1(\ln P_r)}$

$Z_c = 0.25$:
$$\frac{H° - H}{T_c} = \frac{6.5 + 4.48(-\ln P_r)^{0.3952}}{1.0 - 0.00185(\ln P_r)}$$

$Z_c = 0.23$:
$$\frac{H° - H}{T_c} = \frac{7.0 + 4.5688(-\ln P_r)^{0.333}}{1.0 + 0.004(\ln P_r)}$$

Subcooled liquid, $P_r = 0.01$–30, $T_r = 0.5$–4.0

$Z_c = 0.29$:
$$\frac{H° - H}{T_c} = -0.09572107(P_r - 4.2) - 9.501235(T_r - 0.77)$$
$$- 17.30389(T_r - 0.77)^2 - 0.3195707(P_r - 4.2)(T_r - 0.77)$$
$$+ 1.368092 \ln P_r + 4.227096(\ln P_r)(\ln T_r)$$
$$+ 3.181639(\ln P_r)(\ln T_r)^2 + 9.707447$$

$Z_c = 0.27$:
$$\frac{H° - H}{T_c} = -0.1368774(P_r - 4.664) - 14.56975(T_r - 0.79749)$$
$$- 7.812724(T_r - 0.79749)^2 - 0.1642482(T_r - 0.79749)(P_r - 4.664)$$
$$+ 1.036851 \ln P_r + 4.463472(\ln P_r)(\ln T_r) + 4.525831(\ln P_r)(\ln T_r)^2$$
$$+ 10.86085$$

$Z_c = 0.25$:
$$\frac{H° - H}{T_c} = -0.1074635(P_r - 4.2) - 15.80132(T_r - 0.77)$$
$$- 15.18611(T_r - 0.77)^2 - 0.1476876(P_r - 4.2)(T_r - 0.77)$$
$$+ 0.7800774 \ln P_r + 3.154058(\ln P_r)(\ln T_r)$$
$$+ 2.988533(\ln P_r)(\ln T_r)^2 + 12.28618$$

$Z_c = 0.23$:
$$\frac{H° - H}{T_c} = -0.08644293(P_r - 4.2) - 12.93889(T_r - 0.77)$$
$$- 10.81311(T_r - 0.77)^2 - 0.1568094(P_r - 4.2)(T_r - 0.77)$$
$$+ 0.7466842 \ln P_r + 3.17422(\ln P_r)(\ln T_r)$$
$$+ 2.930566(\ln P_r)(\ln T_r)^2 + 12.72429$$

$f(T_r, P_r, Z_c)$. This correlation is shown in Table 5-8 and Figs. 5-2 to 5-5; values of $(H° - H)/T_c$ are given for discrete values of Z_c. Interpolation is required if Z_c lies between values of 0.23, 0.25, 0.27, or 0.29, but extrapolation to Z_c values less than 0.23 or greater than 0.29 should *not* be made, as serious errors may result. When applied to mixtures, Eqs. (4-2.1) and (4-2.2) can be used to estimate T_{c_m} and P_{c_m}† and a mole-

†Yen and Alexander used a mole-fraction average P_{c_m}.

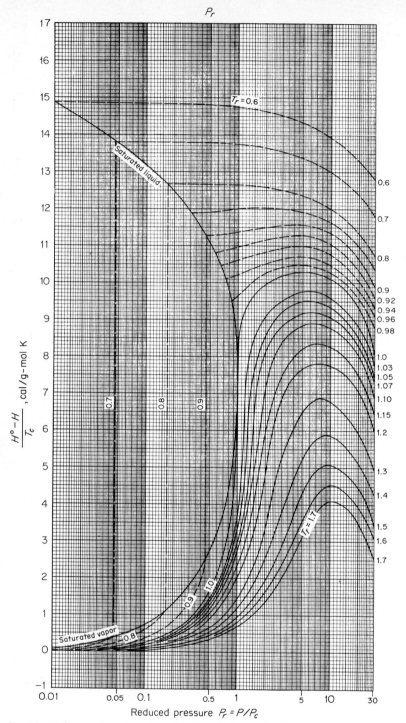

Fig. 5-2 Enthalpy departure from ideal-gas behavior for fluids, $Z_c = 0.23$. (*From Ref.* 103.)

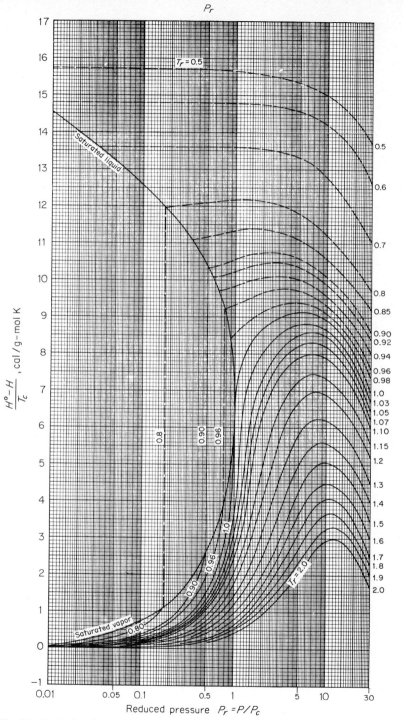

Fig. 5-3 Enthalpy departure from ideal-gas behavior for fluids, $Z_c = 0.25$. (*From Ref.* 103.)

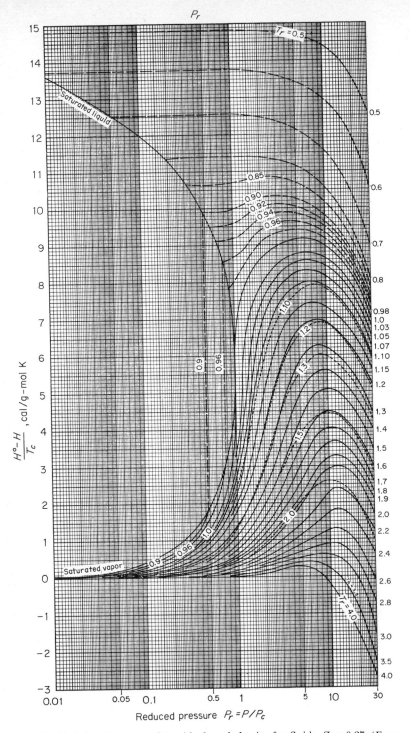

Fig. 5-4 Enthalpy departure from ideal-gas behavior for fluids, $Z_c = 0.27$. (*From Ref*. 103.)

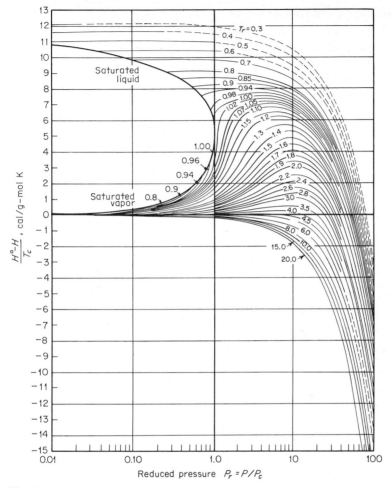

Fig. 5-5 Enthalpy departure from ideal-gas behavior for fluids, $Z_c = 0.29$; $--$ liquid, — vapor. (*Modified from Ref.* 28.)

fraction average for Z_{c_m} unless He, H₂, or Ne is present, when special rules devised by Gunn et al. [35] should be used.

Discussion. In testing the departure functions presented in this chapter, only estimated values of $H° - H$ can be compared with experimental data, and even here, most reliable data are limited to simple hydrocarbons and the *permanent* light gases. Also, it is convenient to summarize at this point the methods for estimating molal volumes of gas mixtures.†

†Recommended methods for estimating the molar volumes of *pure* gases are given in Sec. 3-12 and for pure liquids in Sec. 3-16.   Liquid mixtures are treated in Sec. 4-10.   The Lee-Kesler and the Benedict-Webb-Rubin equations have also been used to calculate molar volumes of pure liquids and liquid mixtures of light hydrocarbons, but the methods described in Sec. 4-10 are generally more accurate.

Molar Volumes for Gas Mixtures   No comprehensive test has been made of all estimation techniques.   Authors of individual methods have made limited comparisons, as have the authors of this book.   Generally errors were found to be less than 2 to 3 percent, except near the critical point or for mixtures containing highly polar components in significant concentrations.

To estimate the molar volume of *hydrocarbon* gas mixtures (including those with components associated with natural gas, e.g., CO_2, H_2S), any of the four relationships shown below are accurate, and are recommended:

	Equations	Tables
Benedict-Webb-Rubin	(3-8.1), (4-6.1)	3-6 to 3-8
Lee-Erbar-Edmister	(3-10.1)	3-10, 4-4
Soave modification of Redlich-Kwong	(3-5.11), (3-5.15)	4-1, 4-2
Lee-Kesler	(3-9.1)–(3-9.4), (4-6.3)–(4-6.7)	3-1, 3-2, 3-9

The Soave, Lee-Kesler, and the Benedict-Webb-Rubin equations have been extensively tested.   Sass et al. [76] found that the Benedict-Webb-Rubin relation yields poor results when applied to high-pressure ethylene-CO_2 mixtures.   Difficulty was also reported in the application of the Benedict-Webb-Rubin equation to *n*-butane [80], but this was refuted by Barner and Adler [5].   Wolfe [102] illustrates the Benedict-Webb-Rubin form for natural-gas mixtures.

To estimate the molar volumes of gas mixtures containing nonhydrocarbons the following equations are recommended:

	Equations	Tables
Barner-Adler	(3-6.1), (4-4.1)–(4-4.7)	3-4, 4-3
Sugie-Lu	(4-5.1)–(4-5.4)	3-5
Lee-Kesler	(3-9.1)–(3-9.4), (4-6.3)–(4-6.7)	3-1, 3-2, 3-9

None of these equations has been critically evaluated for nonhydrocarbons; in most cases, binary interaction parameters are required for high accuracy.

The equations of state noted above for hydrocarbon and nonhydrocarbon mixtures are complex and normally require a computer for efficient utilization.   The Redlich-Kwong equation of state may be slightly less accurate, but it is simpler in form, and the truncated virial equation is the easiest of all equations of state to use, although for mixtures its accuracy may be poor unless care is taken to estimate the cross-coefficient B_{12}.

Enthalpy Departures for Pure Gases and Gas Mixtures   Of all methods for estimating enthalpy departures, that of Yen-Alexander has been the most extensively tested [60, 79, 94, 100, 104].   Garcia-Rangel and Yen [28] compared calculated values of $H° - H$ with over 7000 literature values for both pure gases and gas mixtures.   Enthalpy departures for nonpolar gases were usually predicted to within 2 to 3 cal/g.   Even for polar-gas mixtures, errors were normally less than 6 cal/g.

For hydrocarbon gas mixtures, the Soave modification of the Redlich-Kwong equation, the Lee-Kesler, and the Lee-Erbar-Edmister equation generally predict enthalpy departures within 1 cal/g [17, 92, 97].   The Benedict-Webb-Rubin equation is also quite accurate, although the best results are obtained when C_0 is allowed to vary with temperature† and interaction parameters are used in mixture combining rules.   For the methane-propane system over a wide temperature range, Starling and Powers [86] found they had to modify the Benedict-Webb-Rubin equation by adding three additional constants (total of 11) to fit experimental values of $H° - H$ within 1 cal/g.

The Barner-Adler and Sugie-Lu equations of state have not been tested extensively in calculating enthalpy departures.   The few comparisons made by the authors indicate errors in the same range as those found for the Yen-Alexander correlation.   Somewhat higher errors are to be expected when the original Redlich-Kwong or truncated virial equation is employed.   The latter, however, is capable of high accuracy if the interaction virial B_{12} can be accurately estimated as a function of temperature.

Though only tested extensively with hydrocarbons, the Lee-Kesler relation for estimating enthalpy departures appears to be more accurate [92] than that of Yen and Alexander.

Recommendations   For hydrocarbon gas mixtures (including light gases such as N_2, CO_2, and H_2S), calculate $H° - H$ from the Soave modification of the Redlich-Kwong equation, the Lee-Kesler equation, or from the Lee-Erbar-Edmister equation.   If the mixture contains hydrogen, the Soave equation should not be used.   Even over wide temperature and pressure ranges, errors of only a few calories per gram are to be expected.

For gas mixtures containing nonhydrocarbons, the Lee-Kesler correlation [Eqs. (5-4.2) to (5-4.5) or Tables 5-2 and 5-3], the Yen-Alexander correlation (Table 5-8 or Figs. 5-2 to 5-5) or the equation-of-state methods of Barner and Adler or Sugie and Lu are recommended to estimate $H° - H$.   Though errors vary, for nonpolar-gas mixtures,

---

†Ruf et al. [71] allowed both C_0 and γ to vary with temperature and obtained good results for mixtures of helium, nitrogen, methane, and propane.

differences between calculated and experimental values of $H° - H$ should be only a few calories per gram.

The Redlich-Kwong or truncated virial equations of state are simple to employ and may be useful where extensive iterative calculations must be performed. However, the truncated virial equation can be used only at low or moderate densities.

All the correlations noted above may be used up to the saturated-vapor envelope.† As this envelope is approached, and especially in the critical region, errors are expected to increase. At very high pressures, the correlation of Breedveld and Prausnitz [10] can be used.

Enthalpy Departures for Pure Liquids and Liquid Mixtures   To estimate the enthalpy departure of *pure* liquids, it is generally preferable to break the computation into several steps, i.e.,

$$H^L - H° = (H^L - H^{SL}) + (H^{SL} - H^{SV}) + (H^{SV} - H°) \quad (5\text{-}4.12)$$

where H^L = liquid enthalpy at T and P
   $H°$ = ideal-gas enthalpy at T and $P°$
   H^{SL} = saturated liquid enthalpy at T and P_{vp}
   H^{SV} = saturated vapor enthalpy at T and P_{vp}

The vapor contribution $H^{SV} - H°$ can be estimated by methods described earlier in this chapter. The term $H^{SL} - H^{SV}$ is simply $-\Delta H_v$ and can be obtained from enthalpy of vaporization correlations given in Chap. 6. Finally, $H^L - H^{SL}$ represents the effect of pressure on liquid enthalpy. It is normally small relative to the other two terms. The Yen-Alexander corresponding-states correlation given in Table 5-8 can be employed to estimate $H^L - H^{SL}$, that is, for a given Z_c,

$$\frac{H^L - H^{SL}}{T_c} = \left(\frac{H° - H}{T_c}\right)_{SL} - \left(\frac{H° - H}{T_c}\right)_{SCL} \quad (5\text{-}4.13)$$

where SL stands for saturated liquid and SCL for subcooled liquid. Equation (5-4.13) is not accurate near the critical point. In fact, in using the equations given in Table 5-8, $[(H° - H)/T_c]_{SL} - [(H° - H)/T_c]_{SCL}$ does not vanish when applied to the critical point, i.e., when $P_r = T_r = 1$.

Alternatively, the Lee-Kesler correlation [Eqs. (5-4.2) to (5-4.5) or Tables 5-2 and 5-3] may be applied to the liquid phase and differences taken between $(H° - H)_{SL}$ and $(H° - H)_{SCL}$. Lu et al. [53] also tabulate residual enthalpies for the subcooled liquid. These are extensions of an earlier set [65, 66, 98].

In the Yen-Alexander and Lee-Kesler methods, the enthalpy depar-

---

†Except the truncated virial, which is limited in range to about one-half the critical density.

ture term is $H° - H$, that is, the *difference* between the enthalpy of the fluid in an ideal-gas state at T and that of the fluid at P, T (liquid or gas). It is not generally recommended that enthalpies of *liquids* be calculated directly from this difference. The corresponding-states methods are usually not sufficiently accurate to estimate ΔH for a phase change. It is preferable to determine phase-change ΔH values separately, as indicated in Eq. (5-4.12), and to use other methods for $H^L - H^{SL}$ and $H^{SV} - H°$.

When enthalpy departures are desired for liquid mixtures, no completely satisfactory recommendations can be formulated. One method now in wide use is simply to use the Yen-Alexander method shown in Table 5-8 with the mixing rules noted earlier. Errors vary, but for nonpolar liquid mixtures Garcia-Rangel and Yen [28] found errors normally less than 6 cal/g. Excellent results were also found by Tully and Edmister [94] when this correlation was employed to estimate integral heats of vaporization for high-pressure methane-ethylene mixtures.

Another approach employs a modified form of Eq. (5-4.12). $H^{SV} - H°$ is calculated as described earlier for gas mixtures, while $H^{SL} - H^{SV}$ is set equal to the mole-fraction average of the *pure*-component values of $-\Delta H_v$,

$$H^{SL} - H^{SV} = -\sum_j x_j \, \Delta H_{v_j} \qquad (5\text{-}4.14)$$

Finally, $H^L - H^{SL}$ can be neglected at low pressure, or it can be estimated from Tables 5-2 and 5-3 with pseudocritical constants determined from Eqs. (4-2.1) to (4-2.3) or (4-6.3) to (4-6.7). This approach is useful only if all components are subcritical. Further, it neglects any heat of mixing in the liquid phase, an assumption which is often warranted unless the liquid phase contains polar components [87].

Good results have been reported when the Soave modification of the Redlich-Kwong equation of state has been used to calculate liquid-mixture enthalpy departures for hydrocarbon mixtures not containing hydrogen [97]. This relation is preferred to the Benedict-Webb-Rubin or Lee-Erbar-Edmister equation or to other modified Redlich-Kwong equations [41, 101]. A theoretical equation-of-state method applicable for cryogenic mixtures is also available [62].

If Tables 5-2 and 5-3 are used directly to calculate liquid-mixture enthalpy departures, good results are often reported [28, 40, 92, 101].

Mixture enthalpies of vaporization can also be obtained from phase-equilibrium measurements [50, 91], but very accurate data are needed to yield reasonable enthalpy values.

To estimate enthalpies (and densities) of hydrocarbon mixtures containing hydrogen, the generalized method of Chueh and Deal [13] is recommended. Huang and Daubert [39] illustrate how one can predict enthalpies of liquid (and vapor) mixtures of petroleum fractions, and Ghormley and Lenoir treat saturated-liquid and saturated-vapor enthalpies for aliphatic-hydrocarbon mixtures [29].

Departure Entropies and Fugacity Coefficients   To estimate departure functions for entropy and fugacity coefficients of *pure gases or mixtures,* where possible, follow the recommendations made earlier for enthalpies.   (The Yen-Alexander correlation is applicable only to enthalpy.)   Table 5-1 may be used for $S° - S$ if an analytical equation of state is selected and $\ln(f/P)$ determined from Eq. (5-3.11).   If the Lee-Kesler method is chosen, use Eqs. (5-4.6) to (5-4.11) or Tables 5-4 to 5-7.

Finally, it should be pointed out that enthalpy departures, entropy departures, and fugacity coefficients are not independent and are related by

$$\frac{H° - H}{RT} = \frac{S° - S}{R} - \ln \frac{f}{P°} \tag{5-4.15}$$

where $P°$ is the pressure in the ideal-gas reference state (see Sec. 5-3).

Example 5-2   Estimate the enthalpy and entropy departures for propylene at 125°C and 10 MPa.†   Bier et al. [7] report experimental values as

$$H° - H = 244.58 \text{ J/g} \qquad S° - S = 1.4172 \text{ J/g K}$$

(The ideal-gas reference pressure for entropy $P°$ is 0.1 MPa.)

   solution   From Appendix A, $M = 42.081$, $T_c = 365.0$ K, $P_c = 45.6$ atm, $Z_c = 0.275$, and $\omega = 0.148$.   Thus $T_r = (125 + 273.2)/365.0 = 1.09$, and $P_r = (10)(9.87)/45.6 = 2.16$.

   Lee-Kesler Method.   Equations (5-4.4) and (5-4.8) will be used with Tables 5-2 to 5-5

$$\left(\frac{H° - H}{RT_c}\right)^{(0)} = 3.11 \qquad \left(\frac{H° - H}{RT_c}\right)^{(1)} = 1.61$$

$$\left(\frac{S° - S}{R}\right)^{(0)} = 2.17 \qquad \left(\frac{S° - S}{R}\right)^{(1)} = 1.57$$

Then

$$H° - H = \frac{RT_c}{M} \left[\left(\frac{H° - H}{RT_c}\right)^{(0)} + \omega \left(\frac{H° - H}{RT_c}\right)^{(1)}\right]$$

$$= \frac{(8.314)(365.0)}{42.081} [3.11 + (0.148)(1.61)] = 242 \text{ J/g}$$

$$S° - S = \frac{R}{M} \left[\left(\frac{S° - S}{R}\right)^{(0)} + \omega \left(\frac{S° - S}{R}\right)^{(1)} - \ln \frac{P°}{P}\right]$$

$$= \frac{8.314}{42.081} \left[2.17 + (0.148)(1.57) - \ln \frac{0.1}{10}\right]$$

$$= 1.39 \text{ J/g K}$$

This method underpredicts $H° - H$ by 2 J/g and $S° - S$ by 0.03 J/g K.

†1 MPa = 1 megapascal = 10^6 N/m² = 9.87 atm = 145 psia = 10 bars

Yen-Alexander Corresponding-States Method. With Table 5-8, since $Z_c = 0.275$ lies between Z_c of 0.27 and 0.29, we shall calculate $(H° - H)/T_c$ for both and linearly interpolate:

$$\frac{H° - H}{T_c} = \frac{mP_r(1 - P_r/X_0)}{\exp\left[-C_1(P_r)^2\right]\{\ \} + C_4 + C_5P_r + C_6P_r^2}$$

where

$$\{\ \} = 1 - C_2 - C_4 - C_5P_r + C_2\left[\frac{1}{\pi}\arctan(C_3 - C_3P_r) + 0.50\right]^2$$

	Z_c	
	0.27	0.29
C_1	0.296	0.616
C_2	0.340	0.168
C_3	6.587	4.040
C_4	0.161	0.153
C_5	0.170	0.183
C_6	8.113×10^{-4}	8.831×10^{-4}
X_0	97.41	104.75
m	1.722	1.662
$(H° - H)/T_c$	6.433	6.185
$H° - H$, cal/g-mol K	2348	2259
J/g K	233	225

For $Z_c = 0.275$: $H° - H = 231$ J/g K

The entropy departure cannot be predicted by this method.

Equation-of-State Method. To illustrate the use of one equation of state, let us select the Soave modification of the Redlich-Kwong equation. Using Table 5-1 for a pure material gives

$$P = \frac{RT}{V - b} - \frac{\Omega_a}{\Omega_b}\frac{RTbF}{V(V + b)}$$

$$A - A° = RT\left(-\ln\frac{V - b}{V} - \frac{\beta}{b}\ln\frac{V + b}{V} - \ln\frac{V}{V°}\right)$$

$$S - S° = R\left(\ln\frac{V - b}{V} - \frac{\gamma}{b}\ln\frac{V + b}{V} + \ln\frac{V}{V°}\right)$$

$$\Omega_a = 0.42748 \qquad \Omega_b = 0.08664$$

$$R = 82.04\ \text{cm}^3\ \text{atm/g-mol K}$$

$$b = \frac{\Omega_b RT_c}{P_c} = \frac{(0.08664)(82.04)(365.0)}{45.6}$$

$$= 56.92\ \text{cm}^3/\text{g-mol}$$

$$P = 10\ \text{MPa} = 98.7\ \text{atm} \qquad T = 125°C = 398.2\ \text{K}$$

$$F = \frac{1}{T_r}[1 + (0.480 + 1.57\omega - 0.176\omega^2)(1 - T_r^{0.5})]^2$$

$$= \frac{1}{1.09}\{1 + [0.480 + (1.57)(0.148) - (0.176)(0.148)^2][1 - (1.09)^{0.5}]\}^2$$

$$= 0.861$$

Since pressure and temperature are known in this case, the volume is found to be 147.5 cm³/g-mol and $Z = 0.446$. For the $A - A°$ calculation, let us select a reference pressure of 0.1 MPa. Then $V/V° = ZP°/P = (0.446)(0.1)/10 = 4.46 \times 10^{-3}$. Also, for a pure component, from Table 5-1,

$$\beta = \frac{\Omega_a}{\Omega_b} bF = \frac{0.42748}{0.08664} (56.92)(0.861) = 241.8 \text{ cm}^3/\text{g-mol}$$

Thus

$$\frac{A - A°}{RT} = -\ln \frac{147.5 - 56.92}{147.5} - \frac{241.8}{56.92} \ln \frac{147.5 + 56.92}{147.5} - \ln (4.46 \times 10^{-3})$$

$$= 4.515$$

For $S - S°$,

$$\gamma = -\frac{1}{2} \frac{\Omega_a}{\Omega_b} bF \frac{2f_\omega}{F^{1/2}} = -\frac{\Omega_a}{\Omega_b} bF^{1/2} f_\omega$$

$$= -\frac{0.42748}{0.08664} (56.92)(0.861)^{1/2}[0.480 + (1.574)(0.148) - (0.176)(0.148)^2]$$

$$= 185.6 \text{ cm}^3/\text{g-mol}$$

and

$$\frac{S - S°}{R} = \ln \frac{147.5 - 56.92}{147.5} - \frac{185.6}{56.92} \ln \frac{147.5 + 56.92}{147.5} + \ln (4.46 \times 10^{-3})$$

$$= -6.965$$

$$S - S° = -\frac{(8.314)(6.965)}{42.081} = -1.376 \text{ J/g K}$$

Then for Eq. (5-3.8),

$$H° - H = -[(A - A°) + T(S - S°) + RT(Z - 1)]$$

$$= -(8.314)(398.2) \frac{4.515 - 6.965 + (0.446 - 1)}{42.081}$$

$$= 236 \text{ J/g}$$

All three methods illustrated yield similar values for $H° - H$ and $S° - S$. Though all are slightly smaller than experimental values, all are satisfactory for engineering use.

5-5 Derivative Properties

Thermodynamic partial derivatives (other than those with respect to composition or mole numbers) can be expressed in terms of P, V, T, $(\partial P/\partial V)_T$, $(\partial P/\partial T)_V$, C_p (or $C_p°$), and (sometimes) a reference molar entropy. Thus the basis of all such derivatives is an equation of state along with $C_p°$ and $S°$ [59]. Any of the equations of state introduced in Chap. 3 can be used to obtain $(\partial P/\partial V)_T$ and $(\partial P/\partial T)_V$. For example, using the Lee-Kesler modification of the Benedict-Webb-Rubin equation [Eqs. (3-9.1) to (3-9.4) and Table 3-9], gives

$$\left(\frac{\partial P_r}{\partial T_r}\right)_{V_r} = \left[\left(\frac{\partial P_r}{\partial T_r}\right)_{V_r}\right]^{(0)} + \frac{\omega}{\omega^R} \left\{\left[\left(\frac{\partial P_r}{\partial T_r}\right)_{V_r}\right]^{(R)} - \left[\left(\frac{\partial P_r}{\partial T_r}\right)_{V_r}\right]^{(0)}\right\} \quad (5\text{-}5.1)$$

$$\left(\frac{\partial P_r}{\partial V_r}\right)_{T_r} = \left[\left(\frac{\partial P_r}{\partial V_r}\right)_{T_r}\right]^{(0)} + \frac{\omega}{\omega^R} \left\{\left[\left(\frac{\partial P_r}{\partial V_r}\right)_{T_r}\right]^{(R)} - \left[\left(\frac{\partial P_r}{\partial V_r}\right)_{T_r}\right]^{(0)}\right\} \quad (5\text{-}5.2)$$

The functions $[\ ]^{(R)}$ and $[\ ]^{(0)}$ are found from Eqs. (5-5.3) and (5-5.4) following a procedure analogous to that described in Sec. 5-4 for enthalpy departure functions determined with the Lee-Kesler correlation:

$$\left[\left(\frac{\partial P_r}{\partial T_r}\right)_{V_r^{(0)}}\right]^{(0)} = \frac{1}{V_r^{(0)}}\left[1 + \frac{b_1 + b_3/T_r^2 + 2b_4/T_r^3}{V_r^{(0)}}\right.$$

$$+ \frac{c_1 - 2c_3/T_r^3}{(V_r^{(0)})^2} + \frac{d_1}{(V_r^{(0)})^5}$$

$$\left. - \frac{2c_4}{T_r^3(V_r^{(0)})^2}\left\{\left[\beta + \frac{\gamma}{(V_r^{(0)})^2}\right]\exp\left[-\frac{\gamma}{(V_r^{(0)})^2}\right]\right\}\right. \tag{5-5.3}$$

A similar expression can be written for $[(\partial P_r/\partial T_r)_{V_r}]^{(R)}$ with the reference-fluid constants in Table 3-9; ω^R is the acentric factor for the reference fluid, n-octane, equal to 0.3978:

$$\left[\left(\frac{\partial P_r}{\partial V_r}\right)_{T_r}\right]^{(0)} = -\frac{T_r}{(V_r^{(0)})^2}\left\{1 + \frac{2B}{V_r^{(0)}} + \frac{3C}{(V_r^{(0)})^2} + \frac{6D}{(V_r^{(0)})^5}\right.$$

$$+ \frac{c_4}{T_r^3(V_r^{(0)})^2}\left[3\beta + \left\{5 - 2\left[\beta + \frac{\gamma}{(V_r^{(0)})^2}\right]\right\}\frac{\gamma}{(V_r^{(0)})^2}\right]$$

$$\left.\exp\left[-\frac{\gamma}{(V_r^{(0)})^2}\right]\right\} \tag{5-5.4}$$

Again $[(\partial P_r/\partial V_r)_{T_r}]^{(R)}$ is given by a similar expression with the reference-fluid constants (Table 3-9).

An alternate method for estimating these derivatives employs the *derivative compressibility factor*. We define two functions

$$Z_p \equiv Z - P_r\left(\frac{\partial Z}{\partial P_r}\right)_{T_r} \tag{5-5.5}$$

$$Z_T \equiv Z + T_r\left(\frac{\partial Z}{\partial T_r}\right)_{P_r} \tag{5-5.6}$$

Then it is easily shown that

$$\left(\frac{\partial P_r}{\partial T_r}\right)_{V_r} = \frac{P_r/Z_p}{T_r/Z_T} \tag{5-5.7}$$

$$\left(\frac{\partial P_r}{\partial V_r}\right)_{T_r} = -\frac{P_r/Z_p}{V_r/Z} \tag{5-5.8}$$

Edmister first suggested correlating Z_p and Z_T with reduced properties [21, 22] and a later correlation [69] related Z_p and Z_T to T_r, P_r, and ω,

$$Z_p = Z_p^{(0)} + \omega Z_p^{(1)} \tag{5-5.9}$$

$$Z_T = Z_T^{(0)} + \omega Z_T^{(1)} \tag{5-5.10}$$

The simple-fluid functions $Z_p^{(0)}$ and $Z_T^{(0)}$ are shown in Figs. 5-6 and 5-8 and the deviation functions $Z_p^{(1)}$ and $Z_T^{(1)}$ in Figs. 5-7 and 5-9.

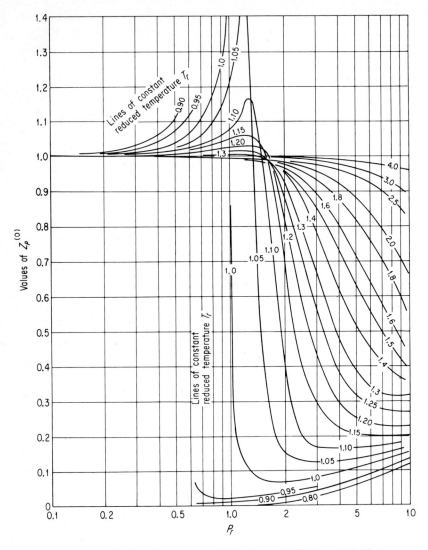

Fig. 5-6 Simple-fluid derivative compressibility factor $Z_p^{(0)}$. (*From Ref. 22a.*)

To illustrate the utility of the functions Z_p and Z_T, a Bridgman table is given in Table 5-9. For example, if the derivative $(\partial H/\partial P)_T$ were desired, $(\partial H)_T = (RT/P)(Z_T - Z)$ and $(\partial P)_T = -1$. Thus $(\partial H/\partial P)_T = (RT/P)(Z - Z_T)$.

To use derivative compressibility factors for gas mixtures, pseudocritical constants may be estimated as shown in Eqs. (4-2.1) to (4-2.3). Also, Z_p and Z_T for liquids have recently been determined by Hsi and Lu[38].

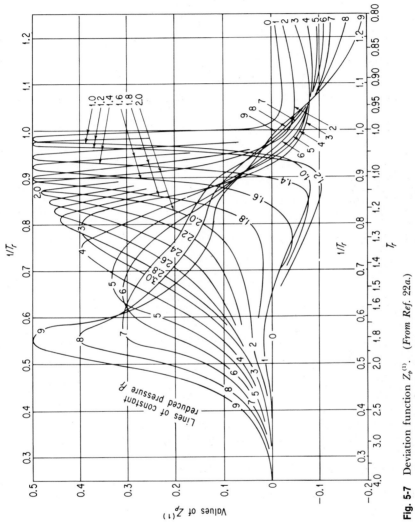

Fig. 5-7 Deviation function $Z_p^{(1)}$. (*From Ref. 22a.*)

TABLE 5-9 Bridgman Tables in Terms of Z_T and Z_p [69]

$$(\partial T)_P = -(\partial P)_T = 1$$

$$(\partial V)_P = -(\partial P)_V = \frac{RZ_T}{P}$$

$$(\partial S)_P = -(\partial P)_S = \frac{C_p}{T}$$

$$(\partial U)_P = -(\partial P)_U = C_p - RZ_T$$

$$(\partial H)_P = -(\partial P)_H = C_p$$

$$(\partial G)_P = -(\partial P)_G = -S$$

$$(\partial A)_P = -(\partial P)_A = -(S + RZ_T)$$

$$(\partial V)_T = -(\partial T)_V = \frac{Z_p RT}{P^2}$$

$$(\partial S)_T = -(\partial T)_S = \frac{RZ_T}{P}$$

$$(\partial U)_T = -(\partial T)_U = \frac{RZ_T T}{P} - \frac{RZ_p T}{P} = \frac{RT}{P}(Z_T - Z_p)$$

$$(\partial H)_T = -(\partial T)_H = -\frac{ZRT}{P} + \frac{Z_T RT}{P} = \frac{RT}{P}(Z_T - Z)$$

$$(\partial G)_T = -(\partial T)_G = -V = -\frac{ZRT}{P}$$

$$(\partial A)_T = -(\partial T)_A = -\frac{Z_p RT}{P}$$

$$(\partial S)_V = -(\partial V)_S = \frac{R}{P^2}(-C_p Z_p + RZ_T^2)$$

$$(\partial U)_V = -(\partial V)_U = -\frac{C_p Z_p RT}{P^2} + \frac{R^2 TZ_T^2}{P^2} = \frac{RT}{P^2}(-C_p Z_p + RZ_T^2)$$

$$(\partial H)_V = -(\partial V)_H = \frac{RT}{P^2}(-C_p Z_p + RZ_T^2 - RZZ_T)$$

$$(\partial G)_V = -(\partial V)_G = -\frac{RT}{P^2}(-SZ_p + RZZ_T)$$

$$(\partial A)_V = -(\partial V)_A = \frac{SZ_p RT}{P^2}$$

$$(\partial U)_S = -(\partial S)_U = \frac{R}{P}(-C_p Z_p + RZ_T^2)$$

$$(\partial H)_S = -(\partial S)_H = -\frac{ZRC_p}{P}$$

$$(\partial G)_S = -(\partial S)_G = \frac{R}{P}(SZ_T - C_p Z)$$

$$(\partial A)_S = -(\partial S)_A = \frac{R}{P}(-C_p Z_p + RZ_T^2 + SZ_T)$$

$$(\partial H)_U = -(\partial U)_H = \frac{RT}{P}(-ZC_p + RZZ_T + C_p Z_p - RZ_T^2)$$

$$(\partial G)_U = -(\partial U)_G = \frac{RT}{P}(RZ_T Z - ZC_p + SZ_T - SZ_p)$$

$$(\partial A)_U = -(\partial U)_A = \frac{RT}{P}(-C_p Z_p + RZ_T^2 + SZ_T - SZ_p)$$

$$(\partial G)_H = -(\partial H)_G = \frac{RT}{P}(-SZ - C_p Z + SZ_T)$$

$$(\partial A)_H = -(\partial H)_A = \frac{RT}{P}[(S + RZ_T)(Z_T - Z) - C_p Z_p]$$

$$(\partial A)_G = -(\partial G)_A = -\frac{RT}{P}(ZS - Z_p S + RZZ_T)$$

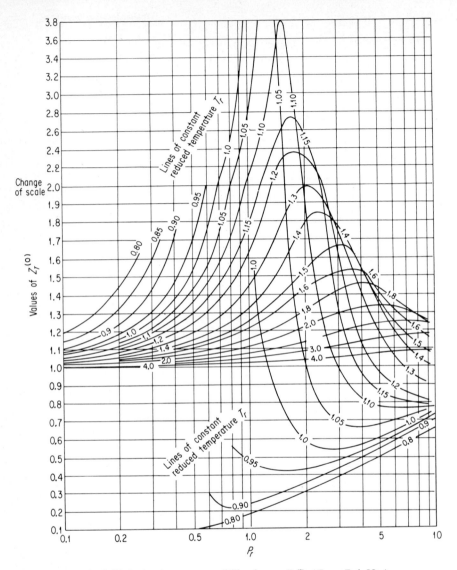

Fig. 5-8 Simple-fluid derivative compressibility factor $Z_T^{(0)}$. (*From Ref. 22a.*)

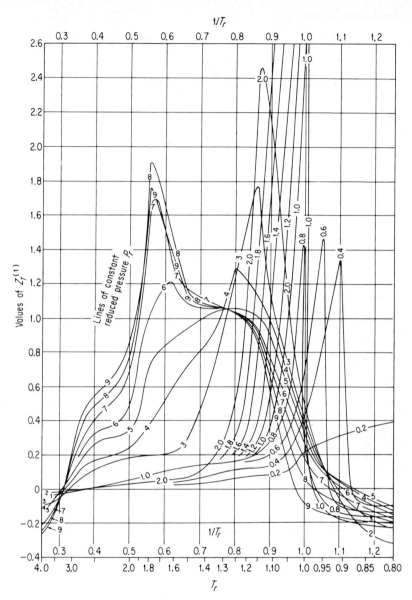

Fig. 5-9 Deviation function $Z_T^{(1)}$. (*From Ref.* 22a.)

Example 5-3 Estimate V, $(\partial V/\partial P)_T$, and $(\partial V/\partial T)_P$ for propane at 60°C and 84 atm. Experimental values are 2.15×10^{-3} m³/kg, -1.93×10^{-6} m³/kg atm, and 6.92×10^{-6} m³/kg K, respectively[73].

solution From Appendix A, for propane, $M = 44.097$, $T_c = 369.8$ K, $P_c = 41.9$ atm, and $\omega = 0.152$. Thus $T_r = (60 + 273.2)/369.8 = 0.90$, and $P_r = 84/41.9 = 2.0$. From Eq. (3-3.1) and Tables 3-1 and 3-2, $Z^{(0)} = 0.311$ and $Z^{(1)} = -0.106$. Thus

$$Z = Z^{(0)} + \omega Z^{(1)} = 0.311 + (0.152)(-0.106) = 0.296$$

$$V = \frac{ZRT}{PM} = \frac{(0.296)(82.04)(60 + 273.2)}{(84)(44.097)}$$

$$= 2.18 \text{ cm}^3/\text{g} = 2.18 \times 10^{-3} \text{ m}^3/\text{kg}$$

With Figs. 5-6 to 5-9, $Z_p^{(0)} = 0.025$, $Z_p^{(1)} = -0.045$, $Z_T^{(0)} = 0.36$, and $Z_T^{(1)} = -0.10$. Then, with Eqs. (5-5.9) and (5-5.10),

$$Z_p = 0.025 + (0.152)(-0.045) = 0.018$$

$$Z_T = 0.36 + (0.152)(-0.10) = 0.345$$

With Table 5-9,

$$\left(\frac{\partial V}{\partial P}\right)_T = \frac{-RTZ_p}{P^2 M} = -\frac{(82.04)(60 + 273.2)(0.018)}{(84^2)(44.097)}$$

$$= -1.6 \times 10^{-3} \text{ cm}^3/\text{atm g} = -1.6 \times 10^{-6} \text{ m}^3/\text{atm kg}$$

$$\left(\frac{\partial V}{\partial T}\right)_P = \frac{RZ_T}{PM} = \frac{(82.04)(0.345)}{(84)(44.097)}$$

$$= 7.6 \times 10^{-3} \text{ cm}^3/\text{g K} = 7.6 \times 10^{-6} \text{ m}^3/\text{kg K}$$

The errors are 1, -17, and 10 percent, respectively.

Many other thermodynamic relations can be expressed as simple partial derivatives and evaluated in terms of Z, Z_T, and Z_p. Two of the more common are the difference $C_p - C_v$ and the Joule-Thompson coefficient, i.e.,

$$C_p - C_v = T\left(\frac{\partial P}{\partial T}\right)_V \left(\frac{\partial V}{\partial T}\right)_P = -T\left(\frac{\partial P}{\partial V}\right)_T \left(\frac{\partial V}{\partial T}\right)_P^2 = \frac{RZ_T^2}{Z_P} \quad (5\text{-}5.11)$$

and

$$\mu \equiv \left(\frac{\partial T}{\partial P}\right)_H = \frac{T(\partial V/\partial T)_P - V}{C_p}$$

and

$$\mu C_p = \frac{RT}{P}(Z_T - Z) \quad (5\text{-}5.12)$$

Edmister [20–22] has developed graphs of μC_p and $C_p - C_v$ as well as isentropic coefficients as functions of T_r and P_r without introducing a third parameter such as ω.

As further illustration of the use of the derivative compressibility factors, the generalized sonic-velocity correlation of Sherwood [82] for compressed gases may be expressed as

$$V_s^2 = [aTZ^2(C_p^\circ + \Delta C_p)][Z_P(C_p^\circ + \Delta C_p) - RZ_T^2]^{-1} \quad (5\text{-}5.13)$$

where V_s = velocity of sound, m/s

a = dimensional constant = $8308/M$ m²/s² K

M = molecular weight

$\Delta C_p = C_p - C_p^\circ$ (see Sec. 5-6)

and the remainder of the terms have been defined previously. This equation is illustrated in Example 5-4.

Example 5-4 Estimate the velocity of sound in ethane gas at 305 K and 116 atm. The experimental value reported by Sherwood is 583 m/s, and the heat capacity at zero pressure C_p° is 12.8 cal/g-mol K.

solution From Appendix A, $T_c = 305.4$ K, $P_c = 48.2$ atm, $\omega = 0.098$. Thus, $T_r = 305/305.4 = 1.0$; $P_r = 116/48.2 = 2.4$. From Tables 3-1 and 3-2 and Figs. 5-6 to 5-9, $Z^{(0)} = 0.374$, $Z^{(1)} = -0.094$, $Z_p^{(0)} = 0.072$, $Z_p^{(1)} = -0.05$, $Z_T^{(0)} = 0.55$, $Z_T^{(1)} = 0.43$; thus

$$Z = Z^{(0)} + \omega Z^{(1)} = 0.374 + (0.098)(-0.094) = 0.365$$

$$Z_p = Z_p^{(0)} + \omega Z_p^{(1)} = 0.072 + (0.098)(-0.05) = 0.067$$

$$Z_T = Z_T^{(0)} + \omega Z_T^{(1)} = 0.55 + (0.098)(0.43) = 0.595$$

From Tables 5-10 and 5-11, with Eq. (5-6.4),

$$\left(\frac{C_p - C_p^\circ}{R}\right)^{(0)} = 4.64 \quad \text{and} \quad \left(\frac{C_p - C_p^\circ}{R}\right)^{(1)} = 12.095$$

Then

$$C_p - C_p^\circ = (1.987)[4.64 + (0.098)(12.095)] = 11.57 \text{ cal/g-mol K}$$

at $T_r = 1.0$, $P_r = 2.4$. Thus, with $M = 30.070$,

$$V_s^2 = \frac{(8308)(305)(0.365)^2(12.8 + 11.57)}{(30.070)[(0.067)(12.8 + 11.57) - (1.987)(0.595)^2]} = 295,000 \text{ m}^2/\text{s}^2$$

$$V_s = 543 \text{ m/s}$$

$$\text{Error} = \frac{543 - 583}{583} \times 100 = -6.9\%$$

5-6 Heat Capacities of Real Gases

In Chap. 7 we present methods for estimating the heat capacity of pure gases in the ideal-gas state as a function of temperature. Also, in this ideal-gas state, for a gas mixture,

$$C_{p_m}^\circ = \sum_j y_j C_{p_j}^\circ \tag{5-6.1}$$

The heat capacity of a real gas is related to the value in the ideal-gas state, at the same temperature and composition:

$$C_p = C_p^\circ + \Delta C_p \tag{5-6.2}$$

where this relation applies to either a pure gas or gas mixture at constant composition. ΔC_p is a residual heat capacity; it can be determined by taking the partial derivative of the enthalpy departure at constant pressure and composition

$$\Delta C_p = \frac{\partial}{\partial T}(H - H^\circ)_{T,\text{comp}} \tag{5-6.3}$$

TABLE 5-10 Residual Heat Capacities [51]

Simple fluid $\left(\dfrac{C_p - C_p^\circ}{R}\right)^{(0)}$

	P_r						
T_r	0.010	0.050	0.100	0.200	0.400	0.600	0.800
0.30	2.805	2.807	2.809	2.814	2.830	2.842	2.854
0.35	2.808	2.810	2.812	2.815	2.823	2.835	2.844
0.40	2.925	2.926	2.928	2.933	2.935	2.940	2.945
0.45	2.989	2.990	2.990	2.991	2.993	2.995	2.997
0.50	3.006	3.005	3.004	3.003	3.001	3.000	2.998
0.55	0.118	3.002	3.000	2.997	2.990	2.984	2.978
0.60	0.089	3.009	3.006	2.999	2.986	2.974	2.963
0.65	0.069	0.387	3.047	3.036	3.014	2.993	2.973
0.70	0.054	0.298	0.687	3.138	3.099	3.065	3.033
0.75	0.044	0.236	0.526	3.351	3.284	3.225	3.171
0.80	0.036	0.191	0.415	1.032	3.647	3.537	3.440
0.85	0.030	0.157	0.336	0.794	4.404	4.158	3.957
0.90	0.025	0.131	0.277	0.633	1.858	5.679	5.095
0.93	0.023	0.118	0.249	0.560	1.538	4.208	6.720
0.95	0.021	0.111	0.232	0.518	1.375	3.341	9.316
0.97	0.020	0.104	0.217	0.480	1.240	2.778	9.585
0.98	0.019	0.101	0.210	0.463	1.181	2.563	7.350
0.99	0.019	0.098	0.204	0.447	1.126	2.378	6.038
1.00	0.018	0.095	0.197	0.431	1.076	2.218	5.156
1.01	0.018	0.092	0.191	0.417	1.029	2.076	4.516
1.02	0.017	0.089	0.185	0.403	0.986	1.951	4.025
1.05	0.016	0.082	0.169	0.365	0.872	1.648	3.047
1.10	0.014	0.071	0.147	0.313	0.724	1.297	2.168
1.15	0.012	0.063	0.128	0.271	0.612	1.058	1.670
1.20	0.011	0.055	0.113	0.237	0.525	0.885	1.345
1.30	0.009	0.044	0.089	0.185	0.400	0.651	0.946
1.40	0.007	0.036	0.072	0.149	0.315	0.502	0.711
1.50	0.006	0.029	0.060	0.122	0.255	0.399	0.557
1.60	0.005	0.025	0.050	0.101	0.210	0.326	0.449
1.70	0.004	0.021	0.042	0.086	0.176	0.271	0.371
1.80	0.004	0.018	0.036	0.073	0.150	0.229	0.311
1.90	0.003	0.016	0.031	0.063	0.129	0.196	0.265
2.00	0.003	0.014	0.027	0.055	0.112	0.170	0.229
2.20	0.002	0.011	0.021	0.043	0.086	0.131	0.175
2.40	0.002	0.009	0.017	0.034	0.069	0.104	0.138
2.60	0.001	0.007	0.014	0.028	0.056	0.084	0.112
2.80	0.001	0.006	0.012	0.023	0.046	0.070	0.093
3.00	0.001	0.005	0.010	0.020	0.039	0.058	0.078
3.50	0.001	0.003	0.007	0.013	0.027	0.040	0.053
4.00	0.000	0.002	0.005	0.010	0.019	0.029	0.038

			P_r				
1.000	1.200	1.500	2.000	3.000	5.000	7.000	10.000
2.866	2.878	2.896	2.927	2.989	3.122	3.257	3.466
2.853	2.861	2.875	2.897	2.944	3.042	3.145	3.313
2.951	2.956	2.965	2.979	3.014	3.085	3.164	3.293
2.999	3.002	3.006	3.014	3.032	3.079	3.135	3.232
2.997	2.996	2.995	2.995	2.999	3.019	3.054	3.122
2.973	2.968	2.961	2.951	2.938	2.934	2.947	2.988
2.952	2.942	2.927	2.907	2.874	2.840	2.831	2.847
2.955	2.938	2.914	2.878	2.822	2.753	2.720	2.709
3.003	2.975	2.937	2.881	2.792	2.681	2.621	2.582
3.122	3.076	3.015	2.928	2.795	2.629	2.537	2.469
3.354	3.277	3.176	3.038	2.838	2.601	2.473	2.373
3.790	3.647	3.470	3.240	2.931	2.599	2.427	2.292
4.677	4.359	4.000	3.585	3.096	2.626	2.399	2.227
5.766	5.149	4.533	3.902	3.236	2.657	2.392	2.195
7.127	6.010	5.050	4.180	3.351	2.684	2.391	2.175
10.011	7.451	5.785	4.531	3.486	2.716	2.393	2.159
13.270	8.611	6.279	4.743	3.560	2.733	2.395	2.151
21.948	10.362	6.897	4.983	3.641	2.752	2.398	2.144
******	13.281	7.686	5.255	3.729	2.773	2.401	2.138
22.295	18.967	8.708	5.569	3.821	2.794	2.405	2.131
13.184	31.353	10.062	5.923	3.920	2.816	2.408	2.125
6.458	20.234	16.457	7.296	4.259	2.891	2.425	2.110
3.649	6.510	13.256	9.787	4.927	3.033	2.462	2.093
2.553	3.885	6.985	9.094	5.535	3.186	2.508	2.083
1.951	2.758	4.430	6.911	5.710	3.326	2.555	2.079
1.297	1.711	2.458	3.850	4.793	3.452	2.628	2.077
0.946	1.208	1.650	2.462	3.573	3.282	2.626	2.068
0.728	0.912	1.211	1.747	2.647	2.917	2.525	2.038
0.580	0.719	0.938	1.321	2.016	2.508	2.347	1.978
0.475	0.583	0.752	1.043	1.586	2.128	2.130	1.889
0.397	0.484	0.619	0.848	1.282	1.805	1.907	1.778
0.336	0.409	0.519	0.706	1.060	1.538	1.696	1.656
0.289	0.350	0.443	0.598	0.893	1.320	1.505	1.531
0.220	0.265	0.334	0.446	0.661	0.998	1.191	1.292
0.173	0.208	0.261	0.347	0.510	0.779	0.956	1.086
0.140	0.168	0.210	0.278	0.407	0.624	0.780	0.917
0.116	0.138	0.172	0.227	0.332	0.512	0.647	0.779
0.097	0.116	0.144	0.190	0.277	0.427	0.545	0.668
0.066	0.079	0.098	0.128	0.187	0.289	0.374	0.472
0.048	0.057	0.071	0.093	0.135	0.209	0.272	0.350

TABLE 5-11 Residual Heat Capacities [51]

Deviation function $\left(\dfrac{C_p - C_p^\circ}{R}\right)^{(1)}$

T_r	P_r						
	0.010	0.050	0.100	0.200	0.400	0.600	0.800
0.30	8.462	8.445	8.424	8.381	8.281	8.192	8.10?
0.35	9.775	9.762	9.746	9.713	9.646	9.568	9.49?
0.40	11.494	11.484	11.471	11.438	11.394	11.343	11.29?
0.45	12.651	12.643	12.633	12.613	12.573	12.532	12.49?
0.50	13.111	13.106	13.099	13.084	13.055	13.025	12.99?
0.55	0.511	13.035	13.030	13.021	13.002	12.981	12.96?
0.60	0.345	12.679	12.675	12.668	12.653	12.637	12.62?
0.65	0.242	1.518	12.148	12.145	12.137	12.128	12.11?
0.70	0.174	1.026	2.698	11.557	11.564	11.563	11.55?
0.75	0.129	0.726	1.747	10.967	10.995	11.011	11.01?
0.80	0.097	0.532	1.212	3.511	10.490	10.536	10.56?
0.85	0.075	0.399	0.879	2.247	9.999	10.153	10.24?
0.90	0.058	0.306	0.658	1.563	5.486	9.793	10.18?
0.93	0.050	0.263	0.560	1.289	3.890	******	10.28?
0.95	0.046	0.239	0.505	1.142	3.215	9.389	9.99?
0.97	0.042	0.217	0.456	1.018	2.712	6.588	******
0.98	0.040	0.207	0.434	0.962	2.506	5.711	******
0.99	0.038	0.198	0.414	0.911	2.324	5.027	******
1.00	0.037	0.189	0.394	0.863	2.162	4.477	10.51?
1.01	0.035	0.181	0.376	0.819	2.016	4.026	8.43?
1.02	0.034	0.173	0.359	0.778	1.884	3.648	7.04?
1.05	0.030	0.152	0.313	0.669	1.559	2.812	4.67?
1.10	0.024	0.123	0.252	0.528	1.174	1.968	2.91?
1.15	0.020	0.101	0.205	0.424	0.910	1.460	2.04?
1.20	0.016	0.083	0.168	0.345	0.722	1.123	1.52?
1.30	0.012	0.058	0.116	0.235	0.476	0.715	0.93?
1.40	0.008	0.042	0.083	0.166	0.329	0.484	0.62?
1.50	0.006	0.030	0.061	0.120	0.235	0.342	0.43?
1.60	0.005	0.023	0.045	0.089	0.173	0.249	0.31?
1.70	0.003	0.017	0.034	0.068	0.130	0.187	0.23?
1.80	0.003	0.013	0.027	0.052	0.100	0.143	0.18?
1.90	0.002	0.011	0.021	0.041	0.078	0.111	0.14?
2.00	0.002	0.008	0.017	0.032	0.062	0.088	0.11?
2.20	0.001	0.005	0.011	0.021	0.040	0.057	0.07?
2.40	0.001	0.004	0.007	0.014	0.028	0.039	0.04?
2.60	0.001	0.003	0.005	0.010	0.020	0.028	0.03?
2.80	0.000	0.002	0.004	0.008	0.014	0.021	0.02?
3.00	0.000	0.001	0.003	0.006	0.011	0.016	0.02?
3.50	0.000	0.001	0.002	0.003	0.006	0.009	0.01?
4.00	0.000	0.001	0.001	0.002	0.004	0.006	0.00?

138

			P_r				
1.000	1.200	1.500	2.000	3.000	5.000	7.000	10.000
8.011	7.920	7.785	7.558	7.103	6.270	5.372	4.020
9.430	9.360	9.256	9.080	8.728	8.013	7.290	6.285
11.240	11.188	11.110	10.980	10.709	10.170	9.625	8.803
12.451	12.409	12.347	12.243	12.029	11.592	11.183	10.533
12.964	12.933	12.886	12.805	12.639	12.288	11.946	11.419
12.939	12.917	12.882	12.823	12.695	12.407	12.103	11.673
12.589	12.574	12.550	12.506	12.407	12.165	11.905	11.526
12.105	12.092	12.060	12.026	11.943	11.728	11.494	11.141
11.553	11.536	11.524	11.495	11.416	11.208	10.985	10.661
11.024	11.022	11.013	10.986	10.898	10.677	10.448	10.132
10.583	10.590	10.587	10.556	10.446	10.176	9.917	9.591
10.297	10.321	10.324	10.278	10.111	9.740	9.433	9.075
10.349	10.409	10.401	10.279	9.940	9.389	8.999	8.592
10.769	10.875	10.801	10.523	9.965	9.225	8.766	8.322
11.420	11.607	11.387	10.865	10.055	9.136	8.621	8.152
13.001	******	12.498	11.445	10.215	9.061	8.485	7.986
******	******	******	11.856	10.323	9.037	8.420	7.905
******	******	******	12.388	10.457	9.011	8.359	7.826
******	******	******	13.081	10.617	8.990	8.293	7.747
******	******	******	******	10.805	8.973	8.236	7.670
******	******	******	******	11.024	8.960	8.182	7.595
7.173	2.277	******	******	11.852	8.939	8.018	7.377
3.877	4.002	3.927	******	******	8.933	7.759	7.031
2.587	2.844	2.236	7.716	12.812	8.849	7.504	6.702
1.881	2.095	1.962	2.965	9.494	8.508	7.206	6.384
1.129	1.264	1.327	1.288	3.855	6.758	6.365	5.735
0.743	0.833	0.904	0.905	1.652	4.524	5.193	5.035
0.517	0.580	0.639	0.666	0.907	2.823	3.944	4.289
0.374	0.419	0.466	0.499	0.601	1.755	2.871	3.545
0.278	0.312	0.349	0.380	0.439	1.129	2.060	2.867
0.212	0.238	0.267	0.296	0.337	0.764	1.483	2.287
0.164	0.185	0.209	0.234	0.267	0.545	1.085	1.817
0.130	0.146	0.166	0.187	0.217	0.407	0.812	1.446
0.085	0.096	0.110	0.126	0.150	0.256	0.492	0.941
0.058	0.066	0.076	0.089	0.109	0.180	0.329	0.644
0.042	0.048	0.056	0.066	0.084	0.137	0.239	0.466
0.031	0.036	0.042	0.051	0.067	0.110	0.187	0.356
0.024	0.028	0.033	0.041	0.055	0.092	0.153	0.285
0.015	0.017	0.021	0.026	0.038	0.067	0.108	0.190
0.010	0.012	0.015	0.019	0.029	0.054	0.085	0.146

The enthalpy departure can be calculated from the equations of state shown in Table 5-1 or from the Lee-Kesler method by Eqs. (5-4.1) to (5-4.5):

$$C_p - C_p^\circ = \Delta C_p = (\Delta C_p)^{(0)} + \omega (\Delta C_p)^{(1)} \qquad (5\text{-}6.4)$$

The simple-fluid contribution $(\Delta C_p)^{(0)}$ is given in Table 5-10 and the deviation function $(\Delta C_p)^{(1)}$ in Table 5-11 as a function of T_r and P_r. If Eq. (5-6.4) and Tables 5-10 and 5-11 are employed for mixtures, the pseudocritical rules given in Eqs. (4-6.3) to (4-6.7) should be used. These rules have been developed primarily from hydrocarbon-mixture data, but they should be satisfactory unless highly polar components are present. Tables 5-10 and 5-11 differ somewhat from an earlier correlation by Edmister [24], especially in the critical region, where high accuracy is difficult to achieve in any case. The use of Eq. (5-6.4) with Tables 5-10 and 5-11 was illustrated in Example 5-4.

5-7 True Critical Points of Mixtures

In Chap. 4, emphasis was placed upon the estimation of pseudocritical constants for mixtures. Such constants are necessary if one is to use most corresponding-states correlations to estimate mixture P-V-T or derived properties. However, these pseudocritical constants often differ considerably from the true critical points for mixtures. Estimation techniques for the latter can be evaluated by comparison with experimental data; for the former, evaluation is indirect since the pseudocritical state does not exist in a physical sense.

In this section, we briefly discuss methods of estimating the true critical properties of mixtures. Most techniques are limited to hydro-carbon mixtures or to mixtures of hydrocarbons with CO_2, H_2S, CO, and the permanent gases. Estimation procedures for the cricondenbar or cricondentherm points are poorly developed [83] and are not covered.

Mixture Critical Temperatures The true mixture critical temperature is usually *not* a linear *mole-fraction* average of the pure-component critical temperatures. Li [52] has suggested that if the composition is expressed as

$$\phi_j = \frac{y_j V_{c_j}}{\sum_i y_i V_{c_i}} \qquad (5\text{-}7.1)$$

the true mixture critical temperature can be estimated by

$$T_{c_T} = \sum_j \phi_j T_{c_j} \qquad (5\text{-}7.2)$$

where y_j = mole fraction of component j
V_{c_j} = critical volume of j
T_{c_j} = critical temperature of j
T_{cr} = true mixture critical temperature

Chueh and Prausnitz [14] have proposed a similar technique. By defining a *surface fraction* θ_j,

$$\theta_j = \frac{y_j V_{c_j}^{2/3}}{\sum_i y_i V_{c_i}^{2/3}} \tag{5-7.3}$$

they then relate θ_j and T_{cr} by

$$T_{cr} = \sum_j \theta_j T_{c_j} + \sum_i \sum_j \theta_i \theta_j \tau_{ij} \tag{5-7.4}$$

where τ_{ij} is an interaction parameter. τ_{ii} is considered to be zero, and τ_{ij} $(i \neq j)$ can be estimated for several different binary types by

$$\psi_T = A + B\delta_T + C\delta_T^2 + D\delta_T^3 + E\delta_T^4 \tag{5-7.5}$$

where

$$\psi_T = \frac{2\tau_{ij}}{T_{c_i} + T_{c_j}} \tag{5-7.6}$$

and

$$\delta_T = \left| \frac{T_{c_i} - T_{c_j}}{T_{c_i} + T_{c_j}} \right| \tag{5-7.7}$$

The coefficients for Eq. (5-7.5) are shown below for a few binary types [84], where $0 \leqslant \delta_T \leqslant 0.5$:

Binary	A	B	C	D	E
Containing aromatics	-0.0219	1.227	-24.277	147.673	-259.433
Containing H_2S	-0.0479	-5.725	70.974	-161.319	
Containing CO_2	-0.0953	2.185	-33.985	179.068	-264.522
Containing C_2H_2	-0.0785	-2.152	93.084	-722.676	
Containing CO	-0.0077	-0.095	-0.225	3.528	
All other systems	-0.0076	0.287	-1.343	5.443	-3.038

In Fig. 5-10 the critical temperature of the methane-n-pentane binary is shown plotted as a function of mole fraction, surface fraction, and volume fraction. It is clear that the use of a volume fraction provides essentially a linear relationship between ϕ_j and T_{cr}, as predicted by Eq. (5-7.2); when the surface fraction is employed, T_{cr} is slightly nonlinear with θ_j and the interaction term in Eq. (5-7.4) compensates for this nonlinearity.

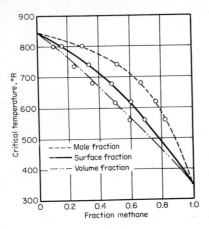

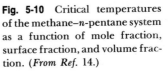

Fig. 5-10 Critical temperatures of the methane–n-pentane system as a function of mole fraction, surface fraction, and volume fraction. (*From Ref.* 14.)

Spencer et al. [85] reviewed and evaluated a number of proposed methods for estimating true critical temperatures for mixtures. They recommended either the Li or Chueh-Prausnitz correlations described above; 135 binary hydrocarbon mixtures were tested, and the average deviation noted for both methods was less than 4 K. For multicomponent hydrocarbon systems, larger errors were found (average deviation about 11 K). When hydrocarbon-nonhydrocarbon mixtures were evaluated, the Chueh-Prausnitz method yielded a smaller average deviation [77].

Of the two methods recommended by Spencer et al. [85], the Li relation (5-7.2) is the easier to use, and unless one of the components is a nonhydrocarbon, it is slightly more accurate.

Example 5-5   Estimate the true critical temperature for a mixture of methane, ethane, and n-butane with mole fractions:

$$y_{C_1} = 0.193$$
$$y_{C_2} = 0.470$$
$$y_{C_4} = \underline{0.337}$$
$$\phantom{y_{C_4} = }1.000$$

Li [52] indicates that the experimental value is 354 K.
   solution   From Appendix A:

	T_c, K	V_c, cm^3/g-mol
Methane	190.6	99.0
Ethane	305.4	148
n-Butane	425.2	255

Li Method. Using Eq. (5-7.1),

	Volume fraction ϕ_j
Methane	0.110
Ethane	0.398
n-Butane	0.492
	1.000

With Eq. (5-7.2)

$$T_{cT} = (0.110)(190.6) + (0.398)(305.4) + (0.492)(425.2) = 352 \text{ K}$$

Deviation = 2 K

Chueh-Prausnitz Method. First, surface fractions θ_j are determined with Eq. (5-7.3):

	Surface fraction θ_j
Methane	0.134
Ethane	0.427
n-Butane	0.439
	1.000

Next, τ_{ij} is found from Eq. (5-7.5) using $A = -0.0076$, $B = 0.287$, $C = -1.343$, $D = 5.443$, and $E = -3.038$:

i	j	δ_T	ψ_T	τ_{ij}
Methane	Ethane	0.231	0.045	11.2
Methane	n-Butane	0.381	0.144	44.3
Ethane	n-Butane	0.164	0.025	9.1

Then, with Eq. (5-7.4),

$$T_{cT} = (0.134)(190.6) + (0.427)(305.4) + (0.439)(425.2)$$
$$+ (2)(0.134)(0.427)(11.2) + (2)(0.134)(0.439)(44.3)$$
$$+ (2)(0.427)(0.439)(9.1)$$
$$= 353 \text{ K}$$

Deviation = 1 K

Mixture Critical Volumes   Only a few experimental values are available for mixture critical volumes.   Thus the range and accuracy of estimation methods are not clearly established.   Grieves and Thodos [34] have suggested an approximate graphical method for hydrocarbon mixtures, but an analytical technique by Chueh and Prausnitz [14], modified by Schick and Prausnitz [77], appears to be more accurate.   When the surface fraction θ_j is defined as in Eq. (5-7.3), the mixture critical volume is given by a relation analogous to Eq. (5-7.4)

$$V_{c_T} = \sum_j \theta_j V_{c_j} + \sum_i \sum_j \theta_i \theta_j \nu_{ij} \qquad (5\text{-}7.8)$$

V_{c_j} is the critical volume of j, and ν_{ij} is an interaction parameter such that $\nu_{ii} = 0$ and ν_{ij} $(i \neq j)$ can be estimated as follows:

$$\psi_v = A + B\delta_v + C\delta_v^2 + D\delta_v^3 + E\delta_v^4 \qquad (5\text{-}7.9)$$

$$\psi_v = \frac{2\nu_{ij}}{V_{c_i} + V_{c_j}} \qquad (5\text{-}7.10)$$

$$\delta_v = \left| \frac{V_{c_i}^{2/3} - V_{c_j}^{2/3}}{V_{c_i}^{2/3} + V_{c_j}^{2/3}} \right| \qquad (5\text{-}7.11)$$

The coefficients for Eq. (5-7.9) are given below for a few binary types [84] when $0 \leq \delta_v \leq 0.5$.

Binary	A	B	C	D	E
Aromatic-aromatic	0	0	0	0	0
Containing at least one cycloparaffin	0	0	0	0	0
Paraffin-aromatic	0.0753	−3.332	2.220	0	0
System with CO_2 or H_2S	−0.4957	17.1185	−168.56	587.05	−698.89
All other systems	0.1397	−2.9672	1.8337	−1.536	0

Spencer et al. [85] evaluated Eq. (5-7.8) for 23 binary hydrocarbon mixtures and 8 binaries consisting of a hydrocarbon and a nonhydrocarbon. They report an average error of 10.5 percent, with particularly poor results for ethane-cyclohexane, ethylene-propylene, and several of the systems containing nonhydrocarbons. However, this large error may in part be explained by experimental inaccuracies. It is much more difficult to measure V_{c_T} than to measure T_{c_T} or P_{c_T}. Spencer et al. [85] note that when V_{c_T} is correlated by

$$V_{c_T} = y_1 V_{c_1} + y_2 V_{c_2} + V_c^{EX} \qquad (5\text{-}7.12)$$

for a binary of 1 and 2, V_c^{EX} values for some systems are positive while for others they are negative and no generalized methods are available to estimate either the sign or magnitude.

Example 5-6   Estimate the true critical volume of a mixture of toluene and n-hexane containing 50.5 mole percent n-hexane.   The experimental value is 325 cm^3/g-mol [96].

solution   Equation (5-7.8) will be used.   From Appendix A, V_c (n-hexane) = 370 cm^3/g-mol and V_c (toluene) = 316 cm^3/g-mol.   Thus with Eq. (5-7.3),

$$\theta_{n\text{-}C_6} = \frac{(0.505)(370^{2/3})}{(0.505)(370^{2/3}) + (0.495)(316^{2/3})} = 0.531$$

$$\theta_{tol} = 0.469$$

With Eq. (5-7.11)

$$\delta_v = \left| \frac{370^{2/3} - 316^{2/3}}{370^{2/3} + 316^{2/3}} \right| = 0.0525$$

Then with Eq. (5-7.9),

$$\psi_v = A + B\delta_v + C\delta_v^2 + D\delta_v^3 + E\delta_v^4$$
$$= 0.0753 + (-3.332)(0.0525) + (2.220)(0.0525)^2$$
$$= -0.094$$
$$= \frac{2\nu_{12}}{370 + 316}$$
$$\nu_{12} = -32$$

Then, with Eq. (5-7.8),

$$V_{cT} = (0.531)(370) + (0.469)(316) + (2)(-32)(0.531)(0.469)$$
$$= 328 \text{ cm}^3/\text{g-mol}$$
$$\text{Error} = \frac{328 - 325}{325} \times 100 = 1\%$$

Mixture Critical Pressure   The dependence of mixture critical pressures on mole fraction is often nonlinear, and estimation of P_{cT} is often unreliable.   Two approaches are illustrated below.

Kreglewski and Kay [49] derived an approximate expression for P_{cT} using conformal solution theory. T_{cT} is required in the calculation, and Kreglewski and Kay also suggest how this property may be determined.   However, Spencer et al. [85] found better results in determining P_{cT} if T_{cT} is found from Eq. (5-7.2).   To find P_{cT}, the liquid molal volumes for each component at $T_r = 0.6$, V_i^*, must first be obtained.   Kreglewski [48] tabulates this volume for many pure liquids; alternatively, it can be estimated by methods presented in Chap. 3.   The sequential set of equations to employ then follow for a binary of 1 and 2:

$$V_{12}^* = \frac{[(V_1^*)^{1/3} + (V_2^*)^{1/3}]^3}{8} \tag{5-7.13}$$

$$V^* = V_1^* y_1 + V_2^* y_2 + (2V_{12}^* - V_1^* - V_2^*) y_1 y_2 \tag{5-7.14}$$

where y_1 and y_2 are mole fractions.   Next, *surface* fractions are defined as

$$\theta_1 = \frac{y_1 V_1^{*2/3}}{y_1 V_1^{*2/3} + y_2 V_2^{*2/3}} \tag{5-7.15}$$

$$\theta_2 = 1 - \theta_1 \tag{5-7.16}$$

$$T_{12}^* = \frac{2 V_{12}^{*1/3}}{V_1^{*1/3}/T_{c_1} + V_2^{*1/3}/T_{c_2}} \tag{5-7.17}$$

$$T^* = V^{*1/3}\left[\frac{T_{c_1}\theta_1}{V_1^{*1/3}} + \frac{T_{c_2}\theta_2}{V_2^{*1/3}} + \left(\frac{2T_{12}^*}{V_{12}^{*1/3}} - \frac{T_{c_1}}{V_1^{*1/3}} - \frac{T_{c_2}}{V_2^{*1/3}}\right)\theta_1\theta_2\right] \tag{5-7.18}$$

$$\omega_{12} = \frac{2}{1/\omega_1 + 1/\omega_2} \tag{5-7.19}$$

$$\omega = \omega_1\theta_1 + \omega_2\theta_2 + (2\omega_{12} - \omega_1 - \omega_2)\theta_1\theta_2 \tag{5-7.20}$$

where ω is the acentric factor.   Finally,

$$P^* = \frac{T^*}{V^{*1/3}} \frac{P_{c_1}\theta_1 + P_{c_2}\theta_2}{T_{c_1}\theta_1/V_1^{*1/3} + T_{c_2}\theta_2/V_2^{*1/3}} \tag{5-7.21}$$

and

$$P_{c_T} = P^*\left[1 + (5.808 + 4.93\omega)\left(\frac{T_{c_T}}{T^*} - 1\right)\right] \tag{5-7.22}$$

with T_{c_T} from Eq. (5-7.2) as noted before.

Spencer et al. [84, 85] evaluated this approach and found an average deviation of about 1 atm when calculated values of P_{c_T} were compared against experimental data; 967 mixture data points were tested. Methane systems were not included, as all available correlations give significant error for such mixtures. The method is illustrated in Example 5-7. In most cases, Eq. (5-7.22) can be simplified by approximating P^*, T^*, and ω by mole-fraction averages of the pure-component properties P_{c_i}, T_{c_i}, and ω_i [43, 84, 85]. Average errors increase to only about 1.3 atm.

Example 5-7 Estimate the critical pressure for a mixture of ethane and benzene which contains 39.2 mole percent ethane. The true critical pressure and temperature are reported to be 83.8 atm and 225.9°C [44].

solution Properties of the pure components from Appendix A are:

Property	Ethane	Benzene
T_c, K	305.4	562.1
P_c, atm	48.2	48.3
ω	0.098	0.212
V_c, cm³/g-mol	148	259

From Kreglewski [48] V^* (ethane) = 54.87 cm³/g-mol and V^* (benzene) = 93.97 cm³/g-mol. Following Eqs. (5-7.13) to (5.7-21), with ethane = 1, and benzene = 2, we have

$$V_{12}^* = \frac{[(54.87)^{1/3} + (93.97)^{1/3}]^3}{8} = 72.7 \text{ cm}^3/\text{g-mol}$$

$$V^* = (54.87)(0.392) + (93.97)(0.608) + [(2)(72.7) - 54.87$$
$$- 93.97](0.392)(0.608) = 77.8 \text{ cm}^3/\text{g-mol}$$

$$\theta_1 = \frac{(0.392)(54.87)^{2/3}}{(0.392)(54.87)^{2/3} + (0.608)(93.97)^{2/3}} = 0.310$$

$$\theta_2 = 1 - 0.310 = 0.690$$

$$T_{12}^* = \frac{(2)(72.7)^{1/3}}{(54.87)^{1/3}/305.4 + (93.97)^{1/3}/562.1} = 406 \text{ K}$$

$$T^* = (77.8)^{1/3} \left\{ \frac{(305.4)(0.310)}{(54.87)^{1/3}} + \frac{(562.1)(0.690)}{(93.97)^{1/3}} \right.$$

$$+ \left. \left[\frac{(2)(406)}{(72.7)^{1/3}} - \frac{305.4}{(54.87)^{1/3}} - \frac{562.1}{(93.97)^{1/3}} \right] (0.310)(0.690) \right\}$$

$$= 462 \text{ K}$$

$$\omega_{12} = \frac{2}{1/0.098 + 1/0.212} = 0.134$$

$$\omega = (0.098)(0.310) + (0.212)(0.690) + [(2)(0.134)$$

$$- 0.098 - 0.212](0.310)(0.690)$$

$$= 0.167$$

$$P^* = \frac{462}{(77.8)^{1/3}} \frac{(48.2)(0.310) + (48.3)(0.690)}{\dfrac{(305.4)(0.310)}{(54.87)^{1/3}} + \dfrac{(562.1)(0.690)}{(93.97)^{1/3}}}$$

$$= 47.4 \text{ atm}$$

Before using Eq. (5-7.22), T_{cT} must be estimated. With Eqs. (5-7.1) and (5-7.2),

$$\theta_1 = \frac{(0.392)(148)}{(0.392)(148) + (0.608)(259)} = 0.269$$

$$T_{cT} = (0.269)(305.4) + (0.731)(562.1) = 493 \text{ K}$$

Then, with Eq. (5-7.22),

$$P_{cT} = 47.4 \left\{ 1 + [5.808 + (4.93)(0.167)] \left(\frac{493}{462} - 1 \right) \right\} = 68.5 \text{ atm}$$

$$\text{Error} = \frac{68.5 - 83.8}{83.8} \times 100 = -18\%$$

Spencer et al. [84] report an average deviation of -13 percent when the Kreglewski and Kay method is applied to this very nonlinear system. Tests in most other systems led to much less error.

An alternate method for predicting the critical pressures for mixtures was developed by Chueh and Prausnitz [14]. P_{c_m} was related to T_{c_m} and V_{c_m} by a modified Redlich-Kwong equation of state (see Secs. 3-5 and 4-3).

$$P_{c_T} = \frac{R T_{c_T}}{V_{c_T} - b_m} - \frac{a_m}{T_{c_T}^{1/2} V_{c_T} (V_{c_T} + b_m)} \tag{5-7.23}$$

where T_{c_T} and V_{c_T} are calculated from methods described earlier in this section. The mixture coefficients for determining P_{c_T} are defined as

$$b_m = \sum_j y_j b_j = \sum_j \frac{y_j \Omega_{b_j}^* R T_{c_j}}{P_{c_j}} \tag{5-7.24}$$

$$a_m = \sum_i \sum_j y_i y_j a_{ij} \tag{5-7.25}$$

with
$$\Omega_{b_j}^* = 0.0867 - 0.0125\omega_j + 0.011\omega_j^2 \tag{5-7.26}$$

$$a_{ii} = \frac{\Omega_{a_i}^* R^2 T_{c_i}^{2.5}}{P_{c_i}} \tag{5-7.27}$$

$$a_{ij} = \frac{(\Omega_{a_i}^* + \Omega_{a_j}^*) R T_{c_{ij}}^{1.5} (V_{c_i} + V_{c_i})}{4[0.291 - 0.04(\omega_i + \omega_j)]} \tag{5-7.28}$$

$$T_{c_{ij}} = (1 - k_{ij}) \sqrt{T_{c_i} T_{c_j}} \tag{5-7.29}$$

$$\Omega_{a_j}^* = \left(\frac{R T_{c_j}}{V_{c_j} - b_j} - P_{c_j} \right) \frac{P_{c_j} V_{c_j} (V_{c_j} + b_j)}{(R T_{c_j})^2} \tag{5-7.30}$$

The interaction parameter k_{ij} usually ranges from 0.1 to 0.01. Values for a large number of binary systems have been tabulated [14].

Spencer et al. [85] reported that the average deviation between P_{c_T} estimated from the Chueh-Prausnitz correlation and experimental values was 2 atm unless methane was one of the components. In such cases, the deviation was often much larger. There have been other techniques suggested to estimate P_{c_m} for systems containing methane [1, 34, 89], but they are either limited to aliphatic hydrocarbons or are graphical with a trial-and-error solution.

Example 5-8 Repeat Example 5-7 using the Chueh-Prausnitz method to estimate P_{c_T}.
  solution The properties of pure ethane and benzene are given in Example 5-7. Using the Chueh-Prausnitz methods to calculate T_{c_T} and V_{c_T} for this mixture gives $T_{c_T} = 493$ K and $V_{c_T} = 184$ cm³/g-mol; $k_{12} = 0.03$ for this binary [14]. Then

$$\Omega_{b_1}^* = 0.0867 - (0.0125)(0.098) + (0.011)(0.098)^2 = 0.0856$$

$$\Omega_{b_2}^* = 0.0846$$

$$b_1 = \frac{(0.0856)(82.04)(305.4)}{48.2} = 44.51 \text{ cm}^3/\text{g-mol}$$

$$b_2 = 80.80 \text{ cm}^3/\text{g-mol}$$

$$b_m = (0.392)(44.51) + (0.608)(80.80) = 66.57 \text{ cm}^3/\text{g-mol}$$

$$\Omega_{a_1}^* = \left(\frac{R T_{c_1}}{V_{c_1} - b_1} - P_{c_1} \right) \frac{P_{c_1} V_{c_1} (V_{c_1} + b_1)}{(R T_{c_1})^2}$$

$$= \left[\frac{(82.04)(305.4)}{148 - 44.51} - 48.2 \right] \left[\frac{(48.2)(148)(148 + 44.51)}{[(82.04)(305.4)]^2} \right]$$

$$= 0.424$$

$$\Omega_{a_2}^* = 0.421$$

$$a_1 = \frac{\Omega_{a_1}^* R^2 T_{c_1}^{2.5}}{P_{c_1}} = \frac{(0.424)(82.04)^2 (305.4)^{2.5}}{48.2}$$

$$= 96.57 \times 10^6 \text{ cm}^6 \text{ atm K}^{0.5}/(\text{g-mol})^2$$

$$a_2 = 439.4 \times 10^6 \text{ cm}^6 \text{ atm K}^{0.5}/(\text{g-mol})^2$$

$$T_{c_{12}} = [(305.4)(562.1)]^{1/2}(1 - 0.03) = 401.9 \text{ K}$$

$$a_{12} = \frac{(\frac{1}{4})(0.424 + 0.421)(82.04)(401.9)^{1.5}(148 + 259)}{0.291 - (0.04)(0.098 + 0.212)}$$

$$= 203.9 \times 10^6 \text{ cm}^6 \text{ atm K}^{0.5}/(\text{g-mol})^2$$

$$a_m = 10^6[(0.392)^2(96.57) + (0.608)^2(439.4)$$

$$+ (2)(0.392)(0.608)(203.9)]$$

$$= 274.4 \times 10^6 \text{ cm}^6 \text{ atm K}^{0.5}/(\text{g-mol})^2$$

Then

$$P_{cT} = \frac{RT_{cT}}{V_{cT} - b_m} - \frac{a_m}{T_{cT}^{1/2} V_{cT}(V_{cT} + b_m)}$$

$$= \frac{(82.04)(493)}{184 - 66.57} - \frac{274.4 \times 10^6}{(493^{1/2})(184)(184 + 66.57)}$$

$$= 76.5 \text{ atm}$$

$$\text{Error} = \frac{76.5 - 83.8}{83.8} \times 100 = -9\%$$

Recommendations   (1) To estimate the true critical temperature of a hydrocarbon mixture, use the method of Li [Eq. (5-7.2)]; if the mixture contains nonhydrocarbons, the Chueh-Prausnitz correlation is preferred [Eq. (5-7.4)], though the interaction parameter τ_{ij} can be evaluated only for relatively simple binary pairs.   (2) To estimate the true critical volume of a mixture, Eq. (5-7.8) is recommended.   However, as above, the interaction parameter ν_{ij} can be determined only for a limited number of binary types.   (3) For the true critical pressure of a mixture, either the Kreglewski-Kay or the Chueh-Prausnitz method may be used.   Neither is particularly applicable for systems containing methane.   Errors found are usually considerably larger for P_{cT} estimations than for T_{cT}.

In all cases, the correlations have been developed and tested primarily with hydrocarbon mixtures.   When applying these estimation methods to nonhydrocarbon mixtures, no reliable estimate of the error can be given.

5-8   Heat Capacities of Liquids

There are *three* liquid heat capacities in common use, C_{P_L}, C_{σ_L}, and C_{sat_L}.   The first represents the change in enthalpy with temperature at constant pressure; the second shows the variation in enthalpy of a *saturated* liquid with temperature; the third indicates the energy required to effect a temperature change while maintaining the liquid in a saturated state.   The three heat capacities are related as follows:

$$C_{\sigma_L} = \frac{dH_{\sigma_L}}{dT} = C_{P_L} + \left[V_{\sigma_L} - T\left(\frac{\partial V}{\partial T}\right)_p\right]\left(\frac{dP}{dT}\right)_{\sigma_L}$$

$$= C_{\text{sat}_L} + V_{\sigma_L}\left(\frac{dP}{dT}\right)_{\sigma_L} \tag{5-8.1}$$

The term $(dP/dT)_{\sigma_L}$ represents the change in vapor pressure with temperature.   Except at high reduced temperatures, all three forms of the liquid heat capacity are in close numerical agreement.   Most estimation techniques yield either C_{p_L} or C_{σ_L}, although C_{sat_L} is often the quantity measured experimentally.

Liquid heat capacities are not strong functions of temperature except above $T_r = 0.7$ to 0.8.   In fact, a shallow *minimum* is often reported at temperatures slightly below the normal boiling point.   At high reduced temperatures, liquid heat capacities are large and are strong functions of temperature.   The general trend is illustrated in Fig. 5-11 for propylene.

Near the normal boiling point, most liquid organic compounds have heat capacities between 0.4 and 0.5 cal/g K.   In this temperature range, there is essentially no effect of pressure [27].

Experimentally reported liquid heat capacities of hydrocarbons have been correlated in a nomograph [27] and expressed in analytical form [36, 37, 90].   A tabulation of available data is given by San Jose [75].

Estimation methods applicable for liquid heat capacities fall into four

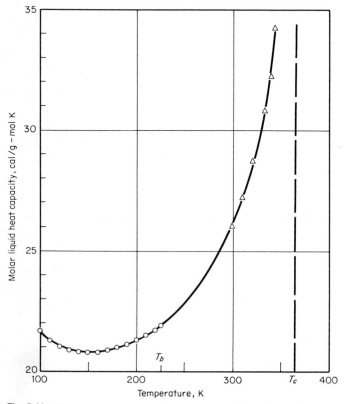

Fig. 5-11   Heat capacity of liquid propylene; ○ Ref. 67, △ Ref. 4.

general categories: theoretical, group-contribution, corresponding-states, or Watson's thermodynamic cycle. Methods from each category except the first are described below, and recommendations are presented at the end of the section. Theoretical methods are based on the estimation of liquid heat capacities at *constant volume* by considering each mode of energy storage separately. Reliable estimation procedures have not yet been developed for engineering use, although Bondi [8, 9] has suggested some useful approximations that are particularly valuable for high-molecular-weight liquids and polymers. A similar, earlier treatment was published by Sakiadis and Coates [74].

Group-Contribution Methods The assumption is made that various groups in a molecule contribute a definite value to the total molar heat capacity independent of other groups present. Johnson and Huang [42], Shaw [81], and Chueh and Swanson [11] have all proposed values for different molecular groups to estimate C_{p_L} at room temperature. Shaw's method is applicable at 25°C, while the others are to be used at 20°C. Both the Shaw and Chueh-Swanson methods are accurate, but the latter is more general and group contributions for this technique are shown in Table 5-12. Another additive method in which structural increments are given from -25 to 100°C has been suggested by Missenard [58], and group contributions for this scheme are given in Table 5-13. The Chueh-Swanson and Missenard methods are illustrated in Examples 5-9 and 5-10. Missenard's method cannot be used for compounds with double bonds, and neither should be used if the temperature corresponds to a reduced temperature in excess of 0.75. When $T_r < 0.75$, the estimated value may be considered to be either C_{p_L}, C_{σ_L}, or C_{sat_L}, as these are essentially identical at low reduced temperatures. Errors for the Chueh-Swanson method rarely exceed 2 to 3 percent and for Missenard's ± 5 percent.

For *hydrocarbons*, Luria and Benson [53a] have proposed an accurate group-contribution method applicable *below* the normal boiling point. For each carbon atom in the hydrocarbon, one must specify the other atoms to which it is covalently bonded. A carbon with four bonds is simply written as C; a carbon double-bonded to another carbon is noted as C_d, and it is only necessary to show the atoms which are attached to the *two* free bonds. Similarly, a carbon triply bonded to another carbon is C_t and it possesses only one free bond, while an aromatic carbon is C_B and the allene group ($>C=C=C<$) is C_a. Table 5-14 shows contributions for many carbon types to allow one to calculate the polynomial constants and estimate C_p as a function of temperature. Several correction terms are also given. The method is generally accurate to less than 1 cal/g-mol K and is illustrated in Example 5-11. The Benson method for estimating ideal-gas properties is similar in nature and is described in Sec. 7-3.

TABLE 5-12 Group Contributions for Molar Liquid Heat Capacity at 20°C [11]

Group	Value†	Group	Value†
Alkane		$-C=O$ (with H below)	12.66
$-CH_3$	8.80		
$-CH_2-$	7.26	$-C-OH$ (with O double bond above)	19.1
$-CH-$	5.00		
$-C-$	1.76	$-C-O-$ (with O double bond above)	14.5
Olefin		$-CH_2OH$	17.5
$=CH_2$	5.20	$-CHOH$	18.2
$=C-H$	5.10	$-COH$	26.6
$=C-$	3.80	$-OH$	10.7
Alkyne		$-ONO_2$	28.5
$-C≡H$	5.90	**Nitrogen**	
$-C≡$	5.90	$H-N-$ (with H above)	14.0
In a Ring			
$-CH-$	4.4	$-N-$ (with H above)	10.5
$-C=$ or $-C-$	2.9	$-N-$	7.5
$-CH=$	5.3	$-N=$ (in a ring)	4.5
$-CH_2-$	6.2	$-C≡N$	13.9
Oxygen		**Sulfur**	
$-O-$	8.4	$-SH$	10.7
$C=O$	12.66	$-S-$	8.0

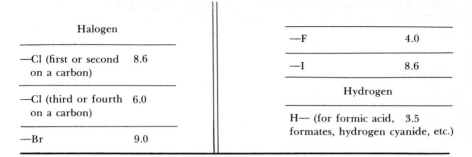

Halogen			—F	4.0
—Cl (first or second on a carbon)	8.6		—I	8.6
—Cl (third or fourth on a carbon)	6.0		Hydrogen	
—Br	9.0		H— (for formic acid, formates, hydrogen cyanide, etc.)	3.5

†Add 4.5 for any carbon group which fulfills the following criterion: a carbon group which is jointed by a single bond to a carbon group connected by a double or triple bond with a third carbon group. In some cases a carbon group fulfills the above criterion in more ways than one. In these cases 4.5 should be added each time the group fulfills the criterion.

Exceptions to the above 4.5 addition rule:

1. No such extra 4.5 additions for —CH$_3$ groups.
2. For a —CH$_2$— group fulfilling the 4.5 addition criterion add 2.5 instead of 4.5. However, when the —CH$_2$— group fulfills the addition criterion in more ways than one, the addition should be 2.5 the first time and 4.5 for each subsequent addition (see Example 5-9).
3. No such extra addition for any carbon group in a ring.

Example 5-9 Estimate the liquid heat capacity of 1,4-pentadiene at 20°C using the Chueh-Swanson group-contribution method.
   solution From Table 5-12,

$$C_{p_L}\ (20°C) = 2(CH_2\!\!=\!) + 2(—CH\!\!=\!) + —CH_2—$$
$$+ \text{corrections noted in Table 5-12}$$
$$= (2)(5.20) + (2)(5.10) + 7.26 + 2.5 + 4.5$$
$$= 34.9\ \text{cal/g-mol K}$$

Tamplin and Zuzic [90] indicate that $C_{p_L} = 35.1$ cal/g-mol K at 20°C.

Example 5-10 Using Missenard's group-contribution method, estimate the liquid heat capacity of isopropyl alcohol at 0°C.
   solution With Table 5-13,

$$C_{p_L}\ (0°C) = 2(—CH_3) + \overset{|}{\underset{|}{—CH}} + —OH$$
$$= (2)(9.55) + 5.7 + 8.0$$
$$= 32.8\ \text{cal/g-mol K}$$

The experimental value is 32.46 cal/g-mol K [32].

TABLE 5-13 Group Contributions for Missenard Method, cal/g-mol K [58]

| Group | Temperature, °C | | | | | |
	−25	0	25	50	75	100
—H	3.0	3.2	3.5	3.7	4.0	4.5
—CH₃	9.2	9.55	9.95	10.4	10.95	11.55
—CH₂—	6.5	6.6	6.75	6.95	7.15	7.4
∣ —CH—	5.0	5.7	5.95	6.15	6.35	6.7
∣ —C— ∣	2.0	2.0	2.0	2.0	2.0	
—C≡C—	11.0	11.0	11.0	11.0		
—O—	6.9	7.0	7.1	7.2	7.3	7.4
—CO— (ketone)	10.0	10.2	10.4	10.6	10.8	11.0
—OH	6.5	8.0	10.5	12.5	14.75	17.0
—COO— (ester)	13.5	13.8	14.1	14.6	15.1	15.5
—COOH	17.0	17.7	18.8	20.0	21.5	22.5
—NH₂	14.0	14.0	15.0	16.0		
—NH—	12.2	12.2	12.2			
∣ —N—	2.0	2.0	2.0			
—CN	13.4	13.5	13.6			
—NO₂	15.4	15.5	15.7	16.0	16.3	
—NH—NH—	19.0	19.0	19.0			
C₆H₅— (phenyl)	26.0	27.0	28.0	29.5	31.0	32.5
C₁₀H₇— (naphthyl)	43.0	44.0	45.0	47.0	49.0	51.0
—F	5.8	5.8	6.0	6.2	6.45	6.75
—Cl	6.9	7.0	7.1	7.2	7.35	7.5
—Br	8.4	8.5	8.6	8.7	8.9	9.1
—I	9.4	9.5	9.65	9.8		
—S—	8.9	9.0	9.2	9.4		

Example 5-11 Estimate the liquid heat capacity of 1,1-dimethylcyclopentane at 300 K using the group method of Luria and Benson.

solution The groups constituting 1,1-dimethylcyclopentane are:

Number	Type
4	C—(C)₂(H)₂
1	C—(C)₄
2	C—(C)(H)₃

TABLE 5-14 Polynomial Coefficients for the Luria and Benson Liquid-Heat Capacity Equation [53a]

$$C_p = A + BT + CT^2 + DT^3 \quad \text{cal/g-mol K}$$

Group	A	B	C	D
C—(C)(H)$_3$	8.459	2.113E−3	−5.605E−5	1.723E−7
C—(C)$_2$(H)$_2$	−1.383	7.049E−2	−2.063E−4	2.269E−7
C—(C)$_3$(H)	2.489	−4.617E−2	3.181E−4	−4.565E−7
C—(C)$_4$	9.116	−2.354E−1	1.287E−3	−1.906E−6
C$_d$—(H)$_2$	8.153	1.776E−2	−1.526E−4	2.542E−7
C$_d$—(C)(H)	5.792	−1.228E−2	6.036E−5	−1.926E−8
C$_d$—(C)$_2$	8.005	−9.456E−2	4.620E−4	−6.547E−7
C$_d$—(C$_d$)(H)	8.127	−7.171E−2	3.894E−4	−5.462E−7
C$_a$	13.756	−1.338E−1	6.553E−4	−9.447E−7
C$_d$—(C$_B$)(C)	5.745	−1.085E−1	5.898E−4	−6.983E−7
C—(C$_d$)$_2$(H)$_2$	9.733	−1.100E−1	5.522E−4	−6.852E−7
C—(C$_d$)(C)(H)$_2$	3.497	−1.568E−2	1.808E−4	−3.277E−7
C—(C$_d$)(C)$_2$(H)	−2.232	−1.773E−2	2.812E−4	−4.199E−7
C—(C$_B$)(C)(H)$_2$	30.192	−2.812E−1	1.002E−3	−1.115E−6
C$_t$—(H)	30.122	−2.081E−1	5.945E−4	−3.430E−7
C$_t$—(C)	−10.407	1.662E−1	−5.679E−4	6.667E−7
C$_B$—(H)	−1.842	5.778E−2	−1.716E−4	1.995E−7
C$_B$—(C)	28.807	−2.824E−1	9.779E−4	−1.103E−6
C$_B$—(C$_B$)	−3.780	2.563E−2	1.190E−5	−9.774E−8

Corrections

	A	B	C	D
Cis with double bond	14.299	−1.646E−1	6.069E−4	−7.716E−7
Cyclopropane ring	28.469	−2.696E−1	6.534E−4	−1.636E−7
Cyclobutane ring	6.060	−3.114E−2	−2.461E−4	8.349E−7
Cyclopentane ring	34.261	−3.803E−1	1.161E−3	−1.118E−6
Spiropentane ring	32.469	−1.991E−1	1.820E−4	4.090E−7
Cyclohexane ring	13.021	−1.468E−1	2.802E−4	−3.185E−8
Cycloheptane ring	210.72	−2.344	8.235E−3	−9.500E−6
Cyclooctane ring	1691.9	−1.680E+1	5.523E−2	−6.033E−5
Cyclopentene ring	13.650	−1.126E−1	7.257E−5	3.400E−7
Cyclohexene ring	5.360	−3.456E−2	−2.232E−4	7.324E−7
Cycloheptatriene ring	−22.158	3.985E−1	−2.059E−3	2.863E−6
Cyclooctatetraene ring	−1.060	2.739E−1	−1.972E−3	3.167E−6
Decahydronaphthalene ring (cis and trans)	141.85	−1.510	4.773E−3	−4.872E−6
1,2,3,4-Tetrahydro-naphthalene	−212.65	2.203	−7.571E−3	8.565E−6

With Table 5-14 and the correction for the cyclopentane ring,

$$C_p = 4(-1.383 + 7.049 \times 10^{-2}T - 2.063 \times 10^{-4}T^2$$
$$+ 2.269 \times 10^{-7}T^3 + (9.116 - 2.354 \times 10^{-1}T$$
$$+ 1.287 \times 10^{-3}T^2 - 1.906 \times 10^{-6}T^3) + (2)(8.459$$
$$+ 2.113 \times 10^{-3}T - 5.605 \times 10^{-5}T^2 + 1.723 \times 10^{-7}T^3)$$
$$+ (34.261 - 3.803 \times 10^{-1}T + 1.161 \times 10^{-3}T^2$$
$$- 1.118 \times 10^{-6}T^3)$$
$$= 54.763 - 0.330T + 1.511 \times 10^{-3}T^2 - 1.772 \times 10^{-6}T^3$$

At 300 K, $C_p = 43.9$ cal/g-mol K.   Luria and Benson [53a] indicate that the experimental value is 44.8 cal/g-mol K.

Corresponding-States Methods   Several corresponding-states methods for liquid heat-capacity estimation have been cast in the form of Eq. (5-6.4).†   Bondi [8] has reviewed many forms.   He modified one suggested originally by Rowlinson [70] to give

$$\frac{C_{p_L} - C_p^\circ}{R} = 2.56 + 0.436(1 - T_r)^{-1}$$
$$+ \omega[2.91 + 4.28(1 - T_r)^{1/3}T_r^{-1} + 0.296(1 - T_r)^{-1}] \quad (5\text{-}8.2)$$

and noted a similar relation, attributed to Sternling and Brown,

$$\frac{C_{p_L} - C_p^\circ}{R} = (0.5 + 2.2\omega)[3.67 + 11.64(1 - T_r)^4 + 0.634(1 - T_r)^{-1}]$$
$$(5\text{-}8.3)$$

Whereas Eqs. (5-8.2) and (5-8.3) apply to C_{p_L}, Yuan and Stiel [105] proposed the following correlations for C_{σ_L}:

Nonpolar liquids:
$$C_{\sigma_L} - C_p^\circ = (\Delta C_\sigma)^{(0)} + \omega(\Delta C_\sigma)^{(1)} \quad (5\text{-}8.4)$$

Polar liquids:
$$C_{\sigma_L} - C_p^\circ = \Delta C_\sigma^{(0p)} + \omega(\Delta C_\sigma)^{(1p)} + X(\Delta C_\sigma)^{(2p)}$$
$$+ X^2(\Delta C_\sigma)^{(3p)} + \omega^2(\Delta C_\sigma)^{(4p)} + X\omega(\Delta C_\sigma)^{(5p)} \quad (5\text{-}8.5)$$

X is the Stiel polar factor, discussed in Sec. 2-6.

The functions $(\Delta C_\sigma)^{(0)}$, etc., are tabulated as a function of reduced temperature in Table 5-15.   Should one wish to obtain C_{p_L} instead of C_{σ_L}, an approximate relation is

$$\frac{C_{p_L} - C_{\sigma_L}}{R} = (1 + \omega)^{0.85} \exp(-0.7074 - 31.014T_r + 34.361T_r^2)$$
$$(5\text{-}8.6)$$

---

†For example, with Tables 5-10 and 5-11, one can estimate the heat-capacity departure function $C_{p_L} - C_p^\circ$ for liquids as well as for gases.   Good results have also been reported using an analytical form of the Lee-Kesler heat-capacity departure function [51] for calculating liquid heat capacities for hydrocarbons [16].

TABLE 5-15   Yuan and Stiel Deviation Functions for Saturated-Liquid Heat Capacity, cal/g-mol K [105]

Reduced temperature	$(\Delta C_\sigma)^{(0)}$	$(\Delta C_\sigma)^{(1)}$	$(\Delta C_\sigma)^{(0p)}$	$(\Delta C_\sigma)^{(1p)}$	$(\Delta C_\sigma)^{(2p)}$	$(\Delta C_\sigma)^{(3p)}$ $\times 10^{-2}$	$(\Delta C_\sigma)^{(4p)}$	$(\Delta C_\sigma)^{(5p)}$
0.96	14.87	37.0						
0.94	12.27	29.2	12.30	29.2	-126	†	†	†
0.92	10.60	27.2	10.68	27.4	-123	†	†	†
0.90	9.46	26.1	9.54	25.9	-121	†	†	†
0.88	8.61	25.4	8.67	24.9	-117.5	†	†	†
0.86	7.93	24.8	8.00	24.2	-115	†	†	†
0.84	7.45	24.2	7.60	23.5	-112.5	†	†	†
0.82	7.10	23.7	7.26	23.0	-110	†	†	†
0.80	6.81	23.3	7.07	22.6	-108	†	†	†
0.78	6.57	22.8	6.80	22.2	-107	†	†	†
0.76	6.38	22.5	6.62	21.9	-106	†	†	†
0.74	6.23	22.2	6.41	22.5	-105	-0.69	-4.22	-29.5
0.72	6.11	21.9	6.08	23.6	-107	0.15	-7.20	-30.0
0.70	6.01	21.7	6.01	24.5	-110	1.31	-10.9	-29.1
0.68	5.91	21.6	5.94	25.7	-113	2.36	-15.2	-22.8
0.66	5.83	21.8	5.79	27.2	-118	3.06	-20.0	-7.94
0.64	5.74	22.2	5.57	29.3	-124	3.24	-25.1	14.8
0.62	5.64	22.8	5.33	31.8	-132	2.87	-30.5	43.0
0.60	5.54	23.5	5.12	34.5	-141	1.94	-36.3	73.1
0.58	5.42	24.5	4.92	37.6	-151	0.505	-42.5	102
0.56	5.30	25.6	4.69	41.1	-161	-1.37	-49.2	128
0.54	5.17	26.9	4.33	45.5	-172	-3.58	-56.3	149
0.52	5.03	28.4	3.74	50.9	-184	-6.02	-64.0	165
0.50	4.88	30.0	2.87	57.5	-198	-8.56	-72.1	179
0.48	4.73	31.7	1.76	65.0	-213	-11.1	-80.6	192
0.46	4.58	33.5	0.68	72.6	-229	-13.3	-89.4	206
0.44	4.42	35.4	0.19	78.5	-244	-15.0	-98.2	221
0.42	4.26	37.4						
0.40	4.08	39.4						

†Data not available for $(\Delta C_\sigma)^{(3p)}$ to $(\Delta C_\sigma)^{(5p)}$ above $T_r = 0.74$; assume zero.

which is valid for $0.85 \leq T_r \leq 0.99$.   Below $T_r = 0.8$, it may be assumed that $C_{p_L} \approx C_{\sigma_L}$.

In another corresponding-states method proposed by Lyman and Danner [55], the *radius of gyration* was introduced as a pure-component correlating parameter. Originally suggested by Thompson [93], it has also been used in a vapor-pressure correlation [64] and in estimation methods for second virial coefficients [61] and dense-fluid transport properties [6]. The Lyman-Danner liquid heat-capacity method is empirical and may be written

$$C_{\text{sat}_L} - C_p^\circ = A_1 + (A_2 + A_3 \bar{R}) T_r + (A_4 + A_5 \bar{R}) T_r^5$$
$$+ A_6 \bar{R}^2 / T_r^2 + A_7 \bar{R} / T_r^3 + A_8 / T_r^5$$
$$+ \kappa (B_1 + B_2 T_r^2 + B_3 T_r^5) + \kappa^2 (B_4 + B_5 T_r^2) \quad \text{cal/g-mol K} \quad (5\text{-}8.7)$$

where C_{sat_L} = saturation heat capacity [see Eq. (5-8.1)]

C_p^o = ideal-gas heat capacity

$\bar{R}$ = radius of gyration of fluid

κ = an association factor defined as

$$\kappa = \frac{-\ln P_c - C_n(1 - T_{b_r}^{-1}) + D_n \ln T_{b_r} - 0.4218\left(\dfrac{1}{P_c T_{b_r}^2} - 1\right)}{1 - T_{b_r}^{-1} - \ln T_{b_r}} \quad (5\text{-}8.8)$$

with P_c in atmospheres and $T_{b_r} = T_b / T_c$. The parameters C_n and D_n are

$$C_n = 4.6773 + 1.8324\bar{R} - 0.03501\bar{R}^2$$
$$D_n = 0.7751 C_n - 2.6354 \quad\quad (5\text{-}8.9)$$

The Lyman-Danner constants are shown in Table 5-16.

TABLE 5-16   Lyman-Danner Constants in Eq.
(5-8.7) [55]

A_1	10.1273	B_1	0.31446
A_2	−15.3546	B_2	2.5346
A_3	3.2008	B_3	−2.0242
A_4	19.7302	B_4	−0.07055
A_5	−0.8949	B_5	0.07264
A_6	−0.01489		
A_7	0.2241		
A_8	−0.04342		

To use the Lyman-Danner method, the normal boiling point as well as the critical temperature and pressure must be known.   Also, the ideal-gas heat capacity is necessary, as is the radius of gyration.   It is this last parameter which is the most difficult to determine.   It can be obtained from the molecular structure if the moment of inertia is known for each axis [93].   Alternatively, Passut and Danner have reported $\bar{R}$ and κ for some 250 compounds, and these are given in Table 5-17.

Below $T_r \approx 0.8$, C_{sat_L} is essentially equal to C_{σ_L}.   From $0.8 < T_r < 0.99$, within an experimental error of about 0.2 cal/g-mol K, for all fluids

$$\frac{C_{\sigma_L} - C_{sat_L}}{R} = \exp(8.655 T_r - 8.385) \quad\quad (5\text{-}8.10)$$

The four corresponding-states estimation methods for liquid heat capacity are illustrated in Examples 5-12 to 5-14.   Calculated and experimental values are compared in Table 5-18.   The accuracy and applicability of these techniques are discussed after the Watson thermodynamic-cycle method has been described.

TABLE 5-17   Radius of Gyration and Association Factors [64]

Compound	Radius of gyration, Å	κ
Methane	1.1234	0.0000
Ethane	1.8314	0.0000
Propane	2.4255	0.0000
n-Butane	2.8885	0.0000
2-Methylpropane	2.8962	− 0.6884
n-Pentane	3.3850	0.0000
2-Methylbutane	3.3130	− 0.7643
2,2-Dimethylpropane	3.1530	− 1.1136
n-Hexane	3.8120	0.0000
2-Methylpentane	3.8090	− 0.8930
3-Methylpentane	3.6797	− 0.6343
2,2-Dimethylbutane	3.4846	− 1.1611
2,3-Dimethylbutane	3.5209	− 0.8695
n-Heptane	4.2665	0.0000
2-Methylhexane	4.2779	− 0.8844
3-Methylhexane	4.1454	− 0.6597
2,2-Dimethylpentane	4.0001	− 1.2464
2,3-Dimethylpentane	3.9210	− 0.7324
2,4-Dimethylpentane	3.9634	− 0.6858
3,3-Dimethylpentane	3.7952	− 1.1610
2,2,3-Trimethylbutane	3.6960	− 1.3293
n-Octane	4.6804	0.0000
2-Methylheptane	4.7401	− 0.9121
3-Methylheptane	4.5932	− 0.6792
4-Methylheptane	4.5581	− 0.5431
2,2-Dimethylhexane	4.4956	− 1.3428
2,3-Dimethylhexane	4.4084	− 0.8089
2,4-Dimethylhexane	4.3463	− 0.7135
2,5-Dimethylhexane	4.5932	− 1.1169
3,3-Dimethylhexane	4.3197	− 1.2865
3,4-Dimethylhexane	4.4000	− 1.0135
2,2,3-Trimethylpentane	4.1618	− 1.4562
2,2,4-Trimethylpentane	4.1714	− 1.3313
2,3,3-Trimethylpentane	4.0859	− 1.4284
2,3,4-Trimethylpentane	4.2052	− 1.0783
n-Nonane	5.1263	0.0000
2,2,3,3-Tetramethylpentane	4.1556	− 1.8930
n-Decane	5.5390	0.0000
n-Undecane	5.9867	0.0000
n-Dodecane	6.4321	0.0000

TABLE 5-17   Radius of Gyration and Association Factors   (*Continued*)

Compound	Radius of gyration, Å	κ
n-Tetradecane	7.3578	0.0000
n-Pentadecane	7.8387	0.0000
n-Hexadecane	8.3180	0.0000
Cyclohexane	3.2605	−0.9985
Methylcyclohexane	3.7467	−1.8288
1,1-Dimethylcyclohexane	4.0925	−2.8029
cis-1,2-Dimethylcyclohexane	4.0612	−2.6532
trans-1,2-Dimethylcyclohexane	4.1814	−2.9308
cis-1,3-Dimethylcyclohexane	4.0549	−2.5495
trans-1,3-Dimethylcyclohexane	4.1462	−2.9371
cis-1,4-Dimethylcyclohexane	4.1446	−2.9658
trans-1,4-Dimethylcyclohexane	4.1670	−2.8887
Ethene	1.5382	0.9391
Propene	2.2283	0.3590
1-Butene	2.7458	−0.0103
cis-2-Butene	2.7765	0.2014
trans-2-Butene	2.7123	0.7144
2-Methylpropene	2.8281	−0.1428
1-Pentene	3.1956	1.1055
cis-2-Pentene	3.2763	−0.2598
trans-2-Pentene	3.2826	−0.3495
2-Methyl-1-Butene	3.2239	−0.2859
2-Methyl-2-Butene	3.2301	1.0668
1-Hexene	3.6472	−0.2243
trans-2-Hexene	3.6964	−1.2781
1-Heptene	4.0971	0.6549
1-Octene	4.5342	−0.0195
1-Nonene	4.9687	−0.0299
1-Decene	5.4017	0.6240
1-Undecene	5.8389	−0.0256
1-Dodecene	6.2966	−0.2172
1-Tridecene	6.7514	−0.2574
1-Tetradecene	7.2244	−0.2025
1-Pentadecene	7.6969	−0.4251
1-Hexadecene	8.1872	−0.7231
1,2-Butadiene	2.7497	1.4910
2,3-Pentadiene	3.0552	0.5219
Ethyne (acetylene)	1.10945	4.8845
Propyne	1.8864	3.2221
1-Butyne	2.7130	−2.2156

Benzene	3.0037	− 0.2179
Methylbenzene	3.4431	− 0.1628
Ethylbenzene	3.8211	− 0.2047
1,2-Dimethylbenzene	3.7889	− 0.0213
1,3-Dimethylbenzene	3.8966	0.1918
1,4-Dimethylbenzene	3.7962	0.3475
Isopropylbenzene	4.1870	− 0.6791
1-Methyl-2-ethylbenzene	4.1296	− 1.2412
1-Methyl-3-ethylbenzene	4.2845	− 0.9775
1-Methyl-4-ethylbenzene	4.1662	− 0.6550
1,2,3-Trimethylbenzene	4.0996	0.7296
1,2,4-Trimethylbenzene	4.1678	0.8416
1,3,5-Trimethylbenzene	4.3408	0.9603
1-Methyl-3-isopropylbenzene	4.5790	− 3.0710
1-Methyl-4-isopropylbenzene	4.5231	− 0.3729
1,2,3,4-Tetramethylbenzene	4.3779	7.1270
1,2,3,5-Tetramethylbenzene	4.4920	3.8842
1,2,4,5-Tetramethylbenzene	4.4542	1.8030
Fluoromethane	1.4186	4.2800
Trifluoromethane	2.3194	2.1131
Tetrafluoromethane	2.6350	− 0.0475
Fluoroethane	2.1758	2.3345
1,1-Difluoroethane	2.4976	2.6225
1,1,1-Trifluoroethane	2.6718	1.8538
Perfluoroethane	3.3797	0.2252
Perfluoro-n-butane	4.4430	0.1000
Perfluoro-n-hexane	5.5208	0.4932
Perfluoro-n-heptane	6.0216	2.2991
Perfluorocyclohexane	4.6834	2.2770
Perfluoromethylcyclohexane	4.8941	2.1612
Perfluoroethene	3.1498	− 0.5233
Fluorobenzene	3.3454	− 0.3441
Perfluorobenzene	4.6577	0.0623
Fluorodichloromethane	2.8801	0.0378
Difluorochloromethane	2.5693	0.7926
Difluorodichloromethane	3.0259	− 1.1524
Trifluorochloromethane	2.8099	− 0.8004
Fluorotrichloromethane	3.3037	− 2.3227
1,2-Difluoro-1,1,2,2-tetrachloroethane	3.9935	− 1.2653
1,1,2-Trifluoro-1,2,2-trichloroethane	3.8158	− 1.6572
1,1,2,2-Tetrafluoro-1,2-dichloroethane	3.6751	− 1.3001
Chloropentafluoroethane	3.4819	− 0.0921
Chlorotrifluoroethene	3.3448	− 0.4045
Chloromethane	1.4500	2.7997
Dichloromethane	2.3423	1.4060

TABLE 5-17   Radius of Gyration and Association Factors   (*Continued*)

Compound	Radius of gyration, Å	κ
Trichloromethane	3.1779	-0.4105
Tetrachloromethane	3.4581	-2.0378
Chloroethane	2.2812	1.3946
1,1-Dichloroethane	2.9945	0.4817
1,2-Dichloroethane	2.8510	1.9778
1,1,1-Trichloroethane	3.3566	2.4867
Chloroethene	2.1220	0.4357
cis-1,2-Dichloroethene	3.0132	-1.1799
trans-1,2-Dichloroethene	5.0132	-0.2465
Trichloroethene	3.7592	-1.6978
Chlorobenzene	3.5684	-0.8499
Bromomethane	1.1796	1.9341
Chloroethane	1.9877	3.2628
Bromobenzene	3.4714	-0.5914
Iodomethane	1.0458	2.7737
Iodobenzene	3.2688	-0.0752
Dimethyl ether	2.1274	2.0857
Methyl ethyl ether	2.6409	1.7172
Diethyl ether	3.1395	1.3295
Ethyl propyl ether	3.5468	1.2872
Diisopropyl ether	3.6805	1.2304
Vinyl ethyl ether	2.9902	1.3107
Ethylene oxide	1.9005	2.7162
1,2-Dimethoxy ethane	3.3998	3.1514
Methanol	1.5360	14.2961
Ethanol	2.2495	14.7937
1-Propanol	2.7359	13.1172
2-Propanol	2.7264	14.7128
2-Methyl-1-propanol	3.1816	10.6592
2-Methyl-2-propanol	3.0190	12.1625
1-Butanol	3.2250	10.5692
2-Butanol	3.1507	10.4651
Cyclohexanol	3.4341	7.8767
Phenol	3.5496	4.3831
o-Cresol	3.7164	3.7663
m-Cresol	3.8349	4.1207
p-Cresol	3.7360	5.9420
Formaldehyde	1.2124	6.2518
Acetone	2.7404	3.1677
Pentafluorochloroacetone	4.0101	0.1284

Perfluoroacetone	3.8076	1.6516
Methyl ethyl ketone	3.1395	2.4878
Methyl n-propyl ketone	3.6265	1.5783
Methyl isopropyl ketone	3.4148	0.2187
Diethyl ketone	3.4817	2.3917
Methyl isobutyl ketone	3.5722	3.0128
Acetic Acid	2.5950	7.2091
Propionic Acid	3.0500	8.8116
Butyric Acid	3.5505	11.9043
Isobutyric Acid	3.3418	10.7015
Acetic anhydride	3.5727	20.6471
Methyl formate	2.3604	2.8389
Ethyl formate	2.8697	2.0680
Propyl formate	3.4192	1.4367
Isobutyl formate	3.5561	3.2871
Methyl acetate	2.8616	3.2788
Ethyl acetate	3.3479	2.9482
Propyl acetate	3.7785	2.4937
Methyl propionate	3.3035	2.8068
Ethyl propionate	3.7124	2.7627
Methyl butyrate	3.7720	2.0266
Methyl isobutyrate	3.6271	2.0875
Ethyl isobutyrate	4.0484	2.7152
Dimethyl oxalate	3.8714	7.2279
Dimethyl sulfide	2.3719	1.1055
Methyl ethyl sulfide	2.8091	2.5851
Diethyl sulfide	3.2067	1.3356
Methyl propyl sulfide	3.3174	− 0.6659
Methyl isopropyl sulfide	3.2385	0.2168
Ethyl butyl sulfide	4.0643	− 0.6389
Methanethiol	1.6114	2.3417
Ethanethiol	2.3408	1.7729
2-Methyl-2-propanethiol	3.1096	0.8852
1-Pentanethiol	3.7914	0.1439
Hydrogen cyanide	0.6496	11.5466
Acetonitrile	1.8213	6.1861
Propionitrile	2.3618	4.6585
Butryonitrile	3.2094	3.5995
Benzonitrile	3.7432	1.5804
Monomethylamine	1.6623	5.8615
Dimethylamine	2.2644	4.3322
Trimethylamine	2.7356	0.2762
Ethylamine	2.3085	3.9599
Diethylamine	3.1606	1.8686
Propylamine	2.7889	3.0940

TABLE 5-17   Radius of Gyration and Association Factors   (Continued)

Compound	Radius of gyration, Å	κ
Aniline	3.3926	3.3293
Nitromethane	2.3063	5.0823
Hydrogen	0.3708	− 3.0010
Nitrogen	0.5471	2.8266
Oxygen	0.6037	2.1474
Fluorine	0.7140	3.0189
Chlorine	0.9873	2.3329
Water	0.6150	9.4339
Chlorine dioxide	1.7115	5.3359
Nitric oxide	0.5302	18.9561
Nitrous oxide	1.1907	3.9052
Nitrogen dioxide	1.4286	23.7186
Carbon monoxide	0.5582	3.1931
Carbon dioxide	0.9918	7.3589
Sulfur dioxide	1.6739	4.8707
Sulfur trioxide	2.2027	9.9043
Hydrogen sulfide	0.6384	3.8149
Carbon disulfide	1.4241	1.7156
Hydrogen fluoride	0.2006	11.8167
Hydrogen chloride	0.2989	5.4689
Hydrogen bromide	0.1568	4.8699
Hydrogen iodide	0.1418	4.0720
Ammonia	0.8533	6.9221
Chlorine trifluoride	2.3066	4.4848
Boron trifluoride	2.3559	7.2179
Oxygen difluoride	1.8057	− 0.3173
Nitrogen trifluoride	2.2624	− 0.1056
Phosphorous trifluoride	2.3574	1.5503
Nitrosyl chloride	1.8458	4.8125
Boron trichloride	3.2794	− 2.4123
Boron tribromide	3.6718	− 0.8700
Carbonyl chloride	2.8269	− 0.1005
Carbonyl sulfide	1.1760	2.2985
Cyanogen	1.4232	6.3992
Hydrazine	1.5119	6.6724
Helium	0.8077	− 13.5387
Neon	0.8687	− 0.0228
Argon	1.0760	0.1792
Krypton	1.1376	0.0435
Xenon	1.2956	− 0.2990

Example 5-12 Estimate the liquid heat capacity of *cis*-2-butene at 76.6°C (349.8 K) using the Rowlinson-Bondi and Sternling-Brown corresponding-states correlations.

solution From Appendix A, $T_c = 435.6$ K, $\omega = 0.202$, CPVAP A $= 0.105$, CPVAP B $= 7.054 \times 10^{-2}$, CPVAP C $= -2.431 \times 10^{-5}$, CPVAP D $= -0.147 \times 10^{-9}$. At 349.8 K,

$$C_p^\circ = 0.105 + (7.054 \times 10^{-2})(349.8) + (-2.431 \times 10^{-5})(349.8)^2$$
$$+ (-0.147 \times 10^{-9})(349.8)^3 = 21.80 \text{ cal/g-mol K}$$

The reduced temperature is $349.8/435.6 = 0.803$.

Rowlinson-Bondi [Eq. (5-8.2)]

$$\frac{C_{p_L} - C_p^\circ}{R} = 2.56 + \frac{0.436}{1 - 0.803}$$
$$+ (0.202)\left[2.91 + \frac{(4.28)(1 - 0.803)^{1/3}}{0.803} + \frac{0.296}{1 - 0.803}\right]$$
$$= 6.29$$
$$C_{p_L} = (1.98)(6.29) + 21.80 = 34.3 \text{ cal/g-mol K}$$

Since the experimental value is 36.5 cal/g-mol K [78],

$$\text{Error} = \frac{34.3 - 36.5}{36.5} \times 100 = -6.0\%$$

Sternling-Brown [Eq. (5-8.3)]

$$\frac{C_{p_L} - C_p^\circ}{R} = [0.5 + (2.2)(0.202)]\left[3.67 + (11.64)(1 - 0.803)^4 + \frac{0.634}{1 - 0.803}\right]$$
$$= 6.52$$
$$C_{p_L} = (1.98)(6.52) + 21.80 = 34.7 \text{ cal/g-mol K}$$
$$\text{Error} = \frac{34.7 - 36.5}{36.5} \times 100 = -4.9\%$$

Example 5-13 Estimate the liquid heat capacity of ethyl mercaptan at 42°C (315.2 K) using the Yuan-Stiel corresponding-states method. The experimental value is 28.7 cal/g-mol K [56].

solution From Appendix A, $T_c = 499$ K, $\omega = 0.190$, CPVAP A $= 3.564$, CPVAP B $= 5.615 \times 10^{-2}$, CPVAP C $= -3.239 \times 10^{-5}$, CPVAP D $= 7.552 \times 10^{-9}$. At 315.2 K,

$$C_p^\circ = 3.564 + (5.615 \times 10^{-2})(315.2) + (-3.239 \times 10^{-5})(315.2)^2$$
$$+ (7.552 \times 10^{-9})(315.2)^3 = 18.28 \text{ cal/g-mol K}$$

Since ethyl mercaptan is a slightly polar fluid, Eq. (5-8.5) will be used. X is found from Table 2-5 as 0.004. $T_r = 315.2/499 = 0.632$. From Table 5-15, using linear interpolation, $(\Delta C_\sigma)^{(0p)} = 5.47$, $(\Delta C_\sigma)^{(1p)} = 30.3$, $(\Delta C_\sigma)^{(2p)} = -127$, $(\Delta C_\sigma)^{(3p)} = 309$, $(\Delta C_\sigma)^{(4p)} = -27.3$, and $(\Delta C_\sigma)^{(5p)} = 26.1$. Then, with Eq. (5-8.5),

$$C_{\sigma_L} - C_p^\circ = 5.47 + (0.190)(30.3) + (0.004)(-127) + (0.004)^2(309) + (0.190)^2(-27.3)$$
$$+ (0.004)(0.190)(26.1) = 9.75 \text{ cal/g-mol K}$$
$$C_{\sigma_L} = 9.75 + 18.28 = 28.0 \text{ cal/g-mol K}$$
$$\text{Error} = \frac{28.0 - 28.7}{28.7} \times 100 = -2.3\%$$

Example 5-14 Repeat Example 5-13 using the Lyman-Danner estimation method.

solution   From Table 5-17, for ethyl mercaptan (ethanethiol), $\bar{R} = 2.3408$ Å and $\kappa = 1.7729$. With T_r and C_p° from Example 5-13, using Eq. (5-8.7) and Table 5-16,

$$
\begin{aligned}
C_{\text{sat}_L} - C_p^{\circ} = {} & 10.1273 + [-15.3546 + (3.2008)(2.3408)](0.632) \\
& + [19.7302 + (-0.8949)(2.3408)](0.632)^5 \\
& + (-0.01489)(2.3408)^2/(0.632)^2 \\
& + (0.2241)(2.3408)/(0.632)^3 + (-0.04342)/(0.632)^5 \\
& + (1.7729)[0.31446 + (2.5346)(0.632)^2 + (-2.0242)(0.632)^5] \\
& + (1.7729)^2[-0.07055 + (0.07264)(0.632)^2] \\
= {} & 10.24 \text{ cal/g-mol K}
\end{aligned}
$$

$$
C_{\text{sat}_L} = 10.24 + 18.28 = 28.52 \text{ cal/g-mol K}
$$

$$
\text{Error} = \frac{28.52 - 28.7}{28.7} \times 100 = -0.6\%
$$

Note that at this low temperature, $C_{\text{sat}_L} \approx C_{\sigma_L} \approx C_{pL}$.

Watson Thermodynamic Cycle

Watson [95] suggested that liquid heat capacities could be estimated by calculating enthalpy changes in a thermodynamic cycle as follows: (1) a saturated liquid at temperature T_1 is heated to T_2 maintaining saturated conditions, (2) liquid at T_2 is vaporized and expanded isothermally to a low-pressure ideal-gas state, (3) as an ideal gas, the material is cooled from T_2 to T_1, and (4) the fluid is compressed isothermally at T_1 to a saturated vapor and condensed.   The sum of enthalpy terms for the cycle is zero.   If T_2 is allowed to approach T_1,

$$
\frac{C_{\sigma_L} - C_p^{\circ}}{R} = -\frac{d}{dT_r}\left(\frac{\Delta H_v}{RT_c}\right) - \frac{d}{dT_r}\left(\frac{H^{\circ} - H_{\text{sv}}}{RT_c}\right) \tag{5-8.11}
$$

H° and C_p° represent the ideal-gas enthalpy and heat capacity. H_{sv} is the enthalpy of the saturated vapor and C_{σ_L} the saturated-liquid heat capacity (dH_{σ_L}/dT).   The estimation of each term on the right-hand side of Eq. (5-8.11) is considered separately.

The derivative, $(d/dT_r)[(H^{\circ} - H_{\text{sv}})/RT_c]$, reflects the change, with reduced temperature, of the reduced enthalpy departure along the *saturation-vapor* curve.   Any of the enthalpy-departure equations presented earlier in this chapter may be employed, with an appropriate vapor-pressure correlation, to obtain $(H^{\circ} - H_{\text{sv}})/RT_c$, and then numerical or analytical differentiation can be used to calculate the derivative.

Alternatively, the enthalpy departure can be expanded in terms of reduced temperature and pressure to give

$$
\frac{d}{dT_r}\left(\frac{H^{\circ} - H_{\text{sv}}}{RT_c}\right) = \frac{\partial}{\partial T_r}\left(\frac{H^{\circ} - H_{\text{sv}}}{RT_c}\right)_{P_r} + \frac{\partial}{\partial P_r}\left(\frac{H^{\circ} - H_{\text{sv}}}{RT_c}\right)_{T_r}\frac{dP_r}{dT_r} \tag{5-8.12}
$$

The two partial derivatives have been expressed in graphical form using both a two-parameter form of the law of corresponding states [95] and a three-parameter form (with Z_c) [68].   More recently, Chueh and Swanson

[11] have proposed that the derivative can be obtained in an analytical form using the equations in Table 5-8 that apply to a saturated vapor. In this case,

$$\frac{H° - H_{sv}}{RT_c} = \frac{1}{R} \frac{DP_r^E}{1 + F(-\ln P_r)^G} \tag{5-8.13}$$

where the constants D, E, F, and G are shown for different values of Z_c in Table 5-8. P_r is the reduced vapor pressure at the T_r of interest; for a pure material, P_r is a function of T_r.

With Eq. (5-8.13),

$$\frac{d}{dT_r}\left(\frac{H° - H_{sv}}{RT_c}\right) = P_r \psi(P_r, Z_c) \frac{d \ln P_r}{dT_r} \tag{5-8.14}$$

where

$$\psi(P_r, Z_c) = \frac{DP_r^{E-1}}{R[1 + F(-\ln P_r)^G]}\left[E + \frac{FG(-\ln P_r)^{G-1}}{1 + F(-\ln P_r)^G}\right] \tag{5-8.15}$$

To use Eq. (5-8.15), Z_c for the pure component (or mixture) is found and values of D, E, F, and G are determined from Table 5-8. P_r and $(d \ln P_r)/dT_r$ are calculated from a vapor-pressure correlation (Chap. 6) appropriate to the liquid considered.

Interpolation to obtain the parameters D, E, F, and G at Z_c values *between* those tabulated in Table 5-8 is not easily accomplished, as those values which are listed do not plot as simple curves when expressed as a function of Z_c. It is therefore recommended that $(d/dT_r)[H° - H_{sv})/RT_c]$ be calculated at the two values of Z_c which bracket the appropriate Z_c and that the derivative be found by linear interpolation.

Values of $(d/dT_r)[H° - H_{sv})/RT_c]$ are positive and increase with temperature. At low pressures, where the gas-phase behavior approximates an ideal gas, this derivative is quite small and essentially negligible.

The other term in Eq. (5-8.11), $(d/dT_r)(\Delta H_v/RT_c)$, is negative and is normally the more important. It is difficult to estimate with high accuracy. One of the easiest ways to calculate this derivative is to assume that the enthalpy of vaporization varies with temperature, as given by the Watson equation (6-16.1),

$$\Delta H_v = \Delta H_{v_1}\left(\frac{1 - T_r}{1 - T_{r_1}}\right)^n \tag{5-8.16}$$

ΔH_{v_1} is the enthalpy of vaporization at some reference temperature T_1. The exponent n is a function of the material and probably of temperature. Often, however, it is assumed constant at a value of 0.38. With this value,

$$\frac{d}{dT_r}\left(\frac{\Delta H_v}{RT_c}\right) = -0.38 \frac{\Delta H_{v_1}/RT_c}{(1 - T_{r_1})^{0.38}}(1 - T_r)^{-0.62} \tag{5-8.17}$$

Equation (5-8.16) often yields reasonable values of $(d/dT_r)(\Delta H_v/RT_c)$ at high reduced temperatures, but it is unsatisfactory below the normal boiling point.   Chueh and Deal [12] discuss this problem and suggest that instead of Eq. (5-8.16) a more satisfactory form for ΔH_v is

$$\Delta H_v = A (1 - T_r)^n + B (1 - T_r)^6 \qquad (5\text{-}8.18)$$

Chueh and Deal obtained n from a correlation of Fishtine [26] and A and B from experimental data for ΔH_v and C_{p_L} at low temperatures.   (B is often close to 39 cal/g.)

A somewhat similar approach for obtaining ΔH_v as a function of temperature was presented by Chueh and Swanson [11].   Equation (5-8.16) is modified to

$$\Delta H_v = \Delta H_{v_1} \left(\frac{1 - T_r}{1 - T_{r_1}}\right)^{0.38 + \beta(1 - T_r)} \qquad (5\text{-}8.19)$$

β is a parameter which must be determined from experimental data.   Then, *at a reference state T_{r_1},*

$$\frac{d}{dT_r} \left(\frac{\Delta H_v}{RT_c}\right)_{\text{at } T_{r_1}} = -\frac{\Delta H_{v_1}}{RT_c} \frac{0.38 + \beta(1 - T_{r_1})}{1 - T_{r_1}} \qquad (5\text{-}8.20)$$

The reference state is selected at a temperature where a liquid heat-capacity datum point is available either from an experiment or as determined from a group-contribution method noted earlier in this section.   Then, with this C_{σ_L} at T_1 and with C_p° at T_1, Eq. (5-8.11) may be written at T_1,

$$\frac{\Delta H_{v_1}}{RT_c} \frac{0.38 + \beta(1 - T_{r_1})}{1 - T_{r_1}} = \left(\frac{C_{\sigma_L} - C_p^\circ}{R}\right)_{\text{at } T_{r_1}} + \frac{d}{dT_r} \left(\frac{H^\circ - H_{sv}}{RT_c}\right)_{\text{at } T_{r_1}} \qquad (5\text{-}8.21)$$

The second term on the right-hand side of Eq. (5-8.21) is calculated at T_{r_1}, as described earlier in this section. ΔH_{v_1} is usually not known, but ΔH_{v_b} is readily estimated (Sec. 6-15) or can be found in Appendix A.   Thus, Eq. (5-8.19) is applied to calculate ΔH_{v_1}, given ΔH_{v_b}.   Equation (5-8.21) can then be solved for the parameter β by a trial-and-error procedure.   When β is known,

$$\frac{d}{dT_r} \left(\frac{\Delta H_v}{RT_c}\right) = -\left[\frac{0.38}{1 - T_r} + \beta \left(1 + \ln \frac{1 - T_r}{1 - T_{r_1}}\right)\right]$$
$$\times \left[\frac{\Delta H_{v_1}}{RT_c} \left(\frac{1 - T_r}{1 - T_{r_1}}\right)^{0.38 + \beta(1 - T_r)}\right] \qquad (5\text{-}8.22)$$

Equation (5-8.22) reduces to Eq. (5-8.17) if $\beta = 0$.   Note that β depends upon the choice of T_{r_1}.   The use of a low-temperature liquid heat capacity significantly improves the predictive accuracy of the Watson method [11].   The method is illustrated in Example 5-15.

Example 5-15 Estimate the saturated-liquid heat capacity of p-xylene, $C_{\sigma_L} = dH_{\sigma_L}/dT$, at 300°C.   Corruccini and Ginnings [15] report a value of 70.2 cal/g-mol K for C_{σ_L}.

solution   We choose a reference state of 20°C and estimate C_{σ_L} from Table 5-12.

$$C_{\sigma_L}(20°C) = 4(\overset{|}{=}CH) + 2(\overset{|}{=}C-) + 2(-CH_3)$$

$$= (4)(5.3) + (2)(2.9) + (2)(8.8)$$

$$= 44.6 \text{ cal/g-mol K}$$

From Appendix A, $M = 106.168$, $T_b = 411.5$ K, $T_c = 616.2$ K, $P_c = 34.7$ atm, $Z_c = 0.260$, CPVAP A $= -5.993$, CPVAP B $= 1.443 \times 10^{-1}$, CPVAP C $= -8.058 \times 10^{-5}$, CPVAP D $= 1.629 \times 10^{-8}$, and $\Delta H_{v_b} = 8600$ cal/g-mol.

1. Calculate C_p° at 20°C (293 K) and 300°C (573 K)

$$C_{P293}^\circ = -5.993 + (1.443 \times 10^{-1})(293) + (-8.058 \times 10^{-5})(293^2)$$

$$+ (1.629 \times 10^{-8})(293^3) = 29.8 \text{ cal/g-mol K}$$

In a similar way,

$$C_{P573}^\circ = 53.3 \text{ cal/g-mol K}$$

2. Determine $(d \ln P_r)/dT_r$ and P_r at 20 and 300°C.   At 20°C, the Antoine vapor-pressure equation (6-3.1) may be used, together with Antoine constants in Appendix A,

$$\ln P = 16.0963 - \frac{3346.65}{293 - 57.84} = 1.865$$

$$P = 6.5 \text{ mm Hg} \quad \text{and} \quad P_r = \frac{6.5}{(760)(34.7)} = 2.46 \times 10^{-4}$$

$$\frac{d \ln P_r}{dT_r} = \frac{BT_c}{(T+C)^2} = \frac{(3346.65)(616.2)}{(293 - 57.84)^2}$$

$$= 37.3$$

At 300°C, any of the wide-range vapor-pressure equations in Chap. 6 can be used to obtain P and $(d \ln P_r)/dT_r$.   We use the Harlacher form [Eq. (6-6.1)] and use the constants in Appendix A:

$$\ln P = 56.175 + \frac{-6673.70}{573} + (-5.543) \ln 573 + 6.19 \frac{P}{573^2}$$

$$P = 14,820 \text{ mm Hg} \qquad P_r = \frac{14,820}{(760)(34.7)} = 0.562$$

Differentiating Eq. (6-6.1) gives

$$\frac{d \ln P_r}{dT_r} = T_c \frac{-B/T^2 + C/T - 2DP/T^3}{1 - DP/T^2}$$

With the Harlacher constants we have

$$\frac{d \ln P_r}{dT_r} = 8.27$$

3. Determine ψ in Eq. (5-8.15).   At 20°C, $P_r = 2.46 \times 10^{-4}$, as calculated above.   With the constants in Table 5-8,

$$\psi(P_r = 2.46 \times 10^{-4}, Z_c = 0.25) = 15.03$$

$$\psi(P_r = 2.46 \times 10^{-4}, Z_c = 0.27) = 8.57$$

Thus

$$\psi(P_r = 2.46 \times 10^{-4}, Z_c = 0.26) = 11.8$$

Similarly

$$\psi(P_r = 0.562, Z_c = 0.26) = 2.27$$

4. Calculate, from Eq. (5-8.14),

$$\left(\frac{d}{dT_r}\right)\left(\frac{H^\circ - H_{sv}}{RT_c}\right) = (2.46 \times 10^{-4})(11.8)(37.3) = 0.108 \text{ at } 20°C$$

$$= (0.562)(2.27)(8.27) = 10.6 \text{ at } 300°C$$

5. Estimate β with Eq. (5-8.21). $T_1 = 20°C = 293$ K, and $T_{r_1} = 0.475$:

$$\frac{\Delta H_{v293}}{(1.987)(616.2)} \frac{0.38 + \beta(1-0.475)}{1-0.475} = \frac{44.6 - 29.8}{1.987} + 0.108$$

But ΔH_{v293} is not known. Using $\Delta H_{v_b} = 8600$ cal/g-mol at $T_b = 411.5$ K ($T_{b_r} = 0.668$), with Eq. (5-8.19), we get

$$8600 = \Delta H_{v293}\left(\frac{1-0.668}{1-0.475}\right)^{0.38+\beta(1-0.668)}$$

Then β can be solved by a trial-and-error solution between these last two equations; if this is done, $\beta = 0.16$ and $\Delta H_{v293} = 10490$ cal/g-mol.

6. At 300°C ($T_r = 0.930$), from Eq. (5-8.22),

$$\frac{d}{dT_r}\frac{\Delta H_v}{RT_c} = -\left[\frac{0.38}{1-0.93} + 0.16\left(1 + \ln\frac{1-0.93}{1-0.475}\right)\right.$$

$$\left.\times \left[\frac{10490}{(1.987)(616.2)}\left(\frac{1-0.93}{1-0.475}\right)^{0.38+0.16(1-0.93)}\right]\right]$$

$$= -20.5$$

7. Calculate C_{σ_L} with Eq. (5-8.11):

$$\frac{C_{\sigma_L} - 53.3}{1.987} = -(-20.5) - (10.2)$$

$$C_{\sigma_L} = 73.9 \text{ cal/g-mol K}$$

$$\text{Error} = \frac{73.9 - 70.2}{70.2} \times 100 = 5.3\%$$

Discussion and Recommendations Five techniques for estimating liquid heat capacities as a function of temperature have been described. All methods require C_p° at the temperature of interest.

The simple Rowlinson-Bondi and Sternling-Brown corresponding-states relations [Eqs. (5-8.2) and (5-8.3)] require, in addition, only the critical temperature and acentric factor as input. Their accuracy is surprisingly good, as indicated by the values shown in Table 5-18 (which is only a subset of a larger test program [68]). Although they are not satisfactory for polar compounds at low temperatures, they are generally accurate to within 5 to 10 percent.

The Yuan-Stiel method [Eqs. (5-8.4) and (5-8.5)] also requires T_c, ω, and (if the fluid is polar) the Stiel polar factor. They report errors usually less than 5 percent, and this result has been confirmed by our testing. The ΔC_p functions in Table 5-15 are not readily expressed in analytical form, and a *table look-up* procedure with a five-point polynomial interpolation technique was used in our tests. The method is not applicable for reduced temperatures less than 0.4 or greater than about 0.95.

The corresponding-states method of Lyman and Danner [Eq. (5-8.7)] yields liquid heat capacities with the same range of accuracy as that of Yuan and Stiel, i.e., usually less than 5 percent.   To use this method, the radius of gyration must be available (Table 5-17) as well as the critical temperature.   With $\bar{R}$, the method is easily programmed for computer calculations.

The final method described is the Chueh-Swanson modification of the Watson thermodynamic-cycle approach [Eq. (5-8.11)].   It is more difficult to employ, and testing showed erratic results.   The critical temperature and pressure must be known as well as one liquid heat capacity and one enthalpy of vaporization.   A vapor-pressure–temperature correlation is also necessary.   In the application of this method, the required liquid heat capacity was always found from the Chueh-Swanson group-contribution method at 20°C (Table 5-12) and the required heat of vaporization from Appendix A.   Comparisons show that the method yields poor results at low temperatures ($<20°C$).   Chueh and Swanson [11] in their tests indicate errors were generally found to be less than 5 percent (3 percent for hydrocarbons).

It is difficult to evaluate the accuracy of liquid heat-capacity correlations.   Experimental data over wide temperature ranges are sparse, and data from different investigators often do not agree.

From these studies, it is *recommended* that the Yuan-Stiel or Lyman-Danner estimation methods be employed over the entire temperature range.   Errors should be less than 5 percent except at very high reduced temperatures.   These methods were developed for pure liquids.   For *liquid mixtures*, no specific correlations have been suggested, and few data are available.   In most cases, one is forced to assume that the mixture molar heat capacity is a mole-fraction average of the pure-component values [18].   This neglects any contribution due to the effect of temperature on heats of mixing.

For hydrocarbons only, the correlations of Hadden [36, 37] and Luria and Benson [53a] (Table 5-14) are normally accurate.

5-9 Vapor-Phase Fugacity of a Component in a Mixture

From thermodynamics, the chemical potential or fugacity can be related to the Gibbs or Helmholtz energies.   Using the latter gives

$$\mu_i \equiv \left(\frac{\partial A}{\partial N_i} \right)_{T,V,N_j[i]} \tag{5-9.1}$$

The subscripts on the partial derivative indicate that the temperature, *total* system volume, and all mole numbers (except i) are to be held constant.   Thus one may take the Helmholtz energy function for a

TABLE 5-18 Comparison between Calculated and Experimental Values of Liquid Heat Capacity

Compound	T, K	C_{σ_L} (exp.),[†] cal/g-mol K	Ref.	Rowlinson and Bondi, Eq. (5-8.2)	Sternling and Brown, Eq. (5-8.3)	Yuan and Stiel, Eqs. (5-8.4) and (5-8.5)	Lyman and Danner, Eq. (5-8.7)	Chueh and Swanson, Eq. (5-8.11)
Methane	102.3	13.1	99	16	5.6	2.8	1.1	
	140.5	14.3		16	0.5	1.0	−2.2	
	180.9	22.2		17	−8.3	−2.2	−2.3	
Propane	100	20.3	47	11	18	...	−24	
	150	21.0		7.8	7.8	6.7	11	
	200	22.3		6.8	3.8	3.8	3.3	−27
	305.3	27.5	72	9.8	9.0	4.7	6.7	9.2
	344.7	35.9		4.6	6.9	−0.9	0.1	12.3
n-Pentane	150	33.8	57	−5.2	−2.5	...	4.9	−23
	200	34.5		−1.4	−0.9	1.7	3.7	−11
	250	36.7		0.1	−0.2	0.7	0.3	−3.9
	300	40.1		1.2	1.3	0.1	−0.3	−0.2
	363	46.2	2	0.8	2.7	−0.2	0.2	−0.5
	443	65.3		−7.1	1.8	−8.2	−10	−8.6
n-Heptane	200	48.1	19, 57	−7.1	−5.6	...	2.4	−8.7
	300	53.9		−0.7	0	0.6	−0.1	2.8
	400	64.6	19	−0.3	2.4	0.1	0.1	2.3
	503	85.0	2	−4.5	5.6	−4.0	−4.9	−4.6
n-Decane	250	71.1	57	−6.2	−4.9	0.4	0.4	
	320	77.8		−2.1	−0.8	1.0	−0.3	
Cyclohexane	305.3	37.4	4	−0.5	−1.4	−1.1	−3.0	4.1
	360.9	43.0		0.6	0.1	−0.9	−3.2	4.6
cis-2-Butene	210.9	26.6	78	8.9	7.9	9.2	−1.0	−30

	305.3	31.4		8.3	7.8	5.8	−6.1	−3.9
	349.8	36.5		3.6	4.9	1.3	−7.7	1.7
Isopropylbenzene	305.3	50.7	78	−1.2	−0.7	1.7	2.3	9.8
	333.1	53.9		−1.4	−0.9	0.3	0.4	8.2
	366.4	58.1		−2.4	−1.8	−1.8	−1.6	5.9
Ethyl alcohol	208	21.8	45	69	74	. . .	−4.7	−32
	294	26.3		38	46	8.9	8.4	−7.7
	383	38.0	25	1.8	15	7.8	4.3	−7.7
Acetone	180	28.0	46	2.3	5.0	. . .	−0.4	−8.9
	209	28.2		2.0	3.4	. . .	−0.1	−4.5
	297	29.8		2.8	3.6	−5.6	0.3	2.6
Ethylene oxide	170	19.7	30	8.1	11	. . .	−2.0	6.7
	230	19.7		6.6	4.9	4.5	1.6	11
	280	20.7		4.3	2.1	0	−0.1	8.9
Ethyl ether	186	36.3	63	−8.5	−7.5	. . .	−4.8	−7.5
	214	37.6		−8.6	−8.3	−3.6	−7.1	−5.1
	290	40.8		−5.5	−5.0	−5.1	−6.5	−0.9
Ethyl chloride	150	23.0	33	−6.8	−3.3	. . .	−5.8	1.0
	200	22.9		−8.4	−8.7	. . .	0.9	3.5
	250	23.6		−11	−13	−7.7	−1.8	1.7
	290	24.7		−11	−13	−11	−1.7	−0.6
Ethyl mercaptan	154	27.0	56	−7.0	−3.4	. . .	−15	−32
	208	26.7		−3.8	−3.7	−2.7	−2.3	−19
	275	27.6		−0.9	−2.7	−2.3	−2.1	−7.7
	315	28.7		0.9	−0.6	−2.3	−0.6	−2.2
Chlorine	200	15.9	31	2.2	−2.6	−8.2	−7.7	
	240	15.7		4.9	−2.7	1.1	−5.2	

†When C_{pL} or C_{sat} was reported, it was converted to $C_{\sigma L}$. Likewise if an estimation method did not predict $C_{\sigma L}$ directly, the estimated value was also converted to $C_{\sigma L}$.

‡Percent error = [(calc. − exp.)/exp.] × 100.

TABLE 5-19   Fugacity Coefficient Expressions

$$\phi_i \equiv \frac{\hat{f}_i}{y_i P}$$

Original Redlich-Kwong [Eq. (3-5.1), Table 5-1, Sec. 4-3]

$$\ln \phi_i = \ln \frac{V}{V-b} + \frac{b_i}{V-b} - \ln Z + \frac{ab_i}{b^2 RT^{1.5}} \left(\ln \frac{V+b}{V} - \frac{b}{V+b} \right) - \frac{2 \sum_j y_j a_{ij}}{bRT^{1.5}} \ln \frac{V+b}{V}$$

Soave Modification of Redlich-Kwong [Eqs. (3-5.11) and (3-5.15), Tables 5-1 and 4-1]

$$\ln \phi_i = \frac{b_i}{b}(Z-1) - \ln Z + \ln \frac{V}{V-b} + \frac{a'(T)}{bRT}$$

$$\times \left\{ \frac{b_i}{b} - 2 \sum_j \frac{(1-\bar{k}_{ij})[a_i'(T)a_j'(T)]^{0.5}}{a'(T)} y_j \right\} \ln \frac{V+b}{V}$$

Barner-Adler

Not convenient in explicit form

Sugie-Lu [Tables 3-5 and 5-1, Sec. 4-5]

$$\ln \phi_i = \ln \frac{f}{P} + \frac{(b-c)\beta_i - 0.08 R\gamma_i \epsilon}{V-b+c} + \frac{a(\beta_i - \alpha_i)}{bRT^{1.5}} \ln \frac{V+b+c}{V+c}$$

$$+ \frac{a}{RT^{1.5}(V+b+c)} \left(\frac{c\beta_i + 0.08 R\gamma_i \epsilon}{V+c} - \beta_i \right)$$

$$+ \frac{1}{RT} \sum_{j=1}^{10} \frac{1}{jV^j} \left\{ \left(j\beta_i + \gamma_i \frac{d_{\omega j}^*}{d_j} \right) d_j T \right.$$

$$+ \left[\alpha_i + (j-1)\beta_i + \gamma_i \frac{e_{\omega j}^*}{e_j^*} \right] e_j T^{-0.5} \right\}$$

where

$$\alpha_i = \frac{2}{\delta} \sum_k y_k \delta_{ik} - \delta \qquad \beta_i = \frac{\epsilon_i - \epsilon}{\epsilon}$$

$$\gamma_i = \omega_i - \omega \qquad \delta = \sum_k y_i y_k \delta_{ik}$$

$$\delta_{ik} = \left(\frac{T_{c_i}^{2.5}}{P_{c_i}} \frac{T_{c_k}^{2.5}}{P_{c_k}} \right)^{0.5} (1 - k_{ik})$$

$$\epsilon = \sum_k y_k \epsilon_k \qquad \epsilon_k = \frac{T_{c_k}}{P_{c_k}}$$

$$d_{\omega j}^*, e_{\omega j}^* = \text{slopes of } d^* \text{ and } e^* \text{ functions in Table 3-5 with respect to } \omega$$

Lee-Erbar-Edmister [Tables 3-10, 4-4, and 5-1, Sec. 4-7]

$$\ln \phi_i = \frac{1}{bRT} \left[(2A_i' - aB_i' - bRT) \ln \frac{V-b}{V} \right.$$

$$+ \left(\frac{cB_i'}{2} - C_i' \right) \ln \left(1 - \frac{b^2}{V^2} \right) \right] + B_i'(Z-1) - \ln Z$$

where

$$B'_i = \frac{b_i}{b}$$

$$A'_i = a_i^{0.5} \sum_{k=1}^{n} y_k \alpha_{ik} a_k^{0.5}$$

$$C'_i = c_i^{0.5} \sum_{k=1}^{n} y_k \beta_{ik} c_k^{0.5}$$

Benedict-Webb-Rubin [Eq. (3-8.1), Sec. 4-6, Table 5-1]

$$\ln \phi_i = -\ln Z + \frac{1}{RT} [(B_0 - B_{0_i})RT - 2(A_0 A_{0_i})^{0.5} - 2(C_0 C_{0_i})^{0.5} T^{-2}] V^{-1}$$

$$+ \frac{3}{2RT} [RT(b^2 b_i)^{1/3} - (a^2 a_i)^{1/3}] V^{-2} + \frac{3}{5RT} \left[a(\alpha^2 \alpha_i)^{1/3} \right.$$

$$\left. + \alpha (a^2 a_i)^{1/3} \right] V^{-5} + \frac{1}{RT} 3 V^{-2} (c^2 c_i)^{1/3} T^{-2} \left(\frac{1 - e^{-\gamma V^{-2}}}{\gamma V^{-2}} - \frac{e^{-\gamma V^{-2}}}{2} \right)$$

$$- \frac{1}{RT} \frac{2 V^{-2} c}{T^2} \frac{\gamma_i}{\gamma} \left(\frac{1 - e^{-\gamma V^{-2}}}{\gamma V^{-2}} - e^{-\gamma V^{-2}} \frac{1 + \gamma V^{-2}}{2} \right)$$

Virial equation (only including B) [Secs. 3-11 and 4-8, Table 5-1]

$$\ln \phi_i = \left(2 \sum_j y_j B_{ij} - B \right) (P/RT)$$

particular equation of state, as shown in Table 5-1, and find μ_i by differentiation. Since the functions so given are expressed as the difference in *specific* Helmholtz energy between the real state and the chosen reference state, one must multiply the entire expression by N, the total moles, before differentiating. Then

$$\mu_i - \mu_i^\circ = \frac{\partial}{\partial N_t} (\underline{A} - \underline{A}^\circ)_{T,V,N_{j[i]}} \qquad (5\text{-}9.2)$$

But this difference in chemical potential is related to fugacity by

$$\mu_i - \mu_i^\circ = RT \ln \frac{\hat{f}_i}{\hat{f}_i^\circ} \qquad (5\text{-}9.3)$$

The fugacity $\hat{f}_i$ refers to the value for component i in the mixture and $\hat{f}_i^\circ$ to the reference state at T, P°, V°, N. However, the reference state was previously chosen as an ideal-gas state, and for such cases, as a part of the definition of fugacity

$$\hat{f}_i^\circ = P^\circ y_i \qquad (5\text{-}9.4)$$

Thus

$$RT \ln \frac{\hat{f}_i}{P^\circ y_i} = \frac{\partial}{\partial N_i} (\underline{A} - \underline{A}^\circ)_{T,V,N_{j[i]}} \tag{5-9.5}$$

and from Eq. (5-3.5), multiplying by N, we have

$$\underline{A} - \underline{A}^\circ = -\int_\infty^V \left(P - \frac{NRT}{V} \right) dV - NRT \ln \frac{V}{V^\circ} \tag{5-9.6}$$

With Eqs. (5-9.5) and (5-9.6), noting that $\underline{V} = ZNRT/P$, $\underline{V}^\circ = NRT/P^\circ$, we have

$$RT \ln \frac{\hat{f}_i}{Py_i} = -\int_\infty^V \left[\left(\frac{\partial P}{\partial N_i} \right)_{T,V,N_{j[i]}} - \frac{RT}{V} \right] dV - RT \ln Z \tag{5-9.7}$$

$$= RT \ln \phi_i \tag{5-9.8}$$

ϕ_i is the fugacity coefficient of i in the gas mixture.

To obtain a usable relation for ϕ_i, Eq. (5-9.7) must be integrated, but before this can be done, the derivative of P with respect to N_i must be found. Thus any pressure-explicit equation of state is convenient provided that the composition dependence of all the parameters can be expressed in analytical form.

For the analytical equations of state covered in Chap. 3, mixture combining rules are given in Chap. 4; thus evaluation of the integral in Eq. (5-9.7) is possible. For example, the original Redlich-Kwong equation is given as Eq. (3-5.1), and this same relation expressed in terms of total volume would be

$$P = \frac{NRT}{V - Nb} - \frac{aN^2}{T^{1/2}V(V + Nb)} \tag{5-9.9}$$

In the differentiation indicated in Eq. (5-9.7), the variables are N, a, and b, where the parameters a and b are shown as functions of composition in Eqs. (4-3.1), (4-3.2), (3-5.7), and (3-5.8). The final result is

$$\ln \phi_i = \ln \frac{V}{V - b} + \frac{b_i}{V - b} - \ln Z + \frac{ab_i}{RT^{3/2}b^2} \left(\ln \frac{V + b}{V} - \frac{b}{V + b} \right)$$

$$- \frac{2 \sum_j y_j a_{ij}}{RT^{3/2}b} \ln \frac{V + b}{V} \tag{5-9.10}$$

For all the analytical equations of state, the working equations for $\ln \phi_i$ are given in Table 5-19.

It is difficult to evaluate these expressions since fugacity coefficients themselves are difficult to determine. Presumably, if the mixture equation of state is a valid representation, the derived property ϕ_i should also be accurate. This assumes, of course, that the equation of state not only yields accurate predictions of volumetric properties but also that it yields accurate derivatives of pressure with respect to mole numbers.   These two attributes are not necessarily compatible.

Vapor-phase fugacity expressions are of value in vapor-liquid equilibrium calculations, as described in Chap. 8.

NOTATION

a = Redlich-Kwong constant, Eq. (3-5.1)

A = Helmholtz energy, cal/g-mol

$\underline{A}$ = Helmholtz energy, cal

b = Redlich-Kwong constant, Eq. (3-5.1)

C = heat capacity, cal/g-mol K; C_p, at constant pressure; C_v, at constant volume; $C_{\sigma L}$, variation of saturated liquid enthalpy with temperature; C_{sat}, $(dQ/dT)_{SL}$

f = fugacity, atm; $\hat{f}_i$, fugacity of i in a mixture

G = Gibbs energy, cal/g-mol

H = enthalpy, cal/g-mol

ΔH_v = enthalpy of vaporization, cal/g-mol

M = molecular weight

N = total moles

P = pressure, usually atm; P_c, at the critical point; P_r, P/P_c

Q = heat, cal

R = gas constant, 82.04 atm cm³/g-mol K; 8314 J/kg-mol K; 1.986 cal/g-mol K

$\bar{R}$ = radius of gyration, Å

S = entropy, cal/g-mol K

T = temperature, K; T_c, at the critical point; T_r, T/T_c; T_b, at the normal boiling point

U = internal energy, cal/g-mol

V = volume, cm³/g-mol; V_c, at the critical point; V_r, V/V_c

$\underline{V}$ = volume, cm³

V^* = Kreglewski volume, cm³/g-mol

V_s = velocity of sound, cm/s or m/s

x_i = mole fraction i

y_i = mole fraction i

Z = compressibility factor; Z_c, at the critical point; Z_p, Z_T, derivative compressibility factors in Eqs. (5-5.5) and (5-5.6)

Greek

β = parameter in Eq. (5-8.19)

δ_T, δ_v = parameters in Eqs. (5-7.7) and (5-7.11)

θ = surface fraction, Eq. (5-7.3)

κ = association parameter, Eq. (5-8.8)

μ = chemical potential or Joule-Thompson coefficient $(\partial T/\partial P)_H$

ν_{ij} = interaction parameter in Eq. (5-7.8)

τ_{ij} = interaction parameter in Eq. (5-7.4)

ϕ = volume fraction defined in Eqs. (5-7.1) and (5-7.15)

ϕ_i = fugacity coefficient, $\hat{f}_i/Py_i$

ψ_T, ψ_v = parameters in Eqs. (5-7.6) and (5-7.10)

ω = Pitzer acentric factor

Superscripts

$°$ = reference state or an ideal-gas state

(0) = simple fluid function
(R) = simple fluid function for the reference fluid
R = reference fluid
SL = saturated liquid
SV = saturated vapor
L = liquid
EX = excess function
(1) = deviation function

Subscripts

b = normal boiling point
c = critical state
SCL = subcooled liquid
SL = saturated liquid
SV = saturated vapor
m = mixture
T = true critical property of mixture
σ_L = saturated-liquid state

REFERENCES

1. Akiyama, T., and G. Thodos: *Can. J. Chem. Eng.*, **48**: 311 (1970).
2. Amirkhanov, Kh. I., B. G. Alibekov, D. I. Vikhrov, V. A. Mirskaya, and L. N. Levina: *Teplofiz. Vys. Temp.*, **9**: 1211 (1971).
3. Amirkhanov, Kh. I., B. G. Alibekov, D. I. Vikhrov, V. A. Mirskaya, and L. N. Levina: *Teplofiz. Vys. Temp.*, **9**: 1310 (1971).
4. Auerbach, C. E., B. H. Sage, and W. N. Lacey: *Ind. Eng. Chem.*, **42**: 110 (1950).
5. Barner, H. E., and S. B. Adler: *Ind. Eng. Chem.*, **59**(7): 60 (1967).
6. Bauer, H., *Ind. Eng. Chem. Fundam.*, **13**: 286 (1967).
7. Bier, K., G. Ernst, J. Kunze, and G. Maurer: *J. Chem. Thermodyn.*, **6**: 1039 (1974).
8. Bondi, A.: *Ind. Eng. Chem. Fundam.*, **5**: 443 (1966).
9. Bondi, A.: "Physical Properties of Molecular Crystals, Liquids and Glasses," Wiley, New York, 1968.
10. Breedveld, G. J. F., and J. M. Prausnitz: *AIChE J.*, **19**: 783 (1973).
11. Chueh, C. F., and A. C. Swanson: (*a*) *Chem. Eng. Prog.*, **69**(7): 83 (1973); (*b*) *Can. J. Chem. Eng.*, **51**: 596 (1973).
12. Chueh, P. L., and C. H. Deal: "Thermophysical Properties of Pure Chemical Compounds," paper presented at *65th Ann. AIChE Meet., New York, November 1972.*
13. Chueh, P. L., and C. H. Deal: *AIChE J.*, **19**: 138 (1973).
14. Chueh, P. L., and J. M. Prausnitz: *AIChE J.*, **13**: 1099 (1967).
15. Corruccini, R. J., and D. C. Ginnings: *J. Am. Chem. Soc.*, **69**: 2291 (1947).
16. Daubert, T. E.: private communication, 1975.
17. Dillard, D. D., W. C. Edmister, J. H. Erbar, and R. L. Robinson, Jr.: *AIChE J.*, **14**: 923 (1968).
18. Dimoplon, W.: *Chem. Eng.*, **74**(22): 64 (1972).
19. Douglas, T. B., G. T. Furukawa, R. E. McCoskey, and A. F. Ball: *J. Res. Bur. Stand.*, **53**: 139 (1954).
20. Edmister, W. C.: *Petrol. Refiner*, **27**(11): 609 (1948).
21. Edmister, W. C.: *Petrol. Refiner*, **28**(1): 128 (1949).
22. Edmister, W. C.: *Petrol. Refiner*, **37**(7): 153 (1957).
22a. Edmister, W. C.: *Hydrocarbon Process.*, **46**(3): 155 (1967).
23. Edmister, W. C.: *Hydrocarbon Process.*, **46**(4): 165 (1967).
24. Edmister, W. C.: *Hydrocarbon Process.*, **46**(5): 187 (1967).

25. Fick, E. F., D. C. Ginnings, and W. B. Holten: *J. Res. Bur. Stand.*, **6**: 881 (1931).
26. Fishtine, S. H.: *Ind. Eng. Chem.*, **55**(4): 20, **55**(5): 49, **55**(6): 47 (1963); *Hydrocarbon Process. Pet. Refiner*, **42**(10): 143 (1963).
27. Gambill, W. R.: *Chem. Eng.*, **64**(5): 263, **64**(6): 243, **64**(7); 263; **64**(8): 257 (1957).
28. Garcia-Rangel, S., and L. C. Yen: Evaluation of Generalized Correlations for Mixture Enthalpy Predictions, paper presented at *159th Natl. Meet., Am. Chem. Soc., Houston, Tex., February 1970.*
29. Ghormley, E. L., and J. M. Lenoir: *Can. J. Chem. Eng.*, **50**: 89 (1972).
30. Giauque, W. F., and J. Gordon: *J. Am. Chem. Soc.*, **71**: 2176 (1949).
31. Giauque, W. F., and T. M. Powell: *J. Am. Chem. Soc.*, **61**: 1970 (1939).
32. Ginnings, D. C., and R. J. Corruccini: *Ind. Eng. Chem.*, **40**: 1990 (1948).
33. Gordon, J., and W. F. Giauque: *J. Am. Chem. Soc.*, **70**: 1506 (1948).
34. Grieves, R. B., and G. Thodos: *AIChE J.*, **9**: 25 (1963).
35. Gunn, R. D., P. L. Chueh, and J. M. Prausnitz: *AIChE J.*, **12**: 937 (1966).
36. Hadden, S. T.: *Hydrocarbon Process Pet. Refiner*, **45**(7): 137 (1966).
37. Hadden, S. T.: *J. Chem. Eng. Data*, **15**: 92 (1970).
38. Hsi, C., and B. C.-Y. Lu: *AIChE J.*, **20**: 616 (1974).
39. Huang, P. K., and T. E. Daubert: *Ind. Eng. Chem. Process Des. Dev.*, **13**: 359 (1974).
40. Joffe, J.: *Ind. Eng. Chem. Fundam.*, **12**: 259 (1973).
41. Joffe, J., and D. Zudkevitch: paper presented at *159th Natl. Meet., Am. Chem. Soc., Houston, Tex., February 1970.*
42. Johnson, A. I., and C. J. Huang: *Can. J. Technol.*, **33**: 421 (1955).
43. Kay, W. B., D. W. Hissong, and A. Kreglewski: *Ohio State Univ. Res. Found. Rep. 3, API Proj.* PPC 15.8, 1968.
44. Kay, W. B., and T. D. Nevens: *Chem. Eng. Prog. Symp. Ser.*, **48**(3): 108 (1952).
45. Kelley, K. K.: *J. Am. Chem. Soc.*, **51**: 770 (1929).
46. Kelley, K. K.: *J. Am. Chem. Soc.*, **51**: 1145 (1929).
47. Kemp, J. D., and C. J. Egan: *J. Am. Chem. Soc.*, **60**: 1521 (1938).
48. Kreglewski, A.: *J. Phys. Chem.*, **73**: 608 (1969).
49. Kreglewski, A., and W. B. Kay: *J. Phys. Chem.*, **73**: 3359 (1969).
50. Lee, B. I., and W. C. Edmister: *AIChE J.*, **15**: 615 (1969).
51. Lee, B. I., and M. G. Kesler: *AIChE J.*, **21**: 510 (1975).
52. Li, C. C.: *Can. J. Chem. Eng.*, **19**: 709 (1971).
53. Lu, B. C.-Y., C. Hsi, and D. P. L. Poon: *Chem. Eng. Prog. Symp. Ser.*, **70**(140): 56 (1974).
53a. Luria, M., and S. W. Benson: *J. Chem. Eng. Data* **21**, (1976).
54. Lydersen, A. L., R. A. Greenkorn, and O. A. Hougen: Generalized Thermodynamic Properties of Pure Fluids, *Univ. Wisconsin Coll. Eng., Eng. Exp. Stn. Rep.* 4, Madison, 1955.
55. Lyman, T. J., and R. P. Danner: *AIChE J.*, **22**: 759 (1976).
56. McCullough, J. P., D. W. Scott, H. L. Finke, M. E. Gross, K. D. Williamson, R. E. Pennington, G. Waddington, and H. M. Huffman: *J. Am. Chem. Soc.*, **74**: 2801 (1952).
57. Messerly, J. F., G. B. Guthrie, S. S. Todd, and H. L. Finke: *J. Chem. Eng. Data*, **12**: 338 (1967).
58. Missenard, F.-A.: *C. R.*, **260**: 5521 (1965).
59. Modell, M., and R. C. Reid: "Thermodynamics and Its Applications in Chemical Engineering," chap. 5, Prentice-Hall, Englewood Cliffs, N.J., 1974.
60. Nathan, D. I.: *Br. Chem. Eng.*, **12**(2): 223 (1967).
61. O'Connell, J. P.: private communication, 1975.
62. Orentlicher, M., and J. M. Prausnitz: *Can. J. Chem. Eng.*, **45**: 78 (1967).
63. Parks, G. S., and H. M. Huffman: *J. Am. Chem. Soc.*, **48**: 2788 (1926).
64. Passut, C. A., and R. P. Danner: *Chem. Eng. Prog. Symp. Ser.*, **70**(140): 30 (1974).
65. Pitzer, K. S., and R. F. Curl: *J. Am. Chem. Soc.*, **77**: 3427 (1955).
66. Pitzer, K. S., D. Z. Lippman, R. F. Curl, C. M. Huggins, and D. E. Petersen: *J. Am. Chem. Soc.*, **77**: 3433 (1955).

67. Powell, T. M., and W. F. Giauque: *J. Am. Chem. Soc.*, **61**: 2366 (1939).
68. Reid, R. C., and J. E. Sobel: *Ind. Eng. Chem. Fundam.*, **4**: 328 (1965).
69. Reid, R. C., and J. R. Valbert: *Ind. Eng. Chem. Fundam.*, **1**: 292 (1962).
70. Rowlinson, J. S.: "Liquids and Liquid Mixtures," 2d ed., Butterworth, London, 1969.
71. Ruf, J. F., F. Kurata, and T. F. McCall: *Ind. Eng. Chem. Process Des. Dev.*, **12**: 1 (1973).
72. Sage, B. H., and W. N. Lacey: *Ind. Eng. Chem.*, **27**: 1484 (1935).
73. Sage, B. H., J. G. Schaafsma, and W. N. Lacey: *Ind. Eng. Chem.*, **26**: 1218 (1934).
74. Sakiadis, B. C., and J. Coates: *AIChE J.*, **2**: 88 (1956).
75. San Jose, J.: Ph.D. thesis, Department of Chemical Engineering, Massachusetts Institute of Technology, Cambridge, Mass., 1975.
76. Sass, A., B. F. Dodge, and R. H. Bretton: *J. Chem. Eng. Data*, **12**: 169 (1967).
77. Schick, L. M., and J. M. Prausnitz: *AIChE J.*, **14**: 673 (1968).
78. Schlinger, W. G., and B. H. Sage: *Ind. Eng. Chem.*, **44**: 2454 (1952).
79. Sehgal, I. J. S., V. F. Yesavage, A. E. Mather, and J. E. Powers: *Hydrocarbon Process.*, **47**(8): 137 (1968).
80. Shah, K. K., and G. Thodos: *Ind. Eng. Chem.*, **57**(3): 30 (1965).
81. Shaw, R.: *J. Chem. Eng. Data*, **14**: 461 (1969).
82. Sherwood, T. K.: *J. Chem. Eng. Data*, **7**: 47 (1962).
83. Silverman, E. D., and G. Thodos: *Ind. Eng. Chem. Fundam.*, **1**: 299 (1962).
84. Spencer, C. F., T. E. Daubert, and R. P. Danner: "Technical Data Book," chap. 4, Critical Properties, American Petroleum Institute, OP72, 539, Xerox University Microfilm.
85. Spencer, C. F., T. E. Daubert, and R. P. Danner: *AIChE J.*, **19**: 522 (1973).
86. Starling, K. E., and J. E. Powers, Enthalpy of Mixtures by Modified BWR Equation, paper presented at *150th Meet., Am. Chem. Soc., Houston, Tex., February 1970*.
87. Stein, F. P., and J. J. Martin: *Chem. Eng. Prog. Symp. Ser.*, **59**(44): 112 (1963).
88. Stipp, G. K., S. D. Bai, and L. I. Stiel: *AIChE J.*, **19**: 1227 (1973).
89. Sutton, J. R.: "Advances in Thermophysical Properties at Extreme Temperatures and Pressures," p. 76, *ASME*, New York, 1965.
90. Tamplin, W. S., and D. A. Zuzic: *Hydrocarbon Process.*, **46**(8): 145 (1967).
91. Tao, L. C.: *AIChE J.*, **15**: 362, 469 (1969).
92. Tarakad, R., and T. E. Daubert: API-5-74, Pennsylvania State University, University Park, Sept. 23, 1974.
93. Thompson, W. H.: A Molecular Association Factor for Use in the Extended Theory of Corresponding States, Ph.D. thesis, Pennsylvania State University, University Park, 1966.
94. Tully, P. C., and W. C. Edmister: *AIChE J.*, **13**: 155 (1967).
95. Watson, K. M.: *Ind. Eng. Chem.*, **35**: 398 (1943).
96. Watson, L. M., and B. F. Dodge: *Chem. Eng. Prog. Symp. Ser.*, **48**(3): 73 (1952).
97. West, E. W., and J. H. Erbar: An Evaluation of Four Methods of Predicting Properties of Light Hydrocarbon Systems, paper presented at *NGPA 52d Ann. Meet., Dallas, Tex., March 1973*.
98. White, M. G., R. A. Greenkorn, and K. C. Chao: *Ind. Eng. Chem. Process Des. Dev.*, **13**: 453 (1974).
99. Wiebe, R., and M. J. Brevoort: *J. Am. Chem. Soc.*, **52**: 622 (1930).
100. Wiener, L. D.: *Hydrocarbon Process.*, **46**(4): 131 (1967).
101. Wilson, G. M.: *Adv. Cryog. Eng.*, **11**: 392 (1966).
102. Wolfe, J. F.: *J. Pet. Technol.*, March **1966**: 364.
103. Yen, L. C., and R. E. Alexander: *AIChE J.*, **11**; 334 (1965).
104. Yesavage, V. F., A. E. Mather, D. L. Katz, and J. E. Powers: *Ind. Eng. Chem.*, **59**(11): 35 (1967).
105. Yuan, T.-F., and L. I. Stiel: *Ind. Eng. Chem. Fundam.*, **9**: 393 (1970).

Chapter Six

Vapor Pressures and Enthalpies of Vaporization of Pure Fluids

6-1 Scope

This chapter covers methods for estimating and correlating vapor pressures of pure liquids. Since enthalpies of vaporization are often derived from vapor-pressure–temperature data, the estimation of this property is also included.

6-2 Theory and Corresponding-States Correlations

When the vapor phase of a pure fluid is in equilibrium with its liquid phase, the equality of chemical potential, temperature, and pressure in both phases leads to the Clausius-Clapeyron equation

$$\frac{dP_{vp}}{dT} = \frac{\Delta H_v}{T \, \Delta V_v} = \frac{\Delta H_v}{(RT^2/P_{vp}) \, \Delta Z_v} \qquad (6\text{-}2.1)$$

or

$$\frac{d \ln P_{vp}}{d(1/T)} = - \frac{\Delta H_v}{R \, \Delta Z_v} \qquad (6\text{-}2.2)$$

Most vapor-pressure estimation and correlation equations stem from an integration of Eq. (6-2.2).   When this is done, an assumption must be made regarding the dependence of the group $\Delta H_v / \Delta Z_v$ on temperature, and in the integration a constant is obtained which must be evaluated using one vapor-pressure–temperature point.

The simplest approach is to assume that the group $\Delta H_v / R \, \Delta Z_v$ is constant and independent of temperature.   Then, with the constant of integration denoted as A, Eq. (6-2.2) becomes

$$\ln P_{\rm vp} = A - \frac{B}{T} \qquad \text{where } B = \Delta H_v / R \, \Delta Z_v \qquad (6\text{-}2.3)$$

Equation (6-2.3) is sometimes called the *Clapeyron equation*.   Surprisingly, it is a fairly good relation for approximating vapor pressure over small temperature intervals.   Except near the critical point, ΔH_v and ΔZ_v are both weak functions of temperature, and since both decrease with an increase in temperature, the result is a compensatory effect.   However, over large temperature ranges, Eq. (6-2.3) normally represents vapor-pressure data poorly.   This is shown in Fig. 6-1.   The ordinate is the ratio $[P \text{ (exp.)} - P \text{ (calc.)}]/P \text{ (exp.)}$ and the abscissa $T_r = T/T_c$. P (calc.) is obtained from Eq. (6-2.3), where the constants A and B are found from experimental data at $T_r = 0.7$ and 1.0.   If Eq. (6-2.3) were an exact correlating equation, a horizontal line of value zero would be found.   At high reduced temperatures, the fit is reasonably good for oxygen and a typical hydrocarbon, 2,2,4-trimethylpentane, but poor for an associating liquid, n-butanol. Ambrose [2] points out the complexity of the curves in this figure and notes that to represent the changes in curvature that are evident, at least a four-constant vapor-pressure equation would be necessary.   Also, it is important to note that there is usually a change in curvature between T_r of 0.8 to 0.85; this fact is utilized in several later developments.

Extending our consideration of Eq. (6-2.3) one step further, a common practice is to use both the normal boiling point and the critical point to obtain generalized constants.   Expressing pressures in atmospheres and temperatures on the absolute scale (kelvins or degrees Rankine), with $P = P_c$, $T = T_c$ and $P = 1$, $T = T_b$, Eq. (6-2.3) becomes

$$\ln P_{\rm vp_r} = h \left(1 - \frac{1}{T_r} \right) \qquad (6\text{-}2.4)$$

$$h = T_{b_r} \frac{\ln P_c}{1 - T_{b_r}} \qquad (6\text{-}2.5)$$

As noted in Fig. 6-1, the linear form of $\ln P_{\rm vp}$ vs. $1/T$ is not satisfactory for associating materials.   Equation (6-2.3) generally overpredicts vapor pressures below T_b (see Fig. 6-1 or Table 6-1).

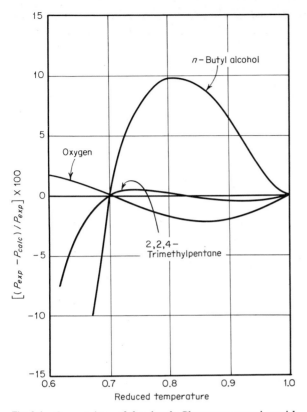

Fig. 6-1   Comparison of the simple Clapeyron equation with experimental vapor-pressure data. (*Adapted from Ref. 2.*)

Equation (6-2.4) is an example of a two-parameter corresponding-states correlation for vapor pressure. To achieve more accuracy, several investigators have proposed three-parameter forms. The Pitzer expansion is one of the more successful:

$$\ln P_{vp_r} = f^{(0)}(T_r) + \omega f^{(1)}(T_r) \tag{6-2.6}$$

The functions $f^{(0)}$ and $f^{(1)}$ have been tabulated over wide ranges in reduced temperature [15, 29, 36, 68] and expressed in analytical form by Lee and Kesler [43] as follows:

$$f^{(0)} = 5.92714 - \frac{6.09648}{T_r} - 1.28862 \ln T_r + 0.169347 T_r^6 \tag{6-2.7}$$

$$f^{(1)} = 15.2518 - \frac{15.6875}{T_r} - 13.4721 \ln T_r + 0.43577 T_r^6 \tag{6-2.8}$$

Values of the acentric factor ω are tabulated in Appendix A for many fluids, but when Eq. (6-2.6) is employed, it is recommended that Eq. (2-3.4) be used to compute ω. The latter equation was obtained from Eq. (6-2.6) with $T_r = T_{b_r}$. The Lee-Kesler form of the Pitzer equation generally predicts vapor pressures within 1 to 2 percent between T_b and T_c. Below T_b, it may underpredict P_{vp} by several percent.

As can readily be verified, Eqs. (6-2.6) to (6-2.8) satisfy the definition of the acentric factor [Eq. (2-3.1)].

Example 6-1   Estimate the vapor pressure of ethylbenzene at 74.1 and 186.8°C using both Eq. (6-2.4) and the Lee-Kesler relations.   Experimental values are 100 mm Hg [94] and 2494 mm Hg [3], respectively.

solution   From Appendix A, $T_b = 409.3$ K, $T_c = 617.1$ K, and $P_c = 35.6$ atm.

Equation (6-2.4).   First h is determined from Eq. (6-2.5), with $T_{b_r} = 409.3/617.1 = 0.663$,

$$h = 0.663 \frac{\ln 35.6}{1 - 0.663} = 7.028$$

Then

$$\ln P_{vp_r} = 7.028(1 - T_r^{-1})$$

Lee-Kesler.   With $T_{b_r} = 0.663$, from Eq. (2-3.4), $\omega = 0.299$.   Then, with Eq. (6-2.6),

$$\ln P_{vp_r} = f^{(0)}(T_r) + 0.299 f^{(1)}(T_r)$$

For the cases considered,

		Eq. (6-2.4)			Lee-Kesler	
T, °C	T_r	P (exp.), mm Hg	P (calc.), mm Hg	% error	P (calc.), mm Hg	% error
74.1	0.563	100	116	16	99.2	−0.8
186.8	0.745	2494	2441	−2.1	2515	0.8

The error was calculated as $[(\text{calc.} - \text{exp.})/\text{exp.}] \times 100$.

6-3   Antoine Vapor-Pressure Correlation

Antoine [7] proposed a simple modification of Eq. (6-2.3) which has been widely used,

$$\ln P_{vp} = A - \frac{B}{T + C} \tag{6-3.1}$$

When $C = 0$, Eq. (6-3.1) reverts to the Clapeyron equation (6-2.3).

Simple rules have been proposed [23, 87] to relate C to the normal boiling point for certain classes of materials, but these rules are not reliable and the only way to obtain values of the constants is to regress experimental data [13, 38, 44, 49, 73, 86].

We tabulate values of A, B, and C for many materials in Appendix A for P_{vp} in millimeters of mercury and T in kelvins.   The applicable temperature range is not large and in most instances corresponds to a pressure interval of about 10 to 1500 mm Hg.   The equation should *not* be used outside the stated temperature limits.

Cox [19] suggested a graphical correlation in which the ordinate, representing P_{vp}, is a log scale, and a straight line (with a positive slope) is drawn.   The sloping line is taken to represent the vapor pressure of water (or some other reference substance).   Since the vapor pressure of water is accurately known as a function of temperature, the abscissa scale can be marked in temperature units.   When the vapor-pressure and temperature scales are prepared in this way, vapor pressures for other compounds are found to be nearly straight lines.   Calingaert and Davis [14] have shown that the temperature scale on this *Cox chart* is nearly equivalent to the function $(T + C)^{-1}$, where C is approximately -43 K for many materials boiling between 0 and 100°C.   Thus the Cox chart closely resembles a plot of the Antoine vapor-pressure equation.   Also, for homologous series, a useful phenomenon is often noted on Cox charts.   The straight lines for each member of the homologous series usually converge to a point when extrapolated.   This point, called the *infinite point*, is useful for providing one value of vapor pressure for a new member of the series.   Dreisbach [21] presents a tabulation of these infinite points for many homologous series.

Example 6-2   Estimate the vapor pressure of acetone at 273.4 K using the Antoine equation.

   solution   From   Appendix   A,   ANTA $= 16.6513$,   ANTB $= 2940.46$,   and   ANTC $= -35.93$.   With Eq. (6-3.1),

$$\ln P_{vp} = 16.6513 - \frac{2940.46}{273.4 - 35.93} = 71.4 \text{ mm Hg}$$

The experimental value is 71.2 mm Hg [6] and

$$\text{Error} = \frac{71.4 - 71.2}{71.2} \times 100 = 0.3\%$$

Usually, in the range of 10 to 1500 mm Hg, the Antoine equation provides an excellent correlating equation for vapor pressures; above 1500 mm Hg, the equation normally underpredicts vapor pressures.

6-4   Integration of the Clausius-Clapeyron Equation

In Sec. 6-2 we emphasized that most vapor-pressure estimation equations result from an integration of Eq. (6-2.2).   The options one has in this integration are limited, but the results available in the literature vary widely as each author normally introduces correction terms to obtain more accuracy.

Of the two terms ΔH_v and ΔZ_v which must be expressed as a function of temperature (or pressure), let us consider the latter first.   With the subscripts SV for saturated vapor and SL for saturated liquid, ΔZ_v is defined as

$$\Delta Z_v = Z_{SV} - Z_{SL}$$

$$= \frac{P}{RT}(V_{SV} - V_{SL}) \tag{6-4.1}$$

ΔZ_v is often assumed equal to unity, is approximated by a simple analytical function, or is estimated from an equation of state that is assumed to apply to both the saturated-liquid and vapor phases (Chap. 3).

One of the more often quoted estimation methods for ΔZ_v is due to Haggenmacher [27]. An equation of state (similar to that of van der Waals) is assumed, and both saturated-liquid and vapor volumes are calculated. Generalizing the equation with the aid of Eqs. (3-4.1) and (3-4.2), it can be shown that

$$\Delta Z_v = \left(1 - \frac{P_r}{T_r^3}\right)^{1/2} \tag{6-4.2}$$

Equation (6-4.2) provides a good approximation at reduced temperatures near or below T_{b_r}, but it should be used with caution when $T_r > T_{b_r}$.

To express ΔH_v as a function of temperature, either one of two approaches is normally used. ΔH_v is expressed as a linear function of temperature, or the Watson correlation [Eq. (6-16.1)] is assumed. The first technique is only approximate and not valid at high temperatures. However, if ΔZ_v is set equal to a constant and

$$\Delta H_v = a + bT \tag{6-4.3}$$

then, with Eq. (6-2.2),

$$\ln P_{vp} = A + \frac{B}{T} + C \ln T \tag{6-4.4}$$

The $\ln T$ term arises from the assumed linear dependence of ΔH_v on T. While Eq. (6-4.4) is of little direct use, the general form appears in a number of correlations, frequently modified by the addition of other terms. Equation (6-4.4) is called the *Rankine* or *Kirchhoff vapor-pressure equation*.

6-5 Riedel's Vapor-Pressure Equation

Riedel [77] proposed a vapor-pressure correlation based upon a modification of Eq. (6-4.4) [69]:

$$\ln P_{vp} = A + \frac{B}{T} + C \ln T + DT^6 \tag{6-5.1}$$

The T^6 term is presumably included to account for the fact that ΔZ_v is not unity at high temperatures (nor is ΔH_v a linear function of temperature in that region). The power 6 is not significant, and other similar values could be used [1, 39] without affecting the accuracy, although the values of the constants are necessarily different.

To determine the constants in Eq. (6-5.1), Riedel defined a parameter α

$$\alpha \equiv \frac{d \ln P_{vp_r}}{d \ln T_r} \tag{6-5.2}$$

and Plank and Riedel [69, 70] showed, from a study of experimental vapor-pressure data, that

$$\frac{d\alpha}{dT_r} = 0 \qquad \text{at } T_r = 1 \tag{6-5.3}$$

Let α_c stand for α at the critical point.   If one assumes that the constant D is related to α_c, using the critical state as a datum, Riedel found that

$$\ln P_{vp_r} = A^+ - \frac{B^+}{T_r} + C^+ \ln T_r + D^+ T_r^6 \tag{6-5.4}$$

where $\qquad A^+ = -35Q \qquad B^+ = -36Q \qquad C^+ = 42Q + \alpha_c$
$$D^+ = -Q \qquad Q = 0.0838(3.758 - \alpha_c) \tag{6-5.5}$$

Knowing values of P_c, T_c, and α_c, one can use Eqs. (6-5.4) and (6-5.5) to estimate vapor pressures.   Note that the normal boiling point was *not* used.   However, since it is not easy (or even desirable) to determine α_c by its defining equation at the critical point, α_c is usually found from Eqs. (6-5.4) and (6-5.5) by inserting $P = 1$ atm, $T = T_b$ and calculating α_c.   When this is done, we obtain

$$\alpha_c = \frac{0.315\psi_b + \ln P_c}{0.0838\psi_b - \ln T_{r_b}} \tag{6-5.6}$$

$$\psi_b = -35 + \frac{36}{T_{b_r}} + 42 \ln T_{b_r} - T_{b_r}^6 \tag{6-5.7}$$

Example 6-3   Repeat Example 6-1 using the Riedel correlation.
  solution   As given in Example 6-1, for ethylbenzene, $P_c = 35.6$ atm, $T_b = 409.3$ K, and $T_c = 617.1$ K.   Thus $T_{b_r} = 409.3/617.1 = 0.663$.   From Eq. (6-5.7),

$$\psi_b = -35 + \frac{36}{0.663} + 42 \ln 0.663 - (0.663)^6 = 1.9525$$

Then, with Eq. (6-5.6),

$$\alpha_c = \frac{(0.315)(1.9525) + \ln 35.6}{(0.0838)(1.9525) - \ln 0.663}$$

$$= 7.2880$$

The constants for Riedel's vapor-pressure correlation are found from Eqs. (6-5.5):

$$Q = (0.0838)(3.758 - 7.2880) = -0.2958$$
$$A^+ = -35Q = 10.353 \qquad B^+ = -36Q = 10.649$$
$$C^+ = 42Q + \alpha_c = -5.136 \qquad D^+ = -Q = 0.2958$$
$$\ln P_{vp_r} = 10.353 - \frac{10.649}{T_r} - 5.136 \ln T_r + 0.2958 T_r^6$$

At 74.1 and 186.8°C,

T, °C	T_r	P (calc.), mm Hg	P (exp.), mm Hg	$\dfrac{P\text{ (calc.)} - P\text{ (exp.)}}{P\text{ (exp.)}} \times 100$
74.1	0.563	99.9	100	−0.1
186.8	0.745	2509	2494	0.6

Riedel's equation is generally more accurate for $T_b > T_c$ (see Table 6-1).

6-6 Frost-Kalkwarf-Thodos Vapor-Pressure Equation

Frost and Kalkwarf[25] also integrated Eq. (6-2.2) assuming ΔH_v to be given by Eq. (6-4.3); but instead of using $\Delta Z_v \approx 1.0$ they estimated ΔZ_v from van der Waals' equation of state. The result is only slightly different from the Riedel form, Eq. (6-5.1):

$$\ln P_{vp} = A + \frac{B}{T} + C \ln T + \frac{DP_{vp}}{T^2} \tag{6-6.1}$$

D is related to the van der Waals constant a and also to the critical properties:

$$D = \frac{a}{R^2} = \frac{27}{64} \frac{T_c^2}{P_c} \tag{6-6.2}$$

Thodos and coworkers [9, 30, 66, 75, 76, 80, 81] have examined the behavior of Eq. (6-6.1) in detail. They propose that

$$C = 0.7816B + 2.67 \tag{6-6.3}$$

With Eqs. (6-6.2) and (6-6.3) and the use of the critical point as a datum, Eq. (6-6.1) becomes

$$\ln P_{vp_r} = B\left(\frac{1}{T_r} - 1\right) + C \ln T_r + \frac{27}{64}\left(\frac{P_{vp_r}}{T_r^2} - 1\right) \tag{6-6.4}$$

B is found by applying Eq. (6-6.4) at the normal boiling point, that is, $P = 1$ atm, $T = T_b$:

$$B = \frac{\ln P_c + 2.67 \ln T_{b_r} + \frac{27}{64}[(1/P_c T_{b_r}^2) - 1]}{1 - 1/T_{b_r} - 0.7816 \ln T_{b_r}} \tag{6-6.5}$$

Additive methods for determining the constant B are available for hydrocarbons [12], and alternative relations to Eq. (6-6.4) have been proposed [50]. For general use, however, Eq. (6-6.4) is preferable; it is, nevertheless, somewhat cumbersome because it is not explicit in P_{vp}. The Frost-Kalkwarf equation was also used by Passut and Danner [67], although they introduced the radius of gyration of a molecule (see Sec. 5-8) as a correlating parameter to improve the accuracy.

Example 6-4   Repeat Example 6-1 using the generalized Frost-Kalkwarf-Thodos vapor-pressure correlation.

solution   For ethylbenzene, with $P_c = 35.6$ atm and $T_{b_r} = 0.663$, Eq. (6-6.5) can be used to determine B:

$$B = \frac{\ln 35.6 + 2.67 \ln 0.663 + \dfrac{27}{64}\left[\dfrac{1}{(35.6)(0.663)^2} - 1\right]}{1 - \dfrac{1}{0.663} - 0.7816 \ln 0.663}$$

$$= -11.120$$

Then, with Eqs. (6-6.3) and (6-6.4),

$$\ln P_{vp_r} = -11.120\left(\frac{1}{T_r} - 1\right) - 6.021 \ln T_r + \frac{27}{64}\left(\frac{P_{vp_r}}{T_r^2} - 1\right)$$

Vapor pressures calculated by a trial-and-error procedure at both 74.1 and 186.8°C are:

T, °C	T_r	P (calc.), mm Hg	P (exp.), mm Hg	$\dfrac{P\text{ (calc.)} - P\text{ (exp.)}}{P\text{ (exp.)}} \times 100$
74.1	0.563	101	100	1
186.8	0.745	2491	2494	−0.1

Harlacher and Braun [31] also adopt Eq. (6-6.4). Using experimentally reported data for 242 substances, they developed and tabulated values of B and C for each fluid. An approximate correlation between these constants and the parachor† and acentric factor was also suggested. To employ the Harlacher-Braun constants to estimate vapor pressures in Eq. (6-6.4), one must have the critical constants, T_c and P_c. These constants are also given by Harlacher and Braun, but a few of them differ from those shown in Appendix A. We have used all the Harlacher-Braun constants (B, C, T_c, P_c) in Eq. (6-6.4) and obtained a set of constants A, B, C, D applicable for Eq. (6-6.1). As this is not a reduced equation, the problem of which critical properties to use ceases to exist. We have tabulated these constants for many fluids in Appendix A. When these constants are used, T is in kelvins and P_{vp} is in millimeters of mercury.

Example 6-5   Repeat Example 6-1 using the Harlacher-Braun constants for ethylbenzene.

solution   From Appendix A, HARA = 58.100, HARB = −6792.54, HARC = −5.802, and HARD = 5.75.

T, °C	T, K	P (calc.), mm Hg	P (exp.), mm Hg	$\dfrac{P\text{ (calc.)} - P\text{ (exp.)}}{P\text{ (exp.)}} \times 100$
74.1	347.3	99.7	100	−0.3
186.8	460.0	2510	2494	0.6

†The parachor is defined in terms of surface tension and is discussed in Chap. 12.

For ethylbenzene, the percent error is less than 1 percent between 300 and 500 K.   The error above the latter temperature is larger (3 percent was the largest) because Harlacher and Braun reported a critical pressure of about 27,900 mm Hg whereas, in Appendix A, a value of 27,060 mm Hg is listed [3].

6-7   Riedel-Plank-Miller Vapor-Pressure Equation

Miller has published extensively on the subject of vapor-pressure correlations [49–51, 54].   Only one of his equations is given here. Beginning with

$$\ln P_{vp} = A + \frac{B}{T} + CT + DT^3 \qquad (6\text{-}7.1)$$

Miller used the critical point, the normal boiling point, and the Riedel restriction, Eq. (6-5.3).   By relating α_c to h [see Eq. (6-2.5)] he obtained [54] the reduced-vapor-pressure relation

$$\ln P_{vp_r} = -\frac{G}{T_r}[1 - T_r^2 + k(3 + T_r)(1 - T_r)^3] \qquad (6\text{-}7.2)$$

$$G = 0.4835 + 0.4605h \qquad (6\text{-}7.3)$$

k is found by applying Eq. (6-7.2) at the normal boiling point:

$$k = \frac{h/G - (1 + T_{b_r})}{(3 + T_{b_r})(1 - T_{b_r})^2} \qquad (6\text{-}7.4)$$

Example 6-6   Repeat Example 6-1 using the Riedel-Plank-Miller vapor-pressure correlation.
   solution   For ethylbenzene, as given in Example 6-1, $P_c = 35.6$ atm, $T_{b_r} = 0.663$, and $h = 7.028$.   Then, with Eq. (6-7.3),

$$G = 0.4835 + (0.4605)(7.028) = 3.720$$

and, with Eq. (6-7.4),

$$k = \frac{7.028/3.720 - (1 + 0.663)}{(3 + 0.663)(1 - 0.663)^2} = 0.544$$

The Riedel-Plank-Miller vapor-pressure equation for ethylbenzene is, by Eq. (6-7.2),

$$\ln P_{vp_r} = \frac{-3.720}{T_r}[1 - T_r^2 + 0.544(3 + T_r)(1 - T_r)^3]$$

Then

T, °C	T_r	P (calc.), mm Hg	P (exp.), mm Hg	$\dfrac{P\,(\text{calc.}) - P\,(\text{exp.})}{P\,(\text{exp.})} \times 100$
74.1	0.563	101.9	100	1.9
186.8	0.745	2479	2494	−0.6

6-8 Thek-Stiel Vapor-Pressure Equation

In a major effort to improve earlier attempts to integrate the Clausius-Clapeyron equation, Thek and Stiel [84] proposed the use of Eq. (6-16.1) to relate ΔH_v to temperature. An exponent of 0.375 was selected and the Watson function expanded as a power-series polynomial in T_r. Any effect of temperature on ΔZ_v was included as a correction term in the ΔH_v expansion. This correction term was devised to possess several characteristics: it should vanish at low temperatures, where $\Delta Z_v \approx 1.0$; it should lead to a minimum in the predicted $\Delta H_v / \Delta Z_v$ at around $T_r = 0.8$ (see Sec. 6-16); and it should be of such a form that the vapor-pressure equation leads to the appropriate value of α at the critical point [see Eq. (6-5.2)]. With these conditions the final equation can be written

$$
\begin{aligned}
\ln P_{\mathrm{vp}_r} = A \Bigg(& 1.14893 - \frac{1}{T_r} - 0.11719 T_r \\
& - 0.03174 T_r^2 - 0.375 \ln T_r \Bigg) \\
& + (1.042 \alpha_c - 0.46284 A) \\
& \times \left[\frac{T_r^{5.2691 + 2.0753 A - 3.1738 h} - 1}{5.2691 + 2.0753 A - 3.1738 h} \right] \\
& + 0.040 \left(\frac{1}{T_r} - 1 \right) \Bigg]
\end{aligned}
\tag{6-8.1}
$$

where

$$
A = \frac{\Delta H_{v_b}}{R T_c (1 - T_{r_b})^{0.375}}
\tag{6-8.2}
$$

and h is defined by Eq. (6-2.5). Equation (6-8.1) contains only a single constant, α_c, which can be found from the conditions $P = 1$ atm, $T = T_b$.

Thek and Stiel claim that the major advantage of their correlation, relative to those presented earlier, is improved accuracy for estimating the vapor pressures for polar and hydrogen-bonded substances at low temperatures (below T_b). Note that to use the Thek-Stiel form, T_b, T_c, P_c, and ΔH_{v_b} must be known. The last property especially reflects the polar character of the material; it is, however, often difficult to obtain, and in practical cases it must often be estimated from correlations of the type discussed in Sec. 6-15.

When tested by Thek and Stiel for some 69 compounds, both polar and nonpolar, average errors were less than 1 percent and rarely did the maximum error exceed 5 percent over wide temperature ranges. Less extensive testing by the authors confirms this degree of accuracy although the calculated value of P_{vp} was found to be particularly sensitive to the value of ΔH_{v_b} chosen. At all temperatures, except near T_c, estimation of P_{vp} was also sensitive to the selected value of T_c. Less sensitivity was noted for the dependence of P_{vp} on the chosen value of P_c, that is, $\Delta P_{\mathrm{vp}} / \Delta P_c \approx 1$ for high T_r and ≈ 0.5 for $T < T_b$.

Example 6-7 Repeat Example 6-1 using the Thek-Stiel vapor-pressure estimation equation.

solution   For ethylbenzene, $P_c = 35.6$ atm, $T_{b_r} = 0.663$, and $T_c = 617.1$ K.   In addition to these constants, the enthalpy of vaporization at T_b is also necessary.   Wilhoit and Zwolinski [94] quote a value of 8500 cal/g-mol.   Then, with Eq. (6-8.2),

$$A = \frac{8500}{(1.987)(617.1)(1 - 0.663)^{0.375}} = 10.423$$

From Example 6-1, $h = 7.028$.   With $P_{vp_r} = 1/P_c$, $T_r = T_{b_r}$, solving Eq. (6-8.1), one obtains a value $\alpha_c = 7.418$.   Note that this is slightly different from the α_c calculated by Riedel's equation in Example 6-3.

With these parameters, Eq. (6-8.1) becomes

$$\ln P_{vp_r} = 10.423 \left(1.14893 - \frac{1}{T_r} - 0.11719 T_r - 0.03174 T_r^2 - 0.375 \ln T_r \right)$$

$$+ 0.6298(T_r^{4.607} - 1) + 0.1161 \left(\frac{1}{T_r} - 1 \right)$$

At 74.1 and 186.8°C, the calculated vapor pressures for ethylbenzene are:

T, °C	T_r	P (calc.), mm Hg	P (exp.), mm Hg	$\dfrac{P\ (\text{calc.}) - P\ (\text{exp.})}{P\ (\text{exp.})} \times 100$
74.1	0.563	101	100	1.1
186.8	0.745	2487	2494	−0.3

6-9   Correlation Equations

Most vapor-pressure equations presented in the previous sections have been introduced primarily as *estimating* equations.   That is, given some data such as a boiling point and the critical properties, it is possible to develop the constants so that vapor pressures can be estimated as a function of temperature.

Upon occasion, however, one is favored with experimental vapor-pressure data over a wide range of temperatures, and a means of storing this information in an analytical form is wanted.   Using standard regression techniques, one can determine the best values of the constants employing any one of the equation forms introduced earlier in this chapter.   This is routinely done to obtain the Antoine constants in Eq. (6-3.1).

Ambrose et al. [4] recommend the use of Chebyshev polynomials as particularly appropriate correlating equations for vapor pressures.   In this case

$$T \ln P_{vp} = \frac{A_0}{2} + \sum_{i=1}^{n} A_i E_i(x) \tag{6-9.1}$$

The variable x defines the temperature within the range of applicability between T_{max} and T_{min}:

$$x = \frac{2T - (T_{max} + T_{min})}{T_{max} - T_{min}} \qquad (6\text{-}9.2)$$

where x varies from -1 to $+1$. $E_i(x)$ is the Chebyshev polynomial, where

$$
\begin{aligned}
E_0(x) &= 1 \\
E_1(x) &= x \\
E_{i+1}(x) - 2xE_i(x) + E_{i-1}(x) &= 0
\end{aligned}
\qquad (6\text{-}9.3)
$$

The advantage of Eq. (6-9.1) is that successive terms contribute in a rapidly decreasing fashion and the addition of higher terms has little effect on the leading coefficients. This is not the case for most other vapor-pressure equations presented in this chapter; in the latter cases, omission of the last (or any other) term usually leads to a completely wrong answer. Regression techniques are straightforward [17], and examples of the use of such polynomials are given by Ambrose and Sprake [5].

6-10   Discussion and Recommendations for Vapor-Pressure Estimation and Correlation

Starting from the Clausius-Clapeyron equation, Eq. (6-2.2), we have shown only a few of the many vapor-pressure equations which have been published. We have emphasized those which appear to be most accurate and general. In almost all cases, only one boiling point (usually the normal boiling point) and the critical temperature and pressure are required. It is amazing how well most of the techniques predict vapor pressures over wide ranges with this little input. We show in Table 6-1 a detailed comparison between calculated and experimental vapor pressures for acetone for the seven estimation techniques described in this chapter. The range shown is from 32 mm Hg to the critical point, 35,250 mm Hg. The least accurate is, as expected, the Clapeyron equation, especially at the lower temperatures.

The Antoine equation should not be used above 1500 to 2000 mm Hg when the constants are obtained from experimental data below this pressure. In the range for which the constants are applicable, it is very accurate.

Of the five other methods shown, i.e., Lee-Kesler, Riedel, Frost-Kalkwarf-Thodos, Riedel-Plank-Miller, and Thek-Stiel, it is hard to justify any real advantage of one over another. The Frost-Kalkwarf-Thodos form requires an iterative solution since the vapor pressure occurs on both sides of the equation. The other four can readily be handled with a small desk calculator.

TABLE 6-1 Comparison between Calculated and Experimental Vapor Pressures for Acetone

T, K	T_r	P_{vp}(exp.), mm Hg[a]	Clapeyron, Eq. (6-2.4)	Antoine, Eq. (6-3.1)[b]	Riedel, Eq. (6-5.4)	Frost, Kalkwarf, and Thodos, Eq. (6-6.4)	Riedel, Plank, and Miller, Eq. (6-7.2)	Thek and Stiel, Eq. (6-8.1)	Lee and Kesler, Eq. (6-2.6)
						Percent error[c]			
259.2	0.510	32.0	24	1.6	−7.2	−6.2	−3.0	1.3	−8.3[e]
273.4	0.538	71.2	15	0.3	−5.6	−4.5	−3.0	0.5	−6.3
290.1	0.571	161.4	8.3	−0.1	−3.3	−2.3	−1.4	0.7	−3.7
320.5	0.631	558.4	1.5	−0.7	−0.3	0	0.2	0.4	−0.2
350.9	0.691	1,512	−1.0	−0.6	1.6	1.1	0.8	0.2	1.9
390.3	0.768	4,241	−1.5	0.1	1.8	0.9	−0.1	−0.8	2.2
446.4	0.878	13,260	−0.3	−0.5	0.4	−0.2	−1.3	−1.7	0.7
470.6	0.926	19,970	0.5	−1.5[d]	0.2	−0.1	−1.0	−1.2	0.3
508.1	1.00	35,250		−4.5[d]					

[a] Experimental data from Ref. 6.
[b] ln P_{vp} = 16.6513 − 2940.45/(T − 35.93).
[c] Percent error = [(calc. − exp.)/exp.] × 100.
[d] Constants not applicable at high reduced temperature.
[e] ω calculated from Eq. (2-3.4) (ω = 0.301) rather than obtained from Appendix A (ω = 0.309).

Tests similar to the one shown in Table 6-1 were made for other fluids. Comparable results were obtained, although the Clapeyron equation becomes less accurate as the polarity of the compound increases. For very polar substances, the Thek-Stiel method was usually (but not always) more accurate; note that in this case a value of ΔH_{v_b} must be supplied along with T_b and T_c, P_c.

Since the Clapeyron, Lee-Kesler, Riedel, Frost-Kalkwarf-Thodos, and Riedel-Plank-Miller equations can be viewed as being of the form

$$P_{vp_r} = f(T_{b_r}, T_r, P_c)$$

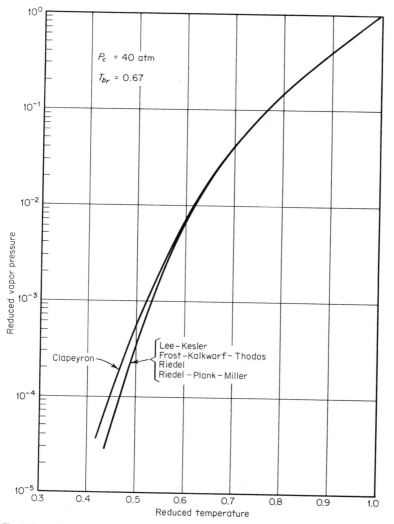

Fig. 6-2 Reduced vapor-pressure correlations.

it is interesting to compare the predictions of these five correlations with one another.

After a number of *hypothetical* fluids with different values of P_c and T_{b_r} had been chosen, values of P_{vp_r} were determined as a function of T_r. Except for the Clapeyron equation, all agree well, and this reemphasizes the difficulty of selecting any one as being *more accurate* than another. All methods must agree when $T_r = T_{b_r}$ or $T_r = 1$, as these values were chosen as data points. In Fig. 6-2, we show one illustrative case where $P_c = 40$ atm and $T_{b_r} = 0.67$. Within the thickness of a fine line, all methods yield similar results, except for the Clapeyron form, which predicts a higher P_{vp_r} at low reduced temperatures.

Othmer and his coworkers have published several papers that show how vapor pressures of one substance can be correlated with those of a reference substance when both are compared at the same temperature or sometimes at the same reduced temperature [59, 60, 63]. More recently, this approach was used in computer calculations [62, 64]. Similar comparisons can be made for boiling temperatures at equal vapor pressures, and Thomas and Smith [85] have demonstrated that if a nonassociated substance such as benzene is chosen as a reference material, correlations of high accuracy are possible.

Narsimhan [57] introduced the parameter $\Delta H_v (T \to 0)$ as a correlating third parameter to develop a simple reduced-vapor-pressure correlation.

Miller [52] has published a detailed error analysis of several vapor-pressure estimation relations. For cases where the necessary input data are T_b, P_c, T_c, his recommendations are similar to those given below. Miller also evaluated several vapor-pressure correlations which do not require P_c or even T_c as input. In such cases, however, values of ΔH_v must be known. In only a few instances do we have experimental values of ΔH_v when no critical properties are known or can be estimated with confidence (see Sec. 2-2). Thus, we have not considered such estimation methods in this chapter.

Recommendations in Priority Form

1. If $P_{vp} < 10$ mm Hg, none of the methods described in this chapter yields accurate predicted values.

2. If $P_{vp} < 1500$ mm Hg and the constants are available (as in Appendix A), the Antoine equation (6-3.1) is recommended.

3. For 10 mm Hg $< P < P_c$, if the Harlacher-Braun constants are available in Appendix A, vapor pressures should be estimated with these constants and Eq. (6-6.1). To estimate vapor pressures near the critical point, a check should be made to verify that the tabulated constants do, in fact, yield a reliable estimate of the critical pressure given in Appendix A.

4. If the fluid is nonpolar, the Lee-Kesler equation (6-2.6), the Riedel equation (6-5.4), the Frost-Kalkwarf-Thodos equation (6-6.4), the Riedel-Plank-Miller equation (6-7.2), and the Thek-Stiel equation (6-8.1) all yield accurate estimates of the vapor pressure. For the first three methods, only T_b, T_c, and P_c are required. For the Thek-Stiel method, these constants as well as a value of ΔH_{vb} are necessary.

5. For polar liquids, especially hydrogen-bonded substances, all methods yield results of lower accuracy. The Thek-Stiel method is probably the most accurate in the majority of cases.

The methods described in this chapter were developed for fluids with normal molecular weight. For very heavy hydrocarbons, where experimental data are scarce, a correlation has been established, based on Prigogine's theory of fluids containing polysegmented molecules. This correlation [80a] is useful for estimating vapor pressures of very high-boiling hydrocarbons, including those whose molecules contain naphthenic or aromatic rings.

6-11 Enthalpy of Vaporization of Pure Compounds

The enthalpy of vaporization ΔH_v is sometimes referred to as the *latent heat of vaporization*. It is the difference between the enthalpy of the saturated vapor and that of the saturated liquid at the same temperature.

Because of the forces of attraction between the molecules of the liquid, the molecules escaping are those of higher than average energy. The average energy of the remaining molecules in the liquid is reduced, and energy must be supplied to maintain the temperature constant. This is the *internal energy of vaporization* ΔU_v. Work is done on the vapor phase as vaporization proceeds, since the vapor volume increases if the pressure is maintained constant at P_{vp}. This work is $P_{vp}(V_g - V_L)$. Thus

$$\Delta H_v = \Delta U_v + P_{vp}(V_g - V_L) = \Delta U_v + RT(Z_g - Z_L)$$
$$= \Delta U_v + RT\,\Delta Z_v \tag{6-11.1}$$

Many estimation methods for ΔH_v can be traced to Eq. (6-2.2), where it was shown that ΔH_v is related to the slope of the vapor-pressure–temperature curve. Other methods are based on the law of corresponding states. We review the more accurate techniques in Secs. 6-12 to 6-16; recommendations are presented in Sec. 6-17.

It is often difficult to trace the origin of many "experimental" enthalpies of vaporization. A few have been determined from calorimetric measurements, but in a large number of cases, the so-called experimental value was obtained directly from Eq. (6-2.2). Some technique was employed to determine ΔZ_v separately, and also $(d \ln P_{vp})/dT$ was found by numerical differentiation of experimental vapor-pressure data or by differentiating analytically some P_{vp}-T correlation. An example of this latter approach can be found in the recent reissue of the API Tables [94]. Enthalpies of vaporization were determined using Eq. (6-2.2), where dP_{vp}/dT was found from the Antoine vapor-pressure equation (Sec. 6-3); the saturated-vapor compressibility factor was estimated from a virial equation of state (Sec. 3-11), and experimental data were employed for saturated-liquid compressibility factors.

An element of uncertainty is introduced in using any analytical vapor-pressure–temperature equation to obtain accurate values of slopes dP_{vp}/dT. The constants in the equation may be optimum for correlating vapor pressures, but it does not necessarily follow that these same constants give the best fit for computing slopes.

Since so few calorimetric measurements of ΔH_v are available, there is little that can be done to rectify the problem. A critical survey of reported ΔH_v values would nevertheless be of value since one would like to avoid the logical pitfalls of comparing estimated values of ΔH_v with values found (estimated?) by other approximate methods and then making recommendations based on such a comparison.

6.12 Estimation of ΔH_v from the Clausius-Clapeyron Equation and Experimental Vapor Pressures

Equation (6-2.1) can be employed to obtain ΔH_v directly from vapor-pressure data by using any one of various methods to obtain the slope dP_{vp}/dT. For example, the Douglass-Avakian method of differentiation is convenient where P_{vp} is tabulated over equal temperature intervals. Applied to the Clausius-Clapeyron equation, this gives

$$\Delta H_v = T\,\Delta V_v \left(\frac{397 \sum nP}{1512 \delta_T} - \frac{7 \sum n^3 P}{216 \delta_T} \right) \qquad (6\text{-}12.1)$$

where δ_T is the temperature interval and n is the number of the point in question, varying from -3 to $+3$. ΔH_v is obtained at the midpoint of the seven temperatures ($n = 0$). To use Eq. (6-12.1), a value of ΔV_v or ΔZ_v is necessary.

Example 6-8 Given the following vapor pressures of water

Temp., °C	15	20	25	30	35	40	45
Vapor pressure, mm Hg	12.79	17.54	23.76	31.82	42.18	55.32	71.88

evaluate ΔH_v at 30°C.

solution Following the indicated procedure for numerical differentiation, $n = -3$ at $T = 15$, -2 at $20, \ldots,$ $+3$ at $45°$. $\Sigma\, nP$ is 271.27, and $\Sigma\, n^3 P$ is 1916.22. The product $T\, \Delta V_v$ is $(\Delta Z_v\, RT^2)/P$. As an approximation, take $\Delta Z_v = 1.0$. Then

$$\Delta H_v = \frac{RT^2}{P}\frac{(397)(271.27)}{(1512)(5)} - \frac{(7)(1916.22)}{(216)(5)}$$

$$= \frac{(1.987)(303^2)}{31.82}(14.244 - 12.420) = 10,445 \text{ cal/g-mol}$$

The experimental value is 10,452 cal/g-mol.

6-13 Estimation of ΔH_v from the Law of Corresponding States

Equation (6-2.1), in reduced form, becomes

$$d \ln P_{\mathrm{vp}_r} = \frac{-\Delta H_v}{RT_c\, \Delta Z_v}\, d\,\frac{1}{T_r} \tag{6-13.1}$$

The reduced enthalpy of vaporization $-\Delta H_v/RT_c$ is a function of $(d \ln P_{\mathrm{vp}_r})/d(1/T_r)$ and ΔZ_v; both these parameters are commonly assumed to be functions of T_r or P_{vp_r} and some third parameter such as α_c, ω, Z_c, or h. A number of correlations based on this approach have been suggested.

Riedel-Factor Correlation [78] Riedel tabulated $\Delta H_v/T_c$ as a function of T_r and α_c, the Riedel factor defined by Eq. (6-5.2) at the critical point. These tables provide reliable estimates for nonpolar liquids. For example, when they were tested with 33 hydrocarbons at over 478 temperatures, the average error in estimating ΔH_v was only about 1.8 percent [88]. However, Pitzer's method described below is slightly more convenient and generally more accurate.

Lydersen, Greenkorn, and Hougen Correlation [46] Enthalpies of the saturated vapor and liquid for various reduced temperatures have been related to T_r and Z_c by Lydersen, Greenkorn, and Hougen. From these values, ΔH_v can be obtained, but the accuracy is poor. A similar technique was suggested by Hobson and Weber [34].

Pitzer Acentric-Factor Correlation [68] Pitzer et al. have shown that ΔH_v can be related to T, T_r, and ω by an expansion similar to that used to estimate compressibility factors, Eq. (3-3.1), i.e.,

$$\frac{\Delta H_v}{T} = \Delta S_v^{(0)} + \omega\, \Delta S_v^{(1)} \tag{6-13.2}$$

where $\Delta S_v^{(0)}$ and $\Delta S_v^{(1)}$ are expressed in entropy units, for example, cal/g-mol K, and are functions only of T_r.   Multiplying Eq. (6-13.2) by T_r/R gives

$$\frac{\Delta H_v}{RT_c} = \frac{T_r}{R}(\Delta S_v^{(0)} + \omega \; \Delta S_v^{(1)}) \qquad (6\text{-}13.3)$$

Thus $\Delta H_v/RT_c$ is a function only of ω and T_r.   From the tabulated $\Delta S_v^{(0)}$ and $\Delta S_v^{(1)}$ functions given by Pitzer et al. and extended to low reduced temperatures by Carruth and Kobayashi [15], Fig. 6-3 was constructed.   For a close approximation, an analytical representation of this correlation for $0.6 < T_r \leqslant 1.0$ is

$$\frac{\Delta H_v}{RT_c} = 7.08(1 - T_r)^{0.354} + 10.95\omega(1 - T_r)^{0.456} \qquad (6\text{-}13.4)$$

The effect of temperature on ΔH_v is similar to that suggested by Watson (see Sec. 6-16).

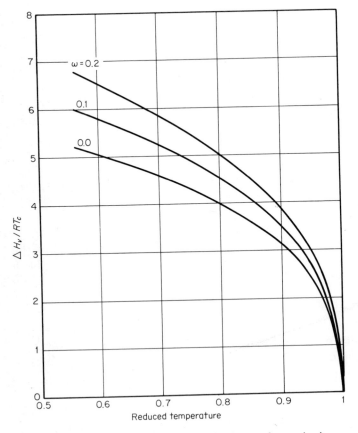

Fig. 6-3   Plot of Pitzer et al. correlation for heats of vaporization.

Example 6-9 Using the Pitzer et al. corresponding-states correlation, estimate the enthalpy of vaporization of propionaldehyde at 48°C. Repeat for n-octane at 171°C. The literature values are 6760 and 7254 cal/g-mol, respectively [18, 48].

  solution

	T_c, K	ω	T, K	T_r	$\dfrac{\Delta H_v}{RT_c}$
Propionaldehyde	496	0.313	321	0.647	7.008
n-Octane	568.6	0.394	444.2	0.781	6.294

where $\Delta H_v/RT_c$ was calculated from Eq. (6-13.4). Therefore

Propionaldehyde:
$$\Delta H_v = (7.008)(1.987)(496) = 6907 \text{ cal/g-mol}$$
$$\text{Error} = \frac{6907 - 6760}{6760} \times 100 = 2.1\%$$

n-Octane:
$$\Delta H_v = (6.294)(1.987)(568.8) = 7110 \text{ cal/g-mol}$$
$$\text{Error} = \frac{7110 - 7254}{7254} \times 100 = -2.0\%$$

6-14 Estimation of ΔH_v from Vapor-Pressure Equations

The vapor-pressure correlations covered in Secs. 6-2 to 6-8 can be used to estimate enthalpies of vaporization. From Eq. (6-13.1), we can define a dimensionless group ψ as

$$\psi \equiv \frac{\Delta H_v}{RT_c \, \Delta Z_v} = \frac{-d \ln P_{\text{vp}_r}}{d(1/T_r)} \qquad (6\text{-}14.1)$$

Differentiating the vapor-pressure equations discussed earlier, we can obtain various expressions for ψ. These are shown in Table 6-2. To use these expressions, one must refer back to the reference vapor-pressure equation for the definition of the various parameters.

TABLE 6-2 ψ Values from Vapor-Pressure Equations

Vapor-pressure equation	$\psi \equiv \dfrac{\Delta H_v}{RT_c \, \Delta Z_v}$	
Clapeyron, Eq. (6-2.4)	h	(6-14.2)
Lee-Kesler, Eq. (6-2.6)	$6.09648 - 1.28862\,T_r + 1.016\,T_r^7$ $+ \omega(15.6875 - 13.4721\,T_r + 2.615\,T_r^7)$	(6-14.3)
Antoine, Eq. (6-3.1)	$\dfrac{B}{T_c}\left(\dfrac{T_r}{T_r - C/T_c}\right)^2$	(6-14.4)
Riedel, Eq. (6-5.4)	$B^+ + C^+ T_r + 6D^+ T_r^7$	(6-14.5)

TABLE 6-2 ψ **Values from Vapor-Pressure Equations** *(Continued)*

Vapor-pressure equation	$\psi \equiv \dfrac{\Delta H_v}{RT_c\,\Delta Z_v}$	
Frost-Kalkwarf-Thodos, Eq. (6-6.4)	$\dfrac{-B + CT_r - (\frac{27}{32})(P_r/T_r)}{1 - (\frac{27}{64})(P_r/T_r^2)}$	(6-14.6)
Riedel-Plank-Miller, Eq. (6-7.2)	$G[1 + T_r^2 + 3k(1 - T_r^2)^2]$	(6-14.7)
Thek-Stiel, Eq. (6-8.1)	$A(1 - 0.375T_r - 0.11719T_r^2 - 0.06348T_r^3)$ $+ (1.042\alpha_c - 0.46248A)(T_r^{\epsilon+1} - 0.040)$ where $\epsilon \equiv 5.2691 + 2.0753A - 3.1738h$	(6-14.8)

In Fig. 6-4, we show experimental values of ψ for propane. These were calculated from several reported values of ΔH_v and saturated-phase volumetric properties. The agreement between the various investigators is reasonable, and differences are less than 1 percent, except close to $T_r = 1$. Note the pronounced minimum in the curve

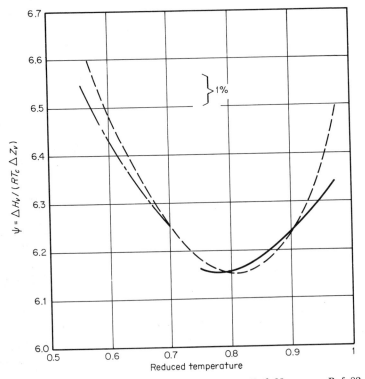

Fig. 6-4 Literature values of ψ for propane; —— Ref. 32, −−−− Ref. 82, −−− Ref. 8.

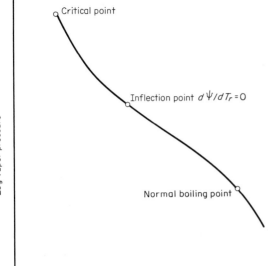

Fig. 6-5 Schematic vapor-pressure plot.

around $T_r = 0.8$.   Since

$$\psi = \frac{-d \ln P_{vp_r}}{d(1/T_r)} \tag{6-14.9}$$

we have

$$\frac{d\psi}{dT_r} = \frac{1}{T_r^2} \frac{d^2 \ln P_{vp_r}}{d(1/T_r)^2} \tag{6-14.10}$$

At low values of T_r, $d\psi/dT_r < 0$ so that $(d^2 \ln P_{vp_r})/d(1/T_r)^2$ is also < 0.   At high values of T_r, the signs reverse.   When $d\psi/dT_r = 0$, an inflection point results.   Thus the general (though exaggerated) shape of a log vapor-pressure–inverse-temperature curve is that shown in Fig. 6-5.

To illustrate how the various vapor-pressure equations can predict the shape of Fig. 6-4, we have drawn Fig. 6-6 (pp. 204–205).   Except for the Clapeyron equation (where ψ is a constant equal to h), the other equations show a remarkable fit to the value of ψ calculated from literature data for propane.   The Antoine equation does not predict the ψ-T_r minimum, and the calculated curve was drawn only up to a T_r value corresponding to 1500 mm Hg.

Other comparisons also produced good results.   One is therefore tempted to recommend these vapor-pressure correlations to predict ψ,

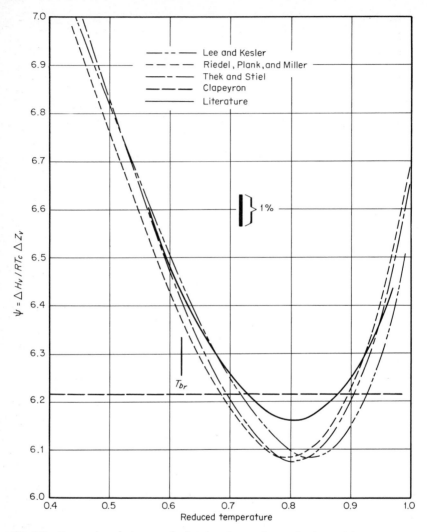

Fig. 6-6 Comparison between calculated and experimental values of $\Delta H_v/(RT_c\ \Delta Z_v)$ for propane.

and thus ΔH_v.   However, accurate values of ΔZ_v must be available.   To date, specific correlations for ΔZ_v [e.g., see Eq. (6-4.2)] are not accurate. ΔZ_v is determined best from P-V-T estimation techniques covered in Chap. 3 for both the saturated vapor and liquid.

6-15 ΔH_v at the Normal Boiling Point

A pure-component constant that is occasionally used in property correlations is the enthalpy of vaporization at the normal boiling point

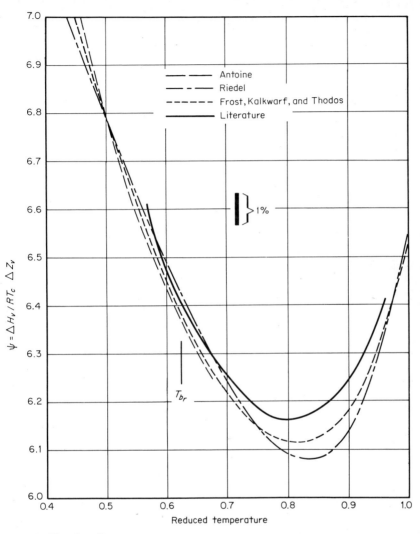

Fig. 6-6 *(Continued)*

ΔH_{v_b}.   Any one of the correlations discussed in Secs. 6-12 to 6-14 can be used for this state where $T = T_b$, $P = 1$ atm.   We discuss some of the techniques below.   In addition, several special estimation methods are suggested.

ΔH_{v_b} from Vapor-Pressure Relations   In Table 6-2, we show equations for $\psi \equiv \Delta H_v/(RT_c \, \Delta Z_v)$ as determined from a number of the more accurate vapor-pressure equations.   Each can be used to determine

$\psi(T_b)$ except the Thek-Stiel form, where ΔH_{v_b} is required as an input variable.   With $\psi(T_b)$ and $\Delta Z_v(T_b)$, ΔH_{v_b} can be determined.

The Antoine correlation employs specific constants (such as those shown in Appendix A).   For the Clapeyron, Lee-Kesler, Riedel, Frost-Kalkwarf-Thodos, and Riedel-Plank-Miller relations, ψ is given in Table 6-2; when $\psi(T_b)$ is desired, it can be shown that

$$\psi(T_b) = f(T_{b_r}, P_c) \qquad (6\text{-}15.1)$$

In Figs. 6-7 and 6-8, we show $\psi(T_b)$ for these vapor-pressure forms for two cases.   In Fig. 6-7, $\psi(T_b)$ is given as a function of T_{b_r} for a critical pressure of 40 atm.   In Fig. 6-8, we show a cross-plot where $\psi(T_b)$ is related to P_c when $T_{b_r} = 0.66$.   Except for the Clapeyron equation, the vapor-pressure relations yield similar results.   When ΔZ_{v_b} is determined

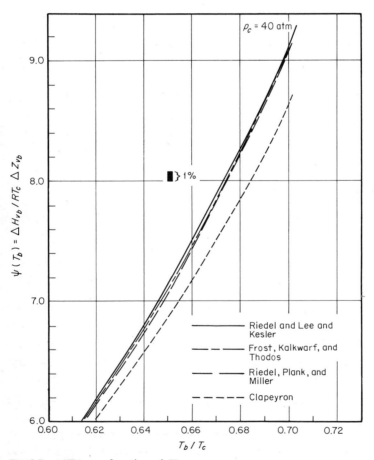

Fig. 6-7 $\psi(T_b)$ as a function of T_{b_r}.

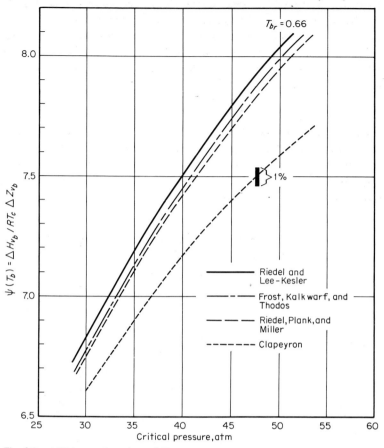

Fig. 6-8 $\psi(T_b)$ as a function of P_c.

from $P\text{-}V\text{-}T$ relations discussed in Chap. 3, the Riedel, Frost-Kalkwarf-Thodos, or Riedel-Plank-Miller vapor-pressure relations allow an accurate determination of ΔH_{v_b}.

Special mention should be made when the Clapeyron equation is used to calculate ψ [see Eq. (6-14.2) in Table 6-2]. Here, ψ is equal to h regardless of T_r, that is,

$$\psi(T_r) = \psi(T_b) = T_{b_r} \frac{\ln P_c}{1 - T_{b_r}} \tag{6-15.2}$$

and

$$\Delta H_{v_b} = RT_c \Delta Z_{v_b} \left(T_{b_r} \frac{\ln P_c}{1 - T_{b_r}} \right) \tag{6-15.3}$$

Equation (6-15.3) has been widely employed to make rapid estimates of ΔH_{v_b}, and usually, in such cases, ΔZ_{v_b} is set equal to unity.   In this form, it has been called the *Giacalone equation* [26]; extensive testing of this simplified form indicates that it normally overpredicts ΔH_{vb} by a few percent. Correction terms have been suggested [23, 41] to improve the accuracy of the Giacalone equation, but better results are obtained with other relations, noted below.

Riedel Method.   Riedel [78] modified Eq. (6-15.3) slightly and proposed that

$$\Delta H_{vb} = 1.093RT_c\left[T_{b_r}\frac{(\ln P_c - 1)}{0.930 - T_{b_r}}\right] \tag{6-15.4}$$

When tested as shown in Table 6-3, errors are almost always less than 2 percent.

Chen Method.   Chen [16] used Eq. (6-13.3) and a similar expression proposed by Pitzer et al. to correlate vapor pressures so that the acentric factor is eliminated. He obtained a relation between ΔH_v, $P_{vp,}$, and T_r. When applied to the normal boiling point,

$$\Delta H_{vb} = RT_cT_{b_r}\frac{3.978T_{b_r} - 3.938 + 1.555\ln P_c}{1.07 - T_{b_r}} \tag{6-15.5}$$

This correlation was tested, with the results shown in Table 6-3. The accuracy is similar to that of the Riedel equation (6-15.4). In a more complete test, Chen compared estimated ΔH_{vb} with literature values for 169 materials and found an average error of 2.1 percent.

Vetere Method.   By starting with the Frost-Kalkwarf vapor-pressure relation [Eq. (6-6.1)] and regressing ΔH_{vb} data to obtain the constants,

TABLE 6-3   Comparison between Calculated and Experimental Values of ΔH_{vb}

Class of compounds	Number of compounds	Average absolute percent error			
		Giacalone, Eq. (6-15.3)	Riedel, Eq. (6-15.4)	Chen, Eq. (6-15.5)	Vetere, Eq. (6-15.6)
Saturated hydrocarbons	22	2.9	0.9	0.4	0.4
Unsaturated hydrocarbons	8	2.4	1.4	1.2	1.2
Cycloparaffins and aromatics	12	1.1	1.3	1.2	1.1
Alcohols	7	3.6	4.0	4.0	3.8
Nitrogen and sulfur compounds (organic)	10	1.6	1.7	1.7	1.9
Halogenated compounds	10	1.3	1.6	1.5	1.5
Inert gases	5	8.4	2.1	2.2	2.5
Nitrogen and sulfur compounds (inorganic)	4	3.0	2.7	2.7	2.1
Inorganic halides	4	0.6	1.4	1.4	0.9
Oxides	6	6.9	4.4	4.9	4.6
Other polar compounds	6	2.2	1.5	1.8	1.6
Total	94	2.8	1.8	1.7	1.6

Vetere [89] proposed a relation similar to the one suggested by Chen:

$$\Delta H_{v_b} = RT_c T_{b_r} \frac{0.4343 \ln P_c - 0.68859 + 0.89584 T_{b_r}}{0.37691 - 0.37306 T_{b_r} + 0.14878 P_c^{-1} T_{b_r}^{-2}} \qquad (6\text{-}15.6)$$

where P_c is in atmospheres and T_c in kelvins.

As shown in Table 6-3, this empirical equation is capable of providing a good estimate of ΔH_{v_b}; errors are normally less than 2 percent.

Other Methods. A number of other methods have been proposed for estimating ΔH_{v_b}. None appears to offer any significant advantages over those given above. Miller [53] used an earlier version of the Riedel-Plank-Miller vapor-pressure equation to determine ψ and then, with $\Delta Z_{v_b} \approx 1 - 0.97/P_c T_{b_r}$, he obtained a relation for ΔH_{v_b} in terms of T_{b_r} and P_c. The final result is more complex than Eqs. (6-15.4) to (6-15.6) and yields similar results.

Ibrahim and Kuloor [37] present two correlations for ΔH_{v_b} in which the molecular weight or liquid volume is employed as the independent variable. Specific constants are required for various homologous series. Ogden and Lielmezs [58] proposed a similar type of relation, but in this case the Altenburg quadratic mean radius was used. The latter term reflects the mass distribution in a molecule. Procopio and Su [72] regressed ΔH_{v_b} values reported in the literature to determine the best values of K and Y in the expression

$$\Delta H_{v_b} = KRT_c T_{b_r} \frac{(\ln P_c)(1 - P_c^{-1})^Y}{1 - T_{b_r}} \qquad (6\text{-}15.7)$$

P_c is in atmospheres, T_c in kelvins, and values of $K = 1.024$ and $Y = 1.0$ were recommended. Equation (6-15.7) is similar to several other correlations noted earlier if ΔZ_{v_b} is approximated as $1 - P_c^{-1}$. The accuracy of Eq. (6-15.7) is similar to that of the Riedel form, Eq. (6-15.4). Viswanath and Kuloor [91] have used essentially the same relation and suggested $Y = 0.69$ and $K = 1.02$. If $Y = 0$ and $K = 1.0$, the Giacalone form is obtained.

Narsimhan [56] relates ΔH_{v_b} to a density function, and McCurdy and Laidler [47] and Fedors [22] propose additive group methods.

Comparison of Estimated with Literature Values of ΔH_{v_b}. Table 6-3 compares calculated and experimental values of ΔH_{v_b} using the estimation methods described in this section. The Riedel, Pitzer-Chen, and Vetere relations [Eqs. (6-15.4) to (6-15.6)] are convenient and are generally accurate. In every case, T_b, T_c, and P_c must be known or estimated.

Example 6-10 Estimate the enthalpy of vaporization of propionaldehyde at the boiling point. The experimental value is 6760 cal/g-mol [18].

solution From Appendix A, $T_b = 321$ K, $T_c = 496$ K, and $P_c = 47.0$ atm. Thus $T_{b_r} = 0.647$.

Riedel Method. With Eq. (6-15.4),

$$\Delta H_{v_b} = (1.093)(1.987)(496)\frac{(0.647)(\ln 47.0 - 1)}{0.930 - 0.647}$$

$$= 7020 \text{ cal/g-mol}$$

$$\text{Error} = \frac{7020 - 6760}{6760} \times 100 = 3.8\%$$

Giacalone Method. With Eq. (6-15.3) and assuming $\Delta Z_{v_b} = 1.0$,

$$\Delta H_{v_b} = (1.987)(496)\left(\frac{0.647 \ln 47.0}{1 - 0.647}\right)$$

$$= 6950 \text{ cal/g-mol}$$

$$\text{Error} = \frac{6950 - 6760}{6760} \times 100 = 2.9\%$$

Chen Method. From Eq. (6-15.5),

$$\Delta H_{v_b} = \frac{(1.987)(496)(0.647)[(3.978)(0.647) - 3.938 + 1.555 \ln 47.0]}{1.07 - 0.647}$$

$$= 6970 \text{ cal/g-mol}$$

$$\text{Error} = \frac{6970 - 6760}{6760} \times 100 = 3.1\%$$

Vetere Method. From Eq. (6-15.6),

$$\Delta H_{v_b} = (1.987)(496)(0.647)$$

$$\times \frac{0.4343 \ln 47.0 - 0.68859 + (0.89584)(0.647)}{0.37691 - (0.37306)(0.647) + (0.14878)(47.0)^{-1}(0.647)^{-2}}$$

$$= 6960 \text{ cal/g-mol}$$

$$\text{Error} = \frac{6960 - 6760}{6760} \times 100 = 3.0\%$$

6-16 Variation of ΔH_v with Temperature

The latent heat of vaporization decreases with temperature and is zero at the critical point. Typical data are shown in Fig. 6-9 for a few compounds. The shapes of these curves agree with most other enthalpy-of-vaporization data.

The variation of ΔH_v with temperature could be determined from any of the ψ relations shown in Table 6-2 although the variation of ΔZ_v with temperature would also have to be specified.

A widely used correlation between ΔH_v and T is the Watson relation [75]

$$\Delta H_{v_2} = \Delta H_{v_1} \left(\frac{1 - T_{r_2}}{1 - T_{r_1}}\right)^n \tag{6-16.1}$$

where the subscripts 1 and 2 refer to temperatures 1 and 2. A common choice for n is 0.375 or 0.38 [88]. Silverberg and Wenzel [79] found that

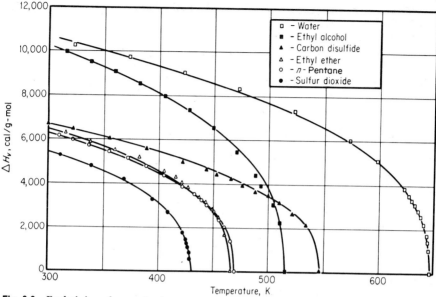

Fig. 6-9 Enthalpies of vaporization.

n varies from substance to substance.   For 44 fluids, the average value was 0.378; for parahydrogen, however, $n = 0.237$, and for acetaldehyde, $n = 0.589$.   Attempts were made to correlate n with T_c, P_c, the acentric factor, the solubility parameter, and many other parameters, but no trend could be observed.   Viswanath and Kuloor [92] also studied this problem.

It is recommended that n be chosen as a constant and equal to 0.38.   Between T_b and T_c, this value is generally satisfactory for engineering calculations.   Below T_b, ΔH_v increases more rapidly with decreasing temperature than predicted with a constant n.†·

The Watson equation predicts that ΔH_v decreases as temperature increases.   We have plotted $\psi \equiv \Delta H_v / (RT_c \Delta Z_v)$ for ethylbenzene in Fig. 6-10 to illustrate how the Watson equation [(6-16.1) with $n = 0.38$] agrees with values obtained from several vapor-pressure equations described earlier in this chapter (see Table 6-2).   For the Watson form, we estimated ΔZ_v by the Haggenmacher relation, Eq. (6-4.2), which is especially crude at high reduced temperatures.   Also the Watson-Haggenmacher base point was chosen at $T_{br} = 0.663$ to match the Frost-Kalkwarf-Thodos ψ value.   Though no definite conclusions can be drawn, it is clear that all relations except the Clapeyron and Antoine equations predict a minimum in ψ at about $T_r = 0.84$.   The Clapeyron

---

†In Sec. 5-8, this point is discussed further, and a technique is suggested for relating low-temperature enthalpies of vaporization to liquid heat capacities.

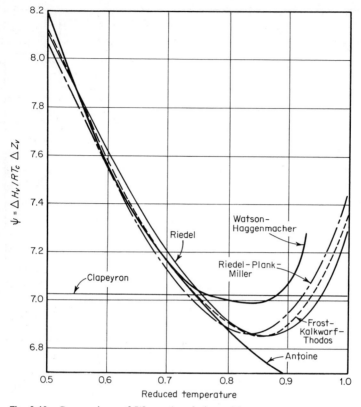

Fig. 6-10 Comparison of Watson's relation with vapor-pressure equations for estimating ψ.

equation predicts a constant ψ, and the Antoine equation is not applicable above a reduced temperature of about 0.75.

Fish and Lielmezs [22a] suggest a different way to correlate ΔH_v with T:

$$\Delta H_v = \Delta H_{v_b} \frac{T_r}{T_{b_r}} \frac{\chi + \chi^q}{1 + \chi^p} \qquad (6\text{-}16.2)$$

where

$$\chi = \frac{T_{b_r}}{T_r} \frac{1 - T_r}{1 - T_{b_r}} \qquad (6\text{-}16.3)$$

and the parameters q and p are

	q	p
Liquid metals	0.20957	−0.17467
Quantum liquids†	0.14543	0.52740
Inorganic and organic liquids	0.35298	0.13856

†Helium, hydrogen, deuterium, neon.

Compared with Eq. (6-16.1), with $n = 0.38$, Eq. (6-16.2) predicts slightly smaller values of ΔH_v when $T < T_b$ and somewhat higher when $T > T_b$. For liquid metals and quantum liquids, Eq. (6-16.2) is more accurate than Eq. (6-16.1), but for inorganic and organic liquids, the average errors reported by Fish and Lielmezs for Eqs. (6-16.1) and (6-16.2) were not significantly different.

6-17 Discussion and Recommendations for Enthalpy of Vaporization

Three techniques for estimating enthalpies of vaporization of pure liquids have been proposed. The first is based on Eq. (6-2.1) and requires finding dP_{vp}/dT either from a vapor-pressure–temperature correlation (Sec. 6-14) or from actual vapor-pressure data (Sec. 6-12). In both cases, a separate estimation of ΔZ_v must be made before ΔH_v can be obtained. This procedure is inherently accurate, especially if ΔZ_v is obtained from reliable P-V-T correlations discussed in Chap. 3. Any number of modifications could be programmed for use in a machine computation system We recommend the Lee-Kesler, Riedel, Frost-Kalkwarf-Thodos, or Riedel-Plank-Miller vapor-pressure equations for use over the entire liquid range, although the accuracy is lower near the freezing point and the critical point.

The second method is characteristic of many from the law of corresponding states. We selected the Pitzer et al. form as one of the most accurate and convenient. In an analytical form, this equation for ΔH_v is approximated by Eq. (6-13.4). Thompson and Braun [88] also recommended the Pitzer et al. form for hydrocarbons. The critical temperature and acentric factor are required.

The third method is to estimate ΔH_{v_b} as recommended in Sec. 6-15 and then scale with temperature with the Watson or Fish-Lielmezs functions discussed in Sec. 6-16.

All three of these techniques are satisfactory, and all yield approximately the same error when averaged over many types of fluids and over large temperature ranges. Fishtine [24], in a separate review of predictive methods for ΔH_v, did not consider the first of the methods discussed above but recommended the latter two as being of comparable accuracy.

All the methods noted above can be improved in accuracy for special cases. For example, Halm and Stiel [28, 29] have modified the Pitzer et al. correlation for polar fluids at lower reduced temperatures.

Finally, for all correlations discussed, T_c and P_c are required either directly or indirectly. Although these constants are available for many fluids—and can be estimated for most others—there are occasions when one would prefer not to use critical properties. (For example, for some high-molecular-weight materials or for polyhydroxylated compounds, it

TABLE 6-4 Vetere's Modification of the Kistiakowsky Equation

$\Delta H_{vb}/T_b = \Delta S_{vb}$, cal/g-mol K; T_b in kelvins; M = molecular weight

Type of compound	Correlation
Alcohols, acids, methylamine	$\Delta S_{vb} = 19.388 + 3.1269 \log T_b - 6.1589\, \dfrac{T_b}{M} + 0.035021\, \dfrac{T_b^2}{M} - 5.1056 \times 10^{-5}\, \dfrac{T_b^3}{M}$
Other polar compounds†	$\Delta S_{vb} = 10.604 + 3.664 \log T_b + 0.09354\, \dfrac{T_b}{M} + 1.035 \times 10^{-3}\, \dfrac{T_b^2}{M} - 1.345 \times 10^{-6}\, \dfrac{T_b^3}{M}$
Hydrocarbons	$\Delta S_{vb} = 13.91 + 3.27 \log M + 1.55\, \dfrac{[T_b - (263M)^{0.581}]^{1.037}}{M}$

†For esters, multiply the calculated value of ΔS_{vb} by 1.03.

is difficult to assign reliable values to T_c and P_c.) In such cases, one may have to use an approximate rule for the entropy of vaporization at the normal boiling point and then scale with temperature with the Watson relation described in Sec. 6-16. Several entropy-of-vaporization rules are described in an earlier edition of this book [74]. One of the more famous of such rules was suggested by Kistiakowsky [40].

$$\frac{\Delta H_{v_b}}{T_b} = \Delta S_{v_b} = 8.75 + R \ln T_b \qquad (6\text{-}17.1)$$

where T_b is in kelvins and ΔS_{v_b} in cal/g-mol K. Fishtine [23] improved the accuracy of Eq. (6-17.1) by employing a multiplicative correction factor that is a function of the compound class. More recently, Vetere [90] proposed a form wherein ΔS_{v_b} is correlated as a function of T_b and M, and his relations are given in Table 6-4. ΔH_{v_b} predicted from these equations is almost always within ± 5 percent of the experimental value, and for most cases the error is below 3 percent.

Example 6-11 Repeat Example 6-10 using the Vetere equation for ΔS_{v_b}.

 solution From Appendix A, $T_b = 321$ K and $M = 58.08$. Using the second equation in Table 6-4 gives

$$\Delta S_{v_b} = 10.604 + 3.664 \log 321 + 0.09354 \frac{321}{58.08}$$

$$+ 1.035 \times 10^{-3} \frac{321^2}{58.08} - 1.345 \times 10^{-6} \frac{321^3}{58.08}$$

$$= 21.37 \text{ cal/g-mol K}$$

$$\Delta H_{v_b} = (21.37)(321) = 6861 \text{ cal/g-mol}$$

$$\text{Error} = \frac{6861 - 6760}{6760} \times 100 = 1.5\%$$

Finally, we have not mentioned any of the literature dealing with estimation methods where ΔH_v is correlated relative to a reference material whose ΔH_v-T functionality is well established. Othmer has published widely on this method [61, 65], and Lu [45] and Das and Kuloor [20] have contributed. Such comparative methods, though convenient and often accurate, appear to offer no real advantages over other estimation techniques described earlier in this chapter.

6-18 Enthalpy of Fusion

The enthalpy change on fusion or melting is commonly referred to as the *latent heat of fusion*. It depends in part on the crystal form of the solid phase, and attempts to obtain general correlations have been unsuccessful. The Clausius-Clapeyron equation is applicable, but its use to calculate ΔH_m requires data on the variation of melting point with pressure—information which is seldom available.

TABLE 6-5 Sample Tabulation of Some Enthalpies and Entropies of Fusion for Simple Hydrocarbons†

Compound	T_m °C	T_m K	M	ΔH_m cal/g	ΔH_m cal/g-mol	$\Delta S_m = \dfrac{\Delta H_m}{T}$, cal/g-mol K	ΔH_{v_b}, cal/g	$\dfrac{\Delta H_m}{\Delta H_{v_b}}$
Methane	−182.5	90.7	16.04	14.03	225	2.48	122	0.12
Ethane	−183.3	89.9	30.07	22.73	683	7.60	117	0.19
Propane	−187.7	85.5	44.09	19.10	842	9.85	102	0.19
n-Butane	−138.4	134.8	58.12	19.17	1,114	8.26	92.1	0.21
Isobutane	−159.6	113.6	58.12	18.67	1,085	9.55	87.6	0.21
n-Pentane	−129.7	143.5	72.15	27.81	2,006	13.98	85.4	0.33
Isopentane	−159.9	113.3	72.15	17.06	1,231	10.86	81.0	0.21
Neopentane	−16.6	256.6	72.15	10.79	779	3.03	75.4	0.14
n-Hexane	−95.4	177.8	86.17	36.14	3,114	17.51	80.0	0.45
2-Methylpentane	−153.7	119.5	86.17	17.41	1,500	12.55	77.1	0.23
2,2-Dimethylbutane	−99.9	173.3	86.17	1.61	139	0.80	73.0	0.02
2,3-Dimethylbutane	−128.5	144.7	86.17	2.25	194	1.34	75.7	0.03
n-Heptane	−90.6	182.6	100.2	33.47	3,350	18.37	75.6	0.44
2-Methylhexane	−118.3	155.0	100.2	21.91	2,195	14.16	73.1	0.30
3-Ethylpentane	−118.6	154.6	100.2	22.78	2,283	14.77	73.8	0.31
2,2-Dimethylpentane	−123.8	149.4	100.2	13.89	1,392	9.33	69.6	0.20
2,4-Dimethylpentane	−119.2	154.0	100.2	16.32	1,635	10.62	70.4	0.23
3,3-Dimethylpentane	−134.5	138.7	100.2	16.86	1,689	12.18	70.7	0.24
2,2,3-Trimethylbutane	−24.9	248.3	100.2	5.39	540	2.18	69.0	0.08
n-Octane	−56.8	216.4	114.2	43.40	4,956	22.90	71.9	0.60
3-Methylheptane	−120.5	152.7	114.2	23.81	2,719	17.81	71.3	0.33
4-Methylheptane	−120.9	152.3	114.2	22.68	2,590	17.01	70.9	0.32
n-Nonane	−53.5	219.7	128.2	28.83	3,696	16.82	68.4	0.42
n-Decane	−29.6	243.6	142.3	48.24	6,865	28.18	66.0	0.73
n-Dodecane	−9.6	263.6	170.3	51.69	8,803	33.40	61.3	0.84
n-Octadecane	28.2	301.4	254.4	57.65	14,660	48.66	51.5	1.12
n-Nonadecane	32.1	305.3	268.5	40.78	10,950	35.88	50.3	0.81
n-Eicosane	36.8	310.0	282.5	59.11	16,700	53.87	48.8	1.21
Benzene	5.5	278.7	78.1	30.09	2,350	8.43	94.1	0.32
Toluene	−94.9	178.3	92.1	17.1	1,575	8.83	86.8	0.20
Ethylbenzene	−94.9	178.3	106.1	20.63	2,188	12.27	81.0	0.25
o-Xylene	−25.1	248.1	106.1	30.61	3,250	13.10	82.9	0.37
m-Xylene	−47.9	225.3	106.1	26.04	2,760	12.25	82.0	0.32
p-Xylene	13.3	286.5	106.1	38.5	4,080	14.24	81.2	0.47
n-Propylbenzene	−99.5	173.7	120.1	16.97	2,040	11.74	76.0	0.22
Isopropylbenzene	−96.0	177.2	120.1	14.15	1,700	9.59	74.6	0.19
1,2,3-Trimethylbenzene	−25.4	247.8	120.1	16.6	1,990	8.03	79.6	0.21
1,2,4-Trimethylbenzene	−43.8	229.4	120.1	25.54	3,070	13.4	78.0	0.33
1,3,5-Trimethylbenzene	−44.7	228.5	120.1	19.14	2,300	10.1	77.6	0.25
Cyclohexane	6.5	279.7	84.1	7.57	637	2.28	85.4	0.09
Methylcyclohexane	−126.5	146.7	98.1	16.4	1,610	10.97	77.2	0.21
Ethylcyclohexane	−111.3	160.9	112.2	17.73	1,930	12.00	73.9	0.24
1,1-Dimethylcyclohexane	−33.5	239.7	112.2	1.32	148	0.62	70.2	0.02
1,cis-2-Dimethylcyclohexane	−50.0	223.2	112.2	3.50	393	1.76	72.9	0.05
1,trans-2-Dimethylcyclohexane	−88.1	185.1	112.2	22.34	2,507	13.54	71.1	0.31

†Data taken primarily from R. R. Dreisbach, Physical Properties of Chemical Compounds, *Adv. Chem. Ser., ACS Monogr.* 15 and 22, 1955, 1959.

Sutra [83] and Mukherjee [55] have indicated how the prediction of ΔH_m might be approached on the basis of the modern hole theory of liquids, but their calculated values compare poorly with experimental values. Kuczinski [42] presents a theory relating ΔH_m and the modulus of rigidity of the solid. Good agreement between theoretical and experimental values is shown for eight metallic elements. For monatomic substances, ΔS_m is about equal to the gas constant R [33]. The best theoretical treatment has been presented by Bondi [11], who has related the entropy of fusion of molecular crystals to structure.

The difficulty in obtaining a general correlation of ΔH_m in terms of other physical properties is suggested by the selected values tabulated for some hydrocarbons in Table 6-5. It is evident that the simple introduction of a methyl group often has a marked effect, either increasing or decreasing ΔH_m. Results for optical and stereoisomers differ markedly. The variation of ΔH_m and of the entropy of fusion $\Delta H_m / T_m$ is as great as that of the melting points, which have never been correlated with other properties. It appears that there is no simple correlation between ΔH_m and melting point. Table 6-5 shows, for comparison, the relative constancy of the latent heat of vaporization at the normal boiling point. The ratio of $\Delta H_m / \Delta H_{v_b}$, however, varies greatly; i.e., in the table, values are found between 0.02 and 1.12. Since the enthalpy of vaporization increases with decreasing temperature, the ratio of heat effects between melting and vaporization at the melting point probably varies from about 0.1 to about 0.9.

6-19   Enthalpy of Sublimation

Solids vaporize without melting (*sublime*) at temperatures below the triple-point temperature. Sublimation involves an enthalpy increase, or *latent heat of sublimation*. This may be considered to be the sum of a latent heat of fusion and a latent heat of vaporization, even though liquid cannot exist at the pressure and temperature in question.

The latent heat of sublimation ΔH_s is best obtained from solid vapor-pressure data where such exist. For this purpose, the Clausius-Clapeyron equation (6-2.1) is applicable.

In only a very few cases is the sublimation pressure at the melting point known with any accuracy. It can be calculated from a liquid vapor-pressure correlation, with input such as T_b and T_c, by extrapolating to the melting point. Such a method is not recommended generally, as none of the vapor-pressure correlations is accurate in the very low pressure range. Even if P_{vp} at T_m is known, at least one other value of the vapor pressure of the solid is still necessary to calculate ΔH_s from the integrated form of the Clausius-Clapeyron equation. Thus, there appear to be no useful generalized correlations available for vapor pressures and enthalpies of sublimation for solids.

Another difficulty arises from transitions within the solid phase. Whereas the freezing point is a first-order transition[†] between vapor and solid, the solid may have some liquid characteristics such as free rotation. In many cases, at temperatures somewhat below the melting point, there is another first-order solid-solid transition between the crystallinelike solid and the more mobile liquidlike solid. The enthalpy of fusion and enthalpy of sublimation are different for these different solid phases. Ordinarily the literature values for ΔH_m and ΔH_s refer to the liquidlike solid phase. Bondi [10] suggests that ΔH_m or ΔH_s correlations would be much better if the crystalline solid phase were considered; i.e., the correlations would be based on the lowest first-order transition temperature, where there is some definite semblance of order in the solid structure. Solid-phase transitions are also discussed by Preston et al. [71].

In some cases, it is possible to obtain ΔH_s from thermochemical data and standard techniques of calculation by using known values of the heats of formation of solid and vapor. This is hardly a basis of estimation of an unknown ΔH_s, however, since the heats of formation tabulated in the standard references are often based in part on measured values of ΔH_s. If the heats of dissociation of both solid and gas phases are known, it is possible to formulate a cycle involving the sublimation of the solid, the dissociation of the vapor, and the recombination of the elements to form the solid compound.

Bondi [10] has suggested an additive-group technique to estimate ΔH_s at the lowest first-order transition temperature for molecular crystals of organic substances and of inorganic hydrides, perhalides, and percarbonyls. Usually the lowest first-order transition temperature is not much less than the melting point; e.g., for paraffins the two are almost identical.[‡] For some molecules, however, there is a considerable difference. Cyclohexane melts at 6.5°C, but the lowest first-order transition temperature is −87°C. There appear to be no general rules for predicting any of the first-order transition temperatures. Bondi's method, however, is only approximate, and great care must be taken in deciding what contributions are necessary in many cases.

Finally, as a rough engineering rule, one might estimate ΔH_v and ΔH_m

---

[†]In a first-order phase transition, there is a discontinuity in volume, enthalpy, and entropy between the phases in equilibrium. Vaporization and melting are first-order transitions. Second-order transitions are relatively rare and are characterized by the fact that the volume, enthalpy, and entropy are continuous between the phases, but there is a discontinuity in the heat capacities and compressibilities between the phases.

[‡]The exception appears to be n-butane; the usually quoted freezing point is 135 K, whereas the lowest first-order transition temperature is near 107 K.

separately and obtain ΔH_s as the sum. The latent heat of fusion is usually less than one-quarter of the sum; therefore the estimate may be fair even though that for ΔH_m is crude.

NOTATION

a = constant in Eq. (6-4.3); van der Waals constant

A = constant in Eqs. (6-3.1), (6-4.4), (6-5.1), (6-6.1), (6-7.1); parameter in Thek-Stiel correlation, Eq. (6-8.1)

A^+ = constant in Eq. (6-5.4) and defined in Eq. (6-5.5)

b = constant in Eq. (6-4.3)

B = constant in Eqs. (6-3.1), (6-4.4), (6-5.1), (6-6.1), (6-7.1)

B^+ = constant in Eq. (6-5.4) and defined in Eq. (6-5.5)

C = constant in Eqs. (6-3.1), (6-4.4), (6-5.1), (6-6.1), (6-7.1)

C^+ = constant in Eq. (6-5.4) and defined in Eq. (6-5.5)

D = constant in Eqs. (6-5.1), (6-6.1), (6-7.1)

D^+ = constant in Eq. (6-5.4) and defined in Eq. (6-5.5)

$E_i(x)$ = Chebyshev polynomial, Eq. (6-9.3)

$f^{(0)}, f^{(1)}$ = functions of reduced temperature in Eq. (6-2.6) and defined in Eqs. (6-2.7) and (6-2.8)

G = parameter in Eq. (6-7.2) and defined in Eq. (6-7.3)

h = parameter defined in Eq. (6-2.5)

ΔH_m = enthalpy change on melting, cal/g-mol

ΔH_v = enthalpy of vaporization, cal/g-mol; ΔH_{vb}, at T_b

ΔH_s = enthalpy of sublimation, cal/g-mol

k = parameter in Eq. (6-7.2) and defined in Eq. (6-7.4)

K = constant in Eq. (6-15.7)

n = exponent in Eq. (6-16.1); usually chosen as 0.38

P = pressure, atm; $P_r = P/P_c$

P_c = critical pressure, atm

P_{vp} = vapor pressure; $P_{vp_r} = P_{vp}/P_c$

Q = parameter defined in Eq. (6-5.5)

R = gas constant, 1.987 cal/g-mol K

ΔS_v = entropy of vaporization, cal/g-mol K; $\Delta S^{(0)}$ and $\Delta S^{(1)}$, Pitzer parameters in Eq. (6-13.2)

T = temperature, K; $T_r = T/T_c$; T_b, normal boiling point; T_m, melting point

T_c = critical temperature, K

ΔU_v = internal energy of vaporization, cal/g-mol

V = volume, cm^3/g-mol; V_g, saturated vapor; V_L, saturated liquid

V_c = critical volume, cm^3/g-mol

ΔV_v = volume change in vaporization, cm^3/g-mol

x = temperature function in Eq. (6-9.2)

Z = compressibility factor, PV/RT; Z_g or Z_{SV}, saturated vapor; Z_L or Z_{SL}, saturated liquid

$\Delta Z_v = Z_g - Z_L$; ΔZ_{vb}, at normal boiling point

$Z_c = Z$ at the critical point

Greek

α = parameter in Eq. (6-5.2)

α_c = Riedel factor, α at T_c

δ_T = parameter in Eq. (6-12.1)

χ = temperature parameter defined in Eq. (6-16.3)

$\psi = \Delta H_v / (RT_c \, \Delta Z_v)$

ψ_b = parameter defined in Eq. (6-5.7)

ω = acentric factor

REFERENCES

1. Abrams, D. S., H. A. Massaldi, and J. M. Prausnitz: *Ind. Eng. Chem. Fundam.*, **13**: 259 (1974).
2. Ambrose, D.: "Vapor-Pressure Equations," *Nat. Phys. Lab. Rep. Chem.* 19, November 1972.
3. Ambrose, D., B. E. Broderick, and R. Townsend: *J. Chem. Soc.*, **1967A**: 633.
4. Ambrose, D., J. F. Counsell, and A. J. Davenport: *J. Chem. Thermodyn.*, **2**: 283 (1970).
5. Ambrose, D., and C. H. S. Sprake: *J. Chem. Soc.*, **1971A**: 1263.
6. Ambrose, D., C. H. S. Sprake, and R. Townsend: *J. Chem. Thermodyn.*, **6**: 693 (1974).
7. Antoine, C.: *C. R.*, **107**: 681, 836 (1888).
8. "ASHRAE Handbook for Thermodynamic Properties, 1972," table 45; data from York Division of Borg Warner, Inc., unpublished.
9. Bond, D. L., and G. Thodos: *J. Chem. Eng. Data*, **5**: 289 (1960).
10. Bondi, A.: *J. Chem. Eng. Data*, **8**: 371 (1963).
11. Bondi, A.: *Chem. Rev.*, **67**: 565 (1967).
12. Bondi, A., and R. B. McConaughy: Estimation of Vapor Pressures for Pure Hydrocarbons with 5 to 30 Carbon Atoms, paper presented at *27th Midyear Meet., Am. Pet. Inst., San Francisco, May 1962.*
13. Boublík, T., V. Fried, and E. Halá: "The Vapor Pressures of Pure Substances," Elsevier, New York, 1973.
14. Calingaert, G., and D. S. Davis: *Ind. Eng. Chem.*, **17**: 1287 (1925).
15. Carruth, G. F., and R. Kobayashi: *Ind. Eng. Chem. Fundam.*, **11**: 509 (1972).
16. Chen, N. H.: *J. Chem. Eng. Data*, **10**: 207 (1965).
17. Clenshaw, C. W.: *Computer J.*, **2**: 170 (1960); C. W. Clenshaw and J. G. Hayes, *J. Inst. Math. Appl.*, **1**: 164 (1965).
18. Counsell, J. F., and D. A. Lee: *J. Chem. Thermodyn.*, **4**: 915 (1972).
19. Cox, E. R.: *Ind. Eng. Chem.*, **15**: 592 (1923).
20. Das, T. R., and N. R. Kuloor: *Hydrocarbon Process.*, **47**(2): 137 (1968).
21. Dreisbach, R. R.: "Pressure-Volume-Temperature Relationships of Organic Compounds," 3d ed., McGraw-Hill, New York, 1952.
22. Fedors, R. F.: *Polym. Eng. Sci.*, **14**: 147 (1974).
22a. Fish, L. W., and J. Lielmezs: *Ind. Eng. Chem. Fundam.*, **14**: 248 (1975).
23. Fishtine, S. H.: *Ind. Eng. Chem.*, **55**(4): 20 (1963), **55**(5): 49 (1963), **55**(6): 47 (1963); *Hydrocarbon Process. Pet. Refiner*, **42**(10): 143 (1963).
24. Fishtine, S. H.: *Hydrocarbon Process.*, **45**(4): 173 (1966).
25. Frost, A. A., and D. R. Kalkwarf: *J. Chem. Phys.*, **21**: 264 (1953).
26. Giacalone, A.: *Gazz. Chim. Ital.*, **81**: 180 (1951).
27. Haggenmacher, J. E.: *J. Am. Chem. Soc.*, **68**: 1633 (1946).
28. Halm, R. L.: Ph.D. thesis, Syracuse University, Syracuse, N.Y., 1968.
29. Halm, R. L., and L. I. Stiel: *AIChE J.*, **13**: 351 (1967).
30. Hamrin, C. E., Jr., and G. Thodos: *J. Chem. Phys.*, **35**: 899 (1961).
31. Harlacher, E. A., and W. G. Braun: *Ind. Eng. Chem. Process Des. Dev.*, **9**: 479 (1970).
32. Helgeson, N. L., and B. H. Sage: *J. Chem. Eng. Data*, **12**: 47 (1967).
33. Hirschfelder, J. O., C. F. Curtiss, and R. B. Bird: "Molecular Theory of Gases and Liquids," Wiley, New York, 1954.

34. Hobson, M., and J. H. Weber: *AIChE J.*, **2**: 354 (1956).
35. Hobson, M., and J. H. Weber: *Petrol. Process.*, **12**(8): 43 (1957).
36. Hooper, E. D., and J. Joffee: *J. Chem. Eng. Data*, **5**: 155 (1960).
37. Ibrahim, S. H., and N. R. Kuloor: *Chem. Eng.*, **74**(12): 147 (1967).
38. Karapet'yants, M. Kh., and K. Ch'ung: *Ssu Ch'uan Ta Hsueh Hsueh Pao-Tzu Jan K'o Hsueh*, **1958**: 91; *Chem. Abstr.*, **53**, 17613 (1959).
39. King, M. B., and H. Al-Najjar: *Chem. Eng. Sci.*, **29**: 1003 (1974).
40. Kistiakowsky, W.: *Z. Phys. Chem.*, **107**: 65 (1923).
41. Klein, V. A.: *Chem. Eng. Prog.*, **45**: 675 (1949).
42. Kuczinski, G. C.: *J. Appl. Phys.*, **24**: 1250 (1953).
43. Lee, B. I., and M. G. Kesler: *AIChE J.*, **21**: 510 (1975).
44. Lu, B. C.-Y.: *Can. J. Chem. Eng.*, **38**: 33 (1960).
45. Lu, B. C.-Y.: *Can. J. Chem. Eng.*, **42**: 123 (1964).
46. Lydersen, A. L., R. A. Greenkorn, and O. A. Hougen: Generalized Thermodynamic Properties of Pure Fluids, *Univ. Wisconsin, Coll. Eng., Eng. Exp. Stn. Rep.* 4, Madison, Wis., October 1955.
47. McCurdy, K. G., and K. J. Laidler: *Can. J. Chem.*, **41**: 1867 (1963).
48. McKay, R. A., and B. H. Sage: *J. Chem. Eng. Data*, **5**: 21 (1960).
49. Miller, D. G.: *Ind. Eng. Chem. Fundam.*, **2**: 68 (1963).
50. Miller, D. G. (also discussion by G. Thodos): *Ind. Eng. Chem. Fundam.*, **2**: 78, 80 (1963).
51. Miller, D. G.: *Ind. Eng. Chem.*, **56**(3): 46 (1964).
52. Miller, D. G.: *J. Phys. Chem.*, **68**: 1399 (1964).
53. Miller, D. G.: personal communication, April 1964.
54. Miller, D. G.: *Univ. California Rad. Lab. Rep.* 14115-T, Apr. 21, 1965.
55. Mukherjee, N. R.: *J. Chem. Phys.*, **19**: 502, 1431 (1951).
56. Narsimhan, G.: *Br. Chem. Eng.*, **10**(4): 253 (1965), **12**(6): 897 (1967).
57. Narsimhan, G.: *Can. J. Chem. Eng.*, **45**: 230 (1967).
58. Ogden, J. M., and J. Lielmezs: *AIChE J.*, **15**: 469 (1969).
59. Othmer, D. F.: *Ind. Eng. Chem.*, **32**: 841 (1940).
60. Othmer, D. F.: *Ind. Eng. Chem.*, **34**: 1072 (1942).
61. Othmer, D. F., and H.-T. Chen: *Ind. Eng. Chem.*, **60**(4): 39 (1968).
62. Othmer, D. F., and H. N. Huang: *Ind. Eng. Chem.*, **57**(10): 42 (1965).
63. Othmer, D. F., P. W. Maurer, C. J. Molinary, and R. C. Kowalski: *Ind. Eng. Chem.*, **49**: 125 (1957).
64. Othmer, D. F., and E.-S. Yu: *Ind. Eng. Chem.*, **60**(1): 23 (1968).
65. Othmer, D. F., and D. Zudkevitch: *Ind. Eng. Chem.*, **51**: 791 (1959).
66. Pasek, G., and G. Thodos: *J. Chem. Eng. Data*, **7**: 21 (1962).
67. Passut, C. A., and R. P. Danner: *Chem. Eng. Prog. Symp. Ser.*, **70**(140): 30 (1974).
68. Pitzer, K. S., D. Z. Lippmann, R. F. Curl, C. M. Huggins, and D. E. Petersen: *J. Am. Chem. Soc.*, **77**: 3433 (1955).
69. Plank, R., and L. Riedel: *Ing. Arch.*, **16**: 255 (1948).
70. Plank, R., and L. Riedel: *Tex. J. Sci.*, **1**: 86 (1949).
71. Preston, G. T., E. W. Funk, and J. M. Prausnitz: *J. Phys. Chem.*, **75**: 2345 (1971).
72. Procopio, J. M., and G. J. Su: *Chem. Eng.*, **75**(12), 101 (1968).
73. Rehberg, C. E.: *Ind. Eng. Chem.*, **42**: 829 (1950).
74. Reid, R. C., and T. K. Sherwood: "The Properties of Gases and Liquids," 2d ed., pp. 149–153, McGraw-Hill, New York, 1966.
75. Reynes, E. G., and G. Thodos: *AIChE J.*, **8**: 357 (1962).
76. Reynes, E. G., and G. Thodos: *Ind. Eng. Chem. Fundam.*, **1**: 127 (1962).
77. Riedel, L.: *Chem. Ing. Tech.*, **26**: 83 (1954).
78. Riedel, L.: *Chem. Ing. Tech.*, **26**: 679 (1954).
79. Silverberg, P. M., and L. A. Wenzel: *J. Chem. Eng. Data*, **10**: 363 (1965).

80. Smith, C. H., and G. Thodos: *AIChE J.*, **6**: 569 (1960).
80a. Smith, G., J. Winnick, D. S. Abrams, and J. M. Prausnitz: *Can. J. Chem. Eng.*, in press, 1976.
81. Sondak, N. E., and G. Thodos: *AIChE J.*, **2**: 347 (1956).
82. Stearns, W. V., and E. M. George: *Ind. Eng. Chem.*, **35**: 602 (1943).
83. Sutra, G.: *C. R.*, **233**: 1027, 1186 (1951).
84. Thek, R. E., and L. I. Stiel: *AIChE J.*, **12**: 599 (1966), **13**: 626 (1967).
85. Thomas, L. H., and H. Smith: *J. Appl. Chem.*, **20**(1): 33 (1970).
86. Thomson, G. W.: *Chem. Rev.*, **38**: 1 (1946).
87. Thomson, G. W.: "Techniques of Organic Chemistry," A. Weissberger (ed.), 3d ed., vol. I, pt. I, p. 473, Interscience, New York, 1959.
88. Thompson, W. H., and W. G. Braun: *29th Midyear Meet., Am. Pet. Inst., Div. Refining, St. Louis, Mo., May 11, 1964, prepr.* 06–64.
89. Vetere, A.: New Generalized Correlations for Enthalpy of Vaporization of Pure Compounds, Laboratori Ricerche Chimica Industriale, SNAM PROGETTI, San Donato Milanese, 1973.
90. Vetere, A.: Modification of the Kistiakowsky Equation for the Calculation of the Enthalpies of Vaporization of Pure Compounds, Laboratori Ricerche Chimica Industriale, SNAM PROGETTI, San Donato Milanese, 1973.
91. Viswanath, D. S., and N. R. Kuloor: *J. Chem. Eng. Data*, **11**: 69, 544 (1966).
92. Viswanath, D. S., and N. R. Kuloor: *Can. J. Chem. Eng.*, **45**: 29 (1967).
93. Watson, K. M.: *Ind. Eng. Chem.*, **35**: 398 (1943).
94. Wilhoit, R. C., and B. J. Zwolinski: "Handbook of Vapor Pressures and Heats of Vaporization of Hydrocarbons and Related Compounds," Texas A & M University, Thermodynamics Research Center, College Station, 1971.

Chapter Seven

Thermodynamic Properties of Ideal Gases

7-1 Scope and Definitions

The present chapter describes methods for estimating enthalpies of formation, heat capacities, and entropies (or Gibbs energies) of organic compounds in the ideal-gas state.

Enthalpies and heat capacities of ideal gases are not functions of pressure; for these properties, therefore, no reference pressure need be stipulated. For entropies and free energies, the reference pressure employed throughout the chapter is 1 atm unless specifically stated otherwise.

The enthalpy of formation of a compound $\Delta H^{\circ}_{f_T}$ is defined as the isothermal enthalpy change in a synthesis reaction from the elements in their standard states. In such a reaction scheme, the elements are assumed initially to be at reaction temperature, at 1 atm, and in their most stable configuration, e.g., diatomic oxygen as an ideal gas at T, carbon as a solid in the form of graphite, etc. Ordinarily one need not be concerned with enthalpies of formation of the elements since, to obtain a standard heat of reaction, by virtue of the conservation of atoms, the enthalpies of formation of all elements cancel.   For a general reaction

$$aA + bB = cC + dD$$

the standard enthalpy change to form products C and D from A and B in stoichiometric amounts when reactants and products are pure and at T and 1 atm is given by

$$\Delta H_T^\circ = c\ \Delta H_{f_T}^\circ(C) + d\ \Delta H_{f_T}^\circ(D) - a\ \Delta H_{f_T}^\circ(A) - b\ \Delta H_{f_T}^\circ(B) \quad (7\text{-}1.1)$$

No enthalpies of formation of any elements appear (if A, B, C, or D are elements, each $\Delta H_{f_T}^\circ$ is set equal to zero).

Only the enthalpy of formation is normally given at 298 K.   Should one want ΔH_f° at other temperatures,

$$\Delta H_{f_T}^\circ = \Delta H_{f_{298}}^\circ + \int_{298}^{T} \Delta C_p^\circ\, dT \quad (7\text{-}1.2)$$

Here ΔC_p° is the difference in heat capacity for the material and the heat capacities of the elements which constitute the material, each element is in its standard state (as noted earlier), and each is multiplied by the appropriate stoichiometric multiplier.   To determine how standard heats of reaction vary with temperature, combine Eqs. (7-1.1) and (7-1.2):

$$\Delta H_T^\circ = \sum_j \nu_j\ \Delta H_{f_{298}}^\circ(j) + \sum_j \int_{298}^{T} \nu_j C_p^\circ(j)\, dT \quad (7\text{-}1.3)$$

where ν_j is the stoichiometric multiplier for each component (negative for reactants and positive for products).   Thus, to determine ΔH_T°, we need to know $\Delta H_{f_{298}}^\circ$ and C_p°, the ideal-gas heat capacity for all reactants and products.†

The Gibbs energy of formation $\Delta G_{f_T}^\circ$ is defined in a manner analogous to that for $\Delta H_{f_T}^\circ$, and the standard Gibbs energy of reaction ΔG_T° may be written in a form similar to that of Eq. (7-1.1).   Unless otherwise specified, all reactants and all products are pure, ideal gases at temperature T and at 1 atm.

Finally, standard entropies of elements and compounds S_T° are also in the pure, ideal-gas state, at 1 atm and at temperature T.   This entropy is relative to that at absolute zero temperature in a perfectly ordered solid state.   The entropy of formation $\Delta S_{f_T}^\circ$ could also have been introduced, but no estimation schemes yield this property directly.   For use in chemical reactions, since the standard-state entropies of the

---

†If some of the products or reactants are elements, C_p° refers to the heat capacity of the element in its standard state; this state may not be an ideal gas, e.g., carbon.   In such cases, the actual standard-state element heat capacities must be known.

elements cancel, one may employ S_T° directly; i.e., for the general reaction introduced above,

$$\Delta S_T^\circ = c S_T^\circ(C) + d S_T^\circ(D) - a S_T^\circ(A) - b S_T^\circ(B) \qquad (7\text{-}1.4)$$

ΔS_T° may be expressed in terms of values of S_{298}° for reactants and products in a manner similar to Eq. (7-1.3):

$$\Delta S_T^\circ = \sum_j \nu_j S_{298}^\circ(j) + \sum_j \int_{298}^T \nu_j C_p^\circ(j) \, d \ln T \qquad (7\text{-}1.5)$$

Finally, the standard Gibbs energy change for a chemical reaction ΔG_T° can be expressed as

$$\Delta G_T^\circ = \Delta H_T^\circ - T \, \Delta S_T^\circ \qquad (7\text{-}1.6)$$

where ΔH_T° and ΔS_T° are found from Eqs. (7-1.3) and (7-1.5).

7-2 Estimation Methods

Since the properties covered in this chapter are those of an *ideal gas*, intermolecular forces can play no part in their estimation. Also, by the same reasoning, the law of corresponding states, used so widely in other chapters, is inapplicable. All estimation methods for C_p°, ΔH_f°, S°, and ΔG_T° involve some form of group-estimation method based on the structure of the molecule. Benson [4] and Benson and Buss [5] have pointed out a hierarchy of such methods. Most simple would be those which assign contributions based only on the atoms present in a molecule. While exact for molecular weights and even reasonably good for a few other properties, e.g., liquid molal volume at the boiling point, such methods are completely inadequate for the properties discussed in this chapter.

Only slightly more complicated are methods which assign contributions to various chemical bonds. Such techniques are easy to use but yield only approximate results. The most successful method assigns contributions to common molecular groupings, for example, —CH₃, —NH₂, —COOH, and, by simple additivity, one can then estimate ideal-gas properties from a table of group values.

Proceeding to more complicated, and usually more accurate, methods, atoms or molecular groups are chosen, and allowance is made for next-nearest neighbors to this atom or group. The method of Benson and coworkers, discussed later, is illustrative of this approach. There is no limit to the extension of the *neighborhood-environment* technique. Allowance could be made for neighboring atoms (or groups) some distance away, though usually the effect of neighbors separated by more than one atom becomes so small as to be insignificant. An exception would be a gauche interaction between carbon atoms separated by two other carbon atoms. Obvious interactions may become apparent from molecular models.

Most estimation methods described later consist of a mixture of the hierarchial listing discussed above. Authors have formulated specific rules to accompany their group contributions, and it is imperative that these rules be followed to obtain accurate estimates.

7-3   Ideal-Gas Heat Capacity

Before attempting to estimate a heat capacity, one should first ascertain whether or not values are already available.   In Appendix A, constants are given for many compounds to allow the calculation of C_p° with the equation

$$C_p^\circ = A + BT + CT^2 + DT^3 \tag{7-3.1}$$

where C_p° is in cal/g-mol K and T is in kelvins.   Other convenient references for C_p° include the JANAF tables [18] and the book by Stull et al. [33], where C_p° is tabulated as a function of temperature for several hundred organic compounds.   Passut and Danner [22] and Huang and Daubert [17] list constants for a fourth-order polynomial equation for many hydrocarbons.   Thinh et al. [36] also have tabulated A, B, C, and D [in Eq. (7-3.1)] for a large number of compounds.   Yuan and Mok [42] use an exponential temperature function to correlate C_p° with temperature and show constants for many hydrocarbons and non-hydrocarbons.   A nomograph by Tans [34] allows a rapid, rough approximation of C_p° for paraffins.

Theoretical methods are available to calculate C_p° if sufficient spectroscopic data are available to delineate the important vibrational and rotational frequencies in a molecule.   In fact many *experimental* data were not determined calorimetrically but were calculated from experimental spectral data.

Translational and external rotational contributions can readily be found from classical theory; i.e., for translation, $C_p^\circ = \frac{3}{2}R$ and for external rotation, $C_p^\circ = \frac{3}{2}R$ for nonlinear and R for linear molecules.   (R is the gas constant.)   The difficulty of selecting appropriate average vibrational and rotational frequencies over wide temperature ranges has led to the use of more accurate, but empirical, estimation methods based on group or bond contributions.

To illustrate the latter, bond contributions are shown in Table 7-1 for estimating C_p° at 298 K (S_{298}° and ΔH_{f298}° values are discussed later).   Although one may not treat certain types of molecules with the bonds shown, e.g., acetylenes, the estimation of C_p° (298 K) is rapid and usually reasonable.   In Table 7-5, calculated and literature values of C_p° are compared, and the method is illustrated in Example 7-1.

TABLE 7-1   Bond Contributions for C_p°, S°, and ΔH_f° at 298 K [4]

Bond	C_p°, cal/g-mol K	S°, cal/g-mol K	ΔH_f°, kcal/g-mol
C—H	1.74	12.90	-3.83
C—D	2.06	13.60	-4.73
C—C	1.98	-16.40	2.73
C_d—H†	2.6	13.8	3.2
C_d—C	2.6	-14.3	6.7
ϕ—H‡	3.0	11.7	3.25
ϕ—C	4.5	-17.4	7.25
C—F	3.34	16.90	
C_d—F	4.6	18.6	
C—Cl	4.64	19.70	-7.4
C_d—Cl	5.7	21.2	-0.7
C—Br	5.14	22.65	2.2
C_d—Br	6.3	24.1	9.7
C—I	5.54	24.65	14.1
C_d—I	6.7	26.1	21.7
C—O	2.7	-4.0	-12.0
O—H	2.7	24.0	-27.0
O—D	3.1	24.8	-27.9
O—Cl	5.5	32.5	9.1
O—O	4.9	9.1	21.5
H—CO§	4.2	26.8	-13.9
C—CO	3.7	-0.6	-14.4
O—CO	2.2	9.8	-50.5
F—CO	5.7	31.6	
Cl—CO	7.2	35.2	-27.0
C—N	2.1	-12.8	9.3
N—H	2.3	17.7	-2.6
C—S	3.4	-1.5	6.7
S—H	3.2	27.0	-0.8
S—S	5.4	11.6	
(NO_2)—O	$\cdots$	43.1	-3.0
(NO)—O	$\cdots$	35.5	9.0

†C_d represents the tetravalent group, $\rangle C{=}C\langle$.

‡ϕ represents the hexavalent aromatic nucleus.

§H—CO represents a bond between hydrogen and a carbonyl carbon.

Example 7-1   Using Table 7-1, estimate the ideal-gas heat capacity of ethyl acetate at 298 K.
solution

$$C_p^\circ = 8(\text{C—H}) + \text{C—CO} + \text{O—CO}$$
$$+ \text{C—O} + \text{C—C}$$
$$= 8(1.74) + 3.7 + 2.2 + 2.7 + 1.98$$
$$= 24.5 \text{ cal/g-mol K}$$

Stull [33] reports a value of 27.16 cal/g-mol K at this temperature.

TABLE 7-2 Thinh, Duran, and Ramalho Group Contributions to Ideal-Gas Heat Capacity [35]

Group	A	B_1	C_1	n_1	B_2	C_2	n_2
			Aliphatic Hydrocarbon Groups				
—CH$_3$	4.7366	20.4410	1013.8229	1.0489	0	0	0
—CH$_2$	3.0820	136.8702	788.7739	1.0380	120.3019	832.8313	1.0452
—C—H	−0.6214	75.9952	601.8911	0.9953	61.3229	1013.8229	1.0489
—C—	−2.4690	89.6797	733.9538	1.0396	81.7638	1013.8229	1.0489
=CH$_2$	3.4382	15.3919	527.1308	0.9644	0	0	0
—C≡	2.4980	24.0973	1133.5426	1.0774	20.4410	1013.8229	1.0489
≡CH	1.6538	12.4458	21.3585	0.5	0	0	0
=C=	1.5307	36.4231	321.4962	0.9048	30.7839	527.1308	0.9644
H\C=CH$_2$	1.6606	31.4786	134.8699	0.8030	0	0	0
\C=CH$_2$	4.9218	62.5821	652.9594	0.9990	40.8819	1013.8229	1.0489

$\diagdown$C=C$\diagup$ (tetrasubstituted)	−0.0015	102.8316	511.5828	0.9573	81.7638	1013.8229	1.0489
$\diagdown$C=C$\diagdown$H	2.6362	78.5993	874.4157	1.0464	61.3229	1013.8229	1.0489
H$\diagdown$C=C$\diagup$H (cis)	3.1836	62.2851	1110.9532	1.0843	40.8819	1013.8229	1.0489
H$\diagdown$C=C$\diagup$H (trans)	5.9189	60.2442	990.4639	1.0590	40.8819	1013.8229	1.0489
$\diagdown$C=C=CH$_2$	7.5007	66.6656	578.8631	0.9870	40.8819	1013.8229	1.0489
$\diagdown$C=C=CH$_2$ H	8.5582	49.9143	700.3515	1.0163	20.4410	1013.8229	1.0489
$\diagdown$C=C=C$\diagup$H H	8.0606	64.5906	949.4988	1.0596	40.8819	1013.8229	1.0489
Aromatic Hydrocarbon Groups							
HC $\diagup\diagdown$	1.4345	10.8720	1174.9378	1.1387	0	0	0

TABLE 7-2 Thinh, Duran, and Ramalho Group Contributions to Ideal-Gas Heat Capacity [35] *(Continued)*

Group	A	B_1	C_1	n_1	B_2	C_2	n_2
		Aromatic Hydrocarbon Groups *(Continued)*					
$-C\diagup^{\diagdown}_{\diagup}$	1.3635	79.6089	1269.0479	1.1387	74.9947	986.1985	1.0928
$\leftrightarrow C\diagup^{\diagdown}_{\diagup}$	2.3397	47.6491	1587.2948	1.1915	43.4881	1174.9378	1.1387
		Ring-Formation Correction					
Three-membered ring	−1.7223	30.1047	4304.6037	1.3222	32.9162	1237.7501	1.1173
Five-membered ring, cyclopentane	−5.5560	86.9988	1281.9257	1.1221	82.7124	545.5184	0.9912
Five-membered ring, cyclopentene	−2.3485	70.5966	1877.9329	1.1852	70.5846	813.0059	1.0604
Six-membered ring, cyclohexane	−1.8728	95.6365	4371.2289	1.3210	99.2549	545.5184	0.9912
Six-membered ring, cyclohexene	−4.7552	91.0076	972.7163	1.1001	87.0564	756.9254	1.0481
		Correction for Branching in Cycloparaffins					
Branching in five-membered ring:							
Single branching	2.6826	102.9302	1040.8363	1.0906	104.7224	1062.1344	1.1013
Double branching:							
1,1 position	5.4945	115.3182	1362.2334	1.1394	122.2821	951.9636	1.0909
1, cis-2	5.1851	116.5713	1170.5749	1.1148	122.2821	951.9636	1.0909

1, trans-2	5.0336	117.2347	1084.4398	1.1022	122.2821	951.9636	1.0909
1, cis-3	5.0336	117.2347	1084.4398	1.1022	122.2821	951.9636	1.0909
1, trans-3	5.0336	117.2347	1084.4398	1.1022	122.2881	951.9636	1.0909

Branching in six-membered ring:

Single branching	1.5560	112.8656	2470.4850	1.2344	113.1267	3061.5552	1.2706

Double branching:

1,1 position	4.5367	124.6332	3819.4942	1.3044	130.5289	2405.6012	1.2375
cis 1,2	3.6537	126.7920	2640.9182	1.2462	130.5289	2405.6012	1.2375
trans 1,2	2.9896	128.1617	2189.0807	1.2193	130.5289	2405.6012	1.2375
cis 1,3	4.7228	125.9625	3032.2880	1.2661	130.5289	2405.6012	1.2375
trans 1,3	3.8571	127.3311	2403.6929	1.2294	130.5289	2405.6012	1.2375
cis 1,4	3.8571	127.3311	2403.6929	1.2294	130.5289	2405.6012	1.2375
trans 1,4	3.4805	126.8064	2616.9819	1.2473	130.5289	2405.6012	1.2375

Correction for Branching in Aromatics

Double branching:

1,2 position	2.2108	93.1940	1137.6065	1.1076	93.8766	1381.8801	1.1442
1,3 position	0.4109	94.1667	1266.8220	1.1287	93.8766	1381.8801	1.1442
1,4 position	1.0741	93.3095	1414.2979	1.1428	93.8766	1381.8801	1.1442

Triple branching:

1,2,3 position	3.2258	106.7469	1416.3668	1.1311	108.0631	1496.1894	1.1513
1,2,4 position	3.7823	105.6478	1513.4331	1.1414	108.0631	1496.1894	1.1513
1,3,5 position	1.4321	107.5030	1482.6086	1.1443	108.0631	1496.1894	1.1513

Special Correction for First Few $-CH_2$ in Normal Series

For first three $-CH_2$ in normal paraffins:

First	−1.6075	59.4196	653.8562	0.9992	57.8390	779.2506	1.0174
$-CH_2-$							

TABLE 7-2 Thinh, Duran, and Ramalho Group Contributions to Ideal-Gas Heat Capacity [35] (*Continued*)

Group	A	B_1	C_1	n_1	B_2	C_2	n_2
Special Correction for First Few —CH$_2$ in Normal Series (*Continued*)							
First and Second —CH$_2$	−0.5942	74.1066	745.5631	1.0920	74.5838	694.6245	1.0053
First, Second, and Third —CH$_2$	−0.3156	89.9889	750.0491	1.0258	91.2438	652.7659	0.9996
For first three —CH$_2$ in normal alkyl benzenes: First	−0.6231	96.7804	972.3475	1.0980	95.8682	1115.4125	1.1162
First and Second —CH$_2$	0.9122	110.6626	1065.8703	1.1084	112.1937	1014.5118	1.0997
First, Second, and Third —CH$_2$	1.1845	126.5265	1024.6444	1.1005	128.5645	943.4540	1.0870
For first two —CH$_2$ in normal monoolefins: First	1.8693	64.0369	616.9205	0.9973	68.6181	313.8732	0.8983

First and Second $-CH_2$	3.0142	78.5361	721.9341	1.0225	84.9976	352.8265	0.9180

For first two $-CH_2$ in normal acetylenes:

First $-CH_2$	−1.9514	54.0869	333.3507	0.9005	49.9997	504.9147	0.9628
First and Second $-CH_2$	−0.2389	66.9791	533.4160	0.9750	66.5609	510.4336	0.9686

For first $-CH_2$, outside ring, in normal alkyl cyclopentanes:

First $-CH_2$	−7.2831	140.4183	312.1951	0.9010	119.3679	956.9970	1.0777

For first two $-CH_2$, outside ring, in normal alkyl cyclohexanes:

First $-CH_2$	0.1936	128.1843	2165.9117	1.2143	128.8507	2097.6983	1.2081
First and Second $-CH_2$	−0.5250	147.1091	1465.2225	1.1521	144.9198	1839.8190	1.1870

Method of Thinh, Duran, and Ramalho [35]   An additive group method for estimating C_p° for *hydrocarbons*, this method employs the equation

$$C_p^\circ = \sum_i n_i (A + B_1\, e^{-C_1/T^{n_1}} - B_2\, e^{-C_2/T^{n_2}})_i \qquad (7\text{-}3.2)$$

The values of the constants A, B_1, C_1, n_1, B_2, C_2, and n_2 are given in Table 7-2 for a large number of hydrocarbon groups.   The temperature-dependent function in parentheses is determined for each ith group and the resultant is summed.

Equation (7-3.2) is modified from a simpler form suggested earlier by Yuan and Mok [42] and is generally accurate.   Thinh et al. made an extensive comparison with available C_p° data for 42 hydrocarbons [1] and found an average error less than 0.5 percent.

The technique is illustrated in Example 7-2, and calculated values of C_p° are compared with those reported in the literature in Table 7-5. Except in a few cases, errors are well under 1 percent.

Example 7-2   Using the group-contribution method of Thinh et al., estimate the ideal-gas heat capacity of isoprene (2-methyl-1,3-butadiene) at 800 K.
   solution

$$C_p^\circ = -CH_3 + \overset{H}{\underset{\diagup}{\diagdown}}C{=}CH_2 + \overset{\diagdown}{\underset{\diagup}{}}C{=}CH_2$$

For $-CH_3$ at 800 K, from Eq. (7-3.2) and Table 7-2,

$$-CH_3 = 4.7366 + 20.4410 \exp\left(\frac{-1013.8229}{800^{1.0489}}\right)$$

$$= 12.93$$

$$\overset{H}{\underset{\diagup}{\diagdown}}C{=}CH_2 = 1.6606 + 31.4786 \exp\left(\frac{-134.8699}{800^{0.8030}}\right)$$

$$= 18.44$$

$$\overset{\diagdown}{\underset{\diagup}{}}C{=}CH_2 = 4.9218 + 62.5821 \exp\left(\frac{-652.9594}{800^{0.9990}}\right) - 40.8819 \exp\left(\frac{-1013.8229}{800^{1.0489}}\right)$$

$$= 16.04$$

Thus, $C_p^\circ = 12.93 + 18.44 + 16.04 = 47.42$ cal/g-mol K.   The value reported in the literature is 48.0 cal/g-mol K [33].

Method of Rihani and Doraiswamy [26]   Another additive group method which is applicable to many types of organic compounds is based on the equation

$$C_p^\circ = \sum_i n_i a_i + \sum_i n_i b_i T + \sum_i n_i c_i T^2 + \sum_i n_i d_i T^3 \qquad (7\text{-}3.3)$$

where n_i represents the number of groups of type i.   The parameters a_i, b_i, c_i, and d_i can be found from Table 7-3.   Estimates for ring

TABLE 7-3 Rihani and Doraiswamy's Group Contributions to Ideal-Gas Heat Capacities [26]

Group	a	$b \times 10^2$	$c \times 10^4$	$d \times 10^6$
Aliphatic Hydrocarbon Groups				
$-CH_3$	0.6087	2.1433	-0.0852	0.01135
$-CH_2$	0.3945	2.1363	-0.1197	0.002596
$=CH_2$	0.5266	1.8357	-0.0954	0.001950
$-\overset{\mid}{\underset{\mid}{C}}-H$	-3.5232	3.4158	-0.2816	0.008015
$-\overset{\mid}{\underset{\mid}{C}}-$	-5.8307	4.4541	-0.4208	0.012630
$\overset{H}{\diagdown}C=CH_2$	0.2773	3.4580	-0.1918	0.004130
$\diagup C=CH_2$	-0.4173	3.8857	-0.2783	0.007364
$\overset{H}{\diagdown}C=C\overset{H}{\diagup}$	-3.1210	3.8060	-0.2359	0.005504
$\overset{H}{\diagdown}C=C\underset{H}{\diagup}$	0.9377	2.9904	-0.1749	0.003918
$\diagdown C=C\overset{H}{\diagup}$	-1.4714	3.3842	-0.2371	0.006063
$\diagdown C=C\diagup$	0.4736	3.5183	-0.3150	0.009205
$\overset{H}{\diagdown}C=C=CH_2$	2.2400	4.2896	-0.2566	0.005908
$\diagdown C=C=CH_2$	2.6308	4.1658	-0.2845	0.007277

TABLE 7-3 Rihani and Doraiswamy's Group Contributions to Ideal-Gas Heat Capacities [26] (Continued)

Group	a	$b \times 10^2$	$c \times 10^4$	$d \times 10^6$
Aliphatic Hydrocarbon Groups				
$\diagup C{=}C{=}C\diagdown$ (H, H)	-3.1249	6.6843	-0.5766	0.017430
Aromatic Hydrocarbon Groups				
$HC\diagup_\diagdown$	-1.4572	1.9147	-0.1233	0.002985
$-C\diagup_\diagdown$	-1.3883	1.5159	-0.1069	0.002659
$\leftrightarrow C\diagup_\diagdown$	0.1219	1.2170	-0.0855	0.002122
Contributions Due to Ring Formation				
Three-membered ring	-3.5320	-0.0300	0.0747	-0.005514
Four-membered ring	-8.6550	1.0780	0.0425	-0.000250
Five-membered ring:				
Pentane	-12.2850	1.8609	-0.1037	0.002145
Pentene	-6.8813	0.7818	-0.0345	0.000591
Six-membered ring:				
Hexane	-13.3923	2.1392	-0.0429	-0.001865
Hexene	-8.0238	2.2239	-0.1915	0.005473
Oxygen-containing Groups				
$-OH$	6.5128	-0.1347	0.0414	-0.001623
$-O-$	2.8461	-0.0100	0.0454	-0.002728
$-\overset{H}{\underset{\|}{C}}{=}O$	3.5184	0.9437	0.0614	-0.006978
$\diagup C{=}O$	1.0016	2.0763	-0.1636	0.004494
$-\overset{O}{\overset{\|}{C}}-O-H$	1.4055	3.4632	-0.2557	0.006886
$-C\overset{O}{\diagdown}{}_{O-}$	2.7350	1.0751	0.0667	-0.009230

Group	a	$b \times 10^2$	$c \times 10^4$	$d \times 10^6$
$O\diagup\diagdown$	-3.7344	1.3727	-0.1265	0.003789
Nitrogen-containing Groups				
$-C\equiv N$	4.5104	0.5461	0.0269	-0.003790
$-N\equiv C$	5.0860	0.3492	0.0259	-0.002436
$-NH_2$	4.1783	0.7378	0.0679	-0.007310
$\diagdown NH$	-1.2530	2.1932	-0.1604	0.004237
$\diagdown N-$	-3.4677	2.9433	-0.2673	0.007828
$N\diagup\diagdown$	2.4458	0.3436	0.0171	-0.002719
$-NO_2$	1.0898	2.6401	-0.1871	0.004750
Sulfur-containing Groups				
$-SH$	2.5597	1.3347	-0.1189	0.003820
$-S-$	4.2256	0.1127	-0.0026	-0.000072
$S\diagup\diagdown$	4.0824	-0.0301	0.0731	-0.006081
$-SO_3H$	6.9218	2.4735	0.1776	-0.022445
Halogen-containing Groups				
$-F$	1.4382	0.3452	-0.0106	-0.000034
$-Cl$	3.0660	0.2122	-0.0128	0.000276
	2.7605	0.4731	-0.0455	0.001420
$-I$	3.2651	0.4901	-0.0539	0.001782

compounds (including heterocyclic types) can be made, although the technique is inapplicable to acetylenic materials. The method is illustrated in Example 7-3, and calculated values are compared with those from the literature in Table 7-5. Errors are generally less than 2 to 3 percent.

Example 7-3 Estimate the ideal-gas heat capacity of 3-methyl thiophene at 800 K using the method of Rihani and Doraiswamy.

solution   From Table 7-3

	a	$b \times 10^2$	$c \times 10^4$	$d \times 10^6$
—CH$_3$	0.6087	2.1433	−0.0852	0.001135
$\displaystyle \mathop{C=C}$ (trans) (H ... H)	−3.1210	3.8060	−0.2359	0.005504
$\displaystyle \mathop{C=C}$ (H)	−1.4714	3.3842	−0.2371	0.006063
S	4.0824	−0.0301	0.0731	−0.006081
	0.0987	9.3034	−0.4851	0.006621

Thus,
$$C_p^\circ = 0.0987 + (9.3034 \times 10^{-2})T - (0.4851 \times 10^{-4})T^2 + (0.006621 \times 10^{-6})T^3$$

and at 800 K, $C_p^\circ = 46.87$ cal/g-mol K. The value reported by Stull [33] is 45.95 cal/g-mol K.

Method of Benson   For ideal-gas heat capacities an accurate method has been developed by Benson and his colleagues; it is thoroughly described in a book [4] and in a comprehensive review paper [6].   It is a group-contribution method applicable for C_p°, $\Delta H_{f_{298}}^\circ$, and S_{298}°. Contributions are given only for atoms with valences greater than unity. For each group, the key atom is given but followed by a notation specifying other atoms bonded to the key atom. For example, C—(C)(H)$_3$ refers to a carbon atom bonded to another carbon and three hydrogens, that is, —CH$_3$.   The contributions for this method are shown in Table 7-4.

To employ this method, one must become acquainted with the shorthand notation introduced; for example, C$_d$ refers to a carbon atom which also participates in a double bond with another carbon atom; it is assumed to have a valence of 2.   Notes at the bottom of the tables define terms which are not immediately obvious.

Preparation of Table 7-4 was a major undertaking, and work is continuing to include additional groups.   The review article cited above [6] presents many illustrative examples.   In addition, Eigenmann et al. [11] have modified several ΔH_f° values, Shaw [29] has extended the method to cover nitroaromatics, O'Neal and Benson [21] consider polycyclic compounds, and Seaton and Freedman [28] discuss an operational computer system (CHETAH) which employs the Benson method to estimate properties.   Other contributions, included in Table 7-4,

TABLE 7-4 Benson Group Contributions to Ideal-Gas Properties[a]

Group[b]	ΔH°_{f298}, kcal/g-mol	S°_{298}, cal/g-mol K	C_p°, cal/g-mol K at:					
			300 K	400 K	500 K	600 K	800 K	1000 K
		Hydrocarbon Groups						
C—(C)(H)$_3$	−10.08	30.41	6.19	7.84	9.40	10.79	13.02	14.77
C—(C)$_2$(H)$_2$	−4.95	9.42	5.50	6.95	8.25	9.35	11.07	12.34
C—(C)$_3$(H)	−1.90	−12.07	4.54	6.00	7.17	8.05	9.31	10.05
C—(C)$_4$	0.50	−35.10	4.37	6.13	7.36	8.12	8.77	8.76
C$_d$—(H)$_2$	6.26	27.61	5.10	6.36	7.51	8.50	10.07	11.27
C$_d$—(C)(H)	8.59	7.97	4.16	5.03	5.81	6.50	7.65	8.45
C$_d$—(C)$_2$	10.34	−12.7	4.10	4.61	4.99	5.26	5.80	6.08
C$_d$—(C$_d$)(H)	6.78	6.38	4.46	5.79	6.75	7.42	8.35	8.99
C$_d$—(C$_d$)(C)	8.88	−14.6	(4.40)	(5.37)	(5.93)	(6.18)	(6.50)	(6.62)
C$_d$—(C$_d$)$_2$	4.6							
C$_d$—(C$_B$)(H)	6.78	6.4	4.46	5.79	6.75	7.42	8.35	8.99
C$_d$—(C$_B$)(C)	8.64	(−14.6)	(4.40)	(5.37)	(5.93)	(6.18)	(6.50)	(6.62)
C$_d$—(C$_B$)$_2$	8.0							
C$_d$—(C$_t$)(H)	6.78	6.4	4.46	5.79	6.75	7.42	8.35	8.99
C$_d$—(C$_t$)(C)	8.53	...	4.40	5.37	5.93	6.18	6.50	6.62
C—(C$_d$)(H)$_3$	−10.08	30.41	6.19	7.84	9.40	10.79	13.02	14.77
C—(C$_d$)$_2$(H)$_2$	−4.29	(10.2)	(4.7)	(6.8)	(8.4)	(9.6)	(11.3)	(12.6)
C—(C$_d$)$_2$(C)$_2$	1.16	...	3.57	5.98	7.51	8.37	9.00	9.02
C—(C$_d$)(C)$_3$	1.68	(−34.72)	(3.99)	(6.04)	(7.43)	(8.26)	(8.92)	(8.96)
C—(C$_d$)(C)(H)$_2$	−4.76	9.8	5.12	6.86	8.32	9.49	11.22	12.48
C—(C$_d$)(C)$_2$(H)	−1.48	(−11.7)	(4.16)	(5.91)	(7.34)	(8.19)	(9.46)	(10.19)
C—(C$_d$)$_2$(C)(H)	−1.24	...	3.74	5.85	7.32	8.30	9.54	10.31

TABLE 7-4 Benson Group Contributions to Ideal-Gas Properties[a] (Continued)

Group[b]	ΔH°_{f298}, kcal/g-mol	S°_{298}, cal/g-mol K	C°_p, cal/g-mol K at: 300 K	400 K	500 K	600 K	800 K	1000 K
		Hydrocarbon Groups						
C—(C$_t$)(H)$_3$	−10.08	30.41	6.19	7.84	9.40	10.79	13.02	14.77
C—(C$_t$)(C)(H)$_2$	−4.73	10.3	4.95	6.56	7.93	9.08	10.86	12.19
C—(C$_t$)(C)$_2$(H)	−1.72	(−11.2)	(3.99)	(5.61)	(6.85)	(7.78)	(9.10)	(9.90)
C—(C$_B$)(H)$_3$	−10.08	30.41	6.19	7.84	9.40	10.79	13.02	14.77
C—(C$_B$)(C)(H)$_2$	−4.86	9.3	5.84	7.61	8.98	10.01	11.49	12.54
C—(C$_B$)(C)$_2$(H)	−0.98	(−12.2)	(4.88)	(6.66)	(7.90)	(8.75)	(9.73)	(10.25)
C—(C$_B$)(C)$_3$	2.81	(−35.18)	(4.37)	(6.79)	(8.09)	(8.78)	(9.19)	(8.96)
C—(C$_B$)$_2$(C)(H)	−1.24	...	3.74	5.85	7.32	8.30	9.54	10.31
C—(C$_B$)$_2$(C)$_2$	1.16	...	3.57	5.98	7.51	8.37	9.00	9.02
C—(C$_B$)(C$_d$)(H)$_2$	−4.29	(10.2)	(4.7)	(6.8)	(8.4)	(9.6)	(11.3)	(12.6)
C$_t$—(H)	26.93	24.7	5.27	5.99	6.49	6.87	7.47	7.96
C$_t$—(C)	27.55	6.35	3.13	3.48	3.81	4.09	4.60	4.92
C$_t$—(C$_d$)	29.20	(6.43)	(2.57)	(3.54)	(3.50)	(4.92)	(5.34)	(5.50)
C$_t$—(C$_B$)	(29.20)	6.43	2.57	3.54	3.50	4.92	5.34	5.50
C$_B$—(H)	3.30	11.53	3.24	4.44	5.46	6.30	7.54	8.41
C$_B$—(C)	5.51	−7.69	2.67	3.14	3.68	4.15	4.96	5.44
C$_B$—(C$_d$)	5.68	−7.80	3.59	3.97	4.38	4.72	5.28	5.61
C$_B$—(C$_t$)	5.7	−7.80	3.59	3.97	4.38	4.72	5.28	5.61
C$_B$—(C$_B$)	4.96	−8.64	3.33	4.22	4.89	5.27	5.76	5.95
C$_a$	34.20	6.0	3.9	4.4	4.7	5.0	5.3	5.5

Next-Nearest-Neighbor Correction

	ΔH°_{298}, kcal/g-mol	S°_{298},[d] cal/g-mol K	C_p°, cal/g-mol K at:					
			300 K	400 K	500 K	600 K	800 K	1000 K
Alkane gauche	0.80							
Alkene gauche	0.50							
Cis	1.00[c]		−1.34	−1.09	−0.81	−0.61	−0.39	−0.26
Ortho	0.57	−1.61	1.12	1.35	1.30	1.17	0.88	0.66

Corrections to be Applied for Ring Compounds

Ring $(\sigma)^{e,f}$	ΔH°_{298}, kcal/g-mol	S°_{298}, cal/g-mol K	300 K	400 K	500 K	600 K	800 K	1000 K
Cyclopropane (6)	27.6	32.1	−3.05	−2.53	−2.10	−1.90	−1.77	−1.62
Cyclopropene (2)	53.7	33.6						
Cyclobutane (8)	26.2	29.8	−4.61	−3.89	−3.14	−2.64	−1.88	−1.38
Cyclobutene (2)	29.8	29.0	−2.53	−2.19	−1.89	−1.68	−1.48	−1.33
Cyclopentane (10)	6.3	27.3	−6.5	−5.5	−4.5	−3.8	−2.8	−1.9
Cyclopentene (2)	5.9	25.8	−5.98	−5.35	−4.89	−4.14	−2.93	−2.26
Cyclopentadiene	6.0	28.0	−4.3					
Cyclohexane (6)	0	18.8	−5.8	−4.1	−2.9	−1.3	1.1	2.2
Cyclohexene (2)	1.4	21.5	−4.28	−3.04	−1.98	−1.43	−0.29	0.08
Cycloheptane (1)	6.4	15.9						
Cyclooctane (8)	9.9	16.5						
Naphthalene	...	8.1						

TABLE 7-4 Benson Group Contributions to Ideal-Gas Properties[a] (Continued)

Group[b]	$\Delta H^\circ_{f_{298}}$, kcal/g-mol	S°_{298}, cal/g-mol K	C°_p, cal/g-mol K at:					
			300 K	400 K	500 K	600 K	800 K	1000 K
		Oxygen-containing Compounds						
CO—(CO)(H)	−26.0	...	6.72	7.83	8.90	9.89	11.43	12.12
CO—(CO)(C)	−29.2	...	5.46	6.32	7.16	7.87	9.00	9.76
CO—(O)(C$_d$)	−32.5	...	5.97	6.7	7.41	8.02	8.87	9.36
CO—(O)(C$_B$)	−32.5	...	2.18	2.75	3.98	5.03	6.29	7.06
CO—(O)(C)	−35.1	4.78	5.97	6.70	7.40	8.02	8.87	9.36
CO—(O)(H)	−32.1	34.93	7.03	7.87	8.82	9.68	11.16	12.20
CO—(C$_d$)(H)	−31.7	...	7.03	7.87	8.82	9.68	11.16	12.20
CO—(C$_B$)$_2$	−38.1	...	5.26	6.77	7.67	8.48	9.62	9.85
CO—(C$_B$)(C)	−30.9	...	5.68	6.92	7.70	8.36	9.39	9.76
CO—(C$_B$)(H)	−34.6	...	6.40	7.72	8.91	9.85	11.49	12.09
CO—(C)$_2$	−31.4	15.01	5.59	6.32	7.09	7.76	8.89	9.61
CO—(C)(H)	−29.1	34.93	7.03	7.87	8.82	9.68	11.16	12.20
CO—(H)$_2$	−26.0	53.67	8.47	9.38	10.46	11.52	13.37	14.81
O—(C$_B$)(CO)	−32.5	...	2.06	2.70	3.11	3.42	3.88	4.18
O—(CO)$_2$	−50.9	...	−0.41	1.78	3.20	4.00	5.13	5.85
O—(CO)(O)	−19.0	...	3.7	3.7	3.7	3.7	4.2	4.2
O—(CO)(C$_d$)	−46.9	...	1.44	2.98	3.98	4.49	4.97	5.20
O—(CO)(C)	−44.3	8.39	3.90	3.61	4.19	4.62	4.99	4.82
O—(CO)(H)	−58.1	24.52	3.81	4.98	5.80	6.34	7.19	7.75
O—(O)(C)	(−4.5)	(9.4)	(3.7)	(3.7)	(3.7)	(3.7)	(4.2)	(4.2)
O—(O)$_2$	(−19.0)	(9.4)	(3.7)	(3.7)	(3.7)	(3.7)	(4.2)	(4.2)
O—(O)(H)	−16.27	27.85	5.17	5.79	6.28	6.66	7.15	7.51

Group								
O—(Cd)2	−32.8	10.1	3.4	3.7	3.7	3.8	4.4	4.6
O—(Cd)(C)	−31.9	9.7	3.4	3.7	3.7	3.8	4.4	4.6
O—(CB)2	−21.1	...	1.09	1.22	1.50	1.99	2.85	3.51
O—(CB)(C)	−22.6	...	3.4	3.7	3.7	3.8	4.4	4.6
O—(CB)(H)	−37.9	29.1	4.3	4.5	4.8	5.2	6.0	6.6
O—(C)2	−23.7	8.68	3.4	3.7	3.7	3.8	4.4	4.6
O—(C)(H)	−37.9	29.07	4.33	4.45	4.82	5.23	6.02	6.61
Cd—(CO)(O)	9.0	...	5.59	7.00	7.48	7.75	8.02	8.13
Cd—(CO)(C)	9.4	...	3.73	4.48	5.02	5.40	5.95	6.37
Cd—(CO)(H)	8.5	...	3.79	4.90	5.84	6.64	7.80	8.74
Cd—(O)(Cd)	8.9	...	(4.40)	(5.37)	(5.93)	(6.18)	(6.50)	(6.62)
Cd—(O)(C)	10.3	...	4.10	4.61	4.99	5.26	5.80	6.08
Cd—(O)(H)	8.6	...	4.16	5.03	5.81	6.50	7.65	8.45
CB—(CO)	9.7	...	2.67	3.14	3.68	4.15	4.96	5.44
CB—(O)	−0.9	−10.2	3.9	5.3	6.2	6.6	6.9	6.9
C—(CO)2(H)2	−7.6	...	5.60	7.05	8.39	9.68	11.58	12.87
C—(CO)(C)2(H)	−1.8	−12.0	6.21	7.56	8.00	8.21	9.18	9.63
C—(CO)(C)(H)2	−5.2	9.6	6.2	7.7	8.7	9.5	11.1	12.2
C—(CO)(C)3	1.6	...	5.07	6.88	7.81	8.27	8.80	8.62
C—(CO)(H)3	−10.1	30.41	6.19	7.84	9.40	10.79	13.02	14.77
C—(O)2(C)2	−18.6	...	1.59	3.95	6.20	7.39	7.62	8.48
C—(O)2(C)(H)	−16.3	...	5.06	7.28	9.03	9.41	10.31	10.75
C—(O)2(H)2	−15.1	...	2.83	5.06	7.52	9.12	10.32	11.29
C—(O)(CB)(H)2	−8.1	9.7	3.71	6.27	8.28	9.79	11.79	13.20
C—(O)(CB)(C)(H)	−6.08	...	5.14	7.30	8.83	9.43	10.23	10.6
C—(O)(Cd)(H)2	−6.9	...	4.66	6.97	8.65	9.88	11.54	12.73
C—(O)(C)3	−6.60	−33.56	4.33	6.19	7.25	7.70	8.20	8.24
C—(O)(C)2(H)	−7.2	−11.00	4.80	6.64	8.10	8.73	9.81	10.40
C—(O)(C)(H)2	−8.1	9.8	4.99	6.85	8.30	9.43	11.11	12.33
C—(O)(H)3	−10.1	30.41	6.19	7.84	9.40	10.79	13.03	14.77

Table 7-4 Benson Group Contributions to Ideal-Gas Properties[a] *(Continued)*

Strain or rings	ΔH°_{f298}, kcal/g-mol	S°_{298}, cal/g-mol K	C°_p, cal/g-mol K at: 300 K	400 K	500 K	600 K	800 K	1000 K
Strain and Ring Corrections for Oxygen-containing Compounds *(Continued)*								
Ether oxygen, gauche	0.3	⋯	−0.10	−0.89	−1.10	−0.73	−0.60	−0.23
Ditertiary ethers	7.8	⋯	−3.94	−5.64	−7.15	−8.83	−12.04	−14.90
Ethylene oxide, H₂C—CH₂ / O	27.6	31.4	−2.0	−2.8	−3.0	−2.6	−2.3	−2.3
Trimethylene oxide, CH₂—CH₂ / H₂C—O	26.4	27.7	−4.6	−5.0	−4.2	−3.5	−2.6	0.2
Tetrahydrofuran, H₂C—CH₂ / H₂C—O—CH₂	6.7	⋯	−4.25	−4.54	−4.07	−3.55	−3.09	−2.61
Tetrahydropyran, CH₂—CH₂ / H₂C—CH₂ / H₂C—O—CH₂	2.2	⋯	−4.28	−3.04	−1.98	−1.43	−0.29	0.08

1,3-Dioxane,	0.9	...	−2.51	−2.88	−2.28	−1.49	−0.26	0.56
1,4-Dioxane,	5.4	...	−4.16	−4.57	−3.11	−1.88	−1.09	−0.47
1,3,5-Trioxane,	5.1	...	1.79	0.56	−0.61	−0.65	−1.20	−2.43
Furan,	−5.8	...	−4.19	−3.63	−2.92	−2.39	−1.99	−1.72
Dihydropyran,	1.2	...	−4.44	−3.20	−1.56	−0.45	0.42	0.66

TABLE 7-4 Benson Group Contributions to Ideal-Gas Properties[a] (Continued)

Strain or rings	ΔH°_{f298}, kcal/g-mol	S°_{298}, cal/g-mol K	C°_p, cal/g-mol K at:					
			300 K	400 K	500 K	600 K	800 K	1000 K
Strain and Ring Corrections for Oxygen-containing Compounds *(Continued)*								
Cyclopentanone,	5.2	...	−8.53	−7.19	−5.31	−3.72	−2.26	−1.22
Cyclohexanone,	2.2		−8.10	−6.57	−4.24	−1.91	0.70	1.97
Succinic anhydride,	4.5	...	−7.90	−6.02	−4.49	−3.58	−3.35	−3.06
Glutaric anhydride,	0.8	...	−7.93	−6.04	−4.50	−3.59	3.35	−3.07
Maleic anhydride,	3.6	...	−5.12	−3.38	−2.02	−2.19	−0.37	−0.01

Group[g]	ΔH°_{f298}, kcal/g-mol	S°_{298}, cal/g-mol K	C°_p, cal/g-mol K at:					
			300 K	400 K	500 K	600 K	800 K	1000 K
		Nitrogen-containing Compounds						
C—(N)(H)₃	−10.08	30.41	6.19	7.84	9.40	10.79	13.02	14.77
C—(N)(C)(H)₂	−6.6	9.8	5.25	6.90	8.28	9.39	11.09	12.34
C—(N)(C)₂(H)	−5.2	−11.7	4.67	6.32	7.64	8.39	9.56	10.23
C—(N)(C)₃	−3.2	−34.1	4.35	6.16	7.31	7.91	8.49	8.50
N—(C)(H)₂	4.8	29.71	5.72	6.51	7.32	8.07	9.41	10.47
N—(C)₂(H)	15.4	8.94	4.20	5.21	6.13	6.83	7.90	8.65
N—(C)₃	24.4	−13.46	3.48	4.56	5.43	5.97	6.56	6.67
N—(N)(H)₂	11.4	29.13	6.10	7.38	8.43	9.27	10.54	11.52
N—(N)(C)(H)	20.9	9.61	4.82	5.8	6.5	7.0	7.8	8.3
N—(N)(C)₂	29.2	−13.80	1.56	2.50	3.31	3.87	4.62	4.99
N—(N)(C_B)(H)	22.1	⋯	3.28	4.05	4.75	5.31	6.28	6.91
N_I—(H)	(16.3)	(12.3)	2.95	4.58	6.45	7.71	9.13	9.92
N_I—(C)	21.3	⋯	2.48	3.34	3.95	4.29	4.59	4.60
N_I—(C_B)	16.7	⋯	2.60	3.22	3.81	4.22	4.79	5.12
N_A—(H)	25.1	26.8	4.38	4.89	5.44	5.94	6.77	7.42
N_A—(C)	32.5	8.0	2.70	4.10	4.92	5.34	5.69	5.71
N—(C_B)(H)₂	4.8	29.71	5.72	6.51	7.32	8.07	9.41	10.47
N—(C_B)(C)(H)	14.9	⋯	3.82	4.89	5.71	6.28	7.19	7.73
N—(C_B)(C)₂	26.2	⋯	0.62	2.02	3.27	4.13	5.23	5.59
N—(C_B)₂(H)	16.3	⋯	2.16	3.12	4.13	5.10	6.76	7.88
C_B—(N)	−0.5	−9.69	3.95	5.21	5.94	6.32	6.53	6.56
N_A—(N)	23.0	⋯	2.12	4.18	5.51	6.77	6.86	7.05
CO—(N)(H)	−29.6	34.93	7.03	7.87	8.82	9.68	11.16	12.20
CO—(N)(C)	−32.8	16.2	5.37	6.17	7.07	7.66	9.62	11.19
N—(CO)(H)₂	−14.9	24.69	4.07	5.74	7.13	8.29	9.96	11.22
N—(CO)(C)(H)	−4.4	3.9	3.87	5.08	5.95	6.76	6.87	6.54
N—(CO)(C)₂	4.7	⋯	1.83	3.79	5.24	6.19	7.11	7.42
N—(CO)(C_B)(H)	0.4	⋯	3.03	3.91	4.60	5.58	6.23	6.32

TABLE 7-4 Benson Group Contributions to Ideal-Gas Properties[a] *(Continued)*

Group[g]	ΔH°_{f298}, kcal/g-mol	S°_{298}, cal/g-mol K	C°_p, cal/g-mol K at:					
			300 K	400 K	500 K	600 K	800 K	1000 K
Nitrogen-containing Compounds (Continued)								
N—(CO)$_2$(H)	−18.5	...	3.59	5.54	6.70	7.39	7.95	8.19
N—(CO)$_2$(C)	−5.9	...	1.07	3.10	4.31	5.00	5.48	6.47
N—(CO)$_2$(C_B)	−0.5	...	0.98	3.06	4.23	4.85	5.28	5.29
C—(CN)(C)(H)$_2$	22.5	40.20	11.10	13.40	15.50	17.20	19.7	21.30
C—(CN)(C)$_2$(H)	25.8	19.80	11.00	12.70	14.10	15.40	17.30	18.60
C—(CN)(C)$_3$	29.0	−2.80	8.65	11.16	12.89	14.05	15.51	16.19
C—(CN)$_2$(C)$_2$	...	28.40	14.72	17.79	20.00	21.61	23.78	24.96
C_d—(CN)(H)	37.4	36.58	9.80	11.70	13.30	14.50	16.30	17.30
C_d—(CN)(C)	39.15	15.91	9.74	11.28	12.48	13.26	14.45	14.93
C_d—(CN)$_2$	84.1	...	13.60	16.55	18.68	20.25	22.34	23.59
C_d—(NO$_2$)(H)	...	44.4	12.3	15.1	17.4	19.2	21.6	23.2
C_B—(CN)	35.8	20.50	9.8	11.2	12.3	13.1	14.2	14.9
C_t—(CN)	63.8	35.40	10.30	11.30	12.10	12.70	13.60	14.30
C—(NO$_2$)(C)(H)$_2$	−15.1	48.4	12.59	15.82	18.52	20.66	23.79	25.90
C—(NO$_2$)(C)$_2$(H)	−15.8	26.9	11.99	15.21	17.72	19.61	22.18	23.70
C—(NO$_2$)(C)$_3$	...	3.9	9.89	13.34	15.86	17.62	19.41	20.86
C—(NO$_2$)$_2$(C)(H)	−14.9	...	17.32	22.82	27.07	30.21	34.35	36.83
O—(NO)(C)	−5.9	41.9	9.10	10.30	11.2	12.0	13.3	13.9
O—(NO$_2$)(C)	−19.4	48.50	9.54	11.54	13.26	15.60	16.39	17.38

	ΔH°_{f298}, kcal/g-mol	S°_{298}, cal/g-mol K	C°_p, cal/g-mol K at:					
Ring			300 K	400 K	500 K	600 K	800 K	1000 K

Ring Corrections for Nitrogen-containing Compounds

Ethyleneimine, H_2C–CH_2, N, H	27.7	31.6	−2.07	−2.18	−2.17	−2.05	−1.94	−1.88
Azetidine, CH_2, H_2C, CH_2, N, H	26.2	29.3	−4.73	−4.52	−4.08	−3.61	−2.66	0.01
Pyrrolidine, H_2C–CH_2, H_2C, CH_2, N, H	6.8	26.7	−6.17	−5.58	−4.80	−4.00	−2.87	−2.17
Piperidine, CH_2, H_2C, CH_2, H_2C, CH_2, N, H	1.0	...	−0.56	0.37	1.08	1.56	1.71	−0.46
$C_6H_{12}N_2$, CH_2, CH_2, H_2C, N, CH_2, H_2C, N, CH_2	3.4	...	−9.44	−8.70	−7.00	−5.36	−3.21	−1.84
Succinimide, H_2C, CH_2, C=O, C=O, N, H	8.5	...	2.16	4.08	6.14	8.01	9.11	9.77

TABLE 7-4 Benson Group Contributions to Ideal-Gas Properties[a] *(Continued)*

Group	ΔH°_{f298}, kcal/g-mol	S°_{298}, cal/g-mol K	C_p°, cal/g-mol K at:					
			300 K	400 K	500 K	600 K	800 K	1000 K
		Halogen Groups						
C—(F)$_3$(C)	−158.4	42.5	12.7	15.0	16.4	17.9	19.3	20.0
C—(F)$_2$(H)(C)	−109.3	39.1	9.9	12.0	13.7	15.1	16.7	17.8
C—(F)(H)$_2$(C)	−51.5	35.4	8.1	10.0	12.0	13.0	15.2	16.6
C—(F)$_2$(C)$_2$	−97.0	17.8	9.9	11.8	13.5	14.4	16.1	16.6
C—(F)(H)(C)$_2$	−49.0	14.0	7.30	9.04	10.47	11.56	13.10	14.01
C—(F)(C)$_3$	−48.5	...	6.80	8.86	10.20	11.16	12.43	12.72
C—(F)$_2$(Cl)(C)	−106.3	40.5	13.7	16.1	17.5	18.6	19.8	20.4
C—(Cl)$_3$(C)	−20.7	50.4	16.3	18.0	19.1	19.8	20.6	21.0
C—(Cl)$_2$(H)(C)	(−18.9)	43.7	12.1	14.0	15.4	16.5	17.9	18.7
C—(Cl)(H)$_2$(C)	−16.5	37.8	8.9	10.7	12.3	13.4	15.3	16.7
C—(Cl)$_2$(C)$_2$	−22.0	22.4	12.2	14.88	15.95	16.48	16.96	17.02
C—(Cl)(H)(C)$_2$	−14.8	17.6	9.0	9.9	10.5	11.2	13.9	14.6
C—(Cl)(C)$_3$	−12.8	−5.4	9.3	10.5	11.0	11.3	12.4	12.7
C—(Br)$_3$(C)	...	55.7	16.7	18.0	18.8	19.4	19.9	20.3
C—(Br)(H)$_2$(C)	−5.4	40.8	9.1	11.0	12.6	13.7	15.5	16.8
C—(Br)(H)(C)$_2$	−3.4	...	8.93	10.66	11.96	12.84	14.05	14.72
C—(Br)(C)$_3$	−0.4	−2.0	9.3	11.0	11.5	12.3	13.3	13.3
C—(I)(H)$_2$(C)	8.0	43.0	9.2	11.0	12.9	13.9	15.8	17.2
C—(I)(H)(C)$_2$	10.5	21.3	9.2	10.9	12.2	13.0	14.2	14.8
C—(I)(C)(C$_d$)(H)	13.32	...	8.13	10.02	11.82	12.6	14.0	14.9
C—(I)(C$_d$)(H)$_2$	8.19	...	8.82	10.91	12.97	14.04	15.95	17.34
C—(I)(C)$_3$	13.0	0	9.83	11.75	12.92	13.45	13.79	13.60

C—(Cl)(Br)(H)(C)	⋯	45.7	12.4	14.0	15.6	16.3	17.9	19.0
N—(F)₂(C)	−7.8	⋯	8.25	10.13	11.52	12.80	14.37	14.98
C—(Cl)(C)(O)(H)	−21.6	15	9.85	10.41	11.05	11.57	12.45	13.14
C—(I)₂(C)(H)	(26.0)	(54.6)	12.69	14.78	16.21	17.12	18.31	19.03
C—(I)(O)(H)₂	3.8	40.7	8.22	10.49	12.23	13.55	15.35	16.57
Cd—(F)₂	−77.5	37.3	9.7	11.0	12.0	12.7	13.8	14.5
Cd—(Cl)₂	−1.8	42.1	11.4	12.5	13.3	13.9	14.6	15.0
Cd—(Br)₂	⋯	47.6	12.3	13.2	13.9	14.3	14.9	15.2
Cd—(F)(Cl)	⋯	39.8	10.3	11.7	12.6	13.3	14.2	14.7
Cd—(F)(Br)	⋯	42.5	10.8	12.0	12.8	13.5	14.3	14.7
Cd—(Cl)(Br)	⋯	45.1	12.1	12.7	13.5	14.1	14.7	14.7
Cd—(F)(H)	−37.6	32.8	6.8	8.4	9.5	10.5	11.8	12.7
Cd—(Cl)(H)	−1.2	35.4	7.9	9.2	10.3	11.2	12.3	13.1
Cd—(Br)(H)	11.0	38.3	8.1	9.5	10.6	11.4	12.4	13.2
Cd—(I)(H)	24.5	40.5	8.8	10.0	10.9	11.6	12.6	13.3
Cd—(C)(Cl)	−2.1	15.0	8.0	8.4	8.5	9.0	9.2	9.4
Cd—(C)(I)	23.6	⋯	8.9	9.2	9.1	9.4	9.5	9.6
Cd—(Cd)(Cl)	−3.56	⋯	8.3	9.2	9.4	9.9	9.9	9.9
Cd—(Cd)(I)	22.14	⋯	9.2	9.9	10.0	10.3	10.3	10.1
Ct—(Cl)	⋯	33.4	7.9	8.4	8.7	9.0	9.4	9.6
Ct—(Br)	⋯	36.1	8.3	8.7	9.0	9.2	9.5	9.7
Ct—(I)	⋯	37.9	8.4	8.8	9.1	9.3	9.6	9.8
CB—(F)	−42.8	16.1	6.3	7.6	8.5	9.1	9.8	10.2
CB—(Cl)	−3.8	18.9	7.4	8.4	9.2	9.7	10.2	10.4
CB—(Br)	10.7	21.6	7.8	8.7	9.4	9.9	10.3	10.5
CB—(I)	24.0	23.7	8.0	8.9	9.6	9.9	10.3	10.5
C—(CB)(F)₃	−162.7	42.8	12.5	15.3	17.2	18.5	20.1	21.0
C—(CB)(Br)(H)₂	−6.9	⋯	9.29	11.10	12.47	13.69	15.59	16.71
C—(CB)(I)(H)₂	8.4	⋯	9.78	11.56	12.90	14.08	15.88	16.91
C—(Cl)₂(CO)(H)	−17.8	⋯	12.8	14.75	15.85	16.65	17.93	18.56
C—(Cl)₃(CO)	−19.6	⋯	17.0	18.75	19.55	19.95	20.63	20.86
CO—(Cl)(C)	−30.2	⋯	8.87	9.44	10.24	11.08	12.53	13.59

251

TABLE 7-4 Benson Group Contributions to Ideal-Gas Properties^a *(Continued)*

Next nearest neighbors	ΔH°_{f298}, kcal/g-mol	S°_{298}, cal/g-mol K	C_p°, cal/g-mol K at:					
			300 K	400 K	500 K	600 K	800 K	1000 K
Corrections for Next-Nearest-Neighbor Halogen Compounds								
Ortho (F)(F)	5.0	⋯	0	0	0	0	0	0
Ortho (Cl)(Cl)	2.2	⋯	−0.50	−0.44	−0.55	−0.53	−0.28	−0.02
Ortho (alkane) (halogen)	0.6	⋯	0.42	0.44	0.28	0.19	0.12	0.14
Cis (halogen) (halogen)	0.3	⋯	−0.19	−0.01	−0.03	−0.17	0	−0.03
Cis (halogen) (alkane)	−0.8	⋯	−0.97	−0.70	−0.53	−0.47	−0.24	−0.13

Group	ΔH°_{f298}, kcal/g-mol	S°_{298}, cal/g-mol K	C_p°, cal/g-mol K at:					
			300 K	400 K	500 K	600 K	800 K	1000 K
Organosulfur Groups								
C—(H)$_3$(S)	−10.08	30.41	6.19	7.84	9.40	10.79	13.02	14.77
C—(C)(H)$_2$(S)	−5.65	9.88	5.38	7.08	8.60	9.97	12.26	14.15
C—(C)$_2$(H)(S)	−2.64	−11.32	4.85	6.51	7.78	8.69	9.90	10.57
C—(C)$_3$(S)	−0.55	−34.41	4.57	6.27	7.45	8.15	8.72	8.10
C—(C$_B$)(H)$_2$(S)	−4.73	⋯	4.11	6.75	8.70	10.15	11.93	13.10
C—(C$_d$)(H)$_2$(S)	−6.45	⋯	5.00	6.99	8.67	10.07	12.41	14.29
C$_B$—(S)	−1.8	10.20	3.90	5.30	6.20	6.60	6.90	6.90
C$_d$—(H)(S)	8.56	8.0	4.16	5.03	5.81	6.50	7.65	8.45
C$_d$—(C)(S)	10.93	−12.41	3.50	3.57	3.83	4.09	4.41	5.00
S—(C)(H)	4.62	32.73	5.86	6.20	6.51	6.78	7.30	7.71
S—(C$_B$)(H)	11.96	12.66	5.12	5.26	5.57	6.03	6.99	7.84
S—(C)$_2$	11.51	13.15	4.99	4.96	5.02	5.07	5.41	5.73

Group								
S—(C)(C_d)	9.97		4.22	5.08	5.56	5.77	5.87	5.87
S—(C_d)_2	−4.54	16.48	4.79	5.58	5.53	6.29	7.94	9.73
S—(C_B)(C)	19.16	...	3.02	3.39	3.71	4.04	4.62	5.00
S—(C_B)_2	25.90	...	2.00	2.01	2.24	2.74	3.80	4.71
S—(S)(C)	7.05	12.37	5.23	5.42	5.51	5.51	5.38	5.12
S—(S)(C_B)	14.5	...	2.89	3.39	3.72	4.15	4.78	5.10
S—(S)_2	3.04	13.36	4.7	5.0	5.1	5.2	5.3	5.4
C—(SO)(H)_3	−10.08	30.41	6.19	7.84	9.40	10.79	13.02	14.77
C—(C)(SO)(H)_2	−7.72	...	4.55	6.42	7.95	9.16	10.95	12.22
C—(C)_3(SO)	−3.05	...	3.06	4.58	4.84	6.60	7.54	7.96
C—(C_d)(SO)(H)_2	−7.35	...	4.40	6.36	6.94	9.25	10.97	12.25
C_B—(SO)	2.3	...	2.67	3.14	3.68	4.15	4.96	5.44
SO—(C)_2	−14.41	18.10	8.88	10.03	10.50	10.79	10.98	11.17
SO—(C_B)_2	−12.0	...	5.72	9.09	9.70	11.45	11.46	11.25
C—(SO_2)(H)_3	−10.08	30.41	6.19	7.84	9.40	10.79	13.02	14.77
C—(C)(SO_2)(H)_2	−7.68	...	5.38	7.08	8.60	9.97	12.26	14.15
C—(C)_2(SO_2)(H)	−2.62	...	4.42	6.25	7.56	8.48	9.64	10.30
C—(C)_3(SO_2)	−0.61	...	2.32	4.38	5.70	6.49	7.27	7.46
C—(C_d)(SO_2)(H)_2	−7.14	...	5.00	6.99	8.67	10.07	12.41	14.29
C—(C_B)(SO_2)(H)_2	−5.54	...	3.71	6.57	8.28	9.79	11.89	13.20
C_B—(SO_2)	2.3	...	2.67	3.14	3.68	4.15	4.96	5.44
C_d—(H)(SO_2)	12.53	...	3.04	4.67	5.93	6.84	7.87	8.67
C_d—(C)(SO_2)	14.47	...	1.85	3.11	3.98	4.60	5.33	5.67
SO_2—(C_d)(C_B)	−68.58	...	9.89	11.50	13.35	14.61	15.72	15.92
SO_2—(C_d)_2	−73.58	...	11.52	11.97	13.35	14.28	15.38	15.88
SO_2—(C)_2	−69.74	20.90	10.18	11.74	12.92	13.77	15.13	16.0
SO_2—(C)(C_B)	−72.29	...	9.94	11.50	13.45	14.51	15.62	15.92
SO_2—(C_B)_2	−68.58	...	8.36	11.03	13.55	14.94	15.86	15.96
SO_2—(SO_2)(C_B)	−76.25	...	9.81	11.50	13.52	14.73	15.71	16.03
CO—(S)(C)	−31.56	15.43	5.59	6.32	7.09	7.76	8.89	9.61
S—(H)(CO)	−1.41	31.20	7.63	8.09	8.12	8.17	8.50	8.24
C—(S)(F)_3	...	38.9	9.88	13.01	14.83	16.37	18.17	19.11

TABLE 7-4 Benson Group Contributions to Ideal-Gas Properties[a] (Continued)

Group[b]	ΔH°_{f298}, kcal/g-mol	S°_{298}, cal/g-mol K	C_p°, cal/g-mol K at:					
			300 K	400 K	500 K	600 K	800 K	1000 K
CS—(N)$_2$	−31.56	15.43	5.59	6.32	7.09	7.76	8.89	9.61
N—(CS)(H)$_2$	12.78	29.19	6.07	7.28	8.18	8.91	10.09	10.98
S—(S)(N)	−4.90	...	3.7	3.7	3.7	3.7	4.2	4.2
N—(S)(C)$_2$	29.9	...	3.97	5.17	6.21	6.94	7.39	9.24
SO—(N)$_2$	−31.56	...	5.59	6.32	7.09	7.76	8.89	9.61
N—(SO)(C)$_2$	16.0	...	4.20	5.88	6.12	6.53	6.83	8.34
SO$_2$—(N)$_2$	−31.56	...	5.59	6.32	7.09	7.76	8.89	9.61
N—(SO$_2$)(C)$_2$	−20.4	...	6.02	6.35	7.54	8.23	9.03	9.19

Ring Corrections for Sulfur-containing Compounds

Ring $(\sigma)^{e,f}$	ΔH°_{f298}, kcal/g-mol	S°_{298}, cal/g-mol K	C_p°, cal/g-mol K at:					
			300 K	400 K	500 K	600 K	800 K	1000 K
Thiirane (2),	17.7	29.47	−2.85	−2.59	−2.66	−3.02	−4.32	−5.82
Trimethylene sulfide (2),	19.37	27.18	−4.59	−4.18	−3.91	−3.91	−4.60	−5.70

Compound								
Tetrahydrothiophene (2), $H_2C\overset{S}{-}CH_2$, H_2C-CH_2	1.73	23.56	−4.90	−4.67	−3.68	−3.66	−4.41	−5.57
Thiacyclohexane (2), $H_2C\overset{S}{-}CH_2$, $H_2C-CH_2-CH_2$	0	17.46	−6.22	−4.26	−2.24	−0.69	0.86	1.29
Thiacycloheptane (2)	3.89	...	−7.75	−4.92	−1.22	2.59	4.79	4.61
3-Thiocyclopentene (2), $H_2C\overset{S}{-}CH_2$, $HC=CH$	5.07	...	−6.44	−4.24	−4.23	−4.18	−4.80	−5.96
2-Thiocyclopentene (1), $H_2C\overset{S}{-}CH$, H_2C-CH	5.07	...	−6.44	−4.24	−4.23	−4.18	−4.80	−5.96
$C_4H_6SO_2$ (2), $H_2C\overset{O\;\;S\;\;O}{-}CH_2$, $HC=CH$	5.74	...	−4.90	−4.67	−3.68	−3.66	−4.41	−5.57

TABLE 7-4 Benson Group Contributions to Ideal-Gas Properties[a] (Continued)

Ring (σ)[e,f]	ΔH°_{f298}, kcal/g-mol	S°_{298}, cal/g-mol K	C°_p, cal/g-mol K at: 300 K	400 K	500 K	600 K	800 K	1000 K
Ring Corrections for Sulfur-containing Compounds (Continued)								
Thiophene (2),	1.73	23.56	−4.90	−4.67	−3.68	−3.66	−4.41	−5.57

[a] Data were obtained largely from Refs. 4 and 6. Some H°_{f298} values were obtained from Ref. 11. Many C°_p values are from Ref. 20. Grateful acknowledgment is extended to Shell Development Co., Bellaire Research Center, and particularly to Dr. P. Chueh, for supplying missing values as well as additional contributions. Finally, Dr. S. W. Benson provided an up-to-date errata list that showed recent modifications of the original tables.

[b] C_d represents a carbon atom that is joined to another carbon atom by a double bond. It is considered divalent. For example, 2-pentene would have the groups $C—(C_d)(H)_3$, $C_d—(C)(H)$ twice, $C—(C_d)(H)_2$, and $C—(C)(H)_3$. C_t represents a carbon atom that is joined to another carbon atom by a triple bond. It is considered monovalent. For example, propyne would have the groups $C_t—(H)$, $C_t—(C)$, and $C—(C_t)(H_3)$. C_B represents a carbon atom in an aromatic ring. It is considered monovalent. For example, p-ethyl toluene would have the groups $C—(C)(H)_3$, $C—(C_B)(C)(H)_2$, $C_B—(C)(H)_3$, $C_B—(C_B)(H)_2$, $C_B—(C)$ twice, and $C_B—(H)$ four times. C_a represents the allene group, $>C=C=C<$; the end carbons are treated as normal C_d atoms. For example, 1,2-butadiene would have the groups C_a, $C_d—(H)_2$, $C_d—(C)(H)$, and $C—(C_a)(H)_3$.

[c] When one of the groups is t-butyl, the cis correction = 4.0; when both are t-butyl, the cis correction = 10.0; and when there are two cis corrections around one double bond, the total correction = 3.0.

[d] Value is 1.2 for but-2-ene but zero for other dienes and 0.6 for trienes.

[e] The number in parentheses beside each ring is the symmetry number.

[f] For ΔH°_{f298} contributions for other ring structures, see *Chem. Rev.*, **69**: 279 (1969).

[g] N_I represents a double-bonded nitrogen in imines; $N_I—(C_B)$ represents a pyridine nitrogen. N_A represents a double-bonded nitrogen in azo compounds. For ortho or para substitution in pyridine add −1.5 kcal/g-mol per group to ΔH°_{f298}.

have been determined by Shell Development Co. [8] and by Olson [20]. When care is used in its application, the method is accurate, as illustrated in Example 7-4; a comparison between calculated and literature values of C_p° is presented in Table 7-5. Errors are almost always less than 1 percent for a wide range of compounds.

Example 7-4   Using Benson's method, estimate the ideal-gas heat capacity of 2-methyl-2-butanethiol at 800 K.

solution   From Table 7-4

Group	No.	Contribution	
C—(C)(H)$_3$	3	13.02	39.06
C—(C)$_2$(H)$_2$	1	11.07	11.07
C—(C)$_3$(S)	1	8.72	8.72
S—(C)(H)	1	7.30	7.30
			66.15

Thus, $C_p^\circ(800) = 66.15$ cal/g-mol K. Stull et al. [33] report the value 66.28 cal/g-mol K.

Discussion and Recommendations   A comparison between calculated and literature values for C_p° at 298 and at 800 K is shown in Table 7-5. Clearly, where applicable, the Benson method is the most accurate, although for hydrocarbons, Thinh's technique yields comparable results. The bond method is easy to use but is only approximate and is limited to 298 K. The Rihani-Doraiswamy method is applicable to a large variety of compounds and yields a polynomial in temperature. It is, however, generally less accurate, particularly at lower temperatures.

In this section no exhaustive review was given covering the many methods proposed to estimate C_p°. Other methods are reviewed in an earlier edition of this book [23] and in other books dealing with thermochemistry [9, 14, 19, 41]. The methods described here are believed to be the most accurate and general for engineering use.

7-4 Standard Heat of Formation

The standard heat of formation ΔH_f° is defined in Sec. 7-1. In the present section, five estimation methods for ΔH_{f298}° are discussed and illustrated by examples. In Table 7-8, a comparison is presented between estimated and literature values for a number of diverse compounds.

Heats of formation are often found from measured heats of combustion ΔH_c° or from an experimental enthalpy of reaction wherein ΔH_f° values are known for all but the compound of interest. Domalski [10] has critically reviewed literature values of ΔH_{f298}° and ΔH_{c298} for over 700

TABLE 7-5 Comparison of Estimated and Literature Values for Ideal-Gas Heat Capacity

Compound	T, K	$C_p^{° a}$, cal/g-mol K	Percent error[b] calculated by method of:			
			Bond values, Table 7-1	Benson values, Table 7-4	Rihani-Doraiswamy, Table 7-3	Thinh et al., Table 7-2
Propane	298	17.66	1.2	0.7	3.6	-0.5
	800	37.08	...	0.1	-0.5	0.2
n-Heptane	298	39.67	0.1	0	4.3	-0.1
	800	81.43	...	0	0.1	0.8
2,2,3-Trimethylbutane	298	39.33	-4.0	0.8	1.1	0.7
	800	82.73	...	0.5	1.1	1.0
trans-2-Butene	298	20.99	-7.1	-1.8	-0.2	0
	800	41.50	...	-0.4	-0.1	0.1
3,3-Dimethyl-1-butene	298	30.23	-2.7	2.2	5.3	6.6
	800	63.60	...	3.3	4.3	4.3
2-Methyl-1,3-butadiene	298	25.0	-6.3	0.4	-3.3	-2.0
	800	48.0	...	0	-1.5	-1.2
2-Pentyne	298	23.59	c	-0.5	c	-0.3
	800	45.90	...	0.4		0.4
p-Ethyltoluene	298	36.22	1.9	0.2	-1.3	-1.3
	800	77.60	...	0	-0.2	-0.5
2-Methylnaphthalene	298	38.19	-20	-0.5	-2.5	-2.2
	800	82.03	...	0.3	-0.5	-0.4
cis-1,3-Dimethylcyclopentane	298	32.14	19	-5.9	-3.6	0
	800	75.84	...	-1.1	0.3	0.1
2-Butanol	298	27.08	-0.3	-0.8	2.6	
	800	52.68	...	0.5	-1.9	
p-Cresol	298	29.75	c	0.1	1.3	
	800	61.11	...	-0.1	-0.7	
Isopropyl ether	298	37.83	-0.4	-0.8	-2.3	
	800	74.39	...	2.3	1.6	
p-Dioxane	298	22.48	28	-1.0	31	
	800					

258

Compound	T (K)	C_p° [a]				
Methyl ethyl ketone	298	24.59	−5.2	−2.5	−1.7	
	800	46.08		−0.1	0.6	
Ethyl acetate	298	27.16	−9.8	0	−9.4	
	800	51.01		0	−6.3	
Trimethylamine	298	21.93	0.1	0	0	
	800	45.62			0	
Propionitrile	298	17.46	c	−1.4	4.0	
	800	32.14		1.8	−7.6	
2-Nitrobutane	298	29.51	c	0.9	2.0	
	800	59.44		0.2	0	
3-Picoline	298	23.80	c	c	5.0	
	800	53.12		−2.3	−0.2	
1,1-Difluoroethane	298	16.24	27	5.1	−5.3	
	800	29.69		−6.2	0.3	
Octafluorocyclobutane	298	37.32	−7.2	−6.6	−5.9	
	800	58.65	c	2.3	11	
Bromobenzene	298	23.35	c	0.5	−4.9	
	800	47.78		0.1	−1.4	
Trichloroethylene	298	19.17	2.8	−0.1	−9.0	
	800	26.94		0	−4.3	
Butyl methyl sulfide	298	33.64	0	−1.0	2.2	
	800	66.53		0	−3.6	
2-Methyl-2-butanethiol	298	34.30	−1.9	−0.2	−0.3	
	800	66.28		0.2	0.7	
Propyl disulfide	298	44.30	0.4	−0.3	0.9	
	800	83.70		2.0	−4.1	
3-Methylthiophene	298	22.67		−1.9	4.5	
	800	45.95	10		2.0	
Number of compounds			22	27	27	10
Average error [b]			7.1	1.1	3.2	1.1 [d]

[a] Literature values of C_p° from Ref. 33.

[b] [(calc. − lit.)/lit.] × 100.

[c] No contribution available for one or more bonds or groups.

[d] Thinh et al. method application only to hydrocarbons.

259

organic compounds, and this source should be consulted before any *estimation* scheme is employed; also valuable are Refs. 4, 6, 9, 18, and 33. ΔH_c can be estimated directly for both simple and complex organic compounds using a method developed by Handrick [15], discussed in an earlier edition of this book [23]. One must use care in interpreting reported ΔH_c values, especially if sulfur and/or phosphorus is present. For materials containing only carbon, hydrogen, oxygen, and nitrogen, the assumed products of combustion are carbon dioxide, water,† and elemental diatomic nitrogen. Halogens usually are assumed to appear as HX.

Handrick's ΔH_c values for sulfur-containing materials assume SO_2 as a product, whereas Domalski uses (liquid) $H_2SO_4 \cdot 115H_2O$. Also the latter author assumes that reactants containing phosphorous combust to yield crystalline H_3PO_4. The standard heats of formation of the common combustion products are

	ΔH°_{f298}, kcal/g-mol
$CO_2(g)$	-94.051
$H_2O(l)$	-68.315
$HF(g)$	-64.20
$HCl(g)$	-22.10
$HBr(g)$	-8.70
$HI(g)$	6.20
$SO_2(g)$	-71.00
$H_2SO_4 \cdot 115H_2O(l)$	-212.192
$H_3PO_4(c)$	-305.7

Bond Energies   There are a number of methods available for estimating ΔH°_f from bond energies. Most require that one begin with the elements making up the compound and decompose them, if necessary, to gaseous atoms, for example, $O_2(g) \rightarrow 2O(g)$, or to vaporize the elements, e.g., $C(s) \rightarrow C(g)$, and then proceed to synthesize the compound of interest. In each step, there is an enthalpy change, and the net sum is ΔH°_f. The method suffers from the fact that usually large positive and negative enthalpy values have to be added; upon calculating differences between large numbers, high accuracy is difficult to achieve. In this book, a simpler bond-increment method is chosen, i.e., with the values shown in Table 7-1, a rapid, simple additive method is available.

---

†If product water (liquid) is assumed, the *higher* heat of combustion is determined; if product water vapor is chosen, the *lower* heat of combustion is calculated. Higher heats of combustion are those normally reported.

Example 7-5 With Table 7-1, determine the value of $\Delta H^{\circ}_{f_{298}}$ of p-ethyltoluene.

  solution

$$\Delta H^{\circ}_{f_{298}} = 4(\phi{-}H) + 2(\phi{-}C) + C{-}C + 8(C{-}H)$$
$$= (4)(3.25) + (2)(7.25) + 2.73 + 8(-3.83)$$
$$= -0.41 \text{ kcal/g-mol}$$

Stull et al. [33] report a value of -0.78 kcal/g-mol.

Benson Method A group-contribution method proposed by Benson and his colleagues [4, 6, 11] is described in Sec. 7-3, and in Table 7-4 group contributions for determining $\Delta H^{\circ}_{f_{298}}$ are also presented for a wide variety of compounds. The technique is illustrated in Example 7-6, and a comparison between estimated and literature values is shown in Table 7-8. Differences are usually less than 1 kcal/g-mol for the diverse compounds chosen for comparison.

Example 7-6 Estimate the heat of formation of methyl methacrylate at 298 K using Benson's technique.

  solution The structural formula of this compound is

$$CH_2{=}C(CH_3){-}COO{-}CH_3$$

and the contributions from Table 7-4 are

	Contribution
C—(O)(H)$_3$	-10.1
O—(CO)(C)	-44.3
CO—(O)(C$_d$)	-32.5
C$_d$—(C)(CO)	9.4
C$_d$—(H)$_2$	6.26
C—(C$_d$)(H)$_3$	-10.08†
	-81.32

†C$_d$ represents a carbon atom joined by a double bond to another carbon atom; it is considered divalent.

Benson et al. [6] quote a value of $\Delta H^{\circ}_{f_{298}}$ of -79.3 kcal/g-mol.

Methods of Franklin and Verma-Doraiswamy Similar group-contribution methods for estimating ΔH°_f were developed by Franklin [12, 13] and by Verma and Doraiswamy [40]. As originally presented, both methods allowed one to determine ΔH°_f as a function of temperature, but as pointed out in Sec. 7-1, since the ultimate use for ΔH°_f is to determine enthalpies of reaction, these can be found from $\Delta H^{\circ}_{f_{298}}$ and C°_p for the products and reactants as shown in Eq. (7-1.3). Thus, in Table 7-6, the group contributions shown are applicable only at 298 K.

TABLE 7-6   Group Contributions for $\Delta H^{\circ}_{f_{298}}$, kcal/g-mol

	Verma and Doraiswamy [40]	Franklin [12, 13]
Carbon-Hydrogen Groups		
—CH$_3$	−10.25	−10.12
—CH$_2$	−4.94	−4.93
—CH	−1.29	−1.09
—C—	0.62	0.80
=CH$_2$	6.19	6.25
=C=	32.24	33.42
≡C—	27.38	27.34
≡CH	27.10	27.10
H⟩C=CH$_2$	15.02	15.00
⟩C=CH$_2$	20.50	16.89
⟩C=C⟨	30.46	24.57
⟩C=C⟨H	20.10	20.19
H⟩C=C⟨H (cis)	17.96	18.88
⟩C=C⟨H (trans)	17.83	17.83

Structure	Value 1	Value 2
$\diagup^{\diagdown}C{=}C{=}CH_2$	51.30	
$\leftrightarrow CH_2$	...	10.08
$\leftrightarrow C\diagup^{H}_{\diagdown}$	...	12.04
$^{H}\diagdown C{=}C{=}CH_2$	49.47	
$\diagdown_{H}C{=}C{=}C\diagup^{\diagup}_{\diagdown H}$	55.04	
$HC\diagup^{\nearrow}_{\searrow}$	3.27	3.30
$-C\diagup^{\nearrow}_{\searrow}$	5.55	5.57
$\leftrightarrow C\diagup^{\nearrow}_{\searrow}$	4.48	4.28

Ring Correction

C_3 cycloparaffin ring	24.13	24.22
C_4 cycloparaffin ring	18.45	18.4
C_5 cycloparaffin ring	5.44	4.94
C_6 cycloparaffin ring	−0.76	−0.45

Branching in Paraffins

Side chain with two or more C atoms	0.80	0.80
Three adjacent $-\overset{\vert}{\underset{\vert}{C}}H$ groups	−1.2	2.3
Adjacent $-\overset{\vert}{\underset{\vert}{C}}-$ and $-\overset{\vert}{\underset{\vert}{C}}H$ groups	0.6	2.5
Adjacent $-\overset{\vert}{\underset{\vert}{C}}-$ and $-\overset{\vert}{\underset{\vert}{C}}-$ groups	...	5.4

TABLE 7-6 Group Contributions for $\Delta H^\circ_{f_{298}}$, kcal/g-mol (Continued)

	Verma and Doraiswamy [40]	Franklin [12, 13]
Branching in Paraffins (*Continued*)		
$-\overset{\displaystyle\mid}{\underset{\displaystyle\mid}{C}}-$ not adjacent to terminal carbon	. . .	1.7
Branching in Aromatics		
Double branching:		
1,2 position	0.94	
1,3 position	0.38	
1,4 position	0.58	
Triple branching:		
1,2,3 position	1.80	
1,2,4 position	0.44	
1,3,5 position	0.44	
1,2-Dimethyl or 1,3-methylethyl	. . .	0.6
1,2-Methylethyl or 1,2,3-trimethyl	. . .	1.4
Branching in Six-membered Rings		
Single branching	0	
Double branching:		
1,1 position	2.44	
Cis 1,2 position	−0.20	
Trans 1,2 position	−2.69	
Cis 1,3 position	−2.98	
Trans 1,3 position	−0.48	
Cis 1,4 position	−0.48	
Trans 1,4 position	−2.98	
Branching in Five-membered Rings		
Single branching	0	
Double branching:		
1,1 position	0.30	
Cis 1,2 position	0.70	
Trans 1,2 position	−1.10	
Cis 1,3 position	−0.30	
Trans 1,3 position	−0.90	
Oxygen-containing Groups		
—OH (primary alcohol)	−41.2	−41.9
—OH (secondary alcohol)	−43.8	−44.9
—OH (tertiary alcohol)	−47.60	−49.2

—OH (on aromatic group)	− 45.10	− 46.9
—CHO (aldehyde)	− 29.71	− 33.9
$\diagdown$C=O (ketone)	− 31.48	− 31.6
—COOH (acid)	− 94.68	− 94.6
—COO— (ester)	. . .	− 79.8
—O— (ether)	− 24.2	− 27.2

—C=O
    \O
    /O
—C=O
	. . .	− 102.6

O⇗⇘	− 21.62	

Nitrogen and Sulfur-containing Groups

—C≡N	36.82	29.5
—NO₂	− 7.94	− 8.5
—ONO	. . .	− 10.9
—ONO₂	. . .	− 18.4
—N=C	. . .	44.4
Aliphatic: —NH₂	3.21	
>NH	13.47	
>N—	18.94	
Aromatic: —NH₂	− 1.27	− 6.4
>NH	8.50	
>N—	19.21	
—SH	4.60	3.1
—S—	11.17	10.6
S ⇗⇘	. . .	7.8

Example 7-7   Using the group-contribution method suggested by Franklin and Verma and Doraiswamy, estimate ΔH°_{f298} for cis-1,3-dimethylcyclopentane.

solution   From Table 7-6

Group	No.	Franklin	Verma and Doraiswamy
—CH$_2$—	3	(3)(−4.93)	(3)(−4.94)
>CH—	2	(2)(−1.09)	(2)(−1.29)
—CH$_3$	2	(2)(−10.12)	(2)(−10.25)
C$_5$ ring correction		4.94	5.44
Branching in five-membered ring; cis-1,3 position		. . .	−0.30
		−32.27	−32.76

Stull et al. [33] report a value of ΔH°_{f298} of −32.47 kcal/g-mol.

Method of Anderson, Beyer, and Watson   A group-estimation method suggested by Anderson, Beyer, and Watson in 1944 [3, 16] can be applied to many types of materials. S°_{298} can be estimated as well as ΔH°_{f298}. In the former method, as discussed later in Sec. 7-5, an advantage is that symmetry corrections are not necessary.

A compound is considered to be composed of a base group which is modified by the substitution of other groups. These base groups, shown in Table 7-7, comprise methane, cyclopentane, benzene, naphthalene, methylamine, dimethylamine, trimethylamine, ethyl ether, and formamide. For example, for ethane, the base group (methane) is *modified* by the *deletion* of one hydrogen and the *addition* of a methyl group. The first substitution of a methyl group on the base group is termed a *primary methyl substitution*, and values for this substitution are also given in Table 7-7. Except in the case of the base groups benzene, naphthalene, and cyclopentane, any further substitution of methyl groups *in the base group* is called a *secondary methyl substitution*. The increment in this case depends upon the type of carbon atom where the substitution is made as well as upon the type of adjacent carbon atoms. Let the letter A designate the carbon atom where the substitution is made and the letter B the *highest* type number (see below) carbon adjacent to A. The type numbers are as follows:

Type	1	2	3	4	5
	CH$_3$	CH$_2$	CH	C	C†

†C on benzene or naphthalene ring.

Thus a substitution of a hydrogen in propane by CH$_3$ (to form n-butane) involves a secondary methyl substitution of type A = 1,

TABLE 7-7   Anderson-Beyer-Watson Group Contributions for ΔH°_{f298} and S°_{298}†

Base group	ΔH°_{f298}, kcal/g-mol	S°_{298}, cal/g-mol K
	Base-group Properties	
Methane	−17.89	44.50
Cyclopentane	−18.46	70.00
Cyclohexane	−29.43	71.28
Benzene	19.82	64.34
Naphthalene	35.4	80.7
Methylamine	−7.1	57.7
Dimethylamine	−7.8	65.2
Trimethylamine	−10.9	
Dimethyl ether	−46.0	63.7
Formamide	−49.5	

Contribution of Primary CH_3 Substitution Groups Replacing Hydrogen

1. Methane	−2.50	10.35
2. Cyclopentane		
a. First primary substitution	−7.04	11.24
b. Second primary substitution		
To form 1,1	−7.55	4.63
To form cis 1,2	−5.46	6.27
To form trans 1,2	−7.17	6.43
To form cis 1,3	−6.43	6.43
To form trans 1,3	−6.97	6.43
c. Additional substitutions, each	−7.0	
3. Cyclohexane		
a. Enlargement of ring over 6 C, per carbon atom added to ring	−10.97	1.28
b. First primary substitution on ring	−7.56	10.78
c. Second primary substitution on ring		
To form 1,1	−6.27	5.18
To form cis 1,2	−4.16	7.45
To form trans 1,2	−6.03	6.59
To form cis 1,3	−7.18	6.48
To form trans 1,3	−5.21	7.86
To form cis 1,4	−5.23	6.48
To form trans 1,4	−7.13	5.13
d. Additional substitutions on ring, each	−7.0	

TABLE 7-7   Anderson-Beyer-Watson Group Contributions for $\Delta H^{\circ}_{f_{298}}$ and S°_{298}†   (Continued)

Base group	$\Delta H^{\circ}_{f_{298}}$, kcal/g-mol	S°_{298}, cal/g-mol K
Contribution of Primary CH₃ Substitution Groups Replacing Hydrogen (*Continued*)		
4. Benzene		
a. First substitution	−7.87	12.08
b. Second substitution		
To form 1,2	−7.41	7.89
To form 1,3	−7.83	9.07
To form 1,4	−7.66	7.81
c. Third substitution		
To form 1,2,3	−6.83	9.19
To form 1,2,4	−7.87	10.42
To form 1,3,5	−7.96	6.66
5. Naphthalene		
a. First substitution	−4.5	12.0
b. Second substitution		
To form 1,2	−6.3	8.1
To form 1,3	−6.5	9.2
To form 1,4	−8.0	7.8
6. Methylamine	−5.7	
7. Dimethylamine	−6.3	
8. Trimethylamine	−4.1	
9. Formamide		
Substitution on C atom	−9.0	

Contribution of Secondary CH₃ Substitutions Replacing Hydrogen

A	B	$\Delta H^{\circ}_{f_{298}}$, kcal/g-mol	ΔS°_{298}, cal/g-mol K
1	1	−4.75	10.00
1	2	−4.92	9.18
1	3	−4.42	9.72
1	4	−5.0	11.0
1	5	−4.68	10.76
2	1	−6.31	5.57
2	2	−6.33	7.15
2	3	−5.25	6.53
2	4	−3.83	7.46
2	5	−6.18	6.72
3	1	−8.22	2.81
3	2	−7.00	3.87
3	3	−5.19	3.99
3	4	−4.94	1.88
3	5	−9.2	1.3
1	—O— in ester or ether	−7.0	14.4

Substitution of H of OH group to form ester	9.5	16.7

Additional Corrections for Final Structure of Hydrocarbons

	ΔH°_{f298}, kcal/g-mol	ΔS°_{298}, cal/g-mol K
Additional correction for length of each side chain on ring:		
1. More than 2 C on cyclopentane side chain	− 0.45	0.12
2. More than 2 C on cyclohexane side chain	0.32	− 0.39
3. More than 4 C on benzene side chain	− 0.70	− 0.62
Additional correction for double-bond arrangement:		
1. Adjacent double bonds	13.16	− 3.74
2. Alternate double bonds	− 4.28	− 5.12
3. Double bond adjacent to aromatic ring:		
a. Fewer than 5 C in side chain	− 2.0	− 2.65
b. Over 4 C in side chain	− 1.16	− 2.65

Multiple-Bond Contributions Replacing Single Bonds

Type of Bond	ΔH°_{f298}, kcal/g-mol	ΔS°_{298}, cal/g-mol K
1=1	32.88	− 2.40
1=2	30.00	− 0.21
1=3	28.23	− 0.11
2=2 (cis)	28.39	− 1.19
2=2 (trans)	27.40	− 2.16
2=3	26.72	− 0.28
3=3	25.70	− 0.66
1≡1	74.58	− 9.85
1≡2	69.52	− 4.19
2≡2	65.50	− 3.97

Substitution Group Contributions Replacing CH₃ Group

Group	$\Delta^{\circ} H_{f298}$, kcal/g-mol	ΔS°_{298}, cal/g-mol K
—OH (aliphatic, meta, para)	− 32.7	2.6
—OH ortho	− 47.7	
—NO₂	1.2	2.0
—CN	39.0	4.0

TABLE 7-7   Anderson-Beyer-Watson Group Contributions for ΔH°_{f298} and S°_{298}† (Continued)

Base group	ΔH°_{f298}, kcal/g-mol	S°_{298}, cal/g-mol K
Substitution Group Contributions Replacing CH₃ Group (Continued)		
—Cl	0 for first Cl on a carbon; 4.5 for each additional	0
—Br	10.0	3.0‡
—F	−35.0	−1.0‡
—I	24.8	5.0‡
=O, aldehyde	−12.9	−12.3
ketone	−13.2	−2.4
—COOH	−87.0	15.4
—SH	15.8	5.2
—C₆H₅	32.3	21.7
—NH₂	12.3	−4.8

†From Ref. 3 and modified by J. M. Brown, as reported in Ref. 16, chap. 25.

‡Add 1.0 to the calculated entropy contributions of halides for methyl derivatives; e.g., methyl chloride = 44.4 (base) + 10.4 (primary CH_3) − 0.0 (Cl substitution) + 1.0.

B = 2.   A secondary methyl substitution of a hydrogen on the side chain in toluene to form ethyl benzene is of type A = 1, B = 5.

Special secondary substitutions are defined in Table 7-7 to modify a carboxylic acid to methyl and ethyl esters.   Contributions are also provided for multiple bonds and for nonhydrocarbon groups.   In the latter, the group value shown is for the replacement of a methyl group, *not* a hydrogen atom.   The method is illustrated in Example 7-8, and a comparison is shown between literature and estimated values of ΔH°_f in Table 7-8.

Example 7-8   With the method proposed by Anderson, Beyer, and Watson, estimate ΔH°_{f298} of p-cresol.

solution   The base group to choose is obviously benzene.   Then toluene is synthesized in the primary substitution.   p-Xylene is formed next, and then one —CH₃ is replaced by an —OH to form p-cresol.   Thus:

Benzene base group	19.82
First methyl substitution to form toluene	−7.87
Second methyl substitution, para	−7.66
Replacement of —CH₃ by —OH	−32.7
	−28.41

Stull et al. [33] report that ΔH°_{f298} for p-cresol is −29.97 kcal/g-mol.

TABLE 7-8 Comparison between Estimated and Literature Values of Heats of Formation at 298 K

Compound	ΔH_f°, kcal/g-mol [33]	Bond values, Table 7-1	Difference,† kcal/g-mol, as calculated by method of:			
			Benson et al., Table 7-4	Verma and Doraiswamy, Table 7-6	Franklin, Table 7-6	Anderson, Beyer, and Watson, Table 7-7
Propane	−24.82	0.35	0.29	0.62	0.35	0.32
n-Heptane	−44.88	0.02	0.03	0.32	0.01	−0.06
2,2,3-Trimethylbutane	−48.95	−0.95	−0.35	2.37	−0.56	1.95
trans-2-Butene	−2.67	0.51	0.31	0	−0.26	−0.01
3,3-Dimethyl-1-butene	−10.31	20	3.40	4.80	4.25	4.28
2-Methyl-1,3-butadiene	18.10	−0.19	0	7.17	3.67	−1.36
2-Pentyne	30.80	...	−0.59	−1.48	−1.29	−0.28
p-Ethyltoluene	−0.78	−0.37	0.02	−0.10	2.63	−0.39
2-Methylnaphthalene	27.75	−9.24	0.70	−0.60	−0.64	−0.22
cis-1,3-Dimethylcyclopentane	−32.47	2.0	0.04	0.29	−0.20	0.69
2-Butanol	−69.86	−4.58	0.35	0.67	1.30	−0.79
p-Cresol	−29.97	...	0.20	0.62	2.71	−1.56
Isopropylether	−76.20	−22	0.50	−8.42	−6.34	27
p-Dioxane	75.30	−2.12	−1.90	−7.14	−1.18	
Methyl ethyl ketone	−56.97	−0.27	−0.19	0.01	−0.20	−7.40
Ethyl acetate	−105.86	−1.05	−2.28	...	−0.89	−0.97
Methyl methacrylate	−79.3	18	2.22	...	3.85	
Trimethylamine	−5.70	0.87	0.14	6.11	...	5.20
Propionitrile	12.10	...	0.30	9.53	2.35	−3.16
2-Nitrobutane	−39.1	...	1.80	−4.43	−4.34	−3.93

TABLE 7-8 Comparison between Estimated and Literature Values of Heats of Formation at 298 K (Continued)

| | | Difference,† kcal/g-mol, as calculated by method of: | | | | |
Compound	ΔH_f°, kcal/g-mol [33]	Bond values, Table 7-1	Benson et al., Table 7-4	Verma and Doraiswamy, Table 7-6	Franklin, Table 7-6	Anderson, Beyer, and Watson, Table 7-7
3-Picoline‡	25.37					−16
1,1-Difluoroethane	−118	...	1	...	...	
Octafluorocyclobutane	−365.2	...	−3.4	...	...	
Bromobenzene	25.10	...	2.10	...	...	−3.15
Trichloroethylene	−1.40	−0.30	−1.10	...	...	3.75
Butyl methyl sulfide	−24.42	−0.05	−0.22	−0.27	0.01	−0.32
2-Methyl-2-butanethiol	−30.36	−5.05	0.76	0.11	1.03	−1.57
Propyl disulfide	−28.01	1.29	−0.75	−10	−9.25	
3-Methylthiophene	19.79	5.12	0.10	...	17	

†Difference = calculated-literature. ‡None of the methods appears suitable for heterocyclic nitrogen compounds.

Discussion   Of the five methods for estimating ΔH°_{f298} discussed in this section, the method of Benson and his colleagues (Table 7-4) is the most accurate.   It is also the most complete because it is applicable to many types of organic compounds.   The bond-energy technique is easy to use but is less accurate.   The other three methods (Franklin, Verma-Doraiswamy, and Anderson-Beyer-Watson) are reasonably accurate and general.   In Table 7-8, a comparison is made between calculated and literature values of ΔH°_{f298} for different materials.   Wherever applicable, the Benson method is recommended.   None of the five is reliable for heterocyclic nitrogen compounds.

Other methods for estimating heats of formation are not discussed here, as they are too specialized, less accurate, or not thoroughly developed for engineering use.   As examples, Tans [34] presents a nomograph for estimating $\Delta H^\circ_{f_T}$ for alkanes; for the same type of materials, Somayajulu and Zwolinski [32] discuss a detailed group-contribution method which takes into account the detailed structure of the molecules.   Shotte [30] and Boyd et al. [7] discuss more theoretical methods, while Rihani [24] and Van Tiggelen [39] suggest empirical methods, which, however, are less accurate than those presented here.

7-5 Ideal-Gas Entropy

As noted in Sec. 7-1, S°_T is the entropy of a material in the ideal-gas state, at 1 atm pressure, relative to the entropy of the material in a perfectly ordered solid at 0 K.   It is not an entropy of formation in the same sense as ΔH°_f.   The utility of S°_T lies in determining standard entropies of reaction, and this is accomplished by summing S° for the products and reactants, each multiplied by its respective stoichiometric multiplier.   This technique is workable since, thanks to the conservation of atoms in a chemical reaction, the absolute entropies of all elements involved cancel.   Also implied is the assumption that entropy changes in reactions between perfectly ordered solids at 0 K disappear, i.e., the so-called third law of thermodynamics.   A similar approach is not satisfactory for standard enthalpies of reaction since in that case there still remains an enthalpy of reaction even at 0 K.

Experimentally, S° can be determined from heat capacities and from latent enthalpies of phase transformation as a material is heated from 0 K to T.   Or it can be found using relations derived from statistical mechanics if sufficient information is available on molecular structure and spectra.   This is not the case for ΔH°_f.   An experimental heat of reaction or combustion must always be available to establish ΔH°_f at some temperature; this ΔH°_f can then be adjusted to other temperatures by the use of Eq. (7-1.2).   Similarly, if S° is known at one temperature T_1,

$$S_T^\circ = S_{T_1}^\circ + \int_T^{T_1} C_p^\circ \, d \ln T \tag{7-5.1}$$

Equation (7-1.5) gives the comparable relation for the standard entropy of reaction ΔS_T° referred to a base temperature of 298 K. With ideal-gas heat capacities obtained from data or by techniques presented in Sec. 7-3, it then becomes necessary only to obtain values of S_{298}° to determine entropy changes for chemical reactions at any temperature.

In this section, three methods are suggested for calculating S_{298}°. As noted above, with $C_p^\circ(T)$ values and Eq. (7-1.5), ΔS_T° can then be found. Also with values of $\Delta H_{f_{298}}^\circ$ (Sec. 7-4) and C_p°, with Eq. (7-1.3), ΔH_T° can be found. Then, using Eq. (7-1.6), the standard Gibbs energy of reaction ΔG_T° can be calculated and equilibrium constants can be determined, e.g.,

$$\Delta G^\circ = - RT \ln K \tag{7-5.2}$$

Alternatively, ΔG_T° can be estimated directly. Few methods have been proposed to accomplish this task. One reasonably reliable technique is presented in Sec. 7-6.

If one wants to determine the *entropy of formation* of a material, the synthesis reaction is written and ΔS_f° is then the entropy of reaction. The value of S° for the elements which make up the compound must then be known. Stull et al. [33] present tables of S° (as well as C_p°) for all common elements as a function of temperature.

Corrections for Symmetry When determining S_{298}° by group-contribution methods, ordinarily one must make certain corrections for molecular symmetry. This correction has been the source of some difficulty in that care must be taken to note carefully how an author has defined the symmetry number of a molecule.

The correction originates in the fact that, from statistical mechanics, the entropy is given by $R \ln \mathcal{W}$, where $\mathcal{W}$ is the number of distinguishable configurations of a compound. The rotational entropy contribution must be corrected, since by rotating a molecule one often finds indistinguishable configurations and $\mathcal{W}$ must be reduced by this factor, or if σ is the symmetry number (see below for a more exact definition), the rotational entropy is to be corrected by subtracting $R \ln \sigma$ from the calculated value.

Benson [4] defines σ as "the total number of *independent* permutations of identical atoms (or groups) in a molecule that can be arrived at by simple rigid rotations of the entire molecule." Inversion is not allowed.

It is often convenient to separate σ into two parts, σ_{ext} and σ_{int}, and then

$$\sigma = \sigma_{\text{ext}} \sigma_{\text{int}} \tag{7-5.3}$$

For example, propane has two terminal —CH_3 groups; each has a threefold axis of symmetry. Rotation of these *internal* groups yields $\sigma_{int} = (3)(3)$ as the number of permutations. Also, the entire molecule has a single twofold axis of symmetry, so $\sigma_{ext} = 2$. Then $\sigma = (2)(3^2) = 18$. Some additional examples:

	σ_{ext}	σ_{int}	σ
Benzene	(6)(2)	1	12
Methane	4	3	12
p-Cresol	2	3	6
1,3,5-Trimethylbenzene	2	3^4	162
1,2,4-Trimethylbenzene	1	3^3	27
Cyclohexane	6	1	6
Methanol	1	3	3
t-Butyl alcohol	1	3^4	81
Acetone	2	3^2	18
Acetic acid	1	3	3
Aniline	2	1	2
Trimethylamine	3	3^3	81

Benson et al. [6] show many other examples in their comprehensive review paper.

Corrections for Isomers In addition to the symmetry corrections noted above, if a molecule has optical isomers, i.e., contains one or more completely asymmetric carbon atoms (as in 3-methylhexane), the number of spatial orientations is increased and a correction of $+R \ln \eta$ must be added to the calculated absolute entropy, η being the number of such isomers. The number of possible optical isomers is 2^m, where m is the number of asymmetric carbons. However, in some molecules with more than one asymmetric carbon atom, there exist planes of symmetry which negate the optical activity of some forms, e.g., the meso form of tartaric acid.

In a similar manner, Benson [4] indicates that in molecules of the type ROOH and R—OO—R, the O—H and O—R bonds are at approximately right angles and exist in right- and left-hand forms with a higher entropy by $R \ln 2$.

Bond Entropies With Table 7-1, including any symmetry or isomer corrections, S_{298}° can be rapidly estimated.

Example 7-9 Estimate the entropy of 1,2-diiodopropane at 298 K with Table 7-1. This material has a symmetry number of 3 (that is, $\sigma_{ext} = 1$, $\sigma_{int} = 3$) and one asymmetric carbon atom.

solution

$$S^\circ_{298} = 6(C-H) + 2(C-C) + 2(C-I) - R \ln \sigma + R \ln \eta$$
$$= (6)(12.90) + (2)(-16.40) + (2)(24.65) - 1.987 \ln 3 + 1.987 \ln 2$$
$$= 93.1 \text{ cal/g-mol K}$$

The literature value is 94.3 cal/g-mol K [6].

Benson Method   Additive groups are presented in Table 7-4, and the method is described in Sec. 7-3, where heat-capacity estimation techniques are covered.   As used to determine S°_{298}, the method is illustrated in Example 7-10, and a few calculated values are compared with those from the literature in Table 7-9.

TABLE 7-9   Comparison between Estimated and Literature Values of Entropies at 298 K

		Difference,† cal/g-mol K, as calculated by method of:		
Compound	S°_{298}, cal/g-mol K [33]	Bond values, Table 7-1	Benson et al., Table 7-4	Anderson, Beyer, and Watson, Table 7-7
Propane	64.51	0.15	− 0.01	0.34
n-Heptane	102.27	− 0.01	− 0.09	− 0.70
2,2,3-Trimethylbutane	91.61	21	0.18	− 3.63
trans-2-Butene	70.86	− 0.20	0.16	1.01
3,3-Dimethyl-1-butene	82.16	0.41	1.20	0.04
2-Methyl-1,3-butadiene	75.44	1.48	− 0.21	− 0.44
2-Pentyne	79.30	...	0.16	− 0.06
p-Ethyltoluene	95.34	− 2.28	− 0.22	− 0.35
2-Methylnaphthalene	90.83	10	1.22	1.95
cis-1,3-Dimethylcyclopentane	87.67	− 26	0.21	0.71
2-Butanol	85.81	− 1.90	0.49	− 3.61
p-Cresol	83.09	...	1.09	3.74
Isopropyl ether	93.27	3.63	4.95	10
p-Dioxane	71.65	− 19		
Methyl ethyl ketone	80.81	0.43	0.26	− 3.61
Ethyl acetate	86.70	0.94	3.23	15
Trimethylamine	69.02	− 0.05	0.02	
Propionitrile	68.50	...	− 0.10	9.53
2-Nitrobutane	91.62	...	2.53	− 10
1,1-Difluoroethane	67.52	− 14	− 0.22	0.90
Octafluorocyclobutane	95.69	− 30	1.18	
Bromobenzene	77.53	...	0.34	1.89
Trichloroethylene	77.63	0.23	0.13	1.69
Butyl methyl sulfide	98.43	− 0.19	− 0.10	
2-Methyl-2-butanethiol	92.48	2.77	− 0.06	− 4.87
Propyl disulfide	118.30	− 0.44	0.12	
3-Methylthiophene	76.79	...	− 0.74	

†Difference = calculated-experimental.

Example 7-10 Using the Benson group contributions from Table 7-4, estimate the value of S°_{298} for methyl isopropyl ketone.

solution For this material $\sigma_{ext} = 1$, $\sigma_{int} = 3^3$, and $\eta = 1$. Thus

	$2[C—(C)(H)_3] = (2)(30.41) =$	60.82
	$C—(C)_2(H)(CO)$	$= -12.0$
	$CO—(C)_2$	$= 15.01$
	$C—(CO)(H)_3$	$= 30.41$
Symmetry:	$-R \ln \sigma = -R \ln 3^3$	$= -6.55$
		87.7

The experimental value reported is 88.5 cal/g-mol K [2].

Anderson-Beyer-Watson Method

This additive group method is identical to that described in Sec. 7-4 for $\Delta H^{\circ}_{f_{298}}$ and yields S°_{298} in cal/g-mol K. Contributions are shown in Table 7-7. *Neither symmetry nor isomer corrections are required.*

Example 7-11 Repeat Example 7-10 using the method of Anderson, Beyer, and Watson to estimate S°_{298} of methyl isopropyl ketone.

solution The base group chosen from Table 7-7 is methane.

	S°_{298}
Base group, methane	44.50
Primary methyl substitution to form ethane	10.35
Secondary methyl substitution of the type $A = 1$, $B = 1$, to form propane	10.00
Secondary methyl substitution of the type $A = 1$, $B = 2$, to form *n*-butane	9.18
Secondary methyl substitution of the type $A = 2$, $B = 2$, to form 2-methylbutane	7.15
Secondary methyl substitution of the type $A = 2$, $B = 3$, to form 2,3-dimethylbutane	6.53
Replacement of —CH$_3$ by C=O to form methyl isopropyl ketone	-2.4
	85.31

As noted in Example 7-10, the literature value of S°_{298} is 88.5 cal/g-mol K.

Discussion

A comparison between calculated and literature values of S°_{298} is shown in Table 7-9 for the three methods discussed in this section. The most accurate is the one suggested by Benson et al., which employs the group contributions in Table 7-4. Except in a few cases, the difference between the estimated and reported values is less than 0.5 cal/g-mol K. Differences found using the bond values of Table 7-1 varied from small to very large, whereas the Anderson-Beyer-Watson method was more consistent and differences averaged about 1 cal/g-mol K. In this last method, *no* symmetry or isomer corrections need be made.

Other estimation methods for $S°$ have been proposed by Rihani [25] and Rihani and Doraiswamy [27]. The latter is not recommended as the symmetry and isomer corrections have been multiplied by temperature and are therefore dimensionally inconsistent; also, they have been assigned the wrong algebraic sign [31].

7-6 Standard Gibbs Energy of Formation

Few estimation methods have been suggested to estimate $\Delta G_f^°$ directly. In most instances one determines $\Delta H_f^°$ and $\Delta S_f^°$ (or $S°$) separately and then uses Eq. (7-1.6) to determine the Gibbs energy change in a reaction. However, van Krevelen and Chermin [37, 38] have proposed that $\Delta G_f^°$ can be correlated as a linear function of temperature, as shown in Eq. (7-6.1), with the constants A and B determined as additive functions of groups shown in Table 7-10:

$$\Delta G_f^° = A + BT \qquad (7-6.1)$$

Two broad temperature bands are specified in the table, as a single linear equation was not found to be applicable over the entire range of 300 to 1500 K.

Since

$$\Delta G_f^° = \Delta H_f^° - T \Delta S_f^° \qquad (7-6.2)$$

it is tempting to associate $\Delta H_f^°$ with A and $-\Delta S_f^°$ with B, but it is *not* recommended that this be done to obtain individual estimations of $\Delta H_f^°$ and $\Delta S_f^°$.

When employing this method, one must correct for symmetry and isomers. However, the symmetry correction is $+R \ln \sigma_{ext}$. (See Sec. 7-5 for a discussion of symmetry corrections.) This correction is to be added to the parameter B. The symmetry corrections for σ_{int} have already been included in the contributions shown in Table 7-10.

The isomer correction is now $-R \ln \eta$ and is also applied to parameter B; η is the number of optical isomers (see Sec. 7-5).

The method is illustrated in Example 7-12, and a comparison between calculated and literature values of $\Delta G_f^°$ is shown in Table 7-11. In general, $\Delta G_f^°$ can be estimated within 5 kcal/g-mol. With such uncertainty, one can *estimate* the feasibility of a reaction, but one cannot accurately calculate equilibrium concentrations.

TABLE 7-10 Group Contributions for ΔG_f° [37, 38]

Group	300–600 K		600–1500 K	
	A	$B \times 10^2$	A	$B \times 10^2$
Alkane Groups				
CH_4	-18.948	2.225	-21.250	2.596
$-CH_3$	-10.943	2.215	-12.310	2.436
$-CH_2-$	-5.193	2.430	-5.830	2.544
$-\overset{\displaystyle |}{\underset{\displaystyle |}{C}}H$	-0.705	2.910	-0.705	2.910
$-\overset{\displaystyle |}{\underset{\displaystyle |}{C}}-$	1.958	3.735	4.385	3.350
Alkene Groups				
$H_2C{=}CH_2$	11.552	1.545	9.450	1.888
$H_2C{=}C\diagup\diagdown^H$	13.737	1.655	12.465	1.762
$H_2C{=}C\diagup\diagdown$	16.467	1.915	16.255	1.966
$^H\diagdown_{\diagup}C{=}C\diagup^H\diagdown$	17.663	1.965	16.180	2.116
$^H\diagdown_{\diagup}C{=}C\diagup\diagdown_H$	17.187	1.915	15.815	2.062
$^H\diagdown_{\diagup}C{=}C\diagup\diagdown$	20.217	2.295	19.584	2.354
$\diagdown_{\diagup}C{=}C\diagup\diagdown$	25.135	2.573	25.135	2.573
$H_2C{=}C{=}CH_2$	45.250	1.027	43.634	1.311

TABLE 7-10 Group Contributions for ΔG_f° (Continued)

Group	300–600 K		600–1500 K	
	A	$B \times 10^2$	A	$B \times 10^2$
Alkene Groups (Continued)				
$H_2C{=}C{=}C\langle^H$	49.377	1.035	48.170	1.208
$H_2C{=}C{=}C\langle$	51.084	1.474	51.084	1.474
$^H{>}C{=}C{=}C\langle^H$	52.460	1.483	52.460	1.483
Conjugated Alkene Groups				
$H_2C \leftrightarrow$	5.437	0.675	4.500	0.832
$^H{>}C\leftrightarrow$	7.407	1.035	6.980	1.088
$>C\leftrightarrow$	9.152	1.505	10.370	1.308
Alkyne Groups				
$HC{\equiv}$	27.048	-0.765	26.700	-0.704
$-C{\equiv}$	26.938	-0.525	26.555	-0.550
Aromatic Groups				
$HC\langle$	3.047	0.615	2.505	0.706
$-C\langle$	4.675	1.150	5.010	0.988
$\leftrightarrow C\langle$	3.513	0.568	3.998	0.485

Ring Formation				
3-membered ring	23.458	− 3.045	22.915	− 2.966
4-membered ring	10.73	− 2.65	10.60	− 2.50
5-membered ring	4.275	− 2.350	2.665	− 2.182
6-membered ring	− 1.128	− 1.635	− 1.930	− 1.504
Pentene ring	− 3.657	− 2.395	− 3.915	− 2.250
Hexene ring	− 9.102	− 2.045	− 8.810	− 2.071

Branching in Paraffin Chains				
Side chain with two or more C atoms	1.31	0	1.31	0
Three adjacent —CH groups	2.12	0	2.12	0
Adjacent —CH—C— groups	1.80	0	1.80	0
Two adjacent —C— groups	2.58	0	2.58	0

Branching in Cycloparaffins				
Branching in 5-membered ring:				
Single branching	− 1.04	0	− 1.69	0
Double branching:				
1,1 position	− 1.85	0	− 1.190	− 0.160
Cis 1,2 position	− 0.38	0	− 0.38	0
Trans 1,2 position	− 2.55	0	− 0.945	− 0.266
Cis 1,3 position	− 1.20	0	− 0.370	− 0.166
Trans 1,3 position	− 2.35	0	− 0.800	− 0.264
Branching in 6-membered ring:				
Single branching	− 0.93	0	0.230	− 0.192
Double branching:				
1,1 position	0.835	− 0.367	1.745	− 0.556
Cis 1,2 position	− 0.19	0	1.470	− 0.276
Trans 1,2 position	− 2.41	0	0.045	− 0.398
Cis 1,3 position	− 2.70	0	− 1.647	− 0.185
Trans 1,3 position	− 1.60	0	0.260	− 0.290
Cis 1,4 position	− 1.11	0	− 1.11	0
Trans 1,4 position	− 2.80	0	− 0.995	− 0.245
Branching in aromatics:				
Double branching:				
1,2 position	1.02	0	1.02	0
1,3 position	− 0.31	0	− 0.31	0
1,4 position	0.93	0	0.93	0
Triple branching:				
1,2,3 position	1.91	0	2.10	0
1,2,4 position	1.10	0	1.10	0
1,3,5 position	0	0	0	0

TABLE 7-10 Group Contributions for ΔG_f° (Continued)

Group	300–600 K		600–1500 K	
	A	$B \times 10^2$	A	$B \times 10^2$
Oxygen-containing Groups				
H_2O	−58.076	1.154	−59.138	1.316
—OH	−41.56	1.28	−41.56	1.28
—O—	−15.79	−0.85		
O⟨	−18.37	0.80	−16.07	0.40
H_2CO	−29.118	0.653	−30.327	0.854
H —C=O	−29.28	0.77	−30.15	0.83
⟩C=O	−28.08	0.91	−28.08	0.91
O ‖ HC—OH	−87.660	2.473	−90.569	2.958
O ‖ —C—OH	−98.39	2.86	−98.83	2.93
O ‖ —C—O—	−92.62	2.61	−92.62	2.61
$H_2C=C=O$	−14.515	0.295	−14.515	0.295
HC=C=O |	−12.86	0.46	−12.86	0.46
⟩C=C=O	−9.62	0.72	−9.38	0.73
Nitrogen-containing Groups				
HCN	31.179	−0.826	30.874	−0.775
—C≡N	30.75	−0.72	30.75	−0.72
—N=C	46.32	−0.89	46.32	−0.89
NH_3	−11.606	2.556	−12.972	2.784

—NH$_2$	2.82	2.71	−6.78	3.98
⟩NH	12.93	3.16	12.93	3.16
⟩N—	19.46	3.82	19.46	3.82
⟩N (ring)	11.32	1.11	12.26	0.96
—NO$_2$	−9.0	3.70	−14.19	4.38

Sulfur-containing Groups				
H$_2$S	−20.552	1.026	−21.366	1.167
—SH	−10.68	1.07	−10.68	1.07
—S—	−3.32	1.42	−3.32	1.44
S (ring)	−0.97	0.51	−0.65	0.44
⟩SO	−30.19	3.39	−30.19	3.39
⟩SO$_2$	−82.58	5.58	−80.69	5.26

Halogen-containing Groups				
HF	−64.476	−0.145	−64.884	−0.081
—F	−45.10	−0.20		
HCl	−22.100	−0.215	−22.460	−0.156
—Cl	−8.25	0	−8.25	0
HBr	−12.553	−0.234	−13.010	−0.158
—Br	−1.62	−0.26	−1.62	−0.26
HI	−1.330	−0.225	−1.718	−0.176
—I	7.80	0	7.80	0

Ring Formation				
⟩C—O—C⟨ (epoxide ring)	12.86	−0.63	12.86	−0.63
⟩C—O—C⟨ / ⟩C—C⟨ (five-membered ring)	−5.82	0.25	−3.53	−0.16

TABLE 7-11   Comparison between Estimated and Literature Standard Gibbs Energies of Formation

Values in kilocalories per gram mole for the ideal-gas state at 1 atm and 298 K

Compound	Literature value [33] $\Delta G^\circ_{f_{298}}$	Difference† as estimated by method of van Krevelen and Chermin
Propane	−5.61	0.57
n-Heptane	1.91	−0.04
3-Methylhexane	1.10	−0.02
2,4-Dimethylpentane	0.74	0.32
2,2,3-Trimethylbutane	1.02	0.54
Cyclopentane	9.23	−0.37
Cyclohexane	7.59	−0.43
Methylcyclopentane	8.55	−0.50
Ethylene	16.28	0.69
1-Butene	17.04	−0.67
cis-2-Butene	15.74	−0.50
trans-2-Butene	15.05	0.04
1,3-Butadiene	36.01	0.29
Acetylene	50.00	2.23
Methylacetylene	46.47	2.03
Benzene	30.99	−0.26
Ethylbenzene	31.21	−0.59
o-Xylene	29.18	−0.71
m-Xylene	28.41	−1.27
Ethyl mercaptan	−1.12	8.58
Diethylsulfide	4.25	7.55
Thiophene	30.30	−9.80
Aniline	39.84	4.06
Ethylamine	8.91	0.37
Pyridine	45.46	−5.86
Dimethylether	−26.99	−1.09
Acetaldehyde	−31.86	−0.55
Acetone	−36.58	−3.08
Methyl formate	−71.03	18
Acetic acid	−90.03	4.07
Ethyl acetate	−78.25	13
n-Propyl alcohol	−38.95	−0.95
Isopropyl alcohol	−41.49	−3.19
Phenol	−7.86	−2.66
Phosgene	−49.42	−7.52
Methyl chloride	−15.03	−3.13
Methylene chloride	−16.46	−1.12

Chloroform	− 16.38	− 0.28
Carbon tetrachloride	− 13.92	8.38
Ethyl bromide	− 6.29	− 1.62
Dichlorodifluoromethane	− 94.3	0.90
Fluorobenzene	− 16.50	− 4.80
Chlorobenzene	23.70	1.00

†Difference = estimated-literature.

Example 7-12 Estimate the standard Gibbs energy of formation of 1,2-diiodopropane at 500 K. As noted in Example 7-9, $\sigma_{ext} = 1$, $\sigma_{int} = 3$, $\eta = 2$.
  solution The group contributions are as follows, from Table 7-10:

	A	$B \times 10^2$
—CH$_3$	− 10.943	2.215
—CH$_2$	− 5.193	2.430
—C—H	− 0.705	2.910
2 —I	2(7.80)	2(0)
$+ R \ln \sigma_{ext}$	$\cdots$	0
$- R \ln \eta$	$\cdots$	$- R \ln 2$
	− 1.241	6.178

From Eq. (7-6.1),

$$\Delta G^{\circ}_{f500} = - 1.241 + (6.178 \times 10^{-2})(500)$$
$$= 29.65 \text{ kcal/g-mol}$$

Stull [33] reports that $\Delta G^{\circ}_{f500} = 26.9$ kcal/g-mol. At this temperature, $\Delta H^{\circ}_f = -8.06$ and $\Delta S^{\circ}_f = -7.0 \times 10^{-2}$; clearly these values are not represented by parameters A and $-B$.

NOTATION

C°_p = ideal-gas heat capacity, cal/g-mol K
ΔG°_{fT} = standard Gibbs energy of formation at T, kcal/g-mol
ΔH_c = heat of combustion, kcal/g-mol
ΔH°_{fT} = standard heat of formation at T, kcal/g-mol
m = number of asymmetric carbon atoms
n_i = number of groups of type i
R = gas constant, usually 1.987 cal/g-mol K
S°_T = absolute entropy at T, cal/g-mol K

$\Delta S^\circ_{fT} =$ standard entropy of formation at T, cal/g-mol K

$T =$ temperature, K

$\mathscr{W} =$ number of distinguishable configurations of a molecule

Greek

$\eta =$ number of optical isomers

$\nu_j =$ stoichiometric multiplier, positive for products, negative for reactants

$\sigma =$ symmetry number; σ_{ext}, for rigid body rotation; σ_{int}, for rotation of subgroups constituting the molecule

REFERENCES

1. American Petroleum Institute: "Technical Data Book: Petroleum Refining," 2d ed., chap. 7, Washington, D.C., 1971.
2. Anden, R. J. L.: *J. Chem. Soc.*, **1968A**: 1894.
3. Anderson, J. W., G. H. Beyer, and K. M. Watson: *Natl. Pet. News Tech. Sec.*, **36**: R476 (July 5, 1944).
4. Benson, S. W.: "Thermochemical Kinetics," chap. 2, Wiley, New York, 1968.
5. Benson, S. W., and J. H. Buss: *J. Chem. Phys.*, **29**, 546 (1958).
6. Benson, S. W., F. R. Cruickshank, D. M. Golden, G. R. Haugen, H. E. O'Neal, A. S. Rodgers, R. Shaw, and R. Walsh: *Chem. Rev.*, **69**: 279 (1969).
7. Boyd, R. H., S. M. Breitling, and M. Mansfield: *AIChE J.*, **19**: 1016 (1973).
8. Chueh, P. L., Shell Development Co.: private communication, 1974.
9. Cox, J. D., and G. Pilcher: "Thermochemistry of Organic and Organometallic Compounds," Academic, London, 1970.
10. Domalski, E. S.: *J. Phys. Chem. Ref. Data*, **1**: 221 (1972).
11. Eigenmann, H. K., D. M. Golden, and S. W. Benson: *J. Phys. Chem.*, **77**: 1687 (1973).
12. Franklin, J. L.: *Ind. Eng. Chem.*, **41**: 1070 (1949).
13. Franklin, J. L.: *J. Chem. Phys.*, **21**: 2029 (1953).
14. Glasstone, S.: "Theoretical Chemistry," Van Nostrand, Princeton, N.J., 1944.
15. Handrick, G. R.: *Ind. Eng. Chem.*, **48**: 1366 (1956).
16. Hougen, O. A., K. M. Watson, and R. A. Ragatz: "Chemical Process Principles," 2d ed., pt. II, "Thermodynamics," Wiley, New York, 1959.
17. Huang, P. K., and T. E. Daubert: *Ind. Eng. Chem. Process Des. Dev.*, **13**: 193 (1974).
18. JANAF Thermochemical Tables, 2d ed., NSRDS-NBS 37, June 1971.
19. Janz, G. J.: "Estimation of Thermodynamic Properties of Organic Compounds," Academic, New York, 1958.
20. Olson, B. A.: M.S. thesis in Chemical Engineering, Rutgers University, New Brunswick, N.J., August 1973.
21. O'Neal, H. E., and S. W. Benson: *J. Chem. Eng. Data*, **15**: 266 (1970).
22. Passut, C. A., and R. P. Danner: *Ind. Eng. Chem. Process Des. Dev.*, **11**: 543 (1972).
23. Reid, R. C., and T. K. Sherwood: "Properties of Gases and Liquids," 2d ed., chap. 5, McGraw-Hill, New York, 1966.
24. Rihani, D. N.: *Hydrocarbon Process.*, **47**(3): 137 (1968).
25. Rihani, D. N.: *Hydrocarbon Process.*, **47**(4): 161 (1968).
26. Rihani, D. N., and L. K. Doraiswamy: *Ind. Eng. Chem. Fundam.*, **4**: 17 (1965).
27. Rihani, D. N., and L. K. Doraiswamy: *Ind. Eng. Chem. Fundam.*, **7**: 375 (1968).
28. Seaton, W. H., and E. Freedman: Computer Implementation of a Second Order Additivity Method for the Estimation of Chemical Thermodynamic Data, paper presented at *65th Ann. Meet., AIChE, New York, November 1972.*
29. Shaw, R.: *J. Phys. Chem.*, **75**: 4047 (1971).
30. Shotte, W.: *J. Phys. Chem.*, **72**: 2422 (1968).

31. Small, P. A.: *Ind. Eng. Chem. Fundam.*, **8**: 599 (1969).
32. Somayajulu, G. R., and B. J. Zwolinski: *Trans. Faraday Soc.*, **62**: 2327 (1966); *J. Chem. Soc., Faraday Trans.*, (II)**68**: 1971 (1972).
33. Stull, D. R., E. F. Westrum, and G. C. Sinke: "The Chemical Thermodynamics of Organic Compounds," Wiley, New York, 1969.
34. Tans, A. M. P.: *Hydrocarbon Process.*, **47**(4): 169 (1968).
35. Thinh, T.-P., J.-L. Duran, and R. S. Ramalho: *Ind. Eng. Chem. Process Des. Dev.*, **10**: 576 (1971).
36. Thinh, T.-P., J.-L. Duran, R. S. Ramalho, and S. Kaliaguine: *Hydrocarbon Process.*, **50**(1): 98 (1971).
37. van Krevelen, D. W., and H. A. G. Chermin: *Chem. Eng. Sci.*, **1**: 66 (1951).
38. van Krevelen, D. W., and H. A. G. Chermin: *Chem. Eng. Sci.*, **1**: 238 (1952).
39. Van Tiggelen, A.: *Chem. Eng. Sci.*, **20**: 529 (1965).
40. Verma, K. K., and L. K. Doraiswamy: *Ind. Eng. Chem. Fundam.*, **4**: 389 (1965).
41. Wenner, R. R., "Thermochemical Calculations," McGraw-Hill, New York, 1941.
42. Yuan, S. C., and Y. I. Mok, *Hydrocarbon Process.*, **47**(3): 133 (1968); **47**(7): 153 (1968).

Chapter Eight

Fluid-Phase Equilibria in Multicomponent Systems

8-1 Scope

In the chemical process industries, fluid mixtures are often separated into their components by diffusional operations such as distillation, absorption, and extraction. Design of such separation operations requires quantitative estimates of the partial equilibrium properties of fluid mixtures. Whenever possible, such estimates should be based on reliable experimental data for the particular mixture, at conditions of temperature, pressure, and composition corresponding to those of industrial interest. Unfortunately, such data are only rarely available. In typical cases, only fragmentary data are at hand, and it is necessary to reduce and correlate the limited data to make the best possible interpolations and extrapolations. This chapter discusses some techniques toward that end. Attention is restricted to solutions of nonelectrolytes. Emphasis is given to the calculation of fugacities in liquid solutions; fugacities in gaseous mixtures are discussed in Sec. 5-9.

The scientific literature on fluid-phase equilibria goes back well over 100 years and has reached monumental proportions, including thousands of articles and more than 100 books and monographs. Table 8-1 gives authors and titles of some books which are useful for obtaining data and for more detailed discussions. The list in Table 8-1 is not exhaustive; it is restricted to those publications which are likely to be useful to the practicing engineer in the chemical process industries.

There is an important difference between calculating phase-equilibrium compositions and calculating typical volumetric, energetic, or transport properties of fluids of known composition. In the latter case we are interested in the property of the mixture as a whole, whereas in the former we are interested in the *partial* properties of the individual components which constitute the mixture. For example, to find the pressure drop of a liquid mixture flowing through a pipe, we need the viscosity and the density of that liquid mixture at the particular composition of interest. But if we ask for the composition of the vapor which is in equilibrium with that liquid mixture, it is no longer sufficient to know the properties of the liquid mixture at that particular composition; we must now know, in addition, how certain of its properties (in particular the Gibbs energy) *depend* on composition. In phase-equilibrium calculations, we must know *partial* properties, and to find these, we typically differentiate data with respect to composition. Whenever experimental data are differentiated, there is a loss of accuracy, often a serious loss. Since partial, rather than total, properties are needed in phase equilibria, it is not surprising that prediction of phase-equilibrium relations is usually more difficult and less accurate than prediction of other properties encountered in chemical process design.

In one chapter it is not possible to present a complete review of a large subject. Also, since this subject is so wide in its range, it is not possible to recommend to the reader simple, unambiguous rules for obtaining quantitative answers to his particular phase-equilibrium problem. Since the variety of mixtures is extensive, and since mixture conditions (temperature, pressure, composition) cover many possibilities, and, finally, since there are large variations in the availability, quantity, and quality of experimental data, the reader cannot escape the responsibility of using judgment, which, ultimately, is obtained only by experience.

This chapter, therefore, is qualitatively different from the others in this book. It does not give specific advice on how to calculate specific quantities. It provides only an introduction to some (by no means all) of the tools and techniques which may be useful for an efficient strategy toward calculating particular phase equilibria for a particular process design.

TABLE 8-1 Some Useful Books on Fluid-Phase Equilibria

Book	Remarks
Chao, K. C., and R. A. Greenkorn: "Thermodynamics of Fluids," Dekker, New York, 1975	Good introductory survey including an introduction to statistical thermodynamics of fluids; also gives a summary of surface thermodynamics
Francis, A. W.: "Liquid-Liquid Equilibriums," Wiley-Interscience, New York, 1963	Phenomenological discussion of liquid-liquid equilibria with essentially no thermodynamics; extensive data bibliography
Hála, E., and others: "Vapour-Liquid Equilibrium," 2d English ed., trans. George Standart, Pergamon, Oxford, 1967	An excellent comprehensive survey, including a discussion of experimental methods
Hála, E., I. Wichterle, J. Polák, and T. Boublik: "Vapor-Liquid Equilibrium Data at Normal Pressures," Pergamon, Oxford, 1968	A compilation of experimental data for binary mixtures
Hicks, C. P., K. N. Marsh, A. G. Williamson, I. A. McLure, and C. L. Young: "Bibliography of Thermodynamic Studies," Chemical Society, London, 1975	Gives literature references for vapor-liquid equilibria, enthalpies of mixing, and volume changes of mixing for binary systems
Hildebrand, J. H., and R. L. Scott: "Solubility of Nonelectrolytes," 3d ed., Reinhold, New York, 1950 (reprinted by Dover, New York, 1964)	A classic in its field, giving a splendid survey of solution physical chemistry from a chemist's point of view. While out of date, it nevertheless provides physical insight into how molecules "behave" in mixtures
Hildebrand, J. H., and R. L. Scott: "Regular Solutions," Prentice-Hall, Englewood Cliffs, N.J., 1962	Updates some of the material in Hildebrand's 1950 book; primarily for chemists
Hildebrand, J. H., J. M. Prausnitz, and R. L. Scott: "Regular and Related Solutions," Van Nostrand Reinhold, New York, 1970	Further updates some of the material in Hildebrand's book; primarily for chemists
Hirata, M., S. Ohe, and K. Nagahama: "Computer-aided Data Book of Vapor-Liquid Equilibria," Elsevier, Amsterdam, 1975	A compilation of binary experimental data reduced with the Wilson equation and, for high pressures, with a modified Redlich-Kwong equation

Kehiaian, H. V. (Editor-in-Chief), and B. J. Zwolinski (Executive Officer): "International Data Series: Selected Data on Mixtures," Thermodynamics Research Center, Chemistry Department, Texas A and M University, College Station, Tex. 77843 (continuing since 1973)

Presents a variety of measured thermodynamic properties of binary mixtures; these properties are often represented by empirical equations

King, M. B.: "Phase Equilibrium in Mixtures," Pergamon, Oxford, 1969

A good general British text covering a variety of subjects in mixture thermodynamics

Null, H. R.: "Phase Equilibrium in Process Design," Wiley-Interscience, New York, 1970

A useful, engineering-oriented monograph with a variety of numerical examples

Prausnitz, J. M.: "Molecular Thermodynamics of Fluid-Phase Equilibria," Prentice-Hall, Englewood Cliffs, N.J., 1969

A text suitable for first-year graduate students; stresses application of molecular fundamentals for engineering application

Prausnitz, J. M., C. A. Eckert, J. P. O'Connell, and R. V. Orye: "Computer Calculations for Multicomponent Vapor-Liquid Equilibria," Prentice-Hall, Englewood Cliffs, N.J., 1967

Discusses the thermodynamic basis for computer calculations for vapor-liquid equilibria; computer programs are given

Prigogine, I., and R. Defay: "Chemical Thermodynamics," trans. and rev. by D. H. Everett, Longmans, Green, London, 1954

A semiadvanced text from a European chemist's point of view, it offers many examples and discusses molecular principles; although out of date, it contains much useful information not easily available in standard American texts

Renon, H., L. Asselineau, G. Cohen, and C. Raimbault: "Calcul sur ordinateur des équilibres liquide-vapeur et liquide-liquide," Editions Technip, Paris, 1971

Discusses the thermodynamic basis for computer calculations for vapor-liquid and liquid-liquid equilibria (in French); computer programs are given

Rowlinson, J. S.: "Liquids and Liquid Mixtures," 2d ed., Butterworth, London, 1969

An advanced British treatise with a good discussion of applied statistical mechanics

Treybal, R. E.: "Liquid Extraction," 2d ed., chaps. 2 and 3, McGraw-Hill, New York, 1963

Early chapters give good review of classical mixture thermodynamics with engineering applications

Wichterle, I., J. Linek, and E. Hála: "Vapor-Liquid Equilibrium Data Bibliography," Elsevier, Amsterdam, 1973

A thorough compilation of literature sources for binary and multicomponent data; includes many references to the East European literature

8-2 Basic Thermodynamics of Vapor-Liquid Equilibria†

We are concerned with a liquid mixture which, at temperature T and pressure P, is in equilibrium with a vapor mixture at the same temperature and pressure. The quantities of interest are the temperature, the pressure, and the compositions of both phases. Given some of these quantities, the task is to calculate the others.

For every component i in the mixture, the condition of thermodynamic equilibrium is given by

$$f_i^V = f_i^L \qquad (8\text{-}2.1)$$

where f = fugacity
V = vapor
L = liquid

The fundamental problem is to relate these fugacities to mixture compositions, since it is the latter which are of concern in process design. In the subsequent discussion we neglect effects due to surface forces, gravitation, electric or magnetic fields, semipermeable membranes, or any other special conditions rarely encountered in the chemical process industries.

The fugacity of a component in a mixture depends on the temperature, pressure, and composition of that mixture. In principle, any measure of composition can be used. For the vapor phase, the composition is nearly always expressed by the mole fraction y. To relate f_i^V to temperature, pressure, and mole fraction, it is useful to introduce the fugacity coefficient ϕ_i

$$\phi_i = \frac{f_i^V}{y_i P} \qquad (8\text{-}2.2)$$

which can be calculated from vapor-phase $PVTy$ data, usually given by an equation of state, as discussed in Sec. 5-9. For a mixture of ideal gases $\phi_i = 1$.

The fugacity coefficient ϕ_i depends on temperature and pressure and, in a multicomponent mixture, on *all* mole fractions in the vapor phase, not just y_i. The fugacity coefficient is, by definition, normalized such that as $P \to 0$, $\phi_i \to 1$ for all i. At low pressures, therefore, it is usually a good assumption to set $\phi_i = 1$. But just what "low" means depends on the composition and temperature of the mixture. For typical mixtures of nonpolar (or slightly polar) fluids at a temperature near or above the normal boiling point of the least volatile component, "low" pressure means a pressure less than a few atmospheres. However, for mixtures containing a strongly associating carboxylic acid, e.g., acetic acid–water at 25°C, fugacity coefficients may differ appreciably from unity at

†For a detailed discussion, see, for example, Refs. 56, 66, and 81.

pressures much less than 1 atm.† For mixtures containing one component of very low volatility and another of high volatility, e.g., decane-methane at 25°C, the fugacity coefficient of the light component may be close to unity for pressures up to 10 or 20 atm while at the same pressure the fugacity coefficient of the heavy component is much less than unity. A more detailed discussion is given in chap. 5 of Ref. 66.

The fugacity of component i in the liquid phase is related to the composition of that phase through the activity coefficient γ_i. In principle, any composition scale may be used; the choice is strictly a matter of convenience. For some aqueous solutions, frequently used scales are molality (moles of solute per 1000 g of water) and molarity (moles of solute per liter of solution); for polymer solutions, a useful scale is the volume fraction, discussed briefly in Sec. 8-14. However, for typical solutions containing nonelectrolytes of normal molecular weight (including water), the most useful measure of concentration is the mole fraction x. The activity coefficient γ_i is related to x_i and to the standard-state fugacity f_i° by

$$\gamma_i \equiv \frac{a_i}{x_i} = \frac{f_i^L}{x_i f_i^\circ} \tag{8-2.3}$$

where a_i is the activity of component i. The standard-state fugacity f_i° is the fugacity of component i at the temperature of the system, i.e., the mixture, and at some arbitrarily chosen pressure and composition. The choice of standard-state pressure and composition is dictated only by convenience, but it is important to bear in mind that the numerical values of γ_i and a_i have no meaning unless f_i° is clearly specified.

While there are some important exceptions, activity coefficients for most typical solutions of nonelectrolytes are based on a standard state, where for every component i, f_i° is the fugacity of *pure* liquid i at system temperature and pressure; i.e., the arbitrarily chosen pressure is the total pressure P, and the arbitrarily chosen composition is $x_i = 1$. Frequently, this standard-state fugacity refers to a hypothetical state since it may well happen that component i cannot physically exist as a pure liquid at system temperature and pressure. Fortunately, for many common mixtures it is possible to calculate this standard-state fugacity by modest extrapolations with respect to pressure, and since liquid-phase properties remote from the critical region are not sensitive to pressure (except at high pressures), such extrapolation introduces little uncertainty. In some mixtures, however, namely, those which contain supercritical components, extrapolations with respect to temperature

†A useful correlation for fugacity coefficients at modest pressures, including carboxylic acids, is given by J. G. Hayden and J. P. O'Connell, *Ind. Eng. Chem. Process Des. Dev.*, **14**: 3 (1975).

are required, and these, when carried out over an appreciable temperature region, may lead to large uncertainties. We briefly return to this problem in Sec. 8-12.

Whenever the standard-state fugacity is that of the pure liquid at system temperature and pressure, we obtain the limiting relation that $\gamma_i \to 1$ as $x_i \to 1$.

8-3 Fugacity of a Pure Liquid

To calculate the fugacity of a pure liquid at a specified temperature and pressure, we require two primary thermodynamic properties: the saturation (vapor) pressure, which depends only on temperature, and the liquid density, which depends primarily on temperature and to a lesser extent on pressure. Unless the pressure is very large, it is the vapor pressure which is by far the more important of these two quantities. In addition, we require volumetric data (equation of state) for pure vapor i at system temperature, but unless the vapor pressure is high, this requirement is of minor, often negligible, importance.

The fugacity of pure liquid i at temperature T and pressure P is given by

$$f_i^L(T, P, x_i = 1) = P_{vp_i}(T)\phi_i^s(T) \exp \int_{P_{vp_i}}^{P} \frac{V_i^L(T, P)}{RT} dP \quad (8\text{-}3.1)$$

where P_{vp} is the vapor pressure (see Chap. 6) and superscript s stands for saturation. The fugacity coefficient ϕ_i^s is calculated from vapor-phase volumetric data, as discussed in Sec. 5-9; for typical nonassociated liquids at temperatures well below the critical, ϕ_i^s is close to unity.

The molar liquid volume V_i^L is the ratio of the molecular weight to the density, where the latter is expressed in units of mass per unit volume.† At a temperature well below the critical, a liquid is nearly incompressible. In that case the effect of pressure on liquid-phase fugacity is not large unless the pressure is very high or the temperature is very low. The exponential term in Eq. (8-3.1) is called the *Poynting factor*.

To illustrate Eq. (8-3.1), the fugacity of liquid water is shown in Table 8-2. Since ϕ^s for a pure liquid is always less than unity, the fugacity at saturation is always lower than the vapor pressure. However, at pressures well above the saturation pressure, the product of ϕ^s and the Poynting factor may easily exceed unity, and then the fugacity is larger than the vapor pressure.

At 500 and 600°F, the vapor pressure exceeds 600 psia, and therefore pure liquid water cannot exist at these temperatures and 600 psia.

---

†For volumetric properties of liquids, see Sec. 3-1̃.̃.̃.̃.̃.

TABLE 8-2 Fugacity of Liquid Water, psia

Temp., °F	P_{vp}	Fugacity		
		Saturation	600 psia	5000 psia
100	0.9492	0.949	0.978	1.21
300	67.01	64.0	65.3	77.2
500	680.8	569	568†	663
600	1543	1159	1119†	1314

†Hypothetical because $P < P_{vp}$.

Nevertheless, the fugacity can be calculated by a mild extrapolation: in the Poynting factor we neglect the effect of pressure on molar liquid volume.

Table 8-2 indicates that the vapor pressure is the primary quantity in Eq. (8-3.1). When data are not available, the vapor pressure can be estimated, as discussed in Chap. 6. Further, for nonpolar (or weakly polar) liquids, the ratio of fugacity to pressure can be estimated from a generalized (corresponding-states) table for liquids, as discussed in Sec. 5-4.

8-4 Simplifications in the Vapor-Liquid Equilibrium Relation

Equation (8-2.1) gives the rigorous, fundamental relation for vapor-liquid equilibrium. Equations (8-2.2), (8-2.3), and (8-3.1) are also rigorous, without any simplifications beyond those indicated in the paragraph following Eq. (8-2.1). Substitution of Eqs. (8-2.2), (8-2.3), and (8-3.1) into Eq. (8-2.1) gives

$$y_i P = \gamma_i x_i P_{vp_i} \mathcal{F}_i \tag{8-4.1}$$

where

$$\mathcal{F}_i = \frac{\phi_i^s}{\phi_i} \exp \int_{P_{vp_i}}^{P} \frac{V_i^L \, dP}{RT} \tag{8-4.2}$$

With few exceptions, the correction factor $\mathcal{F}_i$ is near unity when the total pressure P is sufficiently low. However, even at moderate pressures, we are nevertheless justified in setting $\mathcal{F}_i = 1$ if only approximate results are required and, as happens so often, if experimental information is sketchy, giving large uncertainties in γ.

If, in addition to setting $\mathcal{F}_i = 1$, we assume that $\gamma_i = 1$, Eq. (8-4.1) reduces to the familiar relation known as *Raoult's law*.

In Eq. (8-4.1), γ_i depends on temperature, composition, and pressure. However, remote from critical conditions, and unless the pressure is large, the effect of pressure on γ_i is usually small. [See Eq. (8-12.1).]

8-5 Activity Coefficients; The Gibbs-Duhem Equation and Excess Gibbs Energy

In typical mixtures, Raoult's law provides no more than a rough approximation; only when the components in the liquid mixture are similar, e.g., a mixture of n-butane and isobutane, can we assume that γ_i is essentially unity for all components at all compositions. The activity coefficient, therefore, plays a key role in the calculation of vapor-liquid equilibria.

Classical thermodynamics has little to tell us about the activity coefficient; as always, classical thermodynamics does not give us the experimental quantity we desire but only relates it to another experimental quantity. Thus thermodynamics relates the effect of pressure on the activity coefficient to the partial molar volume, and it relates the effect of temperature on the activity coefficient to the partial molar enthalpy, as discussed in any thermodynamics text (see, for example, chap. 6 of Ref. 66). These relations are of limited use because good data for the partial molar volume and for the partial molar enthalpy are rare.

However, there is one thermodynamic relation which provides a useful tool for correlating and extending limited experimental data: the Gibbs-Duhem equation. This equation is not a panacea, but, given some experimental results, it enables us to use these results efficiently. In essence, the Gibbs-Duhem equation says that in a mixture, the activity coefficients of the individual components are not independent of one another but are related by a differential equation. In a binary mixture the Gibbs-Duhem relation is

$$x_1\left(\frac{\partial \ln \gamma_1}{\partial x_1}\right)_{T,P} = x_2\left(\frac{\partial \ln \gamma_2}{\partial x_2}\right)_{T,P} \qquad (8\text{-}5.1)\dagger$$

Equation (8-5.1) has several important applications.

1. If we have experimental data for γ_1 as a function of x_1, we can integrate Eq. (8-5.1) and calculate γ_2 as a function of x_2. That is, in a binary mixture, activity-coefficient data for one component can be used to predict the activity coefficient of the other component.

2. If we have extensive experimental data for *both* γ_1 and γ_2 as a function of composition, we can test the data for thermodynamic consistency by determining whether or not the data obey Eq. (8-5.1). If

---

†Note that the derivatives are taken at constant temperature T and constant pressure P. In a binary, two-phase system, however, it is not possible to vary x while holding *both* T and P constant. At ordinary pressures γ is a very weak function of P, and therefore it is often possible to apply Eq. (8-5.1) to isothermal data while neglecting the effect of changing pressure. This subject has been amply discussed in the literature; see, for example, chap. 6 and app. IV in Ref. 66.

the data show serious inconsistencies with Eq. (8-5.1), we may conclude that they are unreliable.

3. If we have limited data for γ_1 and γ_2, we can integrate the Gibbs-Duhem equation and obtain thermodynamically consistent equations which relate γ_1 and γ_2 to x. These equations contain a few adjustable parameters, which can then be determined from the limited data. It is this application of the Gibbs-Duhem equation which is of particular use to chemical engineers. However, there is no *unique* integrated form of the Gibbs-Duhem equation; many forms are possible. To obtain a particular relation between γ and x, we must assume some model which is consistent with the Gibbs-Duhem equation.

For practical work, the utility of the Gibbs-Duhem equation is best realized through the concept of excess Gibbs energy, i.e., the observed Gibbs energy of a mixture above and beyond what it would be for an ideal solution at the same temperature, pressure, and composition. By definition, an ideal solution is one where all $\gamma_i = 1$. The *total* excess Gibbs energy G^E for a binary solution, containing n_1 moles of component 1 and n_2 moles of component 2, is defined by

$$G^E = RT(n_1 \ln \gamma_1 + n_2 \ln \gamma_2) \qquad (8\text{-}5.2)$$

Equation (8-5.2) gives G^E as a function of *both* γ_1 and γ_2. Upon applying the Gibbs-Duhem equation, we can relate the *individual* activity coefficients γ_1 or γ_2 to G^E by differentiation

$$RT \ln \gamma_1 = \left(\frac{\partial G^E}{\partial n_1}\right)_{T,P,n_2} \qquad (8\text{-}5.3)$$

$$RT \ln \gamma_2 = \left(\frac{\partial G^E}{\partial n_2}\right)_{T,P,n_1} \qquad (8\text{-}5.4)$$

Equations (8-5.2) to (8-5.4) are useful because they enable us to interpolate and extrapolate limited data with respect to composition. To do so we must first adopt some mathematical expression for G^E as a function of composition. Second, we fix the numerical values of the constants in that expression from the limited data. Third, we calculate activity coefficients at any desired composition by differentiation, as indicated by Eqs. (8-5.3) and (8-5.4).

To illustrate, consider a simple binary mixture. Suppose that we need activity coefficients for a binary mixture over the entire composition range at a fixed temperature T. However, we have experimental data for only one composition, say $x_1 = x_2 = \frac{1}{2}$. From that one datum we calculate $\gamma_1(x_1 = \frac{1}{2})$ and $\gamma_2(x_2 = \frac{1}{2})$; for simplicity, let us assume symmetrical behavior, that is, $\gamma_1(x_1 = \frac{1}{2}) = \gamma_2(x_2 = \frac{1}{2})$.

We must adopt an expression relating G^E to the composition subject to the conditions that at fixed composition G^E is proportional to $n_1 + n_2$

and that $G^E = 0$ when $x_1 = 0$ or $x_2 = 0$. The simplest expression we can construct is

$$G^E = (n_1 + n_2)g^E = (n_1 + n_2)Ax_1x_2 \qquad (8\text{-}5.5)$$

where g^E is the excess Gibbs energy per mole of mixture and A is a constant depending on temperature. The mole fraction x is simply related to mole number n by

$$x_1 = \frac{n_1}{n_1 + n_2} \qquad (8\text{-}5.6)$$

$$x_2 = \frac{n_2}{n_1 + n_2} \qquad (8\text{-}5.7)$$

The constant A is found from substituting Eq. (8-5.5) into Eq. (8-5.2) and using the experimentally determined γ_1 and γ_2 at the composition midpoint:

$$A = \frac{RT}{(\frac{1}{2})(\frac{1}{2})} [\tfrac{1}{2} \ln \gamma_1(x_1 = \tfrac{1}{2}) + \tfrac{1}{2} \ln \gamma_2(x_2 = \tfrac{1}{2})] \qquad (8\text{-}5.8)$$

Upon differentiating Eq. (8-5.5) as indicated by Eqs. (8-5.3) and (8-5.4), we find

$$RT \ln \gamma_1 = Ax_2^{\,2} \qquad (8\text{-}5.9)$$

$$RT \ln \gamma_2 = Ax_1^{\,2} \qquad (8\text{-}5.10)$$

With these relations we can now calculate activity coefficients γ_1 and γ_2 at any desired x even though experimental data were obtained only at one point, namely, $x_1 = x_2 = \tfrac{1}{2}$.

This simplified example illustrates how the concept of excess functions, coupled with the Gibbs-Duhem equation, can be used to interpolate or extrapolate experimental data with respect to composition. Unfortunately the Gibbs-Duhem equation tells us nothing about how to interpolate or extrapolate experimental data with respect to temperature or pressure.

Equations (8-5.2) to (8-5.4) indicate the intimate relation between activity coefficients and excess Gibbs energy G^E. Many expressions relating g^E (per mole of mixture) to composition have been proposed, and a few are given in Table 8-3. All these expressions contain adjustable constants which, at least in principle, depend on temperature. That dependence may in some cases be neglected, especially if the temperature interval is not large. In practice, the number of adjustable constants per binary is typically two or three; the larger the number of constants, the better the representation of the data but, at the same time, the larger the number of reliable experimental data points required to determine the constants. Extensive and highly accurate experimental data are required to justify more than three empirical constants for a binary mixture at a fixed temperature.†

For moderately nonideal binary mixtures, all equations for g^E containing two (or more) binary parameters give good results; there is little reason to choose one over another except that the older ones (Margules, van Laar) are mathematically easier to handle than the newer ones (Wilson, NRTL, UNIQUAC). The two-suffix (one-parameter) Margules equation is applicable only to simple mixtures where the components are similar in chemical nature and in molecular size.

For strongly nonideal binary mixtures, e.g., solutions of alcohols with hydrocarbons, the equation of Wilson is probably the most useful because, unlike the NRTL equation, it contains only two adjustable parameters and it is mathematically simpler than the UNIQUAC equation. For such mixtures, the three-suffix Margules equation and the van Laar equation are likely to represent the data with significantly less success, especially in the region dilute with respect to alcohol, where the Wilson equation is particularly suitable.

The four-suffix (three-parameter) Margules equation has no significant advantages over the three-parameter NRTL equation.

Numerous articles in the literature use the Redlich-Kister expansion [see Eq. (8-9.20)] for g^E. This expansion is mathematically identical to the Margules equation.

The Wilson equation is not applicable to a mixture which exhibits a miscibility gap; it is inherently unable, even qualitatively, to account for phase splitting. Nevertheless, Wilson's equation may be useful even for those mixtures where miscibility is incomplete provided attention is strictly confined to the one-phase region.

Unlike Wilson's equation, the NRTL and UNIQUAC equations are applicable to *both* vapor-liquid and liquid-liquid equilibria.‡ Therefore, mutual-solubility data [see Sec. 8-10] can be used to determine NRTL or UNIQUAC parameters but not Wilson parameters. While UNIQUAC is mathematically more complex than NRTL, it has three advantages: (1) it has only two (rather than three)

---

†The models shown in Table 8-3 are not applicable to solutions of electrolytes; such solutions are not considered here. Thermodynamic properties of solutions containing strong electrolytes are discussed in R. A. Robinson and R. H. Stokes, "Electrolyte Solutions" (2d ed., Butterworth, London, 1959; reprinted 1965), and in chaps. 22–26 of Ref. 49. Some references useful for engineering work are Bromley [16] and Meissner et al. [54]. For dilute aqueous solutions of volatile electrolytes see Edwards et al. [24].

‡Wilson [92] has given a three-parameter form of his equation which is applicable also to liquid-liquid equilibria; the molecular significance of the third parameter has been discussed by Renon and Prausnitz [76]. The three-parameter Wilson equation has not received much attention, primarily because it is not readily extended to multicomponent systems.

TABLE 8-3 Some Models for the Excess Gibbs Energy and Subsequent Activity Coefficients for Binary Systems[a]

Name	g^E	Binary parameters	$\ln \gamma_1$ and $\ln \gamma_2$
Two-suffix[b] Margules	$g^E = A x_1 x_2$	A	$RT \ln \gamma_1 = A x_2^2$
			$RT \ln \gamma_2 = A x_1^2$
Three-suffix[b] Margules	$g^E = x_1 x_2 [A + B(x_1 - x_2)]$	A, B	$RT \ln \gamma_1 = (A + 3B) x_2^2 - 4B x_2^3$
			$RT \ln \gamma_2 = (A - 3B) x_1^2 + 4B x_1^3$
van Laar	$g^E = \dfrac{A x_1 x_2}{x_1 (A/B) + x_2}$	A, B	$RT \ln \gamma_1 = A \left(1 + \dfrac{A}{B} \dfrac{x_1}{x_2} \right)^{-2}$
			$RT \ln \gamma_2 = B \left(1 + \dfrac{B}{A} \dfrac{x_2}{x_1} \right)^{-2}$
Wilson	$\dfrac{g^E}{RT} = -x_1 \ln (x_1 + \Lambda_{12} x_2) - x_2 \ln (x_2 + \Lambda_{21} x_1)$	$\Lambda_{12}, \Lambda_{21}$	$\ln \gamma_1 = -\ln (x_1 + \Lambda_{12} x_2) + x_2 \left(\dfrac{\Lambda_{12}}{x_1 + \Lambda_{12} x_2} - \dfrac{\Lambda_{21}}{\Lambda_{21} x_1 + x_2} \right)$
			$\ln \gamma_2 = -\ln (x_2 + \Lambda_{21} x_1) - x_1 \left(\dfrac{\Lambda_{12}}{x_1 + \Lambda_{12} x_2} - \dfrac{\Lambda_{21}}{\Lambda_{21} x_1 + x_2} \right)$
Four-suffix[b] Margules	$g^E = x_1 x_2 [A + B(x_1 - x_2) + C(x_1 - x_2)^2]$	A, B, C	$RT \ln \gamma_1 = (A + 3B + 5C) x_2^2 - 4(B + 4C) x_2^3 + 12C x_2^4$
			$RT \ln \gamma_2 = (A - 3B + 5C) x_1^2 + 4(B - 4C) x_1^3 + 12C x_1^4$

NRTL[c]

$$\frac{g^E}{RT} = x_1 x_2 \left(\frac{\tau_{21} G_{21}}{x_1 + x_2 G_{21}} + \frac{\tau_{12} G_{12}}{x_2 + x_1 G_{12}} \right)$$

where $\tau_{12} = \dfrac{\Delta g_{12}}{RT}$ $\qquad \tau_{21} = \dfrac{\Delta g_{21}}{RT}$

$\ln G_{12} = -\alpha_{12} \tau_{12}$ $\qquad \ln G_{21} = -\alpha_{12} \tau_{21}$

$$\ln \gamma_1 = x_2^2 \left[\tau_{21} \left(\frac{G_{21}}{x_1 + x_2 G_{21}} \right)^2 + \frac{\tau_{12} G_{12}}{(x_2 + x_1 G_{12})^2} \right]$$

$$\ln \gamma_2 = x_1^2 \left[\tau_{12} \left(\frac{G_{12}}{x_2 + x_1 G_{12}} \right)^2 + \frac{\tau_{21} G_{21}}{(x_1 + x_2 G_{21})^2} \right]$$

$\Delta g_{12}, \Delta g_{21}, \alpha_{12}$ [d]

UNIQUAC[e]

$g^E = g^E \text{(combinatorial)} + g^E \text{(residual)}$

$$\frac{g^E \text{(combinatorial)}}{RT} = x_1 \ln \frac{\Phi_1}{x_1} + x_2 \ln \frac{\Phi_2}{x_2} + \frac{z}{2} \left(q_1 x_1 \ln \frac{\theta_1}{\Phi_1} + q_2 x_2 \ln \frac{\theta_2}{\Phi_2} \right)$$

$$\frac{g^E \text{(residual)}}{RT} = -q_1 x_1 \ln [\theta_1 + \theta_2 \tau_{21}] - q_2 x_2 \ln [\theta_2 + \theta_1 \tau_{12}]$$

$\Phi_1 = \dfrac{x_1 r_1}{x_1 r_1 + x_2 r_2}$ $\qquad \theta_1 = \dfrac{x_1 q_1}{x_1 q_1 + x_2 q_2}$

$\ln \tau_{21} = -\dfrac{\Delta u_{21}}{RT}$ $\qquad \ln \tau_{12} = -\dfrac{\Delta u_{12}}{RT}$

r and q are pure-component parameters and coordination number $z = 10$

Δu_{12} and Δu_{21} [f]

$$\ln \gamma_i = \ln \frac{\Phi_i}{x_i} + \frac{z}{2} q_i \ln \frac{\theta_i}{\Phi_i} + \Phi_j \left(\ell_i - \frac{r_i}{r_j} \ell_j \right)$$
$$- q_i \ln (\theta_i + \theta_j \tau_{ji}) + \theta_j q_i \left(\frac{\tau_{ji}}{\theta_i + \theta_j \tau_{ji}} - \frac{\tau_{ij}}{\theta_j + \theta_i \tau_{ij}} \right)$$

where $i = 1$ $\quad j = 2$ $\quad$ or $\quad i = 2$ $\quad j = 1$

$\ell_i = \dfrac{z}{2}(r_i - q_i) - (r_i - 1)$

$\ell_j = \dfrac{z}{2}(r_j - q_j) - (r_j - 1)$

[a] Reference 66 discusses the Margules, van Laar, Wilson, and NRTL equations. The UNIQUAC equation is discussed in Ref. 3.
[b] Two-suffix signifies that the expansion for g^E is quadratic in mole fraction. Three-suffix signifies a third-order, and four-suffix signifies a fourth-order equation.
[c] NRTL = Non Random Two Liquid.
[d] $\Delta g_{12} = g_{12} - g_{22}$; $\Delta g_{21} = g_{21} - g_{11}$.
[e] UNIQUAC = Universal Quasi Chemical. Parameters q and r can be calculated from Eq. (8-10.42).
[f] $\Delta u_{12} = u_{12} - u_{22}$; $\Delta u_{21} = u_{21} - u_{11}$.

adjustable parameters; (2) because of its better theoretical basis, UNIQUAC's parameters often have a smaller dependence on temperature; and (3) because the primary concentration variable is a surface fraction (rather than mole fraction), UNIQUAC is applicable to solutions containing small or large molecules, including polymers.

Simplifications: One-Parameter Equations   It frequently happens that experimental data for a given binary mixture are so fragmentary that it is not possible to determine two (or three) *meaningful* binary parameters; limited data can often yield only one significant binary parameter. In that event, it is tempting to use the two-suffix (one-parameter) Margules equation, but this is usually an unsatisfactory procedure because activity coefficients in a real binary mixture are rarely symmetric with respect to mole fraction.   In most cases better results are obtained by choosing the van Laar, Wilson, NRTL, or UNIQUAC equation and reducing the number of adjustable parameters through reasonable physical approximations.

To reduce the van Laar equation to a one-parameter form, for mixtures of nonpolar fluids, the ratio A/B can often be replaced by the ratio of molar liquid volumes: $A/B = V_1^L/V_2^L$. This simplification, however, is not valid for binary mixtures containing one (or two) polar components.

To simplify the Wilson equation, we first note that

$$\Lambda_{ij} = \frac{V_j^L}{V_i^L} \exp\left(-\frac{\lambda_{ij} - \lambda_{ii}}{RT}\right) \qquad (8\text{-}5.11)$$

where V_i^L is the molar volume of pure liquid i and λ_{ij} is an energy parameter characterizing the interaction of molecule i with molecule j.

The Wilson equation can be reduced to a one-parameter form by assuming that $\lambda_{ij} = \lambda_{ji}$† and that

$$\lambda_{ii} = -\beta(\Delta H_{v_i} - RT) \qquad (8\text{-}5.12)$$

where β is a proportionality factor and ΔH_{v_i} is the enthalpy of vaporization of pure component i at T.   A similar equation is written for λ_{jj}.   When β is fixed, the only adjustable binary parameter is λ_{ij}.

Tassios [86] set $\beta = 1$, but from theoretical considerations it makes more sense to assume that $\beta = 2/z$, where z is the coordination number (typically, $z = 10$).   This assumption, used by Wong and Eckert [96] and Schreiber and Eckert [77], gives good estimates for a variety of binary mixtures.   Hiranuma and Honma [40] have had some success in cor-

---

†The simplifying assumption that cross-parameter $\lambda_{ij} = \lambda_{ji}$ (or, similarly, $g_{ij} = g_{ji}$ or $u_{ij} = u_{ji}$) is empirical.   In some models it may lead to theoretical inconsistencies.

relating λ_{ij} with energy contributions from intermolecular-dispersion and dipole-dipole forces.

Ladurelli et al. [48] have suggested that $\beta = 2/z$ for component 2, having the smaller molar volume, while for component 1, having the larger molar volume, $\beta = (2/z)(V_2^L/V_1^L)$. This suggestion follows from the notion that a larger molecule has a larger area of interaction; parameters λ_{ii}, λ_{jj}, and λ_{ij} are considered as interaction energies per segment rather than per molecule. In this particular case the unit segment is that corresponding to one molecule of component 2.

Using similar arguments, Bruin and Prausnitz [17] have shown that it is possible to reduce the number of adjustable binary parameters in the NRTL equation by making a reasonable assumption for α_{12} and by substituting NRTL parameter g_{ii} for Wilson parameter λ_{ii} in Eq. (8-5.12). Bruin gives some correlations for g_{ij}, especially for aqueous systems.

Finally, Abrams and Prausnitz [3] have shown that the UNIQUAC equation can be simplified by assuming that

$$u_{11} = \frac{-\Delta U_1}{q_1} \quad \text{and} \quad u_{22} = \frac{-\Delta U_2}{q_2} \tag{8-5.13}$$

and that
$$u_{12} = u_{21} = (u_{11}u_{22})^{1/2}(1 - c_{12}) \tag{8-5.14}\dagger$$

where, remote from the critical temperature, energy ΔU_i is given very nearly by $\Delta U_i \approx \Delta H_{v_i} - RT$. The only adjustable binary parameter is c_{12}, which, for mixtures of nonpolar liquids, is positive and small compared with unity. For some mixtures containing polar components, however, c_{12} is of the order of 0.5, and when the unlike molecules in a mixture are attracted more strongly than like molecules, c_{12} may be negative, e.g., acetone-chloroform.

For mixtures of nonpolar liquids, a one-parameter form (van Laar, Wilson, NRTL, UNIQUAC) often gives results nearly as good as those obtained using two, or even three parameters. However, if one (or both) components are polar, significantly better results are usually obtained using two parameters, provided that the experimental data used to determine these parameters are of sufficient quantity and quality.

8-6 Calculation of Binary Vapor-Liquid Equilibria

First consider the isothermal case. At some constant temperature T, we wish to construct two diagrams: y vs. x and P vs. x. We assume that

†See footnote on page 302.

since the pressure is low, we can use Eq. (8-4.1) with $\mathscr{F}_i = 1$.   The steps toward that end are:

1. Find the pure-liquid vapor pressures P_{vp_1} and P_{vp_2} at T.

2. Suppose a few experimental points for the mixture are available at temperature T.   Arbitrarily, to fix ideas, suppose there are five points; i.e., for five values of x there are five corresponding experimental equilibrium values of y and P.   For each of these points calculate γ_1 and γ_2 according to

$$\gamma_1 = \frac{y_1 P}{x_1 P_{vp_1}} \qquad (8\text{-}6.1)$$

$$\gamma_2 = \frac{y_2 P}{x_2 P_{vp_2}} \qquad (8\text{-}6.2)$$

3. For each of the five points, calculate the molar excess Gibbs energy g^E:

$$g^E = RT(x_1 \ln \gamma_1 + x_2 \ln \gamma_2) \qquad (8\text{-}6.3)$$

4. Choose one of the equations for g^E given in Table 8.3.   Adjust the constants in that equation to minimize the deviation between g^E calculated from the equation and g^E found from experiment in step 3.

5. Using Eqs. (8-5.3) and (8-5.4), find γ_1 and γ_2 at arbitrarily selected values of x_1 from $x_1=0$ to $x_1=1$.

6. For each selected x_1 find the corresponding y_1 and P by solving Eqs. (8-6.1) and (8-6.2) coupled with the mass-balance relations $x_2=1-x_1$ and $y_2=1-y_1$.   The results obtained give the desired y-vs.-x and P-vs.-x diagrams.

The simple steps outlined above provide a rational, thermodynamically consistent procedure for interpolation and extrapolation with respect to composition.   The crucial step is 4.   Judgment is required to obtain the best, i.e., the most representative, constants in the expression chosen for g^E.   To do so it is necessary to decide on how to weight the five individual experimental data; some of these may be more reliable than others.   For determining the constants, the experimental points which give the most information are those at the ends of the composition scale, that is, y_1 when x_1 is small and y_2 when x_2 is small.   Unfortunately, however, these experimental data are those which, in a typical case, are the most difficult to measure.   Thus it frequently happens that the data which are potentially most valuable are also the ones which are likely to be least accurate.

Now let us consider the more complicated isobaric case.   At some constant pressure P, we wish to construct two diagrams: y vs. x and T vs. x.   Assuming that the pressure is low, we again use Eq. (8-4.1) with $\mathscr{F}_i = 1$.   The steps toward construction of these diagrams are:

1. Find pure-component vapor pressures P_{vp_1} and P_{vp_2}.   Prepare

plots (or obtain analytical representation) of P_{vp_1} and P_{vp_2} vs. temperature in the region where $P_{vp_1} \approx P$ and $P_{vp_2} \approx P$. (See Chap. 6.)

2. Suppose there are available a few experimental data points for the mixture at pressure P or at some other pressure not far removed from P or, perhaps, at some constant temperature such that the total pressure is in the general vicinity of P. As in the previous case, to fix ideas, we arbitrarily set the number of such experimental points at five. By experimental point we mean, as before, that for some value of x_1 we have the corresponding experimental equilibrium values of y_1, T, and total pressure.

For each of the five points, calculate activity coefficients γ_1 and γ_2 according to Eqs. (8-6.1) and (8-6.2). For each point the vapor pressures P_{vp_1} and P_{vp_2} are evaluated at the experimentally determined temperature for that point. In these equations, the experimentally determined total pressure is used for P; the total pressure measured is not necessarily the same as the pressure for which we wish to construct the equilibrium diagrams.

3. For each of the five points, calculate the molar excess Gibbs energy according to Eq. (8-6.3).

4. Choose one of the equations for g^E given in Table 8-3. As in step 4 of the previous (isothermal) case, find the constants in that equation which give the smallest deviation between calculated values of g^E and those found in step 3. When the experimental data used in Eq. (8-6.3) are isobaric rather than isothermal, it may be advantageous to choose an expression for g^E which contains the temperature as one of the explicit variables. Such a choice, however, complicates the calculations in step 6.

5. Find γ_1 and γ_2 as functions of x by differentiation according to Eqs. (8-5.3) and (8-5.4).†

6. Select a set of arbitrary values for x_1 for the range $x_1 = 0$ to $x_1 = 1$. For each x_1, by iteration, solve simultaneously the two equations of phase equilibrium [Eqs. (8-6.1) and (8-6.2)] for the two unknowns, y_1 and T. In these equations the total pressure P is now the one for which the equilibrium diagrams are desired.

Simultaneous solution of Eqs. (8-6.1) and (8-6.2) requires trial and error because at a given x, both y and T are unknown and both P_{vp_1} and P_{vp_2} are strong, nonlinear functions of T. In addition, γ_1 and γ_2 may also vary with T (as well as x), depending on which expression for g^E has been chosen in step 4. For simultaneous solution of the two equilibrium equations, the best procedure is to assume a reasonable temperature for each selected value of x_1. Using this assumed tempera-

---

†Some error is introduced here because Eqs. (8-5.3) and (8-5.4) are based on the isobaric *and* isothermal Gibbs-Duhem equation. For most practical calculations this error is not serious. See chap. 6 and app. IV of Ref. 66.

ture, calculate y_1 and y_2 from Eqs. (8-6.1) and (8-6.2). Then check if $y_1 + y_2 = 1$. If not, repeat the calculation assuming a different temperature. In this way, for fixed P and for each selected value of x, find corresponding equilibrium values y and T.

Calculation of isothermal or isobaric vapor-liquid equilibria can be efficiently performed with a computer. Further, it is possible in such calculations to include the correction factor $\mathscr{F}_i$ [Eq. (8-4.1)] when necessary. In that event, the calculations are more complex in detail but not in principle. Computer programs for such calculations have been published [68].

When the procedures outlined above are followed, the accuracy of any vapor-liquid equilibrium calculation depends primarily on the extent to which the expression for g^E accurately represents the behavior of the mixture at the particular conditions (temperature, pressure, composition) for which the calculation is made. This accuracy of representation often depends not so much on the algebraic form of g^E as on the reliability of the constants appearing in that expression. This reliability, in turn, depends on the quality and quantity of the experimental data used to determine the constants.

While some of the expressions for g^E shown in Table 8-3 have a better theoretical foundation than others, all have a strong empirical flavor. Experience has indicated that the more recent equations for g^E (Wilson, NRTL, and UNIQUAC) are more consistently reliable than the older equations, in the sense that they can usually reproduce accurately even highly nonideal behavior, using only two or three adjustable parameters.

The oldest equation for g^E, that of Margules, is a power series in mole fraction. With a power series it is always possible to increase accuracy of representation by including higher terms, where each term is multiplied by an empirically determined coefficient. (The van Laar equation, as shown by Wohl [95], is also a power series in effective volume fraction, but in practice this series is almost always truncated after the quadratic term.) However, inclusion of higher-order terms in g^E is dangerous because subsequent differentiation to find γ_1 and γ_2 can then lead to spurious maxima or minima. Also, inclusion of higher-order terms in binary-data reduction often leads to serious difficulties when binary data are used to estimate multicomponent phase equilibria.

It is desirable to use an equation for g^E which is based on a relatively simple model and which contains only two (or at most three) adjustable binary parameters. Experimental data are then used to find the "best" binary parameters. Since experimental data are always of limited accuracy, it often happens that several sets of binary parameters may equally well represent the data within experimental uncertainty [3]. Only in rare cases, where experimental data are both plentiful and highly accurate, is there any justification in using more than three adjustable binary parameters.

8-7 Effect of Temperature on Vapor-Liquid Equilibria

A particularly troublesome question is the effect of temperature on the molar excess Gibbs energy g^E. This question is directly related to s^E, the molar excess entropy of mixing, about which little is known. In practice, either one of two approximations is frequently used.

Athermal Solution. This approximation sets $g^E = -Ts^E$, which assumes that the components mix at constant temperature without change of enthalpy ($h^E = 0$). This assumption leads to the conclusion that, at constant composition, $\ln \gamma_i$ is independent of T or, its equivalent, that g^E/RT is independent of temperature.

Regular Solution. This approximation sets $g^E = h^E$, which is the same as assuming that $s^E = 0$. This assumption leads to the conclusion that, at constant composition, $\ln \gamma_i$ varies as $1/T$ or, its equivalent, that g^E is independent of temperature.

Neither one of these extreme approximations is valid, although the second one is often better than the first. Good experimental data for the effect of temperature on activity coefficients are rare, but when such data are available, they suggest that, for a moderate temperature range, they can be expressed by an empirical equation of the form

$$(\ln \gamma_i)_{\substack{\text{constant} \\ \text{composition}}} = c + \frac{d}{T} \tag{8-7.1}$$

where c and d are empirical constants that depend on composition. In most cases constant d is positive. It is evident that when $d = 0$, Eq. (8-7.1) reduces to assumption a and when $c = 0$, it reduces to assumption b. Unfortunately, in typical cases c and d/T are of comparable magnitude.

Thermodynamics relates the effect of temperature on γ_i to the partial molar enthalpy $\bar{h}_i$:

$$\left[\frac{\partial \ln \gamma_i}{\partial (1/T)}\right]_{x,P} = \frac{\bar{h}_i - h_i^\circ}{R} \tag{8-7.2}$$

where h_i° is the enthalpy of liquid i in the standard state, usually taken as pure liquid i at system temperature and pressure. Sometimes (but rarely) experimental data on $\bar{h}_i - h_i^\circ$ may be available, and if so, they can be used to provide information on how the activity coefficient changes with temperature. However, even if such data are at hand, Eq. (8-7.2) must be used with caution because $\bar{h}_i - h_i^\circ$ depends on temperature and often strongly so.

Some of the expressions for g^E shown in Table 8-3 contain T as an explicit variable.   However, one should not therefore conclude that the constants appearing in those expressions are independent of temperature.   The explicit temperature dependence indicated provides only an approximation.   This approximation is usually, but not always, better than approximations a or b, but, in any case, it is not exact.

Fortunately, the primary effect of temperature on vapor-liquid equilibria is contained in the pure-component vapor pressures or, more precisely, in the pure-component liquid fugacities [Eq. (8-3.1)].   While the activity coefficients depend on temperature as well as composition, the temperature dependence of the activity coefficient is usually small when compared with the temperature dependence of the pure-liquid vapor pressures.   In a typical mixture, a rise of 10°C increases the vapor pressures of the pure liquids by a factor of 1.5 or 2, but the change in activity coefficient is likely to be only a few percent, often less than the experimental uncertainty.   Therefore, unless there is a large change in temperature, it is frequently satisfactory to neglect the effect of temperature on g^E when calculating vapor-liquid equilibria.   However, in calculating liquid-liquid equilibria, vapor pressures play no role at all, and therefore the effect of temperature on g^E, while small, may seriously affect liquid-liquid equilibria.   Even small changes in activity coefficients can have a large effect on multicomponent liquid-liquid equilibria, as briefly discussed in Sec. 8.13.

8-8   Binary Vapor-Liquid Equilibria: Examples

To introduce the general ideas, we present first two particularly simple methods for reduction of vapor-liquid equilibria.   These are followed by a brief introduction to more accurate, but also mathematically more complex, procedures.

Example 8-1   Given five experimental vapor-liquid equilibrium data for the binary system methanol (1)–1,2-dichloroethane (2) at 50°C, calculate the P-y-x diagram at 50°C and predict the P-y-x diagram at 60°C.

Experimental Data at 50°C [90]

$100x_1$	$100y_1$	P, mm Hg
30	59.1	483.8
40	60.2	493.2
50	61.2	499.9
70	65.7	501.4
90	81.4	469.7

solution   To interpolate in a thermodynamically consistent manner, we must choose an algebraic expression for the molar excess Gibbs energy.   For simplicity, we choose the van Laar equation (see Table 8-3).   To evaluate the van Laar constants A' and B', we rearrange the van Laar equation in a linear form†

$$\frac{x_1 x_2}{g^E/RT} = D + C(2x_1 - 1) \qquad \text{where} \qquad \begin{array}{l} A' = (D - C)^{-1} \\ B' = (D + C)^{-1} \end{array} \qquad (8\text{-}8.1)$$

Constants D and C are found from a plot of $x_1 x_2 (g^E/RT)^{-1}$ vs. x_1. The intercept at $x_1 = 0$ gives $D - C$, and the intercept at $x_1 = 1$ gives $D + C$.

The molar excess Gibbs energy is calculated from the definition

$$\frac{g^E}{RT} = x_1 \ln \gamma_1 + x_2 \ln \gamma_2 \qquad (8\text{-}8.2)$$

For the five available experimental points, activity coefficients γ_1 and γ_2 are calculated from Eq. (8-4.1) with $\mathscr{F}_i = 1$ and from pure-component vapor-pressure data.

Table 8-4 gives $x_1 x_1 (g^E/RT)^{-1}$ as needed to obtain van Laar constants. Figure 8-1 shows the linearized van Laar equation. The results shown are obtained with Antoine constants given in Appendix A.

From Fig. 8-1 we obtain the van Laar constants

$$A' = 1.94 \qquad B' = 1.61 \qquad \frac{A'}{B'} = 1.20$$

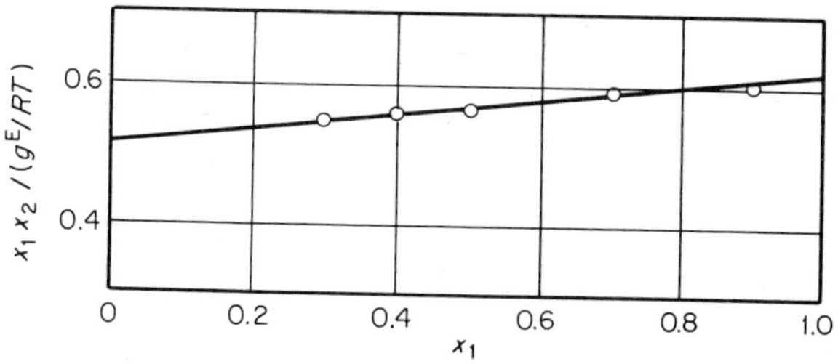

Fig. 8-1 Determination of van Laar constants for the system methanol (1)–1,2-dichloroethane (2) at 50°C.

TABLE 8-4 Experimental Activity Coefficients for Linearized van Laar Plot, Methanol (1)–1,2-Dichloroethane (2) at 50°C

x_1	γ_1	γ_2	$\dfrac{x_1 x_2}{g^E/RT}$
0.3	2.29	1.21	0.550
0.4	1.78	1.40	0.555
0.5	1.47	1.66	0.560
0.7	1.12	2.46	0.601
0.9	1.02	3.75	0.604

$y_i = \gamma_i x_i p_i^{pat}$

We can now calculate γ_1 and γ_2 at any mole fraction:

$$\ln \gamma_1 = 1.94 \left(1 + 1.20 \frac{x_1}{x_2}\right)^{-2} \qquad (8\text{-}8.3)$$

$$\ln \gamma_2 = 1.61 \left(1 + \frac{x_2}{1.20 x_1}\right)^{-2} \qquad (8\text{-}8.4)$$

†From Table 8-3, $A' = A/RT$ and $B' = B/RT$.

Using Eqs. (8-8.3) and (8-8.4) and the pure-component vapor pressures, we can now find y_1, y_2, and the total pressure P. There are two unknowns: y_1 (or y_2) and P. To find them, we must solve simultaneously the two equations of vapor-liquid equilibrium

$$y_1 = \frac{x_1 \gamma_1 P_{vp1}}{P} \qquad (8\text{-}8.5)$$

$$1 - y_1 = \frac{x_2 \gamma_2 P_{vp2}}{P} \qquad (8\text{-}8.6)$$

Calculated results at 50°C are shown in Table 8-5.

TABLE 8-5   Calculated Vapor-Liquid Equilibria in the System Methanol (1)–1,2-Dichloroethane (2) at 50 and 60°C

$100x_1$	γ_1 50°C	γ_1 60°C	γ_2 50°C	γ_2 60°C	$100y_1$ 50°C	$100y_1$ 60°C	P, mm Hg 50°C	P, mm Hg 60°C	$100\Delta y$ 50°C	$100\Delta y$ 60°C	ΔP, mm Hg 50°C	ΔP, mm Hg 60°C
5	5.56	5.28	1.01	1.01	34.1	33.9	339.6	494.2				
10	4.53	4.33	1.02	1.02	46.9	46.8	402.8	587.3	−0.9	0.4	−1.4	1.1
20	3.15	3.04	1.09	1.09	56.4	56.5	465.9	682.7	0.2	0.9	8.3	15.4
40	1.82	1.79	1.37	1.36	61.3	62.1	495.1	732.2	1.1	2.2	1.9	19.8
60	1.28	1.27	1.95	1.91	63.8	65.0	502.0	743.6	1.3	1.8	−1.3	17.2
80	1.06	1.06	3.01	2.91	71.6	73.1	493.9	736.2	0.5	1.2	1.1	26.1
90	1.01	1.01	3.85	3.70	80.9	82.1	468.7	702.7	−0.5	−0.1	−1.0	22.7
									Σ 1.7	6.4	7.6	102.3

$\Delta y = y$ (calc.) $- y$ (exp.) $\Delta P = P$ (calc.) $- P$ (exp.)

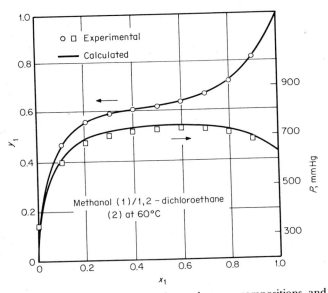

Fig. 8-2 Calculated and experimental vapor compositions and total pressures.

To predict vapor-liquid equilibria at 60°C, we assume that the effect of temperature on activity coefficients is given by the regular-solution approximation (see Sec. 8-7):

$$\frac{\ln \gamma_i \ (60°C)}{\ln \gamma_i \ (50°C)} = \frac{273 + 50}{273 + 60} \tag{8-8.7}$$

Pure-component vapor pressures at 60°C are found from the Antoine relations. The two equations of equilibrium [Eqs. (8-8.5) and (8-8.6)] are again solved simultaneously to obtain y and P as a function of x. Calculated results at 60°C are shown in Table 8-5 and in Fig. 8-2.

Predicted y's are in good agreement with experiment [90], but predicted pressures are too high. This suggests that Eq. (8-8.7) is not a good approximation for this system.

Equation (8-8.7) corresponds to approximation b in Sec. 8-7. If approximation a had been used, the predicted pressures would have been even higher.

Example 8-2   Given five experimental vapor-liquid equilibrium data for the binary system propanol (1)–water (2) at 760 mm Hg, predict the T-y-x diagram for the same system at 1000 mm Hg.

Experimental Data at 760 mm Hg
[55]

$100x_1$	$100y_1$	T, °C
7.5	37.5	89.05
17.9	38.8	87.95
48.2	43.8	87.80
71.2	56.0	89.20
85.0	68.5	91.70

solution   To represent the experimental data, we choose the van Laar equation, as in Example 8-1. Since the temperature range is small, we neglect the effect of temperature on the van Laar constants.

As in Example 8-1, we linearize the van Laar equation as shown in Eq. (8-8.1). To obtain the van Laar constants A' and B' we need, in addition to the data shown above, vapor-pressure data for the pure components.

Activity coefficients γ_1 and γ_2 are calculated from Eq. (8-4.1) with $\mathscr{F}_i = 1$, and g^E/RT is calculated from Eq. (8-8.2). Antoine constants are from Appendix A. Results are given in Table 8-6. The linearized van Laar plot is shown in Fig. 8-3.

TABLE 8-6   Experimental Activity Coefficients for Linearized van Laar Plot, n-Propanol (1)–Water (2) at 760 mm Hg

$100x_1$	T, °C	γ_1	γ_2	$\dfrac{x_1 x_2}{g^E/RT}$
7.50	89.05	6.84	1.01	0.446
17.9	87.95	3.10	1.17	0.448
48.2	87.80	1.31	1.71	0.615
71.2	89.20	1.07	2.28	0.720
85.0	91.70	0.99	2.85	0.848

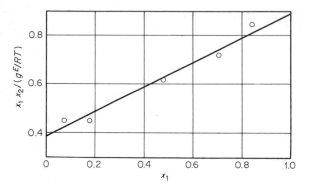

Fig. 8-3   Determination of van Laar constants for the system in n-propanol (1)–water (2) at 760 mm Hg.

From the intercepts in Fig. 8-3 we obtain

$$A' = 2.60 \qquad B' = 1.13 \qquad \frac{A'}{B'} = 2.30 \tag{8-8.8}$$

Activity coefficients γ_1 and γ_2 are now given by the van Laar equations

$$\ln \gamma_1 = 2.60\left(1 + 2.30\frac{x_1}{x_2}\right)^{-2} \tag{8-8.9}$$

$$\ln \gamma_2 = 1.13\left(1 + \frac{x_2}{2.30x_1}\right)^{-2} \tag{8-8.10}$$

To obtain the vapor-liquid equilibrium diagram at 1000 mm Hg, we must solve simultaneously the two equations of equilibrium

$$y_1 = \frac{\gamma_1 x_1 P_{vp_1}(T)}{1000} \tag{8-8.11}$$

$$1 - y_1 = y_2 = \frac{\gamma_2 x_2 P_{vp_2}(T)}{1000} \tag{8-8.12}$$

In this calculation we assume that γ_1 and γ_2 depend only on x (as given by the van Laar equations) and not on temperature. However, P_{vp_1} and P_{vp_2} are strong functions of temperature.

The two unknowns in the equations of equilibrium are y_1 and T. To solve for these unknowns, it is also necessary to use the Antoine relations for the two pure components.

The required calculations contain the temperature as an implicit variable; solution of the equations of equilibrium must be attained by iteration.

While iterative calculations are best performed with a computer, in this example it is possible to obtain results rapidly by hand calculations. Dividing one of the equations of equilibrium by the other, we obtain

$$y_1 = \left(1 + \frac{\gamma_2}{\gamma_1}\frac{x_2}{x_1}\frac{P_{vp_2}}{P_{vp_1}}\right)^{-1} \tag{8-8.13}$$

Although P_{vp_2} and P_{vp_1} are strong functions of temperature, the ratio P_{vp_2}/P_{vp_1} is a much weaker function of temperature.

For a given x_1, find γ_2/γ_1 from the van Laar equations. Choose a reasonable temperature and find P_{vp_2}/P_{vp_1} from the Antoine relations. Equation (8-8.13) then gives a first estimate for y_1. Using this estimate, find P_{vp_1} from

$$P_{vp_1} = \frac{1000y_1}{x_1\gamma_1} \tag{8-8.14}$$

The Antoine relation for component 1 then gives a first estimate for T. Using this T, find the ratio P_{vp_2}/P_{vp_1} and, again using Eq. (8-8.13), find the second estimate for y_1. This second estimate for y_1 is then used with the Antoine relation to find the second estimate for T. Repeat until there is negligible change in the estimate for T.

It is clear that Eq. (8-8.14) for component 1 could be replaced with the analogous equation for component 2. Which one should be used? In principle, either one may be used, but for components of comparable volatility, convergence is likely to be more rapid if Eq. (8-8.14) is used for $x_1 > \frac{1}{2}$ and the analogous equation for component 2 is used when $x_1 < \frac{1}{2}$. However, if one component is much more volatile than the other, the equation for that component is likely to be more useful.

Table 8-7 presents calculated results at 1000 mm Hg. Unfortunately no experimental results at this pressure are available for comparison.

TABLE 8-7 Calculated Vapor-Liquid Equilibria for n-Propanol (1)–Water (2) at 1000 mm Hg

$100x_1$	γ_1	γ_2	T, °C	$100y_1$
5	7.92	1.01	98.4	31.6
10	5.20	1.05	96.0	37.9
20	2.85	1.16	95.3	40.5
40	1.50	1.51	95.0	42.2
50	1.27	1.73	95.2	44.9
60	1.14	1.98	95.5	48.8
80	1.02	2.51	98.2	64.6
90	1.01	2.79	100.6	78.5

The two simple examples above illustrate the essential steps for calculating vapor-liquid equilibria from limited experimental data. Because of their illustrative nature, these examples are intentionally simplified, and for more accurate results it is desirable to replace some of the details by more sophisticated techniques. For example, it may be worthwhile to include corrections for vapor-phase nonideality and perhaps the Poynting factor, i.e., to relax the simplifying assumption $\mathscr{F}_i = 1$ in Eq. (8-4.1). At the modest pressures encountered here, however, such modifications are likely to have a small effect. A more important change would be to replace the van Laar equation with a better equation for the activity coefficients, e.g., the Wilson equation or the UNIQUAC equation. If this is done, the calculational procedure is the same but the details of computation are more complex. Because of algebraic simplicity, the van Laar equations can easily be linearized, and therefore a convenient graphical procedure can be used to find the van

Laar constants.† Equations like UNIQUAC or that of Wilson cannot easily be linearized, and therefore, for practical application, it is necessary to use a computer for data reduction to find the binary constants which appear in these equations.

In Examples 8-1 and 8-2 we have not only made simplifications in the thermodynamic relations but have also neglected to take into quantitative consideration the effect of experimental error.

It is beyond the scope of this chapter to discuss in detail the highly sophisticated statistical methods now available for optimum reduction of vapor-liquid equilibrium data. Nevertheless, a very short discussion may be useful as an introduction for readers who want to obtain the highest possible accuracy from the available data.

A particularly effective data-reduction method is described by Fabries and Renon [25], who base their analysis on the principle of maximum likelihood while taking into account probable experimental errors in all experimentally determined quantities.

A similar technique by Abrams and Prausnitz [3], based on a numerical method described by Britt and Luecke [15], defines a calculated pressure (constraining function) by

$$P^c = \exp\left(x_1 \ln \frac{\gamma_1 x_1 f^L_{\text{pure 1}}}{y_1 \phi_1} + x_2 \ln \frac{\gamma_2 x_2 f^L_{\text{pure 2}}}{y_2 \phi_2}\right) \qquad (8\text{-}8.15)$$

where $f^L_{\text{pure }i}$ is at system temperature and pressure. The most probable values of the parameters (appearing in the function chosen for g^E) are those which minimize the function I:

$$I = \sum_i \left(\frac{(x_i^0 - x_i^M)^2}{\sigma_{x_i}^2} + \frac{(y_i^0 - y_i^M)^2}{\sigma_{y_i}^2} + \frac{(P_i^0 - P_i^M)^2}{\sigma_{P_i}^2} + \frac{(T_i^0 - T_i^M)^2}{\sigma_{T_i}^2}\right) \qquad (8\text{-}8.16)$$

Superscript M indicates a measured value, and superscript 0 indicates an estimate of the true value of the variable. The σ^2's are estimates of the variances of the measured values, i.e., an indication of the probable experimental uncertainty. These may vary from one point to another but need not.

Using experimental P-T-x-y data and the UNIQUAC equation with estimated parameters $u_{12} - u_{22}$ and $u_{21} - u_{11}$, Abrams and Prausnitz obtain estimates of x_i^0, y_i^0, T_i^0, and P_i^0; the last of these is found from Eq. (8-8.15). They then evaluate I, having previously set average variances σ_x^2, σ_y^2, σ_P^2, and σ_T^2 from a critical inspection of the data's quality. Upon changing the estimate of UNIQUAC parameters, they calculate a new I; with a suitable computer program, they search for

---

†The three-suffix Margules equation is also easily linearized, as shown by H. C. Van Ness, "Classical Thermodynamics of Nonelectrolyte Solutions," p. 129, Pergamon, New York, 1964.

those parameters which minimize I.   Convergence is achieved when, from one iteration to the next, the relative change in I is less than 10^{-5}.   After the last iteration, the variance of fit σ_F^2 is given by

$$\sigma_F^2 = \frac{I}{D-L} \tag{8-8.17}$$

where D is the number of data points and L is the number of adjustable parameters.

Since all experimental data have some experimental uncertainty, and since any equation for g^E can provide only an approximation to the experimental results, it follows that the parameters obtained from data reduction are not unique; there are many sets of parameters which can equally well represent the experimental data within experimental uncertainty.   To illustrate this lack of uniqueness, Fig. 8-4 shows results of data reduction for the binary mixture ethanol (1)–water at 70°C.   Experimental data reported by Mertl [55] were reduced using the UNIQUAC equation with the variances

$$\sigma_x = 10^{-3} \qquad \sigma_y = 10^{-2} \qquad \sigma_p = 0.5 \text{ mm Hg} \qquad \sigma_T = 0.1 \text{ K}$$

For this binary system, the fit is very good; $\sigma_F^2 = 5 \times 10^{-4}$.

The ellipse in Fig. 8-4 clearly shows that while parameter $u_{21} - u_{11}$ is strongly correlated with parameter $u_{12} - u_{22}$, there are many sets of these

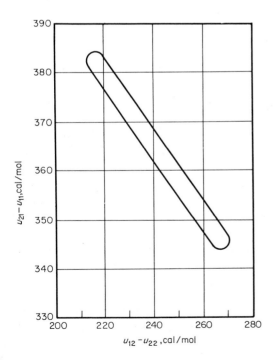

Fig. 8-4 The 99 percent confidence ellipse for UNIQUAC parameters in the system ethanol (1)–water (2) at 70°C.

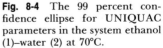

parameters that can equally well represent the data. The experimental data used in data reduction are not sufficient to fix a unique set of "best" parameters. Realistic data reduction can determine only a region of parameters.†

While Fig. 8-4 pertains to the UNIQUAC equation, similar results are obtained when other equations for g^E are used; only a region of acceptable parameters can be obtained from P-T-y-x data. For a two-parameter equation this region is represented by an area; for a three-parameter equation it is represented by a volume. If the equation for g^E is suitable for the mixture, the region of acceptable parameters shrinks as the quality and quantity of the experimental data increase. However, considering the limits of both theory and experiment, it is unreasonable to expect this region to shrink to a single point.

As indicated by numerous authors, notably Abbott and Van Ness [2], experimental errors in vapor composition y are usually larger than those in experimental pressure P, temperature T, and liquid-phase composition x. Therefore, a relatively simple fitting procedure is provided by reducing only P-x-T data; y data, even if available, are not used.‡ The essential point is to minimize the deviation between calculated and observed pressures.

The pressure is calculated according to

$$P \text{ (calc.)} = y_1 P + y_2 P = \gamma_1 x_1 P_{vp_1} \mathscr{F}_1 + \gamma_2 x_2 P_{vp_2} \mathscr{F}_2 \qquad (8\text{-}8.18)$$

where $\mathscr{F}_i$ is given by Eq. (8-4.2).

Thermodynamically consistent equations are now chosen to represent γ_1 and γ_2 as functions of x (and perhaps T); some are suggested in Table 8-3. These equations contain a number of adjustable binary parameters. With a computer, these parameters can be found by minimizing the deviation between calculated and measured pressures.

At low pressures, we can assume that $\mathscr{F}_1 = \mathscr{F}_2 = 1$. However, at higher pressures, correction factors $\mathscr{F}_1$ and $\mathscr{F}_2$ are functions of pressure, temperature, and vapor compositions y_1 and y_2; these compositions are calculated from

$$y_1 = \frac{\gamma_1 x_1 P_{vp_1} \mathscr{F}_1(P, T, y)}{P} \quad \text{and} \quad y_2 = \frac{\gamma_2 x_2 P_{vp_2} \mathscr{F}_2(P, T, y)}{P} \qquad (8\text{-}8.19)\S$$

---

†Instead of the restraint given by Eq. (8-8.15), it is sometimes preferable to use instead two restraints; first, Eq. (8-8.18), and second,

$$y_1 = \frac{x_1 \gamma_1 f_1^0/\phi_1}{x_1 \gamma_1 f_1^0/\phi_1 + x_2 \gamma_2 f_2^0/\phi_2}$$

or the corresponding equation for y_2.

‡This technique is commonly referred to as *Barker's method*.

§If the Lewis fugacity rule is used to calculate vapor-phase fugacity coefficients, $\mathscr{F}_1$ and $\mathscr{F}_2$ depend on pressure and temperature but are independent of y. The Lewis rule provides mathematical simplification, but unfortunately it is a poor rule and if a computer is available, there is no need to use it.

The data-reduction scheme, then, is iterative; to get started, it is necessary first to assume an estimated y for each x. After the first iteration, a new set of estimated y's is found from Eq. (8-8.19). Convergence is achieved when, following a given iteration, the calculated y's differ negligibly from those calculated after the previous iteration and when the pressure deviation is minimized.

8-9 Multicomponent Vapor-Liquid Equilibria

The equations required to calculate vapor-liquid equilibria in multicomponent systems are, in principle, the same as those required for binary systems. In a system containing N components, we must solve simultaneously N equations: Eq. (8-4.1) for each of the N components. We require the saturation (vapor) pressure of each component, as a pure liquid, at the temperature of interest. If all pure-component vapor pressures are low, the total pressure is also low. In that event, the factor $\mathcal{F}_i$ [Eq. (8-4.2)] can often be set equal to unity.

Activity coefficients γ_i are found from an expression for the excess Gibbs energy, as discussed in Sec. (8-5). For a mixture of N components, the excess Gibbs energy G^E is defined by

$$G^E = RT \sum_{i=1}^{N} n_i \ln \gamma_i \qquad (8\text{-}9.1)$$

where n_i is the number of moles of component i. The molar excess Gibbs energy g^E is simply related to G^E by

$$g^E = \frac{G^E}{n_T} \qquad (8\text{-}9.2)$$

where n_T, the total number of moles, is equal to $\sum\limits_{i=1}^{N} n_i$.

Individual activity coefficients can be obtained from G^E upon introducing the Gibbs-Duhem equation for a multicomponent system at constant temperature and pressure. That equation is

$$\sum_{i=1}^{N} n_i \, d \ln \gamma_i = 0 \qquad (8\text{-}9.3)$$

The activity coefficient γ_i is found by a generalization of Eq. (8-5.3):

$$RT \ln \gamma_i = \left(\frac{\partial G^E}{\partial n_i} \right)_{T,P,n_j} \qquad (8\text{-}9.4)$$

where n_j indicates that all mole numbers (except n_i) are held constant in the differentiation.

The key problem in calculating multicomponent vapor-liquid equilibria is to find an expression for g^E which provides a good approximation for the properties of the mixture. Toward that end, the expressions for g^E for binary systems, shown in Table 8-3, can be extended to multicomponent systems. A few of these are shown in Table 8-8.

The excess-Gibbs-energy concept is particularly useful for multicomponent mixtures because in many cases, to a good approximation, extension from binary to multicomponent systems can be made in such a way that only binary parameters appear in the final expression for g^E. When that is the case, a large saving in experimental effort is achieved, since experimental data are then required only for the mixture's constituent binaries, not for the multicomponent mixture itself. For example, activity coefficients in a ternary mixture (components 1, 2, and 3) can often be calculated with good accuracy using only experimental data for the three binary mixtures: components 1 and 2, components 1 and 3, and components 2 and 3.

Many physical models for g^E for a binary system consider only two-body intermolecular interactions, i.e., interactions between two (but not more) molecules. Because of the short range of molecular interaction between nonelectrolytes, it is often permissible to consider only interactions between molecules that are first neighbors and then to sum all the two-body, first-neighbor interactions. A useful consequence of these simplifying assumptions is that extension to ternary (and higher) systems requires only binary, i.e., two-body, information; no ternary (or higher) constants appear. However, not all physical models use this simplifying assumption, and those which do not, often require additional simplifying assumptions if the final expression for g^E is to contain only constants derived from binary data.

To illustrate with the simplest case, consider the two-suffix Margules relation for g^E (Table 8-3). For a binary mixture, this relation is given by Eq. (8-5.5), leading to activity coefficients given by Eqs. (8-5.9) and (8-5.10). The generalization to a system containing N components is

$$g^E = \frac{1}{2} \sum_{i=1}^{N} \sum_{j=1}^{N} A_{ij} x_i x_j \qquad (8\text{-}9.5)$$

where the factor $\frac{1}{2}$ is needed to avoid counting molecular pairs twice. The coefficient A_{ij} is obtained from data for the ij binary. [In the summation indicated in Eq. (8-9.5), $A_{ii} = A_{jj} = 0$ and $A_{ij} = A_{ji}$.] For a ternary system Eq. (8-9.5) becomes

$$g^E = A_{12} x_1 x_2 + A_{13} x_1 x_3 + A_{23} x_2 x_3 \qquad (8\text{-}9.6)$$

Activity coefficients are obtained by differentiating Eq. (8-9.6) according to Eq. (8-9.4), remembering that $x_i = n_i / n_T$, where n_T is the total number of moles. Upon performing this differentiation, we obtain for component k

$$RT \ln \gamma_k = \sum_{i=1}^{N} \sum_{j=1}^{N} (A_{ik} - \tfrac{1}{2} A_{ij}) x_i x_j \qquad (8\text{-}9.7)$$

TABLE 8-8 Three Expressions for the Molar Excess Gibbs Energy and Activity Coefficients of Multicomponent Systems Using Only Pure-Component and Binary Parameters

Symbols defined in Table 8-3; the number of components is N

Name	Molar excess Gibbs energy	Activity coefficient for component i
Wilson	$$\frac{g^E}{RT} = -\sum_i^N x_i \ln\left(\sum_j^N x_j \Lambda_{ij}\right)$$	$$\ln \gamma_i = -\ln\left(\sum_j^N x_j \Lambda_{ij}\right) + 1 - \sum_k^N \frac{x_k \Lambda_{ki}}{\sum_j^N x_j \Lambda_{kj}}$$
NRTL	$$\frac{g^E}{RT} = \sum_i^N x_i \frac{\sum_j^N \tau_{ji} G_{ji} x_j}{\sum_k^N G_{ki} x_k}$$	$$\ln \gamma_i = \frac{\sum_j^N \tau_{ji} G_{ji} x_j}{\sum_k^N G_{ki} x_k} + \sum_j^N \frac{x_j G_{ij}}{\sum_k^N G_{kj} x_k}\left(\tau_{ij} - \frac{\sum_k^N x_k \tau_{kj} G_{kj}}{\sum_k^N G_{kj} x_k}\right)$$
UNIQUAC†	$$\frac{g^E}{RT} = \sum_i^N x_i \ln \frac{\Phi_i}{x_i} + \frac{z}{2}\sum_i^N q_i x_i \ln \frac{\theta_i}{\Phi_i} - \sum_i^N q_i x_i \ln\left(\sum_j^N \theta_j \tau_{ji}\right)$$	$$\ln \gamma_i = \ln \frac{\Phi_i}{x_i} + \frac{z}{2} q_i \ln \frac{\theta_i}{\Phi_i} + l_i - \frac{\Phi_i}{x_i}\sum_j^N x_j l_j - q_i \ln\left(\sum_j^N \theta_j \tau_{ji}\right) + q_i - q_i \sum_j^N \frac{\theta_j \tau_{ij}}{\sum_k^N \theta_k \tau_{kj}}$$

where

$$\Phi_i = \frac{r_i x_i}{\sum_k^N r_k x_k} \qquad \text{and} \qquad \theta_i = \frac{q_i x_i}{\sum_k^N q_k x_k}$$

†Parameters q and r can be calculated from Eq. (8-10.42).

For a ternary system, Eq. (8-9.7) becomes

$$RT \ln \gamma_1 = A_{12}x_2^2 + A_{13}x_3^2 + (A_{12} + A_{13} - A_{23})x_2x_3 \qquad (8\text{-}9.8)$$

$$RT \ln \gamma_2 = A_{12}x_1^2 + A_{23}x_3^2 + (A_{12} + A_{23} - A_{13})x_1x_3 \qquad (8\text{-}9.9)$$

$$RT \ln \gamma_3 = A_{13}x_1^2 + A_{23}x_2^2 + (A_{13} + A_{23} - A_{12})x_1x_2 \qquad (8\text{-}9.10)$$

All constants appearing in these equations can be obtained from binary data; no ternary data are required.

Equations (8-9.8) to (8-9.10) follow from the simplest model for g^E. This model is adequate only for nearly ideal mixtures, where the molecules of the constituent components are similar in size and chemical nature, e.g., benzene-cyclohexane-toluene. For most mixtures encountered in the chemical process industries, more elaborate models for g^E are required.

First it is necessary to choose a model for g^E. Depending on the model chosen, some (or possibly all) of the constants in the model may be obtained from binary data. Second, individual activity coefficients are found by differentiation, as indicated in Eq. (8-9.4).

Once we have an expression for the activity coefficients as a function of liquid-phase composition and temperature, we can then obtain vapor-liquid equilibria by solving simultaneously *all* the equations of equilibrium. For every component i in the mixture

$$y_i P = \gamma_i x_i P_{\text{vp}_i} \mathscr{F}_i \qquad (8\text{-}9.11)$$

where $\mathscr{F}_i$ is given by Eq. (8-4.2).

Since the equations of equilibrium are highly nonlinear, simultaneous solution is almost always achieved only by iteration. Such iterations can be efficiently performed with a computer [68].

Example 8-3   A simple example illustrating how binary data can be used to predict ternary equilibria is provided by Steele, Poling, and Manley [84], who studied the system 1-butene (1)–isobutane (2)–1,3-butadiene (3) in the range 40 to 160°F.

solution   Steele et al. measured isothermal total pressures of the three binary systems as a function of liquid composition. For the three pure components, the pressures are given as a function of temperature by the Antoine equation

$$\ln P_{\text{vp}} = a + b(c + t)^{-1} \qquad (8\text{-}9.12)$$

where P_{vp} is in pounds per square inch and t is in degrees Fahrenheit. Pure-component constants a, b, and c are shown in Table 8-9.

For each binary system the total pressure P is given by

$$P = \sum_{i=1}^{2} y_i P = \sum_{i=1}^{2} x_i \gamma_i P_{\text{vp}_i} \exp \frac{(V_i^L - B_{ii})(P - P_{\text{vp}_i})}{RT} \qquad (8\text{-}9.13)$$

where γ_i is the activity coefficient of component i in the liquid mixture, V_i^L is the molar volume of pure liquid i, and B_{ii} is the second virial coefficient of pure vapor i, all at system temperature T. Equation (8-9.13) assumes that vapor-phase imperfections are described by the (volume-explicit) virial equation truncated after the second term (see Sec.

TABLE 8-9 Antoine Constants for 1-Butene (1)–Isobutene (2)–1,3-Butadiene (3) at 40 to 160°F [Eq. (8-9.12)] [84]

	a	$-b$	c
(1)	12.0502	4067.25	413.786
(2)	12.1465	4170.45	430.786
(3)	12.1118	4126.44	414.039

3-11). Also, since the components are chemically similar, and since there is little difference in molecular size, Steele et al. used the Lewis fugacity rule $B_{ij} = (\frac{1}{2})(B_{ii} + B_{jj})$. For each pure component, the quantity $(V_i^L - B_{ii})/RT$ is shown in Table 8-10.

For the molar excess Gibbs energy of the binary-liquid phase, a one-parameter (two-suffix) Margules equation was assumed:

$$\frac{g_{ij}^E}{RT} = A'_{ij}x_ix_j \tag{8-9.14}$$

From Eq. (8-9.14) we have

$$\ln \gamma_i = A'_{ij}x_j^2 \quad \text{and} \quad \ln \gamma_j = A'_{ij}x_i^2 \tag{8-9.15}$$

Equation (8-9.15) is used at each temperature to reduce the binary, total-pressure data yielding the Margules constant A'_{ij}. For the three binaries studied, Margules constants are shown in Table 8-11.

To predict ternary phase equilibria, Steele et al. assume that the molar excess Gibbs energy is given by

$$\frac{g^E}{RT} = A'_{12}x_1x_2 + A'_{13}x_1x_3 + A'_{23}x_2x_3 \tag{8-9.16}$$

Activity coefficients γ_1, γ_2, and γ_3 are then found by differentiation. [See Eqs. (8-9.8) to (8-9.10), noting that $A'_{ij} = A_{ij}/RT$.]

Vapor-liquid equilibria are found by writing for each component

$$y_iP = \gamma_ix_iP_{vp_i}\mathscr{F}_i \tag{8-9.17}$$

TABLE 8-10 Pure-Component Parameters for 1-Butene (1)–Isobutane (2)–1,3-Butadiene (3) [84]

Temp., °F	$10^4(V_i^L - B_{ii})/RT$, psia^{-1}		
	(1)	(2)	(3)
40	24.22	26.63	23.39
70	22.78	22.78	21.96
100	19.51	19.87	19.03
130	16.79	17.25	16.08
160	14.65	15.25	14.02

TABLE 8-11 Margules Constants A'_{ij} for Three Binary Mixtures formed by 1-Butene (1), Isobutane (2), and 1,3-Butadiene (3) [84]

Temp., °F	$10^3 A'_{12}$	$10^3 A'_{13}$	$10^3 A'_{23}$
40	73.6	77.2	281
70	60.6	64.4	237
100	52.1	54.8	201
130	45.5	47.6	172
160	40.7	42.4	147

where, consistent with earlier assumptions,

$$\mathscr{F}_i = \exp \frac{(V_i^L - B_{ii})(P - P_{vp_i})}{RT} \tag{8-9.18}$$

Steele and coworkers find that predicted ternary vapor-liquid equilibria are in excellent agreement with their ternary data.

Example 8-4 A simple procedure for calculating multicomponent vapor-liquid equilibria from binary data is to assume that for the multicomponent mixture

$$g^E = \sum_{\substack{\text{all} \\ \text{binary} \\ \text{pairs}}} g_{ij}^E \tag{8-9.19}$$

solution To illustrate Eq. (8-9.19), we consider the ternary mixture acetonitrile–benzene–carbon tetrachloride studied by Clarke and Missen [21] at 45°C.

The three sets of binary data were correlated by the Redlich-Kister expansion, which is equivalent to the Margules equation:

$$g_{ij}^E = x_i x_j [A + B(x_i - x_j) + C(x_i - x_j)^2 + D(x_i - x_j)^3] \tag{8-9.20}$$

The constants are given in Table 8-12.

When Eq. (8-9.20) for each binary is substituted into Eq. (8-9.19), the excess Gibbs energy of the ternary is obtained. Clarke and Missen compared excess Gibbs energies calculated in this way with those obtained from experimental data for the ternary system according to the definition

$$g^E = RT(x_1 \ln \gamma_1 + x_2 \ln \gamma_2 + x_3 \ln \gamma_3) \tag{8-9.21}$$

Calculated and experimental excess Gibbs energies were in good agreement, as illustrated by a few results shown in Table 8-13. Comparison between calculated and experimental

TABLE 8-12 Redlich-Kister Constants for the Three Binaries Formed by Acetonitrile (1), Benzene (2), and Carbon Tetrachloride (3) at 45°C [see Eq. (8-9.20)] [21]

Binary system		cal/g-mol			
i	j	A	B	C	D
1	2	643.3	−8.1	70	0
2	3	75.9	−0.85	0	0
3	1	1134.3	118.9	162.2	99.5

TABLE 8-13 Calculated and Observed Molar Excess Gibbs Energies for Acetonitrile (1)–Benzene (2)–Carbon Tetrachloride (3) at 45°C [21]

Calculations from Eq. (8-9.19)

Composition		g^E, cal/g-mol	
x_1	x_2	Calc.	Obs.
0.156	0.767	99	103
0.422	0.128	255	254
0.553	0.328	193	185
0.673	0.244	170	164
0.169	0.179	165	173
0.289	0.506	170	169

results for more than 60 compositions showed that the average deviation (without regard to sign) was only 4 cal/mol. Since the uncertainty due to experimental error is about 3 cal/mol, Clarke and Missen conclude that Eq. (8-9.19) provides an excellent approximation for this ternary system.

Since accurate experimental studies on ternary systems are not plentiful, it is difficult to say to what extent the positive conclusion of Clarke and Missen can be applied to other systems. It appears that for mixtures of typical organic fluids, Eq. (8-9.19) usually gives reliable results, although some deviations have been observed, especially for systems with appreciable hydrogen bonding. In many cases the uncertainties introduced by assuming Eq. (8-9.19) are of the same magnitude as the uncertainties due to experimental error in the binary data.

Example 8-5 While the additivity assumption [Eq. (8-9.19)] often provides a good approximation, for strongly nonideal mixtures, there may be noticeable deviations between experimental and calculated multicomponent equilibria. Such deviations, however, are significant only if they exceed experimental uncertainty. To detect significant deviations, data of high accuracy are required, and such data are rare, especially for ternary systems; they are nearly nonexistent for quaternary (and higher) systems. To illustrate, we consider the ternary system chloroform-ethanol-heptane at 50°C studied by Abbott et al. [1]. Highly accurate data were first obtained for the three binary systems. The data were reduced using Barker's method, as explained by Abbott and Van Ness [2] and elsewhere [66]; the essential feature of this method is that it uses only P-x data (at constant temperature); it does not use data for vapor composition y.

solution To represent the binary data, Abbott et al. considered a five-suffix Margules equation and a modified Margules equation

$$\frac{g^E}{RT} = x_1 x_2 [A'_{21}x_1 + A'_{12}x_2 - (\lambda_{21}x_1 + \lambda_{12}x_2)x_1x_2] \tag{8-9.22}†$$

$$\frac{g^E}{RT} = x_1 x_2 \left(A'_{21}x_1 + A'_{12}x_2 - \frac{\alpha_{12}\alpha_{21}x_1x_2}{\alpha_{12}x_1 + \alpha_{21}x_2 + \eta x_1x_2} \right) \tag{8-9.23}†$$

†The α's and λ's are not to be confused with those used in the NRTL and Wilson equations.

If in Eq. (8-9.22), $\lambda_{21} = \lambda_{12} = D$, and if in Eq. (8-9.23) $\alpha_{12} = \alpha_{21} = D$ and $\eta = 0$, both equations reduce to

$$\frac{g^E}{RT} = x_1 x_2 (A'_{21} x_1 + A'_{12} x_2 - D x_1 x_2) \tag{8-9.24}$$

which is equivalent to the four-suffix Margules equation shown in Table 8-3.

If, in addition, $D = 0$, Eqs. (8-9.22) and (8-9.23) reduce to the three-suffix Margules equation.

For the two binaries chloroform-heptane and chloroform-ethanol, experimental data were reduced using Eq. (8-9.22); however, for the binary ethanol-heptane, Eq. (8-9.23) was used. Parameters reported by Abbott et al. are shown in Table 8-14. With these parameters, calculated total pressures for each binary are in excellent agreement with those measured.

For the ternary, Abbott and coworkers expressed the excess Gibbs energy by

$$\frac{g^E_{123}}{RT} = \frac{g^E_{12}}{RT} + \frac{g^E_{13}}{RT} + \frac{g^E_{23}}{RT} + (C_0 - C_1 x_1 - C_2 x_2 - C_3 x_3) x_1 x_2 x_3 \tag{8-9.25}$$

where C_0, C_1, C_2, and C_3 are ternary constants and g_{ij}^E is given by Eq. (8-9.22) or (8-9.23) for the ij binary. Equation (8-9.25) successfully reproduced the ternary data within experimental error (rms $\Delta P = 0.89$ mm Hg).

Abbott et al. considered two simplifications:

Simplification a: $\qquad\qquad\qquad C_0 = C_1 = C_2 = C_3 = 0$

Simplification b: $\qquad\qquad\qquad C_1 = C_2 = C_3 = 0 \qquad C_0 = \frac{1}{2} \sum_{i \neq j} \sum A'_{ij}$

where the A'_{ij}'s are the binary parameters shown in Table 8-14.

Simplification b was first proposed by Wohl in 1953 [94] on semitheoretical grounds.

When calculated total pressures for the ternary system were compared with experimental results, the deviations exceeded the experimental uncertainty.

Simplification	rms ΔP, mm Hg
a	38.8
b	3.3

TABLE 8-14 Binary Parameters in Eq. (8-9.22) or (8.9-23) and rms Deviation in Total Pressure for the Systems Chloroform–Ethanol–n-Heptane at 50°C [1]

	Chloroform (1), ethanol (2)	Chloroform (1), heptane (2)	Ethanol (1), heptane (2)
A'_{12}	0.4713	0.3507	3.4301
A'_{21}	1.6043	0.5262	2.4440
α_{12}		0.1505	11.1950
α_{21}		0.1505	2.3806
η		0	9.1369
λ_{12}	-0.3651		
λ_{21}	0.5855		
rms ΔP, mm Hg	0.56	0.54	0.34

These results suggest that Wohl's approximation (simplification *b*) provides significant improvement over the additivity assumption for g^E (simplification *a*). However, one cannot generalize from results for one system. Abbott et al. made similar studies for another ternary (acetone-chloroform-methanol) and found that for this system simplification *a* gave significantly better results than simplification *b*, although both simplifications produced errors in total pressure beyond the experimental uncertainty.

While the results of Abbott and coworkers illustrate the limits of predicting ternary (or higher) vapor-liquid equilibria for nonelectrolyte mixtures from binary data only, these limitations are rarely serious for engineering work. As a practical matter, it is common that experimental uncertainties in binary data are as large as the errors which result when multicomponent equilibria are calculated with some model for g^E using only parameters obtained from binary data.

While Eq. (8-9.19) provides a particularly simple approximation, the UNIQUAC equation and the Wilson equation can be generalized to multicomponent mixtures without using that approximation but also without requiring ternary (or higher) parameters. Experience has shown that multicomponent vapor-liquid equilibria can usually be calculated with satisfactory engineering accuracy using the Wilson equation, the NRTL equation, or the UNIQUAC equation provided that care is exercised in obtaining binary parameters.

8-10 Estimation of Activity Coefficients

As discussed in Secs. 8-5 and 8-6, activity coefficients in binary liquid mixtures can often be estimated from a few experimental vapor-liquid equilibrium data for that mixture by using some empirical (or semiempirical) excess function, as shown in Table 8-3. These excess functions provide a thermodynamically consistent method for interpolating and extrapolating limited binary experimental mixture data and for extending binary data to multicomponent mixtures. Frequently, however, few or no mixture data are at hand, and it is necessary to estimate activity coefficients from some suitable correlation. Unfortunately, few such correlations have been established. Theoretical understanding of liquid mixtures is still in an early stage, and while there has been progress for simple mixtures containing small, spherical, nonpolar molecules, e.g., argon-xenon, little theory is available for mixtures containing larger molecules, especially if they are polar or form hydrogen bonds. Therefore, the few available correlations are essentially empirical. This means that predictions of activity coefficients can be made only for systems similar to those used to establish the empirical correlation. Even with this restriction, it must be emphasized that, with few exceptions, the accuracy of prediction is not likely to be high whenever predictions for a binary system do not utilize at least some reliable binary data for that system or for another one that is closely related. In the following sections we summarize a few of the activity-coefficient correlations which are useful for chemical engineering applications.

Regular-Solution Theory Following ideas first introduced by van der Waals and van Laar, Hildebrand and Scatchard, working independently [39], showed that for binary mixtures of nonpolar molecules, activity coefficients γ_1 and γ_2 can be expressed by

$$RT \ln \gamma_1 = V_1^L \Phi_2^2 (c_{11} + c_{22} - 2c_{12}) \qquad (8\text{-}10.1)$$

$$RT \ln \gamma_2 = V_2^L \Phi_1^2 (c_{11} + c_{22} - 2c_{12}) \qquad (8\text{-}10.2)$$

where V_i^L is the liquid molar volume of pure liquid i at temperature T, R is the gas constant, and volume fractions Φ_1 and Φ_2 are defined by

$$\Phi_1 = \frac{x_1 V_1^L}{x_1 V_1^L + x_2 V_2^L} \qquad (8\text{-}10.3)$$

$$\Phi_2 = \frac{x_2 V_2^L}{x_1 V_1^L + x_2 V_2^L} \qquad (8\text{-}10.4)$$

with x denoting the mole fraction.

For pure liquid i, the cohesive energy density c_{ii} is defined by

$$c_{ii} = \frac{\Delta U_i}{V_i^L} \qquad (8\text{-}10.5)$$

where ΔU_i is the energy required isothermally to evaporate liquid i from the saturated liquid to the ideal gas. At temperatures well below the critical

$$\Delta U_i \approx \Delta H_{v_i} - RT \qquad (8\text{-}10.6)$$

where ΔH_{v_i} is the molar enthalpy of vaporization of pure liquid i at temperature T.

Cohesive energy density c_{12} reflects intermolecular forces between molecules of component 1 and 2; this is the key quantity in Eqs. (8-10.1) and (8-10.2). Formally, c_{12} can be related to c_{11} and c_{22} by

$$c_{12} = (c_{11} c_{22})^{1/2} (1 - \ell_{12}) \qquad (8\text{-}10.7)$$

where ℓ_{12} is a binary parameter, positive or negative, but small compared with unity. Upon substitution, Eqs. (8-10.1) and (8-10.2) can be rewritten

$$RT \ln \gamma_1 = V_1^L \Phi_2^2 [(\delta_1 - \delta_2)^2 + 2\ell_{12}\delta_1\delta_2] \qquad (8\text{-}10.8)$$

$$RT \ln \gamma_2 = V_2^L \Phi_1^2 [(\delta_1 - \delta_2)^2 + 2\ell_{12}\delta_1\delta_2] \qquad (8\text{-}10.9)$$

where solubility parameter δ_i is defined by

$$\delta_i = (c_{ii})^{1/2} = \left(\frac{\Delta U_i}{V_i^L}\right)^{1/2} \qquad (8\text{-}10.10)$$

For a first approximation, Hildebrand and Scatchard assume that $\ell_{12} = 0$. In that event, Eqs. (8-10.8) and (8-10.9) contain no binary parameter, and activity coefficients γ_1 and γ_2 can be predicted using only pure-component data.

Although δ_1 and δ_2 depend on temperature, the theory of regular solutions assumes that the excess entropy is zero. It then follows that at constant composition

$$RT \ln \gamma_i = \text{const} \qquad (8\text{-}10.11)$$

Therefore, the right-hand sides of Eqs. (8-10.8) and (8-10.9) may be evaluated at any convenient temperature provided that all quantities are calculated at the same temperature. For many applications the customary convenient temperature is 25°C. A few typical solubility parameters and molar liquid volumes are shown in Table 8-15, and some calculated vapor-liquid equilibria (assuming $\ell_{12} = 0$) are shown in Figs. 8-5 to

TABLE 8-15 Molar Liquid Volumes and Solubility Parameters of Some Nonpolar Liquids†

	V^L, cm³/g-mol	δ, (cal/cm³)$^{1/2}$
Liquefied gases at 90 K		
Nitrogen	38.1	5.3
Carbon monoxide	37.1	5.7
Argon	29.0	6.8
Oxygen	28.0	7.2
Methane	35.3	7.4
Carbon tetrafluoride	46.0	8.3
Ethane	45.7	9.5
Liquid solvents at 25°C		
Perfluoro-n-heptane	226	6.0
Neopentane	122	6.2
Isopentane	117	6.8
n-Pentane	116	7.1
n-Hexane	132	7.3
1-Hexene	126	7.3
n-Octane	164	7.5
n-Hexadecane	294	8.0
Cyclohexane	109	8.2
Carbon tetrachloride	97	8.6
Ethyl benzene	123	8.8
Toluene	107	8.9
Benzene	89	9.2
Styrene	116	9.3
Tetrachloroethylene	103	9.3
Carbon disulfide	61	10.0
Bromine	51	11.5

†More complete tables are given in Refs. 53 and 35.

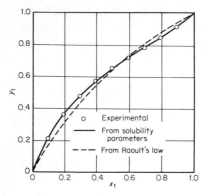

Fig. 8-5 Vapor-liquid equilibria for C_6H_6 (1)–n-C_7H_{16} (2) at 70°C. (*From Ref.* 66.)

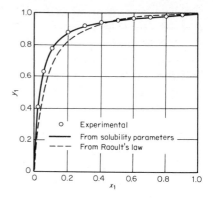

Fig. 8-6 Vapor-liquid equilibria for CO (1)–CH_4 (2) mixtures at 90.7 K. (*From Ref.* 66.)

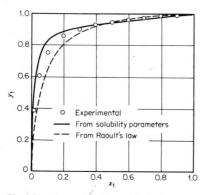

Fig. 8-7 Vapor-liquid equilibria for neo-C_5H_{12} (1)–CCl_4 (2) at 0°C. (*From Ref.* 66.)

8-7.   For typical nonpolar mixtures, calculated results are often in good agreement with experiment.

The regular-solution equations are readily generalized to multicomponent mixtures.   For component k

$$RT \ln \gamma_k = V_k^L \sum_i \sum_j (A_{ik} - \tfrac{1}{2}A_{ij})\Phi_i\Phi_j \qquad (8\text{-}10.12)$$

where

$$A_{ij} = (\delta_i - \delta_j)^2 + 2\ell_{ij}\delta_i\delta_j \qquad (8\text{-}10.13)$$

If all binary parameters ℓ_{ij} are assumed equal to zero, Eq. (8-10.12) simplifies to

$$RT \ln \gamma_k = V_k^L (\delta_k - \bar{\delta})^2 \qquad (8\text{-}10.14)$$

where

$$\bar{\delta} = \sum_i \Phi_i\delta_i \qquad (8\text{-}10.15)$$

where the summation refers to all components, including component k.

The simplicity of Eq. (8-10.14) is striking. It says that in a multicomponent mixture, activity coefficients for all components can be calculated at any composition and temperature using only solubility parameters and molar liquid volumes for the pure components. For mixtures of hydrocarbons, Eq. (8-10.14) often provides a good approximation.

Although binary parameter ℓ_{12} is generally small compared with unity in nonpolar mixtures, its importance may be significant, especially if the difference between δ_1 and δ_2 is small. To illustrate, suppose $T = 300$ K, $V_1^L = 100$ cm^3/mol, $\delta_1 = 7.0$, and $\delta_2 = 7.5$ (cal/cm^3)$^{1/2}$. At infinite dilution ($\Phi_2 = 1$) we find from Eq. (8-10.8) that $\gamma_1^\infty = 1.04$ when $\ell_{12} = 0$. However, if $\ell_{12} = 0.01$, we obtain $\gamma_1^\infty = 1.24$, and if $\ell_{12} = 0.03$, $\gamma_1^\infty = 1.77$. These illustrative results indicate that calculated activity coefficients are often sensitive to small values of ℓ_{12} and that much improvement in predicted results can often be achieved when just one binary datum is available for evaluating ℓ_{12}.

Efforts to correlate ℓ_{12} have met with little success. In their study of binary cryogenic mixtures, Bazúa and Prausnitz [6] found no satisfactory variation of ℓ_{12} with pure-component properties, although some rough trends were found by Cheung and Zander [20] and by Preston and Prausnitz [71]. In many typical cases ℓ_{12} is positive and becomes larger

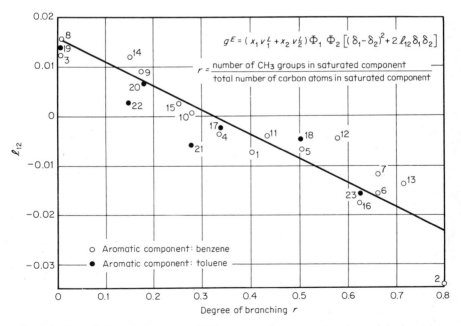

Fig. 8-8 Correlation of the excess Gibbs energy for aromatic-saturated hydrocarbon mixtures at 50°C. Numbers refer to list of binary systems in Ref. 30, table 1. (*From Ref. 30.*)

as the differences in molecular size and chemical nature of the components increase. For example, for carbon dioxide–paraffin mixtures at low temperatures, Preston found that $\ell_{12} = -0.02$ (methane); $+0.08$ (ethane); $+0.08$ (propane); $+0.09$ (butane).

Since ℓ_{12} is an essentially empirical parameter, it depends on temperature. However, for typical nonpolar mixtures over a modest range of temperature, that dependence is usually small.

For mixtures of aromatic and saturated hydrocarbons, Funk and Prausnitz [30] found a systematic variation of ℓ_{12} with the structure of the saturated component, as shown in Fig. 8-8. In this case, a good correlation could be established because experimental data are relatively plentiful and because the correlation is severely restricted with respect to the chemical nature of the components. Figure 8-9 shows the effect of ℓ_{12} on calculating relative volatility in a typical binary system considered by Funk and Prausnitz.

Our inability to correlate ℓ_{12} for a wide variety of mixtures follows

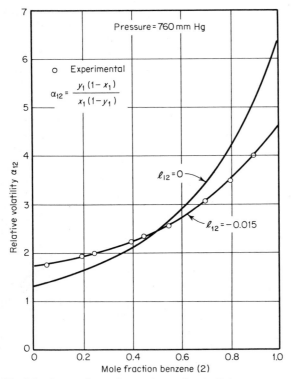

Fig. 8-9 Comparison of experimental volatilities with volatilities calculated by Scatchard-Hildebrand theory for 2,2-dimethylbutane (1)–benzene (2). (*From Ref. 30.*)

from our lack of understanding of intermolecular forces, especially between molecules at short separations.

Several authors have tried to extend regular-solution theory to mixtures containing polar components, but unless the classes of components considered is restricted, such extension has only semiquantitative significance. In establishing these extensions, the cohesive energy density is divided into separate contributions from nonpolar (dispersion) forces and from polar forces:

$$\left(\frac{\Delta U}{V^L}\right)_{total} = \left(\frac{\Delta U}{V^L}\right)_{nonpolar} + \left(\frac{\Delta U}{V^L}\right)_{polar} \qquad (8\text{-}10.16)$$

Equations (8-10.1) and (8-10.2) are used with the substitutions

$$c_{11} = \tau_1^2 + \lambda_1^2 \qquad (8\text{-}10.17)$$

$$c_{22} = \tau_2^2 + \lambda_2^2 \qquad (8\text{-}10.18)$$

$$c_{12} = \lambda_1\lambda_2 + \tau_1\tau_2 + \psi_{12} \qquad (8\text{-}10.19)$$

where λ_i is the nonpolar solubility parameter $[\lambda_i^2 = (\Delta U_i/V_i^L)_{nonpolar}]$ and τ_i is the polar solubility parameter $[\tau_i^2 = (\Delta U_i/V_i^L)_{polar}]$. The binary parameter ψ_{12} is not negligible, as shown by Weimer and Prausnitz [91] in their correlation of activity coefficients at infinite dilution for hydrocarbons in polar non-hydrogen-bonding solvents.

Further extension of the Scatchard-Hildebrand equation to include hydrogen-bonded components makes little sense theoretically since the assumptions of regular-solution theory are seriously in error for mixtures containing such components. Nevertheless, some semiquantitative success has been achieved by Hansen et al. [35] and others [18] interested in establishing criteria for formulating solvents for paints and other surface coatings. Also, Null and Palmer [60] have used extended solubility parameters for establishing an empirical correlation of activity coefficients.

Activity Coefficients at Infinite Dilution Experimental activity coefficients at infinite dilution are particularly useful for calculating the parameters needed in an expression for the excess Gibbs energy (Table 8-3). In a binary mixture, suppose experimental data are available for infinite-dilution activity coefficients γ_1^∞ and γ_2^∞. These can be used to evaluate two adjustable constants in any desired expression for g^E. For example, consider the van Laar equation

$$g^E = Ax_1x_2\left(x_1\frac{A}{B} + x_2\right)^{-1} \qquad (8\text{-}10.20)$$

As indicated in Sec. 8-5, this gives

$$RT \ln \gamma_1 = A\left(1 + \frac{A}{B}\frac{x_1}{x_2}\right)^{-2} \tag{8-10.21}$$

and

$$RT \ln \gamma_2 = B\left(1 + \frac{B}{A}\frac{x_2}{x_1}\right)^{-2} \tag{8-10.22}$$

In the limit, as $x_1 \rightarrow 0$ or as $x_2 \rightarrow 0$, Eqs. (8-10.21) and (8-10.22) become

$$RT \ln \gamma_1^\infty = A \tag{8-10.23}$$

and

$$RT \ln \gamma_2^\infty = B \tag{8-10.24}$$

The calculation of parameters from γ^∞ data is particularly simple for the van Laar equation, but in principle, similar calculations can be made using any two-parameter equation for the excess Gibbs energy. If a three-parameter equation, e.g., NRTL, is used, an independent method must be chosen to determine the third parameter α_{12}.

In recent years, relatively simple experimental methods have been developed for rapid determination of activity coefficients at infinite dilution. These are based on gas-liquid chromatography and on ebulliometry [23, 45, 60, 96, 98, 99].

Schreiber and Eckert [77] have shown that if reliable values of γ_1^∞ and γ_2^∞ are available, either from direct experiment or from a correlation, it is possible to predict vapor-liquid equilibria over the entire range of composition. For completely miscible mixtures the Wilson equation is particularly useful. Parameters Λ_{12} and Λ_{21} are found from simultaneous solution of the relations

$$\ln \gamma_1^\infty = -\ln \Lambda_{12} - \Lambda_{21} + 1 \tag{8-10.25}$$

$$\ln \gamma_2^\infty = -\ln \Lambda_{21} - \Lambda_{12} + 1 \tag{8-10.26}$$

Table 8-16 shows some typical results obtained by Schreiber and Eckert. The average error in vapor composition using γ^∞ data alone is

TABLE 8-16   Fit of Binary Data Using Limiting Activity Coefficients in the Wilson Equation [77]

System and γ^∞	Temp., °C	Average absolute error in calc. $y \times 10^3$	
		All points	γ_1^∞ and γ_2^∞ only
Acetone (1.65)–benzene (1.52)	45	2	4
Carbon tetrachloride (5.66)–acetonitrile (9.30)	45	7	11
Ethanol (18.1)–n-hexane (9.05)	69–79	10	12
Chloroform (2.00)–methanol (9.40)	50	10	28
Acetone (8.75)–water (3.60)	100	10	15

only slightly larger than that obtained when γ data are used over the entire composition range. Schreiber and Eckert also show that reasonable results are often obtained when γ_1^∞ or γ_2^∞ (but not both) are used. When only one γ^∞ is available, it is necessary to use the one-parameter Wilson equation, as discussed earlier. [See Eq. (8-5.12).]

An extensive correlation for γ^∞ data in binary systems has been presented by Pierotti, Deal, and Derr [65]. This correlation can be used to predict γ^∞ for water, hydrocarbons, and typical organic components, e.g., esters, aldehydes, alcohols, ketones, nitriles, in the temperature region 25 to 100°C. The pertinent equations and tables are summarized by Treybal [89] and, with slight changes, are reproduced in Tables 8-17 and 8-18. The accuracy of the correlation varies considerably from one system to another; provided that γ^∞ is not one or more orders of magnitude removed from unity, the average deviation in γ^∞ is about 8 percent.

To illustrate use of Table 8-17, an example, closely resembling one given by Treybal, follows.

Example 8-6   Estimate infinite-dilution activity coefficients for the ethanol-water binary system at 100°C.

solution   First we find γ^∞ for ethanol. Subscript 1 stands for ethanol, and subscript 2 stands for water. From Table 8-17, $\alpha = -0.420$, $\epsilon = 0.517$, $\zeta = 0.230$, $\theta = 0$, and $N_1 = 2$. Using Eq. (a) at the end of Table 8-17, we have

$$\log \gamma^\infty = -0.420 + (0.517)(2) + \frac{0.230}{2} = 0.769$$

$$\gamma^\infty \text{ (ethanol)} = 5.875$$

Next, for water, we again use Table 8-17. Now subscript 1 stands for water and subscript 2 stands for ethanol.

$$\alpha = 0.617 \qquad \epsilon = \zeta = 0 \qquad \theta = -0.280 \qquad N_2 = 2$$

$$\log \gamma^\infty = 0.617 - \frac{0.280}{2} = 0.477$$

$$\gamma^\infty \text{ (water)} = 3.0$$

These calculated results are in good agreement with experimental data of Jones et al. [43].

Azeotropic Data   Many binary systems exhibit azeotropy, i.e., a condition where the composition of a liquid mixture is equal to that of its equilibrium vapor. When the azeotropic conditions (temperature, pressure, composition) are known, activity coefficients γ_1 and γ_2 at that condition are readily found. These activity coefficients can then be used to calculate two parameters in some arbitrarily chosen expression for the excess Gibbs energy (Table 8-3). Extensive compilations of azeotropic data are available [41].

TABLE 8-17 Correlating Constants for Activity Coefficients at Infinite Dilution, Homologous Series of Solutes and Solvents [65]

Solute (1)	Solvent (2)	Temp., °C	α	ϵ	ζ	η	θ	Eq.
n-Acids	Water	25	−1.00	0.622	0.490	⋯	0	(a)
		50	−0.80	0.590	0.290	⋯	0	(a)
		100	−0.620	0.517	0.140	⋯	0	(a)
n-Primary alcohols	Water	25	−0.995	0.622	0.558	⋯	0	(a)
		60	−0.755	0.583	0.460	⋯	0	(a)
		100	−0.420	0.517	0.230	⋯	0	(a)
n-Secondary alcohols	Water	25	−1.220	0.622	0.170	0	⋯	(b)
		60	−1.023	0.583	0.252	0	⋯	(b)
		100	−0.870	0.517	0.400	0	⋯	(b)
n-Tertiary alcohols	Water	25	−1.740	0.622	0.170	⋯	⋯	(c)
		60	−1.477	0.583	0.252	⋯	⋯	(c)
		100	−1.291	0.517	0.400	⋯	⋯	(c)
Alcohols, general	Water	25	−0.525	0.622	0.475	0	⋯	(d)
		60	−0.33	0.583	0.39	0	⋯	(d)
		100	−0.15	0.517	0.34	0	⋯	(d)
n-Allyl alcohols	Water	25	−1.180	0.622	0.558	⋯	0	(a)
		60	−0.929	0.583	0.460	⋯	0	(a)
		100	−0.650	0.517	0.230	⋯	0	(a)
n-Aldehydes	Water	25	−0.780	0.622	0.320	⋯	0	(a)
		60	−0.400	0.583	0.210	⋯	0	(a)
		100	−0.03	0.517	0	⋯	0	(a)
n-Alkene aldehydes	Water	25	−0.720	0.622	0.320	⋯	0	(a)
		60	−0.540	0.583	0.210	⋯	0	(a)
		100	−0.298	0.517	0	⋯	0	(a)

Compound	Solvent							Ref.
n-Ketones	Water	25	−1.475	0.692	0.500	0	...	(b)
		60	−1.040	0.583	0.330	0	...	(b)
		100	−0.621	0.517	0.200	0	...	(b)
n-Acetals	Water	25	−2.556	0.622	0.486	...	...	(e)
		60	−2.184	0.583	0.451	...	...	(e)
		100	−1.780	0.517	0.426	...	...	(e)
n-Ethers	Water	20	−0.770	0.640	0.195	0	...	(b)
n-Nitriles	Water	25	−0.587	0.622	0.760	...	0	(a)
		60	−0.368	0.583	0.413	...	0	(a)
		100	−0.095	0.517	0	...	0	(a)
n-Alkene nitriles	Water	25	−0.520	0.622	0.760	...	0	(a)
		60	−0.323	0.583	0.413	...	0	(a)
		100	−0.074	0.517	0	...	0	(a)
n-Esters	Water	20	−0.930	0.640	0.260	0	...	(b)
n-Formates	Water	20	−0.585	0.640	0.260	...	0	(a)
n-Monoalkyl chlorides	Water	20	1.265	0.640	0.073	...	0	(a)
n-Paraffins	Water	16	0.688	0.642	0	...	0	(a)
n-Alkyl benzenes	Water	25	3.554	0.622	−0.466	...	0	(f)
n-Alcohols	Paraffins	25	1.960	0	0.475	−0.00049	...	(d)
		60	1.460	0	0.390	−0.00057	...	(d)
		100	1.070	0	0.340	−0.00061	...	(d)
n-Ketones	Paraffins	25	0.0877	0	0.757	−0.00049	...	(b)
		60	0.016	0	0.680	−0.00057	...	(b)
		100	−0.067	0	0.605	−0.00061	...	(b)

TABLE 8-17 Correlating Constants for Activity Coefficients at Infinite Dilution, Homologous Series of Solutes and Solvents *(Continued)*

Solute (1)	Solvent (2)	Temp., °C	α	ϵ	ζ	η	θ	Eq.
Water	n-Alcohols	25	0.760	0	0	...	−0.630	(a)
		60	0.680	0	0	...	−0.440	(a)
		100	0.617	0	0	...	−0.280	(a)
Water	sec-Alcohols	80	1.208	0	0	...	−0.690	(c)
Water	n-Ketones	25	1.857	0	0	...	−1.019	(c)
		60	1.493	0	0	...	−0.73	(c)
		100	1.231	0	0	...	−0.557	(c)
Ketones	n-Alcohols	25	−0.088	0.176	0.50	−0.00049	−0.630	(g)
		60	−0.035	0.138	0.33	−0.00057	−0.440	(g)
		100	−0.035	0.112	0.20	−0.00061	−0.280	(g)
Aldehydes	n-Alcohols	25	−0.701	0.176	0.320	−0.00049	−0.630	(h)
		60	−0.239	0.138	0.210	−0.00057	−0.440	(h)
Esters	n-Alcohols	25	0.212	0.176	0.260	−0.00049	−0.630	(g)
		60	0.055	0.138	0.240	−0.00057	−0.440	(g)
		100	0	0.112	0.220	−0.00061	−0.280	(g)
Acetals	n-Alcohols	60	−1.10	0.138	0.451	−0.00057	−0.440	(i)
Paraffins	Ketones	25	...	0.1821	...	−0.00049	0.402	(j)
		60	...	0.1145	...	−0.00057	0.402	(j)
		90	...	0.0746	...	−0.00061	0.402	(j)

Equations

(a) $\log \gamma_1^\infty = \alpha + \epsilon N_1 + \dfrac{\zeta}{N_1} + \dfrac{\theta}{N_2}$

(b) $\log \gamma_1^\infty = \alpha + \epsilon N_1 + \zeta\left(\dfrac{1}{N_1'} + \dfrac{1}{N_1''}\right) + \eta(N_1 - N_2)^2$

(c) $\log \gamma_1^\infty = \alpha + \epsilon N_1 + \zeta\left(\dfrac{1}{N_1'} + \dfrac{1}{N_1''} + \dfrac{1}{N_1'''}\right) + \theta\left(\dfrac{1}{N_2'} + \dfrac{1}{N_2''}\right)$

(d) $\log \gamma_1^\infty = \alpha + \epsilon N_1 + \zeta\left(\dfrac{1}{N_1'} + \dfrac{1}{N_1''} + \dfrac{1}{N_1'''} - 3\right) + \eta(N_1 - N_2)^2$

(e) $\log \gamma_1^\infty = \alpha + \epsilon N_1 + \zeta\left(\dfrac{1}{N_1'} + \dfrac{1}{N_1''} + \dfrac{2}{N_1'''}\right)$

(f) $\log \gamma_1^\infty = \alpha + \epsilon N_1 + \zeta\left(\dfrac{1}{N_1} - 4\right)$

(g) $\log \gamma_1^\infty = \alpha + \epsilon\dfrac{N_1}{N_2} + \zeta\left(\dfrac{1}{N_1'} + \dfrac{1}{N_1''}\right) + \eta(N_1 - N_2)^2 + \dfrac{\theta}{N_2}$

(h) $\log \gamma_1^\infty = \alpha + \epsilon\dfrac{N_1}{N_2} + \dfrac{\zeta}{N_1} + \eta(N_1 - N_2)^2 + \dfrac{\theta}{N_2}$

(i) $\log \gamma_1^\infty = \alpha + \epsilon\dfrac{N_1}{N_2} + \zeta\left(\dfrac{1}{N_1'} + \dfrac{1}{N_1''} + \dfrac{2}{N_1'''}\right) + \eta(N_1 - N_2)^2 + \dfrac{\theta}{N_2}$

(j) $\log \gamma_1^\infty = \epsilon\dfrac{N_1}{N_2} + \eta(N_1 - N_2)^2 + \theta\left(\dfrac{1}{N_2'} + \dfrac{1}{N_2''}\right)$

$N_1, N_2 =$ total number of carbon atoms in molecules 1 and 2, respectively

$N', N'', N''' =$ number of carbon atoms in respective branches of branched compounds, counting the polar grouping; thus, for t-butanol, $N' = N'' = N''' = 2$

TABLE 8-18 Correlating Constants for Activity Coefficients at Infinite Dilution, Homologous Series of Hydrocarbons in Specific Solvents [65]

Temperature, °C	Heptane	Methyl ethyl ketone	Furfural	Phenol	Ethanol	Triethylene glycol	Diethylene glycol	Ethylene glycol
η								
Value of ϵ								
25	−0.00049	0.0455	0.0937	0.0625	0.088	...	0.191	0.275
50	−0.00055	0.033	0.0878	0.0590	0.073	0.161	0.179	0.249
70	−0.00058	0.025	0.0810	0.0586	0.065	...	0.173	0.236
90	−0.00061	0.019	0.0686	0.0581	0.059	0.134	0.158	0.226
Value of θ								
25	0.2105	0.1435	0.1152	0.1421	0.2125	0.181	0.2022	0.275
70	0.1668	0.1142	0.0836	0.1054	0.1575	0.129	0.1472	0.2195
130	0.1212	0.0875	0.0531	0.0734	0.1035	0.0767	0.0996	0.1492
Value of κ								
25	0.1874	0.2079	0.2178	0.2406	0.2425	0.3124	0.3180	0.4147
70	0.1478	0.1754	0.1675	0.1810	0.1753	0.2406	0.2545	0.3516
130	0.1051	0.1427	0.1185	0.1480	0.1169	0.1569	0.1919	0.2772
Value of α								
25	0	0.335	0.916	0.870	0.580	...	0.875	1.208
50	0	0.332	0.756	0.755	0.570	0.72	0.815	1.154
70	0	0.331	0.737	0.690	0.590	...	0.725	1.089
90	0	0.330	0.771	0.620	0.610	0.68	0.72	

Solute (1)	Eq.	ζ
Paraffins	(a)	0
		0
		0
		0

Temp											
	Alkyl cyclohexanes	(a)	−0.260	0.18	0.70	1.26	1.20	1.06	...	1.675	
25			−0.260	0.18	0.70	1.26	1.20	1.06	...	1.675	
50			−0.220	...	0.650	1.120	1.040	1.01	1.46	1.61	2.36
70			−0.195	0.131	0.581	1.020	0.935	0.972	...	1.550	2.22
90			−0.180	0.09	0.480	0.930	0.843	0.925	1.25	1.505	2.08
25	Alkyl benzenes	(a)	−0.466	0.328	0.277	0.67	0.694	1.011	...	1.08	
50			−0.390	0.243	...	0.55	0.580	0.938	0.80	1.00	1.595
70			−0.362	0.225	0.240	0.45	0.500	0.900	...	0.96	1.51
90			−0.350	0.202	0.239	0.44	0.420	0.862	0.74	0.935	1.43
25	Alkyl naphthalenes	(a)	−0.10	0.53	0.169	0.46	0.595	1.06	...	1.00	
50			−0.14	0.53	0.141	0.40	0.54	1.03	0.75	1.00	1.92
70			−0.173	0.53	0.215	0.39	0.497	1.02	...	0.991	1.82
90			−0.204	0.53	0.232	...	0.445	...	0.83	1.01	1.765
25	Alkyl tetralins	(a)	0.28	0.244	0.179	0.652	0.378		...	1.43	
50			0.24	...	...	0.528	0.364		...	1.38	
70			0.21	0.220	0.217	0.447	0.371		1.00	1.33	
90			0.19	...	...	0.373	0.348		0.893	1.28	
25	Alkyl decalins	(a)	−0.43	...	0.871	1.54	1.411		...	2.46	
50			−0.368	...	...	1.367	1.285		...	2.25	
70			−0.355	0.356	0.80	1.253	1.161		1.906	2.07	
90			−0.320	...	...	1.166	1.078		1.68	2.06	
25	Unalkylated aromatics, naphthenes, naphthene aromatics	(b)	1.176† 1.845‡	−1.072	−0.7305	−0.290	−0.383	−0.485	−0.406	−0.377	−0.154
70			0.846† 1.362‡	−0.886	−0.625	−0.080	−0.226	−0.212	−0.186	−0.0775	−0.0174
130			0.544† 0.846‡	−0.6305	−0.504	+0.020	−0.197	+0.47	+0.095	+0.181	+0.229

TABLE 8-18 Correlating Constants for Activity Coefficients at Infinite Dilution, Homologous Series of Hydrocarbons in Specific Solvents (Continued)

Equations

(a) $\log \gamma_1^\infty = \alpha + \epsilon N_p + \dfrac{\zeta}{N_p + 2} + \eta(N_1 + N_2)^2$

(b) $\log \gamma_1^\infty = \alpha + \theta N_a + \kappa N_n + \xi\left(\dfrac{1}{r} - 1\right)$

where N_1, N_2 = total number of carbon atoms in molecules 1 and 2, respectively
N_p = number of paraffinic carbon atoms in solute

N_a = number of aromatic carbon atoms, including $=\!C-$, $=\!CH-$, ring-juncture naphthenic carbons $-\overset{\displaystyle |}{\underset{\displaystyle |}{C}}-H$, and naphthenic carbons in the α position to an aromatic nucleus

N_n = number of naphthenic carbon atoms not counted in N_a
r = number of rings

Examples:

Butyl decalin:	$N_p = 4$	$N_a = 2$	$N_n = 8$	$N_1 = 14$	$r = 2$
Butyl tetralin:	$N_p = 4$	$N_a = 8$	$N_n = 2$	$N_1 = 14$	$r = 2$

†Condensed, naphthalene-like.
‡Tandem, diphenyl-like.

For a binary azeotrope, $x_1 = y_1$ and $x_2 = y_2$. Therefore, Eq. (8-4.1), with $\mathscr{F}_i = 1$, becomes

$$\gamma_1 = \frac{P}{P_{vp_1}} \quad \text{and} \quad \gamma_2 = \frac{P}{P_{vp_2}} \tag{8-10.27}$$

Knowing total pressure P and pure-component vapor pressures P_{vp_1} and P_{vp_2}, we determine γ_1 and γ_2. With these activity coefficients and the azeotropic composition x_1 and x_2 it is now possible to find two parameters A and B by simultaneous solution of two equations of the form

$$RT \ln \gamma_1 = \mathcal{f}_1(x_2, A, B) \tag{8-10.28}$$
$$RT \ln \gamma_2 = \mathcal{f}_2(x_1, A, B) \tag{8-10.29}$$

where, necessarily, $x_1 = 1 - x_2$, and where functions $\mathcal{f}_1$ and $\mathcal{f}_2$ represent thermodynamically consistent equations derived from the choice of an expression for the excess Gibbs energy. Simultaneous solution of Eqs. (8-10.28) and (8-10.29) is simple in principle although the necessary algebra may be tedious if $\mathcal{f}_1$ and $\mathcal{f}_2$ are complex.

Example 8-7   To illustrate, consider an example similar to one given by Treybal [89] for the system ethyl acetate (1)–ethanol (2). This system forms an azeotrope at 760 mm Hg, 71.8°C, and $x_2 = 0.462$.

solution   At 760 mm Hg and 71.8°C, we use Eq. (8-10.27):

$$\gamma_1 = \tfrac{760}{631} = 1.204 \qquad \gamma_2 = \tfrac{760}{581} = 1.307$$

where 631 and 581 mm Hg are the pure-component vapor pressures at 71.8°C.

For functions $\mathcal{f}_1$ and $\mathcal{f}_2$ we choose the van Laar equations shown in Table 8-3. Upon algebraic rearrangement, we obtain an explicit solution for A and B.

$$\frac{A}{RT} = \ln 1.204 \left(1 + \frac{0.462 \ln 1.307}{0.538 \ln 1.204} \right)^2 = 0.93$$

$$\frac{B}{RT} = \ln 1.307 \left(1 + \frac{0.538 \ln 1.204}{0.462 \ln 1.307} \right)^2 = 0.87$$

and $A/B = 1.07$.

At 71.8°C, the activity coefficients are given by

$$\ln \gamma_1 = \frac{0.93}{(1 + 1.07 x_1/x_2)^2}$$

$$\ln \gamma_2 = \frac{0.87}{(1 + x_2/1.07 x_1)^2}$$

Figure 8-10 shows a plot of the calculated activity coefficients. Also shown are experimental results at 760 mm Hg by Furnas and Leighton [31] and by Griswold, Chu, and Winsauer [33]. Since the experimental results are isobaric, the temperature is not constant. However, in this example, the calculated activity coefficients are assumed to be independent of temperature.

Figure 8-10 shows good overall agreement between experimental and calculated activity coefficients. Generally, fair agreement is found if the azeotropic data are accurate, if the binary system is not highly complex and, most important, if the azeotropic composition is in the midrange

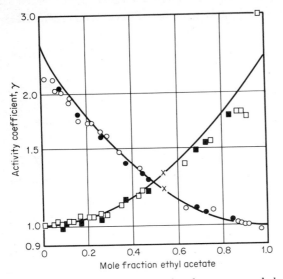

Fig. 8-10 Activity coefficients in the system ethyl acetate–ethanol. Calculated lines from azeotropic data (indicated by x) at 760 mm Hg. Points are experimental [31, 33]. (*From Ref.* 89.)

$0.25 < x_1$ (or x_2) < 0.75. If the azeotropic composition is at either dilute end, azeotropic data are of much less value for estimating activity coefficients over the entire composition range. This negative conclusion follows from the limiting relation $\gamma_1 \rightarrow 1$ as $x_1 \rightarrow 1$. Thus, if we have an azeotropic mixture where $x_2 \ll 1$, the experimental value of γ_1 gives us very little information since γ_1 is necessarily close to unity. For such a mixture, only γ_2 supplies significant information, and therefore we cannot expect to calculate two meaningful adjustable parameters when we have only one significant datum. However, if the azeotropic composition is close to unity, we may, nevertheless, use the azeotropic data to find one activity coefficient, namely, γ_2 (where $x_2 \ll 1$), and then use that γ_2 to determine the single adjustable parameter in any of the one-parameter equations for the molar excess Gibbs energy, as discussed in Sec. 8-5.

Mutual Solubilities of Liquids When two liquids are only partially miscible, experimental data for the two mutual solubilities can be used to estimate activity coefficients over the entire range of composition in the homogeneous regions. Suppose the solubility (mole fraction) of component 1 in component 2 is $x_1^{s'}$ and that of component 2 in component 1 is $x_2^{s''}$, where superscript s denotes saturation and the primes designate the two liquid phases. If $x_1^{s'}$ and $x_2^{s''}$ are known at some temperature T, it is possible to estimate activity coefficients for both components in the homogeneous regions $0 \leqslant x_1' \leqslant x_1^{s'}$ and $0 \leqslant x_2'' \leqslant x_2^{s''}$.

To estimate the activity coefficients, it is necessary to choose some thermodynamically consistent analytical expression which relates activity coefficients γ_1 and γ_2 to mole fraction x. (See Sec. 8-5.) Such an expression contains one or more constants characteristic of the binary system; these constants are generally temperature-dependent, although the effect of temperature is often not large. From the equations of liquid-liquid equilibrium, it is possible to determine two of these constants. The equations of equilibrium are

$$(\gamma_1 x_1)^{s'} = (\gamma_1 x_1)^{s''} \quad \text{and} \quad (\gamma_2 x_2)^{s'} = (\gamma_2 x_2)^{s''} \quad (8\text{-}10.30)$$

Suppose we choose a two-constant expression for the molar excess Gibbs energy g^E. Then, as discussed in Sec. 8-5,

$$RT \ln \gamma_1 = f_1(x_2, A, B) \quad \text{and} \quad RT \ln \gamma_2 = f_2(x_1, A, B) \quad (8\text{-}10.31)$$

where f_1 and f_2 are known functions and the two (unknown) constants are designated by A and B. These constants can be found by simultaneous solution of Eqs. (8-10.30) and (8-10.31) coupled with experimental values for $x_1^{s'}$ and $x_2^{s''}$ and the material balances

$$x_2^{s'} = 1 - x_1^{s'} \quad \text{and} \quad x_1^{s''} = 1 - x_2^{s''} \quad (8\text{-}10.32)$$

In principle, the calculation is simple although the algebra may be tedious, depending on the complexity of the functions f_1 and f_2.

To illustrate, Table 8-19 presents results obtained by Brian [14] for five binary aqueous systems, where subscript 2 refers to water. Calculations are based on both the van Laar equation and the three-suffix (two-parameter) Margules equation (see Table 8-3). Table 8-19 shows the calculated activity coefficients at infinite dilution which are easily related to the constants A and B. [See Eqs. (8-10.23) and (8-10.24).]

Brian's calculations indicate that results are sensitive to the expression

TABLE 8-19 Limiting Activity Coefficients as Calculated from Mutual Solubilities in Five Binary Aqueous Systems [14]

Component (1)	Temp., °C	Solubility limits		$\log \gamma_1^\infty$		$\log \gamma_2^\infty$	
		x_1''	x_2'	van Laar	Margules	van Laar	Margules
Aniline	100	0.01475	0.372	1.8337	1.5996	0.6076	−0.4514
Isobutyl alcohol	90	0.0213	0.5975	1.6531	0.6193	0.4020	−3.0478
1-Butanol	90	0.0207	0.636	1.6477	0.2446	0.3672	−4.1104
Phenol	43.4	0.02105	0.7325	1.6028	−0.1408	0.2872	−8.2901
Propylene oxide	36.3	0.166	0.375	1.1103	1.0743	0.7763	0.7046

arbitrarily chosen for the molar excess Gibbs energy. Brian found that compared with experimental vapor-liquid equilibrium data for the homogeneous regions the Margules equations gave poor results while the van Laar equation gave fair, but not highly accurate, results.

Calculations of this sort can also be made using a three-parameter equation for g^E, but in that event, the third parameter must be estimated independently. A nomogram for such calculations, using the NRTL equation, has been given by Renon and Prausnitz [75].

Generally speaking, mutual-solubility data provide only approximate values of activity coefficients, although such estimates may be better than none at all. These estimates are sensitive not only to the choice of an expression for g^E but also to small errors in the experimental mutual-solubility data.

Group-Contribuion Methods For correlating thermodynamic properties, it is often convenient to regard a molecule as an aggregate of functional groups; as a result, some thermodynamic properties of pure fluids, e.g., heat capacity and critical volume, can be calculated by summing group contributions. Extension of this concept to mixtures was suggested long ago by Langmuir, and several attempts have been made to establish group-contribution methods for heats of mixing and for activity coefficients. Here we mention only two methods, both for activity coefficients, which appear to be particularly useful for making reasonable estimates for those strongly nonideal mixtures where data are sparse or totally absent. The two methods, called ASOG and UNIFAC, are similar in principle but differ in detail.

In any group-contribution method, the basic idea is that whereas there are thousands of chemical compounds of interest in chemical technology, the number of functional groups which constitute these compounds is much smaller. Therefore, if we assume that a physical property of a fluid is the sum of contributions made by the molecule's functional groups, we obtain a possible technique for correlating the properties of a very large number of fluids in terms of a much smaller number of parameters which characterize the contributions of individual groups.

Any group-contribution method is necessarily approximate because the contribution of a given group in one molecule is not necessarily the same as that in another molecule. The fundamental assumption of a group-contribution method is additivity: the contribution made by one group within a molecule is assumed to be independent of that made by any other group in that molecule. This assumption is valid only when the influence of any one group in a molecule is not affected by the nature of other groups within that molecule.

For example, we would not expect the contribution of a carbonyl group in a ketone (say, acetone) to be the same as that of a carbonyl group in an organic acid (say, acetic acid). On the other hand, experience suggests that the contribution of a carbonyl group in, for example, acetone, is close to (although not identical with) the contribution of a carbonyl group in another ketone, say 2-butanone.

Accuracy of correlation improves with increasing distinction of groups; in considering, for example, aliphatic alcohols, in a first approximation no distinction is made between the position (primary or secondary) of a hydroxyl group, but in a second approximation such a distinction is desirable. In the limit, as more and more distinctions are made, we recover the ultimate group, namely, the molecule itself. In that event, the advantage of the group-contribution method is lost. For practical utility, a compromise must be attained. The number of distinct groups must remain small but not so small as to neglect significant effects of molecular structure on physical properties.

Extension of the group-contribution idea to mixtures is extremely attractive because, while the number of pure fluids in chemical technology is already very large, the number of different mixtures is still larger, by many orders of magnitude. Thousands, perhaps millions, of multicomponent liquid mixtures of interest in the chemical industry can be constituted from perhaps 20, 50, or at most 100 functional groups.

ASOG Method The analytical-solution-of-groups (ASOG) method was developed by Derr and Deal [22] and Wilson [92] following earlier work by Redlich, Derr, Pierotti, and Papadopoulos [74]. An introduction to ASOG was presented by Palmer [63].

For component i in a mixture, activity coefficient γ_i consists of a configurational (entropic) contribution due to differences in molecular size and a group-interaction contribution due primarily to differences in intermolecular forces:

$$\ln \gamma_i = \ln \gamma_i^S + \ln \gamma_i^G \qquad (8\text{-}10.33)$$

where superscript S designates size and superscript G designates group.

Activity coefficient γ_i^S depends only on the number of size groups, e.g., CH_2, CO, OH, in the various molecules that constitute the mixture. From Flory-Huggins theory for athermal mixtures of unequal-sized molecules,

$$\ln \gamma_i^S = 1 - \mathscr{R}_i + \ln \mathscr{R}_i \qquad (8\text{-}10.34)$$

where
$$\mathscr{R}_i = \frac{s_i}{\sum_j s_j x_j} \qquad (8\text{-}10.35)$$

where x_j = mole fraction of component j in mixture
s_j = number of size groups in molecule j

Parameter s_j is independent of temperature.   The summation extends over all components, including component i.

To calculate γ_i^G, we need to know all the *group* mole fractions X_k, where subscript k stands for a particular group in molecule j

$$X_k = \frac{\displaystyle\sum_j x_j \nu_{kj}}{\displaystyle\sum_j x_j \sum_k \nu_{kj}} \qquad (8\text{-}10.36)$$

where ν_{kj} is the number of interaction groups k in molecule j.   Activity coefficient γ_i^G is given by

$$\ln \gamma_i^G = \sum_k \nu_{ki} \ln \Gamma_k - \sum_k \nu_{ki} \ln \Gamma_k^* \qquad (8\text{-}10.37)$$

where Γ_k = activity coefficient of group k in the mixture
Γ_k^* = activity coefficient of group k in the standard state
This standard state depends on molecule i.

Activity coefficient Γ_k is given by Wilson's equation

$$\ln \Gamma_k = -\ln \sum_\ell X_\ell A_{k\ell} + \left(1 - \sum_\ell \frac{X_\ell A_{\ell k}}{\sum_m x_m A_{\ell m}}\right) \qquad (8\text{-}10.38)$$

where the summations extend over all groups present in the mixture.

Equation (8-10.38) is also used to find Γ_k^* for component i, but in that case it is applied to a "mixture" of groups as found in pure component i.   For example, if i is water, hexane,† or benzene, there is only one kind of group and $\ln \Gamma_k^*$ is zero.   However, if i is methanol, $\ln \Gamma_k^*$ has a finite value for both hydroxyl and methyl groups.

Parameters $A_{k\ell}$ and $A_{\ell k}$ $(A_{k\ell} \neq A_{\ell k})$ are group-interaction parameters which depend on temperature.   These parameters are obtained from reduction of vapor-liquid equilibria, and a substantial number of such parameters have been reported by Derr and Deal [22].   The important point here is that at a fixed temperature, these parameters depend only on the nature of the groups and, by assumption, are independent of the nature of the molecule.   Therefore, group parameters obtained from available experimental data for some mixtures can be used to predict activity coefficients in other mixtures that contain not the same molecules but the same groups.   For example, suppose we wish to predict activity coefficients in the binary system dibutyl ketone–nitrobenzene.   To do so, we require group-interaction parameters for characterizing interactions between methyl, phenyl, keto, and nitrile

---

†It is assumed here that with respect to group interactions, no distinction is made between groups CH_2 and CH_3.

groups.   These parameters can be obtained from other binary mixtures which contain these groups, e.g., acetone-benzene, nitropropane-toluene, and methyl ethyl ketone–nitroethane.

The UNIFAC Method   The fundamental idea of a solution-of-groups model is to utilize existing phase-equilibrium data for predicting phase equilibria of systems for which no experimental data are available.   In concept, the UNIFAC method follows the ASOG method, wherein activity coefficients in mixtures are related to interactions between structural groups.   The essential features are:

1. Suitable reduction of experimentally obtained activity-coefficient data to yield parameters characterizing interactions between pairs of structural groups in nonelectrolyte systems.

2. Use of these parameters to predict activity coefficients for other systems which have not been studied experimentally but which contain the same functional groups.

The molecular activity coefficient is separated into two parts: one part provides the contribution due to differences in molecular size and the other provides the contribution due to molecular interactions.   In ASOG, the first part is arbitrarily estimated using the athermal Flory-Huggins equation; the Wilson equation, applied to functional groups, is chosen to estimate the second part.   Much of this arbitrariness is removed by combining the solution-of-groups concept with the UNIQUAC equation (see Table 8-3); first, the UNIQUAC model per se contains a combinatorial part, essentially due to differences in size and shape of the molecules in the mixture, and a residual part, essentially due to energy interactions; and second, functional-group sizes and interaction-surface areas are introduced from independently obtained, pure-component molecular-structure data.

The UNIQUAC equation gives good representation of both vapor-liquid and liquid-liquid equilibria for binary and multicomponent mixtures containing a variety of nonelectrolytes such as hydrocarbons, ketones, esters, water, amines, alcohols, nitriles, etc.   In a multicomponent mixture, the UNIQUAC equation for the activity coefficient of (molecular) component i is

$$\ln \gamma_i = \underset{\text{Combinatorial}}{\ln \gamma_i^C} + \underset{\text{Residual}}{\ln \gamma_i^R} \qquad (8\text{-}10.39)$$

where

$$\ln \gamma_i^C = \ln \frac{\Phi_i}{x_i} + \frac{z}{2} q_i \ln \frac{\theta_i}{\Phi_i} + \ell_i - \frac{\Phi_i}{x_i} \sum_j x_j \ell_j \qquad (8\text{-}10.40)$$

and

$$\ln \gamma_i^R = q_i \left[1 - \ln \left(\sum_j \theta_j \tau_{ji} \right) - \sum_j \frac{\theta_j \tau_{ij}}{\sum_k \theta_k \tau_{kj}} \right] \qquad (8\text{-}10.41)$$

TABLE 8-20 Group-Volume and Surface-Area Parameters†

Group no.	Subgroup no.	Group or subgroup	Name	R_k	Q_k	Sample group assignment
1		CH₂	Alkane group			
	1A	CH₃	End group of hydrocarbon chain	0.9011	0.848	Ethane: 2 CH₃
	1B	CH₂	Middle group in hydrocarbon chain	0.6744	0.540	n-Butane: 2 CH₃, 2 CH₂
	1C	CH	Middle group in hydrocarbon chain	0.4469	0.228	Isobutane: 3 CH₃, 1 CH
2		C=C	Olefin group, α-olefin only	1.3454	1.176	α-Butene: 1 C=C, 1 CH₂, 1 CH₃
3		ACH	Aromatic carbon group	0.5313	0.400	Benzene: 6 ACH
4		ACCH₂	Aromatic carbon-alkane group			
	4A	ACCH₂	General case	1.0396	0.660	Ethylbenzene: 5 ACH, 1 ACCH₂, 1 CH₃
	4B	ACCH₃	Toluene group	1.2663	0.968	Toluene: 5 ACH, 1 ACCH₃
5		COH	Alcohol group, includes nearest CH₂			
	5A	COH	General case	1.2044	1.124	Ethanol: 1 CH₃, 1 COH
	5B	MCOH	Methanol	1.4311	1.432	Methanol: 1 MCOH
	5C	CHOH	Secondary alcohol	0.9769	0.812	Isopropanol: 2 CH₃, 1 CHOH
6		H₂O	Water	0.9200	1.400	Water: 1 H₂O
7		ACOH	Aromatic carbon-alcohol group	0.8952	0.680	Phenol: 5 ACH, 1 ACOH
8		CO	Carbonyl group	0.7713	0.640	Acetone: 2 CH₃, 1 CO

No.	Group	Description			Examples
9	CHO	Aldehyde group	0.9980	0.948	Propionaldehyde: 1 CH_3, 1 CH_2, 1 CHO
10	COO	Ester group	1.0020	0.880	Methyl acetate: 2 CH_3, 1 COO
11	O	Ether group	0.2439	0.240	Diethyl ether: 2 CH_3, 2 CH_2, 1 O
12	CNH_2	Primary amine group, includes nearest CH_2			
12A	CNH_2	General case	1.3692	1.236	n-Propylamine: 1 CH_3, 1 CH_2, 1 CN
12B	$MCNH_2$	Methylamine	1.5959	1.544	Methylamine: 1 $MCNH_3$
13	NH	Secondary amine group	0.5326	0.396	Diethylamine: 2 CH_3, 2 CH_2, 1 NH
14	$ACNH_2$	Aromatic carbon-amine group	1.0600	0.816	Aniline: 5 ACH, 1 $ACNH_2$
15	CCN	Nitrile group, includes nearest CH_2			
15A	MCCN	Acetonitrile	1.8701	1.724	Acetonitrile: 1 MCCN
15B	CCN	General case	1.6434	1.416	Propionitrile: 1 CCN, 1 CH_3
16	Cl	Chloride group			
16A	Cl-1	Cl on end carbon	0.7660	0.720	1,2-Dichloroethane: 2 CH_2, 2 Cl-1
16B	Cl-2	Cl on middle carbon	0.8069	0.728	1,2,3-Trichloropropane: 2 CH_2, 1 CH, 2 Cl-1, 1 Cl-2
17	$CHCl_2$	Dichloride group, end group only	2.0672	1.684	1,1-Dichloroethane: 1 CH_3, 1 $CHCl_2$
18	ACCl	Aromatic carbon-chloride group	1.1562	0.844	Chlorobenzene: 5 ACH, 1 ACCl

†An updated and more extensive list of parameters is available upon request from J. M. Prausnitz.

$$\ell_i = \frac{z}{2}(r_i - q_i) - (r_i - 1) \qquad z = 10$$

$$\theta_i = \frac{q_i x_1}{\sum_j q_j x_j} \qquad \Phi_i = \frac{r_i x_1}{\sum_j r_j x_j} \qquad \tau_{ji} = \exp\left(-\frac{u_{ji} - u_{ii}}{RT}\right)$$

In these equations, x_i is the mole fraction of component i and the summations in Eqs. (8-10.40) and (8-10.41) are over all components, including component i; θ_i is the area fraction, and Φ_i is the segment fraction, which is similar to the volume fraction. Pure-component parameters r_i and q_i are, respectively, measures of molecular van der Waals volumes and molecular surface areas.

In UNIQUAC, the two adjustable binary parameters τ_{ij} and τ_{ji} appearing in Eq. (8-10.41) must be evaluated from experimental phase-equilibrium data. No ternary (or higher) parameters are required for systems containing three or more components.

In the UNIFAC method [28], the combinatorial part of the UNIQUAC activity coefficients, Eq. (8-10.40), is used directly. Only pure-component properties enter into this equation. Parameters r_i and q_i are calculated as the sum of the group volume and area parameters R_k and Q_k, given in Table 8-20:

$$r_i = \sum_k \nu_k^{(i)} R_k \qquad \text{and} \qquad q_i = \sum_k \nu_k^{(i)} Q_k \qquad (8\text{-}10.42)$$

where $\nu_k^{(i)}$, always an integer, is the number of groups of type k in molecule i. Group parameters R_k and Q_k are obtained from the van der Waals group volume and surface areas V_{w_k} and A_{w_k}, given by Bondi [10]:

$$R_k = \frac{V_{w_k}}{15.17} \qquad \text{and} \qquad Q_k = \frac{A_{w_k}}{2.5 \times 10^9} \qquad (8\text{-}10.43)$$

The normalization factors 15.17 and 2.5×10^9 are determined by the volume and external surface area of a CH_2 unit in polyethylene.

The residual part of the activity coefficient, Eq. (8-10.41), is replaced by the solution-of-groups concept. Instead of Eq. (8-10.41) we write

$$\ln \gamma_i^R = \sum_{\substack{k \\ \text{all groups}}} \nu_k^{(i)}(\ln \Gamma_k - \ln \Gamma_k^{(i)}) \qquad (8\text{-}10.44)$$

where Γ_k is the group residual activity coefficient and $\Gamma_k^{(i)}$ is the residual activity coefficient of group k in a reference solution containing only molecules of type i. (In UNIFAC, $\Gamma^{(i)}$ is similar to ASOG's Γ^*.) In Eq. (8-10.44) the term $\ln \Gamma_k^{(i)}$ is necessary to attain the normalization that activity coefficient γ_i becomes unity as $x_i \to 1$. The activity coefficient for group k in molecule i depends on the molecule i in which k is situated. For example, $\Gamma_k^{(i)}$ for the COH group† in ethanol refers to a

---

†COH is shortened notation for CH_2OH.

"solution" containing 50 group percent COH and 50 group percent CH_3 at the temperature of the mixture, whereas $\Gamma_k^{(i)}$ for the COH group in n-butanol refers to a "solution" containing 25 group percent COH, 50 group percent CH_2, and 25 group percent CH_3.

The group activity coefficient Γ_k is found from an expression similar to Eq. (8-10.41):

$$\ln \Gamma_k = Q_k \left[1 - \ln \left(\sum_m \theta_m \Psi_{mk} \right) - \sum_m \frac{\theta_m \Psi_{km}}{\sum_n \theta_n \Psi_{nm}} \right] \qquad (8\text{-}10.45)$$

Equation (8-10.45) also holds for $\ln \Gamma_k^{(i)}$. In Eq. (8-10.45), θ_m is the area fraction of group m, and the sums are over all different groups. θ_m is calculated in a manner similar to that for θ_i:

$$\theta_m = \frac{Q_m X_m}{\sum_n Q_n X_n} \qquad (8\text{-}10.46)$$

where X_m is the mole fraction of group m in the mixture. The group-interaction parameter Ψ_{mn} is given by

$$\Psi_{mn} = \exp \left(-\frac{U_{mn} - U_{nn}}{RT} \right) = \exp \left(-\frac{a_{mn}}{T} \right) \qquad (8\text{-}10.47)$$

where U_{mn} is a measure of the energy of interaction between groups m and n. The group-interaction parameters a_{mn} must be evaluated from experimental phase-equilibrium data. Note that a_{mn} has units of kelvins and that $a_{mn} \neq a_{nm}$. Parameters a_{mn} and a_{nm} are obtained from a data base using a wide range of experimental results. Some of these are shown in Table 8-21. Efforts toward updating and extending Table 8-21 are in progress in several university laboratories.

The combinatorial contribution to the activity coefficient [Eq. (8-10.40)] depends only on the sizes and shapes of the molecules present. As the coordination number z increases, for large-chain molecules $q_i / r_i \to 1$ and in that limit, Eq. (8-10.40) reduces to the Flory-Huggins equation used in the ASOG method.

The residual contribution to the activity coefficient [Eqs. (8-10.44) and (8-10.45)] depends on group areas and group interactions. When all group areas are equal, Eqs. (8-10.44) and (8-10.45) are similar to those used in the ASOG method.

The functional groups considered in this work are those given in Table 8-20. Whereas each group listed has its own values of R and Q, the subgroups within the same main group, e.g., subgroups 1A, 1B, and 1C, are assumed to have identical energy-interaction parameters. We present two examples which illustrate (1) the nomenclature and use of Table 8-20 and (2) the UNIFAC method for calculating activity coefficients.

TABLE 8-21   Group Interaction Parameters a_{mn}, K†

	CH₂	C=C	ACH	ACCH₂	COH	H₂O	ACOH	CO	CHO	
CH₂	0	−200.0	32.08	26.78	931.2	1452	1860	1565	685.9	1
C=C	2520	0	651.6	1490	943.3	578.3	x	1400	x	2
ACH	15.26	−144.3	0	167.0	705.9	860.7	1310	651.1	x	3
ACCH₂	−15.84	−309.2	−146.8	0	856.2	3000	740.0	3000	x	4
COH	169.7	254.2	83.5	92.61	0	−320.8	x	462.3	480.0	5
H₂O	657.7	485.4	361.5	385.0	287.5	0	462.6	470.8	234.5	6
ACOH	3000	x	3000	3000	x	−558.2	0	x	x	7
CO	3000	3000	101.8	75.00	−106.5	−532.6	x	0	−49.24	8
CHO	343.2	x	x	x	3000	−226.4	x	39.47	0	9
COO	348.0	x	325.5	3000	167.5	x	−266.5	333.6	x	10
O	2160	x	−75.50	3000	−13.44	x	x	−39.81	x	11
CNH₂	−16.74	90.37	−38.64	x	−109.8	−527.7	x	x	x	12
NH	3000	8.922	37.94	x	−700.0	−882.7	x	x	x	13
ACNH₂	3000	x	3000	3000	x	236.6	x	x	x	14
CCN	27.31	43.03	−66.44	−150.0	337.9	227.0	x	447.7	x	15
Cl	−119.6	242.1	−90.43	52.69	357.0	618.2	x	62.00	x	16
CHCl₂	31.06	−72.88	x	x	x	467.0	x	37.63	x	17
ACCl	121.1	x	1000	x	586.3	1472	x	x	x	18
	1	2	3	4	5	6	7	8	9	

	COO	O	CNH₂	NH	ACNH₂	CCN	Cl	CHCl₂	ACCl	
CH₂	687.5	472.6	422.1	800.0	1330	601.6	523.2	60.45	194.2	1
C=C	x	x	349.9	515.2	x	691.3	253.8	259.5	x	2
ACH	159.1	37.24	179.7	487.2	680.0	290.1	124.0	x	−99.9	3
ACCH₂	110.0	680.0	x	x	640.0	3000	33.84	x	x	4
COH	174.3	−204.6	−166.8	3000	x	79.85	194.6	x	69.97	5
H₂O	x	x	385.3	743.8	−314.6	118.5	158.4	247.2	190.6	6
ACOH	−482.2	x	x	x	x	x	x	x	x	7

	COO	O	CNH$_2$	NH	ACNH$_2$	CCN	Cl	CHCl$_2$	ACCl	
CO	− 180.1	475.5	x	x	x	− 307.4	628.0	874.5	x	8
CHO	x	x	x	x	x	x	x	x	x	9
COO	0	− 26.15	x	x	x	x	x	x	x	10
O	− 290.0	0	x	x	x	x	x	x	x	11
CNH$_2$	x	x	0	x	x	x	x	x	− 10.0	12
NH	x	x	x	0	x	x	x	x	− 60.0	13
ACNH$_2$	x	x	x	x	0	x	x	x	3000	14
CCN	x	x	x	x	x	0	− 100.0	x	25.0	15
Cl	x	x	x	x	x	100.0	0	− 308.5	x	16
CHCl$_2$	x	x	x	x	x	x	790.0	0	x	17
ACCl	x	x	3000	3000	110.0	3000	x	x	0	18
	10	11	12	13	14	15	16	17	18	

†An updated and more extensive list of parameters is available upon request from J. M. Prausnitz. x indicates that data are inadequate or totally absent.

Example 8-9 Consider an equimolar benzene (1)-n-propanol (2) binary mixture.   Benzene has six ACH groups, group 3.   Thus $\nu_3^{(1)} = 6$; $R_3 = 0.5313$, and $Q_3 = 0.400$; $r_1 = (6)(0.5313) = 3.1878$; $q_1 = (6)(0.400) = 2.400$. n-Propanol has one CH$_3$ group (1A), one CH$_2$ group (1B), and one COH group (5A).   Thus $\nu_{1A}^{(2)} = 1$, $\nu_{1B}^{(2)} = 1$, and $\nu_{5A}^{(2)} = 1$; $R_{1A} = 0.9011$, $R_{1B} = 0.6744$, $R_{5A} = 1.2044$,   and $r_2 = (1)(0.9011) + (1)(0.6744) + (1)(1.2044) = 2.7799$;   similarly, $q_2 = (1)(0.848) + (1)(0.540) + (1)(1.124) = 2.512$.   The needed group interaction parameters are obtained from Table 8-21:

$$a_{1,5} = 931.2 \text{ K} \qquad a_{5,1} = 169.7 \text{ K} \qquad a_{1,3} = 32.08 \text{ K} \qquad a_{3,1} = 15.26 \text{ K}$$
$$a_{3,5} = 705.9 \text{ K} \qquad a_{5,3} = 83.50 \text{ K}$$

$$x_1 = x_2 = \tfrac{1}{2}$$

$$X_{1A} = \frac{\frac{1}{2}}{\frac{6}{2} + \frac{1}{2} + \frac{1}{2} + \frac{1}{2}} = \frac{1}{9}$$

$$X_3 = \frac{\frac{6}{2}}{\frac{6}{2} + \frac{1}{2} + \frac{1}{2} + \frac{1}{2}} = \frac{2}{3}$$

Similarly,
$$X_{1B} = X_{5A} = \tfrac{1}{9}$$

At constant temperature the activity coefficient of group k ($k = $ 1A, 1B, 3, or 5A) is a function of the group composition:

$$\Gamma_k = \Gamma(X_{1A}, X_{1B}, X_3, X_{5A})$$

For an equimolar mixture
$$\Gamma_k = \Gamma(\tfrac{1}{9}, \tfrac{1}{9}, \tfrac{2}{3}, \tfrac{1}{9})$$

In pure benzene (1)
$$\Gamma_3^{(1)} = \Gamma(0, 0, 1, 0) = 1$$

In pure n-propanol (2)
$$\Gamma_k^{(2)} = \Gamma(\tfrac{1}{3}, \tfrac{1}{3}, 0, \tfrac{1}{3}) \neq 1 \qquad k = \text{1A, 1B, or 5A}$$

Example 8-10 Obtain activity coefficients for the acetone (1)–n-pentane (2) system at 307 K and $x_1 = 0.047$.

$$r_1 = (2)(0.9011) + (1)(0.7713) = 2.5735 \qquad q_1 = (2)(0.848) + (1)(0.640) = 2.336$$

$$r_2 = (2)(0.9011) + (3)(0.6744) = 3.8254 \qquad q_2 = (2)(0.848) + (3)(0.540) = 3.316$$

$$\Phi_1 = \frac{(2.5735)(0.047)}{(2.5735)(0.047) + (3.8254)(0.953)} = 0.0321 \qquad \Phi_2 = 0.9679$$

$$\theta_1 = \frac{(2.336)(0.047)}{(2.336)(0.047)(3.316)(0.953)} = 0.0336 \qquad \theta_2 = 0.9664$$

$$\ell_1 = (5)(2.5735 - 2.336) - 1.5735 = -0.3860$$

$$\ell_2 = (5)(3.8254 - 3.316) - 2.8254 = -0.2784$$

$$\ln \gamma_1^C = \ln \frac{0.0321}{0.047} + (5)(2.336) \ln \frac{0.0336}{0.0321} - 0.3860$$

$$+ \frac{0.0321}{0.047} [(0.047)(0.3860) + (0.953)(0.2784)] = -0.0403$$

$$CH_3 = 1A \qquad CH_2 = 1B \qquad CO = 8$$

$$a_{1,8} = 1565 \text{ K} \qquad a_{8,1} = 3000 \text{ K}$$

$$\Psi_{1,8} = \exp\left(-\frac{1565}{307}\right) = 0.00611 \qquad \Psi_{8,1} = \exp\left(-\frac{3000}{307}\right) = 0.000057$$

For pure acetone:

$$\ln \Gamma_{1A}^{(1)} = \ln \Gamma_{1A}^{(1)}(X_{1A} = \tfrac{2}{3}, X_8 = \tfrac{1}{3})$$

$$\theta_{1A}^{(1)} = \frac{(\tfrac{2}{3})(0.848)}{(\tfrac{2}{3})(0.848) + (\tfrac{1}{3})(0.640)} = 0.726 \qquad \theta_8^{(1)} = 0.274$$

$$\ln \Gamma_{1A}^{(1)} = 0.848 \left\{ 1 - \ln [0.726 + (0.274)(0.000057)] \right.$$
$$\left. - \left[\frac{0.726}{0.726 + (0.274)(0.000057)} + \frac{(0.274)(0.00611)}{(0.726)(0.00611) + 0.274} \right] \right\} = 0.2664$$

$$\ln \Gamma_8^{(1)} = 0.640 \left\{ 1 - \ln [(0.726)(0.00611) + 0.274] \right.$$
$$\left. - \left[\frac{(0.726)(0.000057)}{0.726 + (0.274)(0.000057)} + \frac{0.274}{(0.726)(0.00611) + 0.274} \right] \right\} = 0.8284$$

For $x_A = 0.047$:

$$X_{1A} = \frac{(0.047)(2) + (0.953)(2)}{(0.047)(3) + (0.953)(5)} = 0.4077 \quad X_{1B} = 0.5828 \qquad X_8 = 0.0096$$

$$\theta_{1A} = 0.5187 \qquad \theta_{1B} = 0.4721 \qquad \theta_8 = 0.0092$$

$$\ln \Gamma_{1A} = 0.848 \left\{ 1 - \ln [0.5187 + 0.4721 + (0.0092)(0.000057)] \right.$$

$$- \left[\frac{0.5187 + 0.4721}{0.5187 + 0.4721 + (0.000057)(0.0092)} \right.$$

$$\left. + \frac{(0.0092)(0.00611)}{(0.5187 + 0.4721)(0.00611 + 0.0092)} \right]\Bigg\}$$

$$= 0.0047$$

$$\ln \Gamma_8 = 0.640 \left\{ 1 - \ln [(0.5187 + 0.4721)(0.00611) + 0.0092] \right.$$

$$- \left[\frac{(0.5187 + 0.4721)(0.000057)}{0.5187 + 0.4721 + (0.0092)(0.000057)} \right.$$

$$\left. + \frac{0.0092}{(0.5187 + 0.4721)(0.00611) + 0.0092} \right]\Bigg\}$$

$$= 2.9310$$

$\ln \gamma_1{}^R = (2)(0.0047 - 0.2664) + (1)(2.9310 - 0.8284) = 1.5792$

$\ln \gamma_1 = \ln \gamma_1{}^C + \ln \gamma_1{}^R = -0.0403 + 1.5792 = 1.5389$ or $\gamma_1 = 4.66$

By following exactly the same procedure for pentane, it is found that $\gamma_2 = 1.02$. The corresponding experimental values of Lo et al. [50] are $\gamma_1 = 4.41$ and $\gamma_2 = 1.11$.

8-11 Solubilities of Gases in Liquids

At modest pressures, most gases are only sparingly soluble in typical liquids. For example, at 25°C and a partial pressure of 1 atm, the (mole-fraction) solubility of nitrogen in cyclohexane is $x = 7.6 \times 10^{-4}$, and that in water is $x = 0.18 \times 10^{-4}$. While there are some exceptions (notably, hydrogen), the solubility of a gas in typical solvents usually falls with rising temperature. However, at higher temperatures, approaching the critical temperature of the solvent, the solubility of a gas usually rises with temperature, as illustrated in Fig. 8-11.

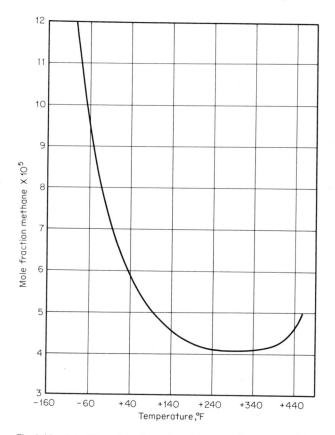

Fig. 8-11 Solubility of methane in n-heptane when vapor-phase fugacity of methane is 0.01 atm. (*From Ref.* 66.)

Experimentally determined solubilities have been reported in the chemical literature for over 100 years, but many of the data are of poor quality. Although no truly comprehensive and critical compilation of the available data exists, a few compilations have been published, as indicated in Table 8-22.

Unfortunately, a variety of units has been employed in reporting gas solubilities. The most common of these are two dimensionless coefficients: *Bunsen coefficient,* defined as the volume (corrected to 0°C and 1 atm) of gas dissolved per unit volume of solvent at system temperature T when the partial pressure of the solute is 1 atm; *Ostwald coefficient,* defined as the volume of gas at system temperature T and partial pressure p dissolved per unit volume of solvent. If the solubility is small and the gas phase is ideal, the Ostwald coefficient is independent of p and these two coefficients are simply related by

$$\text{Ostwald coefficient} = \frac{T}{273} \text{ (Bunsen coefficient)}$$

where T is in kelvins. Friend and Adler [29] have discussed these and other coefficients for expressing solubilities as well as some of their applications for engineering calculations.

These units are often found in older articles. In recent years it has become more common to report solubilities in units of mole fraction or Henry's constants.

When the solubility is small, Henry's law provides a good approximation. In a binary system let subscript 2 refer to the gaseous solute and subscript 1 to the liquid solvent. Henry's law is written

$$f_2 = H_{2,1}^{(P_{vp_1})} x_2 \qquad x_2 \ll 1 \qquad (8\text{-}11.1)$$

where x is the mole fraction, f the fugacity, and H Henry's constant, rigorously defined by

$$H_{2,1}^{(P_{vp_1})} = \lim_{x_2 \to 0} \left(\frac{f}{x}\right)_2 \qquad (8\text{-}11.2)$$

The subscript 2,1 indicates that Henry's constant H is for solute 2 in solvent 1. Superscript P_{vp_1} indicates that the pressure of the system (as

TABLE 8-22 **Some Compilations for Solubilities of Gases in Liquids**

Seidell, A.: "Solubilities of Inorganic and Metal-Organic Compounds," Van Nostrand, New York, 1958, and "Solubilities of Inorganic and Organic Compounds," Van Nostrand, New York, 1952

Stephen, H., and T. Stephen: "Solubilities of Inorganic and Organic Compounds," Pergamon Press, Oxford, and Macmillan, New York, 1963

Battino, R., and H. L. Clever: *Chem. Rev.,* **66**: 395 (1966)

Wilhelm, E., and R. Battino: *Chem. Rev.,* **73**: 1 (1973)

$x_2 \to 0$) is equal to the saturation (vapor) pressure of solvent 1 at temperature T. Henry's constant depends on temperature and often strongly so.

If the gas pressure is large, the effect of pressure on Henry's constant must be taken into consideration. In that event, Eq. (8-11.1) takes the more general form

$$\ln \left(\frac{f}{x}\right)_2 = \ln \left(\frac{\phi y P}{x}\right)_2 = \ln H_{2,1}^{(P_{vp_1})} + \frac{\bar{V}_2^{\infty}(P - P_{vp_1})}{RT} \qquad (8\text{-}11.3)$$

where ϕ = vapor-phase fugacity coefficient

P = system pressure

$\bar{V}_2^{\infty}$ = partial molar volume of solute 2 at infinite dilution in the liquid phase

R = gas constant

[Equation (8-11.3) assumes that $\bar{V}_2^{\infty}$ is independent of pressure in the interval $P - P_{vp_1}$.] An illustration of how it can be used to reduce solubility data is given in Fig. 8-12, which shows the solubility of nitrogen in water. In this case, since the vapor phase is predominantly nitrogen, the fugacity coefficient for nitrogen is calculated using the Lewis fugacity rule.† The truly important assumption in Eq. (8-11.3) is that x_2 is small

†The Lewis fugacity rule assumes that $\phi_i^V(T, P, y_i) = \phi_i^V(T, P, y_i = 1)$, where superscript V refers to the vapor phase.

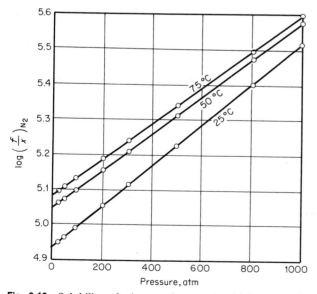

Fig. 8-12 Solubility of nitrogen in water at high pressures. (*From Ref. 66.*)

compared with unity. Just what "small" means depends on the chemical nature of the solute and solvent, as discussed elsewhere [66]. In general, the larger the difference in chemical nature between solute and solvent, the smaller x_2 must be for Eq. (8-11.3) to hold.

In Eq. (8-11.3), the first term on the right-hand side is always dominant; the second term is a correction. Correlations for $\bar{V}_2^\infty$ have been presented by Lyckman et al. [51], by Brelvi and O'Connell [13], and by Tiepel and Gubbins [87]. Some typical values for $\bar{V}_2^\infty$ quoted by Hildebrand and Scott are shown in Table 8-23. At temperatures well below the solvent's critical temperature, $\bar{V}^\infty$ is larger than, but in the vicinity of, the solute's molar volume at its normal boiling temperature. However, $\bar{V}^\infty$ may become much larger at temperatures close to the critical temperature of the solvent [66].

Many attempts have been made to correlate gas solubilities, but success has been severely limited because, on the one hand, a satisfactory theory for gas-liquid solutions has not been established and, on the other, reliable experimental data are not plentiful, especially at temperatures remote from 25°C. Among others, Battino and Wilhelm [5] have obtained some success in correlating solubilities in nonpolar systems near 25°C using concepts from perturbed-hard-sphere theory, but, as yet, these are of limited use for engineering work. A more useful graphical correlation, including polar systems, was prepared by Hayduk et al. [36], and a correlation based on regular-solution theory for nonpolar systems was established by Prausnitz and Shair [70] and, in similar form, by Yen and McKetta [97]. The regular-solution correlation is limited to nonpolar (or weakly polar) systems, and while its accuracy is not high, it has two advantages: it applies over a wide temperature range, and it requires no mixture data. Correlations for nonpolar systems, near 25°C, are given by Hildebrand and Scott [39].

TABLE 8-23   Partial Molal Volumes $\bar{V}^\infty$ of Gases in Liquid Solution at 25°C, cm³/mol†

	H_2	N_2	CO	O_2	CH_4	C_2H_2	C_2H_4	C_2H_6	CO_2	SO_2
Ethyl ether	50	66	62	56	58					
Acetone	38	55	53	48	55	49	58	64	...	68
Methyl acetate	38	54	53	48	53	49	62	69	...	47
Carbon tetrachloride	38	53	53	45	52	54	61	67	...	54
Benzene	36	53	52	46	52	51	61	67	...	48
Methanol	35	52	51	45	52	...	...	...	43	
Chlorobenzene	34	50	46	43	49	50	58	64	...	48
Water	26	40	36	31	37	...	...	...	33	
Molar volume of pure solute at its normal boiling point	28	35	35	28	39	42	50	55	40	45

†J. H. Hildebrand and R. L. Scott, "Solubility of Nonelectrolytes," 3d ed., Reinhold, New York, 1950.

A crude estimate of solubility can be obtained rapidly by extrapolating the vapor pressure of the gaseous solute on a linear plot of $\log P_{vp}$ vs. $1/T$.   The so-called *ideal solubility* is given by

$$x_2 = \frac{y_2 P}{P_{vp_2}}$$ (8-11.4)

where P_{vp_2} is the (extrapolated) vapor pressure of the solute at system temperature T.   The ideal solubility is a function of temperature, but it is independent of the solvent.   Table 8-24 shows that for many typical cases, Eq. (8-11.4) provides an order-of-magnitude estimate.

TABLE 8-24   Solubilities of Gases in Several Liquid Solvents at 25°C and 1 atm Partial Pressure.   Mole Fraction × 10⁴

	Ideal†	n-C_7F_{16}	n-C_7H_{16}	CCl_4	CS_2	$(CH_3)_2CO$
H_2	8	14.01	6.88	3.19	1.49	2.31
N_2	10	38.7	...	6.29	2.22	5.92
CH_4	35	82.6	...	28.4	13.12	22.3
CO_2	160	208.8	121	107	32.8	

†See Eq. (8-11.4).

To find the solubility of a gas in a mixed solvent, a first approximation is provided by the expression

$$\ln H_{2,\text{mix}} = \sum_{\substack{i=1,3,\dots \\ i \neq 2}} x_i \ln H_{2,i}$$ (8-11.5)

where $H_{2,\text{mix}}$ is Henry's constant for solute 2 in the solvent mixture and $H_{2,i}$ is Henry's constant for solute 2 in solvent i, all at system temperature.   The mole fraction x_i is on a solute-free basis.

Equation (8-11.5) is rigorous when the mixture is ideal.   However, even for nonideal mixtures, this equation provides a reasonable approximation.   For more accurate estimates, it is necessary to add to Eq. (8-11.5) terms which depend on the nonideality of the solvent mixture [61, 87].

8-12 Vapor-Liquid Equilibria at High Pressures

Vapor-liquid equilibrium calculations at high pressures are more difficult than those at low or modest pressures for several reasons.

1. The effect of pressure on liquid-phase properties is significant only at high pressures.   At low or modest pressure this effect can often be neglected or approximated; the common approximation is to assume in

Eq. (8-2.3) that the standard-state fugacity depends on pressure (as given by the Poynting factor) but that the activity coefficient is independent of pressure at constant composition and temperature. Since

$$\left(\frac{\partial \ln \gamma_i}{\partial P}\right)_{T,x} = \frac{\bar{V}_i^L - V_{\text{pure } i}^L}{RT} \tag{8-12.1}$$

the assumption that activity coefficient γ_i is independent of pressure is equivalent to assuming that in the liquid phase the partial molar volume $\bar{V}_i^L$ is equal to the molar volume of pure liquid i. At high pressures, especially in the critical region, this assumption can lead to serious error.

2. The vapor-phase fugacity coefficient ϕ_i must be found from an equation of state suitable for high pressures, as discussed in Sec. 5-9. Such equations tend to be complex. By contrast, at low pressures we can often set $\phi_i = 1$, and at modest pressures we can often calculate ϕ_i with the virial equation truncated after the second term.

3. In high-pressure vapor-liquid equilibria we frequently must deal with supercritical components; we are often concerned with mixtures at a temperature which is larger than the critical temperature of one (or possibly more) of the components. In that event, how do we evaluate the standard-state fugacity of the supercritical component? Normally we use as a standard state the pure liquid at system temperature and pressure. For a supercritical component the pure liquid at system temperature is hypothetical, and therefore there is no unambiguous way to calculate its fugacity. The problem of supercritical hypothetical standard states can be avoided by using the unsymmetric convention for normalizing activity coefficients [66, chap. 6], and some correlations for engineering use have been established on that basis [61, 67]. However, there are computational disadvantages in using unsymmetrically normalized activity coefficients, especially in multicomponent systems, and therefore their use in engineering work is not popular.

4. In high-pressure vapor-liquid equilibria we frequently encounter critical phenomena, including retrograde condensation. Since these phenomena are little understood, it is at present not possible to establish simple algebraic equations for representing them.

In the following paragraphs we describe very briefly two well-known and useful methods for calculating high-pressure vapor-liquid equilibria. (For details the reader must consult the original literature.) The first one is based on an extension of the methods used for low-pressure equilibria; it retains the concept of activity coefficient and standard state. The second one avoids the use of activity coefficients (and thereby avoids the problem of standard states for supercritical components) but instead uses an integration of volumetric properties (equation of state) for both the vapor phase and the liquid phase.

High-Pressure Vapor-Liquid Equilibria Using Activity Coefficients for the Liquid Phase

For every component i, the equation of equilibrium is written

$$\phi_i y_i P = \gamma_i x_i f^L_{\text{pure } i} \qquad (8\text{-}12.2)$$

where $f^L_{\text{pure } i}$ is the fugacity of pure liquid i at system temperature and pressure. To use Eq. (8-12.2), we require (1) an equation of state for calculating ϕ_i, (2) an expression for γ_i as a function of temperature, pressure and composition, and (3) a method for calculating $f^L_{\text{pure } i}$ for supercritical as well as subcritical components.

For mixtures of nonpolar components (primarily hydrocarbons), a reasonable procedure is to use the Redlich-Kwong equation for ϕ_i, the Scatchard-Hildebrand (solubility-parameter) equation for γ_i and to back out values of $f^L_{\text{pure } i}$ for supercritical components from binary data. This, essentially, is the procedure suggested by Prausnitz, Edmister, and Chao [69] and developed by Chao and Seader [19].

The vapor-phase fugacity coefficient is given by Eq. (5-9.10). The Redlich-Kwong constants a and b are related to mole fraction y by

$$b = \sum_j y_j b_j \qquad \text{and} \qquad a = \sum_i \sum_j y_i y_j a_{ij} \qquad (8\text{-}12.3)$$

To a first approximation, the mixture constant a_{ij} is given by

$$a_{ij} = (a_i a_j)^{1/2} \qquad (8\text{-}12.4)$$

but better results are obtained by calculating a_{ij} with a binary constant, as discussed in Chap. 4.

The vapor-phase molar volume is found from the equation of state when the pressure, temperature, and composition are known.

The activity coefficient is estimated by the Scatchard-Hildebrand relation (see Sec. 8-10):

$$\ln \gamma_i = \frac{V_i^L (\delta_i - \bar{\delta})^2}{RT} \qquad (8\text{-}12.5)$$

where V_i^L = molar volume
δ_i = solubility parameter of pure liquid i
$\bar{\delta}$ = the average solubility parameter for liquid mixture defined by Eq. (8-10.15)

Equation (8-12.5) is based on regular-solution theory. It gives the activity coefficient as a function of temperature and composition, but it is independent of pressure.

For subcritical components, pure-component data are used to find molar liquid volume V_i^L, solubility parameter δ_i, and standard-state fugacity $f^L_{\text{pure } i}$. For supercritical components these are hypothetical quantities which must be back-calculated from binary vapor-liquid equilibrium data.

In their original correlation, Chao and Seader presented empirical equations for V_i^L as a function of temperature and $f_{\text{pure }i}^L$ as a function of temperature and pressure. A table of solubility parameters was also presented. These equations and tables have been modified and extended by others [32].

The advantage of this type of correlation is that it is relatively simple, and if Eq. (8-12.5) is used for γ and the Redlich-Kwong equation is used for ϕ, no mixture data are needed since only pure-component data are required. Therefore, the Chao-Seader type of correlation is often used with success in the petroleum industry. However, the accuracy of the results is often poor, especially if calculations are performed at conditions remote from those used to establish the hypothetical parameters δ_i, V_i^L, and $f_{\text{pure }i}^L$ for supercritical components. Much improvement can be obtained by introducing binary constants into the expressions for γ and ϕ.

Since the effect of pressure on γ is neglected, the Scatchard-Hildebrand equation is necessarily poor for calculating vapor-liquid equilibria in the critical region.

High-Pressure Vapor-Liquid Equilibria from an Equation of State   While the concepts of activity and activity coefficient are useful for most problems in vapor-liquid equilibria, we can, in principle, do without them provided we have available an equation of state which holds for the mixture of interest in *both* the vapor phase and the liquid phase. Application of an equation of state to calculate vapor-liquid equilibria was indicated by van der Waals in the early years of this century.

Suppose we have an equation of state of the form

$$P = f(T, V_T, n_1, n_2, \ldots) \tag{8-12.6}$$

where V_T is the total volume occupied by n_1 moles of component 1, n_2 moles of component 2, etc. We assume that this equation holds for the liquid phase as well as the vapor phase.

As discussed in Sec. 5-9, we can calculate the fugacity of component i in the vapor phase:

$$RT \ln f_i^V = \int_{V_T^V}^{\infty} \left[\left(\frac{\partial P}{\partial n_i} \right)_{T, V_T, n_j} - \frac{RT}{V_T} \right] dV_T - RT \ln \frac{V^V}{y_i RT} \tag{8-12.7}$$

Similarly, the fugacity of component i in the liquid phase is given by

$$RT \ln f_i^L = \int_{V_T^L}^{\infty} \left[\left(\frac{\partial P}{\partial n_i} \right)_{T, V_T, n_j} - \frac{RT}{V_T} \right] dV_T - RT \ln \frac{V^L}{x_i RT} \tag{8-12.8}$$

where V_T^V = total volume of vapor phase

$\quad\quad V_T^L$ = total volume of liquid phase

The molar volume V is related to total volume by $V = V_T/n_T$, where the total number of moles is given by $n_T = \Sigma_i\, n_i$. Subscript j refers to all components except component i.

If an equation of state is available, Eqs. (8-2.1) and (8-12.6) to (8-12.8) are sufficient to calculate vapor-liquid equilibria. The calculations are tedious, and even for a binary mixture using a relatively simple equation of state trial-and-error (iteration) calculations are required. The equation-of-state method for calculating vapor-liquid equilibria can be reduced to practice only with a computer.

To fix ideas, suppose we have a mixture of N components. Suppose further that we know pressure P and all liquid-phase mole fractions $x_1, x_2, \ldots, x_N$. We want to calculate the equilibrium values of the temperature T and all vapor-phase mole fractions $y_1, y_2, \ldots, y_N$.

The unknowns are

$y_1, y_2, \ldots, y_{N-1}$	$N - 1$ mole fractions†
T	Temperature
V^V, V^L	Molar volumes of equilibrium vapor phase and liquid phase
Total	$N + 2$ unknowns

†Since $\Sigma_{i=1}^{N}\, y_i = 1$, mole fraction y_N is fixed once all other y's are known.

The independent equations are

$f_i^V = f_i^L$	N equations, where fugacities f_i^V and f_i^L are found from Eqs. (8-12.7) and (8-12.8)
$P = f(T, V^L, x_1, \ldots)$	Equation of state for liquid phase
$P = f(T, V^V, y_1, \ldots)$	Equation of state for vapor phase
Total	$N + 2$ independent equations

In principle, therefore, the problem can be solved because the number of independent equations is equal to the number of unknowns.

In practice, however, it is not simple to solve simultaneously a large number of strongly nonlinear equations. Not only is a computer an absolute necessity, but, in addition, economic considerations require an efficient computer program for rapid convergence of the iterative calculations.†

While the computing problem is not trivial, the more difficult problem is to establish an equation of state which can represent volumetric properties of mixtures in both the gas phase and the liquid phase. Van der Waals and his coworkers used his equation of state for calculating vapor-liquid equilibria, and it is surprising how well that relatively simple equation can *qualitatively* represent phase behavior. For quantitative purposes, however, the van der Waals equation is not sufficiently precise. Many empirical equations of state have been proposed for calculating vapor-liquid equilibria; for mixtures of relatively simple molecules, these equations have often produced useful results. The best-known example is the Benedict-Webb-Rubin equation (and its extensions by various authors), which is applicable to mixtures of the lighter hydrocarbons and, to a lesser extent, to mixtures also containing carbon dioxide and hydrogen sulfide. (See Secs. 3-9 and 4-6.)

The term $(\partial P/\partial n_i)_{T,V_{T},n_j}$ in Eqs. (8-12.7) and (8-12.8) indicates that vapor-liquid equilibrium calculations are sensitive to mixing rules, i.e., to the effect of composition on the constants appearing in the equation of state. A small change in mixing rules can produce a large change in vapor-liquid equilibria; for good results it is nearly always necessary to include some binary parameters in the mixing rules for at least one of the equation-of-state constants, as discussed in Chap. 4. For example, consider the Benedict-Webb-Rubin equation as used by Orye [62]. For the constant A_0, Orye writes

$$A_0 = \sum_i x_i^2 A_{0i} + \sum_i \sum_{\substack{j \neq i \\ j > i}} M_{ij} x_i x_j (A_{0i} A_{0j})^{1/2} \qquad (8\text{-}12.9)$$

According to the simple rule, first suggested by Benedict,

$$A_0 = \sum_i (x_i A_{0i}^{1/2})^2 \qquad (8\text{-}12.10)$$

Eq. (8-12.9) is identical to Eq. (8-12.10) if $M_{ij} = 2$ for all ij pairs $(i \neq j)$. Orye relaxes this restrictive assumption by allowing M_{ij} to be an adjustable binary constant depending on temperature but not on density or composition.

To illustrate, Fig. 8-13 shows K factors $(K = y/x)$ for methane in the methane–n-heptane binary system. Lines represent calculations made by Orye using his form of the Benedict-Webb-Rubin equation. Serious error would have resulted if M_{ij} had been set equal to 2.

Another illustration is given in Fig. 8-14 for the system hydrogen

---

†A variety of computer programs for calculating high-pressure vapor-liquid equilibria are available from Gas Processors Association, 1812 First Place, Tulsa, Okla. 74103.

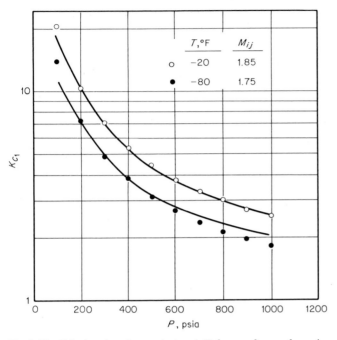

Fig. 8-13 Calculated and experimental K factors for methane in the system methane–n-heptane.   Experimental points from Ref. 46.   (*From Ref. 62.*)

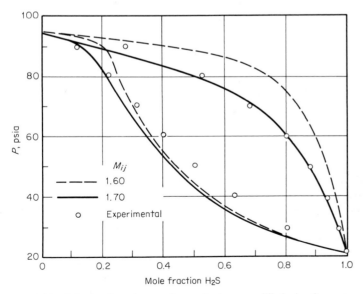

Fig. 8-14 Calculated and experimental phase equilibria in the system carbon dioxide–hydrogen sulfide at $-60°F$.   (*From Ref. 62.*)

sulfide–carbon dioxide.   Results are again sensitive to the value chosen for M_{ij}.   Adjusting the constant M_{ij} much improves the accuracy of the calculations.

An extended Benedict-Webb-Rubin equation of state, developed by Starling and coworkers, uses 11 constants for each pure component. For mixtures, these constants depend on composition according to essentially arbitrary mixing rules; to obtain good agreement with experiment, it is essential that at least one of these mixing rules contain an adjustable binary parameter, as in Orye's method [Eq. (8-12.9)].   Starling and Han [83] present binary parameters for mixtures containing light hydrocarbons (to undecane), nitrogen, carbon dioxide, and hydrogen sulfide.   For multicomponent mixtures, no ternary (or higher) constants are used; only pure-component constants and binary parameters are required, as in Orye's method.   Figures 8-15 and 8-16 indicate the good results obtained by Starling and Han for two binary mixtures.

To obtain high accuracy in representation of fluid-phase properties,

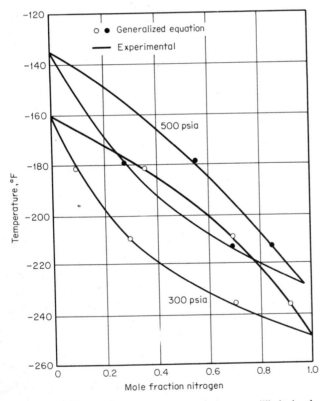

Fig. 8-15 Calculated and experimental phase equilibria in the system methane-nitrogen. (*From Ref.* 83.)

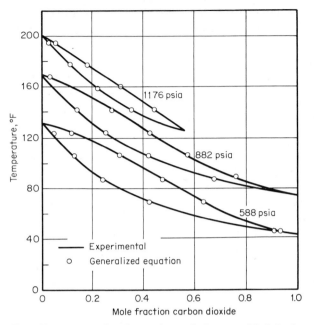

Fig. 8-16   Calculated and experimental phase equilibria in the system carbon dioxide–hydrogen sulfide.   (*From Ref. 83.*)

some workers in this field have used equations of state of increasing complexity requiring additional constants.   However, to obtain the constants, more experimental data are needed.   The multiconstant equation of state is therefore useful only for a limited number of fluids where experimental data are plentiful.   For example, Bender [8] uses a highly complicated equation with 20 constants.   He applies it to nitrogen, oxygen, and argon and then, using additional binary constants, he calculates vapor-liquid equilibria for liquefied air with excellent results.   Unfortunately, however, Bender's method is limited to those very few systems where experimental data are extensive.

Calculating vapor-liquid equilibria from an equation of state is attractive primarily because it avoids the troublesome problem of specifying liquid-phase standard states for supercritical components.   However, the accuracy of such calculations necessarily depends on the equation of state used, and, as yet, no truly satisfactory equation of state has been established.   In practice, empirical equations of state are used, and the danger with such equations is that calculated results are reliable only when they are restricted to that range of temperature, pressure, and composition for which empirical constants were obtained.   For calculations outside that range, it is necessary to extrapolate the empirical equations, and such extrapolations may lead to serious error when the equation of state has no theoretical basis.

Extrapolations are particularly dangerous when the equation of state is complex. If extrapolations must be made, it is safer to use a simple equation of state, which perhaps provides less accuracy in the region used to determine the constants but which exhibits physically reasonable behavior at extreme conditions outside that region. Another advantage of a simple equation of state is that the calculations are then much easier.

It is for these reasons that much attention has been given to modifications of the Redlich-Kwong equation of state, as indicated in Chaps. 3 and 4. One particularly effective modification is that proposed by Soave [82], which often gives good predictions of vapor-liquid equilibria in hydrocarbon mixtures. [See Eq. (3-5.15) and Table 4-1.] This good prediction follows, at least in part, from the use of pure-component vapor-pressure data to determine the constants. As a result, Soave's equation is often reliable for predicting K factors but usually gives erroneous liquid densities.

Instead of using a particular analytical equation of state, it is possible to base phase-equilibrium calculations on the known (experimental) P-V-T properties of a reference substance coupled with the theorem of corresponding states. This approach has been developed by several authors, with particular success by Mollerup [57] who has developed a useful correlation for properties of natural-gas mixtures.

8-13 Liquid-Liquid Equilibria

Many liquids are only partially miscible, and in some cases, e.g., mercury and hexane at normal temperatures, the mutual solubilities are so small that, for practical purposes, the liquids may be considered immiscible. Partial miscibility is observed not only in binary mixtures but also in ternary (and higher) systems, thereby making extraction a possible separation operation. This section introduces some useful thermodynamic relations which, in conjunction with limited experimental data, can be used to obtain quantitative estimates of phase compositions in liquid-liquid systems.

At ordinary temperatures and pressures, it is (relatively) simple to obtain experimentally the compositions of two coexisting liquid phases, and, as a result, the technical literature is rich in experimental results for a variety of binary and ternary systems near 25°C and near atmospheric pressure. However, as temperature and pressure deviate appreciably from those corresponding to normal conditions, the availability of experimental data falls rapidly.

Partial miscibility in liquids is often called *phase splitting*. The thermodynamic criteria which indicate phase splitting are well understood regardless of the number of components, but almost all thermodynamic texts confine discussion to binary systems. Stability analysis shows that for a binary system, phase splitting occurs when

$$\left(\frac{\partial^2 g^E}{\partial x_1{}^2}\right)_{T,P} + RT\left(\frac{1}{x_1} + \frac{1}{x_2}\right) < 0 \tag{8-13.1}$$

where g^E is the molar excess Gibbs energy of the binary mixture (see Sec. 8-5). To illustrate Eq. (8-13.1), consider the simplest nontrivial case. Let

$$g^E = Ax_1x_2 \tag{8-13.2}$$

where A is an empirical coefficient characteristic of the binary mixture. Substituting into Eq. (8-13.1), we find that phase splitting occurs if

$$A > 2RT \tag{8-13.3}$$

In other words, if $A < 2RT$, the two components 1 and 2 are completely miscible; there is only one liquid phase. However, if $A > 2RT$, two liquid phases form because components 1 and 2 are only partially miscible.

The condition where $A = 2RT$ is called *incipient instability*, and the temperature corresponding to that condition is called the *consolute temperature*, designated by T^c; since Eq. (8-13.2) is symmetric in mole fractions x_1 and x_2, the composition at the consolute point is $x_1{}^c = x_2{}^c = 0.5$. In a typical binary mixture, the coefficient A is a function of temperature, and therefore it is possible to have either an upper consolute temperature or a lower consolute temperature, or both, as indicated in Figs. 8-17 and 8-18. Upper consolute temperatures are more common than lower consolute temperatures. Systems with both upper and lower consolute temperatures are rare.†

Stability analysis for ternary (and higher) systems is, in principle, similar to that for binary systems although the mathematical complexity rises with the number of components. (See, for example, Refs. 7 and 73.) However, it is important to recognize that stability analysis can tell us only whether a system can or cannot *somewhere* exhibit phase splitting at a given temperature. That is, if we have an expression for g^E at a particular temperature, stability analysis can determine whether or not there is *some* range of composition where two liquids exist. It does *not* tell us what that composition range is. To find the range of compositions where two liquid phases exist at equilibrium requires a more elaborate calculation. To illustrate, consider again a simple binary

---

†While Eq. (8-13.3) is based on the simple two-suffix (one-parameter) Margules equation, similar calculations can be made using other expressions for g^E. See, for example, Ref. 79.

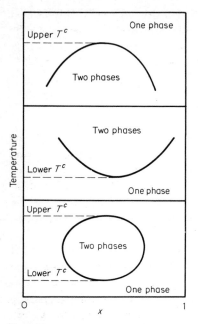

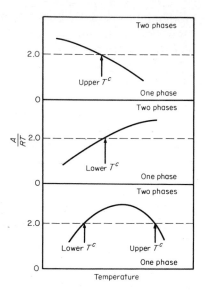

Fig. 8-17 Phase stability in three binary liquid mixtures. (*From Ref.* 66.)

Fig. 8-18 Phase stability in three binary liquid mixtures whose excess Gibbs energy is given by a two-suffix Margules equation. (*From Ref.* 66.)

mixture whose excess Gibbs energy is given by Eq. (8-13.2). If $A > 2RT$, we can calculate the compositions of the two coexisting equations by solving the two equations of phase equilibrium

$$(\gamma_1 x_1)' = (\gamma_1 x_1)'' \quad \text{and} \quad (\gamma_2 x_2)' = (\gamma_2 x_2)'' \tag{8-13.4}$$

where the prime and double prime designate, respectively, the two liquid phases.

From Eq. (8-13.2) we have

$$\ln \gamma_1 = \frac{A}{RT} x_2^{\;2} \tag{8-13.5}$$

and

$$\ln \gamma_2 = \frac{A}{RT} x_1^{\;2} \tag{8-13.6}$$

Substituting into the equation of equilibrium and noting that $x_1' + x_2' = 1$ and $x_1'' + x_2'' = 1$, we obtain

$$x_1' \exp \frac{A(1 - x_1')^2}{RT} = x_1'' \exp \frac{A(1 - x_1'')^2}{RT} \tag{8-13.7}$$

and

$$(1 - x_1') \exp \frac{Ax_1'^2}{RT} = (1 - x_1'') \exp \frac{Ax_1''^2}{RT} \tag{8-13.8}$$

Equations (8-13.7) and (8-13.8) contain two unknowns (x_1' and x_1''), which can be found by iteration.   Mathematically, several solutions of these two equations can be obtained.   However, to be physically meaningful, it is necessary that $0 < x_1' < 1$ and $0 < x_1'' < 1$.

Similar calculations can be performed for ternary (or higher) mixtures. For a ternary system the three equations of equilibrium are

$$(\gamma_1 x_1)' = (\gamma_1 x_1)'' \qquad (\gamma_2 x_2)' = (\gamma_2 x_2)'' \qquad (\gamma_3 x_3)' = (\gamma_3 x_3)'' \quad (8\text{-}13.9)$$

If we have an equation relating the excess molar Gibbs energy g^E of the mixture to the overall composition (x_1, x_2, x_3), we can obtain corresponding expressions for the activity coefficients γ_1, γ_2, and γ_3, as discussed elsewhere [see Eq. (8-9.4)].   The equations of equilibrium [Eq. (8-13.9)], coupled with the material-balance relations ($x_1' + x_2' + x_3' = 1$ and $x_1'' + x_2'' + x_3'' = 1$), can then be solved to obtain the four unknowns (x_1', x_2' and x_1'', x_2'').

Systems containing four or more components are handled in a similar manner.   An expression for g^E for the multicomponent system is used to relate the activity coefficient of each component in each phase to the composition of that phase.   From the equations of equilibrium [$(\gamma_i x_i)' = (\gamma_i x_i)''$ for every component i] the phase compositions x_i' and x_i'' are found by trial and error.

Considerable skill in numerical analysis is required to construct a computer program that finds the equilibrium compositions of a multicomponent liquid-liquid system from an expression for the excess Gibbs energy for that system.   It is difficult to construct a program which always converges to a physically meaningful solution using only a small number of iterations.   This difficulty is especially pronounced in the region near the plait point, where the compositions of the two equilibrium phases become identical.

King [44] has given some useful suggestions for constructing efficient programs toward computation of equilibrium compositions in two-phase systems.

While the thermodynamics of multicomponent liquid-liquid equilibria is, in principle, straightforward, it is difficult to obtain an expression for g^E which is sufficiently accurate to yield reliable results.   Liquid-liquid equilibria are much more sensitive to small changes in activity coefficients than vapor-liquid equilibria.   In the latter, activity coefficients play a role which is secondary to the all-important pure-component vapor pressures.   In liquid-liquid equilibria, however, the activity coefficients are dominant; pure-component vapor pressures play no role at all.   Therefore it has often been observed that good estimates of vapor-liquid equilibria can be made for many systems using only approximate activity coefficients, provided the pure-component vapor pressures are accurately known.   However, in calculating liquid-liquid equilibria, small inaccuracies in activity coefficients can lead to serious errors.

Renon et al. [68] have discussed application of the NRTL equation to liquid-liquid equilibria, and Abrams and Prausnitz have discussed application of the UNIQUAC equation [3]. Regardless of which equation is used, much care must be exercised in determining parameters from experimental data. Whenever possible, such parameters should come from binary mutual-solubility data.

When parameters are obtained from reduction of vapor-liquid equilibrium data, there is always some ambiguity. Unless the experimental data are of very high accuracy, it is usually not possible to obtain a truly unique set of parameters; i.e., in a typical case, there is a range of parameter sets such that any set in that range can equally well reproduce the experimental data within the probable experimental error. (See, for example, Refs. 3 and 25.) When multicomponent vapor-liquid equilibria are calculated, results are not sensitive to which sets of binary parameters are chosen. However, when multicomponent liquid-liquid equilibria are calculated, results are extremely sensitive to the choice of binary parameters. Therefore, it is difficult to establish reliable ternary (or higher) liquid-liquid equilibria using only binary parameters obtained from binary vapor-liquid equilibrium data alone. For reliable results it is usually necessary to utilize at least some multicomponent liquid-liquid equilibrium data.

To illustrate these ideas, we quote some calculations reported by Bender and Block [9], who considered two ternary systems at 25°C:

System I: Water (1), toluene (2), aniline (3)
System II: Water (1), TCE† (2), acetone (3)

To describe these systems, the NRTL equation was used to relate activity coefficients to composition. The essential problem lies in finding the parameters for the NRTL equation. In system I, components 2 and 3 are completely miscible, while components 1 and 2 and components 1 and 3 are only partially miscible. In system II, components 1 and 3 and components 2 and 3 are completely miscible while components 1 and 2 are only partially miscible.

For the completely miscible binaries, Bender and Block set NRTL parameter $\alpha_{ij} = 0.3$. Parameters τ_{ij} and τ_{ji} were then obtained from vapor-liquid equilibria. Since it is not possible to obtain unique values of these parameters from vapor-liquid equilibria, Bender and Block used a criterion suggested by Abrams and Prausnitz [3], namely, to choose those sets of parameters for the completely miscible binary pairs which correctly give the limiting liquid-liquid distribution coefficient for

†1,1,2-Trichloroethane.

the third component at infinite dilution. In other words, NRTL parameters τ_{ij} and τ_{ji} chosen were those which not only represent the ij binary vapor-liquid equilibria within experimental accuracy but which also give the experimental value of K_k^∞ defined by

$$K_k^\infty = \lim_{\substack{w_k \to 0 \\ w'_k \to 0}} \frac{w''_k}{w'_k}$$

where w stands for weight fraction, component k is the third component, i.e., the component *not* in the completely miscible ij binary, and the prime and double prime designate the two equilibrium liquid phases.

For the partially miscible binary pairs, estimates of τ_{ij} and τ_{ji} are obtained from mutual-solubility data following an arbitrary choice for α_{ij} in the region $0.20 \leqslant \alpha_{ij} \leqslant 0.40$. When mutual-solubility data are used, the parameter set τ_{ij} and τ_{ji} depends only on α_{ij}; to find the best α_{ij}, Bender and Block used ternary tie-line data. In other words, since the binary parameters are not unique, the binary parameters chosen were those which gave an optimum representation of the ternary liquid-liquid equilibrium data.

Table 8-25 gives mutual-solubility data for the three partially miscible binary systems. Table 8-26 gives NRTL parameters following the procedure outlined above. With these parameters, Bender and Block obtained good representation of the ternary phase diagrams, essentially within experimental error. Figures 8-19 and 8-20 compare calculated with observed distribution coefficients for systems I and II.

When the NRTL equation is used to represent ternary liquid-liquid equilibria, there are nine adjustable binary parameters; when the UNIQUAC equation is used, there are six. It is tempting to use the ternary liquid-liquid data alone for obtaining the necessary parameters, but this is a dangerous procedure since it is not possible thereby to obtain a set of *meaningful* parameters; in this context "meaningful" indicates those parameters which also reproduce equilibrium data for the binary pairs. As shown by Heidemann and others [37], unusual and bizarre

TABLE 8-25 Mutual Solubilities in Binary Systems at 25°C [9]

Component		Weight fraction	
i	j	i in j	j in i
Water	TCE	0.0011	0.00435
Water	Toluene	0.0005	0.000515
Water	Aniline	0.053	0.0368

TABLE 8-26 NRTL Parameters Used by Bender and Block to Calculate Ternary Liquid-Liquid Equilibria at 25°C

		System I: water (1), toluene (2), aniline (3)		
i	j	τ_{ij}	τ_{ji}	α_{ij}
1	2	7.77063	4.93035	0.2485
1	3	4.18462	1.27932	0.3412
2	3	1.59806	0.03509	0.3
		System II: water (1), TCE (2), acetone (3)		
1	2	5.98775	3.60977	0.2485
1	3	1.38800	0.75701	0.3
2	3	−0.19920	−0.20102	0.3

results can be calculated if the parameter sets are not chosen with care. While experience in this field is not yet plentiful, all indications show that it is always best to use binary data for calculating binary parameters. Since it often happens that binary-parameter sets cannot be determined uniquely, ternary (or higher) data should then be used to

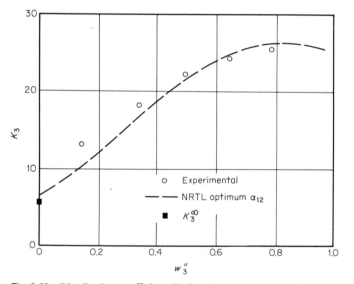

Fig. 8-19 Distribution coefficient K_3 for the system water (1)–toluene (2)–aniline (3) at 25°C. Concentrations are in weight fractions.

$$K_3 = \frac{w_3''}{w_3'} = \frac{\text{weight fraction aniline in toluene-rich phase}}{\text{weight fraction aniline in water-rich phase}}$$

$$K_3^\infty = \frac{\gamma_3'^\infty}{\gamma_3''^\infty} = \frac{\text{activity coefficient of aniline in water-rich phase at infinite dilution}}{\text{activity coefficient of aniline in toluene-rich phase at infinite dilution}}$$

(Activity coefficient γ is here defined using weight fractions.)
(*From Ref. 9.*)

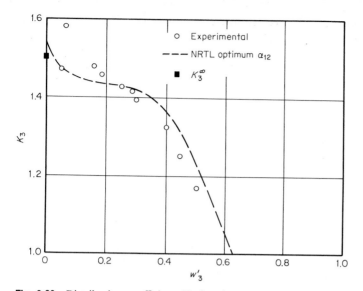

Fig. 8-20 Distribution coefficient K_3 for the system water (1)–TCE (2)–acetone (3) at 25°C. Concentrations are in weight fractions.

$$K_3 = \frac{w''_3}{w'_3} = \frac{\text{weight fraction acetone in TCE-rich phase}}{\text{weight fraction acetone in water-rich phase}}$$

$$K_3^\infty = \frac{\gamma'^\infty_3}{\gamma''^\infty_3} = \frac{\text{activity coefficient of acetone in water-rich phase at infinite dilution}}{\text{activity coefficient of acetone in TCE-rich phase at infinite dilution}}$$

(Activity coefficient γ is here defined using weight fractions.)
(*From Ref. 9.*)

fix the best binary sets from the ranges obtained from the binary data. (For a typical range of binary-parameter sets, see Fig. 8-4.) It is, of course, always possible to add ternary (or higher) terms to the expression for the excess Gibbs energy, thereby introducing ternary (or higher) constants. This is sometimes justified, but it is meaningful only if the multicomponent data are plentiful and of high accuracy.

In calculating multicomponent equilibria, the general rule is to use binary data first. Then use multicomponent data for fine tuning.

8-14 Phase Equilibria in Polymer Solutions

Strong negative deviations from Raoult's law are observed in binary liquid mixtures where one component contains very large molecules (polymers) and the other contains molecules of normal size. For mixtures of normal solvents and amorphous polymers, phase-equilibrium relations are usually described by the Flory-Huggins theory, discussed fully in a book by Flory [26] and in a monograph by Tompa

[88]; a brief introduction is given by Prausnitz [66]. For engineering application, a useful summary is provided by Sheehan and Bisio [80].

There are several versions of the Flory-Huggins theory, and, unfortunately, different authors use different notation. The primary composition variable for the liquid phase is the volume fraction, here designated by Φ and defined by Eqs. (8-10.3) and (8-10.4). In polymer-solvent systems, volume fractions are very different from mole fractions because the molar volume of a polymer is much larger than that of the solvent.

Since the molecular weight of the polymer is often not known accurately, it is difficult to determine the mole fraction. Therefore an equivalent definition of Φ is frequently useful:

$$\Phi_1 = \frac{w_1/\rho_1}{w_1/\rho_1 + w_2/\rho_2} \quad \text{and} \quad \Phi_2 = \frac{w_2/\rho_2}{w_1/\rho_1 + w_2/\rho_2} \quad (8\text{-}14.1)$$

where w_i is the weight fraction of component i and ρ_i is the mass density (*not* molar density) of pure component i.

Let subscript 1 stand for solvent and subscript 2 for polymer. The activity a_1 of the solvent, as given by the Flory-Huggins equation, is

$$\ln a_1 = \ln \Phi_1 + \left(1 - \frac{1}{m}\right)\Phi_2 + \chi \Phi_2{}^2 \quad (8\text{-}14.2)$$

where m is defined by $m = V_2{}^L/V_1{}^L$ and the adjustable constant χ is called the *Flory interaction parameter*. In typical polymer solutions $1/m$ is negligibly small compared with unity, and therefore it may be neglected. While the parameter χ depends on temperature, for polymer-solvent systems where the molecular weight of the polymer is very large, it is nearly independent of polymer molecular weight. In theory, χ is also independent of polymer concentration, but in fact it often varies with concentration, especially in mixtures containing polar molecules, where the Flory-Huggins theory provides only a rough approximation.

In a binary mixture of polymer and solvent at ordinary pressures, only the solvent is volatile; the vapor-phase mole fraction of the solvent is unity, and therefore the total pressure is equal to the partial pressure of the solvent.

In a polymer solution, the activity of the solvent is given by

$$a_1 = \frac{P}{P_{vp1}} \frac{1}{\mathscr{F}_1} \quad (8\text{-}14.3)$$

where factor $\mathscr{F}_1$ is defined by Eq. (8-4.2). At low or moderate pressures, $\mathscr{F}_1$ is equal to unity.

Equation (8-14.2) holds only for those temperatures where the polymer in the pure state is amorphous. If the pure polymer has appreciable crystallinity, corrections to Eq. (8-14.2) are significant, as discussed elsewhere [26].

Equation (8-14.2) is useful for calculating the volatility of a solvent in a polymer solution, provided that the Flory parameter χ is known. Sheehan and Bisio [80] report Flory parameters for a large number of binary systems† and present methods for estimating χ from solubility parameters. Similar data are also given in the "Polymer Handbook" [4]. Table 8-27 shows some χ values reported by Sheehan and Bisio.

A particularly convenient and rapid experimental method for obtaining χ is provided by gas-liquid chromatography [34]. While this experimental technique can be used at finite concentrations of solvent, it is most efficiently used for solutions infinitely dilute with respect to solvent, i.e., at the limit where the volume fraction of polymer approaches unity. Some solvent volatility data obtained from chromatography [59] are shown in Fig. 8-21. From these data, χ can be found by rewriting Eq. (8-14.2) in terms of a weight-fraction activity coefficient Ω

$$\Omega_1 \equiv \frac{a_1}{w_1} = \frac{P}{P_{\mathrm{vp_1}} w_1} \frac{1}{\mathscr{F}_1} \tag{8-14.4}$$

Combining with Eq. (8-14.2), in the limit as $\Phi_2 \to 1$, we obtain

$$\chi = \ln \left(\frac{P}{w}\right)_1^\infty - \left(\ln P_{\mathrm{vp_1}} + \ln \frac{\rho_2}{\rho_1} + 1\right) \tag{8-14.5}$$

where ρ is the mass density (*not* molar density). Equation (8-14.5) also

†Unfortunately, Sheehan and Bisio use completely different notation: v for Φ, x for m, and μ for χ.

TABLE 8-27 Flory χ Parameters for Some Polymer-Solvent Systems near Room Temperature [80]

Polymer	Solvent	χ
Natural rubber	Heptane	0.44
	Toluene	0.39
	Ethyl acetate	0.75
Polydimethyl siloxane	Cyclohexane	0.44
	Nitrobenzene	2.2
Polyisobutylene	Hexadecane	0.47
	Cyclohexane	0.39
	Toluene	0.49
Polystyrene	Benzene	0.22
	Cyclohexane	0.52
Polyvinyl acetate	Acetone	0.37
	Dioxane	0.41
	Propanol	1.2

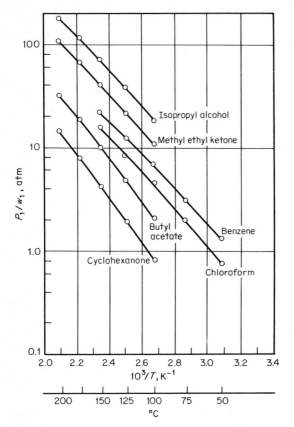

Fig. 8-21 Volatilities of solvents in Lucite 2044 for a small weight fraction of solvent. (*From Ref. 59.*)

assumes that $\mathscr{F}_1 = 1$ and that $1/m \ll 1$. Superscript ∞ denotes that weight fraction w_1 is very small compared with unity. Equation (8-14.5) provides a useful method for finding χ because $(P/w)_1^\infty$ is easily measured by gas-liquid chromatography.

Equation (8-14.2) was derived for a binary system, i.e., one where all polymer molecules have the same molecular weight (monodisperse system). For mixtures containing one solvent and one polymer with a variety of molecular weights (polydisperse system), Eq. (8-14.2) can be used provided m and Φ refer to the polymer whose molecular weight is the number-average molecular weight.

The theory of Flory and Huggins can be extended to multicomponent mixtures containing any number of polymers and any number of solvents. No ternary (or higher) constants are required.

Solubility relations (liquid-liquid equilibria) can also be calculated with the Flory-Huggins theory. Limited solubility is often observed in solvent-polymer systems, and it is common in polymer-polymer systems (incompatibility).   The Flory-Huggins theory indicates that for a solvent-polymer system, limited miscibility occurs when

$$\chi > \frac{1}{2}\left(1 + \frac{1}{m^{1/2}}\right)^2 \qquad (8\text{-}14.6)$$

For large m, the value of χ may not exceed $\frac{1}{2}$ for miscibility in all proportions.

Liquid-liquid phase equilibria in polymer-containing systems are described in numerous articles published in journals devoted to polymer science and engineering.   The thermodynamics of such equilibria is discussed in Flory's book and in articles by Scott and Tompa [78] and, more recently, by Hsu and Prausnitz [42].   A comprehensive review of polymer compatibility and incompatibility is given by Krause [47].

For semiquantitative calculations the three-dimensional solubility-parameter concept [35] is often useful, especially for formulations of paints, coating, inks, etc.

The Flory-Huggins equation contains only one adjustable binary parameter.   For simple nonpolar systems one parameter is often sufficient, but for complex systems, much better representation is obtained by empirical extension of the Flory-Huggins theory using at least two adjustable parameters, as shown by Maron and Nakajima [52] and by Heil and Prausnitz [38].   The latters' extension is a generalization of Wilson's equation.   The UNIQUAC equation with two adjustable parameters is also applicable to polymer solutions [3].

The theory of Flory and Huggins is based on a lattice model which ignores free-volume differences; in general, polymer molecules in the pure state pack more densely than molecules of normal liquids. Therefore, when polymer molecules are mixed with molecules of normal size, the polymer molecules gain freedom to exercise their rotational and vibrational motions; at the same time, the smaller solvent molecules partially lose such freedom.   To account for these effects, an *equation-of-state theory* of polymer solutions has been developed by Flory [27] and Patterson [64] based on ideas suggested by Prigogine [72].   The newer theory is necessarily more complicated, but, unlike the older one, it can at least semiquantitatively describe some forms of phase behavior commonly observed in polymer solutions.   In particular, it can explain the observation that some polymer-solvent systems exhibit lower consolute temperatures as well as upper consolute temperatures, as shown in Fig. 8-17.   Engineering applications of the new theory have been developed only recently.   Introductions are given by Bonner [12], Bondi [10], and Tapavicza and Prausnitz [85].   Application to phase equilibria in the system polyethylene-ethylene at high pressures is discussed by Bonner et al. [12].

8-15 Solubilities of Solids in Liquids

The solubility of a solid in a liquid is determined not only by the intermolecular forces between solute and solvent but also by the melting point and the enthalpy of fusion of the solute. For example, at 25°C, the solid aromatic hydrocarbon phenanthrene is highly soluble in benzene; its solubility is 20.7 mole percent. By contrast, the solid aromatic hydrocarbon anthracene, an isomer of phenanthrene, is only slightly soluble in benzene at 25°C; its solubility is 0.81 mole percent. For both solutes, intermolecular forces between solute and benzene are essentially identical. However, the melting points of the solutes are significantly different: phenanthrene melts at 100°C and anthracene at 217°C. In general, it can be shown that when other factors are held constant, the solute with the higher melting point has the lower solubility. Also, when other factors are held constant, the solute with the higher enthalpy of fusion has the lower solubility.

These qualitative conclusions follow from a quantitative thermodynamic analysis given in several texts. (See, for example, Refs. 39 and 66.)

In a binary system, let subscript 1 stand for solvent and subscript 2 for solute. Assume that the solid phase is pure. At temperature T, the solubility (mole fraction) x_2 is given by

$$\ln \gamma_2 x_2 = -\frac{\Delta h_f}{RT}\left(1 - \frac{T}{T_t}\right) + \frac{\Delta C_p}{R}\left(\frac{T_t - T}{T}\right) - \frac{\Delta C_p}{R}\ln\frac{T_t}{T} \qquad (8\text{-}15.1)$$

where Δh_f is the enthalpy of fusion of the solute at the triple-point temperature T_t and ΔC_p is given by the molar heat capacity of the pure solute:

$$\Delta C_p = C_p \text{ (subcooled liquid solute)} - C_p \text{ (solid solute)} \qquad (8\text{-}15.2)$$

The standard state for activity coefficient γ_2 is pure (subcooled) liquid 2 at system temperature T.

To a good approximation, we can substitute normal melting temperature T_m for triple-point temperature T_t, and we can assume that Δh_f is essentially the same at these two temperatures. In Eq. (8-15.1) the first term on the right-hand side is much more important than the remaining two terms, and therefore a simplified form of that equation is

$$\ln \gamma_2 x_2 = -\frac{\Delta h_f}{RT}\left(1 - \frac{T}{T_m}\right) \qquad (8\text{-}15.3)$$

If we substitute

$$\Delta s_f = \frac{\Delta h_f}{T_m} \qquad (8\text{-}15.4)$$

we obtain an alternate simplified form

$$\ln \gamma_2 x_2 = -\frac{\Delta s_f}{R}\left(\frac{T_m}{T} - 1\right) \qquad (8\text{-}15.5)$$

where Δs_f is the entropy of fusion. A plot of Eq. (8-15.5) is shown in Fig. 8-22.

If we let $\gamma_2 = 1$, we can readily calculate the ideal solubility at temperature T, knowing only the solute's melting temperature and its enthalpy (or entropy) of fusion. This ideal solubility depends only on properties of the solute; it is independent of the solvent's properties. The effect of intermolecular forces between molten solute and solvent are reflected in the activity coefficient γ_2.

To describe γ_2, we can utilize any of the expressions for the excess Gibbs energy, as discussed in Sec. 8-5. However, since γ_2 depends on the mole fraction x_2, solution of Eq. (8-15.5) requires iteration. For

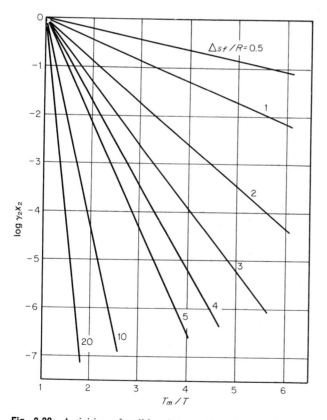

Fig. 8-22 Activities of solid solutes referred to their pure subcooled liquids. (*From Ref. 71.*)

example, suppose that γ_2 is given by a simple one-parameter Margules equation

$$\ln \gamma_2 = \frac{A}{RT}(1 - x_2)^2 \qquad (8\text{-}15.6)$$

where A is an empirical constant. Substitution into Eq. (8-15.5) gives

$$\ln x_2 + \frac{A}{RT}(1 - x_2)^2 = -\frac{\Delta s_f}{R}\left(\frac{T_m}{T} - 1\right) \qquad (8\text{-}15.7)$$

and x_2 must be found by a trial-and-error calculation.

In nonpolar systems, the activity coefficient γ_2 can often be estimated using the Scatchard-Hildebrand equation, as discussed in Sec. 8-10. In that event, since $\gamma_2 \geqslant 1$, the ideal solubility ($\gamma_2 = 1$) is larger than that obtained from regular-solution theory. As shown by Preston and

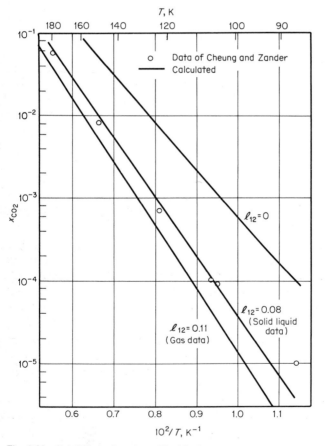

Fig. 8-23 Solubility of carbon dioxide in propane. (*From Ref.* 71.)

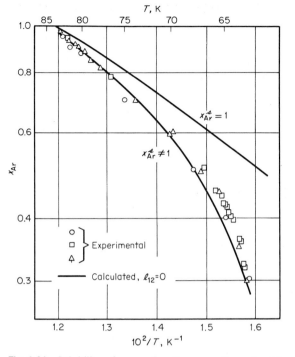

Fig. 8-24 Solubility of argon in nitrogen: effect of solid-phase composition. (*From Ref. 71.*)

Prausnitz [71], and as illustrated in Fig. 8-23, regular-solution theory is useful for calculating solubilities in nonpolar systems, especially when the geometric-mean assumption is relaxed through introduction of an empirical correction ℓ_{12} (see Sec. 8-10).

Figure 8-23 shows three lines: the top line is calculated using the geometric-mean assumption ($\ell_{12} = 0$) in the Scatchard-Hildebrand equation. The bottom line is calculated with $\ell_{12} = 0.11$, a value estimated from gas-phase P-V-T-y data. The middle line is calculated with $\ell_{12} = 0.08$, which is the optimum value obtained from solubility data. Figure 8-23 suggests that even an approximate estimate of ℓ_{12} usually produces better results than assuming that ℓ_{12} is zero. Unfortunately *some* mixture datum is needed to estimate ℓ_{12}. In a few fortunate cases one freezing-point datum, e.g., the eutectic point, may be available to fix ℓ_{12}.

It is important to remember that the calculations outlined above rest on the assumption that the solid phase is pure, i.e., that there is no solubility of the solvent in the solid phase. This assumption is usually a good one, especially if the two components differ appreciably in molecular size and shape. However, many cases are known where the two components are at least partially miscible in the solid phase, and in that event it is necessary to correct for solubility and nonideality in the solid phase as well as in the liquid phase. This complicates the thermodynamic treatment, but, more important, solubility in the solid phase may significantly affect the phase diagram. Figure 8-24 shows results for the solubility of solid argon in liquid nitrogen. The top line presents calculated results assuming that x^δ (argon) = 1, where superscript δ denotes the solid phase. The bottom line takes into account the experimentally known solubility of nitrogen in solid argon $[x^\delta$ (argon) ≠ 1]. In this case it is clear that serious error is introduced by neglecting solubility of the solvent in the solid phase.

8-16 Concluding Remarks

This chapter on phase equilibria has presented no more than a brief introduction to a very broad subject. The variety of mixtures encountered in the chemical industry is extremely large, and, except for general thermodynamic equations, there are no quantitative relations which apply rigorously to all, or even to a large fraction, of these mixtures. Thermodynamics provides only a coarse but reliable framework; the details must be supplied by physics and chemistry, which ultimately rest on experimental data.

For each mixture it is necessary to construct an appropriate mathematical model for representing the properties of that mixture. Whenever possible, such a model should be based on physical concepts, but since our fundamental understanding of fluids is severely limited, any useful model is inevitably influenced by empiricism. While at least some empiricism cannot be avoided, the strategy of the process engineer must be to use enlightened rather than blind empiricism. This means foremost that critical and informed judgment must always be exercised. While such judgment is attained only by experience, we conclude this chapter with a few guidelines.

1. Face the facts: you cannot get something from nothing. Do not expect magic from thermodynamics. If you want reliable results, you will need some reliable experimental data. You may not need many, but you do need some. The required data need not necessarily be for the particular system of interest; sometimes they may come from experimental studies on closely related systems, perhaps represented by a suitable correlation. Only in very simple cases can partial thermodynamic properties in a mixture, e.g., activity coefficients, be found from pure-component data alone.

2. Correlations provide the easy route, but they should be used last, not first. The preferred first step should always be to obtain *reliable* experimental data, either from the literature or from the labora-

tory. Do not at once reject the possibility of obtaining a few crucial data yourself. Laboratory work is more tedious than pushing a computer button, but ultimately, at least in some cases, you may save time by making a few simple measurements instead of a multitude of furious calculations. A small laboratory with a few analytical instruments (especially a chromatograph or a simple boiling-point apparatus) can often save both time and money. If you cannot do the experiment yourself, consider the possibility of having someone else do it for you.

3. It is always better to obtain a few well-chosen and reliable experimental data than to obtain many data of doubtful quality and relevance. Beware of statistics, which may be the last refuge of a poor experimentalist.

4. Always regard published experimental data with some skepticism. Many experimental results are of high quality, but many are not. Just because a number is reported by someone and printed by another, do not automatically assume that it must therefore be correct.

5. When choosing a mathematical model for representing mixture properties, give preference if possible to those which have some physical basis.

6. Seek simplicity; beware of models with many adjustable parameters. When such models are extrapolated even mildly into regions other than those for which the constants were determined, highly erroneous results may be obtained.

7. In reducing experimental data, keep in mind the probable experimental uncertainty of the data. Whenever possible, give more weight to those data which you have reason to believe are most reliable.

8. If you do use a correlation, be sure to note its limitations. Extrapolation outside its domain of validity can lead to large error.

9. Never be impressed by calculated results merely because they come from a computer. The virtue of a computer is speed, not intelligence.

10. Maintain perspective. Always ask yourself: Is this result reasonable? Do other similar systems behave this way? If you are inexperienced, get help from someone who has experience. Phase equilibria in fluid mixtures is not a simple subject. Do not hesitate to ask for advice.

NOTATION

a, b, c = empirical coefficients
a_i = activity of component i
a_{mn} = group interaction parameter, Eq. (8-10.47)
A = empirical constant
B = empirical constant
B_{ij} = second virial coefficient for the ij interaction

c, d = empirical constants in Eq. (8-7.1)

c_{ij} = cohesive energy density for the ij interaction in Sec. 8-10

c_{12} = empirical constant in Eq. (8-5.14)

C = empirical constant

C_p = molar specific heat at constant pressure

D = empirical constant; number of data points, Eq. (8-8.16)

f_i = fugacity of component i

f = a function

$\mathscr{F}_i$ = nonideality factor defined by Eq. (8-4.2)

g_{ij} = empirical constant (Table 8-3)

g^E = molar excess Gibbs energy

G^E = total excess Gibbs energy

G_{ij} = empirical constant (Table 8-3)

h^E = molar excess enthalpy

Δh_f = molar enthalpy of fusion

$\bar{h}_i$ = partial molar enthalpy of component i

H = Henry's constant

ΔH_v = enthalpy of vaporization

I = defined by Eq. (8-8.17)

$K = y/x$ in Sec. 8-12; distribution coefficient in Sec. 8-13

ℓ_i = constant defined in Table 8-3

ℓ_{12} = empirical constant in Sec. 8-10

L = number of adjustable parameters, Eq. (8-8.16)

m = defined after Eq. (8-14.2)

M_{ij} = empirical coefficient

n_i = number of moles of component i

n_T = total number of moles

N = number of components; parameter in Tables 8-17 and 8-18

p = partial pressure

P = total pressure

P_{vp} = vapor pressure

q = molecular surface parameter, an empirical constant (Table 8-3)

Q_k = group surface parameter, Eq. (8-10.43)

r = molecular-size parameter, an empirical constant (Table 8-3); number of rings (Table 8-18)

R = gas constant

R_k = group size parameter, Eq. (8-10.43)

$\mathscr{R}$ = defined by Eq. (8-10.35)

s^E = molar excess entropy

Δs_f = molar entropy of fusion

S_j = number of size groups in molecule j (Sec. 8-10)

t = temperature

T = absolute temperature

T_m = melting-point temperature

T_t = triple-point temperature

u_{ij} = empirical constant (Table 8-3)

ΔU = change in internal energy

V = molar volume

V_T = total volume

w_k = weight fraction of component k

x_i = liquid-phase mole fraction of component i

X_k = group mole fraction for group k

y_i = vapor-phase mole fraction of component i

z = coordination number (Table 8-3)

Greek

α = parameter in Tables 8-17 and 8-18

α_{ij} = empirical constant

β = proportionality factor in Eq. (8-5.12)

γ_i = activity coefficient of component i

Γ_k = activity coefficient of group k (Sec. 8-10)

δ = solubility parameter defined by Eq. (8-10.10)

$\bar{\delta}$ = average solubility parameter defined by Eq. (8-10.15)

ϵ = parameter in Tables 8-17 and 8-18

ζ = parameter in Tables 8-17 and 8-18

η = empirical constant (Table 8-14) and Eq. (8-9.23); empirical constant in Tables 8-17 and 8-18

θ = parameter in Tables 8-17 and 8-18

θ_i = surface fraction of component i (Table 8-3)

λ = nonpolar solubility parameter in Sec. 8-10

λ_{ij} = empirical constant in Eq. (8-5.11) and Table 8-14

Λ_{ij} = empirical constant (Table 8-3)

$\nu_k^{(i)}$ = number of groups of type k in molecule i

ν_{kj} = number of interaction groups k in molecule j (Sec. 8-10)

ρ = density

σ^2 = variance, Eqs. (8-8.16) and (8-8.17)

τ = polar solubility parameter in Sec. 8-10

τ_{ij} = empirical constant (Table 8-3)

ϕ_i = fugacity coefficient of component i

Φ_i = site fraction (or volume fraction) of component i

χ = Flory interaction parameter

Ψ_{mn} = group interaction parameter, Eq. (8-10.47)

ψ_{12} = binary (induction) parameter in Eq. (8-10.19)

Superscripts

c = consolute (Sec. 8-13); calculated quantity, Eq. (8-8.15)

C = configurational

E = excess

G = group (Sec. 8-10)

L = liquid phase

M = measured value, Eq. (8-8.17)

$°$ = standard state as in $f_i^°$

0 = estimated true value, Eq. (8-8.17)

R = residual

s = solid phase

s = saturation

S = size (Sec. 8-10)

V = vapor phase

∞ = infinite dilution

References

1. Abbott, M. M., J. K. Floess, G. E. Walsh, and H. C. Van Ness: *AIChE J.*, **21**: 72 (1975).
2. Abbott, M. M., and H. C. Van Ness: *AIChE J.*, **21**: 62 (1975).

3. Abrams, D. S., and J. M. Prausnitz: *AIChE J.*, **21**: 116 (1975); *Joint Meet. VDI AIChE, Munich, September 1974.*

4. Bandrup, J., and E. H. Immergut: "Polymer Handbook," 2d ed., Wiley, New York, 1975.

5. Battino, R., and E. Wilhelm: *J. Chem. Thermodyn.*, **3**: 379 (1971); L. R. Field, E. Wilhelm, and R. Battino, *J. Chem. Thermodyn.*, **6**: 237 (1974).

6. Bazúa, E. R., and J. M. Prausnitz: *Cryogenics*, **11**: 114 (1971).

7. Beegle, B. L., M. Modell, and R. C. Reid: *AIChE J.*, **20**: 1200 (1974).

8. Bender, E.: *Cryogenics*, **13**: 11 (1973).

9. Bender, E., and U. Block: *Verfahrenstechnik*, **9**: 106 (1975).

10. Bondi, A.: "Physical Properties of Molecular Liquids, Crystals and Glasses," Wiley, New York, 1968.

11. Bonner, D. C., D. P. Maloney, and J. M. Prausnitz: *Ind. Eng. Chem. Process Des. Dev.*, **13**: 198 (1974); D. P. Maloney and J. M. Prausnitz: *Ind. Eng. Chem. Process Des. Dev.*, **15**: 216 (1976) and *AIChE J.*, **22**: 74 (1976).

12. Bonner, D. C., and J. M. Prausnitz: *AIChE J.*, **19**: 943 (1973); Errata, **20**: 206 (1974).

13. Brelvi, S. W., and J. P. O'Connell: *AIChE J.*, **18**: 1239 (1972), **21**: 157 (1975).

14. Brian, P. L. T.: *Ind. Eng. Chem. Fundam.*, **4**: 101 (1965).

15. Britt, H. I., and R. H. Luecke: *Technometrics*, **15**: 233 (1973).

16. Bromley, L. A.: *AIChE J.*, **19**: 313 (1973); *J. Chem. Thermodyn.*, **4**: 669 (1972).

17. Bruin, S., and J. M. Prausnitz: *Ind. Eng. Chem. Process Des. Dev.*, **10**: 562 (1971); S. Bruin, *Ind. Eng. Chem. Fundam.*, **9**: 305 (1970).

18. Burrell, H.: *J. Paint Technol.*, **40**: 197 (1968); J. L. Gardon: *J. Paint Technol.*, **38**: 43 (1966); R. C. Nelson, R. W. Hemwall, and G. D. Edwards: *J. Paint Technol.*, **42**: 636 (1970); "Encyclopedia of Chemical Technology" (Kirk-Othmer), 2d ed., vol. 18, pp. 564–588.

19. Chao, K. C., and G. D. Seader: *AIChE J.*, **7**: 598 (1961).

20. Cheung, H., and E. H. Zander: *Chem. Eng. Prog. Symp. Ser.*, **64**(88): 34 (1968).

21. Clarke, H. A., and R. W. Missen, *J. Chem. Eng. Data*, **19**: 343 (1974).

22. Derr, E. L., and C. H. Deal: *Inst. Chem. Eng. Symp. Ser. Lond.*, **3**(32): 40 (1969).

23. Eckert, C. A., B. A. Newman, G. L. Nicolaides, and T. C. Long: paper presented at *AIChE Meet., Los Angeles, November 1975.*

24. Edwards, T. J., J. Newman, and J. M. Prausnitz: *AIChE J.*, **21**: 248 (1975).

25. Fabries, J. F., and H. Renon: *AIChE J.*, **21**: 735 (1975).

26. Flory, P. J.: "Principles of Polymer Chemistry," Cornell University Press, Ithaca, N.Y., 1953.

27. Flory, P. J.: *Discuss. Faraday Soc.*, **49**: 7 (1970).

28. Fredenslund, A., R. L. Jones, and J. M. Prausnitz: *AIChE J.*, **21**: 1086 (1975).

29. Friend, L., and S. B. Adler: *Chem. Eng. Prog.*, **53**: 452 (1957).

30. Funk, E. W., and J. M. Prausnitz: *Ind. Eng. Chem.*, **62**(9): 8 (1970).

31. Furnas, C. C., and W. B. Leighton: *Ind. Eng. Chem.*, **29**: 709 (1937).

32. Grayson, H. G., and C. W. Streed: *6th World Pet. Congr.*, pap. 20, sec. VII (1963); R. L. Robinson and K. C. Chao: *Ind. Eng. Chem. Process Des. Dev.*, **10**: 221 (1971); B. I. Lee, J. H. Erbar, and W. C. Edmister: *AIChE J.*, **19**: 349 (1973).

33. Griswold, J., P. L. Chu, and W. O. Winsauer: *Ind. Eng. Chem.*, **41**: 2352 (1949).

34. Guillet, J. E.: *Adv. Anal. Chem. Instrum.*, **11**: 187 (1973).

35. Hansen, C. M.: *J. Paint Technol.*, **39**: 104, 505 (1967); C. M. Hansen and K. Skaarup: *J. Paint Technol.*, **39**: 511 (1967); C. M. Hansen and A. Beerbower: "Solubility Parameters" in H. F. Mark, J. J. McKetta, and D. F. Othmer (eds.), "Encyclopedia of Chemical Technology," 2d ed., suppl. vol., Interscience, New York, 1971.

36. Hayduk, W., and S. C. Cheng: *Can. J. Chem. Eng.*, **48**: 93 (1970); W. Hayduk and W. D. Buckley: *Can. J. Chem. Eng.*, **49**: 667 (1971); W. Hayduk and H. Laudie: *AIChE J.*, **19**: 1233 (1973).

37. Heidemann, R. A., and J. M. Mandhane: *Chem. Eng. Sci.*, **28**: 1213 (1973); T. Katayama, M. Kato, and M. Yasuda: *J. Chem. Eng. Jpn.*, **6**: 357 (1973); A. C. Mattelin and L. A. J. Verhoeye: *Chem. Eng. Sci.*, **30**: 193 (1975).

38. Heil, J. F., and J. M. Prausnitz: *AIChE J.*, **12**: 678 (1966).

39. Hildebrand, J. H., and R. L. Scott: "Regular Solutions," Prentice-Hall, Englewood Cliffs, N.J., 1962.

40. Hiranuma, M., and K. Honma: *Ind. Eng. Chem. Process Des. Dev.*, **14**: 221 (1975).

41. Horsley, L. H.: "Azeotropic Data," *Am. Chem. Soc. Adv. Chem. Ser.*, no. 6, 1952; no. 35, 1962; no. 116, 1973.

42. Hsu, C. C., and J. M. Prausnitz: *Macromolecules*, **7**: 320 (1974).

43. Jones, C. A., A. P. Colburn, and E. M. Schoenborn: *Ind. Eng. Chem.*, **35**: 666 (1943).

44. King, C. J.: "Separation Processes," chap. 11, McGraw-Hill, New York, 1971.

45. Kobayashi, R., P. S. Chappelear, and H. A. Deans: *Ind. Eng. Chem.*, **59**: 63 (1967).

46. Kohn, J. P.: *AIChE J.*, **7**: 514 (1961).

47. Krause, S.: *J. Macromol. Sci. Rev. Macromol. Chem.*, **C7**: 251 (1972).

48. Ladurelli, A. J., C. H. Eon, and G. Guiochon: *Ind. Eng. Chem. Fundam.*, **14**: 191 (1975).

49. Lewis, G. N., M. Randall, K. S. Pitzer, and L. Brewer: "Thermodynamics," 2d ed., McGraw-Hill, New York, 1961.

50. Lo, T. C., H. H. Bieber, and A. E. Karr: *J. Chem. Eng. Data*, **7**: 327 (1962).

51. Lyckman, E. W., C. A. Eckert, and J. M. Prausnitz: *Chem. Eng. Sci.*, **20**: 685 (1965).

52. Maron, S. H., and N. Nakajima: *J. Polym. Sci.*, **40**: 59 (1959); S. H. Maron, *J. Polym. Sci.*, **38**: 329 (1959).

53. Martin, R. A., and K. L. Hoy: "Tables of Solubility Parameters," Union Carbide Corp., Chemicals and Plastics, Research and Development Dept., Tarrytown, N.Y., 1975.

54. Meissner, H. P., and C. L. Kusik: *AIChE J.*, **18**: 294 (1972); H. P. Meissner and J. W. Tester, *Ind. Eng. Chem. Process Des. Dev.*, **11**: 128 (1972); *AIChE J.*, **18**: 661 (1972).

55. Mertl, I.: *Coll. Czech. Chem. Commun.*, **37**: 366 (1972).

56. Modell, M., and R. C. Reid: "Thermodynamics and Its Applications," Prentice-Hall, Englewood Cliffs, N.J., 1974.

57. Mollerup, J.: *Adv. Cryogen. Eng.*, **20**: 172 (1975).

58. Murti, P. S., and M. van Winkle: *Chem. Eng. Data Ser.*, **3**: 72 (1958).

59. Newman, R. D., and J. M. Prausnitz: *J. Paint Technol.*, **43**: 33 (1973).

60. Null, H. R., and D. A. Palmer: *Chem. Eng. Prog.*, **65**: 47 (1969); H. R. Null: "Phase Equilibrium in Process Design," Wiley, New York, 1970.

61. O'Connell, J. P.: *AIChE J.*, **12**: 658 (1971).

62. Orye, R. V.: *Ind. Eng. Chem. Process Des. Dev.*, **8**: 579 (1969).

63. Palmer, D. A.: *Chem. Eng.*, June 9, 1975, p. 80.

64. Patterson, D.: *Macromolecules*, **2**: 672 (1969).

65. Pierotti, G. J., C. H. Deal, and E. L. Derr: *Ind. Eng. Chem.*, **51**: 95 (1959).

66. Prausnitz, J. M.: "Molecular Thermodynamics of Fluid-Phase Equilibria," Prentice-Hall, Englewood Cliffs, N.J., 1969.

67. Prausnitz, J. M., and P. L. Chueh: "Computer Calculations for High-Pressure Vapor-Liquid Equilibria," Prentice-Hall, Englewood Cliffs, N.J., 1968.

68. Prausnitz, J. M., C. A. Eckert, R. V. Orye, and J. P. O'Connell: "Computer Calculations for Multicomponent Vapor-Liquid Equilibria," Prentice-Hall, Englewood Cliffs, N.J., 1967; H. Renon, L. Asselineau, G. Cohen, and C. Raimbault: "Calcul sur ordinateur des équilibres liquide-vapeur et liquide-liquide," Editions Technip, Paris, 1971.

69. Prausnitz, J. M., W. C. Edmister, and K. C. Chao: *AIChE J.*, **6**: 214 (1960).

70. Prausnitz, J. M., and F. H. Shair: *AIChE J.*, **7**: 682 (1961).

71. Preston, G. T., and J. M. Prausnitz: *Ind. Eng. Chem. Process Des. Dev.*, **9**: 264 (1970).

72. Prigogine, I.: "The Molecular Theory of Solutions," North-Holland Publishing Co., Amsterdam, 1957.

73. Prigogine, I., and R. Defay: "Chemical Thermodynamics," Longmans, London, 1954.
74. Redlich, O., E. L. Derr, and G. Pierotti: *J. Am. Chem. Soc.*, **81**: 2283 (1959); E. L. Derr and M. Papadopoulous: *J. Amer. Chem. Soc.*, **81**: 2285 (1959).
75. Renon, H., and J. M. Prausnitz: *Ind. Eng. Chem. Process Des. Dev.*, **8**: 413 (1969).
76. Renon, H., and J. M. Prausnitz: *AIChE J.*, **15**: 785 (1969).
77. Schreiber, L. B., and C. A. Eckert: *Ind. Eng. Chem. Process Des. Dev.*, **10**: 572 (1971).
78. Scott, R. L.: *J. Chem. Phys.*, **17**: 279 (1949); H. Tompa: *Trans. Faraday Soc.*, **45**: 1142 (1949).
79. Shain, S. A., and J. M. Prausnitz: *Chem. Eng. Sci.*, **18**: 244 (1963).
80. Sheehan, C. J., and A. L. Bisio: *Rubber Chem. Technol.*, **39**: 149 (1966).
81. Smith, J. M., and H. C. Van Ness: "Introduction to Chemical Engineering Thermodynamics," 3d ed., McGraw-Hill, New York, 1975.
82. Soave, G.: *Chem. Eng. Sci.*, **27**: 1197 (1972).
83. Starling, K. E., and M. S. Han: *Hydrocarbon Process.*, June 1972.
84. Steele, K., B. E. Poling, and D. B. Manley: paper presented at *AIChE Meet.*, *Washington, D.C., December 1974.*
85. Tapavicza, S., and J. M. Prausnitz: *Chem. Ing. Tech.*, **47**: 552 (1975); English translation in *Int. Chem. Eng.*, **16**(2): 329 (April 1976).
86. Tassios, D.: *AIChE J.*, **17**: 1367 (1971).
87. Tiepel, E. W., and K. E. Gubbins: *IEC Fundam.*, **12**: 18 (1973); *Can. J. Chem. Eng.*, **50**: 361 (1972).
88. Tompa, H.: "Polymer Solutions," Butterworths, London, 1956.
89. Treybal, R. E.: "Liquid Extraction," 2d ed., McGraw-Hill, New York, 1963.
90. Udovenko, V. V., and T. B. Frid: *Zh. Fiz. Khim.*, **22**: 1263 (1948).
91. Weimer, R. F., and J. M. Prausnitz: *Hydrocarbon Process. Pet. Refiner*, **44**: 237 (1965).
92. Wilson, G. M.: *J. Am. Chem. Soc.*, **86**: 127, 133 (1964).
93. Wilson, G. M., and C. H. Deal: *Ind. Eng. Chem. Fundam.*, **1**: 20 (1962).
94. Wohl, K.: *Chem. Eng. Prog.*: **49**: 218 (1953).
95. Wohl, K.: *Trans. AIChE*, **42**: 215 (1946).
96. Wong, K. F., and C. A. Eckert: *Ind. Eng. Chem. Fundam.*, **10**: 20 (1971).
97. Yen, L., and J. J. McKetta: *AIChE J.*, **8**: 501 (1962).
98. Yodovich, A., R. L. Robinson, and K. C. Chao: *AIChE J.*, **17**: 1152 (1971).
99. Young, C. L.: *Chromatog. Rev.*, **10**: 129 (1968).

Chapter Nine

Viscosity

9-1 Scope

The first part of this chapter deals with the viscosity of gases and the second with the viscosity of liquids. In each part, methods are recommended for (1) correlating viscosities with temperature, (2) estimating viscosities when no experimental data are available, (3) estimating the effect of pressure on viscosity, and (4) estimating the viscosities of mixtures. The molecular theory of viscosity is considered briefly.

9-2 Definition and Units of Viscosity

If a shearing stress is applied to any portion of a confined fluid, the fluid will move and a velocity gradient will be set up within it with a maximum velocity at the point where the stress is applied. If the shear stress per unit area at any point is divided by the velocity gradient, the ratio obtained is defined as the viscosity of the medium. It can be seen, therefore, that viscosity is a measure of the internal fluid friction which tends to oppose any dynamic change in the fluid motion; i.e., if the friction between layers of fluid is small (low viscosity), an applied shearing force will result in a large velocity gradient. As the viscosity increases, each fluid layer exerts a larger frictional drag on adjacent layers and the velocity gradient decreases.

It is to be noted that viscosity differs in one important respect from the properties discussed previously in this book; namely, viscosity is a dynamic, nonequilibrium property on a macro scale. Density, for example, is a static, equilibrium property. On a micro scale, both properties reflect the effect of molecular motions and interaction. Even though viscosity is ordinarily referred to as a nonequilibrium property, it is a function of the state of the fluid, as are temperature, pressure, and volume, and may be used to define the state of the material.†

The mechanism or theory of gas viscosity has been reasonably well clarified by the application of the kinetic theory of gases, but the theory of liquid viscosity is poorly developed. Brief résumés of both theories will be presented.

Since viscosity is defined as a shearing stress per unit area divided by a velocity gradient, it should have the dimensions of (force)(time)/length^2 or mass/(length)(time). Both dimensional groups are used, although for most scientific work, viscosities are expressed in terms of poises, centipoises, micropoises, etc. A poise (P) denotes a viscosity of 1 dyn s/cm^2 or 1 g-mass/s cm and 1.0 cP = 0.01 P. The following conversion factors apply to the viscosity units:

$$1 \text{ P} = 1.000 \times 10^2 \text{ cP} = 1.000 \times 10^6 \ \mu\text{P} = 6.72 \times 10^{-2} \text{ lb-mass/ft s}$$
$$= 242 \text{ lb-mass/ft h} = 0.1 \text{ N s/m}^2$$

The *kinematic viscosity* is the ratio of the viscosity to the density. With viscosity in poises and the density in grams per cubic centimeter, the unit of kinematic viscosity is the *stokes*, with the units square centimeters per second.

9-3   Theory of Gas and Other Transport Properties

The theory is simply stated but is quite complex to express in equations which can be used directly to calculate viscosities. In simple terms, when a gas undergoes a shearing stress so that there is some bulk motion, the molecules at any one point have the bulk-velocity vector

†This discussion is limited to newtonian fluids, i.e., those fluids in which the viscosity, as defined, is independent of either the magnitude of the shearing stress or velocity gradient (rate of shear). Newtonian fluids include most pure liquids, simple mixtures, and gases. Non-newtonian fluids are characterized by the fact that the viscosity is not independent of the shearing stress or rate of shear and are grouped as being associated with three types of flow: pseudoplastic, dilatant, and plastic. Paper-pulp slurries, paints, polymer solutions, etc., are included in the non-newtonian class.

added to their own random-velocity vector. Molecular collisions cause an interchange of momentum throughout the fluid, and this bulk-motion velocity (or momentum) becomes distributed. Near the source of the applied stress, the bulk-velocity vector is high, but as the molecules move away from the source, they are "slowed down" (in the direction of bulk flow), causing the other sections of the fluid to move in that direction. This random, molecular-momentum interchange is the predominant cause of gaseous viscosity.

Elementary Kinetic Theory If the gas is modeled in the simplest manner, it is possible to show easily the general relationship between viscosity, temperature, pressure, and molecular size. More rigorous treatments will yield similar relationships which contain important correction factors. The elementary gas model assumes all molecules to be noninteracting rigid spheres of diameter σ (with mass m), moving randomly at a mean velocity v. The density is n molecules in a unit volume. Molecules move in the gas, collide, and may transfer momentum or energy if there are velocity or temperature gradients; such processes also result in a transfer of molecular species if a concentration gradient exists. The net flux of momentum, energy, or component mass between two layers is assumed proportional to the momentum, energy, or mass density gradient, i.e.,

$$\text{flux} \propto \left(-\frac{d\rho'}{dz} \right) \qquad (9\text{-}3.1)$$

where the density ρ' decreases in the $+z$ direction and ρ' may be ρ_i, mass density, nmv_y, momentum density, or $C_v nT$, energy density. The coefficient of proportionality for all these fluxes is given by elementary kinetic theory as $vL/3$, where v is the average molecular speed and L the mean free path.

Equation (9-3.1) is also used to define the transport coefficients of diffusivity $\mathscr{D}$, viscosity η, and thermal conductivity λ; that is,

$$\text{Mass flux} = -\mathscr{D}m\frac{dn_i}{dz} = -\frac{vL}{3}\frac{d\rho_i}{dz} \qquad (9\text{-}3.2)$$

$$\text{Momentum flux} = -\eta\frac{dv_y}{dz} = -\frac{vL}{3}mn\frac{dv_y}{dz} \qquad (9\text{-}3.3)$$

$$\text{Energy flux} = -\lambda\frac{dT}{dz} = -\frac{vL}{3}C_v n\frac{dT}{dz} \qquad (9\text{-}3.4)$$

Equations (9-3.2) to (9-3.4) define the transport coefficients $\mathscr{D}$, η, and λ. If the average speed is proportional to $(RT/M)^{1/2}$ and the mean free path to $(n\sigma^2)^{-1}$,

$$\mathscr{D} = \frac{vL}{3} = (\text{const})\frac{T^{3/2}}{M^{1/2}P\sigma^2} \qquad (9\text{-}3.5)$$

$$\eta = \frac{m\rho vL}{3} = (\text{const}) \frac{T^{1/2}M^{1/2}}{\sigma^2} \tag{9-3.6}$$

$$\lambda = \frac{vLC_v n}{3} = (\text{const}) \frac{T^{1/2}}{M^{1/2}\sigma^2} \tag{9-3.7}$$

The constant multipliers in Eqs. (9-3.5) to (9-3.7) are different in each case; the interesting fact to note from these results is the dependency of the various transfer coefficients on T, P, M, and σ. A similar treatment for rigid, noninteracting spheres having a maxwellian velocity distribution yields the same final equations but with slightly different numerical constants.

The viscosity relation [Eq. (9-3.6)] for a rigid, noninteracting sphere model is

$$\eta = 26.69 \frac{\sqrt{MT}}{\sigma^2} \tag{9-3.8}$$

where η = viscosity, μP
M = molecular weight
T = temperature, K
σ = hard-sphere diameter, Å

Analogous equations· for λ and $\mathscr{D}$ are given in Chaps. 10 and 11.

Effect of Intermolecular Forces   If molecules attract or repel one another by virtue of intermolecular forces, the theory of Chapman and Enskog is normally employed [43, 93]. The important assumptions in this development are four: (1) the gas is sufficiently dilute for only binary collisions to occur, (2) the motion of the molecules during a collision can be described by classical mechanics, (3) only elastic collisions occur, and (4) the intermolecular forces act only between fixed centers of the molecules; i.e., the intermolecular potential function is spherically symmetric. With these restrictions, it would appear that the resulting theory should be applicable only to low-pressure, high-temperature monatomic gases. The pressure and temperature restrictions are valid, but for lack of tractable, alternate models, it is very often applied to polyatomic gases, except in the case of thermal conductivity, where a correction for internal energy transfer and storage must be included (see Chap. 10).

The Chapman-Enskog treatment considers in detail the interactions between colliding molecules with a potential energy $\psi(r)$ included. The equations are well known, but their solution is often very difficult. Each choice of an intermolecular potential $\psi(r)$ must be solved separately. In general terms, the solution for viscosity is written

$$\eta = \frac{\frac{5}{16}(\pi MRT)^{1/2}}{(\pi\sigma^2)\Omega_v} = 26.69\frac{\sqrt{MT}}{\sigma^2\Omega_v} \qquad \mu P \qquad (9\text{-}3.9)$$

which is identical to Eq. (9-3.8) except for the inclusion of the *collision integral* Ω_v. Ω_v is unity if the molecules do not attract each other. Given a potential energy of interaction $\psi(r)$, Ω_v can be calculated; results from using the Lennard-Jones and Stockmayer potential functions are given in Sec. 9-4. The former is normally assumed if the molecules are nonpolar. The latter potential is perhaps more reasonable for polar compounds with angle-dependent forces [12, 119, 121].

9-4 Estimation of Low-Pressure Gas Viscosity

Essentially, all gas-viscosity estimation techniques are based either upon the Chapman-Enskog theory or the law of corresponding states. Both approaches are discussed below, and recommendations are presented at the end of the section.

Theoretical Approach The Chapman-Enskog viscosity equation was given as (9-3.9).† To use this relation to estimate viscosities, the collision diameter σ and the collision integral Ω_v must be found. In the derivation of Eq. (9-3.9), Ω_v is obtained as a complex function of a dimensionless temperature T^*. The functionality depends upon the intermolecular potential chosen. As shown in Fig. 9-1, let $\psi(r)$ be the intermolecular energy between two molecules separated by distance r. At large separation distances, $\psi(r)$ is negative; the molecules attract each other.‡ At small distances, repulsion occurs. The minimum in the $\psi(r)$-vs.-r curve, where the forces of attraction and repulsion balance, is termed the *characteristic energy* ϵ. For any potential curve, the dimensionless temperature T^* is related to ϵ by

$$T^* = \frac{kT}{\epsilon} \qquad (9\text{-}4.1)$$

where k is Boltzmann's constant. Referring again to Fig. 9-1, the collision diameter σ is defined as the separation distance when $\psi(r) = 0$.

The relation between $\psi(r)$ and r is called an *intermolecular potential function*. Such a function written using only the parameters ϵ and σ is a two-parameter potential. The Lennard-Jones 12-6 potential given in Eq. (9-4.2) is an example of this type. Many other potential functions with different or additional parameters have also been proposed. The important element, however, is that one must know $\psi(r) = f(r)$ in order

---

†A correction factor, which is essentially unity, was omitted in Eq. (9-3.9).
‡The negative gradient of $\psi(r)$ is the *force* of interaction.

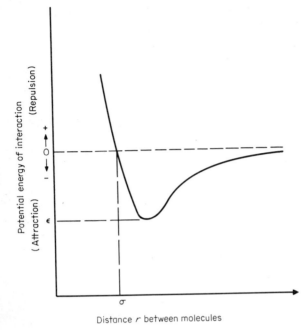

Fig. 9-1   Intermolecular potential relation.

to obtain Ω_v in Eq. (9-3.9).   The working equation for η will, in any case, have as many parameters as used to define the original $\psi(r)$ relation.

   Nonpolar Gases.   The Lennard-Jones 12-6 potential is

$$\psi(r) = 4\epsilon \left[\left(\frac{\sigma}{r} \right)^{12} - \left(\frac{\sigma}{r} \right)^6 \right] \tag{9-4.2}$$

Such a relation is based upon rather tenuous theoretical grounds and has been widely criticized.   However, it is probably one of the more tractable relations for $\psi(r)$, and since Ω_v is relatively insensitive to the exact form of the $\psi(r)$ relation, Eq. (9-4.2) has been extensively used.

   With this potential, the collision integral has been determined by a number of investigators [12, 93, 103, 119, 121, 147].   Recently, Neufeld et al. [150] proposed an empirical equation which is convenient for computer application:

$$\Omega_v = \left(\frac{A}{T^{*B}} \right) + \frac{C}{\exp DT^*} + \frac{E}{\exp FT^*} \tag{9-4.3}$$

where

$$T^* = \frac{kT}{\epsilon} \qquad A = 1.16145 \qquad B = 0.14874$$

$$C = 0.52487 \qquad D = 0.77320 \qquad E = 2.16178 \qquad F = 2.43787$$

Equation (9-4.3) is applicable from $0.3 \leqslant T^* \leqslant 100$ with an average deviation of only 0.064 percent. It is recommended in preference to other suggested analytical relations [37, 133] or nomographs [22, 26, 28].

Ω_v decreases with an increase in T^*, and over the T^* range of interest to most chemical engineers $(0.3 < T^* < 2)$, $\log \Omega_v$ is nearly linear in $\log T^*$, as shown in Fig. 9-2, a fact that will prove later to have interesting consequences.

With values of Ω_v as a function of T^*, a number of investigators have used Eq. (9-3.9) and regressed experimental viscosity-temperature data to find the best values of ϵ/k and σ for many substances. Appendix C lists a number of such sets as reported by Svehla [198]. It should be noted, however, that there appears also to be a number of other quite satisfactory *sets* of ϵ/k and σ for any given compound. For example, with n-butane, Svehla suggested $\epsilon/k = 513.4$ K, $\sigma = 4.730$ Å, whereas Flynn and Thodos [64] recommended $\epsilon/k = 208$ K and $\sigma = 5.869$ Å.

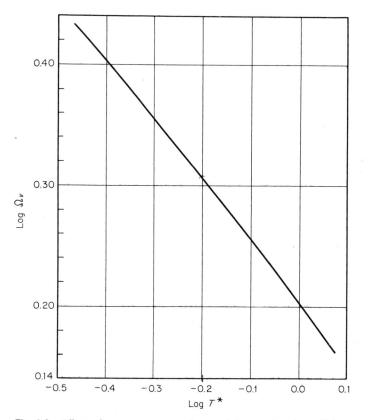

Fig. 9-2 Effect of temperature on Lennard-Jones viscosity collision integral.

Both sets, when used to calculate viscosities, yield almost exactly the same values of viscosity as shown in Fig. 9-3. This interesting paradox has been noted by a number of authors [97, 118, 149, 171] and has been studied by Reichenberg [170]. He suggested that $\log \Omega_v$ is almost a linear function of $\log T^*$ (see Fig. 9-2).†

$$\Omega_v = aT^{*n} \tag{9-4.4}$$

Equation (9-3.9) may then be written

$$\eta = 26.69 M^{1/2} a^{-1} T^{0.5-n} \frac{(\epsilon/k)^n}{\sigma^2} \qquad \mu P \tag{9-4.5}$$

†Kim and Ross [118] do, in fact, propose that $\Omega_v = 1.604 T^{*-0.5}$, where $0.4 < T^* < 1.4$. They note a maximum error of only 0.7 percent.

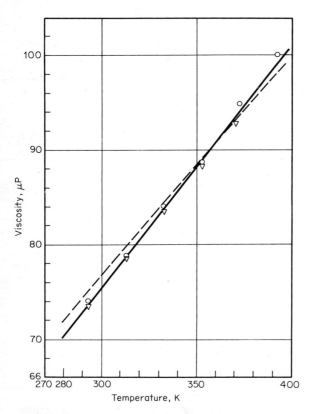

Fig. 9-3 Comparison of calculated and experimental low-pressure gas viscosity of n-butane; $\odot$ Ref. 204, ∇ Ref. 219; Eq. (9-3.9) and the Lennard-Jones potential: $---$ Flynn and Thodos [64] calculated $\sigma = 5.869$ Å and $\epsilon/k = 208$ K; $\underline{\quad\quad}$ Svehla [198] calculated $\sigma = 4.730$ Å and $\epsilon/k = 513.4$ K.

Here, the parameters σ and ϵ/k are combined as a *single* parameter $(\epsilon/k)^n/\sigma^2$. There is then no way of delineating specific values of ϵ/k and σ using experimental viscosity data, at least over the range where Eq. (9-4.4) applies.

The conclusion to be drawn from this discussion is that Eq. (9-3.9) can be used to calculate gas viscosity although the chosen set of ϵ/k and σ may have little relation to molecular properties. There will be an infinite number of acceptable sets as long as the temperature range is not too broad, e.g., if one limits the estimation to the range of reduced temperatures from about 0.3 to 1.2. In using published values of ϵ/k and σ for a fluid of interest, the two values from the same set must be used—never ϵ/k from one set and σ from another.

To use Eq. (9-3.9) to estimate viscosities of nonpolar gases, ϵ/k and σ values can be found for many substances in Appendix C. If they are not in this tabulation, such parameters can be approximated by Eqs. (2-6.2) and (2-6.3) using critical constants and the acentric factor. A number of calculated values of viscosity obtained from Eqs. (9-3.9) and (9-4.3) and Appendix C are compared with experimental values in Table 9-4. In general, the errors are less than 1 percent. Slightly larger errors (1 to 3 percent) are noted when σ and ϵ/k are estimated from Eqs. (2-6.2) and (2-6.3). The technique is illustrated in Example 9-1.

Example 9-1   Estimate the viscosity of n-octane vapor at 37.8°C using the Chapman-Enskog theory and the Lennard-Jones 12-6 potential. Compare the predicted value with the experimental result of 58.2 μP [35].

solution   The ϵ/k and σ values for n-octane are not given in Appendix C and are therefore estimated from Eqs. (2-6.2) and (2-6.3). For this material, from Appendix A, $T_c = 568.8$ K, $P_c = 24.5$ atm, $M = 114.232$, and $\omega = 0.394$. Thus

$$\sigma = \frac{2.3551 - (0.0874)(0.394)}{(24.5/568.6)^{1/3}} = 6.62 \text{ Å}$$

$$\frac{\epsilon}{k} = (568.6)[0.7915 + (0.1693)(0.394)] = 488.0 \text{ K}$$

At 37.8°C

$$T^* = \frac{T}{\epsilon/k} = \frac{37.8 + 273.2}{488.0} = 0.637$$

Then, from Eq. (9-4.3) with $T^* = 0.637$, $\Omega_v = 2.022$. Thus, from Eq. (9-3.9),

$$\eta = \frac{(26.69)(114.232)^{1/2}(311)^{1/2}}{(6.62)^2(2.022)} = 56.9 \ \mu\text{P}$$

$$\text{Error} = \frac{56.9 - 58.2}{58.2} \times 100 = -2.2\%$$

Polar Gases.   For polar molecules, a more suitable potential function would be that suggested by Stockmayer and discussed elsewhere [11, 93, 156]. In essence, this function is identical to the Lennard-Jones 12-6 form except for an additional term to account for permanent dipole-dipole interactions between molecules. If there are no permanent dipoles, the Stockmayer relation simplifies to the Lennard-Jones form. Monchick and Mason [147] have obtained approximate values for Ω_v using this potential function, and these are shown in Table 9-1. To obtain Ω_v, values of ϵ/k and δ are required. δ is a polar

TABLE 9-1   Collision Integrals Ω_v for Viscosity as Calculated by the Stockmayer Potential [147b]

$$\delta = \frac{(\text{dipole moment})^2}{2\epsilon\sigma^3} \qquad T^* = \frac{kT}{\epsilon}$$

$\overset{\delta}{\underset{T^*}{}}$	0	0.25	0.50	0.75	1.0	1.5	2.0	2.5
0.1	4.1005	4.266	4.833	5.742	6.729	8.624	10.34	11.89
0.2	3.2626	3.305	3.516	3.914	4.433	5.570	6.637	7.618
0.3	2.8399	2.836	2.936	3.168	3.511	4.329	5.126	5.874
0.4	2.5310	2.522	2.586	2.749	3.004	3.640	4.282	4.895
0.5	2.2837	2.277	2.329	2.460	2.665	3.187	3.727	4.249
0.6	2.0838	2.081	2.130	2.243	2.417	2.862	3.329	3.786
0.7	1.9220	1.924	1.970	2.072	2.225	2.614	3.028	3.435
0.8	1.7902	1.795	1.840	1.934	2.070	2.417	2.788	3.156
0.9	1.6823	1.689	1.733	1.820	1.944	2.258	2.596	2.933
1.0	1.5929	1.601	1.644	1.725	1.838	2.124	2.435	2.746
1.2	1.4551	1.465	1.504	1.574	1.670	1.913	2.181	2.451
1.4	1.3551	1.365	1.400	1.461	1.544	1.754	1.989	2.228
1.6	1.2800	1.289	1.321	1.374	1.447	1.630	1.838	2.053
1.8	1.2219	1.231	1.259	1.306	1.370	1.532	1.718	1.912
2.0	1.1757	1.184	1.209	1.251	1.307	1.451	1.618	1.795
2.5	1.0933	1.100	1.119	1.150	1.193	1.304	1.435	1.578
3.0	1.0388	1.044	1.059	1.083	1.117	1.204	1.310	1.428
3.5	0.99963	1.004	1.016	1.035	1.062	1.133	1.220	1.319
4.0	0.96988	0.9732	0.9830	0.9991	1.021	1.079	1.153	1.236
5.0	0.92676	0.9291	0.9360	0.9473	0.9628	1.005	1.058	1.121
6.0	0.89616	0.8979	0.9030	0.9114	0.9230	0.9545	0.9955	1.044
7.0	0.87272	0.8741	0.8780	0.8845	0.8935	0.9181	0.9505	0.9893
8.0	0.85379	0.8549	0.8580	0.8632	0.8703	0.8901	0.9164	0.9482
9.0	0.83795	0.8388	0.8414	0.8456	0.8515	0.8678	0.8895	0.9160
10.0	0.82435	0.8251	0.8273	0.8308	0.8356	0.8493	0.8676	0.8901
12.0	0.80184	0.8024	0.8039	0.8065	0.8101	0.8201	0.8337	0.8504
14.0	0.78363	0.7840	0.7852	0.7872	0.7899	0.7976	0.8081	0.8212
16.0	0.76834	0.7687	0.7696	0.7712	0.7733	0.7794	0.7878	0.7983
18.0	0.75518	0.7554	0.7562	0.7575	0.7592	0.7642	0.7711	0.7797
20.0	0.74364	0.7438	0.7445	0.7455	0.7470	0.7512	0.7569	0.7642
25.0	0.71982	0.7200	0.7204	0.7211	0.7221	0.7250	0.7289	0.7339
30.0	0.70097	0.7011	0.7014	0.7019	0.7026	0.7047	0.7076	0.7112
35.0	0.68545	0.6855	0.6858	0.6861	0.6867	0.6883	0.6905	0.6932
40.0	0.67232	0.6724	0.6726	0.6728	0.6733	0.6745	0.6762	0.6784
50.0	0.65099	0.6510	0.6512	0.6513	0.6516	0.6524	0.6534	0.6546
75.0	0.61397	0.6141	0.6143	0.6145	0.6147	0.6148	0.6148	0.6147
100.0	0.58870	0.5889	0.5894	0.5990	0.5903	0.5901	0.5895	0.5885

parameter defined as

$$\delta = \frac{\mu_p^2}{2\epsilon\sigma^3} \qquad (9\text{-}4.6)$$

with μ_p the dipole moment; ϵ and δ are now Stockmayer parameters. δ is dimensionless.† In Table 9-2, values of ϵ/k, σ, and δ are listed for a few representative polar molecules. Halkiadakis and Bowery [86] have published a similar table.

Brokaw [26], in a broad study of the use of the Stockmayer potential, suggests that instead of using Table 9-1, one can closely approximate Ω_v as

$$\Omega_v \text{ (Stockmayer)} = \Omega_v \text{ (Lennard-Jones)} + \frac{0.2\delta^2}{T^*} \qquad (9\text{-}4.7)$$

with Ω_v (Lennard-Jones) from Eq. (9-4.3). Also, if the potential parameters are not shown in Table 9-2, Brokaw recommends they may

---

†μ_p is given in debyes; 1 debye = 10^{-18} esu = 10^{-18} (dyn cm^2)$^{1/2}$ = 3.162×10^{-25} (N m^4)$^{1/2}$. Thus, if ϵ is in joule(newton-meter) and σ in meters, δ is dimensionless.

TABLE 9-2 Stockmayer-Potential Parameters [147b]

	Dipole moment μ_p, debyes	σ, Å	$\frac{\epsilon}{k}$, K	δ
H_2O	1.85	2.52	775	1.0
NH_3	1.47	3.15	358	0.7
HCl	1.08	3.36	328	0.34
HBr	0.80	3.41	417	0.14
HI	0.42	4.13	313	0.029
SO_2	1.63	4.04	347	0.42
H_2S	0.92	3.49	343	0.21
NOCl	1.83	3.53	690	0.4
$CHCl_3$	1.013	5.31	355	0.07
CH_2Cl_2	1.57	4.52	483	0.2
CH_3Cl	1.87	3.94	414	0.5
CH_3Br	1.80	4.25	382	0.4
C_2H_5Cl	2.03	4.45	423	0.4
CH_3OH	1.70	3.69	417	0.5
C_2H_5OH	1.69	4.31	431	0.3
$n\text{-}C_3H_7OH$	1.69	4.71	495	0.2
$i\text{-}C_3H_7OH$	1.69	4.64	518	0.2
$(CH_3)_2O$	1.30	4.21	432	0.19
$(C_2H_5)_2O$	1.15	5.49	362	0.08
$(CH_3)_2CO$	1.20	4.50	549	0.11
CH_3COOCH_3	1.72	5.04	418	0.2
$CH_3COOC_2H_5$	1.78	5.24	499	0.16
CH_3NO_2	2.15	4.16	290	2.3

be estimated as follows:

$$\sigma = \left(\frac{1.585 V_b}{1 + 1.3\delta^2}\right)^{1/3} \tag{9-4.8}$$

$$\frac{\epsilon}{k} = (1.18)(1 + 1.3\delta^2) T_b \tag{9-4.9}$$

$$\delta = \frac{1.94 \times 10^3 \mu_p^2}{V_b T_b} \tag{9-4.10}$$

where σ is in angstroms, ϵ/k and the normal boiling temperature T_b are in kelvins, μ_p is in debyes, and V_b, the liquid molal volume at the normal boiling point, is in cubic centimeters per gram mole.

Table 9-4 shows a few calculated values of polar gas viscosities determined from Eqs. (9-3.9), (9-4.3), and (9-4.7) with the potential parameters found in Table 9-2 or from Eqs. (9-4.8) to (9-4.10). Polar gases are normally defined as those which have $\delta \geqslant 0.1$, though some prefer to include compounds with $\delta \geqslant 0.05$. Again errors found were usually less than 2 percent if σ and ϵ/k were found from Table 9-2. Larger errors were noted when σ and ϵ/k were estimated. The method is illustrated in Example 9-2.

Example 9-2 Ammonia at 220°C and about 1 atm is reported to have a viscosity of 169 μP. How does this compare with the value estimated from the Brokaw modification of the Chapman-Enskog theory? Use Table 9-2 to obtain molecular parameters.

 solution As given in Table 9-2, $\sigma = 3.15$ Å, $\epsilon/k = 358$ K, and $\delta = 0.7$. At 220°C, $T^* = 493.2/358 = 1.378$. $M = 17.031$. The nonpolar portion of the collision integral is given by Eq. (9-4.3):

$$\Omega_v = \frac{1.16145}{1.378^{0.14874}}$$

$$+ \frac{0.52487}{\exp\left[(0.77320)(1.378)\right]}$$

$$+ \frac{2.16178}{\exp\left[(2.43787)(1.378)\right]} = 1.363$$

The polar collision integral is then given by Eq. (9-4.7),

$$\Omega_v = 1.363 + \frac{(0.2)(0.7)^2}{1.378} = 1.434$$

Finally, with Eq. (9-3.9),

$$\eta = \frac{(26.69)(17.031)^{1/2}(493.2)^{1/2}}{(3.15)^2(1.434)} = 172 \ \mu\text{P}$$

$$\text{Error} = \frac{172 - 169}{169} \times 100 = 1.8\%$$

Corresponding-States Methods Several gas-viscosity estimation methods can be traced back to Eq. (9-3.8), where, from simple hard-sphere theory,

$$\eta = K \frac{T^{1/2} M^{1/2}}{\sigma^2} \tag{9-4.11}$$

If σ^3 is associated with V_c and V_c assumed to be proportional to RT_c/P_c, then

$$\eta_r = f(T_r) \tag{9-4.12}$$

where η_r is a dimensionless viscosity,

$$\eta_r = \frac{\eta}{M^{1/2}P_c^{2/3}/(RT_c)^{1/6}} \tag{9-4.13}$$

This reduction was suggested by Trautz in 1931 [206] and discussed in some detail by Golubev in his viscosity handbook [76]. The working equations proposed by Golubev are

$$\eta = \begin{cases} \eta_c^* T_r^{0.965} & T_r < 1 & (9\text{-}4.14) \\ \eta_c^* T_r^{0.71+0.29/T_r} & T_r > 1 & (9\text{-}4.15) \end{cases}$$

where η_c^* is the viscosity at the critical temperature *but at a low pressure*:

$$\eta_c^* = \frac{3.5 M^{1/2} P_c^{2/3}}{T_c^{1/6}} \tag{9-4.16}$$

where M = molecular weight
P_c = critical pressure, atm
T_c = critical temperature, K
η = viscosity, μP

Similar relations were proposed by Thodos and coworkers at about the same time [63, 141, 194]. The Thodos et al. relations (since slightly revised [222]) are:

Nonpolar

$$\eta\xi = 4.610\,T_r^{0.618} - 2.04e^{-0.449T_r} + 1.94e^{-4.058T_r} + 0.1 \tag{9-4.17}$$

Polar Gases. For hydrogen-bonding types, $T_r < 2.0$,

$$\eta\xi = (0.755\,T_r - 0.055)Z_c^{-5/4} \tag{9-4.18}$$

For non-hydrogen-bonding types, $T_r < 2.5$,

$$\eta\xi = (1.90\,T_r - 0.29)^{4/5}Z_c^{-2/3} \tag{9-4.19}$$

where

$$\xi = T_c^{1/6}M^{-1/2}P_c^{-2/3} \tag{9-4.20}$$

Z_c is the compressibility factor at the critical point, and the remaining terms have the same units as shown under Eq. (9-4.16). Malek and Stiel [138] propose alternate forms to Eqs. (9-4.18) and (9-4.19) where $\eta\xi$ is related to T_r, ω, and X. ω is the acentric factor and X the Stiel polar factor; both are discussed in Chap. 2.

The Thodos et al. equations should not be used for hydrogen, helium, or the diatomic halogen gases. This limitation is said not to apply to Eqs. (9-4.14) or (9-4.15). Neither works well with polar gases which associate significantly in the vapor phase.

Reichenberg has suggested a different corresponding-states relation [170]

$$\eta = \frac{a^* T_r}{[1 + 0.36 T_r (T_r - 1)]^{1/6}} \qquad \mu P \qquad (9\text{-}4.21)\dagger$$

In many cases, the parameter a^* is numerically quite similar to η_c^* determined from Eq. (9-4.16). However, for organic compounds, Reichenberg recommends using

$$a^* = \frac{M^{1/2} T_c}{\sum_i n_i C_i} \qquad \mu P \qquad (9\text{-}4.22)$$

where M = molecular weight
T_c = critical temperature, K
n_i = number of atomic groups of ith type

Group contributions for C_i are shown in Table 9-3. At present, only organic compounds can be treated.

Gas viscosity values calculated from these three corresponding-states methods are compared with experimental values in Table 9-4. The methods are illustrated in Examples 9-3 and 9-4.

†In the original form, the denominator of Eq. (9-4.21) was written $[1 + 0.36 T_r (T_r - 1) \times (1 + 4/T_c)]^{1/6}$. However, for most materials, $1 + 4/T_c \approx 1$.

TABLE 9-3 Values of the Group Contributions C_i for the Estimation of a^* in micropoises [170c]

Group	Contribution C_i
—CH$_3$	9.04
$>$CH$_2$ (nonring)	6.47
$>$CH— (nonring)	2.67
$>$C$<$ (nonring)	−1.53
=CH$_2$	7.68
=CH— (nonring)	5.53

$>$C= (nonring)	1.78
≡CH	7.41
≡C— (nonring)	5.24
$>$CH₂ (ring)	6.91
$>$CH— (ring)	1.16
$>$C$<$ (ring)	0.23
=CH— (ring)	5.90
$>$C= (ring)	3.59
—F	4.46
—Cl	10.06
—Br	12.83
—OH (alcohols)	7.96
$>$O (nonring)	3.59
$>$C=O (nonring)	12.02
—CHO (aldehydes)	14.02
—COOH (acids)	18.65
—COO— (esters) or HCOO (formates)	13.41
—NH₂	9.71
$>$NH (nonring)	3.68
=N—(ring)	4.97
—CN	18.13
$>$S (ring)	8.86

TABLE 9-4   Comparison between Calculated and Experimental Values of Low-Pressure Gas Viscosit

Compound	T, °C	Experimental value, μP^b	Theoretical values ϵ/k and σ Available[c]	Est.[d]	Thodos et al., Eqs. (9-4.17)– (9-4.19)	Golubev, Eqs. (9-4.14)– (9-4.16)	Reichenberg, Eq. (9-4.21)
			Nonpolar Gases				
Acetylene	30	102	2.4	−1.0	0.5	2.2	2.4
	101	126	0.8	−2.2	−0.8	3.3	0.7
	200	155	0.6	−2.1	−0.6	12	0.7
Benzene	28	73.2	4.0	−0.3	3.3	6.9	4.4
	100	92.5	2.6	−1.6	1.3	4.0	2.1
	200	117	2.9	−1.2	1.3	3.1	1.5
Isobutane	20	74.4	−0.8	−0.8	1.9	4.3	2.1
	60	84.5	−0.5	−0.5	2.0	3.9	1.8
	120	99.5	−0.2	−0.1	1.8	3.6	1.3
n-Butane	20	73.9	−1.3	−2.4	0.5	3.0	−1.2
	60	83.9	−1.1	−2.0	0.6	2.6	−1.4
	120	99.8	−1.4	−2.5	−0.5	1.2	−2.8
1-Butene	20	76.1	· · ·	−1.6	1.2	3.7	1.8
	60	83.9	· · ·	1.7	4.3	6.4	4.6
	120	99.8	· · ·	1.1	3.2	4.9	3.1
Carbon dioxide	30	151	1.4	−0.9	1.2	2.9	
	100.5	181	2.0	1.2	2.9	7.5	
	200.1	219	2.5	3.3	4.8	19	
Carbon disulfide	30	94.6	6.6	8.0	11	15	
	98.2	119	4.9	6.3	9.2	12	
	200	151	5.5	6.8	9.1	11	
Carbon tetrachloride	125	133	−1.8	−0.6	2.1	4.6	−2.6
	200	156	−1.6	0.7	3.1	5.0	−2.1
	300	190	−1.9	2.3	4.2	6.1	−1.4
Chlorine	20	133	−1.6	−1.0	1.3	3.8	
	100	168	−0.7	−0.1	1.8	3.6	
	200	209	0.3	1.4	2.9	5.8	
Cyclohexane	35	72.3	−0.5	−3.5	0.0	3.4	−4.2
	77.8	81.1	1.1	−1.5	1.6	4.5	−2.8
	100	87.3	−0.3	−2.5	0.4	3.0	−4.1
	200	109	0.1	−0.2	2.2	4.1	−2.9
	300	129	−0.3	1.6	3.5	5.4	−2.0
Ethane	20	90.1	1.8	0.0	1.8	3.5	−2.0
	50	99.8	1.6	0.2	1.8	3.9	−2.2
	100	114		0.2	1.7	6.1	−2.4
	250	153	0.7	0.9	2.4	23	−2.1
Ethylene	−80	71.4	−7.9	−9.7	−7.4	−5.1	−5.4
	0	94.5	−1.2	−3.1	−1.4	0.2	−0.1

	50	111	− 0.9	− 2.7	− 1.2	1.8	− 0.2
	150	141	− 0.8	− 2.5	− 1.1	10	− 0.3
	250	168	− 0.6	− 2.3	− 0.9	26	− 0.4
Methane	20	109	0.7	− 1.9	− 0.5		
	100	133	0.6	− 1.9	− 0.4		
	200	160	0.4	− 2.0	− 0.2		
	300	185	0.4	− 2.0	− 0.1		
	500	227	1.1	− 1.2	0.7		
n-Pentane	125	91.7	0.1	− 1.8	0.7	2.5	− 0.8
	175	103	0.0	− 1.5	0.6	2.3	− 1.1
	225	114	− 0.2	− 1.3	0.6	2.7	− 1.4
	300	130	− 0.6	− 1.2	0.5	4.8	− 1.6
Propane	20	80.6	0.3	− 0.6	1.8	3.8	− 0.6
	60	92.2	− 0.9	− 1.1	0.9	2.7	− 1.7
	125	107	0.9	1.9	3.6	5.9	0.5
	200	125	0.4	2.4	4.0	9.7	0.7
	275	142	− 0.3	2.5	4.1	16.0	0.6
Propylene	20	84.3	− 0.1	− 0.3	2.1	4.1	2.4
	50	93.3	− 0.4	− 0.5	1.5	3.3	1.6
	150	121	− 0.1	− 0.3	1.3	4.6	0.9
	250	147	− 0.1	− 0.3	1.2	11	0.5
Sulfur dioxide	10	120	1.8	− 2.5	0.7	3.4	
	40	135	0.3	− 3.8	− 1.0	1.3	
	100	163	− 0.8	− 4.8	− 2.4	− 0.7	
	300	246	− 1.4	− 4.7	− 3.1	3.4	
	500	315	− 0.5	− 3.4	− 2.0	21	
	700	376	0.0	− 2.5	− 1.2	50	
	900	432	0.2	− 2.1	− 0.8	93	
Toluene	60	78.9		− 7.0	− 3.6	− 0.4	− 2.3
	150	101		− 6.8	− 4.0	− 1.8	− 3.1
	250	123		− 5.0	− 2.7	− 1.0	− 2.2
Average error			1.3	2.2	2.2	8.0	1.9

Polar Gases

Acetone	100	93.3	6.3	4.8	7.1	− 3.0	2.5
	150	108	5.1	3.4	5.4	− 4.9	0.5
	225	128	4.8	2.5	3.9	− 6.2	− 1.0
	325	153	5.1	1.5	2.7	− 5.4	− 2.2
Ammonia	0	90	3.8	6.6	3.7	− 6.8	
	100	131	− 1.0	1.9	0.5	− 13	
	400	251	− 8.5	− 2.8	− 1.8	− 11	
Chloroform	20	100	0.5	− 0.5	− 1.5	7.0	− 1.0
	50	110	1.4	0.5	0.0	7.3	− 0.4
	100	125	2.9	2.0	1.8	8.0	0.6
	200	159	2.5	2.2	1.7	7.1	− 0.1
	350	208	0.5	1.1	0.5	7.8	− 1.4
Ethanol	110	111	0.1	− 8.6	− 1.7	− 11	0.7
	150	123	0.1	− 8.5	− 0.9	− 11	0.2
	200	137	0.6	− 8.1	0.4	− 11	0.1
	300	165	1.4	− 7.1	3.0	− 10	0.0

TABLE 9-4 Comparison between Calculated and Experimental Values of Low-Pressure Gas Viscosity (Continued)

Compound	T, °C	Experimental value, μP^b	Theoretical values ϵ/k and σ Available[c]	Est.[d]	Thodos et al., Eqs. (9-4.17)–(9-4.19)	Golubev, Eqs. (9-4.14)–(9-4.16)	Reichenberg, Eq. (9-4.21)
			Polar Gases	(Continued)			
Ethylacetate	125	101	0.4	0.4	4.0	− 0.7	− 1.2
	175	114	0.3	0.3	3.4	− 1.4	− 2.0
	250	133	0.4	0.0	2.5	− 2.2	− 3.0
	325	153	0.1	− 0.8	1.3	− 1.9	− 4.1
Ethyl ether	125	99.1	1.1	− 0.1	4.2	1.9	0.7
	175	112	0.7	− 0.5	3.4	1.2	− 0.1
	225	124	0.6	− 0.5	3.2	1.6	− 0.4
	300	141	0.9	− 0.1	3.4	4.2	− 0.3
Isopropanol	120	103	− 0.1	− 5.5	4.9	− 5.2	4.2
	160	114	− 0.6	− 6.1	5.1	− 6.3	3.0
	220	130	− 0.2	− 5.9	6.3	− 6.7	2.5
	300	150	1.0	− 5.2	8.6	− 5.4	2.4
Methanol	35	101	− 0.7	− 13	− 5.2	− 22	2.8
	65	111	− 0.4	− 13	− 4.2	− 22	2.5
	120	129	0.6	− 12	− 2.1	− 22	2.6
	160	143	0.4	− 12	− 1.7	− 22	1.8
	240	169	0.6	− 12	− 0.2	− 23	1.0
	320	195	0.7	− 11	1.4	− 22	0.5
Methyl chloride	20	106	− 0.5	3.5	− 0.1	− 0.2	4.1
	50	119	− 2.1	1.9	2.1	− 2.6	1.8
	80	129	− 0.3	3.8	− 0.8	− 1.5	2.9
	130	147	0.1	4.2	− 1.5	− 2.1	2.1
Methyl ether	20	90.9	− 2.1	− 1.1	− 7.9	0.6	2.2
	60	104	− 2.6	− 1.9	− 8.9	− 0.9	0.8
	120	123	− 1.8	− 1.7	− 9.2	− 1.1	0.4
Methylene chloride	20	98.5	3.3	5.5	3.2	7.0	7.8
	100	127	3.1	5.3	3.5	5.2	6.5
	200	160	4.1	5.8	3.3	4.6	5.9
	300	193	5.0	5.6	2.8	5.5	5.3
n-Propanol	125	104	0.2	− 6.7	1.7	− 5.2	0.8
	200	124	− 0.2	− 7.1	2.6	− 6.5	− 0.6
	275	144	− 0.2	− 7.3	3.6	− 7.1	− 1.6
Average error			1.7	4.6	3.1	7.5	2.0

[a] Percent error = [(calc. − exp.)/exp.] × 100.

[b] Most experimental values were obtained from P. M. Craven and J. D. Lambert, *Proc. R. Soc. Lond.*, **A205**: 439 (1951) and Ref. 123.

[c] For nonpolar gases, values were taken from Appendix C; for polar gases, from Table 9-2.

[d] For nonpolar gases, estimates of σ and ϵ/k were obtained from Eqs. (2-6.2) and (2-6.3); for polar gases, Eqs. (9-4.8) to (9-4.10) were used.

Example 9-3   Use corresponding-states correlations to estimate the low-pressure viscosity of sulfur dioxide gas at 40°C.   The critical properties are $T_c = 430.8$ K, $P_c = 77.8$ atm, and $Z_c = 0.268$. M is 64.063.   The experimental value is 135 μP.

   solution   *Thodos   Method.* $\xi = T_c^{1/6} M^{-1/2} P_c^{-2/3} = 0.0188$.   With   Eq.   (9-4.17), $T_r = (40 + 273.2)/430.8 = 0.727$,

$$\eta \xi = (4.610)(0.727)^{0.618} - 2.04 e^{-(0.449)(0.727)} + 1.94 e^{-(4.058)(0.727)} + 0.1$$

$$\eta = \frac{2.516}{0.0188} = 134 \ \mu\text{P}$$

$$\text{Error} = \frac{134 - 135}{135} \times 100 = -0.7\%$$

Gobulev Method.   With Eq. (9-4.16),

$$\eta_c^* = (3.5)(64.063)^{1/2} \frac{(77.8)^{2/3}}{(430.8)^{1/6}} = 186 \ \mu\text{P}$$

From Eq. (9-5.14),

$$\eta = (186)(0.727)^{0.965} = 136 \ \mu\text{P}$$

$$\text{Error} = \frac{136 - 135}{135} \times 100 = 0.7\%$$

Example 9-4   From Reichenberg's correlation [Eq. (9-4.21)] estimate the viscosity of acetylene at 101°C.   The experimental value is 126 μP.

   solution   With Table 9-3, $\sum n_i C_i = (2)(7.41) = 14.82$.   Then, with $M = 26.038$ and $T_c = 308.3$ K

$$a^* = \frac{M^{1/2} T_c}{\sum_i n_i C_i} = (26.038)^{1/2} \frac{308.3}{14.82} = 106 \ \mu\text{P}$$

Since $T_r = (101 + 273)/308.3 = 1.21$, from Eq. (9-4.21),

$$\eta = \frac{(106)(1.21)}{[1 + (0.36)(1.21)(1.21 - 1)]^{1/6}} = 127 \ \mu\text{P}$$

$$\text{Error} = \frac{127 - 126}{126} \times 100 = 0.7\%$$

Recommendations for Estimating Low-Pressure Viscosity of Pure Gases   *Nonpolar Gases.*   If values of ϵ/k and σ are available from Appendix C, use these constants with Eq. (9-3.9) to calculate the viscosity.   Errors are seldom greater than 1 percent.   Should such values not be found, use the Thodos et al. corresponding-states form [Eq. (9-4.17)] or the Reichenberg group-contribution–corresponding-states form [Eq. (9-4.21)].   The expected error is still low and normally lies between 1 and 3 percent.

   Polar Gases.   The most accurate method is Eq. (9-3.9) with the collision integral obtained from Eq. (9-4.7)—providing ϵ/k, σ, and δ values are shown in Table 9-2 (or can be obtained from other similar listings). The expected error lies between 0.5 and 1.5 percent.   If the compound is not shown in Table 9-2, the Reichenberg relation (9-4.21) with Table 9-3 is recommended.   Errors vary, but they are usually less than 4 percent.

9-5 Viscosities of Gas Mixtures at Low Pressures

Unfortunately, the viscosity of a gas mixture is seldom a linear function of composition, as can be noted in Fig. 9-4. There may even be a maximum where the mixture viscosity exceeds the value for either pure component, e.g., system 4, ammonia-hydrogen. No cases of a viscosity minimum have been reported, however. Trends like those shown by

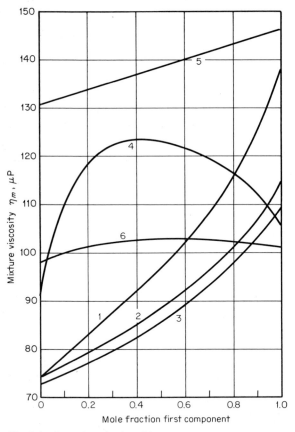

Fig. 9-4 Gas-mixture viscosities.

No.	System	T, K	Ref.
1	Hydrogen sulfide–ethyl ether	331	158
2	Hydrogen sulfide–ammonia	331	158
3	Methane–n-butane	293	117
4	Ammonia-hydrogen	306	157
5	Ammonia-methylamine	423	31
6	Ethylene-ammonia	293	208

ammonia-hydrogen occur most often in polar-nonpolar mixtures [94, 177], where the pure-component viscosities are not greatly different; the maxima are more pronounced as the molecular-weight ratio differs from unity.

The rigorous kinetic theory of Chapman-Enskog can be extended to determine the viscosity of a low-pressure multicomponent gas mixture [23–25, 27, 43, 93]. The final expressions are relatively complicated and consist of the ratio of two determinants which contain elements involving mole fractions, molecular weights, pure-component viscosities, temperature, and various collision integrals. Neglecting second-order effects, the rigorous solution can be approximated in a series as

$$\eta_m = \sum_{i=1}^{n} \frac{y_i \eta_i}{\sum_{j=1}^{n} y_j \phi_{ij}} \tag{9-5.1}$$

In the next several sections, methods for estimating the parameters ϕ_{ij} and ϕ_{ji} are presented. It is also readily shown that after such terms are found, certain systems will show a viscosity maximum. For a binary system of 1 and 2, where $\eta_1 > \eta_2$, if $(\eta_1/\eta_2)\phi_{12}\phi_{21} < 1$, there is a maximum viscosity at some composition between pure 1 and pure 2.

Wilke's Approximation of ϕ_{ij} Wilke [218] used Sutherland's kinetic-theory model to yield

$$\phi_{ij} = \frac{[1 + (\eta_i/\eta_j)^{1/2}(M_j/M_i)^{1/4}]^2}{[8(1 + M_i/M_j)]^{1/2}} \tag{9-5.2}$$

ϕ_{ji} is found by interchanging subscripts or by

$$\phi_{ji} = \frac{\eta_j}{\eta_i} \frac{M_i}{M_j} \phi_{ij} \tag{9-5.3}$$

For a binary system of 1 and 2, with Eqs. (9-5.1) to (9-5.3),

$$\eta_m = \frac{y_1 \eta_1}{y_1 + y_2 \phi_{12}} + \frac{y_2 \eta_2}{y_2 + y_1 \phi_{21}} \tag{9-5.4}$$

where η_m = viscosity of mixture
η_1, η_2 = pure-component viscosities
y_1, y_2 = mole fractions

and
$$\phi_{12} = \frac{[1 + (\eta_1/\eta_2)^{1/2}(M_2/M_1)^{1/4}]^2}{\{8[1 + (M_1/M_2)]\}^{1/2}}$$

$$\phi_{21} = \phi_{12} \frac{\eta_2}{\eta_1} \frac{M_1}{M_2}$$

Equation (9-5.1) with ϕ_{ij} from (9-5.2) has been extensively tested. Wilke [218] compared calculated values with data on 17 binary systems

TABLE 9-5 Comparison of Calculated and Experimental Binary-Gas-Mixture Viscosities

Pressures near atmospheric

System	T, °C	Mole fraction first component	Viscosity (exp.), μP	Ref.	Percent deviation† calculated by method of:			
					Wilke	Herning and Zipperer	Dean and Stiel	Brokaw
Hydrogen-nitrogen	100	0.0	210.1	157, 207				
		0.202	205.8		1.4	−1.2	‡	2.5
		0.490	190.3		5.6	−1.0		8.6
		0.800	152.3		12.0	−2.0		16.0
		1.0	104.2					
Methane-propane	25	0.0	81.0	14			3.1	
		0.2	85.0		−0.4	−0.2	3.9	−0.6
		0.4	89.9		−0.8	−0.7	3.3	−0.7
		0.6	95.0		−0.4	−0.2	3.0	1.4
		0.8	102.0		−0.6	−0.5	1.0	0.8
		1.0	110.0				−1.8	
	225	0.0	131.0	14			1.5	
		0.2	136.0		0.0	0.0	2.1	0.6
		0.4	142.0		0.0	−0.5	3.5	1.0
		0.6	149.0		0.0	−0.6	4.0	1.2
		0.8	157.0		0.0	−0.3	3.3	1.1
		1.0	167.0				0.0	
Sulfur hexachloride–carbon tetrachloride	30	0.0	176.7	167			12.7	
		0.246	164.3		4.8	4.9	4.4	4.8
		0.509	161.5		3.4	3.9	5.9	3.8
		0.743	159.9		2.3	2.3	5.7	2.2
		1.0	159.0				5.2	

System							
Nitrogen–carbon dioxide	20	0.0	146.6	−1.3	−1.1	0.2	−0.9
		0.213	153.5	−1.8	−1.5	−1.9	−1.1
		0.495	161.8	−2.7	−2.6	−3.0	−2.3
		0.767	172.1			−3.4	
		1.0	175.8			−0.1	
Hydrogen-ammonia	33	0.0	105.9	−9.7	−13.0	‡	−4.5
		0.323	120.0	−10.9	−16.0		−4.1
		0.464	122.4	−12.0	−19.0		−4.2
		0.601	123.8	−10.8	−18.0		−3.7
		0.805	118.4				
		1.0	90.6				
Ethyl ether–hydrogen sulfide	58	0.0	137.9	−0.3	4.2	−4.3	2.8
		0.198	116.0	−2.8	3.2	−2.1	1.1
		0.500	96.8	−3.5	0.1	1.0	−1.0
		0.796	83.5			5.4	
		1.0	74.2			11.0	
Ammonia-methylamine	150	0.0	130.7	−0.3	−0.6	‡	0.0
		0.25	134.5	−0.3	−0.6		0.2
		0.75	142.2				
		1.0	146.0				
	400	0.0	204.8	−0.7	−0.8	‡	−0.3
		0.25	212.8	−0.7	−0.7		−0.1
		0.75	228.3				
		1.0	236.0				

†Percent deviation = [(calc. − exp.)/(exp.)] × 100.
‡Method not applicable for systems containing hydrogen or polar compounds.

and reported an average deviation of less than 1 percent; several cases in which η_m passed through a maximum were included.   Many other investigators have tested this method [5, 28, 45, 51, 71, 165, 179, 180, 196, 210, 221].   In most cases, only nonpolar mixtures were compared, and very good results obtained.   For some systems containing hydrogen as one component, less satisfactory agreement was noted.   In Table 9-5, Wilke's method predicted mixture viscosities that were larger than experimental for the H_2–N_2 system, but for H_2–NH_3, it underestimated the viscosities.   Gururaja et al. [84] found that this method also overpredicted in the H_2–O_2 case but was quite accurate for a H_2–CO_2 system. Wilke's approximation has proved reliable even for polar-polar gas mixtures of aliphatic alcohols [172].   The principal reservation appears to lie in those cases where $M_i \gg M_j$ and $\eta_i \gg \eta_j$ [40].   Omitting such cases, a comparison of experimental and calculated values in Table 9-5 indicates that the error is usually less than 2 percent.   For mixtures involving hydrogen or helium as one component, Reichenberg has developed a more accurate, albeit more complex, correlation [169].

Example 9-5   Kestin and Yata [117] report that the viscosity of a mixture of methane and n-butane is 93.35 μP at 20°C when the mole fraction of n-butane is 0.303.   Compare this result with the value estimated by Wilke's method.   For pure methane and n-butane, these same authors report viscosities of 109.4 and 72.74 μP.

    solution   Let 1 refer to methane and 2 to n-butane. $M_1 = 16.043$, and $M_2 = 58.124$.

$$\phi_{12} = \frac{[1 + (109.4/72.74)^{1/2}(58.124/16.043)^{1/4}]^2}{\{8[1 + 16.043/58.124]\}^{1/2}} = 2.268$$

$$\phi_{21} = 2.268 \frac{72.74}{109.4} \frac{16.043}{58.124} = 0.416$$

$$\eta_m = \frac{(0.697)(109.4)}{0.697 + (0.303)(2.268)} + \frac{(0.303)(72.74)}{0.303 + (0.697)(0.146)} = 92.26 \ \mu\text{P}$$

$$\text{Error} = \frac{92.26 - 93.35}{93.35} \times 100 = -1.2\%$$

Herning and Zipperer Approximation of ϕ_{ij}   As an approximate expression for ϕ_{ij} the following is proposed [92]:

$$\phi_{ij} = \left(\frac{M_j}{M_i}\right)^{1/2} = \frac{1}{\phi_{ji}} \tag{9-5.5}$$

In Fig. 9-5, $(M_j/M_i)^{1/2}$ is plotted against the ϕ_{ij} values determined from Eq. (9-5.2) using the compounds shown in Table 9-5 as well as the CH_4–n-C_4H_{10} system of Example 9-5.   Although the dashed line indicates that the ϕ_{ij} in Eq. (9-5.5) is a poor approximation for Wilke's ϕ_{ij}, there is a rather remarkable fit if one examines the solid line drawn through the

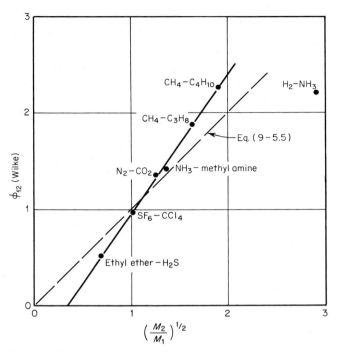

Fig. 9-5 Effect of molecular weight on ϕ_{12}.

data. This line may be expressed as

$$\phi_{12}\,(\text{Wilke}) = 1.45\left(\frac{M_2}{M_1}\right)^{1/2} - 0.505 \qquad (9\text{-}5.6)$$

This simple relationship is *not* applicable for systems containing hydrogen, for example, H_2–NH_3 or H_2–N_2 [which lies off the graph, $(M_{N_2}/M_{H_2})^{1/2} = 3.72$, $\phi_{H_2\text{-}N_2} = 1.90$], but these are just the systems for which neither the Wilke nor the Herning and Zipperer approximations are particularly accurate. In the rest of the test cases shown in Table 9-5, Eq. (9-5.5) proves to be quite reliable. Apparently, multiple sets of ϕ_{ij}-ϕ_{ji} work satisfactorily in Eq. (9-5.1).

Example 9-6 Repeat Example 9-5 using the Herning and Zipperer approximation for ϕ_{ij}.
  solution As before, with 1 as methane and 2 as n-butane,

$$\phi_{12} = \left(\frac{58.124}{16.043}\right)^{1/2} = 1.903 \qquad \phi_{21} = \phi_{12}^{-1} = 0.525$$

$$\eta_m = \frac{(0.697)(109.4)}{0.697 + (0.303)(1.903)} + \frac{(0.303)(72.74)}{0.303 + (0.697)(0.525)} = 92.82\ \mu\text{P}$$

$$\text{Error} = \frac{92.82 - 93.35}{93.35} \times 100 = -0.6\%$$

Brokaw Approximation of ϕ_{ij}   In a particularly complete study of gas-mixture viscosities, Brokaw [25, 26] suggested that ϕ_{ij} be obtained from

$$\phi_{ij} = \left(\frac{\eta_i}{\eta_j}\right)^{1/2} S_{ij} A_{ij} \tag{9-5.7}$$

and ϕ_{ji} be determined by interchanging subscripts.   The term S_{ij} is set equal to unity if i and j are nonpolar.   For polar gases, the determination of S_{ij} is discussed later. A_{ij} is a function only of the molecular-weight ratios, i.e.,

$$A_{ij} = m_{ij}M_{ij}^{-1/2}\left[1 + \frac{M_{ij} - M_{ij}^{0.45}}{2(1 + M_{ij}) + \dfrac{(1 + M_{ij}^{0.45})m_{ij}^{-1/2}}{1 + m_{ij}}}\right] \tag{9-5.8}$$

with

$$m_{ij} = \left[\frac{4}{(1 + M_{ij}^{-1})(1 + M_{ij})}\right]^{0.25} \tag{9-5.9}$$

$$M_{ij} = \frac{M_i}{M_j} \tag{9-5.10}$$

A_{ij} is readily determined with machine computation.   It is also plotted as a function of M_i/M_j in Figs. 9-6 and 9-7, or it can be read from monographs shown by Brokaw [25, 26]. A_{ij} can be approximated by

$$A_{ij} = \left(\frac{M_i}{M_j}\right)^{-0.37} \qquad 0.4 < \frac{M_i}{M_j} < 1.33 \tag{9-5.11}$$

with less than 1 percent error over this range of molecular-weight ratios.

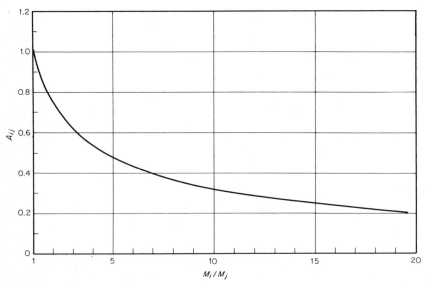

Fig. 9-6   Brokaw values of A_{ij} for $1 < M_i/M_j < 20$.

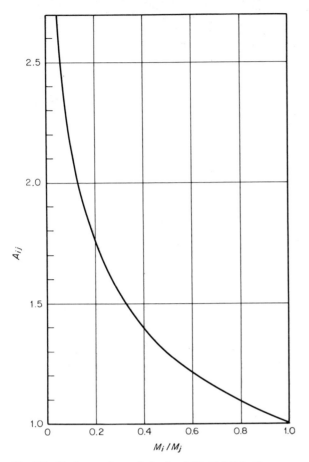

Fig. 9-7 Brokaw values of A_{ij} for $0.05 < M_i/M_j < 1$.

For mixtures containing polar gases, the parameters δ and T^* are required. δ was defined in Eq. (9-4.6), and values can be found in Table 9-2 or estimated from Eq. (9-4.10). T^* equals kT/ϵ, and, again, ϵ is shown for many polar compounds in Table 9-2. An estimation equation is also given in Eq. (9-4.9).

$$S_{ij} = S_{ji} = \frac{1 + (T_i^* T_j^*)^{1/2} + (\delta_i \delta_j/4)}{[1 + T_i^* + (\delta_i^2/4)]^{1/2}[1 + T_j^* + (\delta_j^2/4)]^{1/2}} \qquad (9\text{-}5.12)$$

Equation (9-5.12) should only be used if either δ_i or $\delta_j > 0.1$. If both are less than 0.1, $S_{ij} = S_{ji} = 1.0$.

Shown in Table 9-5 are percent deviations in calculated values of viscosity for several mixtures. Errors are similar to those found by the Wilke method, though for mixtures containing polar components,

Brokaw's method of approximating ϕ_{ij} is definitely superior. Many more systems were compared by Brokaw, and, in general, errors less than 2 percent were found for a wide variety of nonpolar and polar mixtures. An example calculation for a ternary mixture is given below.

Example 9-7 Estimate the viscosity of a ternary mixture containing 0.494 mole fraction methyl chloride, 0.262 mole fraction sulfur dioxide, and 0.244 mole fraction dimethyl ether. The temperature is 80°C. The experimental value is 131.9 μP [41].

solution The viscosities of the pure components at this temperature are shown below [41]. Also, the values of δ and ϵ/k from Table 9-2 are given and T^* calculated from $T^* = kT/\epsilon$, with $T = 353.2$ K.

Component	M	δ	$\frac{\epsilon}{k}$, K	T^*	η, μP
Methyl chloride	50.488	0.5	414	0.853	127.8
Sulfur dioxide	64.063	0.42	347	1.018	152.3
Dimethyl ether	46.069	0.19	432	0.817	109.8

With these values, the S_{ij} factors are found from Eq. (9-5.12). For example, if i = methyl chloride and j = sulfur dioxide,

$$S_{ij} = \frac{1 + [(0.853)(1.018)]^{1/2} + [(0.5)(0.42)/4]}{[1 + 0.853 + (0.5)^2/4]^{1/2}[1 + 1.018 + (0.42)^2/4]^{1/2}} = 1.00$$

Other values are shown below. A_{ij} values are found from Eq. (9-5.8).

Component		$\dfrac{M_i}{M_j}$	A_{ij}	S_{ij}	$\left(\dfrac{\eta_i}{\eta_j}\right)^{1/2}$	ϕ_{ij}
i	j					
Methyl chloride	Sulfur dioxide	0.788	1.095	1.00	0.916	1.013
Sulfur dioxide	Methyl chloride	1.269	0.909	1.01	1.092	1.003
Methyl chloride	Dimethyl ether	1.096	0.964	0.993	1.079	1.033
Dimethyl ether	Methyl chloride	0.912	1.036	0.993	0.927	0.954
Sulfur dioxide	Dimethyl ether	1.390	0.875	0.995	1.178	1.026
Dimethyl ether	Sulfur dioxide	0.719	1.133	0.995	0.849	0.957

From Eq. (9-5.1),

$$\eta_m = \frac{(0.494)(127.8)}{(0.494) + (0.262)(1.013) + (0.244)(1.033)} + \frac{(0.262)(152.3)}{(0.262) + (0.494)(1.003) + (0.244)(1.026)}$$

$$+ \frac{(0.244)(109.8)}{(0.244) + (0.494)(0.954) + (0.262)(0.957)} = 130 \ \mu P$$

$$\text{Error} = \frac{130 - 132}{132} \times 100 = -1.5\%$$

Corresponding-States Correlations Any one of the several corresponding-states methods described in Sec. 9-4 for estimating pure-gas viscosities can also be used for gas mixtures.   However, rules must be made to find M, P_c, and T_c for the mixture.   For example, Dean and Stiel [53] suggested a relation essentially the same as Eq. (9-4.17):

$$\eta_m \xi_m = \begin{cases} (3.40) T_{r_m}^{8/9} & T_r \leqslant 1.5 \\ (16.68)(0.1338 T_r - 0.0932)^{5/9} & T_r > 1.5 \end{cases} \qquad (9\text{-}5.13)$$

where η = mixture viscosity, μP

$$\xi_m = \frac{T_{c_m}^{1/6}}{P_{c_m}^{2/3} \left(\sum_i y_i M_i \right)^{1/2}}$$

$$T_{r_m} = \frac{T}{T_{c_m}} \qquad (9\text{-}5.14)$$

T_{c_m} and P_{c_m} are to be determined from the Prausnitz and Gunn mixing rules, Eqs. (4-2.1) and (4-2.2).   For the nonpolar systems shown in Table 9-5, Dean and Stiel's method yielded good results but with errors normally greater than either Wilke's or Brokaw's technique. There is also no way of employing pure-component viscosity values, should they be available.   In an analogous manner, other corresponding-states methods could be used to predict mixture viscosities. Both Yoon and Thodos [222] and Hattikudur and Thodos [90] have suggested other ways of finding ξ_m for both nonpolar and polar-nonpolar gas mixtures.   None of these corresponding-states methods, however, appears to be as accurate and as general as those employing Eq. (9-5.1) with a reliable estimation of ϕ_{ij}.

Other Methods for Estimating Low-Pressure Gas-Mixture Viscosity Strunk et al. [196, 197] have proposed that Eq. (9-5.1) be used.   To determine σ and $\Omega_v(T_m^*)$ as a function of composition, particular combining rules are specified.   This technique is simple and appealing, although the method as proposed is limited to nonpolar mixtures and the reported accuracy is similar to that found by Wilke's form. Saxena and coworkers [83, 178, 205] also take Eq. (9-5.1) as a starting point for developing a correlation, but at least one viscosity of the mixture is required to obtain other estimates; good results are normally found if such a mixture viscosity is introduced.   Other modifications of Eq. (9-5.1) are available [32, 65, 111].

Recommendations: Viscosities of Gases at Low Pressures   Equation (9-5.1) should be used with the parameter ϕ_{ij} from either Wilke's or Brokaw's equations, (9-5.2) or (9-5.7), respectively.   The latter is probably preferable if any of the components are polar; it is also somewhat more complex.   For polar components, the dipole moment must be known.   For nonpolar mixtures, errors less than 2 to 3 percent are customarily found.   For polar-polar or nonpolar-polar mixtures, errors seldom exceed 3 to 4 percent.   The only systems which present occasional problems are those with a pronounced maximum in the viscosity of the mixture.   Such maxima are found in binary systems with (η_1/η_2) $\phi_{12}\phi_{21} < 1$ and where $\eta_1 > \eta_2$.

9-6   Effect of Pressure on the Viscosity of Pure Gases

The viscosity of gases is a strong function of pressure only in certain regions of pressure and temperature.   Usually, pressure variations are not significant at very high reduced temperatures or low reduced pressures.   In Fig. 9-8, we show experimental viscosities of several gases as reported by Kestin and Leidenfrost [113].   For gases where the reduced temperature is well above unity, there is little effect of pressure.   There is a significant increase in the viscosity of xenon with pressure at 25°C ($T_r = 1.03$).   For CO_2 ($T_r = 0.96$), the highest pressure for which data are reported is 20 atm, where $P_r = 0.27$; this is still a low reduced pressure.   At somewhat higher pressures, the viscosity would be expected to increase sharply.   In Fig. 9-9, viscosity data for n-butane are shown [55].   It is clear that in the vicinity of the saturated vapor and the critical point, pressure effects are quite important.

The effect of pressure is perhaps best seen by examination of Fig. 9-10, which, although only approximate, shows the trends of viscosity with both temperature and pressure [209].   This figure will be discussed later; for the present, it is only necessary to note that the viscosity has been reduced by dividing by the value at the critical point.   At low

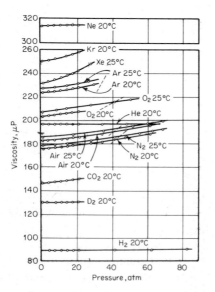

Fig. 9-8   Viscosity of gases at low reduced pressure or high reduced temperature. (*From Ref.* 113.)

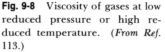

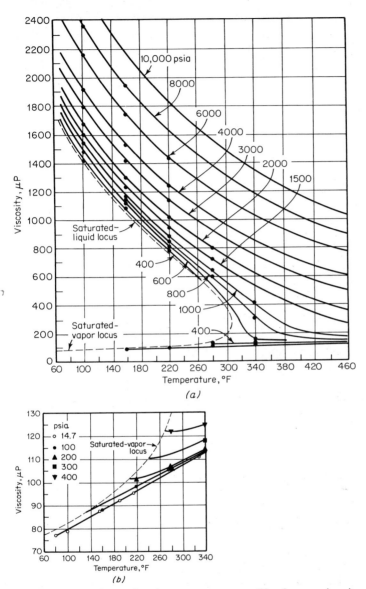

Fig. 9-9 (a) n-Butane viscosity vs. temperature; (b) n-butane viscosity vs. temperature at low pressures. (*From Ref.* 55.)

reduced pressures, we can see that except near the saturated-vapor state there is little effect of pressure. The lower limit of the P_r curves would be indicative of a dilute-gas state, as described in Sec. 9-4. In such a state, η increases with temperature. At high reduced pressures, we see that there is a wide range of temperatures where η decreases with

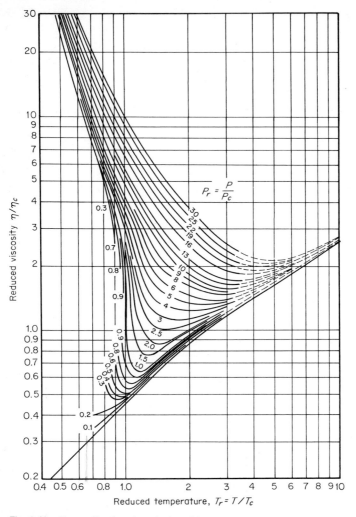

Fig. 9-10 Generalized reduced viscosities. (*From Ref.* 209.)

temperature.   In this region, the viscosity behavior more closely simu-
lates a liquid state, and, as will be shown in Sec. 9-12, an increase
in temperature results in a *decrease* in viscosity. Finally, at very
high reduced temperatures, there again results a condition where there
is but little effect of pressure on gas viscosity and $\partial\eta/\partial T > 0$.   This last
region is representative of many of the "permanent" gases shown in Fig.
9-8.

Childs and Hanley [46] have deduced criteria which indicate whether
or not the pressure effect is significant.   Their results are summarized
in an approximate fashion in Fig. 9-11.   For any given reduced temper-

ature and pressure, one can determine whether the gas is "dilute" (so that the equations given in Sec. 9-4 are suitable) or dense (so that a pressure correction should be applied).   The dividing line is located so that the necessary dense-gas correction is 1 percent or less.   Galloway and Sage [67] also discuss the effect of pressure near atmospheric pressure.

There is also a lower pressure limit, below which flow equations must be modified when the "low-pressure" gas viscosities discussed in Sec. 9-4 are employed.   This is because the assumption of no slip at the walls of the conduit no longer holds at very low pressure, not because of a change in the viscosity itself.

For example, the Poiseuille equation for isothermal gas flow in a round tube may be written

$$q = F\frac{\pi r_w{}^4(P_1{}^2 - P_2{}^2)}{16\eta L P_0} \qquad (9\text{-}6.1)$$

where r_w = radius of tube of length L, cm

P_1, P_2 = pressures at the two ends, dyn/cm^2

q = volumetric flow rate as measured at any pressure P_0, cm^3/s

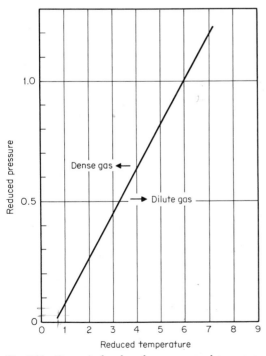

Fig. 9-11   Ranges of reduced pressure and temperature for separating dilute and dense gases.

The factor F is unity at ordinary pressures but increases with increase in the ratio of the molecular mean free path to tube radius and may reach values of 1000 or more for flow in small capillaries at very low pressure. The correction factor F appears because of the effect of the conduit walls, though it might be said the viscosity is decreased to an effective viscosity η/F. F is less than 1.1 if $(\eta/r_w P_m)(M/RT)^{-1/2}$ is less than about 0.3 [29]. P_m is the arithmetic mean of P_1 and P_2. With the units noted above, R is 8.314×10^7; 1 μm Hg = 1.33 dyn/cm^2.

Enskog Dense-Gas Theory   One of the very few theoretical efforts to predict the effect of pressure on the viscosity of gases is due to Enskog and is treated in detail by Chapman and Cowling [43]. The theory has also been applied to dense-gas diffusion coefficients, bulk viscosities, and, for monatomic gases, to thermal conductivities. The assumption is made that the gas consists of dense, hard spheres and behaves like a low-density hard-sphere system except that all events occur at a faster rate due to the higher rate of collision [2, 3]. The increase in collision rate is proportional to the radial distribution function χ. The Enskog equation for shear viscosity is

$$\frac{\eta}{\eta^\circ} = \chi^{-1} + 0.8 b_0 \rho + 0.761 \chi (b_0 \rho)^2 \qquad (9\text{-}6.2)$$

where η = viscosity, μP
$\quad \eta^\circ$ = low-pressure viscosity, μP
$\quad b_0$ = excluded volume = $\frac{2}{3}\pi N_0 \sigma^3$, cm^3/g-mol
$\quad N_0$ = Avogadro's number
$\quad \sigma$ = hard-sphere diameter, Å
$\quad \rho$ = molal density, g-mol/cm^3
$\quad \chi$ = radial distribution function

In the Enskog model, there is no correlation between successive hard-sphere collisions, although this simplification apparently introduces little error except at high densities [4].

When χ and b_0 are obtained from experimental data, the Enskog equation usually correlates dense-gas viscosities moderately well [57, 143]. The key to using Eq. (9-6.2) is to obtain values of χ. Often, an equation of state is invoked which employs the Enskog modulus $b_0 \rho \chi$ [88], and then, by simple differentiation, the modulus can be expressed in terms of the *thermal pressure* $(\partial P/\partial T)_v$. The viscosity ratio η/η° is then correlated directly with the thermal pressure [52, 76, 124, 125].

In other cases, η/η° is simply expanded in a power series with density as the independent variable, e.g.,

$$\frac{\eta}{\eta^\circ} = 1 + b\rho + c\rho^2 + \cdots \qquad (9\text{-}6.3)$$

$b, c, \ldots$ are temperature-dependent coefficients. Estimation techniques for these coefficients have been suggested [89, 175, 185]. They normally show only a weak dependence on temperature but are functions of composition [215]. The correlation by Coremans and Beenakker [49] is perhaps the most convenient to use. It is discussed elsewhere [173].

Residual-Viscosity Correlations   Whereas the Enskog theory [Eq. (9-6.2)] suggests that the viscosity ratio $\eta/\eta°$ be correlated with density and, perhaps, temperature, it has now been shown that a preferable method employs the residual viscosity function $\eta - \eta°$. η is the dense-gas viscosity and $\eta°$ the dilute-gas viscosity at the same temperature. $\eta°$ would be found from low-pressure data or by estimation techniques described in Sec. 9-4. To illustrate the usual manner of introducing the residual viscosity function, consider Fig. 9-12. Here, the residual viscosity for n-butane is plotted as a function of density [55]. The single curve shown contains essentially all the data plotted earlier in Fig. 9-9. Note that there does not appear to be any specific effect of temperature over the range shown. At the highest density, 0.6 g/cm³, the reduced density ρ/ρ_c is 2.63. Similar plots for many other sub-

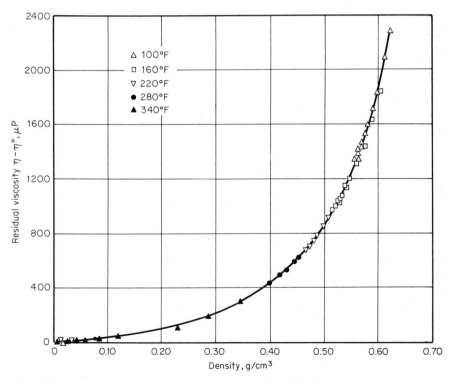

Fig. 9-12   Residual n-butane viscosity vs. density. (*From Ref. 55.*)

stances are available, for example, He, air, O_2, N_2, CH_4 [114]; ammonia [36, 182]; inert gases [181]; diatomic gases [20]; sulfur dioxide [183]; CO_2 [110]; steam [115]; and various hydrocarbons [34, 56, 73, 191, 192]. Other authors have also shown the applicability of a residual viscosity-density correlation [57, 76, 89, 115, 174, 190]. With the success attained for pure materials, it was logical to attempt to generalize this type of correlation. The most accurate method yet suggested is summarized in Eqs. (9-6.4) to (9-6.8).

Nonpolar Gases [108]

$$[(\eta - \eta°)\xi + 1]^{0.25} = 1.0230 + 0.23364\rho_r + 0.58533\rho_r^2 - 0.40758\rho_r^3$$
$$+ 0.093324\rho_r^4 \qquad (9\text{-}6.4)$$

where η = dense-gas viscosity, μP
$\quad\eta°$ = low-pressure gas viscosity, μP
$\quad\rho_r$ = reduced gas density $\rho/\rho_c = V_c/V$
$\quad\xi$ = group $T_c^{1/6}/M^{1/2}P_c^{2/3}$, where T_c is in kelvins and P_c in atmospheres
$\quad M$ = the molecular weight

This relation is reported by Jossi, Stiel, and Thodos to be applicable in the range $0.1 \leqslant \rho_r < 3$. Values of the critical volume V_c are tabulated in Appendix A.

Polar Gases [195]

$$(\eta - \eta°)\xi = 1.656\rho_r^{1.111} \qquad \rho_r \leqslant 0.1 \qquad (9\text{-}6.5)$$
$$(\eta - \eta°)\xi = 0.0607(9.045\rho_r + 0.63)^{1.739} \qquad 0.1 \leqslant \rho_r \leqslant 0.9 \qquad (9\text{-}6.6)$$
$$\log\{4 - \log[(\eta - \eta°)\xi]\} = 0.6439 - 0.1005\rho_r - \Delta \qquad 0.9 \leqslant \rho_r < 2.6$$
$$(9\text{-}6.7)$$

where $\quad \Delta = \begin{cases} 0 & 0.9 \leqslant \rho_r < 2.2 \\ (4.75)(10^{-4})(\rho_r^3 - 10.65)^2 & 2.2 < \rho_r < 2.6 \end{cases} \qquad (9\text{-}6.8)$

Also, $(\eta - \eta°)\xi = 90.0$ and 250 at $\rho_r = 2.8$ and 3.0, respectively. The notation used in Eqs. (9-6.5) to (9-6.8) is defined under Eq. (9-6.4).

Discussion   The relations shown in Eqs. (9-6.4) to (9-6.8) are shown in Figs. 9-13 and 9-14. Although there is some scatter, the results are remarkably good. It is easily seen that the viscosity group $(\eta - \eta°)\xi$ is very sensitive to the reduced density at high densities. The exclusion of any specific temperature effect has been criticized by Rogers and Brickwedde [174], who suggest that as the temperature of a gas decreases toward saturation, $(\eta - \eta°)\xi$ becomes a function of both temperature and density. Starling and Ellington also found this to be true for ethane [191], but other ethane data reported by Carmichael and Sage [34] show no such temperature separation. Certainly, these correla-

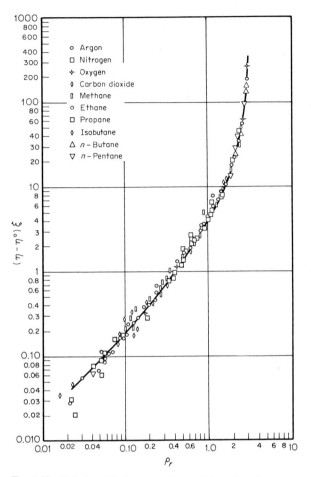

Fig. 9-13 Relationship between $(\eta - \eta^\circ)\xi$ and ρ_r for non-polar substances. (*From Ref.* 108.)

tions are only approximate, but they are reasonably accurate and simple to use.   Density data would usually be estimated as shown in Chap. 3.   Presumably, $(\eta - \eta^\circ)\xi$ could be related to some other density function, but less success has been achieved [73, 125] than when using reduced density.

One should note that if viscosity correlations as shown in Figs. 9-12 to 9-14 are representative, the lack of temperature dependence would indicate that the Enskog theory [Eq. (9-6.2)] could not also apply [73].

We show in Table 9-6 a few values of dense-gas viscosities calculated by Eqs. (9-6.4) to (9-6.7).   The agreement between experimental and estimated viscosities is reasonable.   Generally errors normally do not exceed 10 to 15 percent.

TABLE 9-6 Comparison between Calculated and Experimental Dense-Gas Viscosities

Substance	ξ	T, °C	P, atm	η(exp.), μP	Ref.	$\eta°$, μP	Percent error†	
							Jossi, Stiel, and Thodos	Reichenberg
Ammonia	0.0285	171	13.6	157	36	157	0.7	0.6
			40.8	157			2.4	3.2
			68.0	162			1.4	3.4
			136.1	197			−2.5	−2.5
			204.1	317			−18	−16
			340.2	500			−18	−15
n-Butane	0.0321	171	6.8	113	55	112	1.6	1.5
			34	134			4.7	5.0
			68	405			−7.4	7.6
			340	938			−13	−8.7
			680	1330			−9.1	−12
Carbon dioxide	0.0224	50	41	170	116, 143	161	3.9	5.9
			70	190			7.4	7.6
			90	237			5.1	4.6
			100	295			3.8	1.1
			117	425			−1.5	−1.1
Nitrogen	0.0407	25	35	185	176	178	−0.8	−1.9
			69	190			0.5	−0.8
			137	208			2.6	0.3
			341	286			4.2	1.8
			682	415			0	1.7
Sulfur dioxide	0.0189	200	18	228	183	206	−8.0	−7.9
			52	250			−11	−10
			69	263			−12	−11
			137	346			11	2.0
Average							6.2	5.3

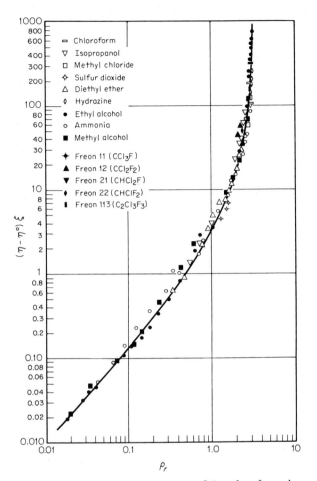

Fig. 9-14 Relationship between $(\eta - \eta^\circ)\xi$ and ρ_r for polar substances. (*From Ref.* 195.)

Example 9-8 Estimate the viscosity of ammonia at 171°C and 136 atm. The experimental value is 197 μP [36].

solution Ammonia is a polar gas, and Eqs. (9-6.5) to (9-6.7) must be used. First, however, the reduced density must be determined. From Appendix A, $T_c = 405.6$ K, $P_c = 111.3$ atm, $V_c = 72.5$ cm^3/g-mol, and $\omega = 0.250$. Thus, $T_r = 444/405.6 = 1.09$, $P_r = 136/111.3 = 1.22$. The compressibility factor Z is calculated from Pitzer's correlation [Eq. (3-3.1)] with $Z^{(0)}$ and $Z^{(1)}$ determined from Tables 3-1 and 3-2 as $Z^{(0)} = 0.558$, $Z^{(1)} = 0.096$. Then $Z = Z^{(0)} + \omega Z^{(1)} = 0.558 + (0.250)(0.096) = 0.582$, and

$$\rho_r = \frac{\rho}{\rho_c} = \rho V_c = \frac{PV_c}{ZRT} = \frac{(136)(72.5)}{(0.558)(82.07)(444)} = 0.485$$

Equation (9-6.6) should be used. ξ is determined as $T_c^{1/6}/M^{1/2}P_c^{2/3} = (405.6)^{1/6}/$

$(17.0)^{1/2}(111.3)^{2/3} = 0.0285.$ $\eta°$ is $157\ \mu$P. Then

$$(\eta - 157)(0.0285) = 0.0607[(9.045)(0.485) + 0.63]^{1.739}$$

$$\eta = 192\ \mu\text{P}$$

$$\text{Error} = \frac{192 - 197}{197} \times 100 = -2.5\%$$

Viscosity-Ratio Correlations Many investigators have suggested that the dense-gas viscosity be expressed as η/η_c, $\eta/\eta_c°$, or $\eta/\eta°$ and this ratio correlated with either T_r and P_r or with other reduced volumetric-type properties [9, 13, 20, 22, 28, 38, 39, 47–49, 68, 76, 81, 131, 132, 176, 181]. In such ratios, η_c is the true viscosity at the critical point, $\eta_c°$ is the low-pressure viscosity at T_c, and $\eta°$ is the low-pressure viscosity at the system temperature. Figure 9-10 is an example of such a correlation [209]. To use this figure, a value of η_c is required. From Eq. (9-4.17), at $T_r = 1$, $\eta_c°\xi = 3.44$. Also, from Eq. (9-6.4), at the critical point, $\rho_r = 1$, $(\eta_c - \eta_c°)\xi = 4.45$. Thus, $\eta_c\xi = 3.44 + 4.45 = 7.89$, which is very close to the value of 7.7 recommended by Uyehara and Watson [209] in 1944.†

Reichenberg [170a] has developed a more accurate correlation for $\eta/\eta°$ which is said to apply to both polar and nonpolar gases. The estimation equation is

$$\frac{\eta}{\eta°} = 1 + (1 - 0.45q)\frac{AP_r^{1.5}}{BP_r + (1 + CP_r^D)^{-1}} \tag{9-6.9}$$

The constants A, B, C, and D are functions of the reduced temperature T_r as shown below:

$$A = \frac{\alpha_1}{T_r}\exp\alpha_2 T_r^{-\alpha_3} \qquad B = A(\beta_1 T_r - \beta_2)$$

$$C = \frac{\gamma_1}{T_r}\exp\gamma_2 T_r^{-\gamma_3} \qquad D = \frac{\delta_1}{T_r}\exp\delta_2 T_r^{-\delta_3}$$

$$\alpha_1 = 1.9824 \times 10^{-3} \qquad \alpha_2 = 5.2683 \qquad \alpha_3 = 0.5767$$

$$\beta_1 = 1.6552 \qquad \beta_2 = 1.2760$$

$$\gamma_1 = 0.1319 \qquad \gamma_2 = 3.7035 \qquad \gamma_3 = 79.8678$$

$$\delta_1 = 2.9496 \qquad \delta_2 = 2.9190 \qquad \delta_3 = 16.6169$$

and

$$q = \frac{668(\mu_p)^2 P_c}{T_c^2}$$

with μ_p in debyes, P_c in atmospheres, and T_c in kelvins. For non-polar materials, $q = 0$. Although this simple relation has not been extensively tested, it appears to yield errors in the same range as those in

---

†Jossi et al. [108] in their correlation indicate that $(\eta_c - \eta_c°)\xi$ is closer to 4.3 than 4.45, as found from Eq. (9-6.4). Thus the agreement is almost exact.

the Jossi, Stiel, and Thodos method as shown in Table 9-6.   Note that the density need not be determined.

Example 9-9   Using the Uyehara and Watson dense-gas viscosity correlation, Fig. 9-10, estimate the viscosity of n-butane at 400°F and 2000 psia.   The experimental value is 462 μP [55].

   solution   For n-butane, from Appendix A, $T_c = 425.2$ K, $P_c = 37.5$ atm, $M = 58.12$. Since 400°F = 477.6 K, $T_r = 477.6/425.2 = 1.12$. $P_r = 2000/(14.696)(37.5) = 3.63$,   and $\xi = T_c^{1/6}/M^{1/2}P_c^{2/3} = (425.5)^{1/6}/(58.12)^{1/2}(37.5)^{2/3} = 0.0321$.   From the relation $\eta_c\xi = 7.7$, $\eta_c = 7.7/0.0321 = 240$ μP.   From Fig. 9-10 at $T_r = 1.12$, $P_r = 3.63$, $\eta/\eta_c = 2.0$, and $\eta = (2.0)(240) = 480$ μP.   The error is about 4 percent.

Example 9-10   Repeat Example 9-8 using the Reichenberg method.
   solution   In this case, $T_r = 1.09$ and $P_r = 1.22$.   With the definitions of A, B, C, and D given under Eq. (9-6.9), at $T_r = 1.09$, $A = 0.273$, $B = 0.144$, $C = 0.121$, and $D = 5.434$.   From Appendix A, $\mu_p = 1.5$ debyes, $P_c = 111.3$ atm, $T_c = 405.6$ K.   Thus, $q = (668)(1.5)^2(111.3)/(405.6)^2 = 1.01$.   From Eq. (9-6.9),

$$\frac{\eta}{\eta^\circ} = 1 + [1 - (0.45)(1.01)]\frac{(0.273)(1.22)^{1.5}}{(0.144)(1.22) + [1 + (0.121)(1.22)^{5.434}]^{-1}}$$

$$= 1.22$$

with $\eta^\circ = 157$ μP, $\eta = 192$ μP.

$$\text{Error} = \frac{192 - 197}{197} \times 100 = -2.5\%$$

Recommendations: Viscosity of Dense Gases

A rapid but approximate estimate of the dense-gas viscosity can be obtained from Fig. 9-10 with η_c found from the relation $\eta_c\xi = 7.7$.

   A more accurate estimation of dense-gas viscosity can be obtained with Eqs. (9-6.4) to (9-6.7).   In this case, a gas density is required, and if no experimental data are available, P-V-T correlations given in Chap. 3 should be used.   Equation (9-6.9) also allows one to estimate high-pressure pure-gas viscosities using only the reduced temperature and pressure.   Errors for both methods usually do not exceed about 10 percent except at very high pressures.

9-7 Effect of Pressure on the Viscosity of Gas Mixtures

As noted in the first part of Sec. 9-6, at moderate pressures and temperatures, a pressure correction to gas viscosities is not necessary.   Figure 9-11 may still be used as a rough guide to indicate whether or not a correction for pressure should be applied.   If such a correction is not warranted, the estimation methods outlined in Sec. 9-5 should be used to obtain gas-mixture viscosities.

   However, to estimate viscosities for dense-gas mixtures, the most accurate method at present appears to be that suggested by Dean and

Stiel [53]. In a manner similar to the residual-viscosity correlations given in Sec. 9-6, they propose

$$(\eta_m - \eta_m^\circ)\xi_m = (1.08)[\exp 1.439\rho_{r_m} - \exp(-1.111\rho_{r_m}^{1.858})] \qquad (9\text{-}7.1)$$

where η_m = high-pressure mixture viscosity, μP
$\quad \eta_m^\circ$ = low-pressure mixture viscosity, μP
$\quad \rho_{r_m}$ = pseudoreduced mixture density, ρ_m/ρ_{c_m}
$\quad \rho_m$ = mixture density, $g\ mol/cm^3$
$\quad \rho_{c_m}$ = pseudocritical mixture density, $g\ mol/cm^3 = P_{c_m}/Z_{c_m}RT_{c_m}$
$\quad \xi_m = T_{c_m}^{1/6}/M_m^{1/2}P_{c_m}^{2/3}$

The mixture molecular weight M_m is a mole-fraction average. The pseudocritical mixture parameters Z_{c_m}, T_{c_m}, and P_{c_m} must be calculated from some assumed pseudocritical-constant rule. Dean and Stiel chose the modified Prausnitz and Gunn rules (Sec. 4-1), i.e.,

$$T_{c_m} = \sum_i y_i T_{c_i} \qquad (9\text{-}7.2)$$

$$Z_{c_m} = \sum_i y_i Z_{c_i} \qquad (9\text{-}7.3)$$

$$V_{c_m} = \sum_i y_i V_{c_i} \qquad (9\text{-}7.4)$$

$$P_{c_m} = \frac{Z_{c_m}RT_{c_m}}{V_{c_m}} \qquad (9\text{-}7.5)$$

These pseudocritical values are then used to calculate ρ_{c_m} and ξ_m.

Equation (9-7.1) is to be used only for nonpolar mixtures; it is said to be applicable both to gases at high pressure and liquids at high temperature, but the accuracy for liquids with reduced densities greater than about 2 is expected to be poor. The equation has never been tested in any detail in the liquid region. When it was tested on nine gas mixtures at various densities (1396 different data points), the average error found was 3.7 percent; most mixtures were composed of light hydrocarbons or hydrocarbons and inert gases. A graph of Eq. (9-7.1) is shown in Fig. 9-15, and for the simple systems shown, the agreement is remarkable. The technique is illustrated in Example 9-11. A similar correlation has been independently proposed by Giddings [73]. In this case, different rules were chosen for determining the pseudocritical constants. A good correlation was obtained for light-hydrocarbon mixtures; it was also found that the correlation could be improved if the mole-fraction molecular weight were employed as a third correlating parameter.

Gambill [68] has reviewed estimation methods for η_m and in several papers discusses techniques to correlate η_m with composition for light-hydrocarbon systems [33, 54, 55, 77].

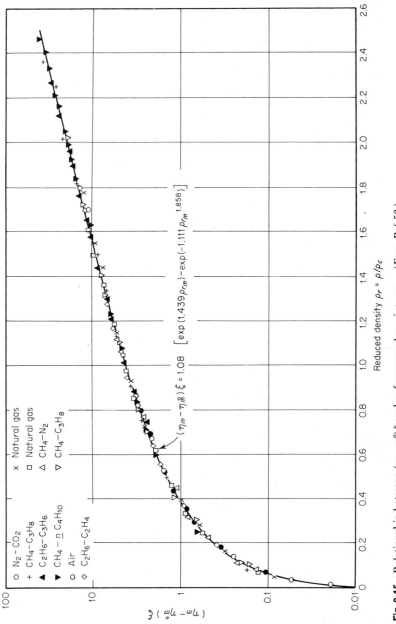

$$(\eta_m - \eta_m^\circ)\,\xi = 1.08\left[\exp(1.439\rho_{rm}) - \exp(-1.111\rho_{rm}^{1.858})\right]$$

Reduced density $\rho_r = \rho/\rho_c$

Fig. 9-15 Relationship between $(\eta - \eta^\circ)\xi$ and ρ_r for nonpolar mixtures. *(From Ref. 53.)*

Legend:

- ○ $N_2 - CO_2$
- + $CH_4 - C_3H_8$
- ▲ $C_2H_6 - C_3H_6$
- ▶ $CH_4 - \underline{n}\, C_4H_{10}$
- ○ Air
- ◇ $C_2H_6 - C_2H_4$
- × Natural gas
- □ Natural gas
- △ $CH_4 - N_2$
- ▽ $CH_4 - C_3H_8$

Example 9-11 Estimate the viscosity of a mixture of 18.65 mole percent ethylene and 81.35 mole percent ethane at 150°C and 120 atm. Dean and Stiel report the experimental value to be 188.2 μP.

solution Letting ethylene be component 1 and ethane component 2, we have, from Appendix A, $T_{c_1} = 282.4$ K, $T_{c_2} = 305.4$ K, $M_1 = 28.054$, $M_2 = 30.070$, $Z_{c_1} = 0.276$, $Z_{c_2} = 0.285$, $V_{c_1} = 129$ cm³/g-mol, $V_{c_2} = 148$ cm³/g-mol, $\omega_1 = 0.085$, and $\omega_2 = 0.098$.

$$T_{c_m} = (0.1865)(282.4) + (0.8135)(305.4) = 301.1 \text{ K}$$

$$M_m = (0.1865)(28.054) + (0.8135)(30.070) = 29.69$$

$$V_{c_m} = (0.1865)(129) + (0.8135)(148) = 144.5 \text{ cm}^3/\text{g-mol}$$

$$Z_{c_m} = (0.1865)(0.276) + (0.8135)(0.285) = 0.283$$

$$\omega_m = (0.1865)(0.085) + (0.8135)(0.098) = 0.096$$

$$P_{c_m} = \frac{Z_{c_m} R T_{c_m}}{V_{c_m}} = \frac{(0.283)(82.07)(301.1)}{144.5} = 48.4 \text{ atm}$$

$$\xi_m = \frac{T_{c_m}^{1/6}}{M_m^{1/2} P_{c_m}^{2/3}} = \frac{(301.1)^{1/6}}{(29.69)^{1/2}(48.4)^{2/3}} = 0.0358$$

To determine ρ_m, Pitzer's correlation [Eq. (3-3.1)], Tables 3-1 and 3-2 may be used with $T_{r_m} = 423/301 = 1.40$.

$$P_{r_m} = \frac{120}{48.4} = 2.48$$

$$Z_m^{(0)} = 0.747 \qquad Z_m^{(1)} = 0.214$$

$$Z_m = Z_m^{(0)} + \omega Z_m^{(1)} = 0.747 + (0.096)(0.214) = 0.767$$

$$V_m = \frac{Z_m R T}{P} = \frac{(0.767)(82.07)(423)}{120} = 222 \text{ cm}^3/\text{g-mol}$$

$$\rho_{r_m} = \frac{V_{c_m}}{V_m} = \frac{144.5}{222} = 0.651$$

The low-pressure mixture viscosity can be determined by any of the techniques discussed in Sec. 9-5. Since Dean and Stiel's high-pressure method is illustrated here, for consistency, their low-pressure value is employed; i.e., from Eq. (9-5.13), $\eta_m^\circ = 128.1 \, \mu$P. Then, from Eq. (9-7.1),

$$(\eta_m - \eta_m^\circ)\xi_m = (\eta_m - 128.1)(0.0358)$$

$$= 1.08(e^{(1.439)(0.651)} - (e^{(-1.111)(0.651)^{1.858}})$$

$$\eta_m = 187 \, \mu\text{P}$$

$$\text{Error} = \frac{187 - 188}{188} \times 100 = -0.5\%$$

Recommendations: Viscosities of Gas Mixtures at Elevated Pressures

To estimate the viscosities of nonpolar gas mixtures, use Eq. (9-7.1) with pseudocritical constants determined from the modified Prausnitz and Gunn rules. Errors are expected to be less than 10 percent for mixtures of low-molecular-weight gases. For high-molecular-weight nonpolar gas mixtures and for mixtures containing one or more polar components, no satisfactory method has yet been devised. Equation (9-7.1) may still be used but with the realization that larger errors can be expected.

9-8   Liquid Viscosity

Gas viscosities at low pressure can be estimated by techniques based on sound theory, but there is no comparable theoretical basis for the estimation of viscosity of liquids.   Certainly the viscosities of liquids are considerably different from those of gases; i.e., they are much larger numerically, and they decrease sharply with an increase in temperature.   The phenomenon of low-pressure gas viscosity is due primarily to the transfer of momentum by individual collisions between molecules moving randomly between layers with different velocities.   A similar momentum transfer may also exist in liquids, although it is usually overshadowed by the interacting force fields between the close-packed liquid molecules.   The densities of liquids are such that the average molecular-separation distance is not greatly different from the effective range of such force fields.

In general, the prevailing theories of liquid viscosity can be divided somewhat arbitrarily into those based on a gaslike liquid and those based on a solidlike liquid.   In the former, the liquid is considered ordered in a short-range sense but disordered in a long-range sense.

In the latter type of theory, the liquid is assumed to exist as a regular lattice, momentum transfer resulting from molecules vibrating within the lattice structure, moving into nearby holes, or combinations of these two events.   The lattices chosen have varied greatly from cubic to types resembling parallel tunnels.   In one well-known theory, the movement from a lattice site to a hole has been considered analogous to an activated chemical reaction.

No theory, as yet, reduces to a simple form which allows liquid viscosities to be calculated a priori, and empirical estimation techniques must be used.   These methods do not conflict with theory: they merely allow some of the unknown or incalculable theoretical constants to be approximated empirically from structure or other physical properties.

The reader interested in probing deeper into some of the quantitative aspects of liquid viscosity can easily do so.   Brush [30] has published a well-written, thoroughly documented review on the theories of liquid viscosity.   Other important references are 8, 72, 93, 127.

The emphasis in the following sections is placed upon ways of calculating liquid viscosities for pure materials and mixtures at various temperatures and pressures; most methods have no definite liquid model as a basis but were suggested from a study of experimental data.

9-9   Effect of High Pressure on Low-Temperature Liquid Viscosity

The viscosity of liquids below the normal boiling point is not particularly affected by moderate pressures, but under very high pressures, large increases have been noted.† The pressure effect is illustrated for some simple liquids in Fig. 9-16; it is obvious that under large pressures order-of-magnitude increases in liquid viscosity are possible. It seems to be a rather general rule that the more complex the molecular structure the larger the effect of pressure [10, 79, 106, 122, 130, 168]. For example, from the detailed experiments of Bridgman [21] carried out to pressures of about 12,000 atm, the fractional increase in the viscosity of liquid mercury was only about 0.32, for isobutyl alcohol 790, and for the complicated molecule eugenol, about 10^7. Figure 9-16 illustrates similar increases. Water presents an anomaly, increasing only a little over twofold from 1 to 10,000 atm. Most of Bridgman's results indicate that a viscosity-pressure graph is linear to a few thousand atmospheres, but at higher pressures, it would be nearly linear if $\ln \eta_L$ were plotted against pressure.

There seems to be no reliable way of estimating low-temperature high-pressure liquid viscosities. Andrade [8] suggested a relationship involving the ratios of the specific volumes and adiabatic-compressibility factors for the compressed and uncompressed liquids, but the relationship is only approximate in the linear portion of the η_L-vs.-P curve and does not even approximate the true situation at high pressures. This formula is also discussed briefly by Gambill in a review article [70].

---

†The discussion in this section is limited to low temperatures, since practically all experimental very-high-pressure data have been obtained in this temperature region; also, high-pressure, high-temperature liquid viscosities are estimated by different types of correlations, i.e., those discussed in Sec. 9-12.

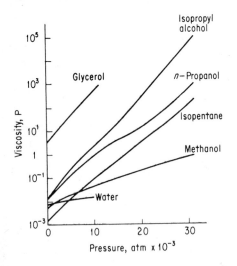

Fig. 9-16 Approximate variation of liquid viscosity with pressure at room temperature. (*From Ref. 19.*)

9-10   Effect of Temperature on Liquid Viscosity

The viscosity of liquids decreases with temperature.   To illustrate, the logarithm of the viscosity of benzene is plotted vs. reciprocal temperature in Fig. 9-17.   Over a wide range in temperatures from somewhat above the normal boiling point to near the freezing point, the plot is linear, i.e.,

$$\eta_L = A e^{B/T} \tag{9-10.1}$$

This equation is commonly known as the *Andrade correlation,* as Andrade first suggested this form from an analysis of the theory of liquid viscosity [6, 7].   In the intervening years since Eq. (9-10.1) was first proposed, literally hundreds of other viscosity-temperature relations have been suggested.   Many can be written in a more general way

$$\eta_L f_1(V) = A \exp \frac{B f_2(V)}{T} \tag{9-10.2}$$

where the functions f_1 and f_2 vary widely in form [82, 187, 189, 220].   Other variations correlate η_L with T_f/T [136] or allow for one or more additional constants [15, 50, 60, 74, 85, 98, 140, 144–146, 200, 214].   One form [58] which has been shown to be quite accurate at low temperatures resembles the Antoine equation (see Chap. 6),

$$\eta_L = A \exp \frac{B}{(T + C)} \tag{9-10.3}$$

Regardless of the many suggested modifications, Eq. (9-10.1) is still the most widely used correlation equation for showing the effect of temperature on liquid viscosities.   However, it often fails at low temperatures, as liquids sometimes show a sharp increase in viscosity as

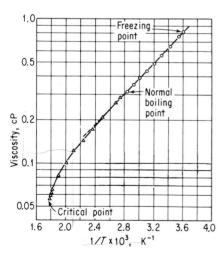

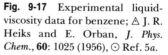

Fig. 9-17 Experimental liquid-viscosity data for benzene; △ J. R. Heiks and E. Orban, *J. Phys. Chem.,* **60**: 1025 (1956), ⊙ Ref. 5a.

the freezing point is approached.   Nor is Eq. (9-10.1) suitable at reduced temperatures above 0.7 to 0.75.   In this region, the viscosity decreases quite rapidly with temperature.   The high-temperature region is discussed in Sec. 9-12.

For the Andrade equation, various attempts have been made to relate the constants A and B to vapor pressure, latent heat of vaporization, or other physical properties [61, 104, 105, 153], but they have not been particularly successful.

Deviations from the exponential function can be allowed for by the use of a graph similar to the Cox chart of vapor pressures.   Irany [100–102] describes such a graph in which viscosity is represented on a scale distorted in such a way as to force the η_L-vs.-T relation to be linear for some substances for which viscosity data are available.   Graphs of η_L vs. T for similar compounds are then found to be linear when the special η_L scale prepared in this way is used.   This is the basis of the ASTM viscosity graph paper [216] widely employed in the petroleum industry for representing the viscosity of oils as a function of temperature.

Equation (9-10.1) requires two viscosity-temperature data points to fix values of the two constants.   If only one is available, one of the few ways to extrapolate this value is to employ the approximate Lewis-Squires chart [127], which is based on the empirical fact that the variation of viscosity with temperature seems to depend primarily upon the value of the viscosity.   This chart, shown in Fig. 9-18, can be used by locating the known value of viscosity on the ordinate and then extending the abscissa by the required number of degrees to find the new viscosity.   For example, if the viscosity is 0.7 cP at 0°C, then at 100°C it

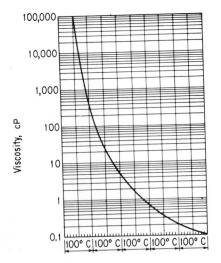

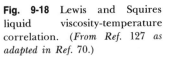

Fig. 9-18 Lewis and Squires liquid viscosity-temperature correlation. (*From Ref.* 127 *as adapted in Ref.* 70.)

will be about 0.2 cP, etc. Gambill [70] mentions several other approximate one-datum-point extrapolation formulas; the estimation techniques discussed in the next section may also be so used by employing a single viscosity point to yield the structural constant.

In summary, from the freezing point to somewhat above the normal boiling point, Eq. (9-10.1) is the best simple temperature-liquid viscosity function.   Two data points are required.   If only one datum point is known, a rough approximation of the viscosity at other temperatures can be obtained from Fig. 9-18.

Above the normal boiling point, no particularly good temperature-viscosity correlations are available.   The technique of Letsou and Stiel, discussed in Sec. 9-12, is probably the best available.

9-11   Estimation of Low-Temperature Liquid Viscosity

A very large number of papers have been published to suggest ways to estimate low-temperature (T_r less than 0.75 to 0.80) liquid viscosity when no experimental data are available.   Of all the prediction methods proposed, four of the best and most general are presented here: none is particularly reliable, and all are empirical.   Estimated values of liquid viscosities are compared with experimental values in Table 9-12.   In many cases, large errors are noted, and it is stressed that, wherever possible, experimental viscosity values should be sought before resorting to an estimation method.   We list in Appendix A many values of the constants B and T_0 for use in Eq. (9-11.5).   These constants can be used to calculate liquid viscosities with reasonable confidence from about 20 to 30°C above the freezing point to a value of $T_r \approx 0.75$.

Orrick and Erbar Method [154]   This method employs a group-contribution technique to estimate A and B in Eq. (9-11.1)

$$\ln \frac{\eta_L}{\rho_L M} = A + \frac{B}{T} \tag{9-11.1}$$

where η_L = liquid viscosity, cP
ρ_L = liquid density at 20°C, g/cm^3
M = molecular weight
T = temperature, K

The group contributions for obtaining A and B are given in Table 9-7.   For liquids that have a normal boiling point below 20°C, use the value of ρ_L at this temperature; for liquids whose freezing point is above 20°C, ρ_L at the melting point should be employed.   Compounds containing nitrogen or sulfur cannot be treated.   Orrick and Erbar

TABLE 9-7 Orrick and Erbar Group Contributions for A and B in Eq. (9-11.1)

Group	A	B
Carbon atoms†	$-(6.95 + 0.21n)$	$275 + 99n$
R—C—R (with R below)	-0.15	35
R—C—R (with R above and below)	-1.20	400
Double bond	0.24	-90
Five-membered ring	0.10	32
Six-membered ring	-0.45	250
Aromatic ring	0	20
Ortho substitution	-0.12	100
Meta substitution	0.05	-34
Para substitution	-0.01	-5
Chlorine	-0.61	220
Bromine	-1.25	365
Iodine	-1.75	400
—OH	-3.00	1600
—COO—	-1.00	420
—O—	-0.38	140
—C=O	-0.50	350
—COOH	-0.90	770

†n = number, not including those in groups shown above.

tested this method for 188 organic liquids. The errors varied widely, but they reported an average deviation of 15 percent. This is close to the average value of 16 percent shown in Table 9-12 for a more limited test.

Example 9-12 Estimate the viscosity of liquid n-butyl alcohol at 120°C.   The experimental value is 0.394 cP.

solution   From Table 9-7,

$$A = -6.95 - (0.21)(4) - 3.00 = -10.79$$
$$B = 275 + (99)(4) + 1600 = 2271$$

From Appendix A, at 20°C, $\rho_L = 0.809$ and $M = 74.12$.   Then, with Eq. (9-11.1)

$$\ln \frac{\eta_L}{(0.809)(74.12)} = -10.79 + \frac{2271}{T}$$

at $T = 120°C = 393$ K, $\eta_L = 0.399$ cP

$$\text{Error} = \frac{0.399 - 0.394}{0.394} \times 100 = 1.3\%$$

Thomas' Method [203]   Thomas has suggested that liquid viscosities at temperatures below the normal boiling point can be calculated from the empirical expression

$$\log\left(8.569 \frac{\eta_L}{\rho_L^{1/2}}\right) = \theta\left(\frac{1}{T_r} - 1\right) \tag{9-11.2}$$

where η_L = viscosity of liquid, cP
ρ_L = density, g/cm^3
θ = viscosity constant calculated from values in Table 9-8
T_r = reduced temperature = T/T_c

Experimental values of liquid viscosity are compared with calculated values in Table 9-12.   The errors are quite variable but generally indicate that aromatics (except benzene), monohalogenated compounds, unsaturates, and high-molecular-weight n-paraffins can be treated with errors usually less than 15 percent.   This method should not be used for alcohols, acids, naphthenes, heterocyclics, amines, aldehydes, or multihalogenated compounds.   The method is illustrated in Example 9-13.

TABLE 9-8   Structural Contributions to Calculation of θ in Eq. (9-11.2) [203]

C	−0.462	Double bond	0.478
H	0.249	C$_6$H$_5$	0.385
O	0.054	S	0.043
Cl	0.340	CO	0.105†
Br	0.326	CN	0.381‡
I	0.335		

†Ketones and esters.
‡Nitriles.

Example 9-13 Estimate the viscosity of chloroform at 60°C by Thomas' method.   The density at this temperature is 1.413 g/cm^3.

solution   The critical temperature of chloroform is 536.4 K, whence T_r is $(273.1 + 60)/536.4 = 0.621$.   Referring to Table 9-8, we see that θ is $(-0.462) + 3(0.340) + 0.249 = 0.807$.   Substituting in Eq. (9-11.2) gives

$$\log \frac{8.569\eta_L}{1.413^{1/2}} = 0.807\left(\frac{1}{0.621} - 1.000\right) = 0.492$$

whence $\eta_L = 0.431$ cP, which is 11 percent greater than the experimental value of 0.390 cP [123].

Morris' Method [148]   Another group-contribution method for estimating liquid viscosities was presented by Morris, who suggested

$$\log \frac{\eta_L}{\eta^+} = J\left(\frac{1}{T_r} - 1\right) \tag{9-11.3}$$

The parameter η^+ is a constant for each compound class, and values are listed in Table 9-9.   From Eq. (9-11.3) it would appear that η^+ should be the viscosity at the critical point, but since this relation is not applicable at high reduced temperatures, η^+ is best interpreted as an empirical constant.

T_r is the reduced temperature T/T_c, and J is a function of the structure. J is calculated from

$$J = \left[0.0577 + \sum_i (b_i n_i)\right]^{1/2} \tag{9-11.4}$$

where b_i = group contribution as determined from Table 9-10
       n_i = number of times group appears in molecule

Morris tested his method with 70 organic compounds and reported an average error of 12 percent.   A less extensive test is shown on Table 9-12, and it is clear that the errors vary widely.   Normally less accuracy is found at the lowest temperatures and for the lower members of a homologous series.   Too few tests were made with branched structures to indicate the usefulness of the method in such cases; it is illustrated in Example 9-14.

TABLE 9-9 Pseudocritical Viscosity η^+ [148]
η_L in centipoises

Hydrocarbons	0.0875
Halogenated hydrocarbons	0.148
Benzene derivatives	0.0895
Halogenated benzene derivatives	0.123
Alcohols	0.0819
Organic acids	0.117
Ethers, ketones, aldehydes, acetates	0.096
Phenols	0.0126
Miscellaneous	0.10

TABLE 9-10 **Structural Contributions for J** [148]

$$J = \sqrt{0.0577 + \sum_i (b_i n_i)}$$

Group	b_i	Group	b_i
CH₃, CH₂, CH	0.0825	CH₂ as saturated ring member	0.1707
Halogen-substituted CH₃	0.0	CH₃, CH₂, CH adjoining ring	0.0520
Halogen-substituted CH₂	0.0893	NO₂ adjoining ring	0.4170
Halogen-substituted CH	0.0667	NH₂ adjoining ring	0.7645
Halogen-substituted C	0.0	F, Cl adjoining ring	0.0
Br	0.2058	OH for alcohols	2.0446
Cl	0.1470	COOH for acids	0.8896
F	0.1344	C=O for ketones	0.3217
I	0.1908	O=C—O for acetates	0.4369
Double bond	−0.0742	OH for phenols	3.4420
C₆H₄ benzene ring	0.3558	—O— for ethers	0.1090
Additional H in ring	0.1446		

Example 9-14 Using Morris' method, estimate the liquid viscosity of n-propylbenzene at 0, 25, and 50°C. Experimental viscosities are 1.178, 0.796, and 0.586 cP.

solution From Table 9-9, η^+ for a benzene derivative is 0.0895. Appendix A indicates that $T_c = 638.3$ K, and J is calculated from Eq. (9-11.4) and Table 9-10.

$$\sum_i b_i n_i = \text{C}_6\text{H}_4 \text{ benzene ring} + \text{additional hydrogen in ring}$$

$$+ (\text{CH}_2 + \text{CH}_2 + \text{CH}_3)_{\text{adjoining ring}}$$

$$= 0.3558 + 0.1446 + (3)(0.0520) = 0.6564$$

$$J = (0.0577 + 0.6564)^{1/2} = 0.8450$$

Then, with Eq. (9-11.3),

$$\log \frac{\eta_L}{0.0895} = 0.8450 \left(\frac{638.3}{T} - 1 \right)$$

			η, cP		
$T°$, C	T, K	T_r	Calc.	Exp.	% error
0	273.2	0.428	1.206	1.178	1.8
25	298.2	0.467	0.824	0.796	3.5
50	323.2	0.506	0.597	0.586	1.8

Van Velzen, Cardozo, and Langenkamp's Method In an unusually detailed study of the effect of structure on liquid viscosities, van Velzen et al. [211, 212] proposed a modification of Eq. (9-10.1),

$$\log \eta_L = B \left(\frac{1}{T} - \frac{1}{T_0} \right) \qquad (9\text{-}11.5)$$

where η_L = liquid viscosity, cP
T = temperature, K

TABLE 9-11 Van Velzen, Cardozo, and Langenkamp Contributions for Liquid Viscosity [211]

Structures or functional group	ΔN_i	ΔB_i	Example: Compound	Example: N^*	Example: B	Example: T_0	Remarks
n-Alkanes	0	0	n-Hexane	6.00	377.86	209.21	
Isoalkanes	$1.389 - 0.238N$	15.51	2-Methylbutane	5.20	351.95	189.83	
Saturated hydrocarbons with two methyl groups in iso position	$2.319 - 0.298N$	15.51	2,3-Dimethylbutane	6.89	437.37	229.29	
n-Alkenes	$-0.152 - 0.042N$	$-44.94 + 5.410N^*$	1-Octene	7.51	446.89	242.41	For any additional CH₃ groups in iso position, increase ΔN by $1.389 - 0.238N$
n-Alkadienes	$-0.304 - 0.084N$	$-44.94 + 5.410N^*$	1,3-Butadiene	3.36	211.21	140.15	
Isoalkenes	$1.237 - 0.280N$	$-36.01 + 5.410N^*$	2-Methyl-2-butene	4.84	307.40	180.68	
Isoalkadienes	$1.085 - 0.322N$	$-36.01 + 5.410N^*$	2-Methyl-1,3-butadiene	4.48	285.89	171.26	
Hydrocarbon with one double bond and two methyl groups in iso position	$2.626 - 0.518N$	$-36.01 + 5.410N^*$	2,3-Dimethyl-1-butene	5.52	347.07	197.74	For any additional CH₃ groups in iso position, increase ΔN by $1.389 - 0.238N$
Hydrocarbon with two double bonds and two methyl groups in iso position	$2.474 - 0.560N$	$-36.01 + 5.410N^*$	2,3-Dimethyl-1,3-butadiene	5.11	323.30	187.57	
Cyclopentanes	$0.205 + 0.069N$	$-45.96 + 2.224N^*$	n-Butylcyclopentane	9.83	527.3	285.7	$N \leqslant 16$; not recommended for $N = 5, 6$
	$3.971 - 0.172N$	$-339.67 + 23.135N^*$	Tridecylcyclopentane	18.87	889.40	392.45	$N \geqslant 16$
Cyclohexanes	1.48	$-272.85 + 25.041N^*$	Ethylcyclohexane	9.48	501.80	279.72	$N < 17$; not recommended for $N = 6, 7$
	$6.517 - 0.311N$	$-272.85 + 25.041N^*$	Dodecylcyclohexane	18.92	994.10	392.87	$N \geqslant 17$
Alkyl benzenes	0.60	$-140.04 + 13.869N^*$	o-Xylene	9.11	563.09	273.20	$N < 16$; not recommended for $N = 6, 7$ [a,c,f]
	$3.055 - 0.161N$	$-140.04 + 13.869N^*$					$N \geqslant 16$ [a,c,f]
Polyphenyls	$-5.340 + 0.815N$	$-188.40 + 9.558N^*$	m-Terphenyl	27.44	1008.7	462.58	[a]
Alcohols:							
Primary	$10.606 - 0.276N$	$-589.44 + 70.519N^*$	1-Pentanol	14.23	1113.0	347.12	[b]
Secondary	$11.200 - 0.605N$	497.58	Isopropyl alcohol	12.38	1141.35	324.12	[b]
Tertiary	$11.200 - 0.605N$	928.83	2-Methyl butanol-2	13.42	1699.1	337.49	[b]
Diols (correction)	See remarks	557.77	Propylene glycol	22.66	1399.71	423.55	For ΔN, use alcohol contributions and add $N - 2.50$
Phenols (correction)	$16.17 - N$	213.68					[a,c,d]
—OH on side chain to aromatic ring (correction)	-0.16	213.68					
Acids	$6.795 + 0.365N^*$	$-249.12 + 22.449N^*$	n-Butyric acid	12.25	665.40	322.36	$N < 11$; not recommended for $N = 1, 2$
	10.71	$-249.12 + 22.449N^*$					$N \geqslant 11$

Group	See remarks		Compound	ΔN	ΔB		Remarks
Iso acids	See remarks	$-249.12 + 22.449N*$	Isobutyric acid	12.01	652.02	319.06	Calculate ΔB as for straight-chain acid; calculate ΔN for straight-chain acid but reduce ΔN by 0.24 for each methyl group in iso position
Acids with aromatic nucleus in structure (correction)	4.81	$-188.40 + 9.558N*$	Phenylacetic acid	22.52	1123.29	422.41	
Esters	$4.337 - 0.230N$	$-149.13 + 18.695N*$	Ethyl valerate	9.73	580.16	284.01	If hydrocarbon groups have iso configuration, see footnote e
Esters with aromatic nucleus in structure (correction)	$-1.174 + 0.376N$	$-140.04 + 13.869N*$	Benzyl benzoate	19.21	1133.17	395.31	Add to values of ΔN, ΔB calculated for ester
Ketones	$3.265 - 0.122N$	$-117.21 + 15.781N*$	Methyl n-butyl ketone	8.53	514.53	262.53	If hydrocarbon groups have iso configuration, see footnote e
Ketones with aromatic nucleus in structure (correction)	2.70	$-760.65 + 50.478N*$	Acetophenone	12.99	645.92	322.10	Add to values of ΔN, ΔB calculated for ketone
Ethers	$0.298 + 0.209N$	$-9.39 + 2.848N*$	Ethyl hexyl ether	9.97	575.96	288.04	If hydrocarbon groups have iso configuration, see footnote e
Aromatic ethers	$11.5 - N$	$-140.04 + 13.869N*$	Propyl phenyl ether	11.5	656.83	311.82	^{c}The ΔN value is not a correction to regular ether value, but the ΔB value is a correction to regular ether
Amines: Primary	$3.581 + 0.325N$	$25.39 + 8.744N*$	Propylamine	7.56	545.01	243.44	If hydrocarbon groups have iso configuration, see footnote e
Primary amine in side chain of aromatic compound (correction)	-0.16	0	Benzylamine	12.70	790.47	328.36	Corrections to be added to amine calculation^c
Secondary	$1.390 + 0.461N$	$25.39 + 8.744N*$	Ethylpropylamine	8.69	605.44	265.53	If hydrocarbon groups have iso configuration, see footnote e
Tertiary	3.27	$25.39 + 8.744N*$		...	...	...	If hydrocarbon groups have iso configuration, see footnote e
Primary amines with NH_2 group on aromatic nucleus	$15.04 - N$	0	m-Toluidine	15.04	904.08	356.13	^{a,c}The ΔN value is not a correction to regular amine value; to find ΔB use primary amine value
Secondary or tertiary amine with at least one aromatic group attached to amino nitrogen	f	f	Benzylphenylamine	21.58	1041.18	414.74	
Nitro compounds: 1-nitro	$7.812 - 0.296N$	$-213.14 + 18.330N*$	Nitromethane	8.57	442.82	263.28	Note alkene contribution is necessary
2-nitro	5.84	$-213.14 + 18.330N*$	2-Nitro-2-pentene	10.48	567.43	296.33	
3-nitro	5.56	$-338.01 + 25.086N*$					
4-nitro; 5-nitro	5.36	$-338.01 + 25.086N*$					
Aromatic nitro-compounds	$7.812 - 0.296N$	$-213.14 + 18.330N*$	Nitrobenzene	13.00	728.79	332.23	For aromatic correction see footnote f

TABLE 9-11 Van Velzen, Cardozo, and Langenkamp Contributions for Liquid Viscosity (Continued)

Structures or functional group	ΔN_i	ΔB_i	Example Compound	N^*	B	T_0	Remarks
Halogenated compounds:							
Fluoride	1.43	5.75	Ethyl chloride	5.21	319.94	190.08	e,f
Chloride	3.21	-17.03	1-Bromo-2-methyl propane	8.15	435.85	255.24	e,f
Bromide	4.39	$-101.97 + 5.954N^*$	Iodobenzene	12.36	589.18	323.85	e,f
Iodide	5.76	-85.32					
Special configurations (corrections):							
$C(Cl)_x$	$1.91 - 1.459x$	-26.38					
—CCl—CCl—	0.96	0					
—C(Br)$_x$—	0.50	$81.34 - 86.850x$					
—CBr—CBr—	1.60	-57.73					
CF_3, in alcohols	-3.93	341.68					
In other compounds	-3.93	25.55					
Aldehydes	3.38	$146.45 - 25.11N^*$	Propionaldehyde	6.38	383.16	217.97	
Aldehydes with an aromatic nucleus in structure (correction)	2.70	$-760.65 + 50.478N^*$	Benzaldehyde	13.08	391.19	333.25	
Anhydrides	$7.97 - 0.50N$	-33.50	Propionic anhydride	10.79	554.87	301.19	
Anhydrides with an aromatic nucleus in structure (correction)	2.70	$-760.65 + 50.478N^*$					
Amides	$13.12 + 1.49N$	$524.63 - 20.72N^*$	Acetamide	18.10	931.07	385.79	
Amides with an aromatic nucleus in structure (correction)	2.70	$-760.65 + 50.478N^*$					

For substitutions on an aromatic nucleus in more than one position, additional corrections are required:

Ortho:	$\Delta N = 0.51$	$\Delta B = \begin{cases} -571.94 & \text{with —OH} \\ 54.84 & \text{without —OH} \end{cases}$
Meta:	$\Delta N = 0.11$	$\Delta B = 27.25$
Para:	$\Delta N = -0.04$	$\Delta B = -17.57$

[b] For alcohols, if there is a methyl group in the iso position, increase ΔN by 0.24 and ΔB by 94.23.

[c] If the compound has an aromatic —OH or —NH$_2$, or if it is an aromatic ether, use ΔN contribution in table but neglect other substituents on the ring such as halogen, CH$_3$, NO$_2$, etc. For the calculation of ΔB, however, such substituents must be taken into account.

[d] For aromatic alcohols and compounds with an —OH on a side chain, the alcohol contribution (primary, etc.) must be included. For example, o-chlorophenol:

$$\Delta B = \Delta B \text{ (primary alcohol)} + \Delta B \text{ (chlorine)} + \Delta B \text{ (phenol)} + \Delta B \text{ (ortho correction}^a)$$

With $N^* = 16.17^c$:

$$\Delta B = (-589.44 + 70.519 \times 16.17) + (-17.03) + (213.68) + (-571.94) = 175.56$$

$$B_a = 745.94 \qquad B = B_a + \Delta B = 921.50$$

2-Phenylethanol:

$$N = 8 \qquad \Delta N = \Delta N \text{ (primary alcohol)} + \Delta N \text{ (correction)} = [10.606 - (0.276)(8)] + (-0.16) = 8.24$$

$$N^* = N + \Delta N = 8 + 8.24 = 16.24$$

$$\Delta B = \Delta B \text{ (primary alcohol)} + \Delta B \text{ (correction)} = [-589.44 + (70.519)(16.24)] + 213.68 = 769.47$$

$$B_a = 747.43 \qquad B = B_a + \Delta B = 1516.9$$

[e] For esters, alkylbenzenes, halogenated hydrocarbons, and ketones, if the hydrocarbon chain has a methyl group in an iso position, decrease ΔN by 0.24 and increase ΔB by 8.93 for each such grouping. For ethers and amines, decrease ΔN by 0.50 and increase ΔB by 8.93 for each iso group.

[f] For alkylbenzenes, nitrobenzenes, halogenated benzenes, and for secondary or tertiary amines where at least one aromatic group is connected to an amino nitrogen, add the following corrections for each aromatic nucleus. If $N < 16$, increase ΔN by 0.60; if $N \geq 16$, increase ΔN by $3.055 - 0.161N$ for each aromatic group. For any N, increase ΔB by $(-140.04 + 13.869N^*)$.

and B and T_0 are related to structure. To determine these parameters, one must first find the *equivalent chain length* N^*, where

$$N^* = N + \sum_i \Delta N_i \qquad (9\text{-}11.6)$$

N is the actual number of carbon atoms in the molecule, and ΔN represents structural contributions from Table 9-11. If the structural or functional group ΔN_i appears n_i times in the molecule, $n_i\,\Delta N_i$ corrections must be added. In the tabulation of ΔN_i contributions, some entries are to be used every time the functional group appears; other entries represent additional *corrections* to be used to modify the basic group contribution. A few examples given below illustrate the technique of calculating N^*.

Example 9-15 Calculate N^* for benzophenone, chloroform, N-methyldiphenylamine, N,N-diethylaniline, allyl alcohol, and m-nitrotoluene.

   solution For benzophenone, $C_6H_5COC_6H_5$, $N = 13$ (number of carbon atoms). There is a ΔN_i contribution for ketones, that is, $3.265 - 0.122N = 3.265 - (0.122)(13) = 1.68$. Also, there is a *correction* for aromatic ketones of 2.70 for each aromatic ring; thus

$$N^* = 13 + 1.68 + (2)(2.70) = 20.08$$

For chloroform, $N = 1$. The three chlorine atoms each have a ΔN_i contribution of 3.21. Also, there is a *correction* term for the $C(Cl)_x$ structure of $1.91 - 1.459x$ where, in this case, $x = 3$. Thus,

$$N^* = 1 + (3)(3.21) + 1.91 - (3)(1.459) = 8.16$$

For N-methyldiphenylamine, $(C_6H_5)_2N(CH_3)$, $N = 13$. This is a tertiary amine so that $\Delta N_i = 3.27$. Also, there is a correction for each of the aromatic groups (see note f of Table 9-11) of 0.6. Then

$$N^* = 13 + 3.27 + (2)(0.6) = 17.47$$

For N,N-diethylaniline, $(C_6H_5)N(C_2H_5)_2$, $N = 10$. Again, a tertiary-amine contribution of 3.27 is required, but only one aromatic group is present with a correction of 0.6:

$$N^* = 10 + 3.27 + 0.6 = 13.87$$

For allyl alcohol, $CH_2{=}CHCH_2OH$, $N = 3$. As a primary alcohol, there is a ΔN_i of $10.606 - 0.276N = 10.606 - (3)(0.276) = 9.778$. In addition, as an alkene, another ΔN_i of $-0.152 - 0.042N = -0.152 - (3)(0.042) = 0.278$ is necessary. Thus,

$$N^* = 3 + 9.778 - 0.278 = 12.50$$

Finally, for m-nitrotoluene, $C_6H_4NO_2(CH_3)$, $N = 7$. The ΔN_i for the aromatic nitro compound is $7.812 - 0.236N = 7.812 - (7)(0.236) = 6.16$. There is no contribution for an alkyl benzene. For the meta correction (see note a of Table 9-11), $\Delta N_i = 0.11$. Thus,

$$N^* = 7 + 6.16 + 0.11 = 13.27$$

The value of N^* is then used to determine B and T_0 in Eq. (9-11.5). For T_0,

$$T_0 = \begin{cases} 28.86 + 37.439N^* - 1.3547N^{*2} + 0.02076N^{*3} & N^* \leq 20 \quad (9\text{-}11.7) \\ 8.164N^* + 238.59 & N^* > 20 \quad (9\text{-}11.8) \end{cases}$$

For B,

$$B = B_a + \sum_i \Delta B_i \qquad (9\text{-}11.9)$$

where

$$B_a = \begin{cases} 24.79 + 66.885N^* - 1.3173N^{*2} - 0.00377N^{*3} & N^* \leqslant 20 \quad (9\text{-}11.10) \\ 530.59 + 13.740N^* & N > 20 \quad (9\text{-}11.11) \end{cases}$$

and $\sum_i \Delta B_i$ can be determined from summing contributions as shown in Table 9-11. Even though a functional group may appear more than once in a compound, the ΔB_i contribution is applied only a single time.

The B and T_0 values found for any specific compound are then used in Eq. (9-11.5) and the liquid viscosity determined. The units of η_L and T are centipoises and kelvins.

Example 9-16   Obtain the constants B and T_0 for benzophenone and estimate the liquid viscosity at 25, 55, 95, and 120°C.   Experimental values reported are 13.61, 4.67, 1.74, and 1.38 cP.

   solution N^* was calculated in Example 9-15 as 20.08.   To determine T_0, since $N^* > 20$, Eq. (9-11.8) must be used.

$$T_0 = (8.164)(20.08) + 238.59 = 402.52$$

For B, first B_a is determined from Eq. (9-11.11),

$$B_a = 530.59 + (13.740)(20.08) = 806.49$$

The ΔB_i corrections are

$$\Delta B \text{ (ketone)} = -117.21 + (15.781)(20.08) = 199.67$$
$$\Delta B \text{ (aromatic ketone correction)} = -760.65 + (50.478)(20.08) = 252.95$$
$$B = 806.49 + 199.67 + 252.95 = 1259.11$$

Then, from Eq. (9-11.5),

$$\log \eta_L = 1259.11 \left(\frac{1}{T} - \frac{1}{402.52} \right)$$

T, °C	η_L Calc.	Exp.	% error
25	12.44	13.61	-8.7
55	5.11	4.67	9.4
95	1.96	1.74	12
120	1.19	1.38	-14

Van Velzen et al. tested their method on 314 different liquids with nearly 4500 datum points, and a careful statistical evaluation was made. Large errors were often noted for the first members of a homologous series. This point is significant since, in the comparison of calculated and experimental liquid viscosities shown in Table 9-12, many

TABLE 9-12 Comparison of Calculated and Experimental Viscosities of Liquids

			Percent error‡ in liquid viscosity calculated by method of:			
Compound	T, °C	η, (exp.) cP†	Thomas	Orrick and Erbar	Van Velzen, Cardozo, and Langenkamp	Morris
Acetone	−90	2.075	−15	−25	−20	−5.0
	−60	0.892	5.8	−6.7	−0.1	14
	0	0.389	3.3	−8.3	0.1	7.0
	30	0.292	0.7	−9.4	−0.2	4.2
	60	0.226	0.5	−8.3	1.6	4.1
Acetic acid	10	1.45	−51	−22	−34	5.5
	40	0.901	−49	−15	−28	6.1
	80	0.561	−37	−9.5	−24	3.1
	110	0.416	−36	−5.3	−21	2.2
Aniline	−5	13.43	...	...	−52	−53
	20	4.38	...	...	−23	−20
	60	1.52	...	...	−2.8	7.6
	120	0.658	...	...	−11	6.5
Benzene	5	0.826	−41	−45	−42	−22
	40	0.492	−31	−35	−30	−16
	80	0.318	−21	−26	−20	−12
	120	0.219	−13	−46	−8.5	−6.4
	160	0.156	−2.5	−7.1	5.8	1.9
	190	0.121	8.3	5.1	20	12
n-Butane	−90	0.63	8.0	−14	−7.2	−7.7
	−60	0.403	2.7	−20	−9.8	−9.7
	−30	0.282	−0.7	−22	−10.5	−9.3
	0	0.210	−2.4	−23	8.8	−7.5
1-Butene	−110	0.79	19	−22	−6.1	−16
	−80	0.45	18	−20	0	−12
	−40	0.26	15	−18	8.7	−5.7
n-Butyl alcohol	0	5.14	−89	−2.1	5.6	−28
	40	1.77	−78	−1.6	0	−19
	80	0.762	−66	0.5	−2.3	−9.3
	120	0.394	−50	−1.4	−5.3	−2.0
Carbon	0	1.369	−0.1	20	−10	−26
tetrachloride	30	0.856	1.9	22	−6.1	−19
	70	0.534	1.9	20	−4.3	−12
	100	0.404	0	19	−4.0	−9.1
Chlorobenzene	0	1.054	15	1.4	15	12
	40	0.639	4.4	−0.6	7.0	11
	80	0.441	1.4	−0.9	0.2	8.6
	120	0.326	−1.2	−5.1	−5.0	7.4
Chloroform	0	0.700	24	40	11	12
	30	0.502	18	34	7.0	12
	60	0.390	10	27	2.1	9.2
Cyclohexane	5	1.300	...	−51	−51	−28
	30	0.826	...	−45	−40	−24
	60	0.528	...	−38	−27	−19
	80	0.411	...	−34	−19	−17

Cyclopentane	−20	0.72	. . .	−36	−38	15
	20	0.439	. . .	−32	−32	2.6
	50	0.323	. . .	−28	−26	−2.3
2,2-Dimethyl propane	−15	0.431	−37	−3.5	7.2	−24
	0	0.328	−31	−1.2	16	−32
	10	0.281	−26	−0.8	20	−28
Ethane	−175	0.985	54	30	2.6	−13
	−120	0.257	27	−12	8.2	−3.2
	−85	0.162	20	−22	12	1.7
Ethylene bromide	20	1.714	−24	−26	−0.8	−18
	60	0.995	−18	−22	0.4	−14
	120	0.562	−12	−20	−2.0	−12
Ethylene chloride	0	1.123	−19	−43	−4.0	−23
	40	0.644	−13	−35	−4.0	−14
	80	0.417	−8.6	−27	−3.5	−4.8
Ethyl alcohol	0	1.770	−79	27	69	1.2
	40	0.826	−68	3.5	38	−6.1
	75	0.465	−55	−5.4	27	−6.0
Ethyl acetate	20	0.458	−3.5	−4.2	3.1	−0.6
	80	0.246	0.4	0.4	7.2	1.4
	140	0.153	4.5	7.4	14	6.4
	190	0.0998	19	27	34	24
Ethylbenzene	−20	1.24	−5.6	−2.9	−3.8	13
	40	0.535	−1.3	−1.2	−3.3	7.4
	100	0.308	−1.3	−1.7	−4.7	1.7
	140	0.231	−0.4	−1.2	−4.6	−0.2
Ethyl bromide	20	0.395	4.0	27	2.9	11
	60	0.269	9.3	32	10	19
	100	0.199	12	36	16	26
	160	0.126	26	58	38	50
Ethylene	−170	0.70	23	−25	4.0	−41
	−140	0.31	19	−27	40	−23
	−100	0.15	27	−22	92	2.4
Ethyl ether	0	0.289	7.6	0	−4.7	4.3
	20	0.236	7.6	0	−3.9	5.7
	60	0.167	8.4	2	−1.2	9.6
	100	0.118	16	11	8.8	22
Ethyl formate	0	0.507	−5.5	−18	−3.3	
	30	0.362	−5.2	−17	−1.8	
	55	0.288	−2.7	−16	−1.2	
Heptane	−90	3.77	−26	−21	18	−17
	−40	0.965	0.1	−0.5	0.8	1.6
	20	0.418	−1.2	−1.9	−1.8	−1.7
	60	0.287	−4.5	−3.8	−4.4	−4.6
	100	0.209	−1.4	−3.3	−4.4	−4.8
Hexane	−60	0.888	8.5	2.9	4.2	4.5
	0	0.381	3.2	−2.4	−0.9	−0.4
	40	0.262	−0.8	−5.3	−4.0	−3.5
	70	0.205	−2.0	−4.9	−3.8	−3.2
Hydrogen chloride	−115	0.581	−5.5			
	−85	0.458	−24			
Hydrogen sulfide	−80	0.511	−27			
	−65	0.443	−29			

TABLE 9-12 Comparison of Calculated and Experimental Viscosities of Liquids (Continued)

Compound	T, °C	η, (exp.) cP†	Thomas	Orrick and Erbar	Van Velzen, Cardozo, and Langenkamp	Morris
				Percent error‡ in liquid viscosity calculated by method of:		
Isobutane	−80	0.628	−22	−23	−3.5	−19
	−40	0.343	−18	−25	−6.4	−17
	−10	0.239	−15	−24	−5.4	−13
Isopropyl	10	3.319	−90	−24	−2.8	−57
alcohol	30	1.811	−84	−15	−3.4	−49
	50	1.062	−77	−10	−3.6	−40
Methane	−185	0.226	46	60	90	
	−160	0.115	53	23	122	
Methyl alcohol.	5	0.746	−56	59	165	94
	20	0.592	−51	39	136	78
	60	0.350	−40	5.1	85	47
2-Methylbutane	−50	0.55	−7.3	−13	−4.0	−15
	−20	0.353	−3.4	−12	−2.7	−9.8
	30	0.205	0	−10	−1.1	−3.2
Pentane	−120	2.35	8.6	−1.0	−1.7	−3.0
	−80	0.791	16	3.8	5.5	−5.8
	−40	0.428	10	−3.3	0.6	1.2
	0	0.279	3.6	−8.2	−3.6	−2.5
	30	0.216	0	−11	−5.2	−3.7
Phenol	50	3.020	−76	0	22	−17
	100	0.783	−42	37	11	−15
Propane	−190	13.8	−12	−7.3	−40	−49
	−140	0.984	17	−1.5	−8.4	−15
	−80	0.327	1.5	−22	−12	−14
	−40	0.205	−2.9	−25	−9.7	−10
n-Propyl alcohol	10	2.897	−86	−9.1	4.1	−33
	40	1.400	−78	−9.8	1.4	−27
	100	0.443	−56	−6.5	2.0	−13
Toluene	−20	1.07	−18	−19	−19	−7.2
	20	0.587	−9.2	−13	−29	−4.7
	60	0.380	−6.1	−10	−8.7	−4.8
	110	0.249	−3.2	−6.8	−4.6	−4.2
o-Xylene	0	1.108	−16	3.1	−9.7	−29
	40	0.625	−9.8	5.0	−13	−23
	100	0.345	−6.1	3.7	−1.1	−18
	140	0.254	−4.7	3.6	−21	−15
m-Xylene	0	0.808	5.3	1.1	9.9	−11
	40	0.492	16	1.4	3.9	−9.4
	80	0.340	4.7	0.3	−1.9	−9.3
	140	0.218	4.1	1.4	−5.8	−7.1
Average error			21.2	16.0	16.1	14.2

†Refs. 5a and 123.
‡[(calc. − exp.)/exp.] × 100.

of the test compounds shown are first members of a series. This reflects unfairly on the van Velzen et al. method.

Some care must be used in selecting appropriate ΔN_i and ΔB_i contributions. A Euratom report [212] is very helpful in illustrating the rules for complex compounds. Also, the contribution for aldehydes, anhydrides, and amides are only tentative [213].

Other Correlations Many other empirical estimation techniques have been suggested, of which Gambill [69] has reviewed many. Albright and Lohrenz [1] suggested the correlating parameter ZZ_c, and at various times other workers have proposed relations involving η_L, T, and one or more of the following: T_b, T_f, M, vapor density, surface tension, sonic velocity, van der Waals volumes, vapor pressure, rheochor, etc. [17, 59, 62, 66, 78, 82, 95, 105, 120, 128, 136, 139, 151, 152, 155, 159, 163, 184, 186, 188, 191, 199, 201]. Most of these methods are valid only at temperatures near the normal boiling point.

Andrade's [8] technique for estimating viscosities at the melting point was studied by Matsen and Johnson [142], who expressed it as $\eta_L = \beta \rho^{2/3} T^{1/2}/M^{1/6}$, where η_L at the melting point is expressed in centipoises, ρ is in grams per cubic centimeter, T is in kelvins, and M is the molecular weight. The minimum value of β was suggested as 0.051 by Andrade, from theoretical considerations. Matsen and Johnson found that actual values were much higher; they tabulated a set of β values for a large number of compounds. There was a definite correlation of β with structure, the simple, symmetrical monatomic liquids approaching the theoretical value but increasing, often by orders of magnitude, for complex unsymmetrical structures. For example, the β values for the n-alkane series oscillated between 0.058 for methane and 2.78 for propane, although β for most compounds up to eicosane fell in the range 0.3 to 0.7.

For a number of associated liquids, Makhija and Stairs [137] have tabulated the parameters A', B', and T' for use in an equation of the form of (9-10.3):

$$\log \eta_L = A' + \frac{B'}{T - T'} \qquad (9\text{-}11.12)$$

where η_L = liquid viscosity, cP
$\quad T$ = temperature, K

A', B', T' for 10 common polar liquids are given in Table 9-13.

Several studies have been directed at correlating the reduced liquid viscosity in terms of the reduced temperature, where the reduced viscosity is equal to $\eta_L[(m\epsilon)^{1/2}/\sigma^2]$ and the reduced temperature is kT/ϵ; m is the mass of the molecule, and ϵ and σ represent characteristic energies and size parameters from an applicable intermolecular-

TABLE 9-13 Polar-Liquid Viscosities [137]

$$\log \eta_L = A' + \frac{B'}{T - T'} \qquad \eta_L \text{ in centipoises}$$
$$T \text{ in kelvins}$$

Liquid	A'	B'	T'	% error	Range, °C
Ammonia	−1.7520	218.76	50.701	0.76	−69 to +40
Methylamine	−1.3634	126.389	102.886	0.32	−70 to −10
1,2-Diaminopropane	−1.0755	165.754	166.751	1.10	−35 to +50
Methanol	−1.6807	354.876	48.585	2.05	−98 to +50
Water	−1.5668	230.298	146.797	0.51	−10 to +160
Ethanol	−2.4401	774.414	−15.249	2.66	−98 to +70
Glycerine	−2.8834	997.586	128.481	4.50	−42 to +30
Ethylene glycol	−1.5923	438.064	141.617	0.18	+20 to +100
n-Propanol	−2.4907	725.903	37.474	1.10	0 to +70
n-Butanol	−3.0037	1033.306	−4.3828	0.80	−51 to +100

potential law.   Usually, the results are acceptable only for monatomic liquids, e.g., argon [18, 202], although applications have also been made to liquid metals [44] and to cryogenic liquids (and mixtures) [164].

Recommendations for Estimating Low-Temperature Liquid Viscosities Four group-contribution estimation methods have been discussed.   In Table 9-12 calculated liquid viscosities are compared with experimental values for 40 different liquids (usually of simple structure).   Large errors may result, as illustrated for all methods. The results are somewhat misleading since the method of van Velzen et al. is *not* recommended for first members of a homologous series.

The methods of Morris and van Velzen et al. assume that $\log \eta_L$ is linear in $1/T$, whereas the Orrick and Erbar and Thomas methods are slightly modified to include the liquid density.   None is reliable for highly branched structures (van Velzen et al. can only treat iso compounds) or for inorganic liquids, and only the method of Thomas can treat sulfur compounds.   All are limited to a temperature range from somewhat above the freezing point to about $T_r \approx 0.75$.

It is recommended that, wherever possible, the method of van Velzen, Cardozo, and Langenkamp be used to estimate low-temperature liquid viscosities.   The main exception would be first members of a homologous series.   Errors vary widely, but as judged from extensive testing [211], errors should be less than 10 to 15 percent in most instances.   Should the van Velzen method not be applicable, the method of Morris or Orrick and Erbar is recommended.   They are not, however, significantly more accurate than Thomas' method.

In Appendix A, values of B and T_0 are tabulated for many liquids. These parameters have been determined from experimental data [212] and not estimated by the method of van Velzen et al. Obviously, they should be used wherever possible with Eq. (9-11.5), although the range of applicability is still below $T_r \approx 0.75$.

9-12 Estimation of Liquid Viscosity at High Temperatures

For saturated liquids at high temperatures, it would appear possible to use the corresponding-states relations developed in Sec. 9-6 provided the density limits are not exceeded. For example, for nonpolar gases, Eq. (9-6.4) should be applicable for reduced densities less than 3. Such a relation indicates that

$$\eta_L = f(\eta^\circ, \xi, \rho_r) \tag{9-12.1}$$

where η° is the low-pressure gas viscosity $= f(\xi, T_r)$, for example, from Eq. (9-4.17),

$$\xi = \frac{T_c^{1/6}}{M^{1/2}P_c^{2/3}} \qquad \begin{array}{l} T_c \text{ in kelvins} \\ P_c \text{ in atmospheres} \end{array}$$

and ρ_r is the reduced saturated-liquid density at T_r. However, as indicated in Sec. 3-15, ρ_r may be related to T_r and either ω or Z_c. Combining these relations, one should be able to express η_L as

$$\eta_L = f(T_r, \xi, \omega \text{ or } Z_c) \tag{9-12.2}$$

An approximate relation using Z_c has been developed [173]. A more recent correlation due to Letsou and Stiel [126] employs the acentric factor

$$\eta_L \xi = (\eta_L \xi)^{(0)} + \omega(\eta_L \xi)^{(1)} \tag{9-12.3}$$

where the parameters $(\eta_L \xi)^{(0)}$ and $(\eta_L \xi)^{(1)}$ are functions only of reduced temperature. Letsou and Stiel tabulate these functions, but, to close approximation, from $0.76 \le T_r \le 0.98$ they can be expressed as

$$(\eta_L \xi)^{(0)} = 0.015174 - 0.02135 T_r + 0.0075 T_r^2 \tag{9-12.4}$$

$$(\eta_L \xi)^{(1)} = 0.042552 - 0.07674 T_r + 0.0340 T_r^2 \tag{9-12.5}$$

where η_L is in centipoises and ξ is defined under Eq. (9-12.1). In Fig. 9-19, $\eta_L \xi$ is plotted as a function of T_r for several values of ω. Above $T_r = 0.98$, the function $\eta_L \xi$ drops rapidly to a value around 0.00072 to 0.00078, and from $T_r = 0.98$ to 1.0, Eqs. (9-12.4) and (9-12.5) are not applicable. Note that the $\eta_L \xi$ value at the critical point compares favorably with the values found in Sec. 9-6, that is, 0.00077 to 0.00079 when the viscosity is expressed in centipoises.

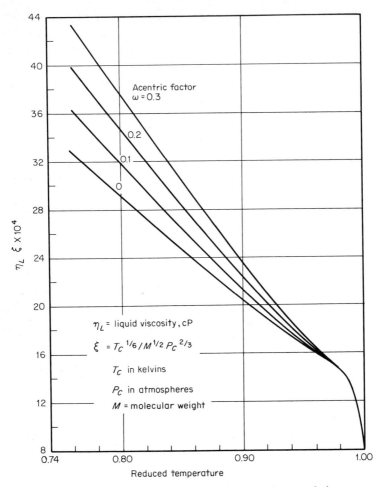

Fig. 9-19 Letsou-Stiel high-temperature liquid-viscosity correlation.

Example 9-17 Estimate the viscosity of liquid n-heptane at 210°C using the Letsou-Stiel correlation.

solution For n-heptane, from Appendix A, $T_c = 540.2$ K, $P_c = 27.0$ atm, $M = 100.205$, and $\omega = 0.351$. Thus, $T_r = (210 + 273.2)/540.2 = 0.894$. From Eqs. (9-12.4) and (9-12.5),

$$(\eta_L \xi)^{(0)} = 0.015174 - (0.02135)(0.894) + (0.0075)(0.894)^2 = 0.00208$$
$$(\eta_L \xi)^{(1)} = 0.042552 - (0.07674)(0.894) + (0.0340)(0.894)^2 = 0.00112$$

Then

$$\eta_L \xi = 0.00208 + (0.351)(0.00112) = 0.00247$$

Since

$$\xi = \frac{T_c^{1/6}}{M^{1/2} P_c^{2/3}} = \frac{(540.2)^{1/6}}{(100.205)^{1/2}(27.0)^{2/3}} = 0.03168$$

we have

$$\eta_L = \frac{0.00247}{0.03168} = 0.078 \text{ cP}$$

The experimental value at this temperature is reported to be 0.085 cP [136].

$$\text{Error} = \frac{0.078 - 0.085}{0.085} \times 100 = -8.2\%$$

Equations (9-12.3) to (9-12.5) represent the best method available for estimating saturated-liquid viscosities at temperatures of $T_r = 0.76$ and higher. It was developed from data on only 14 liquids, mostly simple hydrocarbons, and has not been extensively tested. The authors report average errors of about 3 percent for most materials up to $T_r \approx 0.92$. Larger errors were found as the critical point was approached.

In addition to this correlation, Grunberg and Nissan [80] proposed a nomograph relating η_L, $\eta°$, T_r, and ρ_r. Uyehara and Watson's plot (Fig. 9-10) also shows a liquid region. The Letsou-Stiel correlation, however, is preferable to these methods.

There is a problem in selecting a technique to join low- and high-temperature estimated viscosities. At low temperatures, the methods described in Sec. 9-11 were primarily structural in nature, whereas at high temperature, as the liquid begins to assume some gaslike characteristics, the methods are of a corresponding-states type. It is very unlikely that separate estimates of liquid viscosity in these two temperature regions will lead to a match in the connecting region of $0.74 < T_r < 0.76$. One particularly troublesome problem lies in the fact that around $T_r = 0.76$, it can be seen from Fig. 9-19 that $d\eta_L/dT$ is about constant. This slope is not predicted from the Andrade equation (9-10.1), which forms the basis of most low-temperature estimation methods. Accurate saturated-liquid viscosity data for many liquids over a wide range of temperatures would be very desirable so that one could develop a single liquid viscosity-temperature correlation.

9-13 Liquid-Mixture Viscosity

The effect of composition on low-temperature liquid viscosity cannot be estimated with any accuracy if only the pure-component properties are available. At low temperatures, i.e., below a reduced temperature of about 0.75, the liquid viscosity is quite sensitive to the structure of the liquid, which depends, of course, on composition. We show in Fig. 9-20 a rather extreme case to illustrate this point. A maximum in the solution viscosity is often noted in cases where one component is quite polar or where there can exist some loose association between the materials constituting the mixture. In this particular case, the DMA (N,N-dimethylacetamide) is not usually considered particularly polar, whereas water is strongly associated. The viscosity maximum would normally imply some type of DMA-water association, and Petersen has interpreted this specific case in terms of resonance structures involving

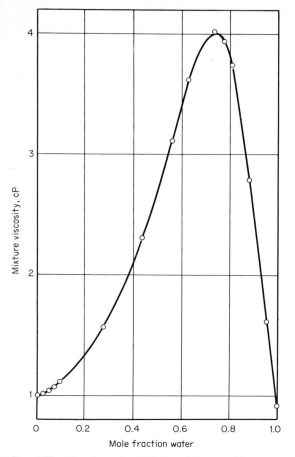

Fig. 9-20 Viscosity   of N,N-dimethylacetamide–water solution; $T = 24°C$.   (*From Ref.* 160.)

the carbonyl bond.   Most other solution viscosity-composition functions that yield maximum or minimum character can also be explained in terms of the properties of the specific substances involved.   However, no general theory exists to allow good a priori predictions.

To discuss the concept of liquid-mixture viscosity qualitatively, one should formulate some picture of the actual molecular processes taking place.   In a binary of A and B, molecular interactions are of both the A-A   and   B-B   types.   There are, in addition, the A-B type of interactions.   The introduction of the A-B type of effects invariably leads to undetermined mixture constants.   A good example of such an approach is that of McAllister [134], who adopted the semitheoretical Eyring approach [16, 75].   In this mechanistic picture, the interaction between layers of molecules in the velocity gradient involves *activated* *jumps* of molecules between layers. A molecule moving in this manner is treated as if it were undergoing a chemical reaction, and the pushing or squeezing of this moving molecule requires that it overcome a potential- (free-) energy barrier ΔG^* in the process. This descriptive mechanism leads, after simplification, to an equation similar to Eq. (9-10.1):

$$\nu_L = \left(\frac{\eta}{\rho}\right)_L = \frac{hN_o}{M}\exp\frac{\Delta G^*}{RT} \tag{9-13.1}$$

$$= \frac{hN_o}{M}\exp\left(-\frac{\Delta S^*}{R}\right)\exp\frac{\Delta H^*}{RT} \tag{9-13.2}$$

The grouping $\rho_L hN_0/M \exp(-\Delta S^*/R)$ is the equivalent of A and $\Delta H^*/R$ of B in Eq. (9-10.1). Here the ΔG^*, ΔH^*, and ΔS^* are the respective Gibbs energies, enthalpies, and entropies of activation in the process of moving a molecule from one layer to another. There is no reliable way at present to determine such energy quantities, but the descriptive picture allows one to postulate a mixture model.

For a mixture of A and B, as molecule A moves over the energy barrier, it may interact predominantly with A, with B, or with some combination of A and B, depending on the local concentration. Moreover, the interaction could be taken as a binary type or, more realistically, as a three-body type, four-body type, etc. Perhaps the simplest to visualize is the two-dimensional, three-body type assumed by McAllister.† The sketch below shows the basic idea:

Here molecule Z is jumping or flowing between molecules X and Y, where, for a binary, Z, X, and Y may be A or B. If all are A, the interaction is the A-A-A type; in general, for a binary, there would be six types, that is, A-A-A, B-B-B, A-B-A, B-A-B, A-A-B, A-B-B. The ΔG^* for the viscous interaction is assumed to be expressed in general as

$$\Delta G^*_m = \sum_{i=1}^{2}\sum_{j=1}^{2}\sum_{k=1}^{2} x_i x_j x_k \, \Delta G^*_{ijk} \tag{9-13.3}$$

where x represents the mole fraction. Furthermore,

$$\Delta G^*_{\text{AAB}} = \Delta G^*_{\text{ABA}} \equiv \Delta G^*_{\text{AB}}$$

$$\Delta G^*_{\text{BBA}} = \Delta G^*_{\text{BAB}} \equiv \Delta G^*_{\text{BA}} \tag{9-13.4}$$

†If the ratio of the molecular diameters is large (>1.5), McAllister suggests that a four-body type or higher interaction model is necessary.

By combining Eqs. (9-13.3) and (9-13.4) into (9-13.1) and defining a mixture kinematic viscosity ν_{AB} (and ν_{BA}) by Eq. (9-13.1),† there results

$$\ln \nu_m = x_A{}^3 \ln \nu_A + 3x_A{}^2 x_B \ln \nu_{AB} + 3x_A x_B{}^2 \ln \nu_{BA} + x_B{}^3 \ln \nu_B + R° \quad (9\text{-}13.5)$$

where

$$R° = x_B{}^3 \ln \frac{M_B}{M_A} + 3x_A x_B{}^2 \ln \frac{1 + 2M_B/M_A}{3}$$

$$+ 3x_A{}^2 x_B \ln \frac{2 + M_B/M_A}{3} - \ln \frac{x_A + x_B M_B}{M_A} \quad (9\text{-}13.6)$$

Equation (9-13.5) contains two undetermined parameters, ν_{AB} and ν_{BA}. They are assumed independent of composition but vary with temperature, as shown in Eq. (9-13.1). Thus, once these constants are established for a given binary, viscosities at other temperatures and compositions can be determined. Expressions of a similar type, but with more undetermined parameters, are obtained if four-body interactions, eight-body interactions, etc., are considered.

The value of $R°$ in Eq. (9-13.6) vanishes if $M_A = M_B$ and is negative except for large or small ratios of M_A/M_B.

McAllister applied Eq. (9-13.5) to the systems benzene-toluene, cyclohexane-heptane, methanol-toluene, and acetone-water at various temperatures. The parameters ν_{AB} and ν_{BA} were fitted by a least-squares technique from the experimental data (actually ΔS^* and ΔH^* values were found for A-A, A-B, B-A, B-B), and with these values the data were well correlated. The most nonideal case, acetone-water, showed the greatest errors, i.e., an average deviation of 6.4 percent and a maximum deviation of 15.8 percent. McAllister showed, however, with a similar type of analysis using a four-body interaction, that the three-parameter result correlated even this system to within 2 to 4 percent.

McAllister's approach has also been shown to apply to ternary systems containing various mixtures of water, ethyl alcohol, ethylene glycol, methyl alcohol, and acetone [42, 109]. The only parameters needed were the ν_{AB}, ν_{BA} values for the three constituent binaries and pure-component viscosities.

Most correlations for liquid viscosity can be generalized to the types

$$f(\eta_m)_L = \sum_i \sum_j x_i x_j f(\eta_{ij})_L \quad (9\text{-}13.7)$$

or

$$f(\eta_m)_L = \sum_i x_i f(\eta_i)_L \quad (9\text{-}13.8)$$

In these relations $f(\eta)_L$ may be η_L, $\ln \eta_L$, $1/\eta_L$, etc., and x_i may be the liquid volume, weight, or mole fraction [135]. Some investigators claim

†Here it is assumed that $M_{AB} = (2M_A + M_B)/3$, $M_{BA} = (2M_B + M_A)/3$.

good results with simple equations of the type of (9-13.8), but it is obvious that such a form cannot accurately account for systems containing maximum, minimum, or both maximum and minimum viscosities as composition is varied. Acetone-water [96], N,N-dimethylacetamide-water [160], and methanol-toluene [87] are good examples of such systems.

Heric and Brewer [91] tested a number of liquid-viscosity correlations, including several of the forms represented by Eqs. (9-13.7) and (9-13.8). They concluded that McAllister's method was the most accurate. In addition to pure-component values, only two binary viscosity data points are necessary at each of two different temperatures to define the entire composition-temperature range over which Eq. (9-13.1) is applicable, although obviously the more data used, the better the resulting correlation. Ratcliff and coworkers [166, 217] have obtained encouraging results in the development of methods for estimating liquid-mixture viscosity from a group-contribution concept. Lobe [129] also tested many liquid-viscosity correlations. He found best results were obtained with an expression

$$\nu_m = \sum_{i=1}^n \phi_i \nu_i \exp\left(\sum_{j=1}^n \frac{\alpha_j \phi_j}{RT}\right) \qquad j \neq i \qquad (9\text{-}13.9)$$

where ν = kinematic viscosity η/ρ, cSt
  ϕ_j = volume fraction of component j
  α_j = characteristic viscosity parameter for j in mixture, cal/g-mol K
  R = gas constant = 1.987 cal/g-mol K
  T = temperature, K

To use Eq. (9-13.9) for an n-component mixture, at least n mixture viscosity values must be known to obtain all α_j values. Equation (9-13.9) will, however, yield relationships between ν_m and composition that allow maximum, minimum, or inflection points.

For a binary system of A and B, Lobe expressed Eq. (9-13.9) as

$$\nu_m = \phi_A \nu_A e^{\phi_B \alpha_B^*} + \phi_B \nu_B e^{\phi_A \alpha_A^*} \qquad (9\text{-}13.10)$$

where $\alpha_A^* = \alpha_A/RT$ and $\alpha_B^* = \alpha_B/RT$, and suggested that if A is chosen as the component with lesser pure-liquid viscosity *and* if the mixture kinematic viscosity varies monotonically with composition, then

$$\alpha_A^* = -1.7 \ln \frac{\nu_B}{\nu_A} \qquad (9\text{-}13.11)$$

$$\alpha_B^* = 0.27 \ln \frac{\nu_B}{\nu_A} + \left(1.3 \ln \frac{\nu_B}{\nu_A}\right)^{1/2} \qquad (9\text{-}13.12)$$

Thus, from only the pure-component kinematic viscosities, it is possible to estimate the binary-liquid-mixture viscosity.

The discussion to this point has been applicable to low-temperature liquid-mixture viscosities. Essentially no experimental data are available, nor have there been correlations suggested for high-temperature liquid-mixture viscosities ($T_r > 0.75$). One might presume that in this region, a corresponding-states approach could be employed where the correlations shown in Sec. 9-12 would be used and the mixture-reduced temperature, pressure, and acentric factor be defined by some mixture rules such as those shown in Eqs. (9-7.2) to (9-7.5). Alternatively, the residual viscosity correlation discussed in Sec. 9-7 might be employed.

Recommendations: Viscosities of Liquid Mixtures

To estimate low-temperature liquid viscosity for mixtures of two components, it is recommended that if sufficient viscosity datum points and mixture volume data are available, Eq. (9-13.5) or (9-13.9) be used. If only one datum point is available at each of two temperatures, Eq. (9-13.7) can be used; a convenient form for $f(\eta)_L$ is $\ln \eta_L$ and x = mole fraction. η_{Lij} is determined at each temperature and assumed to behave as indicated by Eq. (9-10.1).

If no viscosity data are available for the mixture, use Eq. (9-13.8) or Eqs. (9-13.10) to (9-13.12). If Eq. (9-13.8) is employed, use $f(\eta)_L = \ln \eta_L$. Errors to be expected may vary widely for mixtures composed of chemically similar materials but should be less than 15 percent.

Example 9-18 Estimate the viscosity of a liquid mixture of ethyl benzoate and benzyl benzoate at 25°C. The mole fraction of benzyl benzoate is 0.606. The experimental value is 4.95 cP [99].

solution Let the subscripts EB stand for ethyl benzoate and BB for benzyl benzoate. Pure-component data at 25°C are [99]: $\eta_{EB} = 2.01$ cP, $\eta_{BB} = 8.48$ cP, $\rho_{EB} = 1.043$ g/cm³, $\rho_{BB} = 1.112$ g/cm³

(a) Using Eq. (9-13.8), with $f(\eta)_L = \ln \eta_L$,

$$\ln \eta_m = 0.394 \ln 2.01 + 0.606 \ln 8.48$$

$$\eta_m = 4.81 \text{ cP}$$

$$\text{Error} = \frac{4.81 - 4.95}{4.95} \times 100 = -2.8\%$$

(b) With Eqs. (9-13.10) to (9-13.12),

$$\nu_A = \nu_{EB} = \frac{2.01}{1.043} = 1.927 \text{ cST}$$

$$\nu_B = \nu_{BB} = \frac{8.48}{1.112} = 7.626 \text{ cST}$$

$$\alpha_A^* = \alpha_{EB}^* = -1.7 \ln \frac{7.626}{1.927} = -2.338$$

$$\alpha_B^* = \alpha_{BB}^* = 0.27 \ln \frac{7.626}{1.927} + \left(1.3 \ln \frac{7.626}{1.927}\right)^{1/2} = 1.709$$

The molal volumes of EB and BB are 144 and 178 cm^3/g-mol; thus

$$\phi_B = \phi_{BB} = \frac{(0.606)(178)}{(0.606)(178) + (0.394)(144)} = 0.655$$

$$\phi_A = \phi_{EB} = 0.345$$

Then

$$\nu_m = (0.345)(1.927)e^{(0.655)(1.709)} + (0.655)(7.626)e^{(0.345)(-2.338)} = 4.27 \text{ cSt}$$

The density of the solution is 1.091 g/cm^3, so the predicted value of the viscosity is (1.091)(4.27) = 4.66 cP.

$$\text{Error} = \frac{4.66 - 4.95}{4.95} \times 100 = -5.8\%$$

NOTATION

Constants used in specific equations are not usually listed.

A_{ij} = interaction parameter in Eq. (9-5.8)
b_i = group-contribution factor in Morris' method, Table 9-10
b_0 = excluded volume, $\frac{2}{3}\pi N_0 \sigma^3$
B = viscosity parameter in method of van Velzen et al., Eq. (9-11.5) (see Appendix A); activation energy in Eq. (9-10.1)
C_v = heat capacity at constant volume, cal/g-mol K
F = factor in Eq. (9-6.1)
ΔG^* = Gibbs energy of activation for flow, cal/g-mol
h = Planck's constant
ΔH^* = enthalpy of activation
J = Morris' viscosity constant, Table 9-10
k = Boltzmann's constant
l, L = mean free path, length
m = mass of molecule
M = molecular weight
n = number density of molecules
N = number of carbon atoms; N^*, equivalent chain length; ΔN_i, structural contribution in Eq. (9-11.6), Table 9-11
N_o = Avogadro's number
P = pressure, atm; P_c, critical pressure; P_r, reduced pressure, P/P_c; P_{c_m}, pseudocritical pressure of mixture
q = volumetric flow rate; polar parameter in Eq. (9-6.9)
r = distance of separation; r_w, radius
R = gas constant
$R°$ = parameter defined in Eq. (9-13.6)
S_{ij} = polar parameter in Eq. (9-5.12)
ΔS^* = entropy of activation
T = temperature, K; T_c, critical temperature; T_r, reduced temperature, T/T_c; T_{c_m}, pseudocritical temperature of mixture; T_f, freezing temperature; T_0, parameter in method of van Velzen et al., Eq. (9-11.5), also see Appendix A; T_b, normal-boiling-point temperature
$T^* = kT/\epsilon$
v = molecular velocity; v_y, y component of velocity
V = volume, cm^3/g-mol; V_c, critical volume; V_r, reduced volume, V/V_c; V_{c_m}, pseudocritical volume of mixture; V_b, volume of liquid at the normal boiling point; V_m, mixture volume; $\bar{V}_j$, partial molal volume of j

x = mole fraction, liquid
y = mole fraction, vapor
Z = compressibility factor; Z_c, critical compressibility factor; Z_{c_m}, pseudocritical compressibility factor for mixture

Greek

α = parameter in Eq. (9-13.7)
δ = polar parameter in Eq. (9-4.10) and Table 9-2
Δ = correction term in Eq. (9-6.8)
ϵ = energy-potential parameter
θ = Thomas viscosity constant, Table 9-8
η = viscosity (usually micropoises for gas and centipoises for liquid); η°, special notation designating value at pressure around 1 atm; η_L, liquid; η_c, at the critical point; η_m, for the mixture; η_r, reduced, Eq. (9-4.13); η_c° or η_c^*, at critical temperature but 1 atm; $\eta_{L_{ij}}$, interaction liquid-viscosity parameter
λ = thermal conductivity, cal/s K cm
μ_p = dipole moment
ν = kinematic viscosity η/ρ
$\xi = T_c^{1/6} M^{-1/2} P_c^{-2/3}$; ξ_m, for a mixture
ρ = density, g-mol/cm^3; ρ_c, critical density; ρ_r, reduced density ρ/ρ_c; ρ_m, density of mixture; ρ_L, density of liquid; ρ_{c_m}, pseudocritical density of mixture
σ = molecular diameter, Å
ϕ = mixture viscosity parameter in Eq. (9-5.1); volume fraction
χ = factor in Enskog's equation, Eq. (9-6.2)
ψ = intermolecular potential energy as a function of distance r
ω = acentric factor, Sec. 2-3
Ω_v = collision integral for viscosity; Ω_D, collision integral for diffusion

Subscripts

i, j = components i, j
A, B = components A, B
L = liquid

REFERENCES

1. Albright, L. F., and J. Lohrenz: *AIChE J.*, **2**: 290 (1956).
2. Alder, B. J.: Prediction of Transport Properties of Dense Gases and Liquids, UCRL-14891-T, University of California, Berkeley, May 1966.
3. Alder, B. J., and J. H. Dymond: Van der Waals Theory of Transport in Dense Fluids, UCRL-14870-T, University of California, Berkeley, April 1966.
4. Alder, B. J., and T. Wainwright: The Many-Body Problem, pp. 511–522, in J. K. Percus (ed.), *Proc. Symp. Stevens Inst. Tech., 1957*, Interscience, New York, 1963.
5. Amdur, I., and E. A. Mason: *Phys. Fluids*, **1**: 370 (1958).
5a. American Petroleum Institute: "Selected Values of Physical and Thermodynamic Properties of Hydrocarbons and Related Compounds," project 44, Carnegie Press, Pittsburgh, Pa., 1953, and supplements.
6. Andrade, E. N. da C.: *Nature*, **125**: 309 (1930).
7. Andrade, E. N. da C.: *Phil. Mag.*, **17**: 497, 698 (1934).
8. Andrade, E. N. da C.: *Endeavour*, **13**: 117 (1954).
9. Andrussow, L.: *Z. Electrochem.*, **61**: 253 (1957).
10. Babb, S. E., and G. J. Scott: *J. Chem. Phys.*, **40**: 3666 (1964).
11. Bae, J. H., and T. M. Reed III: *Ind. Eng. Chem.*, **10**: 36, 269 (1971).
12. Barker, J. A., W. Fock, and F. Smith: *Phys. Fluids*, **7**: 897 (1964).

13. Baron, J. D., J. G. Root, and F. W. Wells: *J. Chem. Eng. Data*, **4**: 283 (1959).
14. Bicher, L. B.: Ph.D. thesis, University of Michigan, Ann Arbor, 1943.
15. Bingham, E. C., and S. D. Stookey: *J. Am. Chem. Soc.*, **61**: 1625 (1939).
16. Bloomfield, V. A., and R. K. Dewan: *J. Phys. Chem.*, **75**: 3113 (1971).
17. Bondi, A.: *Ind. Eng. Chem. Fundam.*, **2**: 95 (1963); *J. Phys. Chem.*, **68**: 441 (1964); *J. Polym. Sci.*, **(A)2**: 3159 (1964).
18. Boon, J. P., J. C. Legros, and G. Thomaes: *Physica*, **33**: 547 (1967).
19. Bradley, R. S. (ed.): "High Pressure Physics and Chemistry," p. 18, Academic, New York, 1963.
20. Brebach, W. J., and G. Thodos: *Ind. Eng. Chem.*, **50**: 1095 (1958).
21. Bridgman, P. W.: *Proc. Am. Acad. Arts Sci.*, **61**: 57 (1926).
22. Brokaw, R. S.: *NASA Tech. Rep.* R-81, 1961.
23. Brokaw, R. S.: *NASA Tech. Note* D-2502, November 1964.
24. Brokaw, R. S.: *J. Chem. Phys.*, **42**: 1140 (1965).
25. Brokaw, R. S.: *NASA Tech. Note* D-4496, April 1968.
26. Brokaw, R. S.: *Ind. Eng. Chem. Process Des. Dev.*, **8**: 240 (1969).
27. Brokaw, R. S., R. A. Svehla, and C. E. Baker: *NASA Tech. Note* D-2580, January 1965.
28. Bromley, L. A., and C. R. Wilke: *Ind. Eng. Chem.*, **43**: 1641 (1951).
29. Brown, G. P., A. Dinardo, G. K. Cheng, and T. K. Sherwood: *J. Appl. Phys.*, **17**: 802 (1946).
30. Brush, S. G.: *Chem. Rev.*, **62**: 513 (1962).
31. Burch, L. G., and C. J. G. Raw: *J. Chem. Phys.*, **47**: 2798 (1967).
32. Burnett, D.: *J. Chem. Phys.*, **42**, 2533 (1965).
33. Carmichael, L. T., V. Berry, and B. H. Sage: *J. Chem. Eng. Data*, **12**: 44 (1967).
34. Carmichael, L. T., and B. H. Sage: *J. Chem. Eng. Data*, **8**: 94 (1963).
35. Carmichael, L. T., and B. H. Sage: *AIChE J.*, **12**: 559 (1966).
36. Carmichael, L. T., H. H. Reamer, and B. H. Sage: *J. Chem. Eng. Data*, **8**: 400 (1963).
37. Carnahan, N. F., and C. J. Vadovic: *Hydrocarbon Process.*, **49**(5): 159 (1970).
38. Carr, N. L., R. Kobayashi, and D. B. Burroughs: *J. Pet. Technol.*, **6**(10): 47 (1954).
39. Carr, N. L., J. D. Parent, and R. E. Peck: *Chem. Eng. Prog. Symp. Ser.*, **51**(16): 91 (1955).
40. Chakraborti, P. K., and P. Gray: *Trans. Faraday Soc.*, **61**: 2422 (1965).
41. Chakraborti, P. K., and P. Gray: *Trans. Faraday Soc.*, **62**: 1769 (1966).
42. Chandramoule, V. V., and G. S. Laddha: *Indian J. Technol.*, **1**: 199 (1963).
43. Chapman, S., and T. G. Cowling: "The Mathematical Theory of Nonuniform Gases," Cambridge University Press, New York, 1939.
44. Chapman, T. W.: *AIChE J.*, **12**: 395 (1966).
45. Cheung, H.: UCRL Report 8230, University of California, Berkeley, April 1958.
46. Childs, G. E., and H. J. M. Hanley: *Cryogenics*, **8**: 94 (1968).
47. Comings, E. W., and R. S. Egly: *Ind. Eng. Chem.*, **32**: 714 (1940).
48. Comings, E. W., and B. J. Mayland: *Chem. Metall. Eng.*, **42**(3): 115 (1945).
49. Coremans, J. M. J., and J. J. M. Beenakker: *Physica*, **26**: 653 (1960).
50. Cornelissen, J., and H. I. Waterman: *Chem. Eng. Sci.*, **4**: 238 (1955).
51. Dahler, J. S.: "Thermodynamic and Transport Properties of Gases, Liquids, and Solids," pp. 14–24, McGraw-Hill, New York, 1959.
52. Damasius, G., and G. Thodos: *Ind. Eng. Chem. Fundam.*, **2**: 73 (1963).
53. Dean, D. E., and L. I. Stiel: *AIChE J.*, **11**: 526 (1965).
54. Dolan, J. P.: M.S. thesis, Illinois Institute of Technology, Chicago, 1962.
55. Dolan, J. P., K. E. Starling, A. L. Lee, B. E. Eakin, and R. T. Ellington: *J. Chem. Eng. Data*, **8**: 396 (1963).
56. Eakin, B. E., and R. T. Ellington: *J. Pet. Technol.*, **15**: 210 (1963).
57. Ellington, R. T.: Ph.D. thesis, Illinois Institute of Technology, Chicago, 1962.
58. Ertl, H., and F. A. L. Dullien: *AIChE J.*, **19**: 1215 (1973).
59. Ertl, H., and F. A. L. Dullien: *J. Phys. Chem.*, **77**: 3007 (1973).

60. Eversteijn, F. C., J. M. Stevens, and H. I. Waterman: *Chem. Eng. Sci.*, **11**: 267 (1960).
61. Ewell, R. H., and H. Eyring: *J. Chem. Phys.*, **5**: 726 (1937).
62. Eyring, H., and R. P. Marchi: *J. Chem. Ed.*, **40**: 562 (1963).
63. Flynn, L. W., and G. Thodos: *J. Chem. Eng. Data*, **6**: 457 (1961).
64. Flynn, L. W., and G. Thodos: *AIChE J.*, **8**: 362 (1962).
65. Francis, W. E.: *Trans. Faraday Soc.*, **54**: 1492 (1958).
66. Friend, J. N.: *Trans. Faraday Soc.*, **31**: 542 (1935).
67. Galloway, T. R., and B. H. Sage: *J. Chem. Eng. Data*, **12**: 59 (1967).
68. Gambill, W. R.: *Chem. Eng.*, **65**(23): 157 (1958).
69. Gambill, W. R.: *Chem. Eng.*, **66**(1): 127 (1959).
70. Gambill, W. R.: *Chem. Eng.*, **66**(3): 123 (1959).
71. Gandhi, J. M., and S. C. Saxena: *Indian J. Pure Appl. Phys.*, **2**: 83 (1964).
72. Gemant, A.: *J. Appl. Phys.*, **12**: 827 (1941).
73. Giddings, J. D.: Ph.D. thesis, Rice University, Houston, Tex., 1963.
74. Girifalco, L. A.: *J. Chem. Phys.*, **23**: 2446 (1955).
75. Glasstone, S., K. J. Laidler, and H. Eyring: "The Theory of Rate Processes," McGraw-Hill, New York, 1941.
76. Golubev, I. F.: Viscosity of Gases and Gas Mixtures: A Handbook, *Natl. Tech. Inf. Serv.*, TT 70 50022, 1959.
77. Gonzalez, M. H., and A. L. Lee: *AIChE J.*, **14**: 242 (1968).
78. Greet, R. J., and J. H. Magill: *J. Phys. Chem.*, **71**: 1746 (1967).
79. Griest, E. M., W. Webb, and R. W. Schiessler: *J. Chem. Phys.*, **29**: 711 (1958).
80. Grunberg, L., and A. H. Nissan: *Trans. Faraday Soc.*, **44**: 1013 (1948).
81. Grunberg, L., and A. H. Nissan: *Ind. Eng. Chem.*, **42**: 885 (1950).
82. Gubbins, K. E., and M. J. Tham: *AIChE J.*, **15**: 264 (1969).
83. Gupta, G. P., and S. C. Saxena: *AIChE J.*, **14**: 519 (1968).
84. Gururaja, G. J., M. A. Tirunarayanan, and A. Ramachandran: *J. Chem. Eng. Data*, **12**: 562 (1967).
85. Gutman, F., and L. M. Simmons: *J. Appl. Phys.*, **23**: 977 (1952).
86. Halkiadakis, E. A., and R. G. Bowery: *Chem. Eng. Sci.*, **30**: 53 (1975).
87. Hammond, L. W., K. S. Howard, and R. A. McAllister: *J. Phys. Chem.*, **62**: 637 (1958).
88. Hamrin, C. E., and G. Thodos: *J. Chem. Eng. Data*, **13**: 57 (1968).
89. Hanley, H. J. M., R. D. McCarty, and J. V. Sengers: *J. Chem. Phys.*, **50**: 857 (1969).
90. Hattikudur, U. R., and G. Thodos: *AIChE J.*, **17**: 1220 (1971).
91. Heric, E. L., and J. G. Brewer: *J. Chem. Eng. Data*, **12**: 574 (1967).
92. Herning, F., and L. Zipperer: *Gas Wasserfach*, **79**: 49 (1936).
93. Hirschfelder, J. O., C. F. Curtiss, and R. B. Bird: "Molecular Theory of Gases and Liquids," Wiley, New York, 1954.
94. Hirschfelder, J. O., M. H. Taylor, and T. Kihara: *Univ. Wisconsin Theoret. Chem. Lab.*, WIS-OOR-29, Madison, Wis., July 8, 1960.
95. Hogenbloom, D. L., W. Webb, and J. A. Dixon: *J. Chem. Phys.*, **46**: 2586 (1967).
96. Howard, J. S., and R. A. McAllister: *AIChE J.*, **4**: 362 (1958).
97. Hu, A. T., P. S. Chappelear, and R. Kobayashi: *AIChE J.*, **16**: 490 (1970).
98. Innes, K. K.: *J. Phys. Chem.*, **60**: 817 (1956).
99. International Critical Tables, McGraw-Hill, New York, 1928, vol. III, p. 193; vol. V, p. 51.
100. Irany, E. P.: *J. Am. Chem. Soc.*, **60**: 2106 (1938).
101. Irany, E. P.: *J. Am. Chem. Soc.*, **61**: 1734 (1939).
102. Irany, E. P.: *J. Am. Chem. Soc.*, **65**: 1392 (1943).
103. Itean, E. C., A. R. Glueck, and R. A. Svehla: NASA Lewis Research Center, TND-481, Cleveland, Ohio, 1961.
104. Iyer, M. P. V.: *Indian J. Phys.*, **5**: 371 (1930).

105. Jagannathan, T. K., D. S. Viswanath, and N. R. Kuloor: *Hydrocarbon Process.*, **47**(2): 133 (1968).

106. Jobling, A., and A. S. C. Laurence: *Proc. R. Soc. Lond.*, **A206**: 257 (1951).

107. Johnson, D. W., and C. P. Colver: *Hydrocarbon Process.*, **48**(3): 113 (1969).

108. Jossi, J. A., L. I. Stiel, and G. Thodos: *AIChE J.*, **8**: 59 (1962).

109. Kalidas, R., and G. S. Laddha: *J. Chem. Eng. Data*, **9**: 142 (1964).

110. Kennedy, J. T., and G. Thodos: *AIChE J.*, **7**: 625 (1961).

111. Kessel'man, P. M., and A. S. Litvinov: *J. Eng. Phys. USSR*, **10**(3): 385 (1966).

112. Kestin, J., and W. Leidenfrost: *Physica*, **25**: 525 (1959).

113. Kestin, J., and W. Leidenfrost: *Physica*, **25**: 1033 (1959).

114. Kestin, J., and W. Leidenfrost: pp. 321–338, in Y. S. Touloukian (ed.), "Thermodynamic and Transport Properties of Gases, Liquids, and Solids," ASME and McGraw-Hill, New York, 1959.

115. Kestin, J., and J. R. Moszynski: pp. 70–77, in Y. S. Touloukian (ed.), "Thermodynamic and Transport Properties of Gases, Liquids, and Solids," ASME and McGraw-Hill, New York, 1959.

116. Kestin, J., and J. H. Whitelaw, and T. F. Zien: *Physica*, **30**: 161 (1964).

117. Kestin, J., and J. Yata: *J. Chem. Phys.*, **49**: 4780 (1968).

118. Kim, S. K., and J. Ross: *J. Chem. Phys.*, **46**: 818 (1967).

119. Klein, M., and F. J. Smith: *J. Res. Bur. Stand.*, **72A**: 359 (1968).

120. Kreps, S. I., and M. L. Druin: *Ind. Eng. Chem. Fundam.*, **9**: 79 (1970).

121. Krieger, F. A.: Rand Corp. *Rep.* RM-646, Santa Monica, Calif.

122. Kuss, E.: *Z. Angew. Phys.*, **7**: 372 (1955).

123. "Landolt-Börnstein Tabellen," vol. 4, pt. 1, Springer-Verlag, Berlin, 1955.

124. Lefrançois, B.: *Chim. Ind. Paris*, **98**(8): 1377 (1967).

125. Lennert, D. A., and G. Thodos: *AIChE J.*, **11**: 155 (1965).

126. Letsou, A., and L. I. Stiel: *AIChE J.*, **19**: 409 (1973).

127. Lewis, W. K., and L. Squires: *Refiner Nat. Gasoline Manuf.*, **13**(12): 448 (1934).

128. Li, C. C., and T. R. Poole: *J. Chem. Eng. Jpn.*, **6**: 293 (1973).

129. Lobe, V. M.: M.S. thesis, University of Rochester, Rochester, N.Y., 1973.

130. Lowitz, D. A., J. W. Spencer, W. Webb, and R. W. Schliessler: *J. Chem. Phys.*, **30**: 73 (1959).

131. Lucas, K.: *Int. J. Heat Mass Trans.*, **16**: 371 (1973).

132. Lucas, K.: *Chem. Ing. Tech.*, **46**: 157 (1974).

133. Luft, N. W., and O. P. Kharbanda: *Res. Corresp.*, **7**(6): 536 (1954).

134. McAllister, R. A.: *AIChE J.*, **6**: 427 (1960).

135. McLaughlin, E.: *Q. Rev.*, **14**: 236 (1960).

136. Magill, J. H., and R. J. Greet: *Ind. Eng. Chem. Fundam.*, **8**: 701 (1969).

137. Makhija, R. C., and R. A. Stairs: *Can. J. Chem.*, **48**: 1214 (1970).

138. Malek, K. R., and L. I. Stiel: *Can. J. Chem. Eng.*, **50**: 491 (1972).

139. Marks, G. W.: *J. Phys. Chem.*, **43**: 549 (1939).

140. Marschalko, B., and J. Barna: *Acta Tech. Acad. Sci. Hung.*, **19**: 85 (1957).

141. Mathur, G. P., and G. Thodos: *AIChE J.*, **9**: 596 (1963).

142. Matsen, J. M., and E. F. Johnson: *J. Chem. Eng. Data*, **5**: 531 (1960).

143. Michels, A., A. Bolzen, and W. Schuurman: *Physica*, **20**: 1141 (1941).

144. Miller, A. A.: *J. Chem. Phys.*, **38**: 1568 (1963).

145. Miller, A. A.: *J. Phys. Chem.*, **67**: 1031 (1963).

146. Miller, A. A.: *J. Phys. Chem.*, **67**: 2809 (1963).

147. Monchick, L., and E. A. Mason: (*a*) Johns Hopkins University Applied Physics Laboratory, CM-993, February 1961; (*b*) *J. Chem. Phys.*, **35**: 1676 (1961).

148. Morris, P. S.: M.S. thesis, Polytechnic Institute of Brooklyn, Brooklyn, N.Y., 1964.

149. Mueller, C. R., and J. E. Lewis: *J. Chem. Phys.*, **26**: 286 (1957).

150. Neufeld, P. D., A. R. Janzen, and R. A. Aziz: *J. Chem. Phys.*, **57**: 1100 (1972).
151. Nissan, A. H.: *Nature*, **144**: 383 (1939).
152. Nissan, A. H., and L. V. W. Clark: *Nature*, **143**: 722 (1939).
153. Nissan, A. H., L. V. W. Clark, and A. W. Nash: *J. Inst. Pet. Technol.*, **26**: 155 (1940).
154. Orrick, C., and J. H. Erbar: private communication, December 1974.
155. Othmer, D. F., and J. W. Conwell: *Ind. Eng. Chem.*, **37**: 1112 (1945).
156. Pal, A. K., and A. K. Barua: *J. Chem. Phys.*, **48**: 872 (1968).
157. Pal, A. K., and A. K. Barua: *J. Chem. Phys.*, **47**: 216 (1967).
158. Pal, A. K., and P. K. Bhattacharyya: *J. Chem. Phys.*, **51**: 828 (1969).
159. Palit, S. R.: *J. Indian Chem. Soc.*, **40**: 721 (1963).
160. Petersen, R. C.: *J. Phys. Chem.*, **64**: 184 (1960).
161. Powell, R. E., W. E. Rosveare, and H. Eyring: *Ind. Eng. Chem.*, **33**: 430 (1941).
162. Prasad, B.: *Phil. Mag.*, **39**: 884 (1948).
163. Pratap, R., and G. Narsimhan: *Brt. Chem. Eng.*, **7**: 757 (1962).
164. Preston, G. T., T. W. Chapman, and J. M. Prausnitz: *Cryogenics*, **7**: 274 (1967).
165. Ranz, W. E., and H. A. Brodowsky: *Univ. Minn. OOR Proj. 2340 Tech. Rep.* 1, Minneapolis, Minn., March 15, 1962.
166. Ratcliff, G. A., and M. A. Khan: *Can. J. Chem. Eng.*, **49**: 125 (1971).
167. Raw, C. J. G., and H. Tang: *J. Chem. Phys.*, **39**: 2616 (1963).
168. Reamer, H. H., G. Cokelet, and B. H. Sage: *Anal. Chem.*, **31**: 1422 (1959).
169. Reichenberg, D.: The Viscosities of Gas Mixtures at Moderate Pressures, *NPL Rep. Chem.* 29, National Physical Laboratory, Teddington, England, May 1974.
170. Reichenberg, D.: (*a*) *DCS Rep.* 11, National Physical Laboratory, Teddington, England, August 1971; (*b*) *AIChE J.*, **19**: 854 (1973); (*c*) **21**: 181 (1975).
170a. Reichenberg, D.: The Viscosities of Pure Gases at High Pressures, *NPL Rep. Chem.* 38, National Physical Laboratory, Teddington, England, August 1975.
171. Reid, R. C.: *Chem. Eng. Prog. Monogr. Ser.*, no. 5, 1968.
172. Reid, R. C., and L. I. Belenyessy: *J. Chem. Eng. Data*, **5**: 150 (1960).
173. Reid, R. C., and T. K. Sherwood: "Properties of Gases and Liquids," pp. 416, 437–441, 2d ed., McGraw-Hill, New York, 1966.
174. Rogers, J. D., and F. G. Brickwedde: *AIChE J.*, **11**: 304 (1965).
175. Ross, J.: *J. Chem. Phys.*, **42**: 263 (1965).
176. Ross, J. F., and G. M. Brown: *Ind. Eng. Chem.*, **49**: 2026 (1957).
177. Rutherford, R., M. H. Taylor, and J. O. Hirschfelder: *Univ. Wisconsin Theoret. Chem. Lab.*, WIS-OOR-29a, Madison, Wis., August 23, 1960.
178. Saksena, M. P., and S. C. Saxena: *Proc. Natl. Inst. Sci. India*, **31A**(1): 18 (1965).
179. Saxena, S. C., and R. S. Gambhir: *Br. J. Appl. Phys.*, **14**: 436 (1963).
180. Saxena, S. C., and R. S. Gambhir: *Proc. Phys. Soc. Lond.*, **81**: 788 (1963).
181. Shimotake, H., and G. Thodos: *AIChE J.*, **4**: 257 (1958).
182. Shimotake, H., and G. Thodos: *AIChE J.*, **9**: 68 (1963).
183. Shimotake, H., and G. Thodos: *J. Chem. Eng. Data*, **8**: 88 (1963).
184. Shukla, B. P., and R. P. Bhatnagu: *J. Phys. Chem.*, **59**: 988 (1955), **60**: 809 (1956).
185. Singh, Y., and P. K. Bhattacharyya: *J. Phys. B* (*Proc. Phys. Soc.*), (2) **1**: 922 (1968).
186. Smith, A. S., and G. G. Brown: *Ind. Eng. Chem.*, **35**: 705 (1943).
187. Souders, M.: *J. Am. Chem. Soc.*, **59**: 1252 (1937).
188. Souders, M.: *J. Am. Chem. Soc.*, **60**: 154 (1938).
189. Srinivasan, M. K., and B. Prasad: *Phil. Mag.*, **33**: 258 (1942).
190. Starling, K. E.: M.S. thesis, Illinois Institute of Technology, Chicago, 1960.
191. Starling, K. E., and R. T. Ellington: *AIChE J.*, **10**: 11 (1964).
192. Starling, K. E., B. E. Eakin, and R. T. Ellington: *AIChE J.*, **6**: 438 (1960).
193. Staveley, L. A. K., and P. F. Taylor: *J. Chem. Soc. Lond.*, **1956**: 200.
194. Stiel, L. I., and G. Thodos: *AIChE J.*, **7**: 611 (1961), **8**: 229 (1962), **10**: 266 (1964).

195. Stiel, L. I., and G. Thodos: *AIChE J.*, **10**: 275 (1964).
196. Strunk, M. R., W. G. Custead, and G. L. Stevenson: *AIChE J.*, **10**: 483 (1964).
197. Strunk, M. R., and G. D. Fehsenfeld: *AIChE J.*, **11**: 389 (1965).
198. Svehla, R. A.: Estimated Viscosities and Thermal Conductivities at High Temperatures, NASA-TRR-132, 1962.
199. Swift, G. W., J. A. Christy, and F. Kurata: *AIChE J.*, **5**: 98 (1959).
200. Telang, M. S.: *J. Phys. Chem.*, **49**: 579 (1945), **50**: 373 (1946).
201. Telang, M. S.: *J. Chem. Phys.*, **17**: 536 (1949).
202. Tham, M. J., and K. E. Gubbins: *Ind. Eng. Chem. Fundam.*, **8**: 791 (1969).
203. Thomas, L. H.: *J. Chem. Soc.*, **1946**: 573.
204. Titani, T.: *Bull. Inst. Phys. Chem. Res. Tokyo*, **8**: 433 (1929).
205. Tondon, P. K., and S. C. Saxena: *Ind. Eng. Chem. Fundam.*, **7**: 314 (1968).
206. Trautz, M.: *Ann. Phys.*, **11**: 190 (1931).
207. Trautz, M., and P. B. Baumann: *Ann. Phys.*, **5**: 733 (1929).
208. Trautz, M., and R. Heberling: *Ann. Phys.*, **10**: 155 (1931).
209. Uyehara, O. A., and K. M. Watson: *Natl. Pet. News*, **36**: R-714 (1944).
210. Vanderslice, J. T., S. Weissman, E. A. Mason, and R. J. Fallon: *Phys. Fluids*, **5**: 155 (1962).
211. van Velzen, D., R. L. Cardozo, and H. Langenkamp: *Ind. Eng. Chem. Fundam.*, **11**: 20 (1972).
212. van Velzen, D., R. L. Cardozo, and H. Langenkamp: Liquid Viscosity and Chemical Constitution of Organic Compounds: A New Correlation and a Compilation of Literature Data, *Euratom* 4735e, Joint Nuclear Research Centre, Ispra Establishment, Italy, 1972.
213. van Velzen, D.: private communication, September 1973.
214. van Wyk, W. R., J. H. van der Veen, H. C. Brinkman, and W. A. Seeder: *Physica*, **7**: 45 (1940).
215. Wakeham, W. A., J. Kestin, E. A. Mason, and S. I. Sandler: *J. Chem. Phys.*, **57**: 295 (1972).
216. Watson, K. M., F. R. Wien, and G. B. Murphy: *Ind. Eng. Chem.*, **28**: 605 (1936).
217. Wedlake, G. D., and G. A. Ratcliffe: *Can. J. Chem. Eng.*, **51**: 511 (1973).
218. Wilke, C. R.: *J. Chem. Phys.*, **18**: 517 (1950).
219. Wobser, R., and F. Mueller: *Kolloid Beih.*, **52**: 165 (1941).
220. Wright, F. J.: *J. Chem. Eng. Data*, **6**: 454 (1961).
221. Wright, P. G., and P. Gray: *Trans. Faraday Soc.*, **58**: 1 (1962).
222. Yoon, P., and G. Thodos: *AIChE J.*, **16**: 300 (1970).

Chapter Ten

Thermal Conductivity

10-1 Scope

Thermal conductivities of both gases and liquids are considered in this chapter. The theory of thermal conduction in gases is discussed in Sec. 10-2 and of liquids in Sec. 10-8. Estimation techniques for gases at low pressure are covered in Sec. 10-3 and effects of temperature and pressure in Secs. 10-4 and 10-5. Similar topics for liquids are discussed in Secs. 10-9 to 10-11. Thermal conductivities of mixtures of gases and of liquids are covered in Secs. 10-6, 10-7, and 10-12.

The units used for thermal conductivity throughout this chapter are cal/cm s K. To convert these units to Btu/ft h °R, multiply by 241.9; to convert to W/cm K or J/cm s K, multiply by 4.186.

10-2 Theory of Thermal Conduction in Gases

In Sec. 9-3 [Eq. (9-3.7)], we presented a very elementary picture wherein the thermal conductivity of a gas was found to be equal to $vLC_vn/3$, where v is the average velocity of the molecule, L the mean free path, C_v

the heat capacity per molecule, and n the number density of molecules. Similar relations were derived for the viscosity and diffusion coefficients of gases. In the case of the last two properties, this elementary approach yields approximate but reasonable values. For thermal conductivity, it is quite inaccurate. A more detailed treatment is necessary to account for the effect of having a wide spectrum of molecular velocities; also, molecules may store energy in forms other than translational. For monatomic gases, which have no rotational or vibrational degrees of freedom, a more rigorous analysis yields

$$\lambda = \frac{25}{32} (\pi m k T)^{1/2} \frac{C_v/m}{\pi \sigma^2 \Omega_v} \tag{10-2.1}$$

or, written for computational ease,

$$\lambda = \frac{1.989 \times 10^{-4}(T/M)^{1/2}}{\sigma^2 \Omega_v} \tag{10-2.2}$$

where λ = thermal conductivity, cal/cm s K
 T = temperature, K
 m = mass of molecule
 M = molecular weight
 σ = characteristic dimension of molecule, Å
 Ω_v = collision integral, dimensionless

To obtain Eq. (10-2.2) from Eq. (10-2.1), C_v was set equal to $\frac{3}{2}k$, where k is Boltzmann's constant. For a hard-sphere molecule Ω_v is unity; normally, however, it is a function of temperature, and the exact dependence is related to the intermolecular force law chosen. If the Lennard-Jones 12-6 potential [Eq. (9-4.2)] is selected, Ω_v is given by Eq. (9-4.3).

If Eq. (10-2.1) is divided by Eq. (9-3.9),

$$\frac{\lambda M}{\eta C_v} = 2.5 \tag{10-2.3}$$

For $\gamma = C_p/C_v$, the Prandtl number N_{Pr} is

$$N_{Pr} \equiv \frac{C_p \eta}{\lambda M} = \frac{\gamma}{2.5} \tag{10-2.4}$$

Since γ for monatomic gases is close to $\frac{5}{3}$ except at very low temperatures, Eq. (10-2.4) would indicate that $N_{Pr} \approx \frac{2}{3}$, a value close to that found experimentally.

To obtain Eq. (10-2.3), the terms σ^2 and Ω_v have been eliminated, and the result is essentially independent of the intermolecular-potential law chosen. Liley [93] has plotted the group $\lambda M/\eta C_v$ as a function of temperature for several gases, and, as shown in Fig. 10-1, the ratio for

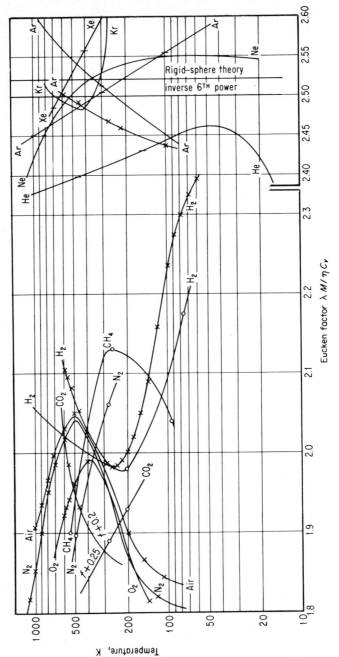

Fig. 10-1 Variation of the Eucken factor with temperature at low pressure. *(From Ref. 93.)*

the monatomic gases is close to the theoretical value of 2.5. The variation from the predicted value is discussed by Kestin et al. [79]; it is concluded that the discrepancies are due principally to experimental error.

It is also apparent from Fig. 10-1 that the group $\lambda M/\eta C_v$, known as the *Eucken factor*, is considerably less than 2.5 for gases other than monatomic. Our discussion so far has considered only energy associated with translational motion; since heat capacities of polyatomic molecules often greatly exceed those for monatomic gases, a substantial fraction of molecular energy is stored in modes other than translational.

10-3   Thermal Conductivity of Polyatomic Gases

Eucken and Modified Eucken Models   Eucken proposed that Eq. (10-2.3) be modified for polyatomic gases by separating the translational- and internal-energy contributions into separate terms:

$$\frac{\lambda M}{\eta} = f_{tr}C_{tr} + f_{int}C_{int} \tag{10-3.1}$$

Thus the translational-energy contribution has been decoupled from any internal-energy interaction [29, 85, 105, 124, 162, 176, 195, 196], although the validity of this step has been questioned [60, 94, 159, 180]. Invariably, f_{tr} is set equal to 2.5 to force Eq. (10-3.1) to reduce to Eq. (10-2.3) for a monatomic gas. C_{tr} is set equal to the classical value of $\frac{3}{2}R$, and C_{int} is conveniently expressed as $C_v - C_{tr}$. Then

$$\frac{\lambda M}{\eta} = \frac{15R}{4} + f_{int}(C_v - \tfrac{3}{2}R) \tag{10-3.2}$$

Eucken chose $f_{int} = 1.0$, whereby Eq. (10-3.2) reduces to

$$\frac{\lambda M}{\eta} = C_v + \frac{9R}{4} = C_v + 4.47 = 4.47 + \frac{C_p}{\gamma} \tag{10-3.3}$$

the well-known *Eucken correlation* for polyatomic gases.

Many of the assumptions leading to Eq. (10-3.3) are open to question, in particular, the choice of $f_{int} = 1.0$. Ubbelohde [186], Chapman and Cowling [22], Hirschfelder [60], and Schäfer [166] have suggested that molecules with excited internal-energy states may be regarded as separate chemical species and the transfer of internal energy is then analogous to a diffusional process. This concept leads to a result that

$$f_{int} = \frac{M\rho\mathcal{D}}{\eta} \tag{10-3.4}$$

where M = molecular weight
    η = viscosity
    ρ = molal density
    $\mathcal{D}$ = diffusion coefficient

Most early theories selected $\mathcal{D}$ to be equivalent to the molecular self-diffusion coefficient, and f_{int} is then the reciprocal of the Schmidt number. With Eqs. (9-3.9) and (11-3.2) it can be shown that $f_{int} \approx 1.32$ and is almost independent of temperature. With this formulation, Eq. (10-3.2) becomes

$$\frac{\lambda M}{\eta} = 1.32 C_v + 3.52 = 3.52 + \frac{1.32 C_p}{\gamma} \qquad (10\text{-}3.5)$$

Equation (10-3.5) is often referred to as the *modified Eucken correlation* and was used by Svehla [180] in his compilation of high-temperature gas properties.

    The modified Eucken relation [Eq. (10-3.5)] predicts larger values of λ than the Eucken form [Eq. (10-3.3)], and the difference becomes greater as C_v increases above the monatomic value of about 3 cal/g-mol K. [Both yield Eq. (10-2.3) when $C_v = \frac{3}{2}R$.] Usually experimental values of λ lie between values calculated by the two Eucken forms, except for polar gases, when both predict λ values that are too high.

    Mason and Monchick Analysis   In a pioneer paper published in 1962 [105], Mason and Monchick employed the formal dynamical treatment of Wang Chang and Uhlenbeck [199] and Taxman [181] to derive an approximation to the thermal conductivity of polyatomic gases. In the formalism of Eq. (10-3.1) they found

$$f_{tr} = \frac{5}{2}\left[1 - \frac{10}{3\pi}\left(1 - \frac{2}{5}\frac{M\rho\mathcal{D}}{\eta}\right)\frac{C_{rot}}{RZ_{rot}}\right] \qquad (10\text{-}3.6)$$

$$f_{int} = \frac{M\rho\mathcal{D}}{\eta}\left[1 + \frac{5}{\pi}\left(1 - \frac{2}{5}\frac{M\rho\mathcal{D}}{\eta}\right)\frac{C_{rot}}{C_{int}Z_{rot}}\right] \qquad (10\text{-}3.7)$$

C_{rot} is the contribution to the heat capacity from rotational storage modes and is obtained from the classical value $F_r R/2$, where F_r is the number of degrees of freedom for external rotation. $C_{int} = C_v - C_{tr}$. Z_{rot}, the collision number, represents the number of collisions required to interchange a quantum of rotational energy with translational energy. [In Eqs. (10-3.6) and (10-3.7) negligible terms involving vibrational-energy contributions have been dropped.] For large values of Z_{rot}, Eqs. (10-3.6) and (10-3.7) reduce to $f_{tr} = 2.5$ and $f_{int} = M\rho\mathcal{D}/\eta$, that is, those used earlier for the modified Eucken relation. If Eqs. (10-3.6) and (10-3.7) are substituted into Eq. (10-3.1) and $M\rho\mathcal{D}/\eta$ assumed to be 1.32,

$$\frac{\lambda M}{\eta} = 1.32 C_v + 3.54 - \frac{0.886 C_{rot}}{Z_{rot}} \qquad (10\text{-}3.8)$$

Equation (10-3.8) should be used only for *nonpolar* polyatomic molecules. The key to its use is the accurate estimation of the rotational collision number. To date this is not possible, although many authors discuss the problem [2, 11, 13, 32, 105, 106, 124, 125, 134, 158–160, 162, 176]. In Fig. 10-2, we show the so-called Eucken factor $\lambda M/\eta C_v$ for hydrogen, nitrogen, and carbon dioxide. As noted earlier, most experimental data show values of $\lambda M/\eta C_v$ between those predicted by the Eucken and modified Eucken relations. In Fig. 10-2, the Z_{rot} shown is that which gave the best fit to the experimental data. Z_{rot} was assumed temperature-independent, although Healy and Storvick [59] clearly show that Z_{rot} increases significantly with temperature and Barua

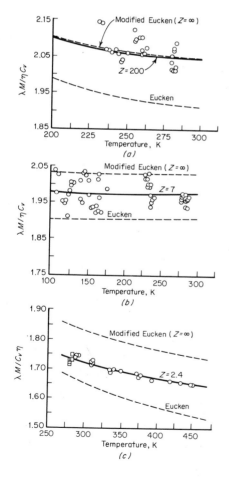

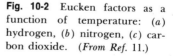

Fig. 10-2 Eucken factors as a function of temperature: (*a*) hydrogen, (*b*) nitrogen, (*c*) carbon dioxide. (*From Ref.* 11.)

[2] suggests that for some polar molecules it may decrease with temperature.

For polar gases, the analysis of Mason and Monchick has been modified by Brokaw [12], but the recommended computational method requires knowledge of the molecule's moment of inertia. The group $M\rho\mathcal{D}/\eta$ is also considered temperature-dependent, and a specific constant is necessary for each polar molecule.

Although Eq. (10-3.8) is probably the best theoretical equation available for estimating the thermal conductivity of a nonpolar polyatomic gas, without some a priori knowledge of Z_{rot} it is not of much practical value. Z_{rot} values normally lie between 1 and 10 and are probably temperature-dependent; widely differing values are reported in the literature. Attempts to relate Z_{rot} to other, more readily available properties of a molecule have not been successful.

Bromley Analysis Bromley [14] started with Eq. (10-3.1) but modified f_{tr} and expanded the term $f_{int}C_{int}$ as follows:

$$\frac{\lambda M}{\eta} = (2.5 - \alpha)C_{tr} + \beta C_{vib} + \psi C_{rot} + C_{ir} \qquad (10\text{-}3.9)$$

C_{tr} and C_{rot} were taken equal to the classical values $\frac{3}{2}R$ and $F_r R/2$, where R is the gas constant and F_r is the number of degrees of freedom for external rotation. C_{vib} was taken to be $C_v - \frac{5}{2}R$ for linear molecules and $C_v - 3R - C_{ir}$ for nonlinear molecules. The internal-rotational heat-capacity contribution C_{ir} was calculated from the magnitude of the potential barrier to internal rotation and the reduced moment of inertia of the rotating groups. Table 10-1 lists values of C_{ir} as a function of temperature for several types of bonds.

The coefficient α allows for the increased interaction on each collision which results from the polar character of the colliding molecules. Although noting that this might better be expressed as a function of a dimensionless group involving the dipole moment, Bromley found it convenient to relate α empirically to the difference between the actual entropy of vaporization per unit volume and the entropy of vaporization given by the Kistiakowsky equation.† The proportionality constant was found empirically to be 3.0,

$$\alpha = 3.0\rho_b(\Delta S_{v_b} - 8.75 - R \ln T_b) \qquad (10\text{-}3.10)$$

where ρ_b = liquid density at normal boiling temperature, g-mol/cm^3
ΔS_{v_b} = actual entropy of vaporization at T_b, cal/g-mol K
R = 1.987 cal/g-mol K
T_b = normal boiling point, K

† $\Delta S_{v_b} = 8.75 + R \ln T_b$; Eq. (6-17.1).

TABLE 10-1   Internal-Rotational Heat-Capacity Contribution† C_{ir}, cal/g-mol K

Type	T, K								
	200	298	400	500	600	700	800	900	1000
—CH₂—CH₂—	2.64	3.18	2.74	2.39	2.12	1.90	1.73	1.60	1.51
CH₃—CH₂—	1.74	2.10	2.17	2.09	1.96	1.82	1.69	1.59	1.51
R—C—H (with vertical bonds)	1.76	2.12	2.19	2.12	2.00	1.85	1.71	1.62	1.55
R—C— (with vertical bonds)	1.60	1.99	2.18	2.23	2.18	2.08	1.96	1.85	1.76
R—CH=‡	2.02	2.02	1.82	1.65	1.50	1.40	1.32	1.26	1.21
R—CH=§	1.48	1.26	1.16	1.11	1.07	1.05	1.04	1.03	1.02
=CH—CH=	2.09	3.02	3.25	2.96	2.67	2.44	2.24	2.09	1.96
≡C—C≡	0.81	0.92	0.76	0.60	0.48	0.39	0.35	0.32	0.31
R—C= (with vertical bond)	1.91	2.15	2.10	1.95	1.79	1.65	1.54	1.46	1.38
CH₃—C=O, R	1.79	1.52	1.34	1.24	1.18	1.10	1.09	1.08	1.06
—CH₂—OH	1.4	1.35	1.25	1.20	1.18	1.16	1.15	1.14	1.13
RCH₂—O—CH₂R	2.24	2.33	2.26	2.09	1.91	1.71	1.66	1.55	1.47

†From M. Souders, C. S. Matthews, and C. O. Hurd, *Ind. Eng. Chem.*, **41**: 1037 (1949).
‡Based on propylene.
§Based on *cis*-2-butene.

α has the units calories per gram mole per kelvin.   Typical values of α are water, 0.90; ammonia, 0.51; methanol, 0.33; *n*-propyl alcohol, 0.24; acetone, 0.06; benzene, 0.   Values of ρ_b can be estimated from Chap. 3 and $\Delta S_{vb} = \Delta H_{vb}/T_b$ from Chap. 6.

Vibrational energy is assumed to be transferred by a diffusional mechanism in the case of linear molecules, and β should be approximately the same as the self-diffusion group $M\rho\mathscr{D}/\eta$, which, as noted previously, is close to a value of 1.3; for nonlinear molecules, β

may be less than 1.30. Bromley attempted to relate β to the group

$$\frac{C_v - 3R - C_{ir}}{3n - 6 - F_{ir}} = Y$$

where F_{ir} is the number of degrees of internal rotation and n is the number of atoms in the molecule. Bromley's plot of this relation shows β constant at about 1.30 for $Y < 0.6$ and then dropping rapidly to 1.0 at $Y > 0.85$. The correlation is poor, however, and it would appear that β might be taken as constant at about 1.30.

The coefficient ψ was found empirically to increase with temperature,

$$\psi = \begin{cases} 1.25 - \dfrac{0.35}{T_r} & \text{for linear molecules} \\[2mm] 1.32 - \dfrac{0.23}{T_r} & \text{for nonlinear molecules} \end{cases}$$

Bromley's correlation may then be summarized:

For monatomic gases: $\qquad \dfrac{\lambda M}{\eta} = 2.5 C_v$

For linear molecules:

$$F_r = 2 \qquad C_{rot} = R \qquad C_{vib} = C_v - \frac{5}{2} R \qquad \beta = 1.30$$

$$\frac{\lambda M}{\eta} = 1.30 C_v + 3.50 - \frac{0.70}{T_r} \tag{10-3.11}$$

For nonlinear molecules:

$$F_r = 3 \qquad C_{rot} = \frac{3R}{2} \qquad C_{vib} = C_v - 3R - C_{ir} \qquad \beta = 1.30$$

$$\frac{\lambda M}{\eta} = 1.30 C_v + 3.66 - 0.3 C_{ir} - \frac{0.69}{T_r} - 3\alpha \tag{10-3.12}$$

where λ = thermal conductivity at low pressure, cal/cm s K
  η = viscosity, P
  C_v = heat capacity at constant volume, cal/g-mol K
  M = molecular weight
  T_r = reduced temperature T/T_c
  α = collision-interaction coefficient from Eq. (10-3.10)
  C_{ir} = internal-rotation heat capacity, cal/g-mol K, obtained from Table 10-1

Equations (10-3.11) and, especially, (10-3.12) bear a striking similarity to Eq. (10-3.8) and may be considered as providing an empirical method for estimating the last term in Eq. (10-3.8), which contains the rotational collision number, i.e.,

$$\frac{0.886C_{\text{rot}}}{Z_{\text{rot}}} \approx \begin{cases} \dfrac{0.70}{T_r} & \text{linear} \\[3mm] 0.3C_{\text{ir}} + 3\alpha + \dfrac{0.69}{T_r} & \text{nonlinear} \end{cases}$$

The semiempirical form of Eq. (10-3.12) also provides a means of treating nonlinear polar molecules.

Empirical Approximations to the Eucken Form   Various writers have attempted to express $\lambda M/\eta$ in a simple fashion to obtain values of this ratio larger than those predicted by the Eucken correlation (10-3.3), but less than the modified Eucken correlation (10-3.5). Some of these proposals are described below to indicate the type of empirical results obtained. Others of a similar nature are given by Liley [93].

	$\lambda M/\eta$
Kinetic theory, monatomic gases	$2.5C_v$
Eucken	$C_v + 4.47$
Modified Eucken	$1.32C_v + 3.52$
Stiel and Thodos [178]	$1.15C_v + 4.04$

Brokaw [9] presents a nomograph of the modified Eucken form, and Svehla [180] calculated values of thermal conductivity for 200 gases from 100 to 5000 K. Figure 10-3 shows some of these relations compared with experimental values for ethyl chloride as given by Vines and Bennett

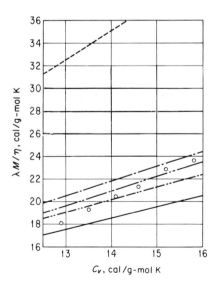

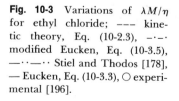

Fig. 10-3 Variations of $\lambda M/\eta$ for ethyl chloride; ――― kinetic theory, Eq. (10-2.3), ―·―· modified Eucken, Eq. (10-3.5), —··—·· Stiel and Thodos [178], — Eucken, Eq. (10-3.3), ○ experimental [196].

[196], who have tabulated such data for many gases.   It is clear that the monatomic kinetic-theory value [Eq. (10-2.3)] yields far too high a ratio; the modified Eucken relation (10-3.5) and the Eucken relation (10-3.3) form the upper and lower bands, with the former better at the higher temperatures and the latter at the lower temperatures.   The temperature range for this plot is about 40 to 140°C.   The empirical form of Stiel and Thodos represents a compromise.   Bromley's method would also have represented such a compromise.   Although not evident from the graph, the dimensionless group $\lambda M / \eta C_v$ is nearly independent of temperature, varying from 1.40 at 40°C to 1.49 at 140°C; this insensitivity to temperature is, over small temperature ranges, typical of most materials, polar and nonpolar [196].   Grilly [55], however, has pointed out exceptions (notably hydrogen) and has studied the temperature effect on $\lambda M / \eta C_v$ over wider temperature ranges.   (See also Fig. 10-2.)

Misic and Thodos Estimation Technique   All methods described above were based on kinetic theory, and the correlating group was $\lambda M / \eta C_v$.   In a different approach, Misic and Thodos [113, 114] proposed an empirical estimation method for λ that is based on dimensional analysis.   By assuming

$$\lambda = f(M^a T_c^b T^c P_c^d V_c^e C_p^f R^g) \tag{10-3.13}$$

it is readily shown by dimensional reasoning that

$$\lambda \Gamma = f\left(T_r^c\left(\frac{C_p}{R}\right)^f Z_c^{5/6-g-f} R^{5/6}\right) \tag{10-3.14}$$

where Γ is defined as

$$\Gamma \equiv \frac{T_c^{1/6} M^{1/2}}{P_c^{2/3}} \tag{10-3.15}$$

The Misic-and-Thodos technique reduces to finding the best functional relationship between $\lambda \Gamma$ and T_r, C_p, and Z_c.   Several equations were obtained, depending upon the type of compound and the reduced temperature.   Only the relations for hydrocarbons are presented here, as the others given by Misic and Thodos are applicable only to simple molecules and are not accurate, in general, for most organic compounds.

For *methane, naphthenes,* and *aromatic hydrocarbons below $T_r = 1$*

$$\lambda = 4.45 \times 10^{-6} T_r \frac{C_p}{\Gamma} \tag{10-3.16}$$

For *all other hydrocarbons* and for other reduced temperature ranges

$$\lambda = (10^{-6})(14.52 T_r - 5.14)^{2/3} \frac{C_p}{\Gamma} \tag{10-3.17}$$

In Eqs. (10-3.15) to (10-3.17),

λ = low-pressure gaseous thermal conductivity, cal/cm s K
T_c = critical temperature, K
P_c = critical pressure, atm
M = molecular weight
C_p = heat capacity at constant pressure, cal/g-mol K

Roy-Thodos Estimation Technique   In an extension of the Misic-Thodos estimation method discussed above, Roy and Thodos [150, 151] separated the $\lambda \Gamma$ product into two parts.   The first, attributed only to translational energy, was obtained from a curve fit of the data for the rare gases [149]; this part varies only with T_r.   In the second, the contribution from rotational, vibrational, etc., interchange was related to the reduced temperature and a specific constant estimated from group contributions.   The fundamental equation may be written

$$\lambda \Gamma = (\lambda \Gamma)_{\text{tr}} + (\lambda \Gamma)_{\text{int}} \tag{10-3.18}$$

where λ = low-pressure gaseous thermal conductivity, cal/cm s K
$\quad \Gamma = T_c^{1/6} M^{1/2} / P_c^{2/3}$
$\quad T_c$ = critical temperature, K
$\quad M$ = molecular weight
$\quad P_c$ = critical pressure, atm

$$(\lambda \Gamma)_{\text{tr}} = 99.6 \times 10^{-6} (e^{0.0464 T_r} - e^{-0.2412 T_r})$$
$$(\lambda \Gamma)_{\text{int}} = \mathscr{C} f(T_r) \tag{10-3.19}$$

Relations for $f(T_r)$ are shown in Table 10-2.   The constant $\mathscr{C}$ is specific for each material and is estimated by a group-contribution technique, as shown below.

TABLE 10-2   Recommended $f(T_r)$ Equations for Roy-Thodos Method

Saturated hydrocarbons†	$-0.152 T_r + 1.191 T_r^2 - 0.039 T_r^3$
Olefins	$-0.255 T_r + 1.065 T_r^2 + 0.190 T_r^3$
Acetylenes	$-0.068 T_r + 1.251 T_r^2 - 0.183 T_r^3$
Naphthenes and aromatics	$-0.354 T_r + 1.501 T_r^2 - 0.147 T_r^3$
Alcohols	$1.000 T_r^2$
Aldehydes, ketones, ethers, esters	$-0.082 T_r + 1.045 T_r^2 + 0.037 T_r^3$
Amines and nitriles	$0.633 T_r^2 + 0.367 T_r^3$
Halides	$-0.107 T_r + 1.330 T_r^2 - 0.223 T_r^3$
Cyclic compounds‡	$-0.354 T_r + 1.501 T_r^2 - 0.147 T_r^3$

†Not recommended for methane.
‡E.g., pyridine, thiophene, ethylene oxide, dioxane, piperidine.

Estimation of Roy-Thodos Constant $\mathscr{C}$. In the discussion to follow, one identifies carbon types as shown:

	H HC— H	H —C— H	H —C— |	| —C— |
Type:	1	2	3	4

Paraffinic hydrocarbons	$\Delta\mathscr{C} \times 10^5$
Base group, methane	0.83
First methyl substitution	2.27
Second methyl substitution	3.62
Third methyl substitution	4.18
Fourth and successive methyl substitutions	5.185

For example, $\mathscr{C}$ for *n*-octane is equal to $[0.83 + 2.27 + 3.62 + 4.18 + 4(5.185)] = 31.64 \times 10^{-5}$.

Isoparaffins are formed by determining the $\mathscr{C}$ for the paraffin with longest possible straight-chain carbon backbone and then making successive substitutions of hydrogen atoms by methyl groups. Values of $\Delta\mathscr{C}$ attributable to such substitutions are shown below:

Type of substitution	$\Delta\mathscr{C} \times 10^5$	Type of substitution	$\Delta\mathscr{C} \times 10^5$
1←2→1	4.14	1←3→1 ↓ 1	3.86
1←2→2	5.36		
1←2→3	6.58	1←3→1 ↓ 2	5.12
2←2→2	6.58		
		1←3→1 ↓ 3	6.38

The type of carbon atom from which the arrow points away is the one involved in the methyl substitution. The arrows point toward the types of adjacent carbon atoms. To calculate $\mathscr{C}$ for an isoparaffin, beginning with the longest straight chain, introduce side chains beginning with the left end and proceed in a clockwise direction. To illustrate with 2,2,4-trimethylpentane,

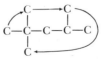

n-pentane $= (0.83 + 2.27 + 3.62 + 4.18 + 5.185)(10^{-5}) = 16.085 \times 10^{-5}$.
Methyl substitutions:

$1 \leftarrow 2 \rightarrow 2$	5.36
$2 \leftarrow 2 \rightarrow 1$	5.36
$1 \leftarrow 3 \rightarrow 2$	5.12
$\downarrow$	
1	

$$\mathscr{C} = \Sigma\,\Delta\mathscr{C} = 31.925 \times 10^{-5}$$

Olefinic and Acetylenic Hydrocarbons. First determine $\mathscr{C}$ for the corresponding saturated hydrocarbon, as described above, then insert the unsaturated bond(s) and employ the following $\Delta\mathscr{C}$ contributions:

		$\Delta\mathscr{C} \times 10^5$
First double bond:	$1 \leftrightarrow 1$	-1.35
	$1 \leftrightarrow 2$	-0.77
	$2 \leftrightarrow 2$	-0.33
Second double bond:	$2 \leftrightarrow 1$	-0.19
Any acetylenic bond		-0.94

Naphthenes. Form the paraffinic hydrocarbon with same number of carbon atoms as in the naphthene ring. Remove two terminal hydrogens and close the ring. $\Delta\mathscr{C} = -1.14 \times 10^{-5}$.

Aromatics. Benzene has a $\mathscr{C}$ value of 15.00×10^{-5}. Methyl-substituted benzenes have a $\mathscr{C}$ value of $(15.00 \times 10^{-5} - 6.0 \times 10^{-5})$(number of methyl substitutions).

Before discussing the estimation of $\mathscr{C}$ for nonhydrocarbons, it is important to note that the simple rules shown above are incomplete and do not cover many types of hydrocarbons. There are, however, no experimental data that can be used to obtain additional $\Delta\mathscr{C}$ contributions. In fact, some of the $\Delta\mathscr{C}$ values quoted above are based on so few data that they should be considered only approximate. For the 27 hydrocarbons studied by Roy and Thodos, one can obtain a rough, but often satisfactory, correlation by plotting $\mathscr{C}$ vs. the molecular weight, as shown in Fig. 10-4. The two aromatic data points (benzene and toluene) fall slightly below the curve, and the $\mathscr{C}$ value for cyclopentane is out of line. (Roy and Thodos note that for this material, the experimental data are in doubt.)

For *nonhydrocarbons*, $\mathscr{C}$ is again estimated by a group-contribution method wherein one mentally synthesizes the final compound by a particular set of rules and employs $\Delta\mathscr{C}$ values for each step.

Alcohols. Synthesize the corresponding hydrocarbon with the same carbon structure and calculate $\mathscr{C}$ as noted above. Replace the

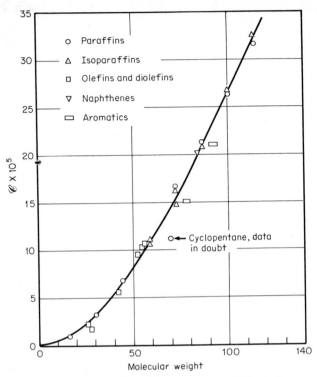

Fig. 10-4   Correlation of Roy-Thodos constant $\mathscr{C}$ for hydrocarbons.

appropriate hydrogen atom by a hydroxyl group and correct $\mathscr{C}$ as noted:

Type of —OH substitution	$\Delta \mathscr{C} \times 10^5$	Type of —OH substitution	$\Delta \mathscr{C} \times 10^5$
On methane	4.31	$3 \leftarrow 1$	4.04
$1 \leftarrow 1$	5.25	$4 \leftarrow 1$	3.45
$2 \leftarrow 1$	4.67	$1 \leftarrow 2 \rightarrow 1$	4.68

The notation is the same as that used earlier, for example, $3 \leftarrow 1$ indicates that the —OH group is replacing a hydrogen atom on a type 1 carbon which is adjacent to a type 3 carbon:

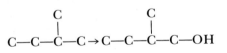

These rules apply only to aliphatic alcohols, and they are incomplete even for these.

Amines.   Estimation of $\mathscr{C}$ for amines is similar to that described above for alcohols.   First, synthesize the corresponding hydrocarbon segment (with the most complex structure) that is finally to be attached to a nitrogen.   For *primary* amines, replace the appropriate terminal hydrogen by a —NH_2 group with the following $\Delta\mathscr{C}$ contributions:

Type of substitution	$\Delta\mathscr{C} \times 10^5$	Type of substitution	$\Delta\mathscr{C} \times 10^5$
On methane	2.96	$2 \leftarrow 2 \rightarrow 1$	8.93
$1 \leftarrow 1$	4.45	$1 \leftarrow 3 \rightarrow 1$	7.39
$1 \leftarrow 2 \rightarrow 1$	5.78	$\downarrow$	
		1	

For *secondary* amines, there are additional $\Delta\mathscr{C}$ values:

	$\Delta\mathscr{C} \times 10^5$
$CH_3—NH_2 \rightarrow CH_3—\overset{\overset{H}{\mid}}{N}—CH_3$	3.76
$—CH_2—NH_2 \rightarrow CH_2—\overset{\overset{H}{\mid}}{N}—CH_3$	5.00

Finally, for *tertiary* amines, Roy and Thodos show three types of corrections applicable to the secondary amines:

	$\Delta\mathscr{C} \times 10^5$
$CH_3—NH—CH_3 \rightarrow (CH_3)_3—N$	2.95
$—CH_2—\overset{\overset{H}{\mid}}{N}—CH_2— \rightarrow —CH_2—\overset{\overset{CH_3}{\mid}}{N}—CH_2—$	3.72
$—CH_2—\overset{\overset{H}{\mid}}{N}—CH_3 \rightarrow —CH_2—N—(CH_3)_2$	3.34

After calculating $\mathscr{C}$ for an amine as noted above, any methyl substitutions for a hydrogen on a side chain increase $\mathscr{C}$ by 5.185×10^{-5} (the same as shown for fourth and successive methyl substitutions in paraffinic hydrocarbons).   For example,

$$—\overset{\mid}{N}—CH_3 \rightarrow —\overset{\mid}{N}—CH_2—CH_3 \qquad \Delta\mathscr{C} = 5.185 \times 10^{-5}$$

Nitriles.   Only three $\Delta\mathscr{C}$ contributions are shown; these were based on thermal-conductivity data for acetonitrile, propionitrile, and acrylic nitrile.

Type of —CN addition	$\Delta \mathscr{C} \times 10^5$
On methane	6.17
$CH_3—CH_3 \rightarrow CH_3—CH_2—CN$	8.10
$—CH{=}CH_2 \rightarrow —CH{=}CH—CN$	7.15

Halides. Suggested contributions are shown below; the order of substitution should be F, Cl, Br, I.

	$\Delta \mathscr{C} \times 10^5$
First halogen substitution on methane:	
Fluorine	0.29
Chlorine	1.57
Bromine	1.77
Iodine	3.07
Second and successive substitutions on methane:	
Fluorine	0.43
Chlorine	2.33
Bromine	3.20
Substitutions on ethane and higher hydrocarbons:	
Fluorine	0.66
Chlorine	3.33

Aldehydes and Ketones. Synthesize the hydrocarbon analog with the same number of carbon atoms and calculate $\mathscr{C}$ as noted above.   Then form the desired aldehyde or ketone by substituting oxygen for two hydrogen atoms:

	$\Delta \mathscr{C} \times 10^5$
$—CH_2—CH_3 \rightarrow —CH_2—CHO$	2.20
$—CH_2—CH_2—CH_2— \rightarrow —CH_2—CO—CH_2—$	3.18

Ethers. Synthesize the primary alcohol with the longest carbon chain on one side of the ether oxygen.   Convert this alcohol to a methyl ether.

$$—CH_2OH \rightarrow —CH_2—O—CH_3 \qquad \Delta \mathscr{C} = 2.80 \times 10^{-5}$$

Extend the methyl chain, if desired, to an ethyl.

$$—CH_2—O—CH_3 \rightarrow —CH_2—O—CH_2—CH_3 \qquad \Delta \mathscr{C} = 4.75 \times 10^{-5}$$

Although Roy and Thodos do not propose extensions beyond the ethyl group, presumably more complex chains could be synthesized using $\Delta \mathscr{C}$ values obtained from the paraffinic and isoparaffinic contributions.

Acids and Esters.   Synthesize the appropriate ether so as to allow the following substitutions:

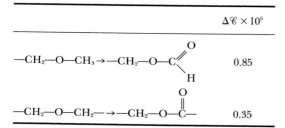

	$\Delta \mathscr{C} \times 10^5$
$-CH_2-O-CH_3 \rightarrow -CH_2-O-C\overset{\displaystyle O}{\diagdown}_H$	0.85
$-CH_2-O-CH_2- \rightarrow -CH_2-O-\overset{\displaystyle O}{\overset{\|}{C}}-$	0.35

Cyclics.   Synthesize the ring, if possible, with the following contributions (not substitutions):

Group	$\delta\mathscr{C} \times 10^5$	Group	$\delta\mathscr{C} \times 10^5$
$-CH_2-$	4.83	$-N=$	3.98
$-CH=$	3.98	$-O-$	4.10
$-NH-$	5.48	$=S=$	7.97

and determine $\mathscr{C}$ as

$$\mathscr{C} = \sum \delta\mathscr{C} - 8.90 \times 10^{-5}$$

These group contributions were obtained from limited data and are averaged values.   Calculations cannot be made for many compounds using the rules given above, but an intelligent guess can often be made for a missing increment.

The Roy-Thodos method can also be used in a different way.   If a single value of λ is available at a known temperature, Eq. (10-3.18) can be used with Table 10-2 to yield a value of $\mathscr{C}$ that can then be employed to determine λ at other temperatures.

Example 10-1   Estimate the thermal conductivity of isopentane vapor at 1 atm and 100°C.   The reported value is 52×10^{-6} cal/cm s K [4].

solution   From Appendix A, $T_c = 460.4$ K, $P_c = 33.4$ atm, $V_c = 306$ cm³/g-mol, $T_b = 301$ K, $Z_c = 0.271$, and $M = 72.151$.

First, the viscosity of isopentane must be estimated.   The Reichenberg correlation, Eq. (9-4.21), is used. With Table 9-3, $\sum n_i C_i = (3)(9.04) + 6.47 + 2.67 = 36.26$.   Then

$$a^* = \frac{M^{1/2} T_c}{\sum n_i C_i} = \frac{(72.15)^{1/2}(460.4)}{36.26} = 107.85 \ \mu P$$

With $T_r = (100 + 273)/460.4 = 0.810$,

$$\eta = \frac{a^* T_r}{[1 + 0.36 T_r (T_r - 1)]^{1/6}}$$

$$= \frac{(107.85)(0.810)}{[1 + (0.36)(0.810)(0.810 - 1)]^{1/6}} = 88.2 \ \mu P$$

[If the Thodos relation, Eq. (9-4.17), is used, η is estimated to be 88.7 μP.]

The heat capacity at constant pressure can be found from the constants shown in Appendix A,

$$C_p^\circ = \text{CPVAP A} + (\text{CPVAP B})(T) + (\text{CPVAP C})(T^2) + (\text{CPVAP D})(T^3)$$
$$= -2.275 + (0.1210)(373) - (6.519 \times 10^{-5})(373^2)$$
$$+ (1.367 \times 10^{-8})(373^3) = 34.5 \text{ cal/g-mol K}$$

C_v° is estimated as $C_p - R = 34.5 - 1.99 = 32.5$ cal/g-mol K.

Eucken Method. With Eq. (10-3.3),

$$\lambda = \frac{\eta}{M}(C_v + 4.47) = \frac{88.2 \times 10^{-6}}{72.15}(32.5 + 4.47)$$
$$= 45.2 \times 10^{-6} \text{ cal/cm s K}$$
$$\text{Error} = \frac{45.2 - 52}{52} \times 100 = -13.1\%$$

Modified Eucken Method. With Eq. (10-3.5),

$$\lambda = \frac{\eta}{M}(1.32C_v + 3.52) = \frac{88.2 \times 10^{-6}}{72.15}[(1.32)(32.5) + 3.52]$$
$$= 56.8 \times 10^{-6} \text{ cal/cm s K}$$
$$\text{Error} = \frac{56.8 - 52}{52} \times 100 = 9.1\%$$

Bromley Method. Since isopentane is a nonlinear molecule, Eq. (10-3.12) is used. To determine α, Eq. (10-3.10) is employed. With the Tyn and Calus method [Eq. (3-14.1)]

$$\rho_b = V_b^{-1} = [(0.285)(V_c)^{1.048}]^{-1} = 8.71 \times 10^{-3} \text{ g-mol/cm}^3$$
$$\Delta S_{v_b} = \frac{\Delta H_{v_b}}{T_b} = \frac{5900}{301} = 19.6 \text{ cal/g-mol K}$$

(ΔH_{v_b} from Appendix A). Then

$$\alpha = 3.0\rho_b(\Delta S_{v_b} - 8.75 - R \ln T_b)$$
$$= (3.0)(8.71 \times 10^{-3})(19.6 - 8.75 - 1.987 \ln 301) = -0.013$$

Let $\alpha \approx 0$. This is reasonable since isopentane is nonpolar.

Next, to determine C_{ir}, we use Table 10-1. Though not exactly comparable, we assume that we have three CH_3—CH_2— bonds and one —CH_2—CH_2— bond. At 373 K these yield an approximate value of C_{ir} as

$$C_{ir} = 3(2.15) + 2.86 = 9.31 \text{ cal/g-mol K}$$

Then, with Eq. (10-3.12),

$$\lambda = \frac{\eta}{M}\left(1.30C_v + 3.66 - 0.3C_{ir} - \frac{0.69}{T_r} - 3\alpha\right)$$
$$= \frac{88.2 \times 10^{-6}}{72.15}\left[(1.30)(32.5) + 3.66 - (0.3)(9.31)\right.$$
$$\left. - \frac{0.69}{0.810} - (3)(0)\right] = 51.6 \times 10^{-6} \text{ cal/cm s K}$$
$$\text{Error} = \frac{51.6 - 52}{52} \times 100 = -0.8\%$$

Misic-Thodos Method. With Eq. (10-3.15),

$$\Gamma = \frac{T_c^{1/6}M^{1/2}}{P_c^{2/3}} = \frac{(460.4)^{1/6}(72.15)^{1/2}}{(33.4)^{2/3}} = 2.276$$

Then, with Eq. (10-3.17),

$$\lambda = 10^{-6}\frac{(14.52T_r - 5.14)^{2/3}C_p}{\Gamma}$$

$$= 10^{-6}\frac{[(14.52)(0.810) - 5.14]^{2/3}(34.5)}{2.276}$$

$$= 53.4 \times 10^{-6}\,\text{cal/cm s K}$$

$$\text{Error} = \frac{53.4 - 52}{52} \times 100 = 2.7\%$$

Roy and Thodos Method.   Equation (10-3.18) is used.   First, $(\lambda\Gamma)_{tr}$ is found by Eq. (10-3.19) when $T_r = 0.810$.

$$(\lambda\Gamma)_{tr} = (99.6 \times 10^{-6})(e^{0.0464T_r} - e^{-0.2412T_r})$$

$$= (99.6 \times 10^{-6})(e^{(0.0464)(0.810)} - e^{-(0.2412)(0.810)})$$

$$= 21.5 \times 10^{-6}$$

To find $(\lambda\Gamma)_{int}$, $\mathscr{C}$ is first determined by synthesizing n-butane with the recommended $\Delta\mathscr{C}$ increments:

$$n\text{-butane} = (0.83 + 2.27 + 3.62 + 4.18) \times 10^{-5} = 10.9 \times 10^{-5}$$

Next, to form isopentane (2-methylbutane), a $1\leftarrow2\rightarrow2$ methyl substitution is required; thus

$$\text{Isopentane} = 10.9 \times 10^{-5} + 5.36 \times 10^{-5} = 16.26 \times 10^{-5}$$

The appropriate $f(T_r)$ is given in Table 10-2 for saturated hydrocarbons:

$$f(T_r) = -0.152T_r + 1.191T_r^2 - 0.039T_r^3$$

$$= -(0.152)(0.810) + (1.191)(0.810)^2 - (0.039)(0.810)^3$$

$$= 0.638$$

Then
$$(\lambda\Gamma)_{int} = \mathscr{C}f(T_r) = (16.26 \times 10^{-5})(0.638) = 103.7 \times 10^{-6}$$

With Eq. (10-3.18), using Γ as determined from the calculations shown with the Misic-Thodos method, we have

$$\lambda\Gamma = (\lambda\Gamma)_{tr} + (\lambda\Gamma)_{int}$$

$$\lambda = \frac{(21.5 + 103.7)(10^{-6})}{2.276} = 55.0 \times 10^{-6}\,\text{cal/cm s K}$$

$$\text{Error} = \frac{55.0 - 52}{52} \times 100 = 5.8\%$$

Example 10-2   Using the Eucken, Bromley, and Roy-Thodos estimation methods, estimate the thermal conductivity of acetone vapor at 1 atm and at temperatures between 0 and 200°C.   In Figs. 10-5 and 10-6, data are shown for the vapor viscosity and heat capacity at constant volume.

solution   *Eucken Method.*   From Eq. (10-3.3),

$$\lambda = \frac{\eta}{M}(C_v + 4.47)$$

With η and C_v values from Figs. 10-5 and 10-6 and $M = 58.08$, estimated values of λ are shown in Table 10-3.

Bromley Method. C_{ir} is estimated from Table 10-1 as $(2 \times CH_3 - C{=}O)$.   From Table

$$\underset{R}{\overset{\displaystyle |}{}}$$

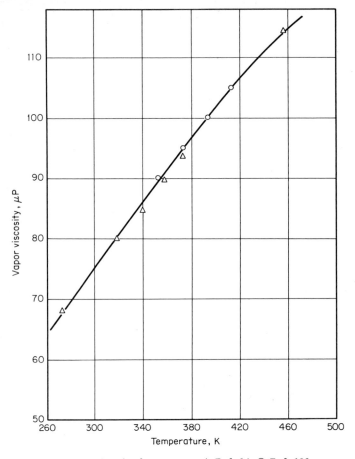

Fig. 10-5   Vapor viscosity for acetone; △ Ref. 14, ⊙ Ref. 196.

TABLE 10-3 Experimental and Estimated Values of the Low-Pressure Vapor Thermal Conductivity of Acetone

	λ, μcal/cm s K			
			Estimated	
T, °C	Exp.	Eucken	Bromley	Roy and Thodos
0	22.8	23.3	24.9	24.3
27	27.2	27.1	29.6	28.8
67	34.8	33.4	37.2	36.0
107	42.8	40.2	45.5	44.0
147	51.4	47.4	54.4	52.5
187	60.0	54.9	63.8	62.3

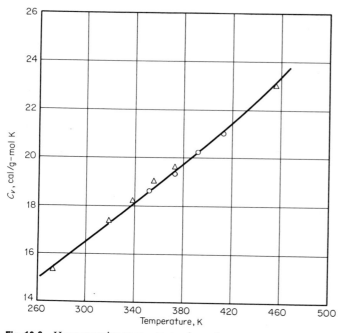

Fig. 10-6 Heat capacity at constant volume for acetone vapor; $\triangle$ Ref. 14, $\odot$ Ref. 196.

3-12, $\rho_b = V_b^{-1} = (77.5)^{-1} = 0.0129$ g-mol/cm^3 and $\Delta S_{v_b} = \Delta H_{v_b}/T_b = 6960/329.4 = 21.1$ cal/g-mol K when $T_b = 329.4$ K and $\Delta H_{v_b} = 6960$ cal/g-mol. From Eq. (10-3.10)

$$\alpha = (3.0\rho_b)(\Delta S_{v_b} - 8.75 - R \ln T_b)$$
$$= (3.0)(0.0129)[21.1 - 8.75 - (1.987) \ln 329.3] = 0.032$$

Then, with Eq. (10-3.12),

$$\lambda = \frac{\eta}{M}\left(1.3 C_v + 3.66 - 0.3 C_{ir} - \frac{0.69}{T_r} - 3\alpha\right)$$

The calculated values of λ are shown in Table 10-3.

Roy-Thodos Method. Γ is defined by Eq. (10-3.15) with

$$T_c = 508.1 \text{ K} \qquad P_c = 46.4 \text{ atm} \qquad M = 58.08$$

$$\Gamma = \frac{T_c^{1/6} M^{1/2}}{P_c^{2/3}} = \frac{(508.2)^{1/6}(58.08)^{1/2}}{(46.4)^{2/3}} = 1.667$$

In Eq. (10-3.19), $T_r = T/T_c = T/508.2$, and $(\lambda \Gamma)_{tr}$ can be determined at various temperatures. To find $(\lambda \Gamma)_{int}$, from Table 10-2, $f(T_r) = -0.082 T_r + 1.045 T_r^2 + 0.037 T_r^3$. $\mathscr{C}$ is computed by a group-contribution method.

Step	$\Delta \mathscr{C} \times 10^5$
Methane	0.83
Replace H by CH$_3$ to form ethane	2.27
Replace H by CH$_3$ to form propane	3.62
Replace two hydrogens by =O to form acetone	3.18
	$\mathscr{C} = 9.90 \times 10^{-5}$

TABLE 10-4 Comparison of Calculated and Experimental Values of the Thermal Conductivity of a Pure Gas at about 1 atm

Compound	T, K	λ(exp.), μcal/cm s K	Ref.	Eucken, Eq. (10-3.3)	Modified Eucken, Eq. (10-3.5)	Bromley, Eqs. (10-3.11) and (10-3.12)	Misic and Thodos, Eqs. (10-3.16) and (10-3.17)	Roy and Thodos, Eq. (10-3.18)
						Percent error†		
Acetone	273	22.8	14	2.0	22	9.2	...	6.9
	300	27.2	196	-0.2	20	8.9	...	5.9
	340	34.8		-4.0	17	6.9	...	3.4
	380	42.8		-6.0	15	6.4	...	2.1
	420	51.4		-8.0	13	5.9	...	2.4
	460	60.0		-8.5	12	6.3	...	3.9
Acetaldehyde	313	30.1	196	9.6	28	16	...	7.5
	333	34.0		6.7	25	14	...	5.2
	353	37.9		4.7	23	13	...	3.9
	373	42.0		2.9	22	12	...	2.7
	393	46.3		1.3	20	11	...	1.6
Acetonitrile	353	29.7	196	13	32	19	...	2.7
	373	32.7		11	31	19	...	2.7
	393	35.8		9.7	29	18	...	2.9
Acetylene	198	28.1	14	4.1	15	1.8	-9.2	10
	273	44.6		3.0	16	7.5	3.0	5.7
	323	57.8		-1.0	13	5.8	3.1	2.7
	373	71.1		-3.8	11	3.9	3.0	1.9

Ammonia							
213	39.3	14	15	26	−3.2		1.0
273	52.9		9.2	20	−4.2		0.1
373	76.4		15	29	7.2		0
							0
Benzene							
353	34.8	196	−14	6.0	1.1	−9.2	
373	39.4		−15	5.4	0.9	−9.0	
393	44.1		−16	4.5	0.3	−9.1	
433	54.0		−18	2.7	−1.0	−9.3	
n-Butane							
273	32.2	39	−9.3	11	−1.9	−13	−2.0
373	58.7		−15	5.5	−3.5	−7.5	−3.7
Carbon dioxide							
200	22.7	14	5.4	15	5.5		
300	39.8		−2.9	7.9	1.8		
473	67.7		−2.1	11	6.8		
598	90.4		−4.8	9.4	4.8		
1273	195		−13	2.4	−0.1		
Carbon tetrachloride							
273	14.2	14	−9.3	9.8	1.9	...	−9.6
373	20.5		−5.5	15.3	9.2	...	5.6
457	26.0		−5.2	16.0	11	...	17
Cyclohexane							
353	38.9	196	−14	7.6	3.1	2.3	−0.6
393	49.2		−8.3	4.9	1.3	0.9	−0.5
433	61.1		−18	2.6	−0.6	−0.8	−1.0
Dichlorodifluoromethane							
273	19.8	14	−13	3.7	−2.9	...	0.7
373	33.0		−18	−1.7	−6.2	...	−0.4
473	46.3		−21	−3.8	−5.5	...	3.8
Ethyl acetate							
319	28.8	14	−5.2	17	8.4	...	3.0
373	38.6		−6.6	17	8.8	...	2.5
457	56.8		−11	12	6.1	...	1.7

TABLE 10-4 Comparison of Calculated and Experimental Values of the Thermal Conductivity of a Pure Gas at about 1 atm (Continued)

Compound	T, K	λ(exp.), μcal/cm s K	Ref.	Eucken, Eq. (10-3.3)	Modified Eucken, Eq. (10-3.5)	Bromley, Eqs. (10-3.11) and (10-3.12)	Misic and Thodos, Eqs. (10-3.16) and (10-3.17)	Roy and Thodos, Eq. (10-3.18)
							Percent error[†]	
Ethyl alcohol	293	35.8	14	−7.6	9.7	−7.1	⋯	−2.9
	375	53.0		−6.7	13	−1.1	⋯	−1.6
	401	59.4		−7.2	12	−0.5	⋯	−1.5
Ethylene	273	41.5	14	−0.5	12	4.0	2.1	−0.3
	323	54.1		−1.9	12	5.5	4.1	−0.2
	373	66.5		−1.4	14	8.5	8.2	3.0
Ethyl ether	273	31.0	14	−13	8.3	−2.3	⋯	−3.2
	373	53.0		−20	−0.6	−8.7	⋯	2.5
	486	83.9	196	−23	−1.6	−8.7	⋯	8.2
n-Hexane	273	26.0	14	−5.4	19	3.3	−10	−2.0
	373	48.0	196	−17	5.0	−5.4	−1.9	−0.8
	433	64.9		−20	2.2	−6.4	−1.6	−1.2
Isopropyl alcohol	304	36.3	14	−19	−1.8	−15	⋯	1.1
	400	59.7		−16	4.0	−4.0	⋯	−1.2
Water	373	56.3	196	28	40	−8.3		
	413	65.9		23	35	−9.3		
Sulfur dioxide	273	19.8	14	8.6	21	13		

†Percent error = [(calc. − exp.)/exp.] × 100.

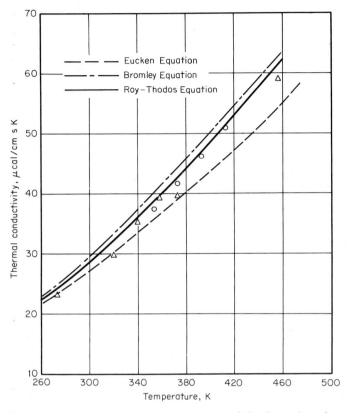

Fig. 10-7 Experimental and estimated values of the thermal conductivity of acetone vapor; △ Ref. 14, ⊙ Ref. 196.

Thus

$$(\lambda \Gamma)_{\text{int}} = 9.90 \times 10^{-5}(-0.082 T_r + 1.045 T_r^2 + 0.037 T_r^3)$$

Calculated values of λ, as determined from Eq. (10-3.18), are shown in Table 10-3.

The results from the three methods are plotted in Fig. 10-7 and compared with data reported in the literature. The Roy-Thodos method provides the best estimate.

Discussion and Recommendations Table 10-4 shows a comparison of the accuracy of five of the thermal-conductivity estimation equations. The simple Eucken relation is surprisingly good and usually more accurate than the modified Eucken form. The most accurate correlation, however, is the Roy-Thodos relation (10-3.18). Group contributions must be available to determine the parameter $\mathscr{C}$, although for hydrocarbons, Fig. 10-4 provides a reasonable correlation of $\mathscr{C}$ as a function of molecular weight. When this latter correlation is inapplicable, either the Eucken equation (10-3.3) or the Bromley correlation [Eq. (10-3.11) or (10-3.12)] should be used. The errors to be expected vary

TABLE 10-5 Coordinates for Nomograph of Thermal Conductivities of Gases

No.	Gas	Temperature range, K	X	Y	No.	Gas	Temperature range, K	X	Y
1	Acetone	250–500	3.7	14.8	27b	Helium	500–5000	15.0	3.0
2	Acetylene	200–600	7.5	13.5	28a	n-Heptane	250–600	4.0	14.8
3a	Air	50–250	12.4	13.9	28b	n-Heptane	600–1000	6.9	14.9
3b	Air	250–1000	14.7	15.0	29	n-Hexane	250–1000	3.7	14.0
3c	Air	1000–1500	17.1	14.5	30a	Hydrogen	50–250	13.2	1.2
4	Ammonia	200–900	8.5	12.6	30b	Hydrogen	250–1000	15.7	1.3
5a	Argon	50–250	12.5	16.5	30c	Hydrogen	1000–2000	13.7	2.7
5b	Argon	250–5000	15.4	18.1	31	Hydrogen chloride	200–700	12.2	18.5
6	Benzene	250–600	2.8	14.2	32	Krypton	100–700	13.7	21.8
7	Boron trifluoride	250–400	12.4	16.4	33a	Methane	100–300	11.2	11.7
8	Bromine	250–350	10.1	23.6	33b	Methane	300–1000	8.5	11.0
9	n-Butane	250–500	5.6	14.1	34	Methyl alcohol	300–500	5.0	14.3
10	i-Butane	250–500	5.7	14.0	35	Methyl chloride	250–700	4.7	15.7
11a	Carbon dioxide	200–700	8.7	15.5	36a	Neon	50–250	15.2	10.2
11b	Carbon dioxide	700–1200	13.3	15.4	36b	Neon	250–5000	17.2	11.0
12a	Carbon monoxide	80–300	12.3	14.2	37	Nitric oxide	100–1000	13.2	14.8
12b	Carbon monoxide	300–1200	15.2	15.2	38a	Nitrogen	50–250	12.5	14.0
13	Carbon tetrachloride	250–500	9.4	21.0	38b	Nitrogen	250–1500	15.8	15.3
14	Chlorine	200–700	10.8	20.1	38c	Nitrogen	1500–3000	12.5	16.5
15a	Deuterium	50–100	12.7	17.3	39a	Nitrous oxide	200–500	8.4	15.0

No.	Substance	Range		
15b	Deuterium	100–400	14.5	19.3
16	Ethane	200–1000	5.4	12.6
17a	Ethyl alcohol	250–350	2.0	13.0
17b	Ethyl alcohol	350–500	7.7	15.2
18	Ethyl ether	250–500	5.3	14.1
19	Ethylene	200–450	3.9	12.3
20a	Fluorine	80–600	12.3	13.8
20b	Fluorine	600–800	18.7	13.8
21	Freon 11	250–500	7.5	19.0
22	Freon 12	250–500	6.8	17.5
23	Freon 13	250–500	7.5	16.5
24	Freon 21	250–450	6.2	17.5
25	Freon 22	250–500	6.5	16.6
26	Freon 113	250–400	4.7	17.0
27a	Helium	50–500	17.0	2.5
39b	Nitrous oxide	500–1000	11.5	15.5
40a	Oxygen	50–300	12.2	13.8
40b	Oxygen	300–1500	14.5	14.8
41	Pentane	250–500	5.0	14.1
42a	Propane	200–300	2.7	12.0
42b	Propane	300–500	6.3	13.7
43	Sulfur dioxide	250–900	9.2	18.5
44	Toluene	250–600	6.4	14.6
45	Xenon	150–700	13.3	25.0

widely, but usually they are less than 10 percent.   With few exceptions, the Roy-Thodos and Bromley methods lead to estimated thermal conductivities larger than found experimentally, while the opposite is true for the Eucken equation.   Not too much faith should be placed in the exact values of the errors, as there is significant disagreement concerning the correct values of the experimental thermal conductivity.   Few reliable values of gaseous thermal conductivity have been reported for complex organic molecules; the lack of such data casts doubt upon the applicability of the estimation methods for such materials.

The nomograph presented by Chang [21], shown in Fig. 10-8 and Table 10-5, is often useful for rapid, approximate estimates of low-pressure gas thermal conductivity.   Other estimation schemes are discussed by Missenard [117].   Data compilations can be found for the rare gases [201] and for a wide variety of polyatomic gases [184].

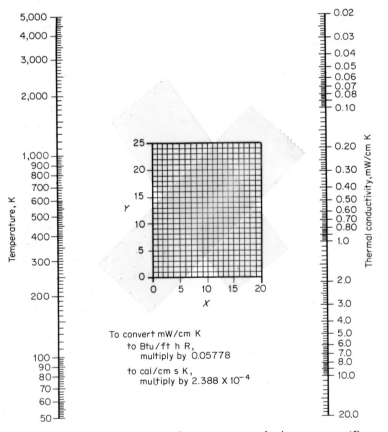

Fig. 10-8 Thermal conductivity of gases at atmospheric pressure. (*From Ref.* 21.)

10-4 Effect of Temperature on the Low-Pressure Thermal Conductivity of Gases

Thermal conductivities of low-pressure gases increase with temperature. Over small temperature ranges, the λ-vs.-T relation is nearly linear, as shown in Fig. 10-9. Values of $d\lambda/dT$ vary from about 0.1 to 0.3 μcal/cm s K^2, and materials with large values of λ usually also have the larger values of $d\lambda/dT$.

Over *wide* temperature ranges, the equations given in Table 10-2, combined with Eq. (10-3.19), indicate that λ increases significantly more rapidly with temperature than implied by a linear function. (See also Fig. 10-7.) If a single thermal-conductivity datum point is available at a particular temperature, it can be used to back-calculate a value of $\mathscr{C}$ in the Roy-Thodos method and then λ can be estimated over wide temperature ranges using the λ-vs.-T_r relations given in this method. Other authors have suggested that

$$\frac{\lambda_{T_2}}{\lambda_{T_1}} = \left(\frac{T_2}{T_1}\right)^n \tag{10-4.1}$$

Owens and Thodos [128] recommend that, except for cyclic compounds, $n = 1.786$, whereas Missenard [120] indicates that n is a temperature-dependent term. Equation (10-4.1), with $n = 1.786$, is shown plotted on the top graph of Fig. 10-9 as dashed lines. There are several such lines with different slopes since the derivative $d\lambda/dT$ depends upon the base point (λ_{T_1}, T_1) chosen.

10-5 Effect of Pressure on the Thermal Conductivity of Gases

The thermal conductivity of all gases increases with pressure, although the effect is relatively small at low and moderate pressures. Three pressure regions are discussed below, in which the effect of pressure is distinctly different.

Very Low Pressure (*below* 1 mm Hg). In this region, often called the *Knudsen domain*, the mean free path of the molecules is large compared with the dimensions of the measuring cell, and there is an important effect of pressure; i.e., at pressures below 0.1 mm Hg, λ is almost proportional to P. In reported thermal-conductivity data, the term *zero-pressure value* is often used; this refers to values extrapolated from higher pressures (above a few millimeters of mercury) and not to measured values in the very-low-pressure region.

Low Pressure. This region extends from approximately 1 mm Hg to perhaps 10 atm and includes the domain discussed in Secs. 10-3 and 10-4. The thermal conductivity increases about 1 percent per

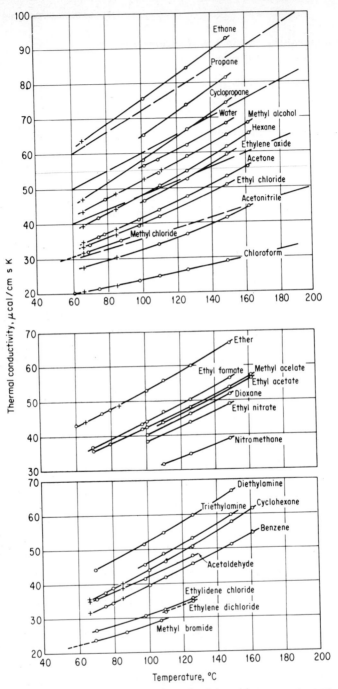

Fig. 10-9 Variation of thermal conductivity with temperature at low pressure. Dashed lines in top graph are from Eq. (10-4.1): $k_{T_2}/k_{T_1} = (T_2/T_1)^{1.786}$. (*From Ref.* 196.)

atmosphere. Such increases are often ignored in the literature, and either the 1-atm value or "zero-pressure" extrapolated value may be referred to as the low-pressure conductivity.

Most of the existing data on λ have been obtained in the low-pressure region, but the experimental variable studied has usually been temperature and not pressure. Vines and Bennett [196] and Vines [194, 195] report data on λ for a variety of organic vapors from 30 to 100°C for the pressure range 10 to 100 mm Hg and record that B, defined as the percent increase in zero-pressure λ per atmosphere, increases with pressure. For polar compounds, B varies from 0.5 to 4.3, with the high values of 10.9 and 4.3 for methanol at 79 and 110°C. For nonpolar compounds, B is less than 2. Data on both polar and nonpolar compounds show that B decreases with rising temperature. Working at subatmospheric pressures, Kannuliuk and Donald [76] found negligible effects of pressure on λ for Ar, He, air, CO_2, N_2O, and CH_4.

A more quantitative picture of the variation of λ with pressure in this region can be obtained from Fig. 10-10. The graphs show the pressure effect on the ratio of λ to $\lambda°$, the low-pressure value of λ at the same temperature. Only for ammonia is λ at $P_r = 0.3$ as much as 8 percent greater than that at low pressure; the effect of pressure for this polar gas is evidently greater at low reduced temperatures.

The data shown in Fig. 10-10 indicate that B, the percent increase per atmosphere, is less than 1.0 in each case. The higher values in a few cases reported by Vines are near saturation conditions.

High Pressure. Many high-pressure thermal-conductivity correlations have been based on the corresponding-states principle, wherein either $\lambda/\lambda°$ or λ/λ_c is plotted as a function of reduced pressure at constant reduced temperature.

The low-pressure value of λ at T is designated by $\lambda°$, and λ_c is the value at T_c and P_c. Plots of $\lambda/\lambda°$ were suggested first by Comings and Mayland [26] and plots of λ/λ_c by Gamson [47].[†] They provide a convenient way of summarizing data for a single substance [27, 91, 92, 112, 177]. However, even for the simple series ethane, propane, and n-butane, Kramer and Comings [84] have shown that the two parameters T_r and P_r are insufficient to correlate the ratio $\lambda/\lambda°$. They suggest the use of $C_p°$ as a third correlating parameter, but this technique does not seem to have been developed.

The thermal conductivity varies significantly with small changes in pressure or temperature near the critical point, with λ showing distinct humps along an isotherm as the pressure (or density) is changed [170]. Figure 10-11 shows data of Guildner for CO_2 [57]; at ρ_c, $(d\lambda/dT)_{\rho_c}$ becomes very large as the critical temperature is approached. The

---

[†]Occasionally the ratio $\lambda/\lambda_c°$ is used; $\lambda_c°$ is the value of λ at low pressure and at T_c.

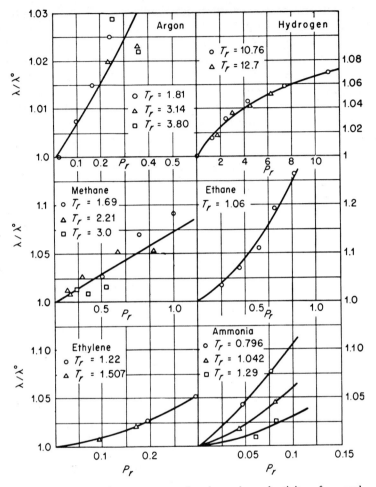

Fig. 10-10   Effect of pressure on the thermal conductivity of several gases.   (*Based on data from Refs. 81 and 82.*)

explanation for this phenomenon is not clear; it may be due to a transition molecular ordering [84] or to a small-scale circulation effect resulting from the migration of small clusters of molecules [89].   In any case, when generalized charts of the effect of pressure on λ are drawn, these irregularities are usually smoothed out and not shown.

A typical graph showing the effect of both temperature and pressure on thermal conductivity is presented in Fig. 10-12 for methane [102].   At high pressures, an increase in temperature results in a decrease in λ, whereas the opposite is true at pressures around 1 atm.

Most correlations describing the effect of pressure on λ have used the remarkably simple correlating technique suggested by Vargaftik

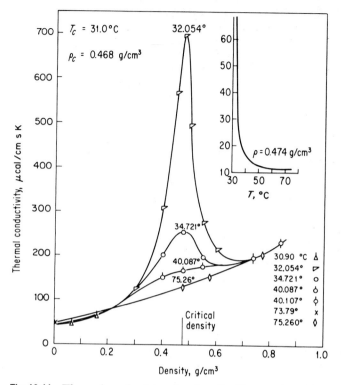

Fig. 10-11 Thermal conductivity of carbon dioxide near the critical point. (*Data from Ref. 57.*)

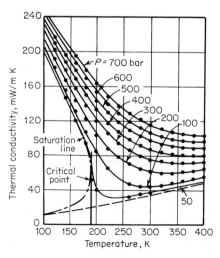

Fig. 10-12 Dependence of thermal conductivity of methane on temperature and pressure; ···· Ref. 102, – – – 1 atm (P_c = 45.0 bars). (*From Ref. 102.*)

[190, 191]. In this scheme, the excess thermal conductivity, $\lambda - \lambda°$, is plotted as a function of density (or reduced density), i.e.,

$$\lambda - \lambda° = f(\rho) \tag{10-5.1}$$

This technique is similar to one developed for correlating the viscosity of dense gases (Sec. 9-6) and has been shown applicable to ammonia [56, 144], ethane [18], n-butane [17, 84], nitrous oxide [143], ethylene [128], methane [19, 102, 127], diatomic gases [115, 165], hydrogen [164], inert gases [126], and carbon dioxide [77]. Temperature and pressure do not enter explicitly, but their effects are included in the parameters $\lambda°$ (temperature only) and density ρ. As an illustration, the methane data shown in Fig. 10-12 are replotted in this fashion in Fig. 10-13.†

Stiel and Thodos [178] have generalized Eq. (10-5.1) by assuming that $f(\rho)$ depends only on T_c, P_c, V_c, M, and ρ. By dimensional analysis they obtain a correlation between $\lambda - \lambda°$, Z_c, Γ, and ρ_r, where Γ is defined in Eq. (10-3.15). From data on 20 nonpolar substances, including inert gases, diatomic gases, CO_2, and hydrocarbons, they established the correlation shown in Fig. 10-14. Approximate analytical expressions for this curve are

$$(\lambda - \lambda°)\Gamma Z_c^5 = (14.0 \times 10^{-8})(e^{0.535\rho_r} - 1) \qquad \rho_r < 0.5 \tag{10-5.2}$$

$$(\lambda - \lambda°)\Gamma Z_c^5 = (13.1 \times 10^{-8})(e^{0.67\rho_r} - 1.069) \qquad 0.5 < \rho_r < 2.0 \tag{10-5.3}$$

$$(\lambda - \lambda°)\Gamma Z_c^5 = (2.976 \times 10^{-8})(e^{1.155\rho_r} + 2.016) \qquad 2.0 < \rho_r < 2.8 \tag{10-5.4}$$

†Rosenbaum and Thodos [147] have also measured the dense-gas thermal conductivity of methane and report that there appears to be a weak effect of temperature other than that included in ρ_r. Large effects of temperature were found for carbon tetrafluoride, and each isotherm had to be plotted separately.

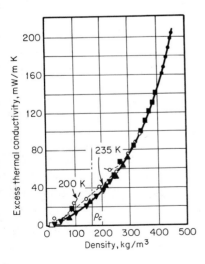

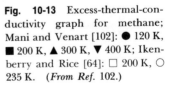

Fig. 10-13 Excess-thermal-conductivity graph for methane; Mani and Venart [102]: ● 120 K, ■ 200 K, ▲ 300 K, ▼ 400 K; Ikenberry and Rice [64]: □ 200 K, ○ 235 K. (*From Ref. 102*.)

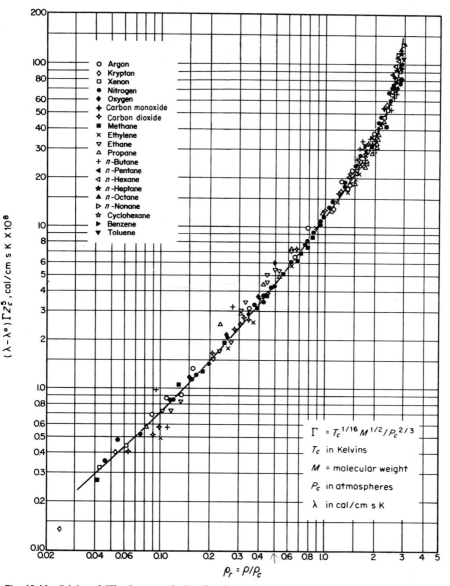

Fig. 10-14 Stiel and Thodos correlation for dense-gas thermal conductivities. (*From Ref. 178.*)

Figure 10-14 should not be used for polar substances or for hydrogen or helium. Its general accuracy is in doubt, and the scatter as shown indicates that errors of ±10 to 20 percent are possible. It appears, however, to be the best generalized correlation at present.

Recommendations Use Fig. 10-14 or Eqs. (10-5.2) to (10-5.4) to determine $\lambda - \lambda°$ for nonpolar materials. Critical constants may be obtained from Appendix A or estimated by the techniques given in Chap. 2. Experimental densities are preferable, but when they are not known, they can be found by the techniques described in Chap. 3. Low-pressure $\lambda°$ values can be determined as shown in Sec. 10-3. The accuracy of this plot is difficult to assess. Near the critical point, the accuracy is probably poor, and in other high-density regions probably no better than 10 to 20 percent accuracy should be expected.

As noted above, Fig. 10-14 should not be employed to estimate high-pressure thermal conductivities of polar compounds, hydrogen, or helium. In the ill-defined range of slightly polar compounds, no guidelines have been established.

Other empirical methods have been suggested, but they do not appear as suitable for general use as Fig. 10-14 [44, 45, 83, 179].

Example 10-3 Estimate the thermal conductivity of nitrous oxide at 105°C and 136 atm. At this pressure and temperature, the experimental value reported is 93.2 μcal/cm s K [143]. At 1 atm and 105°C, $\lambda° = 55.8$ μcal/cm s K [143]. Appendix A lists $T_c = 309.6$ K, $P_c = 71.5$ atm, $V_c = 97.4$ cm³/g-mol, $Z_c = 0.274$, and $M = 44.013$. At 105°C and 136 atm, Z for nitrous oxide is 0.63 [30].

  solution

$$\Gamma = \frac{T_c^{1/6} M^{1/2}}{P_c^{2/3}} = \frac{(309.6)^{1/6}(44.013)^{1/2}}{(71.5)^{2/3}} = 1.00$$

$$V = \frac{ZRT}{P} = \frac{(0.63)(82.07)(378)}{136} = 144 \text{ cm}^3/\text{g-mol}$$

$$\rho_r = \frac{V_c}{V} = \frac{97.4}{144} = 0.676$$

From Fig. 10-14, $(\lambda - \lambda°)\Gamma Z_c^5 = 6.5 \times 10^{-8}$ cal/cm s K or, from Eq. (10-5.3),

$$(\lambda - \lambda°)\Gamma Z_c^5 = (13.1 \times 10^{-8})(e^{(0.67)(0.676)} - 1.069)$$
$$= 6.6 \times 10^{-8} \text{ cal/cm s K}$$

Thus

$$\lambda - \lambda° = \frac{6.6 \times 10^{-8}}{(1.00)(0.274)^5} = 43 \times 10^{-6}$$

$$\lambda = (43 + 55.8)(10^{-6}) = 98.8 \text{ }\mu\text{cal/cm s K}$$

$$\text{Error} = \frac{98.8 - 93.2}{93.2} \times 100 = +6.0\%$$

10-6 Thermal Conductivities of Low-Pressure Gas Mixtures

The thermal conductivity of a gas mixture is not usually a linear function of mole fraction. Generally, if the constituent molecules differ greatly in polarity, the mixture conductivity is larger than would be predicted from a mole-fraction average; for nonpolar molecules, the opposite trend is noted and is more pronounced the greater the difference in molecular weights or sizes of the constituents [51, 113]. Some of these trends are evident in Fig. 10-15, which shows experimental thermal conductivities for four systems. The argon-benzene system typifies a nonpolar case with different molecular sizes, and the methanol–*n*-hexane system is a case representing a significant difference in polarity. The linear systems benzene-hexane and ether-chloroform represent a balance between the effects of size and polarity.

Several reviews have been published in the past few years summarizing various methods of calculating mixture thermal conductivities [12, 38, 45, 48, 88, 93, 110, 111, 121, 157, 163, 169, 185, 188], and a large number of theoretical papers have also appeared discussing the prob-

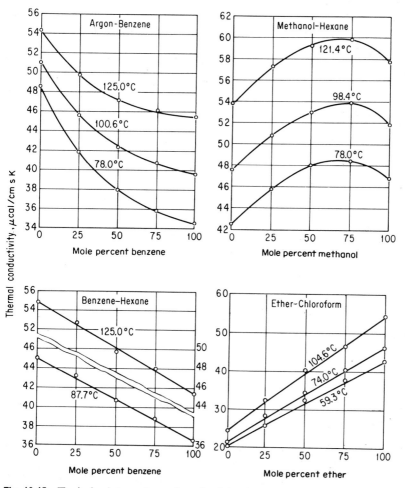

Fig. 10-15 Typical mixture thermal-conductivity data. (*From Ref. 3.*)

lems, approximations, and limitations of the various methods. The theory for calculating the conductivity for *rare-gas* mixtures has been worked out in detail [7, 10, 34, 61, 104, 108, 109, 122]. The more difficult problem, however, is to modify monatomic mixture correlations to apply to polyatomic molecules. Many techniques have been proposed; all are essentially empirical, and most reduce to some form of the Wassiljewa equation.

Wassiljewa Equation   In a form analogous to the theoretical relation for mixture viscosity, Eq. (9-5.1),

$$\lambda_m = \sum_{i=1}^{n} \frac{y_i \lambda_i}{\sum_{j=1}^{n} y_j A_{ij}}$$ (10-6.1)

where λ_m = thermal conductivity of the gas mixture
λ_i = thermal conductivity of pure component i
y_i, y_j = mole fractions of components i and j
A_{ij} = a function, as yet unspecified

This empirical relation was proposed by Wassiljewa in 1904 [200].

Mason and Saxena Modification   By simplifying the more rigorous monatomic gas expression for λ_m and using a suggestion made by Hirschfelder to estimate the Eucken factor for a polyatomic gas mixture, Mason and Saxena [107] suggested that A_{ij} in Eq. (10-6.1) could be expressed as

$$A_{ij} = \kappa \frac{[1 + (\lambda_{tr_i}/\lambda_{tr_j})^{1/2}(M_i/M_j)^{1/4}]^2}{[8(1 + M_i/M_j)]^{1/2}}$$ (10-6.2)

where M = molecular weight
λ_{tr} = monatomic value of thermal conductivity
κ = numerical constant near unity

Mason and Saxena originally suggested a value of κ of 1.065, and Tondon and Saxena later proposed 0.85. As used here, $\kappa = 1.0$.

To determine A_{ij}, if one notes that for monatomic gases $C_{v_i} = C_{v_j} = \frac{3}{2}R$, then from Eq. (10-2.3),

$$\frac{\lambda_{tr_i}}{\lambda_{tr_j}} = \frac{\eta_i}{\eta_j} \frac{M_j}{M_i}$$ (10-6.3)

Substituting Eq. (10-6.3) into (10-6.2) and comparing with Eq. (9-5.2) gives

$$A_{ij} = \phi_{ij}$$ (10-6.4)

where ϕ_{ij} is the interaction parameter for gas-mixture viscosity. This surprising result indicates that the theoretical relation for estimating mixture viscosities is also applicable to thermal conductivities by simply substituting λ for η.   Since both momentum and energy transfer result from different molecular interactions, Eq. (10-6.4) should be viewed as only approximate [100].

The Mason-Saxena form of the Wassiljewa equation (10-6.1) reduces to the use of Eqs. (10-6.4) and (9-5.2) to determine A_{ij}.   An alternate way of proceeding is to obtain the monatomic-gas thermal conductivities from Eq. (10-3.19).   Then

$$\frac{\lambda_{tr_i}}{\lambda_{tr_j}} = \frac{\Gamma_j}{\Gamma_i} \frac{\exp{(0.0464 T_{r_i})} - \exp{(-0.2412 T_{r_i})}}{\exp{(0.0464 T_{r_j})} - \exp{(-0.2412 T_{r_j})}} \qquad (10\text{-}6.5)$$

where Γ is defined by Eq. (10-13.15).   With Eq. (10-6.5), values of A_{ij} become functions of the reduced temperatures of both i and j.   However, with this latter approach, pure-gas viscosities are not required.   Both techniques are illustrated in Example 10-4.

Lindsay and Bromley Modification   Using a Sutherland model of a gas, Lindsay and Bromley [95] proposed,

$$A_{ij} = \tfrac{1}{4}\left\{1 + \left[\frac{\eta_i}{\eta_j}\left(\frac{M_j}{M_i}\right)^{3/4}\frac{T + S_i}{T + S_j}\right]^{1/2}\right\}^2\frac{T + S_{ij}}{T + S_i} \qquad (10\text{-}6.6)$$

where η = pure-gas viscosity
  T = absolute temperature
  S = Sutherland constant†

The parameter A_{ji} is expressed in a similar manner but with all subscripts reversed. S for a pure component can be determined by fitting the Sutherland viscosity equation to experimental data, or it can be estimated from an empirical rule; the one chosen by Lindsay and Bromley is

$$S_i = 1.5 T_{b_i} \qquad (10\text{-}6.7)$$

where T_{b_i} is the normal boiling point of i, K.‡   Other ways of estimating S_i are discussed by Gambill [45], and tables of S for various compounds are available [86, 133].   Although these rules are not accurate, the value of λ_m is relatively insensitive to variations in S; for example, a variation of 20 percent in S affects the calculated value of λ_m by only about 1 percent.

The interaction Sutherland constant S_{ij} is estimated as

$$S_{ij} = S_{ji} = C_s (S_i S_j)^{1/2} \qquad (10\text{-}6.8)$$

where C_s is close to unity unless one of the gases is very polar, when a

---

†The Sutherland viscosity model indicates that η is proportional to $T^{3/2}/(T + S)$.
‡For He, H$_2$, and Ne, S_i should be taken as 79 K [95].

value of 0.73 has been suggested [95]. For nonpolar-polar systems C_s is reported to vary widely [3].

Discussion of the Wassiljewa Form

A number of papers have been published discussing the Wassiljewa equation [23–25] and, in particular, giving some physical significance to the parameters A_{ij} and A_{ji} and to relating A_{ij} to ϕ_{ij} [of Eq. (9-5.1)] [31, 33, 50, 51, 100, 176, 203]. No clear-cut connection between A_{ij} and ϕ_{ij} has been shown, although many of the equations given in this section suggest that they are related.

Reviews by Gray and coworkers [33, 50, 51, 100, 203] relate A_{ij} to the efficiencies by which molecules of one component impede the transport of energy by molecules of the other component. From simple postulates the above authors then can derive the Wassiljewa equation.

Gray et al. [51] have emphasized the mathematical flexibility of the Wassiljewa equation and show that it is capable of representing mixture thermal-conductivity data that have either a maximum or minimum as composition is varied. In general, if $\lambda_1 < \lambda_2$

$$\frac{\lambda_1}{\lambda_2} < A_{12}A_{21} < \frac{\lambda_2}{\lambda_1} \qquad \lambda_m \text{ varies monotonically with composition}$$

$$A_{12}A_{21} \geq \frac{\lambda_2}{\lambda_1} \qquad \lambda_m \text{ has a minimum value below } \lambda_1$$

$$\frac{\lambda_1}{\lambda_2} \geq A_{12}A_{21} \qquad \lambda_m \text{ has a maximum value above } \lambda_2$$

The conditions for points of inflection are also discussed by the same authors.

Brokaw Empirical Method

Brokaw [6] noted that in most nonpolar mixtures λ_m is less than a linear mole-fraction average but larger than a reciprocal average. He then suggested that, for binary mixtures,

$$\lambda_m = q\lambda_{mL} + (1-q)\lambda_{mR}$$

where $\qquad \lambda_{mL} = y_1\lambda_1 + y_2\lambda_2 \qquad$ and $\qquad \dfrac{1}{\lambda_{mR}} = \dfrac{y_1}{\lambda_1} + \dfrac{y_2}{\lambda_2}$ (10-6.9)

TABLE 10-6 Variation of the Brokaw Factor q with Composition of Light Component

Mole fraction light component	Factor q for Eq. (10-6.9)	Mole fraction light component	Factor q for Eq. (10-6.9)
0	0.32	0.6	0.50
0.1	0.34	0.7	0.55
0.2	0.37	0.8	0.61
0.3	0.39	0.9	0.69
0.4	0.42	0.95	0.74
0.5	0.46	1.0	0.80

q is a parameter which is given in Table 10-6 as a function of the mole fraction of the component with lower molecular weight. If q is assumed constant at 0.5, the rule reduces to one similar to that suggested by Burgoyne and Weinberg [16].

Applicability and Accuracy of the Gas-Mixture Estimation Techniques Three principal estimating techniques have been discussed above. Other proposed methods were rejected, either from an error criterion or from a combination of error criterion and the desire to choose simple techniques. Of the three methods, the Lindsay-Bromley method, the oldest, has received the most extensive testing. The authors themselves tested the equation with 16 binary mixtures and found an average error of 1.9 percent. The mixtures included polar (NH_3, H_2O)–nonpolar mixtures but no polar-polar mixtures. Temperatures ranged from 0 to 80°C. Experimental thermal conductivities for the pure components were used in the correlation (and in all the testing described in this section).

Brokaw tested his empirical method with 18 nonpolar mixtures and found an average error of about 2.6 percent, with a maximum of 11.4 percent. Since Brokaw's method cannot predict λ_m larger than the linear mole-fraction average, it cannot be used for mixtures containing a polar component.

The Mason-Saxena method has not been so thoroughly tested, but in the comparison between calculated and experimental results for nine binary and five ternary nonpolar mixtures, average errors of 2 to 3 percent have been reported [107]. For methane-propane mixtures, between 50 and 150°C, good agreement was obtained between experimental and calculated values of λ_m for this technique [175].

In a series of papers by Gray and coworkers [50–52, 99, 100] the Lindsay-Bromley method was shown to yield estimated mixture thermal conductivities within 2 to 4 percent for a wide variety of mixtures. The larger errors often occurred when one component was H_2 or He. Tondon and Saxena [183] also indicated that this method is generally reliable for nonpolar mixtures.

Other tests have been reported, but they do not change any of the relative conclusions that can be reached from the comments given above [28, 36, 53, 58, 70, 123, 134, 161, 182].

Recommendations *Nonpolar Mixtures.* Any of the three methods described are suitable. For the Mason-Saxena form, use Eq. (10-6.1) with $A_{ij} = \phi_{ij}$ in Eq. (9-5.2) or with A_{ij} from Eqs. (10-6.2) and (10-6.5). If the Lindsay-Bromley relation is chosen, Eqs. (10-6.1) and (10-6.6) to (10-6.8) are to be used unless S_i is obtained from experimental viscosity data. Finally, for Brokaw's method, Eq. (10-6.9) is necessary. For all methods one needs pure-component thermal conductivities, and all but

Brokaw's and Mason and Saxena's [with Eq. (10-6.5)] require pure-component viscosities.   Typically, errors do not exceed 1 to 3 percent.

Polar-Nonpolar Mixtures.   The same recommendation is made as for nonpolar mixtures except that the empirical Brokaw method is inapplicable.   The Mason-Saxena method has not been tested as extensively as the Lindsay-Bromley correlation.   Errors rarely exceed 5 percent.

Polar Mixtures.   Use the Lindsay-Bromley method; the accuracy of this technique is in doubt, but with the work of Maczek and Gray [99] as a guide, errors rarely exceed 5 percent even for multicomponent systems.

Example 10-4   Estimate the thermal conductivity of a mixture of methane and propane at 1 atm and 95°C and a mole fraction methane of 0.486.   The experimental mixture value is reported as 7.64×10^{-5} cal/cm s K [24, 175].

   solution   The properties of the pure components which are required in one or all of the estimation techniques illustrated below are as follows:

	Methane	Propane
Molecular weight	16.043	44.097
Pure-gas λ, cal/cm s K [24, 175]	10.49×10^{-5}	6.34×10^{-5}
Pure-gas η, cP [87]	1.32×10^{-2}	1.00×10^{-2}
Normal boiling point, K	111.7	231.1
C_v, cal/g-mol K [1] at 95°C	7.44	19.5
Critical temperature, K	190.6	369.8
Critical pressure, atm	45.4	41.9
Reduced temperature $T_r = T/T_c$	1.93	0.995

   Empirical Brokaw Method.   From Eq. (10-6.9)

$$\lambda_{mL} \times 10^5 = (0.486)(10.49) + (0.514)(6.34) = 8.35$$

$$\lambda_{mR} \times 10^5 = \left(\frac{0.486}{10.49} + \frac{0.514}{6.34}\right)^{-1} = 7.84$$

From Table 10-6, with the mole fraction methane = 0.486, $q = 0.45$; thus

$$\lambda_m = [(0.45)(8.35) + (0.55)(7.84)](10^{-5}) = 8.0 \times 10^{-5} \text{ cal/cm s K}$$

$$\text{Error} = \frac{8.0 - 7.64}{7.64} \times 100 = 4.7\%$$

   Lindsay-Bromley Method.   First the Sutherland constants have to be estimated.   Letting 1 denote methane and 2 propane, we have from Eqs. (10-6.7) and (10-6.8),

$$S_1 = (1.5)(111.7) = 167 \text{ K}$$

$$S_2 = (1.5)(231.1) = 347 \text{ K}$$

$$S_{12} = [(167)(347)]^{1/2} = 240 \text{ K}$$

The A_{12} and A_{21} parameters are determined from Eq. (10-6.6); for example,

$$A_{12} = \frac{1}{4}\left(1 + \left\{\left[\frac{1.32}{1.00}\left(\frac{44.097}{16.043}\right)^{3/4}\right]\frac{1 + 167/368}{1 + 347/368}\right\}^{1/2}\right)^2 \frac{1 + 240/368}{1 + 167/368}$$

$$= 1.71$$

Similarly $A_{21} = 0.61$.   Then, with Eq. (10-6.1),

$$\lambda_m = \frac{y_1\lambda_1}{y_1 + A_{12}y_2} + \frac{y_2\lambda_2}{y_2 + A_{21}y_1} = \frac{(0.486)(10.49 \times 10^{-5})}{0.486 + (1.71)(0.514)}$$

$$+ \frac{(0.514)(6.34 \times 10^{-5})}{(0.514) + (0.61)(0.486)} = 7.75 \times 10^{-5}\ \text{cal/cm s K}$$

$$\text{Error} = \frac{7.75 - 7.64}{7.64} \times 100 = 1.4\%$$

Mason-Saxena Method.   The ϕ_{12} and ϕ_{21} values may be calculated from Eq. (9-5.2) or from Eqs. (10-6.2) and (10-6.5).   With the former,

$$\phi_{12} = \frac{[1 + (1.32/1.00)^{1/2}(44.097/16.043)^{1/4}]^2}{8^{1/2}(1 + 16.04/44.10)^{1/2}} = 1.86$$

Likewise, reversing subscripts, $\phi_{21} = 0.51$.   Then from Eq. (10-6.4)

$$A_{12} = \phi_{12} = 1.86 \qquad A_{21} = \phi_{21} = 0.51$$

Inserting these values into Eq. (10-6.1) gives

$$\lambda_m = \frac{(0.486)(10.49)(10^{-5})}{0.486 + (1.86)(0.514)} + \frac{(0.514)(6.34)(10^{-5})}{0.514 + (0.51)(0.486)} = 7.81 \times 10^{-5}\ \text{cal/cm s K}$$

$$\text{Error} = \frac{7.81 - 7.64}{7.64} \times 100 = 2.2\%$$

If Eqs. (10-6.2) and (10-6.5) are used to determine A_{ij}, pure-component viscosities are not required.   First Γ is calculated for both methane and propane from Eq. (10-3.15):

$$\Gamma = \frac{T_c^{1/6}M^{1/2}}{P_c^{2/3}} \qquad \Gamma_1 = 0.7506 \qquad \Gamma_2 = 1.472$$

With Eq. (10-6.5),

$$\frac{\lambda_{tr1}}{\lambda_{tr2}} = \frac{1.472}{0.7506}\ \frac{e^{(0.0464)(1.93)} - e^{-(0.2412)(1.93)}}{e^{(0.0464)(0.995)} - e^{-(0.2412)(0.995)}}$$

$$= 3.50$$

Substituting this value into Eq. (10-6.2) with $\kappa = 1.0$, gives

$$A_{12} = \frac{[1 + (3.50)^{1/2}(16.04/44.10)^{1/4}]^2}{[8(1 + 16.04/44.10)]^{1/2}} = 1.82$$

Similarly, $A_{21} = 0.521$.   Then, as before, with Eq. (10-6.1)

$$\lambda_m = 7.83 \times 10^{-5}\ \text{cal/cm s K}$$

$$\text{Error} = \frac{7.83 - 7.64}{7.64} \times 100 = 2.5\%$$

10-7 Effect of Temperature and Pressure on the Thermal Conductivity of Gas Mixtures

Temperature Effects   The shape of a mixture thermal-conductivity–composition curve is not usually altered appreciably by changes in temperature.   However, occasionally a mixture which shows a negative deviation at low temperatures may show a positive deviation at higher temperatures, the term *deviation* referring to a mole-fraction linear average curve. N_2–CO_2 seems to show such behavior, as discussed by Brokaw [8].

If one lacks evidence that there is such unusual behavior, the effect of temperature on the mixture thermal conductivity is absorbed in the change of pure-component conductivities.   This variation is discussed in Sec. 10-4; no change in the functional form of the selected mixture correlation is then necessary.

Ulybin [187] has suggested the empirical relation

$$\lambda_m(T_2) = \lambda_m(T_1) \sum_{i=1}^{n} y_i \frac{\lambda_i(T_2)}{\lambda_i(T_1)}$$ (10-7.1)

This has been shown by Saxena and Gupta [161] to work well for mixtures of light gases.

Pressure Effects   Few experimental data are available on the thermal conductivity of high-pressure gas mixtures.   Keyes [80] studied the nitrogen–carbon dioxide system.   Comings and his coworkers reported on ethylene-nitrogen and carbon dioxide–ethylene mixtures [72], rare-gas mixtures [135], and binaries containing carbon dioxide, nitrogen, and ethane [49].   Rosenbaum and Thodos studied methane–carbon dioxide [148] and methane–carbon tetrafluoride [147] binaries.

Although the Lindsay-Bromley method has been suggested as a technique for estimating high-pressure gas thermal conductivities (Sec. 10-5) [49, 70], better results are generally found with the Stiel and Thodos pure-component relations [Eqs. (10-5.2) to (10-5.4)].   The mixture is treated as a hypothetical pure component with pseudocritical properties.   By the use of Prausnitz and Gunn's modified rules [mole fraction average for T_c, V_c, Z_c, and Eq. (4-2.2) for P_c], Table 10-7 has been prepared.   With a few exceptions, this simple technique is reliable.   The use of other pseudocritical rules has been considered, but the results are not significantly affected [148].   In the CH_4–CF_4 case, however, the Stiel and Thodos method yielded the poorest results [147], and it was found necessary to modify Eqs. (10-5.2) to (10-5.4) to make $(\lambda - \lambda°)\Gamma Z_c^5 = f(\rho_r, T, M)$.

More data are necessary to confirm the applicability of the Stiel-Thodos relations for dense-gas-mixture thermal conductivities, but, at present, it is the most reliable general method available.

Example 10-5   Estimate the thermal conductivity of a methane–carbon dioxide gas mixture containing 75.5 mole percent methane at 370.8 K and 172.5 atm.   Rosenbaum and Thodos [148] list an experimental value of 121.3 μcal/cm s K; these same investigators report that, for the same mixture, $\rho = 0.1440$ g/cm^3 and, at 1 atm, $\lambda° = 90.0$ μcal/cm s K.

TABLE 10-7   Comparison between Calculated and Experimental Gas-Mixture Thermal Conductivities at High Pressure

All thermal conductivity values in μcal/cm s K

Mixture	Mol % first component	T, K	P, atm	λ_m (exp.)	λ_m° (exp.) at 1 atm	Ref.	λ (calc.)	% error†
Ethylene–carbon dioxide	55.5	315.1	50.0	63.3	49.6	72	64	1.1
			104.8	137			144	4.9
			155.9	168			191	14
			198.1	192			211	10
Ethylene–carbon dioxide	79.8	315.1	52.1	68.6	51.6	72	70	2.0
			114.4	151			170	14
			155.3	168			195	16
			199.9	198			218	10
Ethylene-nitrogen	67.5	315.1	43.6	64.5	56.3	72	66	2.3
			101.9	88.5			89	0.6
			156.5	119			120	0.8
			196.9	130			135	3.7
Ethylene-nitrogen	38.8	315.1	51.1	67.9	58.2	72	67	−1.3
			102.4	79.5			78	−1.9
			154.9	94.3			95	1.8
			197.8	104			111	6.7
Nitrogen–carbon dioxide	34.1	323.2	19.8	53.0	49.9	80	53	0
			48.8	58.8			58	−1.4
			51.2	59.8			59	−1.3
			81.0	68.5			66	−3.6
Nitrogen–carbon dioxide	52.9	323.2	20.0	56.6	53.7	80	56	−1.0
			42.7	59.8			63	5.3
			68.7	67.0			64	−4.5
			84.3	72.1			69	−4.3
Methane–carbon dioxide	75.5	370.8	33.3	94.9	90.0	148	95.4	0.5
			172.5	121			126	3.9
			415.4	176			195	11
			688.6	228			251	10
Methane–carbon dioxide	24.3	370.7	35.7	68.4	63.0	148	68.8	0.5
			170.8	106			110	3.0
			418.1	172			200	16
			682.2	229			254	11

†Error = [(calc. − exp.)/(exp.)] × 100.

solution   From Appendix A,

	Methane	Carbon dioxide
T_c, K	190.6	304.2
V_c, cm^3/g-mol	99.0	94.0
M	16.043	44.010
Z_c	0.288	0.274

The Prausnitz and Gunn modified pseudocritical rules indicate that T_{c_m}, V_{c_m}, Z_{c_m} should be calculated as linear mole-fraction averages and that P_{c_m} should be determined from $P_{c_m} = Z_{c_m} R T_{c_m} / V_{c_m}$. For example,

$$T_{c_m} = (0.755)(190.6) + (0.245)(304.2) = 218.5 \text{ K}$$

Similarly,

$$V_{c_m} = 97.8 \text{ cm}^3/\text{g-mol} \qquad Z_{c_m} = 0.285 \qquad M_m = 22.895$$

and

$$P_{c_m} = \frac{(0.285)(82.06)(218.5)}{97.8} = 52.24 \text{ atm}$$

Then

$$\rho_{r_m} = \frac{\rho}{\rho_c} = 97.8 \frac{0.1440}{22.895} = 0.615$$

With Eq. (10-5.3), where $\Gamma = (218.5)^{1/6}(22.895)^{1/2}/(52.24)^{2/3} = 0.840$,

$$(\lambda - \lambda^\circ)\Gamma Z_c^5 = (13.1 \times 10^{-8})(e^{(0.67)(0.615)} - 1.069)$$

$$= 5.78 \times 10^{-8}$$

$$\lambda = \frac{5.78 \times 10^{-8}}{(0.840)(0.285)^5} + 90.0 \times 10^{-6}$$

$$= 126 \text{ } \mu\text{cal/cm s K}$$

$$\text{Error} = \frac{126 - 121.3}{121.3} \times 100 = 3.9\%$$

10-8   Thermal Conductivity of Liquids

For many simple organic liquids, the thermal conductivity is between 10 and 100 times as large as that of the low-pressure gas at the same temperature.   There is little pressure dependence, and usually raising the temperature decreases the thermal conductivity.   These characteristics are similar to those noted for liquid viscosity, although the temperature dependence of the latter is pronounced and nearly exponential, whereas that for thermal conductivity is weak and nearly linear.

Values of λ_L for most common organic liquids range between 250 and 400 μcal/cm s K at temperatures below the normal boiling point, but water, NH$_3$, and other highly polar molecules have values 2 to 3 times as large.   Also, in many cases the dimensionless ratio $M\lambda/R\eta$ is nearly constant (for nonpolar liquids) between a value of 2 and 3, so that highly viscous liquids often have a correspondingly larger thermal conductivity.   Liquid metals and some organosilicon compounds have large values of λ_L; the former often are 100 times as large as those for normal organic liquids. The solid thermal conductivity at the melting point is approximately 20 to 40 percent larger than that of the liquid. Some values of liquid thermal conductivities are shown in Table 10-8 to illustrate the range encountered. Liquid thermal-conductivity data have been compiled and evaluated by Jamieson et al. [68].

The difference between transport-property values in the gas phase and those in the liquid phase indicates a distinct change in mechanism of energy (or momentum or mass) transfer, i.e.,

$$\frac{\lambda_L}{\lambda_g} \approx 10 \text{ to } 100 \qquad \frac{\eta_L}{\eta_g} \approx 10 \text{ to } 100 \qquad \frac{\mathscr{D}_L}{\mathscr{D}_g} \approx 10^{-4}$$

In the gas phase, the molecules are relatively free to move about and transfer momentum and energy by a collisional mechanism. The intermolecular force fields, though not insignificant, do not drastically affect the values of λ, η, or $\mathscr{D}$. That is, the intermolecular forces are reflected solely in the collision-integral terms Ω_v and Ω_D, which are really ratios of collision integrals for a real force field and an artificial case where the molecules are rigid, noninteracting spheres. The variation of Ω_v or Ω_D from unity then yields a rough quantitative measure of the importance of intermolecular forces in affecting gas-phase transport coefficients. Reference to Eq. (9-4.3) (for Ω_v) or Eq. (11-3.6) (for Ω_D) shows that Ω values are often near unity. One then concludes that a rigid, noninteracting spherical molecular model yields low-pressure

TABLE 10-8 Thermal Conductivities of Some Liquids at 1 atm

	T, °C	λ_L, μcal/cm s K
Hydrogen	-253	282
Argon	-190	301
n-Hexane	20	291
p-Xylene	20	325
Nitrobenzene	20	361
Acetic acid	20	374
Acetone	20	385
Chloroform	20	277
Glycerol	20	693
Water	20	1,430
Dowtherm A	20	344
Paraffin	25	326
NaCl(aq), sat	20	1,370
Sodium	300	180,000
Mercury	100	22,000
HTS (KNO$_3$–NaNO$_2$–NaNO$_3$)	150	1,360

transport coefficients λ,† η, or $\mathscr{D}$ not greatly different from those computed when intermolecular forces are included.

In the liquid, however, this hypothesis is not even roughly true. The close proximity of molecules to one another emphasizes strongly the intermolecular forces of attraction. There is little wandering of the individual molecules, as evidenced by the low value of liquid diffusion coefficients, and often a liquid is modeled as a lattice with each molecule caged by its nearest neighbors. Energy and momentum are primarily exchanged by oscillations of molecules in the shared force fields surrounding each molecule. Scheffy [167] has pointed out an illuminating description of such processes given by Green [54]: "Imagine the molecules tied together by elastic strings, the tension in which varies in a rather eccentric way so as to simulate the attractive forces. Then, as the molecules move, the elastic energy of the strings will vary and by this means energy can be transported from one part of the assembly to another without actually being carried by the molecules themselves."

The difference between transport mechanisms in a dense gas or liquid and those in a low-pressure gas is discussed in detail by McLaughlin [97].

Theories to date have not led to simple estimating techniques for liquid thermal conductivity [184]; approximate (or frankly empirical) techniques must be used for engineering application.

Only relatively simple organic liquids are considered in the sections to follow. Ho et al. [62] have presented a comprehensive review covering the thermal conductivity of the elements, while Ewing et al. [40] and Gambill [46] consider, respectively, molten metals and molten salt mixtures. Cryogenic liquids are discussed by Preston et al. [137] and Mo and Gubbins [120a].

10-9   Estimation of the Thermal Conductivity of Pure Liquids

There is no scarcity of estimation techniques for calculating the thermal conductivity of a pure liquid. Almost all are empirical, and with only limited examination, some may appear more accurate than they are. As noted above, below the normal boiling point, the conductivities of most organic, nonpolar liquids lie between 250 and 400 μcal/cm s K. With this fact in mind, it is not difficult to devise various schemes for estimating λ_L within this limited domain.

Many estimation methods were tested; three of the best are described in detail below. Others that were considered are noted briefly at the end of the section.

†λ in this case is the monatomic value, *not* including contributions from internal energy-transfer mechanisms.

Method of Robbins and Kingrea   Weber in 1880 [202] suggested that λ_L is proportional to the product $C_p\rho^{4/3}$, and this general type of correlation has been modified and republished several times [132, 173, 174, 191]. The best modification, however, appears to be due to Robbins and Kingrea [146], who have proposed the following relation:

$$\lambda_L = \frac{(88.0 - 4.94H)(10^{-3})}{\Delta S^*} \left(\frac{0.55}{T_r}\right)^N C_p\rho^{4/3} \qquad (10\text{-}9.1)$$

where λ_L = liquid thermal conductivity, cal/cm s K
$\quad T_r$ = reduced temperature T/T_c
$\quad C_p$ = molal heat capacity of the liquid, cal/g-mol K
$\quad \rho$ = molal liquid density, g-mol/cm^3
$\quad \Delta S^* = \Delta H_{vb}/T_b + R \ln (273/T_b) \qquad\qquad\qquad\qquad (10\text{-}9.2)$
$\quad \Delta H_{vb}$ = molal heat of vaporization at the normal boiling point, cal/g-mol
$\quad T_b$ = normal boiling point, K

The parameters H and N are obtained from Table 10-9; H depends upon molecular structure and N upon the liquid density at 20°C. Equation (10-9.1) is relatively easy to use, but ΔH_{vb}, T_b, ρ, C_p values are required. These can be obtained from experimental data or by estimation techniques described earlier in this book.

Robbins and Kingrea tested Eq. (10-9.1) with 70 organic liquids, using 142 data points; some of their test compounds and results are shown in Table 10-10. Rarely do the errors exceed 10 percent. These workers state that the range of applicability is from a T_r of 0.4 to 0.9, but the testing in Table 10-9 was invariably carried out only from about 0.5 to 0.7. Sulfur-containing compounds and inorganics are not included. The abrupt change in the exponent N from zero to unity for compounds with mass densities greater or less than 1.0 g/cm^3 is difficult to accept in many instances; often more reliable results are obtained with $N = 1.0$ even for compounds with $\rho > 1$ g/cm^3.

Boiling-Point Method   Sato [101] suggested that at the normal boiling point,

$$\lambda_{L_b} = \frac{2.64 \times 10^{-3}}{M^{1/2}} \qquad (10\text{-}9.3)$$

where λ_{L_b} = liquid thermal conductivity at the normal boiling point, cal/cm s K
$\quad M$ = molecular weight

To estimate λ at other temperatures, Eqs. (10-9.3) and (10-9.1) can be combined to eliminate terms involving H and ΔS^*. When this is done

[101], assuming $N = 1$,

$$\lambda_L = \frac{2.64 \times 10^{-3}}{M^{1/2}} \frac{C_p}{C_{p_b}} \left(\frac{\rho}{\rho_b}\right)^{4/3} \frac{T_b}{T} \qquad (10\text{-}9.4)$$

where the subscript b represents conditions at the normal boiling point.

TABLE 10-9 *H* and *N* Factors for Eq. (10-9.1) [146a]

Functional group	Number of groups	H†
Unbranched hydrocarbons:		
Paraffins		0
Olefins		0
Rings		0
CH_3 branches	1	1
	2	2
	3	3
C_2H_5 branches	1	2
i-C_3H_7 branches	1	2
C_4H_9 branches	1	2
F substitutions	1	1
	2	2
Cl substitutions	1	1
	2	2
	3 or 4	3
Br substitutions	1	4
	2	6
I substitutions	1	5
OH substitutions	1 (iso)	1
	1 (normal)	-1
	2	0
	1 (tertiary)	5

Oxygen substitutions:		
$-\overset{\displaystyle |}{C}\!\!=\!\!O$ (ketones, aldehydes)		0
$-\overset{\displaystyle O}{\overset{\displaystyle \|}{C}}\!-\!O-$ (acids, esters)		0
$-O-$ (ethers)		2
NH_2 substitutions	1	1
Liquid density, g/cm³	N	
	1	
	0	

†For compounds containing multiple functional groups, the *H*-factor contributions are additive.

TABLE 10-10 Comparison between Calculated and Experimental Values of Liquid Thermal Conductivity†

All values of thermal conductivity in μcal/cm s K

Compound	T, K	λ (exp.)	Method of Robbins and Kingrea			Method of Sato and Riedel	Method of Missenard and Riedel
			H	N	% error‡	% error‡	% error‡
Propane	323	187	0	1	0.5	27	18
n-Pentane	293	272	0	1	-1.8	20	17
	303	265			-3.0	20	17
n-Hexane	293	291	0	1	2.0	12	12
	303	282			3.5	12	12
n-Heptane	273	316	0	1	5.3	6.0	8.9
	293	301			3.0	6.9	9.6
	333	270			1.5	8.5	12
n-Octane	293	308	0	1	4.2	3.2	8.8
	350	267			2.6	5.2	11
n-Nonane	293	313	0	1	3.8	0	8.3
	350	275			-0.7	1.8	11
n-Decane	314	304	0	1	7.9	-2.0	9.5
	349	284			2.8	-1.8	9.8
2-Methylpentane	305	259	1	1	1.5	19	18
	322	247			0	19	18
2,3-Dimethylbutane	305	248	2	1	-2.0	24	20
	322	238			-2.9	23	19
2,2,4-Trimethylpentane	311	231	3	1	-1.3	27	20
	350	201			1.5	23	26
Cyclohexane	298	297	0	1	12	11	3.7

TABLE 10-10 Comparison between Calculated and Experimental Values of Liquid Thermal Conductivity (Continued)
All values of thermal conductivity in μcal/cm s K

Compound	T, K	λ (exp.)	Method of Robbins and Kingrea			Method of Sato and Riedel	Method of Missenard and Riedel
			H	N	% error‡	% error‡	% error‡
Methylcyclopentane	293	289	1	1	1.0	13	3.8
	311	275			1.4	14	4.7
Benzene	293	353	0	1	4.5	−3.4	−5.1
	323	327			2.4	−2.1	−4.0
	389	272			2.6	0	−1.8
Ethylbenzene	293	316	2	1	3.8	2.2	4.4
	353	281			3.6	3.2	5.3
Ethanol	293	394	−1	1	8.1	15	24
	313	363			14	19	28
	333	339			21	21	30
	347	322			28	22	32
n-Butanol	293	367	−1	1	4.1	5.4	20
	313	352			5.7	5.7	21
	333	344			6.7	3.5	18
n-Octanol	293	396	−1	1	−1.5	−19	5.6
	311	383			3.9	−18	6.0
t-Butyl alcohol	311	277	5	1	−2.9	26	77
	350	255			−0.4	22	72
m-Cresol	293	358	2	0	1.1	−3.6	28
	353	347			13	−8.6	21
Ethylene glycol	293	609	0	0	0.3	s	s

Aniline	290	424	1	0	4.7	-15	10
Propionic acid	285	413	0	0	-1.7	-3.4	15
Methylene chloride	253	380	2	0	-4.7	-13	-6.3
	273	373			-7.2	-15	-8.3
	293	353			-5.4	-15	-7.9
Carbon tetrachloride	253	263	3	0	3.4	-0.8	15
	273	256			5.0	-1.9	13
	293	246			5.3	-1.6	14
Ethyl bromide	293	246	4	0	1.6	7.7	-6.9
Chlorobenzene	233	336	1	0	-4.5	0	2.6
	293	305			4.9	0.6	3.3
	353	266			18	4.1	7.1
Iodobenzene	253	254	5	0	-7.9	-0.4	5.1
	293	242			-4.5	-0.4	5.4
	353	224			-0.9	-0.9	4.5
Ethyl acetate	293	351	0	1	2.3	-7.1	3.1
	333	337			-4.7	-12	-2.7
Butyl acetate	293	327	0	1	13	-4.9	9.2
Acetone	273	409	0	1	0.2	-2.2	3.7
	293	385			-0.8	-0.8	5.2
	313	361			-2.4	0.5	6.6
Diethyl ether	293	307	2	1	0	4.5	22
Acetaldehyde	293	454	0	1	-16	-12	-11

†Experimental values of the thermal conductivity were obtained, where possible, from Ref. 69. Most of the calculated values of thermal conductivity for the Robbins-Kingrea method were taken from Ref. 146b. Other sources for thermal conductivity data include Refs. 15, 193, and 35.
‡Percent error = [(calc. − exp.)/exp.] × 100.
§Critical properties not known.

523

Alternatively, a relation between λ_L and T given by Riedel [145] and found to be accurate by Rahalkar et al. [140] and by Martin [103] is

$$\lambda_L = A\,[1 + (\tfrac{20}{3})(1 - T_r)^{2/3}] \tag{10-9.5}$$

With Eqs. (10-9.3) and (10-9.5),

$$\lambda_L = \frac{2.64 \times 10^{-3}}{M^{1/2}} \frac{3 + 20(1 - T_r)^{2/3}}{3 + 20(1 - T_{r_b})^{2/3}} \tag{10-9.6}$$

Equation (10-9.6) was used to estimate λ_L values for Table 10-10. The only data required are T, T_c, T_b, and M. The errors noted vary widely. Poor results were found for low-molecular-weight hydrocarbons and branched hydrocarbons; generally the predicted value was larger than experimental. Better results were obtained for non-hydrocarbons.

Method of Missenard   In his book on thermal conductivity, Missenard [117] suggested that

$$\lambda_{L_0} = \frac{90 \times 10^{-6}(T_b\rho_0)^{1/2}C_{p_0}}{M^{1/2}\mathcal{N}^{1/4}} \tag{10-9.7}$$

where λ_{L_0} = liquid thermal conductivity at 0°C, cal/cm s K
$\quad T_b$ = normal boiling temperature, K
$\quad \rho_0$ = liquid density at 0°C, g-mol/cm³
$\quad C_{p_0}$ = liquid constant pressure heat capacity at 0°C, cal/g-mol K
$\quad M$ = molecular weight
$\quad \mathcal{N}$ = number of atoms in the molecule

When this is combined with Eq. (10-9.5), one obtains

$$\lambda_L = \lambda_{L_0}\frac{3 + 20(1 - T_r)^{2/3}}{3 + 20(1 - 273/T_c)^{2/3}} \tag{10-9.8}$$

Equation (10-9.8) with λ_{L_0} from Eq. (10-9.7) was used in the tests shown in Table 10-10. In almost all cases, the estimated λ_L exceeded the experimentally reported value. The constant 90×10^{-6} in Eq. (10-9.7) should, perhaps, be decreased by 10 to 15 percent. In fact, Missenard later suggested a value of 84×10^{-6} [116].

Other Liquid Thermal-Conductivity Estimation Techniques   Equations (10-5.2) to (10-5.4) (or Fig. 10-14) can be used to determine $\lambda - \lambda^\circ$ as a function of reduced density for the liquid phase when temperatures are well in excess of the normal boiling point. Kanitkar and Thodos [75] suggested a thermal-conductivity correlation for liquid hydrocarbons that relates λ_L and ρ_r, but in most cases the Stiel and Thodos equations (10-5.2) to (10-5.4) appear to be almost as accurate. None of the correlating equations presented earlier in this chapter is believed to be accurate at high reduced temperatures.

Concerning other estimation schemes suggested in the literature, Scheffy and Johnson [167, 168] related λ_L to the difference in temperature between T and the freezing point of the substance. As described elsewhere [141], when tested against experimental data, average errors exceeded 10 percent.

Other recent methods tested and found to be less accurate than those presented above include correlations by Ewing et al. [40], Pachaiyappan [129–131], Puranasamriddhi [138], Djalalian [35], Rahalkar et al. [140], Viswanath [197], and McLaughlin [98]. Earlier estimation methods have been reviewed elsewhere [141].

Discussion and Recommendations   There are often marked differences in reported values of experimental liquid thermal conductivities; such differences frequently are of the same order as the errors reported for the estimation techniques. This is clearly evident in the excellent critical survey of liquid thermal-conductivity data published by the National Engineering Laboratory [69].

Two groups of investigators have contributed notably to the measurement of λ_L, Riedel in Germany [145] and Sakiadis and Coates in the United States [152–155]. Unfortunately, the results of these two groups often do not agree well, Sakiadis and Coates usually reporting values 5 to 10 percent higher. These differences cast some doubt on the acceptability of any estimation method. Challoner and Powell [20] discuss these and other studies and conclude that, in general, Riedel's values are more acceptable, although in the NEL survey noted above, data from both investigators are usually considered reliable. Other recent experimental papers that may prove useful are Refs. 15, 35, 136, 189, and 193. The best recommendations that can be made at the present are as follows.

For organic liquids only:

1. Use the method of Sato and Riedel [Eq. (10-9.6)] if a *quick approximate* value is necessary. The method is not applicable, however, for highly polar liquids, branched-chain and light hydrocarbons, and inorganic liquids or for temperatures well above the normal boiling point. Only the molecular weight, the normal boiling point, and critical temperature need be known.

2. Use the method of Robbins and Kingrea [Eq. (10-9.1)] for a more accurate estimate between reduced temperatures of about 0.4 and 0.8. Liquid densities and heat capacities are required, as are the normal boiling point and latent heat of vaporization. Inorganic or sulfur-containing compounds are not included. Errors are usually less than 5 percent.

3. If the liquid is at a reduced temperature greater than 0.8, refer to the high-pressure-gas thermal-conductivity recommendations in Sec. 10-5.

Example 10-6 Estimate the thermal conductivity of carbon tetrachloride at 20°C. At this temperature, Jamieson and Tudhope [69] list 11 reported values. Six are given a ranking of A and are considered reliable. They range from 244 to 256 μcal/cm s K. Most, however, are close to 246 μcal/cm s K. More recent measurements by Poltz and Jugel [136] appear to substantiate this value.

  solution Data:

	0°C	20°C
Heat capacity, cal/g-mol K [146a]	31.22	31.53
Density, g-mol/cm³ [146a]	0.0106	0.0103
Critical temperature, K		556.4
Normal boiling point, K		349.7
Molecular weight		153.823
Number of atoms $\mathcal{N}$		5
Heat of vaporization at the normal boiling point, cal/g-mol		7170

  Robbins and Kingrea. From Table 10-9, $H = 3$, and since the liquid density $= (0.0103)(153.823) = 1.58 > 1$, $N = 0$. Also with Eq. (10-9.2),

$$\Delta S^* = \frac{7170}{349.7} + 1.987 \ln \frac{273}{349.7} = 20.01 \text{ cal/g-mol K}$$

Then with Eq. (10-9.1),

$$\lambda_L = \left\{ \frac{[88.0 - (4.94)(3)](10^{-3})}{20.01} \right\} (31.53)(0.0103)^{4/3}$$

$$= 259 \times 10^{-6} \text{ cal/cm s K}$$

$$\text{Error} = \frac{259 - 246}{246} \times 100 = 5.3\%$$

  Sato-Riedel. With Eq. (10-9.6), since $T_r = 293/556.4 = 0.527$ and $T_{b_r} = 349.7/556.4 = 0.629$, we have

$$\lambda_L = \frac{2.64 \times 10^{-3}}{(153.84)^{1/2}} \frac{3 + (20)(1 - 0.527)^{2/3}}{3 + (20)(1 - 0.629)^{2/3}}$$

$$= 242 \ \mu\text{cal/cm s K}$$

$$\text{Error} = \frac{242 - 246}{246} \times 100 = -1.6\%$$

  Missenard-Riedel. First, from Eq. (10-9.7), λ_{L_0}, the thermal conductivity at 0°C, is determined:

$$\lambda_{L_0} = (90 \times 10^{-6})[(349.7)(0.0106)]^{1/2} \frac{31.22}{(153.823)^{1/2}} (5)^{1/4}$$

$$= 292 \ \mu\text{cal/cm s K}$$

Then, with Eq. (10-9.8), and $273/556.4 = 0.491$

$$\lambda_L = 292 \times 10^{-6} \frac{3 + (20)(1 - 0.527)^{2/3}}{3 + (20)(1 - 0.491)^{2/3}} = (292 \times 10^{-6})(0.961)$$

$$= 281 \ \mu\text{cal/cm s K}$$

$$\text{Error} = \frac{281 - 246}{246} \times 100 = 14\%$$

As noted in the text, if the constant were reduced from 90×10^{-6} to about 80×10^{-6}, better agreement would normally be obtained between calculated and experimental values.

10-10 Effect of Temperature on the Thermal Conductivity of Liquids

Except for aqueous solutions, water, and multihydroxy molecules, the thermal conductivity of most liquids decreases with temperature.† Below or near the normal boiling point, the decrease is nearly linear and is often represented over a small temperature range by

$$\lambda_L = \lambda_{L_0}[1 + \alpha(T - T_0)] \qquad (10\text{-}10.1)$$

where λ_{L_0} is the thermal conductivity at temperature T_0 and α is a constant for a given compound which varies between -0.0005 and -0.002 K^{-1}. At temperatures near the melting point, the slope $d\lambda/dT$ may decrease [155], and near the critical point the slope becomes large [145].

Over wider temperature ranges, the correlation suggested by Riedel and given as Eq. (10-9.5) is preferable. Although it is not suitable for water, glycerol, glycols, hydrogen, or helium, Jamieson [65] indicates that it represents well the variation of λ_L with T for a wide range of compounds. Figure 10-16 is a graph of the thermal conductivity of liquid chlorotrifluoromethane (F-13) as a function of temperature from $-120°C$ ($T_r \approx 0.5$) to $0°C$ ($T_r \approx 0.9$) [65]. The curve was drawn from Eq. (10-9.5), and the experimental data (not shown) scatter around this curve. Over reasonable temperature ranges, Eq. (10-10.1) is, however, adequate.

Missenard [118, 119] suggests that if the linear λ_L-vs.-T relation [Eq. (10-10.1)] is used, homologous series have a common convergence point. Djalalian [35] and Puranasamriddhi [138] report that $\lambda_L V_L^n$ is a constant, where n is related to $V_{L_{20}}$, the liquid molal volume at 20°C. This relation, however, is only rough and resembles the argument by Vargaftik [191] that λ_L is proportional to $\rho_L^{4/3}$.

Horrocks and McLaughlin [63, 97] argue that $(d \ln \lambda_L)/dT$ should be related to the coefficient of expansion of the liquid, and from their correlation, approximately,

$$\frac{d \ln \lambda_L}{dT} = -2.75 \left(\frac{d \ln V_L}{dT}\right)_p + 0.0015 \qquad (10\text{-}10.2)$$

A graph of this relation is shown in Fig. 10-17, along with data for

†At very high pressures, $d\lambda/dT$ is often positive even for nonpolar liquids [74].

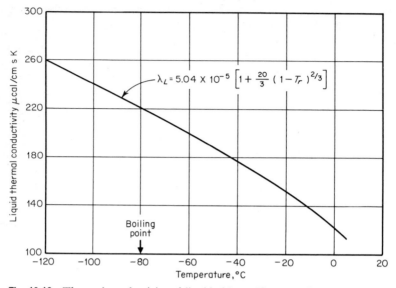

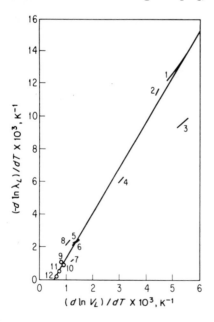

Fig. 10-16   Thermal conductivity of liquid chlorotrifluoromethane.

several compounds.   This correlation indicates that the simple liquids are the more temperature-dependent (they usually have larger values of $[(d \ln V_L)/dT]_p$ and that the particular structure of complex molecules is not of much importance.   For example, as noted by Horrocks and McLaughlin [63], the temperature coefficients of λ_L for

Fig. 10-17   Relation between temperature coefficient of liquid thermal conductivity and thermal coefficient of expansion; $1 =$ carbon monoxide, $2 =$ argon, $3 =$ nitrogen, $4 =$ methane, $5 =$ benzene, $6 =$ cyclohexane, $7 =$ carbon tetrachloride, $8 =$ toluene, $9 =$ diphenyl, $10 = o$-terphenyl, $11 = p$-terphenyl, $12 = m$-terphenyl.   (*From Ref.* 63.)

o-terphenyl and p-terphenyl are almost the same. However, the viscosity temperature coefficients and absolute viscosities are significantly different; for the former the viscosity at the melting point is about 350 mP, while for the latter it is only about 8 mP. Clearly, liquid viscosity is much more sensitive to temperature and molecular structure than thermal conductivity is.

In summary, therefore, the thermal conductivities of liquids usually decrease with temperature; except for highly polar, multihydroxyl and multiamino compounds, increasing T decreases λ_L. The effect is not large, and simple liquids are more temperature-sensitive than complex ones. These comments hold for saturated liquids or subcooled liquids up to pressures around 30 to 40 atm; i.e., over this pressure range the effect of pressure on λ_L is small (except near the critical point, when Fig. 10-13 might be preferable).

10-11 Effect of Pressure on the Thermal Conductivity of Liquids

At moderate pressures, up to 30 to 40 atm, the effect of pressure upon the thermal conductivity of liquids is usually neglected, except near the critical point, where the liquid behaves somewhat like a dense gas (see Sec. 10-5). At lower temperatures, the classical experiments of Bridgman [5] constitute nearly all the available data on the effect of pressure on λ_L. These data show that λ_L increases with pressure.

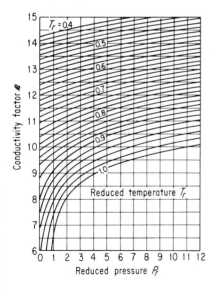

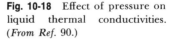

Fig. 10-18 Effect of pressure on liquid thermal conductivities. (*From Ref.* 90.)

A convenient way of estimating the effect of pressure on λ_L is by

$$\frac{\lambda_2}{\lambda_1} = \frac{\ell_2}{\ell_1} \qquad (10\text{-}11.1)$$

where λ_2 and λ_1 refer to liquid thermal conductivities at T and pressures P_2 and P_1 and ℓ_1, ℓ_i are functions of the reduced temperature and pressure, as shown in Fig. 10-18. This correlation was devised by Lenoir and was based primarily on Bridgman's data. Testing with data for 12 liquids, both polar and nonpolar, showed errors of only 2 to 4 percent. The use of Eq. (10-11.1) and Fig. 10-18 is illustrated in Example 10-7 with liquid NO_2, a material *not* used in formulating the correlation.

Example 10-7 Estimate the thermal conductivity of nitrogen dioxide at 311 K and 272 atm. The experimental value quoted is 319 μcal/cm s K [142]. The value of λ_L for saturated liquid at 311 K and 2.09 atm is 296 μcal/cm s K [142].

solution From Appendix A, $T_c = 431.4$ K, $P_c = 100$ atm; thus $T_r = 311/431.4 = 0.721$, $P_{r_1} = 2.09/100 = 0.0209$, $P_{r_2} = 270/100 = 2.72$. From Fig. 10-18, $\ell_2 = 11.75$ *and* $\ell_1 = 11.17$. With Eq. (10-11.1),

$$\lambda\ (272\ \text{atm}) = (296 \times 10^{-6})\frac{11.75}{11.17}$$

$$= 312\ \mu\text{cal/cm s K}$$

$$\text{Error} = \frac{312 - 319}{319} \times 100 = -2.2\%$$

Missenard [119] has proposed a simple correlation for λ_L that extends to much higher pressures. In analytical form

$$\frac{\lambda_L(P_r)}{\lambda_L\ (\text{low pressure})} = 1 + A'P_r^{0.7} \qquad (10\text{-}11.2)$$

$\lambda_L(P_r)$ and λ_L (low pressure) refer to liquid thermal conductivities at high and low, i.e., near saturation, pressure, both at the same reduced temperature. A' is a parameter given in Table 10-11. This correlation is shown in Fig. 10-19.

TABLE 10-11 Values of A' in Eq. (10-11.2)

T_r	P_r					
	1	5	10	50	100	200
0.8	0.036	0.038	0.038	(0.038)	(0.038)	(0.038)
0.7	0.018	0.025	0.027	0.031	0.032	0.032
0.6	0.015	0.020	0.022	0.024	0.025	0.025
0.5	0.012	0.0165	0.017	0.019	0.020	0.020

The correlations of Missenard and Lenoir agree up to a reduced pressure of 12, the maximum value shown for the Lenoir form.

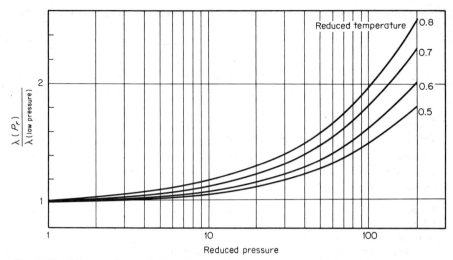

Fig. 10-19 Missenard correlation for liquid thermal conductivities at high pressures. (*From Ref.* 119.)

Example 10-8 Estimate the thermal conductivity of liquid toluene at 6250 atm and 30.8°C. The experimental value at this high pressure is 545 μcal/cm s K [74]. At 1 atm and 30.8°C, $\lambda_L = 307$ μcal/cm s K [74].

solution From Appendix A, $T_c = 591.7$ K and $P_c = 40.6$ atm. Therefore, $T_r = (30.8 + 273.2)/591.7 = 0.514$ and $P_r = 6250/40.6 = 154$. From Table 10-11, $A' = 0.0205$. Then, using Eq. (10-11.2),

$$\lambda_L(P) = (307 \times 10^{-6})[1 + (0.0205)154^{0.7}]$$
$$= 521 \ \mu\text{cal/cm s K}$$
$$\text{Error} = \frac{521 - 545}{545} \times 100 = -4.4\%$$

10-12 Thermal Conductivity of Liquid Mixtures

The thermal conductivities of most mixtures of organic liquids tend to be less than those predicted with a mole- (or weight-) fraction average.† $\lambda_{L_{\text{mix}}}$ values for four typical systems are shown in Fig. 10-20. A review of available data has been published by the National Engineering Laboratory [68], and other, more recent papers presenting new measurements are available in Refs. 66, 67, and 171.

Many correlation techniques have been suggested [37, 78, 96, 97, 137, 139, 156, 193, 204]. Of all the proposed estimation techniques, five are presented here. Each has been extensively tested and shown to be reliable for most mixtures.

†The system aniline–acetic acid may be an exception [42], suggesting that thermal conductivities of other mixtures containing one basic component and one acidic component may also be exceptions.

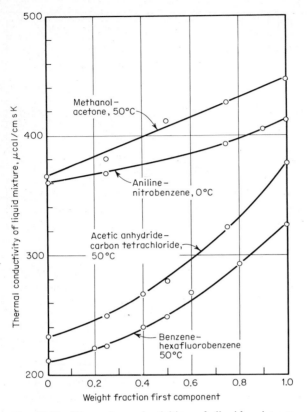

Fig. 10-20 Thermal conductivities of liquid mixtures. (*Data from Ref. 67.*)

NEL Equation

$$\frac{\lambda_m - \lambda_1}{\lambda_2 - \lambda_1} = \alpha w_2^{3/2} + w_2(1 - \alpha) \qquad (10\text{-}12.1)$$

where $\lambda_2 > \lambda_1$. w_2 is the *weight* fraction of component 2, and α is a mixture constant. Values of this constant are given for 60 systems [66], and the range extends from about 0.1 to over 1.0. A value of zero would lead to the prediction that λ_m was a weight-fraction average. Since α is not usually known, a value of unity is often assumed; with this choice, errors still do not usually exceed 4 percent. This relation cannot be used for multicomponent systems.

Filippov Equation Another empirical relation was proposed by researchers in the Soviet Union [41, 43],

$$\frac{\lambda_m - \lambda_1}{\lambda_2 - \lambda_1} = C w_2^2 + w_2(1 - C) \qquad (10\text{-}12.2)$$

It is similar to the NEL form. w_2 is again the weight fraction of component 2, and C is a mixture constant. If experimental values of C are not available, one normally chooses $C = 0.72$. The accuracy of Eq. (10-12.2) is similar to that of the NEL form. It, too, is unsuitable for multicomponent solutions.

Jordan-Coates Equation Shroff [171, 172], after testing many mixture correlations, recommended one suggested by Jordan [71]:

$$\frac{\lambda_m}{\lambda_1^{w_1}\lambda_2^{w_2}} = [\exp(\mathcal{f}|\lambda_2 - \lambda_1|) - 0.5\mathcal{f}(\lambda_2 + \lambda_1)]^{w_1 w_2} \qquad (10\text{-}12.3)$$

In this relation thermal-conductivity values are given in cal/cm s K and $\mathcal{f}$ is a conversion factor equal to 13.80. w is again the weight fraction. If λ is expressed in Btu/ft h °R, $\mathcal{f} = 1.0$. Equation (10-12.3) is not suitable for multicomponent mixtures.

Power-Law Relation To simulate the type of relation between mixture thermal conductivities and composition and to obtain a relation suitable for multicomponent cases, Vredeveld [198] suggests, for a binary,

$$\lambda_m^r = w_1 \lambda_1^r + w_2 \lambda_2^r \qquad (10\text{-}12.4)$$

where w is the weight fraction. The parameter r can be chosen to fit the data best. If it is unity, λ_m is a weight-fraction average. As r decreases below unity, λ_m becomes progressively less than this linear average. The most appropriate value of r appears to depend upon the ratio λ_2/λ_1, where $\lambda_2 > \lambda_1$. In most systems $1 \leqslant \lambda_2/\lambda_1 \leqslant 2$ and in this range, a value of $r = -2$ provides a reasonable agreement between predicted and estimated thermal conductivities. Equation (10-12.4) can readily be generalized for multicomponent mixtures by

$$\lambda_m^r = \sum_{j=1}^{n} w_j \lambda_j^r \qquad (10\text{-}12.5)$$

Li Equation Li [92a] proposed that

$$\lambda_m = \sum_i \sum_j \phi_i \phi_j \lambda_{ij} \qquad (10\text{-}12.6)$$

where

$$\lambda_{ij} = 2(\lambda_i^{-1} + \lambda_j^{-1})^{-1} \qquad (10\text{-}12.7)$$

$$\phi_i = \frac{x_i V_i}{\sum_j x_j V_j} \qquad (10\text{-}12.8)$$

x_i is the mole fraction of component i and ϕ_i is the volume fraction of i. V_i is the molar volume of pure liquid i. For a binary system of 1 and

2, Eq. (10-12.6) becomes

$$\lambda_m = \phi_1^2\lambda_1 + 2\phi_1\phi_2\lambda_{12} + \phi_2^2\lambda_2 \tag{10-12.9}$$

The harmonic mean approximation for λ_{ij} was chosen over a geometric or arithmetic mean after extensive testing and comparison of calculated and experimental values of λ_m. Also, it was found that the V_i terms in Eq. (10-12.8) could be replaced by critical volumes for nonaqueous liquid systems without affecting the results significantly.

Discussion Li [92a] compared the accuracy of his method with the NEL and Filippov equations for a large number of binary liquid mixtures, both aqueous and nonaqueous. Errors for all three methods rarely exceeded 3–4% unless the system exhibited a minimum in λ_m with composition. None of the estimation methods will predict such behavior. Less extensive testing by the present authors indicates that the Jordan-Coates and Power-Law relations are of comparable accuracy to the other three. All methods are illustrated in Example 10-9.

Example 10-9 Estimate the thermal conductivity of a liquid mixture of methanol and benzene at 0°C with a weight fraction methanol of 0.4. At this temperature, the thermal conductivities of pure benzene and methanol are 364 and 501 μcal/cm s K [66], respectively. Compare the estimated value with that determined experimentally, 405 μcal/cm s K.

solution *NEL Equation.* Since λ (methanol) $> \lambda$ (benzene), methanol is designated as component 2. With Eq. (10-12.1)

$$\frac{\lambda_m - 364 \times 10^{-6}}{501 \times 10^{-6} - 364 \times 10^{-6}} = \alpha(0.4)^{3/2} + (0.4)(1 - \alpha)$$

If $\alpha = 1$, $\lambda_m = 399$ μcal/cm s K

$$\text{Error} = \frac{399 - 405}{405} \times 100 = -1.5\%$$

If, as suggested by Jamieson and Hastings [66] for this system, $\alpha = 0.49$,

$$\lambda_m = 409 \ \mu\text{cal/cm s K}$$

$$\text{Error} = \frac{409 - 405}{405} \times 100 = 1.0\%$$

Filippov Equation. With Eq. (10-12.2) with $C = 0.72$,

$$\frac{\lambda_m - 364 \times 10^{-6}}{501 \times 10^{-6} - 364 \times 10^{-6}} = (0.72)(0.4)^2 + (0.4)(1 - 0.72)$$

$$\lambda_m = 395 \ \mu\text{cal/cm s K}$$

$$\text{Error} = \frac{395 - 405}{405} \times 100 = -2.5\%$$

Jordan-Coates Equation. Equation (10-12.3), with $w_1 = 0.6$, $w_2 = 0.4$

$$\frac{\lambda_m}{(364 \times 10^{-6})^{0.6}(501 \times 10^{-6})^{0.4}}$$

$$= \{\exp[(13.80)(501 \times 10^{-6} - 364 \times 10^{-6})]$$

$$- (0.5)(13.80)(501 \times 10^{-6} + 364 \times 10^{-6})\}^{(0.6)(0.4)}$$

$$= 0.9959$$

$$\lambda_m = 412 \ \mu\text{cal/cm s K}$$

$$\text{Error} = \frac{412 - 405}{405} \times 100 = 1.7\%$$

Power-Law Equation.   Employing Eq. (10-12.4) with $r = -2$ and with compositions given in weight fractions,

$$\lambda_m^{-2} = (0.6)(364 \times 10^{-6})^{-2} + (0.4)(501 \times 10^{-6})^{-2}$$

$$\lambda_m = 404 \ \mu\text{cal/cm s K}$$

$$\text{Error} = \frac{404 - 405}{405} \times 100 = -0.2\%$$

Li Equation.   At 0°C, V (methyl alcohol) = 39.6 cm³/g-mol and V (benzene) = 88.9 cm³/g-mol.   If the weight fraction of methanol is 0.4, the mole fraction is 0.619 and, from Eq. (10-12.8), $\phi = 0.420$.   The interaction thermal conductivity λ_{12} is, from Eq. (10-12.7),

$$\lambda_{12} = 2[(501)^{-1} + (364)^{-1}]^{-1} = 422 \ \mu\text{cal/cm s K}$$

Then, from Eq. (10-12.9),

$$\lambda_m = (0.420)^2(501) + (2)(0.420)(0.580)(422)$$

$$+ (0.580)^2(364) = 416 \ \mu\text{cal/cm s K}$$

$$\text{Error} = \frac{416 - 405}{405} \times 100 = 2.7\%$$

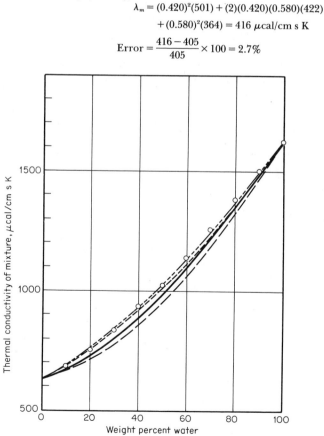

Fig. 10-21   Thermal conductivities of water–ethylene glycol mixtures at 100°C.   — NEL equation (10-12.1), -- Filippov equation (10-12.2), --- Jordan-Coates equation (10-12.3), — --— Li equation (10-12.6).   (*Experimental data from Ref. 69.*)

The five methods illustrated in Example 10-9 are also applicable to water-organic systems. For example, Fig. 10-21 shows experimental and estimated values of λ_{L_m} for the water–ethylene glycol system at 100°C. The power-law estimation is not shown since the λ_2/λ_1 ratio exceeds the range of applicability for which $r = -2$. For this case, where $\lambda_2/\lambda_1 \approx 2.6$, a value of $r \approx 0$ is much more satisfactory. In this special case,

$$\lambda_m = \lambda_1 \left(\frac{\lambda_2}{\lambda_1}\right)^{w_2} \tag{10-12.10}$$

and Eq. (10-12.10) yields values almost exactly the same as those from the Jordan-Coates method. This results since the right-hand side of Eq. (10-12.3) is close to unity, and when this is the case, the Jordan-Coates relation reduces to Eq. (10-12.10), that is, a log-mean average is obtained.

Although all liquid mixture thermal conductivity equations are of comparable accuracy, the Li equation is the only one which may be used for multicomponent mixtures with all terms uniquely determined.

For dilute *ionic* solutions, the mixture thermal conductivity usually decreases with an increase in the concentration of the dissolved salts. To estimate the thermal conductivity of such mixtures, Jamieson and Tudhope [69] recommend the use of an equation proposed

TABLE 10-12 Values of σ_i for Anions and Cations in Eq. (10-12.11) [69]

Anion	$\sigma_i \times 10^5$	Cation	$\sigma_i \times 10^5$
OH^-	20.934	H^+	−9.071
F^-	2.0934	Li^+	−3.489
Cl^-	−5.466	Na^+	0.0000
Br^-	−17.445	K^+	−7.560
I^-	−27.447	NH_4^+	−11.63
NO_2^-	−4.652	Mg^{2+}	−9.304
NO_3^-	−6.978	Ca^{2+}	−0.5815
ClO_3^-	−14.189	Sr^{2+}	−3.954
ClO_4^-	−17.445	Ba^{2+}	−7.676
BrO_3^-	−14.189	Ag^+	−10.47
CO_3^{2-}	−7.560	Cu^{2+}	−16.28
SiO_3^{2-}	−9.300	Zn^{2+}	−16.28
SO_3^{2-}	−2.326	Pb^{2+}	−9.304
SO_4^{2-}	1.163	Co^{2+}	−11.63
$S_2O_3^{2-}$	8.141	Al^{3+}	−32.56
CrO_4^{2-}	−1.163	Th^{4+}	−43.61
$Cr_2O_7^{2-}$	15.93		
PO_4^{3-}	−20.93		
$Fe(CN)_6^{4-}$	18.61		
Acetate$^-$	−22.91		
Oxalate^{2-}	−3.489		

originally by Riedel [145] and tested by Vargaftik and Os'minin [192]:

$$\lambda_m \,(20°C) = \lambda_{H_2O} \,(20°C) + \frac{1}{4.186} \sum \sigma_i c_i \qquad (10\text{-}12.11)$$

where λ_m = thermal conductivity of ionic solution at 20°C, cal/cm s K
λ_{H_2O} = thermal conductivity of water at 20°C, cal/cm s K
c_i = concentration of the electrolyte, g-mol/l
σ_i = coefficient that is characteristic of each ion

Values of σ_i are shown in Table 10-12. To obtain λ_m at other temperatures,

$$\lambda_m(T) = \lambda_m \,(20°C) \frac{\lambda_{H_2O}(T)}{\lambda_{H_2O}\,(20°C)} \qquad (10\text{-}12.12)$$

Except for strong acids and alkalies at high concentrations, Eqs. (10-12.11) and (10-12.12) are usually accurate to within ±5 percent.

NOTATION

A = constant in Eq. (10-9.5)
A' = constant in Eq. (10-11.2)
A_{ij} = parameter in gas-thermal-conductivity mixture equations, Eqs. (10-6.1), (10-6.2)
B = percent increase in gas thermal conductivity per atmosphere change in pressure
c = concentration
C_v = heat capacity at constant volume, cal/g-mol K; C_p, at constant pressure; C_{tr}, for translational energy; C_{rot}, for external rotational energy; C_{int}, for all modes of internal energy; C_{vib}, for vibrational energy; C_{ir}, for internal rotational energy
C = parameter in Eq. (10-12.2)
C_s = constant in Eq. (10-6.8)
$\mathscr{C}$ = Roy-Thodos structural constant in Eq. (10-3.22)
$\mathscr{D}$ = self-diffusion coefficient, cm²/s
ℓ = parameter in Eq. (10-11.1) and Fig. 10-18
f_{tr}, f_{int} = Eucken weighting factors for Eq. (10-3.1)
F_r = degrees of freedom for external rotation; F_{ir}, for internal rotation
H = parameter in Eq. (10-9.1)
ΔH_{vb} = enthalpy of vaporization at the normal boiling point, cal/g-mol
L = mean free path, cm
m = mass of molecule, g
M = molecular weight
n = number density of molecules, cm⁻³; parameter in Eq. (10-4.1)
N = parameter in Eq. (10-9.1)
$\mathscr{N}$ = number of atoms in a molecule
N_{Pr} = Prandtl number $C_p \eta / \lambda M$
P = pressure, atm; P_c, critical pressure; P_r, reduced pressure P/P_c
q = Brokaw parameter in Eq. (10-6.16)
r = exponent in Eq. (10-12.4)
R = gas constant
S = Sutherland constant in Eq. (10-6.6)
ΔS_{vb} = entropy of vaporization at the normal boiling point, cal/g-mol K
ΔS^* = modified entropy of vaporization defined by Eq. (10-9.2)

T = temperature, K; T_b, at normal boiling point; T_c, critical temperature; T_r, reduced temperature, T/T_c

v = average velocity of molecules, cm/s

V = volume, cm³/g-mol; V_L, as a liquid; V_b, at normal boiling point; V_c, critical volume

w = weight fraction

x = mole fraction, usually liquid

y = mole fraction, usually gas

Z_c = critical compressibility factor

Z_{rot} = collision number

Greek

α = polar parameter in Eqs. (10-3.9) and (10-3.10); parameter in Eq. (10-12.1); $d\lambda_L/dT$ for linear form, Eq. (10-10.1)

β = parameter in Eq. (10-3.9)

$\gamma = C_p/C_v$

$\Gamma = T_c^{1/6} M^{1/2}/P_c^{2/3}$

η = viscosity, P

λ = thermal conductivity, cal (or μcal)/cm s K; λ_c, at critical temperature, but usually at low pressure; λ_L, for liquid state; λ_m, for a mixture

ρ = density, usually g-mol/cm³; ρ_b, at normal boiling point; ρ_c, at critical point; ρ_r, ρ/ρ_c; ρ_L, for liquid

σ = characteristic molecular diameter, Å; parameter in Eq. (10-12.11)

ϕ_i = volume fraction

ϕ_{ij} = viscosity parameter for Eq. (9-5.5)

ψ = parameter in Eq. (10-3.9)

Ω_v = collision integral for viscosity and thermal conductivity; Ω_D, for diffusion coefficient

Superscript

° = low-pressure state where material behaves as an ideal gas

Subscripts

i, j = for components i, j

ij = for mixture parameter characteristic of an i-j interaction

b = at normal boiling point

g = gas

L = liquid

m = mixture

0 = 0°C

REFERENCES

1. American Petroleum Institute: "Selected Values of Physical and Thermodynamic Properties of Hydrocarbons and Related Compounds," project 44, Carnegie Press, Pittsburgh, Pa., 1953, and supplements.
2. Barua, A. K., A. Manna, and P. Mukhopadhyay: *J. Chem. Phys.*, **49**: 2422 (1968).
3. Bennett, L. A., and R. G. Vines: *J. Chem. Phys.*, **23**: 1587 (1955).
4. Bretsznajder, S.: "Prediction of Transport and Other Physical Properties of Fluids," p. 251, trans. J. Bandrowski, Pergamon, New York, 1971.
5. Bridgman, P. W.: *Proc. Am. Acad. Arts Sci.*, **59**: 154 (1923).
6. Brokaw, R. S.: *Ind. Eng. Chem.*, **47**: 2398 (1955).

7. Brokaw, R. S.: *J. Chem. Phys.*, **29**: 391 (1958).

8. Brokaw, R. S.: *J. Chem. Phys.*, **31**: 571 (1959).

9. Brokaw, R. S.: Alignment Charts for Transport Properties, Viscosity, Thermal Conductivity, and Diffusion Coefficients for Nonpolar Gases and Gas Mixtures at Low Density, *NASA Tech. Rep.* R-81, Lewis Research Center, Cleveland, Ohio, 1961.

10. Brokaw, R. S.: Approximate Formulas for Viscosity and Thermal Conductivity of Gas Mixtures, *NASA Tech. Note* D-2502, Lewis Research Center, Cleveland, Ohio, 1964.

11. Brokaw, R. S.: *Int. J. Eng. Sci.*, **3**: 251 (1965).

12. Brokaw, R. S.: *Ind. Eng. Chem. Process Des. Dev.*, **8**: 240 (1969).

13. Brokaw, R. S., and C. O'Neal, Jr.: Rotational Relaxation and the Relation between Thermal Conductivity and Viscosity for Some Nonpolar Polyatomic Gases, *9th Int. Symp. Combust.*, p. 725, Academic, New York, 1963.

14. Bromley, L. A.: Thermal Conductivity of Gases at Moderate Pressures, *Univ. California Rad. Lab.* UCRL-1852, Berkeley, Calif., June 1952.

15. Brykov, V. P., G. Kh. Mukhamedzyanov, and A. G. Usmanov: *J. Eng. Phys. USSR*, **18**(1): 62 (1970).

16. Burgoyne, J. H., and F. Weinberg: *4th Symp. Combust.*, p. 294, Williams & Wilkins, Baltimore, 1953.

17. Carmichael, L. T., and B. H. Sage: *J. Chem. Eng. Data*, **9**: 511 (1964).

18. Carmichael, L. T., V. Berry, and B. H. Sage: *J. Chem. Eng. Data*, **8**: 281 (1963).

19. Carmichael, L. T., H. H. Reamer, and B. H. Sage: *J. Chem. Eng. Data*, **11**: 52 (1966).

20. Challoner, A. R., and R. W. Powell: *Proc. R. Soc. Lond.*, **A238**: 90 (1956).

21. Chang, H.-Y.: *Chem. Eng.*, **80**(9): 122 (1973).

22. Chapman, S., and T. G. Cowling: "The Mathematical Theory of Non-uniform Gases," Cambridge University Press, New York, 1961.

23. Cheung, H.: Thermal Conductivity and Viscosity of Gas Mixtures, *Univ. California Rad. Lab.* UCRL-8230, Berkeley, Calif., April 1958.

24. Cheung, H., L. A. Bromley, and C. R. Wilke: Thermal Conductivity and Viscosity of Gas Mixtures, *Univ. California Rad. Lab.* UCRL-8230 rev., Berkeley, Calif., April 1959.

25. Cheung, H., L. A. Bromley, and C. R. Wilke: *AIChE J.*, **8**: 221 (1962).

26. Comings, E. W., and B. J. Mayland: *Chem. Metall. Eng.*, **52**(3): 115 (1945).

27. Comings, E. W., and M. F. Nathan: *Ind. Eng. Chem.*, **39**: 964 (1947).

28. Correla, F. von, B. Schramm, and K. Schaefer: *Ber. Bunsenges. Phys. Chem.*, **72**(3): 393 (1968).

29. Cottrell, T. L., and J. C. McCoubrey: "Molecular Energy Transfer in Gases," Butterworth, London, 1961.

30. Couch, E. J., and K. A. Dobe: *J. Chem. Eng. Data*, **6**: 229 (1961).

31. Cowling, T. G.: *Proc. R. Soc. Lond.*, **A263**: 186 (1961).

32. Cowling, T. G.: *Br. J. Appl. Phys.*, **15**: 959 (1964).

33. Cowling, T. G., P. Gray, and P. G. Wright: *Proc. R. Soc. Lond.*, **A276**: 69 (1963).

34. Devoto, R. S.: *Physica*, **45**: 500 (1970).

35. Djalalian, W. H.: *Kaeltechnik*, **18**(11): 410 (1966).

36. Dognin, A.: *C. R.*, **243**: 840 (1956).

37. Dul'nev, G. N., and Y. P. Zarichnyak: "Thermophysical Properties of Substances and Materials," 3d issue, p. 103, Standards Publications, Moscow, 1971.

38. Dul'nev, G. N., Y. P. Zarichnyak, and B. L. Muratova: *J. Eng. Phys. USSR*, **18**(5): 589 (1970).

39. Ehya, H., F. M. Faubert, and G. S. Springer: *J. Heat Transfer*, **94**: 262 (1972).

40. Ewing, C. T., B. E. Walker, J. A. Grand, and R. R. Miller: *Chem. Eng. Prog. Symp. Ser.*, **53**(20): 19 (1957).

41. Filippov, L. P.: *Vest. Mosk. Univ., Ser. Fiz. Mat. Estestv. Nauk*, (8)**10**(5): 67-69 (1955); *Chem. Abstr.*, **50**: 8276 (1956).

42. Filippov, L. P.: *Int. J. Heat Mass Transfer*, **11**: 331 (1968).

43. Filippov, L. P., and N. S. Novoselova: *Vestn. Mosk. Univ., Ser. Fiz. Mat. Estestv. Nauk,* (3)**10**(2): 37–40 (1955); *Chem. Abstr.,* **49**: 11366 (1955).

44. Franck, E. U.: *Chem. Ing. Tech.,* **25**: 238 (1953).

45. Gambill, W. R.: *Chem. Eng.,* **64**(4): 277 (1957).

46. Gambill, W. R.: *Chem. Eng.,* **66**(16): 129 (1959).

47. Gamson, B. W.: *Chem. Eng. Prog.,* **45**: 154 (1949).

48. Gandhi, J. M., and S. C. Saxena: *Proc. Phys. Soc. Lond.,* **87**: 273 (1966).

49. Gilmore, T. F., and E. W. Comings: *AIChE J.,* **12**: 1172 (1966).

50. Gray, P., S. Holland, and A. O. S. Maczek: *Trans. Faraday Soc.,* **65**: 1032 (1969).

51. Gray, P., S. Holland, and A. O. S. Maczek: *Trans. Faraday Soc.,* **66**: 107 (1970).

52. Gray, P., and P. G. Wright: *Proc. R. Soc. Lond.,* **A263**: 161 (1961).

53. Gray, P., and P. G. Wright: *Proc. R. Soc. Lond.,* **A267**: 408 (1962).

54. Green, H. S.: "The Molecular Theory of Fluids," p. 135, Interscience, New York, 1952.

55. Grilly, E. R.: *Am. J. Phys.,* **20**: 447 (1952).

56. Groenier, W. S., and G. Thodos: *J. Chem. Eng. Data,* **6**: 240 (1961).

57. Guildner, L. A.: *Proc. Natl. Acad. Sci.,* **44**: 1149 (1958).

58. Gutweiler, J., and C. J. Raw: *J. Chem. Phys.,* **48**: 2413 (1968).

59. Healy, R. N., and T. S. Storvick: *J. Chem. Phys.,* **50**: 1419 (1969).

60. Hirschfelder, J. O.: *J. Chem. Phys.,* **26**: 282 (1957).

61. Hirschfelder, J. O., C. F. Curtiss, and R. B. Bird: "Molecular Theory of Gases and Liquids," Wiley, New York, 1954.

62. Ho, C. Y., R. W. Powell, and P. E. Liley: *J. Phys. Chem. Ref. Data,* **1**: 279 (1972).

63. Horrocks, J. K., and E. McLaughlin: *Trans. Faraday Soc.,* **59**: 1709 (1963).

64. Ikenberry, L. D., and S. A. Rice: *J. Chem. Phys.,* **39**: 1561 (1963).

65. Jamieson, D. T.: National Engineering Laboratories, East Kilbride, Glasgow: private communication, March 1971.

66. Jamieson, D. T., and E. H. Hastings: p. 631 in C. Y. Ho and R. E. Taylor (eds.), *Proc. 8th Conf. Thermal Conductivity,* Plenum, New York, 1969.

67. Jamieson, D. T., and J. B. Irving: paper presented at *13th Int. Thermal Conductivity Conf., Univ. Missouri, 1973.*

68. Jamieson, D. T., J. B. Irving, and J. S. Tudhope: "Liquid Thermal Conductivity: A Data Survey to 1973," H.M. Stationery Office, Edinburgh, 1975.

69. Jamieson, D. T., and J. S. Tudhope: *Nat. Eng. Lab. Glasgow, Rep.* 137, March 1964.

70. Johnson, D. W., and C. P. Colver: *Hydrocarbon Process.,* **48**(3): 113 (1969).

71. Jordan, H. B.: M.S. thesis, Louisiana State University, Baton Rouge, 1961.

72. Junk, W. A., and E. W. Comings: *Chem. Eng. Prog.,* **49**: 263 (1953).

73. Kamal, I., and E. McLaughlin: *Trans. Faraday Soc.,* **60**: 809 (1964).

74. Kandiyoti, R., E. McLaughlin, and J. F. T. Pittman: *Chem. Soc. Lond. Faraday Trans.,* **69**: 1953 (1973).

75. Kanitkar, D., and G. Thodos: *Can. J. Chem. Eng.,* **47**: 427 (1969).

76. Kannuliuk, W. G., and H. B. Donald: *Aust. J. Sci. Res.,* **3A**: 417 (1950).

77. Kennedy, J. T., and G. Thodos: *AIChE J.,* **7**: 625 (1961).

78. Kerr, C. P., and J. Coates: *67th Natl. AIChE Meet., Atlanta, Ga.,* Feb. 15–18, 1970, pap. 55c.

79. Kestin, J., S.T. Ro, and W. Wakeham: *Physica,* **58**: 165 (1972).

80. Keyes, F. G.: *Trans. ASME,* **73**: 597 (1951).

81. Keyes, F. G.: *Trans. ASME,* **76**: 809 (1954).

82. Keyes, F. G.: *Trans. ASME,* **77**: 1395 (1955).

83. Kim, S. K., G. P. Flynn, and J. Ross: *J. Chem. Phys.,* **43**: 4166 (1965).

84. Kramer, F. R., and E. W. Comings: *J. Chem. Eng. Data,* **5**: 462 (1960).

85. Lambert, J. D.: in D. R. Bates (ed.), "Atomic and Molecular Processes," Academic, New York, 1962.

86. Lambert, J. D., K. J. Cotton, M. W. Pailthorpe, A. M. Robinson, J. Scrivins, W. R. F. Vale, and R. M. Young: *Proc. R. Soc. Lond.*, **A231**: 280 (1955).
87. "Landolt-Börnstein Tables," vol. 4, pt. 1, Springer-Verlag, Berlin, 1955.
88. Lehmann, H.: *Chem. Technol.*, **9**: 530 (1957).
89. Leng, D. E., and E. W. Comings: *Ind. Eng. Chem.*, **49**: 2042 (1957).
90. Lenoir, J. M.: *Pet. Refiner*, **36**(8): 162 (1957).
91. Lenoir, J. M., and E. W. Comings: *Chem. Eng. Prog.*, **47**: 223 (1951).
92. Lenoir, J. M., W. A. Junk, and E. W. Comings: *Chem. Eng. Prog.*, **49**: 539 (1953).
92a. Li, C. C.: *AIChE J.*, in press, 1976.
93. Liley, P. E.: *Symp. Thermal Prop., Purdue Univ. Lafayette, Ind., Feb. 23–26, 1959*, p. 40.
94. Liley, P. E.: p. 3-216 in R. H. Perry and C. H. Chilton (eds.), "Chemical Engineers' Handbook," 5th ed., McGraw-Hill, New York, 1973.
95. Lindsay, A. L., and L. A. Bromley: *Ind. Eng. Chem.*, **42**: 1508 (1950).
96. Losenicky, Z.: *J. Phys. Chem.*, **72**: 4308 (1968).
97. McLaughlin, E.: *Chem. Rev.*, **64**: 389 (1964).
98. McLaughlin, E.: *Proc. 7th Conf. Thermal Conductivity*, p. 671, *NBS Spec. Publ.* 302, November 1967.
99. Maczek, A. O. S., and P. Gray: *Trans. Faraday Soc.*, **65**: 1473 (1969).
100. Maczek, A. O. S., and P. Gray: *Trans. Faraday Soc.*, **66**: 127 (1970).
101. Maejima, T.: private communication, 1973. Equation (10-9.3) was suggested by Professor K. Sato, of the Tokyo Institute of Technology.
102. Mani, N., and J. E. S. Venart: *Adv. Cryog. Eng.*, **18**: 280 (1973).
103. Martin, C. N. B.: National Engineering Laboratory, East Kilbride, Glasgow: private communication, December 1970.
104. Mason, E. A.: *J. Chem. Phys.*, **28**: 1000 (1958).
105. Mason, E. A., and L. Monchick: *J. Chem. Phys.*, **36**: 1622 (1962).
106. Mason, E. A., and L. Monchick: Theory of Transport Properties of Gases, *9th Int. Symp. Combust.*, Academic, New York, 1963.
107. Mason, E. A., and S. C. Saxena: *Phys. Fluids*, **1**: 361 (1958).
108. Mason, E. A., and S. C. Saxena: *J. Chem. Phys.*, **31**: 511 (1959).
109. Mason, E. A., and H. von Ubisch: *Phys. Fluids*, **3**: 355 (1960).
110. Mather, S., and S. C. Saxena: *Proc. Phys. Soc. Lond.*, **89**: 753 (1966).
111. Mather, S., P. K. Tondin, and S. C. Saxena: *J. Phys. Soc. Jpn.*, **25**: 530 (1968).
112. Michels, A., and A. Botzen: *Physica*, **19**: 585 (1953).
113. Misic, D., and G. Thodos: *AIChE J.*, **7**: 264 (1961).
114. Misic, D., and G. Thodos: *J. Chem. Eng. Data*, **9**: 540 (1963).
115. Misic, D., and G. Thodos: *AIChE J.*, **11**: 650 (1965).
116. Missenard, A.: *C. R.*, **260**(5): 5521 (1965).
117. Missenard, A.: Conductivité thermique des solides, liquides, gaz et de leurs mélanges, Éditions Eyrolles, Paris, 1965.
118. Missenard, A.: *C. R.*, **270**(1): 381 (1970).
119. Missenard, A.: *Rev. Gen. Thermodyn.*, **101**(5): 649 (1970).
120. Missenard, A.: *Rev. Gen. Thermodyn.*, **11**: 9 (1972).
120a. Mo, K. C., and K. E. Gubbins: *Chem. Eng. Comm.*, **1**: 281 (1974).
121. Monchick, L., A. N. G. Pereira, and E. A. Mason: *J. Chem. Phys.*, **42**: 3241 (1965).
122. Muckenfuss, C., and C. F. Curtiss: *J. Chem. Phys.*, **29**: 1273 (1958).
123. Novotny, J. L., and T. F. Irvine, Jr.: *J. Heat Transfer*, **83**: 125 (1961).
124. O'Neal, C., Jr., and R. S. Brokaw: *Phys. Fluids*, **5**: 567 (1962).
125. O'Neal, C., Jr., and R. S. Brokaw: *Phys. Fluids*, **6**: 1675 (1963).
126. Owens, E. J., and G. Thodos: *AIChE J.*, **3**: 454 (1957).
127. Owens, E. J., and G. Thodos: *Proc. Joint Conf. Thermodyn. Transport Prop. Fluids, London, July 1957*, pp. 163–168, Institution of Mechanical Engineers, London, 1958.
128. Owens, E. J., and G. Thodos: *AIChE J.*, **6**: 676 (1960).

129. Pachaiyappan, V.: *Br. Chem. Eng.*, **16**(4/5): 382 (1971).
130. Pachaiyappan, V., S. H. Ibrahim, and N. R. Kuloor: *J. Chem. Eng. Data*, **11**: 73 (1966).
131. Pachaiyappan, V., S. H. Ibrahim, and N. R. Kuloor: *Chem. Eng.*, **74**(4): 140 (1967).
132. Palmer, G.: *Ind. Eng. Chem.*, **40**: 89 (1948).
133. Partington, J.: "An Advanced Treatise on Physical Chemistry," vol. 1, "Fundamental Principles: The Properties of Gases," Longmans, Green, New York, 1949.
134. Pereira, A. N. G., and C. J. G. Raw: *Phys. Fluids*, **6**: 1091 (1963).
135. Peterson, J. N., T. F. Hahn, and E. W. Comings: *AIChE J.*, **17**: 289 (1971).
136. Poltz, H., and R. Jugel: *Int. J. Heat Mass Transfer*, **10**: 1075 (1967).
137. Preston, G. T., T. W. Chapman, and J. M. Prausnitz: *Cryogenics*, **7**(5): 274 (1967).
138. Puranasamriddhi, D.: *Kaeltechnik*, **18**(12): 445 (1966).
139. Rabenovish, B. A.: "Thermophysical Properties of Substances and Materials," 3d ed., Standard, Moscow, 1971.
140. Rahalkar, A. K., D. S. Viswanath, and N. R. Kuloor: *Indian J. Technol.*, **7**(3): 67 (1969).
141. Reid, R. C., and T. K. Sherwood: "Properties of Gases and Liquids," 2d ed., p. 498, McGraw-Hill, New York, 1966.
142. Richter, G. N., and B. H. Sage: *J. Chem. Eng. Data*, **2**: 61 (1957).
143. Richter, G. N., and B. H. Sage: *J. Chem. Eng. Data*, **8**: 221 (1963).
144. Richter, G. N., and B. H. Sage: *J. Chem. Eng. Data*, **9**: 75 (1964).
145. Riedel, L.: *Chem. Ing. Tech.*, **21**: 349 (1949), **23**: 59, 321, 465 (1951).
146. Robbins, L. A., and C. L. Kingrea: (*a*) *Hydrocarbon Proc. Pet. Refiner*, **41**(5): 133 (1962); (*b*) preprint of paper presented at the *Sess. Chem. Eng. 27th Midyear Meet. Am. Pet. Inst., Div. Refining, San Francisco, May 14, 1962.*
147. Rosenbaum, B. M., and G. Thodos: *Physica*, **37**: 442 (1967).
148. Rosenbaum, B. M., and G. Thodos: *J. Chem. Phys.*, **51**: 1361 (1969).
149. Roy, D.: M.S. thesis, Northwestern University, Evanston, Ill., 1967.
150. Roy, D., and G. Thodos: *Ind. Eng. Chem. Fundam.*, 7: 529 (1968).
151. Roy, D., and G. Thodos: *Ind. Eng. Chem. Fundam.*, **9**: 71 (1970).
152. Sakiadis, B. C., and J. Coates: "A Literature Survey of the Thermal Conductivity of Liquids," *Louisiana State Univ., Eng. Exp. Stn. Bull.* 34, 1952.
153. Sakiadis, B. C., and J. Coates: *Louisiana State Univ., Eng. Exp. Stn. Bull.* 46, 1954.
154. Sakiadis, B. C., and J. Coates: *AIChE J.*, **1**: 275 (1955).
155. Sakiadis, B. C., and J. Coates: *AIChE J.*, **3**: 121 (1957).
156. Saksena, M. P., and Harminder: *Ind. Eng. Chem. Fundam.*, **13**: 245 (1974).
157. Saksena, M. P., and S. C. Saxena: *Appl. Sci. Res.*, **17**: 326 (1966).
158. Sandler, S. I.: *Phys. Fluids*, **11**: 2549 (1968).
159. Saxena, S. C., and J. P. Agrawal: *J. Chem. Phys.*, **35**: 2107 (1961).
160. Saxena, S. C., and G. P. Gupta: *Univ. Illinois at Chicago Circle, Dept. Energy Eng., Rep.* TR-E-23, January 1969.
161. Saxena, S. C., and G. P. Gupta: *J. Chem. Eng. Data*, **15**: 98 (1970).
162. Saxena, S. C., M. P. Saksena, and R. S. Gambhir: *Br. J. Appl. Phys.*, **15**: 843 (1964).
163. Saxena, S. C., M. P. Saksena, R. S. Gambhir, and J. M. Gandi: *Physica*, **31**: 333 (1965).
164. Schaefer, C. A., and G. Thodos: *Ind. Eng. Chem.*, **50**: 1585 (1958).
165. Schaefer, C. A., and G. Thodos: *AIChE J.*, **5**: 367 (1959).
166. Schäfer, K.: *Z. Phys. Chem.*, **B53**: 149 (1943).
167. Scheffy, W. J.: Thermal Conduction in Liquids, *Princeton Univ. Project Squid Tech. Rep.* P.R.-85-R, Princeton, N.J., October 1958.
168. Scheffy, W. J., and E. F. Johnson: "Thermal Conductivities of Liquids at High Temperatures," paper presented at *Ann. Meet. Am. Inst. Chem. Eng., St. Paul, Minn., September 1959.*
169. Schramm, V. B., and K. Schaefer: *Ber. Bunsenges. Phys. Chem.*, **69**(2): 110 (1965).
170. Sengers, J. V., and P. H. Keyes: *Phys. Rev. Lett.*, **26**(2): 70 (1971).
171. Shroff, G. H.: Ph.D. thesis, University of New Brunswick, Fredericton, 1968.

172. Shroff, G. H.: p. 643 in C. Y. Ho and R. E. Taylor (eds.), *Proc. 8th. Conf. Thermal Conductivity, Purdue Univ., Oct. 7–10, 1968,* Plenum, New York, 1969.
173. Smith, J. F.: *Ind. Eng. Chem.,* **22**: 1246 (1930).
174. Smith, J. F.: *Trans. ASME,* **58**: 719 (1936).
175. Smith, W. J. S., L. D. Durbin, and R. Kobayashi: *J. Chem. Eng. Data,* **5**: 316 (1960).
176. Srivastava, B. N., and R. C. Srivastava: *J. Chem. Phys.,* **30**: 1200 (1959).
177. Stiel, L. I., and G. Thodos: "Progress in International Research on Thermodynamic and Transport Properties," p. 352, Academic, New York, 1962.
178. Stiel, L. I., and G. Thodos: *AIChE J.,* **10**: 26 (1964).
179. Stolyarov, E. A.: *Zh. Fiz. Khim.,* **24**: 279 (1950); *Chem. Abstr.,* **44**: 6694 (1950).
180. Svehla, R. A.: Estimated Viscosities and Thermal Conductivities of Gases at High Temperatures, *NASA Tech. Rep.* R-132, Lewis Research Center, Cleveland, Ohio, 1962.
181. Taxman, N.: *Phys. Rev.,* **110**: 1235 (1958).
182. Thornton, E.: *Proc. Phys. Soc. Lond.,* **76**: 104 (1960).
183. Tondon, P. K., and S. C. Saxena: *Appl. Sci. Res.,* **19**: 163 (1968).
184. Touloukian, Y. S., and C. Y. Ho (eds.): "Thermal Conductivity," vol. 3, IFI/Plenum Data Corp., New York.
185. Tsederberg, N. V.: "Thermal Conductivity of Gases and Liquids," M.I.T. Press, Cambridge, Mass., 1965.
186. Ubbelohde, A. R.: *J. Chem. Phys.,* **3**: 219 (1935).
187. Ulybin, S. A., V. P. Bugrov, and A. V. Il'in: *Teplofiz. Vys. Temp.,* **4**: 210 (1966).
188. van Dael, W., and H. Cavwenbergh: *Physica,* **40**: 173 (1968).
189. Vanderkool, W. N., D. L. Hildenbrand, and D. R. Stull: *J. Chem. Eng. Data,* **12**: 377 (1967).
190. Vargaftik, N. B.: "Thermal Conductivities of Compressed Gases and Steam at High Pressures," *Izv. Vses. Teplotekh. Inst.,* Nov. 7, 1951; private communication, Professor N. V. Tsederberg, Moscow Energetics Institute.
191. Vargaftik, N. B.: *Proc. Joint Conf. Thermodyn. Transport Prop. Fluids, London, July 1957,* p. 142, Institution of Mechanical Engineers, London, 1958.
192. Vargaftik, N. B., and Y. P. Os'minin: *Teploenergetika,* **3**(7): 11 (1956).
193. Venart, J. E. S., and C. Krishnamurthy: *Proc. 7th Conf. Thermal Conductivity,* p. 659, NBS Spec. Pub. 302, November 1967.
194. Vines, R. G.: *Aust. J. Chem.,* **6**: 1 (1953).
195. Vines, R. G.: *Proc. Joint Conf. Thermodyn. Transport Prop. Fluids, London, July 1957,* pp. 120–123, Institution of Mechanical Engineers, London, 1958.
196. Vines, R. G., and L. A. Bennett: *J. Chem. Phys.,* **22**: 360 (1954).
197. Viswanath, D. S.: *AIChE J.,* **13**: 850 (1967).
198. Vredeveld, D.: private communication, 1973.
199. Wang Chang, C. S., and G. E. Uhlenbeck: Transport Phenomena in Polyatomic Gases, *Univ. Michigan Eng. Res. Rep.* CM-681, Ann Arbor, Mich., 1951.
200. Wassiljewa, A.: *Physik. Z.,* **5**: 737 (1904).
201. Watson, J. T. R.: "The Thermal Conductivity of Gases in Metric Units," H.M. Stationery Office, Edinburgh, 1973.
202. Weber, H. F.: *Wiedemann's Ann., Ann. Phys. Chem.,* **10**: 103 (1880).
203. Wright, P. G., and P. Gray: *Trans. Faraday Soc.,* **58**: 1 (1962).
204. Zimmerling, W., and M. Ratzsch: *Chem. Techn.,* **28**(2): 84 (1976).

Chapter Eleven

Diffusion Coefficients

11-1 Scope

In Sec. 11-2, we discuss briefly several frames of reference from which diffusion can be related and define the diffusion coefficient. Low-pressure binary-gas diffusion coefficients are treated in Secs. 11-3 and 11-4, and recommendations for estimation techniques are given in Sec. 11-5. The pressure and temperature effects on gas-phase diffusion coefficients are covered in Secs. 11-6 and 11-7, respectively. The theory for liquid diffusion coefficients is introduced in Sec. 11-9, and estimation methods for binary-liquid diffusion coefficients are described in Sec. 11-10. Concentration effects are considered in Sec. 11-11 and temperature effects in Sec. 11-12.

Brief comments are given for diffusion in multicomponent mixtures in Secs. 11-8 (gases) and 11-12 (liquids); ionic solutions are covered in Sec. 11-14.

11-2 Basic Concepts and Definitions

The extensive use of the term *diffusion* in the chemical engineering literature is based on an intuitive feel for the concept; i.e., diffusion refers to the net transport of material within a single phase in the absence of mixing (by mechanical means or by convection). Both experiment and theory have shown that diffusion can result from pressure gradients (pressure diffusion), temperature gradients (thermal diffusion), external force fields (forced diffusion), and concentration gradients. Only the last type is considered in this chapter; i.e., the discussion is limited to diffusion in isothermal, isobaric systems with no external-force-field gradients.

Even with this limitation, confusion can easily arise unless care is taken to define clearly diffusion fluxes and diffusion potentials, e.g., driving forces. The proportionality constant between the flux and potential is the *diffusion coefficient,* or *diffusivity.*

Diffusion Fluxes A detailed discussion of diffusion fluxes has been given by Bird, Stewart, and Lightfoot [17]. Various types originate because different reference planes are employed. The most obvious reference plane is fixed on the equipment in which diffusion is occurring. This plane is designated by RR' in Fig. 11-1. Suppose, in a binary mixture of A and B, that A is diffusing to the left and B to the right. If the diffusion rates of these species are not identical, there will be a net depletion or accumulation of molecules in either side of RR'. To maintain the requirements of an isobaric, isothermal system, bulk motion of the mixture occurs. The net movement of A (as measured in the fixed reference frame RR') then results both from diffusion and bulk flow.

Although many reference planes can be delineated [17], a plane of *no net molal flow* is normally used to define a diffusion coefficient in binary mixtures. If $J_A{}^M$ represents a molal flux in a mixture of A and B, $J_A{}^M$ represents the net mole flow of A across the boundaries of a hypothetical (moving) plane such that the total moles of A and B are invariant on

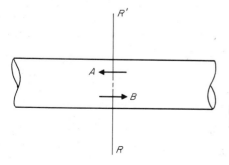

Fig. 11-1 Diffusion across plane RR'.

both sides of the plane. $J_A{}^M$ can be related to fluxes across RR' by

$$J_A{}^M = N_A - x_A(N_A + N_B) \qquad (11\text{-}2.1)$$

where N_A, N_B are the fluxes of A and B across RR' (relative to the fixed plane) and x_A is the mole fraction of A at RR'. Note that $J_A{}^M$, N_A, and N_B are vectorial quantities and a sign convention must be assigned to denote flow directions. Equation (11-2.1) shows that the net flow of A across RR' is due to a diffusion contribution $J_A{}^M$ and a bulk-flow contribution $x_A(N_A + N_B)$. For equimolal counterdiffusion, $N_A + N_B = 0$ and $J_A{}^M = N_A$.

One other flux is extensively used, i.e., one relative to the plane of *no net volume flow*. This plane is less readily visualized. By definition,

$$J_A{}^M + J_B{}^M = 0 \qquad (11\text{-}2.2)$$

and if $J_A{}^V$ and $J_B{}^V$ are vectorial molar fluxes of A and B relative to the plane of no net volume flow, then, by definition,

$$J_A{}^V \bar{V}_A + J_B{}^V \bar{V}_B = 0 \qquad (11\text{-}2.3)$$

where $\bar{V}_A$ and $\bar{V}_B$ are the partial molal volumes of A and B in the mixture. It can be shown that

$$J_A{}^V = \frac{\bar{V}_B}{V} J_A{}^M \qquad \text{and} \qquad J_B{}^V = \frac{\bar{V}_A}{V} J_B{}^M \qquad (11\text{-}2.4)$$

where V is the volume per mole of mixture. Obviously, if $\bar{V}_A = \bar{V}_B = V$, as in an ideal-gas mixture, then $J_A{}^V = J_A{}^M$.

Diffusion Coefficient   Diffusion coefficients for a binary of A and B are defined by

$$J_A{}^M = - c\mathcal{D}_{AB} \nabla x_A \qquad (11\text{-}2.5)$$

$$J_B{}^M = - c\mathcal{D}_{BA} \nabla x_B \qquad (11\text{-}2.6)$$

where c is the total molar concentration $(= V^{-1})$. With Eq. (11-2.2), since $\nabla x_A + \nabla x_B = 0$, we have $\mathcal{D}_{AB} = \mathcal{D}_{BA}$. The diffusion coefficient then represents the proportionality between the flux of A relative to a plane of no net molal flow and the gradient $c \nabla x_A$. From Eqs. (11-2.4) to (11-2.6) and the definition of a partial molal volume it can be shown that for an isothermal, isobaric binary system:

$$J_A{}^V = - \mathcal{D}_{AB} \nabla c_A \qquad \text{and} \qquad J_B{}^V = - \mathcal{D}_{AB} \nabla c_B \qquad (11\text{-}2.7)$$

Thus, when fluxes are expressed in relation to a plane of no net volume flow, the potential is the concentration gradient. $\mathcal{D}_{AB}$ in Eq. (11-2.7) is identical to that defined in Eq. (11-2.5). In many cases $\bar{V}_A = \bar{V}_B = V$ (ideal gases, ideal solutions) and in such instances, $J_A{}^V = J_A{}^M$, $J_B{}^V = J_B{}^M$.

Mutual-, Self-, and Tracer-Diffusion Coefficients The diffusion coefficient $\mathcal{D}_{AB}$ introduced above is termed the *mutual-diffusion coefficient* and refers to the diffusion of one constituent in a binary system. A similar coefficient, $\mathcal{D}_{1m}$, would imply the diffusivity of component 1 in a mixture (see Secs. 11-8 and 11-13).

Tracer-diffusion coefficients (sometimes referred to as *intradiffusion coefficients*) relate to the diffusion of a labeled component within a *homogeneous* mixture. Like mutual-diffusion coefficients, tracer-diffusion coefficients can be a function of composition. If $\mathcal{D}_A^*$ is the tracer diffusivity of A in a mixture of A and B, then as $x_A \rightarrow 1.0$, $\mathcal{D}_A^* \rightarrow \mathcal{D}_{AA}$, where $\mathcal{D}_{AA}$ is the *self-diffusion coefficient* of A in pure A.

In Fig. 11-2, the various diffusion coefficients noted above are shown for a binary mixture of n-octane and n-dodecane at 60°C [218]. In this case, the mutual diffusion of these two hydrocarbons increases as the mixture becomes richer in n-octane. With A as n-octane and B as n-dodecane, as $x_A \rightarrow 1.0$, $\mathcal{D}_{AB} = \mathcal{D}_{BA} \rightarrow \mathcal{D}_{BA}^\circ$, where this notation signifies that this limiting diffusivity represents the diffusion of B in a medium consisting essentially of A, that is, n-dodecane molecules diffusing through almost pure n-octane. Similarly, $\mathcal{D}_{AB}^\circ$ is the diffusivity of A in essentially pure B.

Chemical-Potential Driving Force The mutual-diffusion coefficient $\mathcal{D}_{AB}$ in Eq. (11-2.7) indicates that the flux of a diffusing component is proportional to the actual concentration gradient. Diffusion is affected

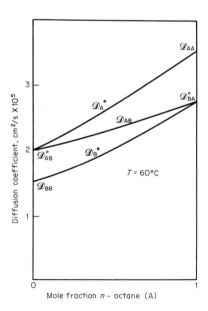

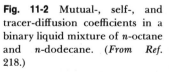

Fig. 11-2 Mutual-, self-, and tracer-diffusion coefficients in a binary liquid mixture of n-octane and n-dodecane. (*From Ref.* 218.)

by more than just the gradient in concentration, e.g., the force fields around molecules [59, 210]. Yet these force fields are some (complex) function of composition as well as of temperature and pressure. Thus fluxes should not be expected to be linear in the concentration gradient [59]. Any inadequacy in the defining equation for $\mathscr{D}_{AB}$ is reflected by a concentration dependence of experimental diffusion coefficients.

Modern theories of diffusion [76] have adopted the premise that if one perturbs the equilibrium composition of a binary system, the subsequent diffusive flow required to attain a new equilibrium state is proportional to the gradient in chemical potential $\nabla \mu_A$. When the chemical potential is related to *activity* a_A by

$$\mu_A = RT \ln a_A + \mu_A(T, P) \tag{11-2.8}$$

all theories lead to an *activity-corrected diffusion coefficient* D_{AB}:

$$D_{AB} = \frac{\mathscr{D}_{AB}}{[(\partial \ln a_A)/(\partial \ln x_A)]_{T,P}} \tag{11-2.9}$$

where D_{AB} is often less sensitive to composition than $\mathscr{D}_{AB}$.

The effect of composition is usually not important for gases, that is, $D_{AB} = \mathscr{D}_{AB}$, but is significant for liquid systems (Sec. 11-9).

11-3 Diffusion Coefficients for Binary Gas Systems at Low Pressures: Prediction from Theory

The theory describing diffusion in binary gas mixtures at low to moderate pressures has been well developed. As noted earlier in Chaps. 9 (Viscosity) and 10 (Thermal Conductivity), the theory results from solving the Boltzmann equation and the results are usually credited both to Chapman and to Enskog, who independently derived the working equation

$$\mathscr{D}_{AB} = \tfrac{3}{16} \left(2\pi k T \frac{M_A + M_B}{M_A M_B} \right)^{1/2} \frac{1}{n \pi \sigma_{AB}^2 \Omega_D} f_D \tag{11-3.1}$$

where, M_A, M_B = molecular weights
$\quad\quad n$ = number density of molecules in mixture
$\quad\quad k$ = Boltzmann's constant
$\quad\quad T$ = absolute temperature

Ω_D, the collision integral for diffusion, is a function of temperature and depends upon the choice of intermolecular-force law between colliding molecules. σ_{AB} is a characteristic length; it also depends upon the intermolecular-force law selected. Finally, f_D is a correction term which is of the order unity. If $M_A \approx M_B$, $1.0 \leqslant f_D \leqslant 1.02$ regardless of composition or intermolecular forces.   Only if the molecular masses are very unequal and the light component is present in trace amounts is the value of f_D significantly different from unity, and even in such cases, f_D is usually between 1.0 and 1.1 [142].

If f_D is chosen as unity and n expressed by the ideal-gas law, Eq. (11-3.1) may be written

$$\mathscr{D}_{AB} = 1.858 \times 10^{-3} T^{3/2} \frac{[(M_A + M_B)/M_A M_B]^{1/2}}{P \sigma_{AB}{}^2 \Omega_D}$$ (11-3.2)

where $\mathscr{D}_{AB}$ = diffusion coefficient, cm^2/s
    T = temperature, K
    P = pressure, atm
    σ = characteristic length, Å
    Ω_D = diffusion collision integral, dimensionless

The key to the use of Eq. (11-3.2) is the selection of an intermolecular-force law and the evaluation of σ_{AB} and Ω_D.

Lennard-Jones 12-6 Potential   As noted earlier [Eq. (9-4.2)], a popular correlation relating the intermolecular energy ψ between two molecules to the distance of separation r is given by

$$\psi(r) = 4\epsilon \left[\left(\frac{\sigma}{r} \right)^{12} - \left(\frac{\sigma}{r} \right)^{6} \right]$$ (11-3.3)

with ϵ and σ as the characteristic Lennard-Jones energy and length, respectively.   Application of the Chapman-Enskog theory to the viscosity of pure gases has led to the determination of many values of ϵ and σ; some of these are given in Appendix C.   Also, methods for estimating ϵ and σ are discussed in Sec. 2-7.

To use Eq. (11-3.2), some rule for obtaining binary value σ_{AB} from σ_A and σ_B must be chosen.   Also, it can be shown that Ω_D is a function only of kT/ϵ_{AB}, where again some rule must be selected to relate ϵ_{AB} to ϵ_A and ϵ_B.   The simple rules shown below are often employed:

$$\epsilon_{AB} = (\epsilon_A \epsilon_B)^{1/2}$$ (11-3.4)

$$\sigma_{AB} = \frac{\sigma_A + \sigma_B}{2}$$ (11-3.5)

Ω_D is tabulated as a function of kT/ϵ for the 12-6 Lennard-Jones potential in several reference sources, e.g., Ref. 101, and various analytical approximations are also available [94, 112, 166].   The accurate relation of Neufeld et al. [166] is

$$\Omega_D = \frac{A}{T^{*B}} + \frac{C}{\exp DT^*} + \frac{E}{\exp FT^*} + \frac{G}{\exp HT^*}$$ (11-3.6)

where $T^* = kT/\epsilon_{AB}$ $A = 1.06036$ $B = 0.15610$

$C = 0.19300$ $D = 0.47635$ $E = 1.03587$

$F = 1.52996$ $G = 1.76474$ $H = 3.89411$

Example 11-1   Estimate the diffusion coefficient for the system N_2–CO_2 at 590 K and 1 atm. The experimental value reported by Ellis and Holsen [63] is 0.575 cm²/s.

   solution   To use Eq. (11-3.2), values of σ_{CO_2}, σ_{N_2}, ϵ_{CO_2}, and ϵ_{N_2} must be obtained. Using the values in Appendix C with Eqs. (11-3.4) and (11-3.5) gives $\sigma_{CO_2} = 3.941$ Å, $\sigma_{N_2} = 3.798$ Å; $\sigma_{CO_2\text{-}N_2} = (3.941 + 3.798)/2 = 3.8695$ Å; $\epsilon_{CO_2}/k = 195.2$ K, $\epsilon_{N_2}/k = 71.4$ K; $\epsilon_{CO_2\text{-}N_2}/k = [(195.2)(71.4)]^{1/2} = 118$ K.   Then $T^* = kT/\epsilon_{CO_2\text{-}N_2} = 590/118 = 5.0$.   With Eq. (11-3.6), $\Omega_D = 0.842$.   Then, with $M_{CO_2} = 44$ and $M_{N_2} = 28$, Eq. (11-3.2) gives

$$\mathscr{D}_{CO_2\text{-}N_2} = \frac{(1.858 \times 10^{-3})(590^{3/2})[(44 + 28)/(44)(28)]^{1/2}}{(1)(3.8695)^2(0.842)}$$

$$= 0.51 \text{ cm}^2/\text{s}$$

The error is 11 percent. Ellis and Holsen recommend values of $\epsilon_{CO_2\text{-}N_2} = 134$ K and $\sigma_{CO_2\text{-}N_2} = 3.660$. With these parameters, the predicted $\mathscr{D}$ is 0.56 cm²/s, a value much closer to that found experimentally. Estimations are insensitive to the ϵ_{AB} value but depend strongly on the σ_{AB} chosen.

Equation (11-3.1) is derived for dilute gases consisting of nonpolar, spherical, monatomic molecules, and the potential function (11-3.3) is essentially empirical, as are the combining rules [Eqs. (11-3.4) and (11-3.5)]. Yet Eq. (11-3.1) gives good results over a wide range of temperatures and provides useful approximate values of $\mathscr{D}_{AB}$ [82, 83]. The general nature of the errors to be expected from this estimation procedure is indicated by the comparison of calculated and experimental values shown in Table 11-2.

   Fortunately, the calculated value of $\mathscr{D}_{AB}$ is relatively insensitive to the value of ϵ_{AB} employed and even to the form of the assumed potential function. Values of ϵ and σ are available from viscosity measurements; values from diffusion data would clearly be preferable, but few are available. No effect of composition is predicted. More detailed theory does indicate that there may be a small effect for those cases where M_A and M_B differ significantly. In a specific study on this effect [239], the binary diffusion coefficient for the system He–$CClF_3$ did vary from about 0.416 to 0.430 cm²/s over the extremes of composition. In another study [163], no effect of concentration was noted for the methyl alcohol–air system, but a small change was observed with chloroform-air.

Low-Pressure Diffusion Coefficients from Viscosity Data

Since the equations for low-pressure gas viscosity [Eq. (9-3.9)] and diffusion [Eq. (11-3.2)] have a common basis in the Chapman-Enskog theory, they can be combined to relate the two gas properties.

The *self-diffusion* coefficient $\mathscr{D}_{AA}$ is easily obtained,

$$\mathscr{D}_{AA} = 1.20 \frac{RT}{MP} \frac{\Omega_v}{\Omega_D} \eta_A \tag{11-3.7}$$

where η is expressed in poises. The ratio of collision integrals Ω_v / Ω_D is a weak function of kT/ϵ and is approximately 1.1 at ordinary temperatures.† This means that the relation between $\mathscr{D}_{AA}$ and η_A is almost independent of the choice of force-law model and potential function. This also explains why the Schmidt number for gases at low pressure is nearly independent of temperature. Equation (11-3.7) has been shown [227, 228] to provide reliable predictions of $\mathscr{D}_{AA}$ not only for simple gases but also for HCl, HBr, BF_3, UF_6, CH_4, CO,' and CO_2.

The same theory also provides a relation between the viscosity and composition of a binary gas mixture. Experimental data on viscosity as a function of composition at constant temperature can therefore be employed as a basis for calculating the *binary* diffusion coefficient $\mathscr{D}_{AB}$ [51, 84, 101, 228]. Weissman and Mason [227, 228] compare the method with a large collection of experimental diffusion data and find excellent agreement. Values of $\mathscr{D}_{AB}$ obtained by this procedure are, in fact, in better agreement with experimental values than those calculated by the use of Eq. (11-3.1). As shown later, Eq. (11-3.1) predicts values of $\mathscr{D}_{AB}$ that are usually several percent low.

Polar Gases. If one or both components of a gas mixture are polar, a modified Lennard-Jones relation, such as the Stockmayer potential, is used. A different collision-integral relation [rather than Eq. (11-3.6)] is then necessary, and Lennard-Jones σ and ϵ values are not sufficient.

Brokaw [19] has suggested an alternate method for estimating diffusion coefficients for binary mixtures containing polar components. (See also Secs. 9-4 and 10-3.) Equation (11-3.1) is still used, but the collision integral Ω_D is now

$$\Omega_D = \Omega_D \text{ [Eq. (11-3.6)]} + \frac{0.19\delta_{AB}^2}{T^*} \tag{11-3.8}$$

where $$T^* = kT/\epsilon_{AB}$$

and $$\delta = \frac{1.94 \times 10^3 \mu_p^2}{V_b T_b} \tag{11-3.9}$$

where μ_p = dipole moment, debyes
V_b = *liquid* molar volume at boiling point, cm^3/g-mol
T_b = normal boiling point, K

$$\frac{\epsilon}{k} = 1.18(1 + 1.3\delta^2) T_b \tag{11-3.10}$$

†The ratio Ω_v / Ω_D is Hirschfelder's $\Omega^{(2,2)*}/\Omega^{(1,1)*}$, tabulated as a function of $T^* = kT/\epsilon$ in Ref. 101, table I-N, p. 1128.

$$\sigma = \left(\frac{1.585 V_b}{1 + 1.3\delta^2}\right)^{1/3} \tag{11-3.11}$$

$$\delta_{AB} = (\delta_A \delta_B)^{1/2} \tag{11-3.12}$$

$$\frac{\epsilon_{AB}}{k} = \left(\frac{\epsilon_A}{k} \frac{\epsilon_B}{k}\right)^{1/2} \tag{11-3.13}$$

$$\sigma_{AB} = (\sigma_A \sigma_B)^{1/2} \tag{11-3.14}$$

Note that, except for the definition of σ_{AB}, Brokaw's modification reduces to the Lennard-Jones relation for mixtures containing one or both nonpolar components ($\mu_p = 0$). The polarity effect is related exclusively to the dipole moment; this may not always be a satisfactory assumption [24].

Example 11-2   Estimate the diffusion coefficient for a mixture of methyl chloride (MC) and sulfur dioxide (SD) at 1 atm and 50°C. Data needed to employ Brokaw's relation are shown below:

	Methyl chloride	Sulfur dioxide
Dipole moment, debyes	1.9	1.6
Liquid molar volume at T_b, cm³/g-mol	50.6	43.8
Normal boiling point, K	249	263

solution   With Eqs. (11-3.9) and (11-3.12)

$$\delta_{MC} = \frac{(1.94 \times 10^3)(1.9)^2}{(50.6)(249)} = 0.55$$

$$\delta_{SD} = \frac{(1.94 \times 10^3)(1.6)^2}{(43.8)(263)} = 0.43$$

$$\delta_{MC\text{-}SD} = [(0.55)(0.43)]^{1/2} = 0.490$$

Also, with Eqs. (11-3.10) and (11-3.13),

$$\frac{\epsilon_{MC}}{k} = 1.18[1 + 1.3(0.55)^2](249) = 409 \text{ K}$$

$$\frac{\epsilon_{SD}}{k} = 1.18[1 + 1.3(0.43)^2](263) = 385 \text{ K}$$

$$\frac{\epsilon_{MC\text{-}SD}}{k} = [(409)(385)]^{1/2} = 397 \text{ K}$$

Finally, with Eqs. (11-3.11) and (11-3.14),

$$\sigma_{MC} = \left[\frac{(1.585)(50.6)}{1 + (1.3)(0.55)^2}\right]^{1/3} = 3.86 \text{ Å}$$

$$\sigma_{SD} = \left[\frac{(1.585)(43.8)}{1 + (1.3)(0.43)^2}\right]^{1/3} = 3.82 \text{ Å}$$

$$\sigma_{MC\text{-}SD} = [(3.86)(3.82)]^{1/2} = 3.84 \text{ Å}$$

To determine Ω_D, $T^* = kT/\epsilon_{AB} = 323/397 = 0.814$. With Eq. (11-3.6), $\Omega_D = 1.598$. Then with Eq. (11-3.8), to correct this value

$$\Omega_D = 1.598 + \frac{(0.19)(0.490)^2}{0.814} = 1.654$$

With Eq. (11-3.2) and $M_{MC} = 50.488$, $M_{SD} = 64.063$,

$$\mathcal{D}_{MC\text{-}SD} = \frac{(1.858 \times 10^{-3})(323^{3/2})\left[\dfrac{50.488 + 64.063}{(50.488)(64.063)}\right]^{1/2}}{(1)(3.84)^2(1.654)}$$

$$= 0.083 \text{ cm}^2/\text{s}$$

The experimental value is $0.077 \text{ cm}^2/\text{s}$ and the estimation error 8 percent.

Discussion   A comprehensive review of the theory and available data for gas-diffusion coefficients is now available [142].   There have been numerous studies covering wide temperature ranges, and the applicability of Eq. (11-3.1) is well verified.   Most investigators select the Lennard-Jones potential for its convenience and simplicity.   The difficult task is to locate appropriate values of σ and ϵ.   A few approximate estimation techniques are described in Sec. 2-7, and a tabulation is shown in Appendix C.   Brokaw also suggests other relations, e.g., Eqs. (11-3.10) and (11-3.11).   Even after the pure-component values of σ and ϵ have been selected, a combination rule is necessary to obtain σ_{AB} and ϵ_{AB}.   Most studies have employed Eqs. (11-3.4) and (11-3.5), as they are simple and theory suggests no particularly better alternatives.

It is important to employ values of σ and ϵ obtained from the same source.   Published values of these parameters differ considerably, but σ and ϵ from a single source will often lead to the same result as the use of a quite different pair from another source.

Besides the few estimation schemes described above, there are off-shoots, partly empirical in nature, that have been suggested for obtaining better agreement with data [143, 152].

11-4   Diffusion Coefficients for Binary Gas Systems at Low Pressures: Empirical Correlations

Several proposed methods of estimating $\mathcal{D}_{AB}$ in low-pressure binary gas systems retain the general form of Eq. (11-3.2), with empirical constants based on experimental data.   These include the equations proposed by Arnold [9], Gilliland [78], Wilke and Lee [234], Slattery and Bird [199], Bailey [11], Chen and Othmer [35], Othmer and Chen [171], and Fuller, Schettler, and Giddings [68, 69].   Values of $\mathcal{D}_{AB}$ estimated by these equations generally agree with experimental values to within 5 to 10 percent, although discrepancies of more than 20 percent are sometimes noted. The method of Fuller, Schettler, and Giddings appears to be slightly more reliable than the earlier equations, probably because it is based on more data.

Fuller, Schettler, and Giddings [68, 69]   The empirical correlation suggested by the form of Eq. (11-3.2) is

$$\mathcal{D}_{AB} = \frac{10^{-3}T^{1.75}[(M_A + M_B)/M_A M_B]^{1/2}}{P[(\Sigma v)_A^{1/3} + (\Sigma v)_B^{1/3}]^2} \qquad (11\text{-}4.1)$$

where T is in kelvins and P in atmospheres. To determine Σv the values of the atomic diffusion volumes shown in Table 11-1 should be used. These values were determined by a regression analysis of 340 experimental diffusion-coefficient values of 153 binary systems. This correlation cannot distinguish between isomers. Nain and Ferron [165] found that the accuracy was poor in polar gas mixtures; they suggest an empirical correction factor to $[(\Sigma v)_A^{1/3} + (\Sigma v)_B^{1/3}]^2$ which allows for a variation in temperature. A similar correction was suggested by Saksena and Saxena for the more general case [188]. Marrero and Mason [143] show that Eq. (11-4.1) is reliable from moderate to high temperatures but is often poor at low temperatures.

TABLE 11-1   Atomic Diffusion Volumes for Use in Estimating $\mathcal{D}_{AB}$ by the Method of Fuller, Schettler, and Giddings†

Atomic and Structural Diffusion-Volume Increments v			
C	16.5	(Cl)	19.5
H	1.98	(S)	17.0
O	5.48	Aromatic ring	−20.2
(N)	5.69	Heterocyclic ring	−20.2

Diffusion Volumes for Simple Molecules Σv			
H_2	7.07	CO	18.9
D_2	6.70	CO_2	26.9
He	2.88	N_2O	35.9
N_2	17.9	NH_3	14.9
O_2	16.6	H_2O	12.7
Air	20.1	(CCl_2F_2)	114.8
Ar	16.1	(SF_6)	69.7
Kr	22.8	(Cl_2)	37.7
(Xe)	37.9	(Br_2)	67.2
		(SO_2)	41.1

†Parentheses indicate that the value listed is based on only a few data points.

Corresponding States   From Eq. (11-3.1), since Ω_D is dimensionless, if $n = kT/P$ and $\sigma \propto V_c^{1/3} \propto (RT_c/P_c)^{1/3}$, one can rewrite the equation using a dimensionless diffusion coefficient, i.e.,

$$\mathscr{D}_{AB_r} \equiv \frac{\mathscr{D}_{AB} M_{AB}^{1/2} P_{c_{AB}}^{1/3}}{(RT_{c_{AB}})^{5/6}} = f(T_r, P_r) \quad \text{with} \quad M_{AB} = \frac{M_A M_B}{M_A + M_B} \quad (11\text{-}4.2)$$

The definitions of $P_{c_{AB}}$ and $T_{c_{AB}}$ are not evident from this manipulation, and no definitive rules can be formulated. Several authors have attempted generalized correlations based on the corresponding-states principle [66, 178, 204]. Mathur and Thodos [145] have employed Eq. (11-4.2) to correlate gas self-diffusion constants at high pressures (see Sec. 11-6).

11-5   Comparison of Estimation Methods for Binary-Gas Diffusion Coefficients at Low Pressures

A number of evaluations of the proposed estimation techniques have been presented in the literature. Lugg [134] measured $\mathscr{D}_{AB}$ for 147 systems in air at 25°C. He also compared his experimental results with several of the estimation techniques referred to above. The Chen-Othmer and Wilke-Lee methods were judged to be the best. The former, however, requires critical properties and could not be used for over half the systems studied. The Wilke-Lee relation is Eq. (11-3.2) with the constant 1.858×10^{-3} replaced by $0.00217 - 0.00050[(M_A + M_B)/M_A M_B]^{1/2}$. This method requires no critical data, but ϵ and σ values must be determined. Lugg adopted empirical rules for the pure-component parameters that were suggested by Wilke and Lee, i.e.,

$$\sigma = 1.18 V_b^{1/3} \qquad \frac{\epsilon}{k} = 1.15 T_b \qquad (11\text{-}5.1)$$

where V_b = Le Bas volume (see Table 3-13)
    T_b = normal boiling point, K

Values for air were chosen as $\sigma = 3.617$ Å, $\epsilon/k = 97.0$ K; these differ slightly from those shown in Appendix C. The use of Eq. (11-5.1) yields results very similar to those found from Brokaw's rules [Eqs. (11-3.10) and (11-3.11)] since, with air as one component, $\delta = 0$. For mixture values, Eqs. (11-3.4) and (11-3.5) were used. An example of the application of the Wilke-Lee relation is given in Example 11-3.

Example 11-3   Estimate the diffusion coefficient of allyl chloride (AC) in air at 25°C and 1 atm. The experimental value reported by Lugg [134] is 0.0975 cm²/s.

solution Employing the method of Fuller, Schettler, and Giddings, values of Σv are first obtained from Table 11.1.

$$\Sigma v \text{ (AC)} = 3(\text{C}) + 5(\text{H}) + \text{Cl}$$
$$= (3)(16.5) + (4)(1.98) + 19.5$$
$$= 78.9$$
$$\Sigma v \text{ (air)} = 20.1$$

With M (AC) = 76.53 and M (air) = 28.8, with Eq. (11-4.1),

$$\mathscr{D}_{\text{AB}} = \frac{(10^{-3})(298^{1.75})[(76.53 + 28.8)/(76.53)(28.8)]^{1/2}}{(1)[(78.9)^{1/3} + (20.1)^{1/3}]^2} = 0.095 \text{ cm}^2/\text{s}$$

Alternatively, using the Wilke-Lee modification of Eq. (11-3.2), from Appendix C, $\sigma_{\text{air}} = 3.711$ Å and $(\epsilon/k)_{\text{air}} = 78.6$ K. Allyl chloride is not given in Appendix C; with Eq. (11-5.1), $T_b = 318$ K, and $V_b = 87.5$ cm^3/g-mol (the Le Bas volume from Table 3-11),

$$\sigma \text{ (AC)} = (1.18)(87.5)^{1/3} = 5.24 \text{ Å}$$

$$\frac{\epsilon}{k} \text{ (AC)} = (1.15)(318) = 366 \text{ K}$$

Then, with Eqs. (11-3.4) and (11-3.5),

$$\left(\frac{\epsilon}{k}\right)_{\text{AC-air}} = [(78.6)(366)]^{1/2} = 170 \text{ K}$$

$$\sigma_{\text{AC-air}} = \frac{3.711 + 5.24}{2} = 4.48 \text{ Å}$$

$$T^* = \frac{T}{(\epsilon/k)_{\text{AC-air}}} = \frac{298}{170} = 1.75$$

and, with Eq. (11-3.6), $\Omega_D = 1.13$. Using the multiplier proposed by Wilke and Lee gives

$$0.00217 - 0.00050 \left[\frac{76.53 + 28.8}{(76.53)(28.8)}\right]^{1/2} = 2.06 \times 10^{-3}$$

Then, with Eq. (11-3.2),

$$\mathscr{D}_{\text{AC-air}} = \frac{(2.06 \times 10^{-3})(298^{1.5})[(76.53 + 28.8)/(76.53)(28.8)]^{1/2}}{(1)(4.48)^2(1.13)}$$

$$= 0.10 \text{ cm}^2/\text{s}$$

The error in each of the two estimation methods illustrated is 2.5 percent.

More recently, Elliott and Watts [62] measured the low-pressure diffusion coefficient of many hydrocarbons in air at 25°C. Again, the Wilke-Lee method was judged the best, although all methods were within about 4 percent when average deviations were computed. The Wilke-Lee calculations were made as outlined above.

An evaluation by Fuller et al. [68, 69] with most low-temperature data available indicated that Eq. (11-4.1) had the least error, although the Wilke-Lee, Chen-Othmer, and Eq. (11-3.2) all show small errors. Gotoh et al. [82, 83] found Eq. (11-3.2) accurate when tested with new data on systems containing H_2, N_2, SF_6, CH_3Cl, C_2H_5Cl, and paraffins from methane to neopentane.

Table 11-2 compares the *experimental* data with values obtained by use

TABLE 11-2   Comparison of Methods of Estimating Gas-Diffusion Coefficients at Low Pressures

System	T, K	$\mathscr{D}_{AB}P$ (obs.), (cm²/s) atm	Ref.	Errors as percent of observed values		
				Theoretical	Wilke and Lee	Fuller, Schettler, and Giddings
Air–carbon dioxide	276.2	0.142	104	−6	4	−3
	317.2	0.177	104	−3	8	−1
Air-ethanol	313	0.145	144	−10	0	−8
Air-helium	276.2	0.624	104	0	3	−7
	317.2	0.765	104	2	5	−3
	346.2	0.902	104	0	3	−4
Air–n-hexane	294	0.0800	31	−6	5	−6
	328	0.0930	31	−1	11	−3
Air–n-heptane	294	0.071	31	15	28	15
	361	0.0985	31	21	35	19
Air-water	313	0.288	31	−18	−11	−5
Ammonia–diethyl ether	288.3	0.0999	202	−23	−16	11
	337.5	0.137	202	−23	−16	7
Argon-ammonia	254.7	0.150	201	3	13	26
	333	0.253	201	3	12	19
Argon–carbon dioxide	276.2	0.133	104	−17	−8	1
Argon-helium	276.2	0.646	104	−1	2	−3
	288	0.696	32	−2	1	−3
	298	0.729	195	−1	3	−2
	418	1.398	32	−10	−8	−7
Argon-hydrogen	242.2	0.562	174	−4	3	−2
	295.4	0.83	230	−9	−1	−7
	448	1.76	230	−13	−6	−9
	628	3.21	230	−15	−8	−10
	806	4.86	230	−15	−9	−8
	958	6.81	230	−19	−13	−11
	1069	8.10	230	−19	−13	−11
Argon-krypton	273	0.119	200	−1	11	3
Argon-methane	298	0.202	32	5	15	6
Argon-neon	273	0.276	200	−3	6	
Argon–sulfur dioxide	263	0.077	144	24	38	25
Argon-xenon	195	0.0518	32	−1	11	2
	194.7	0.0508	2	0	12	4
	329.9	0.137	2	−1	11	−2
	378	0.178	32	−2	10	−5
Carbon dioxide–carbon dioxide	194.8	0.0516	4	−7	1	16
	312.8	0.125	4	−3	5	10
Carbon dioxide–helium	276.2	0.531	104	−2	1	−6
	298	0.612	195	−3	0	−7

TABLE 11-2 Comparison of Methods of Estimating Gas-Diffusion Coefficients at Low Pressures (Continued)

System	T, K	$\mathscr{D}_{AB}P$ (obs.), (cm²/s) atm	Ref.	Errors as percent of observed values		
				Theoretical	Wilke and Lee	Fuller, Schettler, and Giddings
Carbon dioxide–helium	299	0.611	195	−5	−2	−6
(Continued)	346.2	0.765	104	−1	2	−4
	498	1.414	195	−2	0	−1
Carbon dioxide–nitrogen	298	0.167	224	−9	1	−2
	299	0.171	223	−11	−1	−4
Carbon dioxide–nitrous oxide	194.8	0.0531	4	−11	0	3
	312.8	0.128	4	−6	4	−2
Carbon dioxide–oxygen	293.2	0.153	225	−3	7	3
Carbon dioxide–sulfur dioxide	263	0.064	144	5	12	27
	473	0.195	144	8	21	16
Carbon dioxide–water	307.2	0.198	37	−20	−13	12
	328.6	0.257	37	−30	−23	−3
	352.3	0.245	193	−14	−6	11
Carbon monoxide–carbon monoxide	194.7	0.109	3	−14	−6	−13
	373	0.323	3	−7	2	−8
Carbon monoxide–nitrogen	194.7	0.105	3	−9	0	−8
	373	0.318	3	−6	3	−5
Ethane–n-hexane	294	0.0375	31	37	52	80
Ethylene-water	328.4	0.233	193	−7	1	−3
Helium-benzene	423	0.610	195	8	12	−6
Helium–n-butanol	423	0.587	195	10	12	−2
Helium-ethanol	423	0.821	195	3	7	−5
Helium–n-hexanol	423	0.469	195	16	19	1
Helium-methane	298	0.675	32	3	5	−6
Helium-methanol	423	1.032	195	7	10	−1
Helium-neon	242.2	0.792	174	−6	−4	
	341.2	1.405	174	−7	−3	
Helium-nitrogen	298	0.687	195	1	3	1
He trace	298	0.730	231	−5	−3	−5
N₂ trace	298	0.688	231	0	3	1
Helium-oxygen	298	0.729	195	2	5	−2
Helium–n-pentanol	423	0.507	195	16	18	2
Helium-isopropanol	423	0.677	195	7	9	−4
Helium–n-propanol	423	0.676	195	18	20	−3
Helium-water	307.1	0.902	193	−1	2	−3
	352.4	1.121	193	1	3	−4
Hydrogen-acetone	296	0.424	144	0	−3	17

System	T, K	$\mathcal{D}_{AB}P$ (obs.), (cm^2/s) atm	Ref.	Errors as percent of observed values		
				Theoretical	Wilke and Lee	Fuller, Schettler, and Giddings
Hydrogen-ammonia	263	0.57	144	4	1	17
	273	0.745	194	−15	−17	−5
	298	0.783	144	−4	−7	6
	358	1.093	144	−4	−7	4
	473	1.86	144	−5	−8	0
	533	2.149	194	2	−1	7
Hydrogen-benzene	311.3	0.404	107	0	−2	−1
Hydrogen-cyclohexane	288.6	0.319	107	−3	−6	−4
Hydrogen-methane	288	0.694	32	−5	−8	−7
Hydrogen-nitrogen	294	0.763	194	−5	−8	−3
	298	0.784	144	−6	−8	−3
	358	1.052	144	−3	−5	−1
	573	2.417	194	−8	−10	−1
Hydrogen-piperidine	315	0.403	107	−2	−5	−8
Hydrogen-pyridine	318	0.437	107	−5	−7	2
Hydrogen–sulfur dioxide	285.5	0.525	144	−5	−7	−6
	473	1.23	144	0	−2	−3
Hydrogen-thiophene	302	0.400	107	0	−3	6
Hydrogen-water	307.1	0.915	37	−12	−14	1
	328.5	1.121	193	−17	−20	−8
Hydrogen-xenon	341.2	0.751	174	−1	−3	−8
Methane-water	352.3	0.356	193	−11	−5	0
Neon-krypton	273	0.223	200	−2	8	
Nitrogen-ammonia	298	0.230	144	−5	3	10
	358	0.328	144	−6	2	6
Nitrogen-benzene	311.3	0.102	107	−4	6	−2
Nitrogen-cyclohexane	288.6	0.0731	107	1	12	5
Nitrogen-piperidine	315	0.0953	107	−3	7	−2
Nitrogen–sulfur dioxide	263	0.104	144	−3	8	1
Nitrogen-water	307.5	0.256	193	−11	−3	8
	328.9	0.313	37	−17	−11	0
	349.1	0.354	37	−17	−10	−2
	352.1	0.359	193	−17	−10	−2
Oxygen-benzene	311.3	0.101	107	−9	1	−3
Oxygen–carbon tetrachloride	296	0.0749	144	−5	6	35
Oxygen-cyclohexane	288.6	0.0746	107	−7	4	0
Oxygen-piperidine	315	0.0953	107	−7	4	−5
Oxygen-water	352.3	0.352	193	−15	−8	0
Absolute average error				7.5	8.0	6.9

of three of the estimation procedures.   With the exception of several values quoted by Mason and Monchick [144], all the 114 experimental points were obtained from original sources.   No attempt was made to judge the accuracy of the experimental measurements.   Inspection of the table indicates, however, that different experimenters may disagree by several percent.   The two values for helium-nitrogen at 298 K suggest the small effect of mixture composition, which is generally ignored.   Values obtained by four different experimental techniques are included.

Inspection of Table 11-2 suggests that the Fuller-Schettler-Giddings equation does slightly better than the theoretical equation.   The average error by the latter method is 8 percent, but the calculated values average 3.7 percent low.   This may compensate roughly for the fact that experimental values are often high because of failure to eliminate convection.   In systems containing water, the theoretical values are rather consistently low by about 10 percent, suggesting the empirical introduction of a multiplying factor of 1.09 in Eq. (11-3.2) for such cases.   The temperature functions incorporated in the several equations are not adequately tested by Table 11-2; as brought out in an earlier section, that of the theoretical method appears to be good.

Recommendation   From the evaluation shown in Table 11-2 as well as noting other tests described earlier:

1. For simple, nonpolar systems at moderate temperatures, use the Fuller, Schettler, and Giddings correlation (11-4.1); errors should be less than 5 to 10 percent.

2. For systems containing polar components, use the method of Brokaw [Eqs. (11-3.2), (11-3.8) to (11-3.14)]; errors less than 15 percent are usually obtained.   The dipole moments for many materials are tabulated in Appendix A.

11-6   The Effect of Pressure on Diffusion in Gases

At low to moderate pressures, the diffusion coefficient for gases is inversely proportional to pressure or density.   This experimental fact is in agreement with theory, as shown in Eqs. (11-3.1) and (11-3.2).

Most high-pressure diffusion experiments have been limited to the measurement of self-diffusion coefficients.   Dawson, Khoury, and Kobayashi [50] carried out a particularly extensive study with methane from $0.8 < T_r < 1.9$, $0.3 < P_r < 7.4$ and were able to correlate their data with the following equation:

$$\frac{\mathscr{D}\rho}{(\mathscr{D}\rho)^\circ} = 1 + 0.053432\rho_r - 0.030182\rho_r^2 - 0.029725\rho_r^3 \qquad (11\text{-}6.1)$$

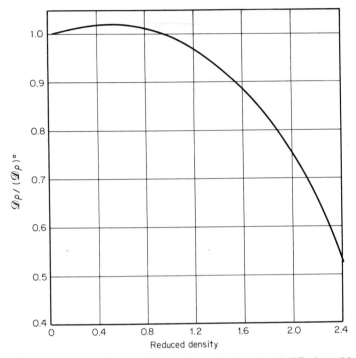

Fig. 11-3   Dawson-Khoury-Kobayashi correlation of self-diffusion with density.

where $\mathscr{D}$ = self-diffusion coefficient at T and ρ
ρ = density
$(\mathscr{D}\rho)^\circ$ = product of $\mathscr{D}$ and ρ evaluated at T but at low pressure
ρ_r = reduced density ρ/ρ_c

Equation (11-6.1) is plotted in Fig. 11-3.   Below a value of $\rho_r = 1$ there is little effect of density, and $\mathscr{D}\rho \approx (\mathscr{D}\rho)^\circ$.   This fact has been confirmed by other experimental self-diffusion data [61, 129, 154, 168, 209], although there is a scatter of about ± 10 percent.   In this same range of reduced densities, Mathur and Thodos [145] have also proposed that

$$\mathscr{D}\rho_r = \frac{10.7 \times 10^{-5} T_r}{\beta} \qquad (11\text{-}6.2)$$

where $\mathscr{D}$ = high-pressure self-diffusion coefficient, cm²/s
ρ_r = reduced density ρ/ρ_c
T_r = reduced temperature T/T_c
$\beta = M^{1/2} P_c^{1/3} / T_c^{5/6}$, with P_c in atmospheres and T_c in kelvins

Equation (11-6.2) would also show that $\mathscr{D}\rho/(\mathscr{D}\rho)^\circ = 1$ for $\rho_r < 1.0$, as indicated in Fig. 11-3.   An alternate correlation leading to the same conclusion was proposed by Stiel and Thodos [204].

Above reduced densities of unity, Eq. (11-6.1) predicts a significant decrease in the $\mathscr{D}\rho$ product. This is shown by Dawson et al. methane data, some self-diffusion data on hydrogen at 36 and 55 K [92], as well as *liquid*-methane data [164]. The latter two sets of data fall slightly above the smooth curve shown in Fig. 11-3. The liquid self-diffusion data for carbon dioxide and propane taken by Robinson and Stewart [186] cannot be used to check Eq. (11-6.1) since no $(\mathscr{D}\rho)°$ value was measured; these data $(1.5 < \rho_r < 2.5, 0.8 < T_r < 0.97)$ do, however, show a decrease in the $\mathscr{D}\rho$ product with increasing density in the appropriate region. Approximate corresponding-states correlations for $\mathscr{D}$ as a function of reduced temperature and reduced pressure have been given by Slattery and Bird [199] and Takahashi [205a].

In summary, for self-diffusion at high pressures, the product $\mathscr{D}\rho$ is essentially constant up to reduced densities of unity. At higher densities, the product $\mathscr{D}\rho$ decreases rapidly with further increases in density. In this region, Eq. (11-6.1) is but a rough approximation.

In contrast to this picture for self-diffusion, the few available data on binary diffusion indicate that $\mathscr{D}_{AB}\rho$ decreases with density even for pseudoreduced densities† less than unity. The use of the density factor itself causes problems. At low densities, ρ, expressed in moles per volume, is independent of composition. However, at high densities the mixtures are usually not ideal, and ρ depends upon composition. The concomitant question whether $\mathscr{D}_{AB}$ depends upon composition at high pressures has not yet been satisfactorily answered, but the evidence indicates that there is an effect.‡

Islam and Stryland [111] studied the argon-methane system at 23°C from 20 to 150 atm. The methane concentration varied from 1 to 1.7 mole percent. $\mathscr{D}_{AB}\rho$ was found to decrease with ρ, as seen in Fig. 11-4.

Berry and Koeller [14] report binary diffusion coefficients for three systems, H_2–N_2, CH_4–C_2H_6, and N_2–C_2H_6, at temperatures from 40 to 77°C and for pressures up to 1000 atm. Below reduced densities of unity, the product $\mathscr{D}_{AB}\rho$ decreased slightly with reduced density; at the higher densities, the decrease was more pronounced. These general trends are in agreement with the study of Islam and Stryland noted above. A slight effect of composition was also found at the high pressures. Figure 11-5 shows the variation of $\mathscr{D}_{AB}$ with density for a 50 percent methane–50 percent ethane mixture at 40°C. The solid line is an extrapolation of a straight line with a slope of -1 to indicate the deviation of the $\mathscr{D}_{AB}\rho$ product from unity at high densities.

It is not clear how one may best estimate the binary diffusion

---

†Since true mixture critical densities are not often known, we have based our discussion on a pseudocritical density $\rho_c = V_c^{-1} = (\Sigma y_i V_{c_i})^{-1}$.

‡The diffusion-coefficient data for the methane-propane system reported by Woessner et al. [238] show a concentration dependence, but these values were *calculated* from self-diffusion coefficients and were not measured.

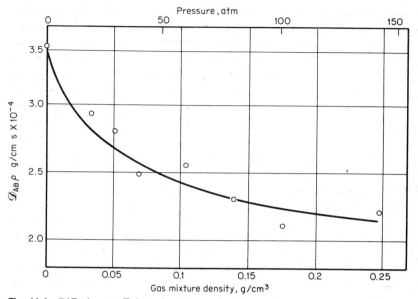

Fig. 11-4 Diffusion-coefficient–density product for methane-argon at 23°C. (*Data from Ref.* 111.)

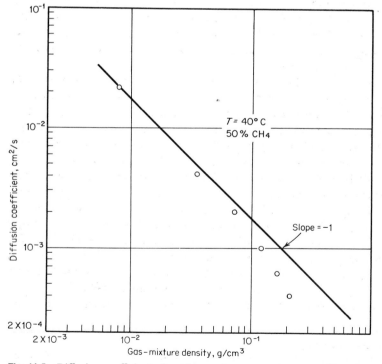

Fig. 11-5 Diffusion coefficient of methane-ethane gas mixtures. (*Data from Ref.* 14.)

coefficient at high pressures.   When Eq. (11-6.1) or (11-6.2) is employed, the estimated value is too high (see Example 11-4 below).   Until more binary-diffusion data become available, no definitive recommendation is possible.

Example 11-4   Estimate the diffusion coefficient for a mixture of methane and ethane at 136.1 atm and 40°C.   The mole fraction methane is 0.8.   At these conditions, the experimental density is 0.135 g/cm^3 and $\mathscr{D}_{AB} = 8.4 \times 10^{-4}$ cm^2/s [14].

solution   If Eq. (11-6.2) is used, pseudocritical values of V_c, P_c, and T_c must be estimated.   Equations (4-2.1) and (4-2.2) are simple rules and probably as good as any for this case.   With T_c (CH$_4$) = 190.6 K, T_c (C$_2$H$_6$) = 305.4 K, V_c (CH$_4$) = 99.0 cm^3/g-mol, V_c (C$_2$H$_6$) = 148 cm^3/g-mol, Z_c (CH$_4$) = 0.288, Z_c (C$_2$H$_6$) = 0.285, M (CH$_4$) = 16.0, M (C$_2$H$_6$) = 30.1 we have

$$T_{cm} = (0.8)(190.6) + (0.2)(305.4) = 213.6 \text{ K}$$

$$V_{cm} = (0.8)(99.0) + (0.2)(148) = 108.8 \text{ cm}^3/\text{g-mol}$$

$$Z_{cm} = (0.8)(0.288) + (0.2)(0.285) = 0.287$$

$$P_{cm} = (0.287)(82.07)\frac{213.6}{108.8} = 46.2 \text{ atm}$$

$$M_m = (0.8)(16.0) + (0.2)(30.1) = 18.85$$

Then,

$$\beta = \frac{M_m^{1/2} P_{cm}^{1/3}}{T_{cm}^{5/6}} = \frac{(18.85)^{1/2}(46.2)^{1/3}}{(213.6)^{5/6}} = 0.178$$

$$\rho_r = \frac{\rho V_{cm}}{M_m} = 0.135 \frac{108.8}{18.85} = 0.779$$

$$T_r = \frac{40 + 273}{213.6} = 1.46$$

From Eq. (11-6.2)

$$\mathscr{D}_{AB} = \frac{(10.7 \times 10^{-5})(1.46)}{(0.178)(0.779)} = 11.3 \times 10^{-4} \text{ cm}^2/\text{s}$$

The estimated value is about 30 percent high.   Of course, Eq. (11-6.2) was not devised for binary diffusion coefficients but it yields approximate, though apparently high, estimated values.

11-7 The Effect of Temperature on Diffusion in Gases

At low pressures, where the ideal-gas law approximation is valid, it is easily seen from Eq. (11-3.2) that

$$\mathscr{D}_{AB} \propto \frac{T^{3/2}}{\Omega_D(T)}; \qquad P = \text{const} \tag{11-7.1}$$

or

$$\left(\frac{\partial \ln \mathscr{D}_{AB}}{\partial \ln T}\right)_P = \frac{3}{2} - \frac{\partial \ln \Omega_D}{\partial \ln T} \tag{11-7.2}$$

Marrero and Mason [142] indicate that in most cases, the term $(\partial \ln \Omega_D)/(\partial \ln T)$ varies from 0 to $-\frac{1}{2}$.   Thus $\mathscr{D}_{AB}$ varies as $T^{3/2}$ to T^2.   This result agrees with the empirical estimation methods referred

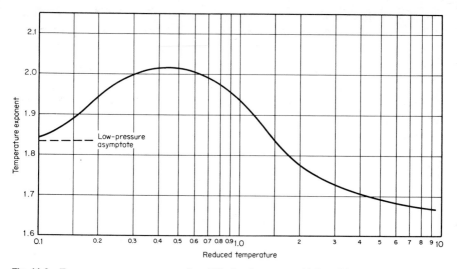

Fig. 11-6 Exponent on temperature for diffusion in gases. (*Adapted from Ref.* 142 *with the approximation that* $\epsilon/k \approx 0.75\,T_c$.)

to in Sec. 11-4, e.g., Fuller-Schettler-Giddings, $\mathscr{D} \propto T^{1.75}$. Over wide temperature ranges, however, the exponent on temperature changes. Figure 11-6 shows the approximate variation of this exponent with reduced temperature. The very fact that the temperature exponent increases and then decreases indicates that most empirical estimation techniques with a constant exponent will be limited in their range of applicability. The theoretical method and the Wilke-Lee method are therefore preferable if wide temperature regions are to be covered.

11-8 Diffusion in Multicomponent Gas Mixtures

A few general concepts of diffusion in multicomponent liquid mixtures presented later (in Sec. 11-13) are also applicable for gas mixtures. One of the problems with diffusion in liquids is that even the binary diffusion coefficients are often very composition-dependent. For multicomponent liquid mixtures, therefore, it is difficult to obtain numerical values of the diffusion coefficients relating fluxes to concentration gradients.

In gases, $\mathscr{D}_{AB}$ is normally assumed independent of composition. With this approximation, multicomponent diffusion in gases is normally described by the Stefan-Maxwell equation

$$\nabla x_i = \sum_{j=1}^{n} \frac{c_i c_j}{c^2 \mathscr{D}_{ij}} \left(\frac{J_j}{c_j} - \frac{J_i}{c_i} \right) \qquad (11\text{-}8.1)$$

where c_i = concentration of i

c = mixture concentration

$J_i, J_j =$ flux of i, j
$\mathscr{D}_{ij} =$ binary diffusion coefficient of ij system
$\nabla x_i =$ gradient in mole fraction of i

This relation is different from the basic binary diffusion relation (11-2.5), but the employment of common binary diffusion coefficients is particularly desirable. Marrero and Mason [142] discuss many of the assumptions behind Eq. (11-8.1).

Few attempts have been made by engineers to calculate fluxes in multicomponent systems. However, one important and simple limiting case is often quoted. If a dilute component i diffuses into a *homogeneous* mixture, then $J_j \approx 0$. With $c_j/c = x_j$, Eq. (11-8.1) reduces to

$$\nabla x_i = -J_i \sum_{\substack{j=1 \\ j \neq i}}^{n} \frac{x_j}{c\mathscr{D}_{ij}} \qquad (11\text{-}8.2)$$

Defining

$$\mathscr{D}_{im} \equiv -\frac{J_i}{\nabla x_i} \qquad (11\text{-}8.3)$$

gives

$$\mathscr{D}_{im} = \left(\sum_{\substack{j=1 \\ j \neq i}}^{n} \frac{x_j}{\mathscr{D}_{ij}} \right)^{-1} \qquad (11\text{-}8.4)$$

This simple relation (sometimes called Blanc's law [18, 142]) was shown to apply to several ternary cases where i was a trace component [146]. Deviations from Blanc's law are discussed by Sandler and Mason [189].

The general theory of diffusion in multicomponent gas systems is covered by Hirschfelder, Curtiss, and Bird [46, 101]. Approximate calculation methods were developed by Wilke [232]. The problem of diffusion in three-component gas systems has been generalized by Toor [208] and checked experimentally by Fairbanks and Wilke [67], Walker, de Haas, and Westenberg [223], and Duncan and Toor [60]. Other relevant articles include Refs. 16, 38, 52, 53, 106, 122, 153, 196, and 205.

11-9   Diffusion in Liquids: Theory

Binary-liquid diffusion coefficients are defined by Eq. (11-2.5) or (11-2.7). Since molecules in liquids are densely packed and strongly affected by force fields of neighboring molecules, values of $\mathscr{D}_{AB}$ for liquids are much smaller than for low-pressure gases, although this does not mean that diffusion rates are necessarily low since concentration gradients can be large.

Liquid-state theories for calculating diffusion coefficients are quite idealized, and none is satisfactory in providing relations for calculating $\mathscr{D}_{AB}$. In several cases, however, the *form* of a theoretical equation has provided the framework for several useful prediction methods. A case in point involves the analysis of large, spherical molecules diffusing in a dilute solution. Hydrodynamic theory [17, 72] then indicates that

$$\mathscr{D}_{AB} = \frac{RT}{6\pi\eta_B r_A} \tag{11-9.1}$$

where η_B is the viscosity of the solvent and r_A the radius of the spherical solute. Equation (11-9.1) is the Stokes-Einstein equation. Although this relation was derived for a very special situation, many authors have used the form $\mathscr{D}_{AB}\eta_B/T = f(\text{solute size})$ as a starting point in developing correlations.

Other theories for modeling diffusion in liquids have been based on kinetic theory [8, 13, 27, 29, 47, 54, 93, 116, 118, 120, 157], absolute-rate theory [44, 65, 72, 79, 125, 128, 170, 184], statistical mechanics [12, 13, 114], and other concepts [21, 56, 105, 117]. Several reviews are available for further consideration [55, 76, 77, 100, 131, 215].

Diffusion in liquid metals is not treated, although estimation techniques are available [173].

11-10 Estimation of Binary-Liquid Diffusion Coefficients at Infinite Dilution

For a binary mixture of solute A in solvent B, the diffusion coefficient $\mathscr{D}_{AB}^{\circ}$ of A diffusing in an infinitely dilute solution of A in B implies that each A molecule is in an environment of essentially pure B. In engineering work, however, $\mathscr{D}_{AB}^{\circ}$ is assumed to be a representative diffusion coefficient even for concentrations of A up to 5, perhaps 10, mole percent.

In this section, several estimation methods for $\mathscr{D}_{AB}^{\circ}$ are discussed; the effect of concentration for mutual diffusion coefficients is covered in Sec. 11-11.

Wilke-Chang Estimation Method [233]   A widely used correlation for $\mathscr{D}_{AB}^{\circ}$, the Wilke-Chang technique is, in essence, an empirical modification of the Stokes-Einstein relation [Eq. (11-9.1)]:

$$\mathscr{D}_{AB}^{\circ} = 7.4 \times 10^{-8} \frac{(\phi M_B)^{1/2} T}{\eta_B V_A^{0.6}} \tag{11-10.1}$$

where $\mathscr{D}_{AB}^{\circ}$ = mutual diffusion coefficient of solute A at very low concentrations in solvent B, cm²/s
$\quad M_B$ = molecular weight of *solvent* B
$\quad T$ = absolute temperature, K
$\quad \eta_B$ = viscosity of *solvent* B, cP

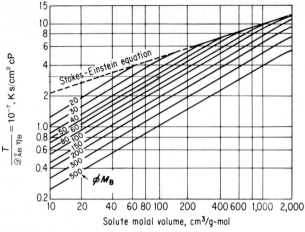

Fig. 11-7 Graphical representation of Wilke-Chang correlation of diffusion coefficients in dilute solutions. (*From Ref.* 233.)

V_A = molal volume of *solute* A at its normal boiling temperature, cm³/g-mol

ϕ = *association factor* of *solvent* B, dimensionless

The value of V_A is best estimated at T_b by methods described in Sec. 3-14; if no experimental data are available, the Le Bas additive volumes given in Table 3-11 may be used.

Wilke and Chang recommend that ϕ be chosen as 2.6 if the solvent is water, 1.9 for methanol, 1.5 for ethanol, and 1.0 for unassociated solvents. When 251 solute-solvent systems were tested by these authors, an average error of about 10 percent was noted. Figure 11-7 is a graphical representation of Eq. (11-10.1) with the dashed line representing Eq. (11-9.1); the latter is assumed to represent the maximum value of the ordinate for any value of V_A.

Diffusion of Gases and Organic Solutes in Water Wise and Houghton [235] measured the diffusion coefficient of a number of sparingly soluble gases in water and reported that the Wilke-Chang correlation predicted values significantly below those determined experimentally. Shrier [197] pointed out that Wise and Houghton had applied the correlation incorrectly, but upon recalculation, their estimated values for $\mathscr{D}_{AW}^{\circ}$ were still some 25 percent low. Yet Himmelblau [100] indicated reasonably good agreement between values from the Wilke-Chang correlation and literature data for many gases in water. Even more recently, Witherspoon and Bonoli [237] made measurements of $\mathscr{D}_{AW}^{\circ}$ for many light hydrocarbons in water from 2 to 60°C and their experimental data differ considerably from those reported earlier by Wise and

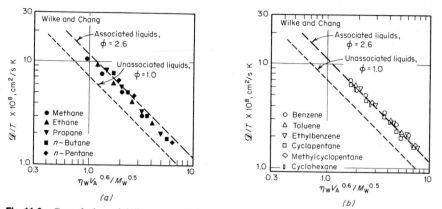

Fig. 11-8 Correlation of diffusion coefficients in water compared with Eq. (11-10.1). (*From Ref.* 237.)

Houghton. The new data agreed well with predictions from Eq. (11-10.1), as indicated in Fig. 11-8a and b. Hayduk and Laudie [98] recently reviewed diffusion coefficient data for 87 different solutes in water. They reported that the Wilke-Chang correlation yielded an average error of 6.9 percent, although if the association factor for water is decreased from 2.6 to 2.26, the average error decreases to only 0.4 percent.

Diffusion of Water in Organic Solvents Olander [169] pointed out that whereas the Wilke-Chang correlation usually is satisfactory for most cases of an organic solute diffusing into water, the opposite situation of water diffusing into an organic solvent is not well predicted. He found the calculated value of $\mathscr{D}_{WB}^{\circ}$ to be about 2.3 times larger than the experimental value. He attributed this error to the fact that since water may diffuse as a tetramer, the value of V_W in Eq. (11-10.1) should be multiplied by 4; $V_W^{0.6}$ will then increase by a factor of 2.3. This explanation is disputed by Lusis [135].

A more thorough study of the diffusion of water in various organic liquids was carried out by Lees and Sarram [124]. Some of the data discussed by these authors is shown on Table 11-3. Conclusions are difficult to draw. In some instances the Wilke-Chang relation is preferable and in others the Olander modification. In a few cases, neither technique is particularly suitable. Where the solvent has a high viscosity (triethylene glycol and glycerol), the Wilke-Chang method is probably a poor choice as an estimating relation.

Diffusion in Organic Systems Caldwell and Babb [25] suggest that for systems containing aromatic hydrocarbons, the Wilke-Chang association

TABLE 11-3   Comparison between Experimental and Estimated Diffusion Constants of Water in Organic Solvents

Temperature $\approx 25°C$, water concentration < 1 mole %

	Diffusion coefficient, $cm^2/s \times 10^5$		
Solvent	Exp.	Wilke and Chang,† Eq. (11-10.1)	Olander (value from Wilke and Chang divided by 2.3)
Methanol	1.75	4.35	1.89
Ethanol	1.24	2.75	1.20
1-Propanol	0.61	1.61	0.50
2-Propanol	0.38	1.18	0.51
n-Butanol	0.56	1.26	0.55
Isobutanol	0.30	0.67	0.29
Benzyl alcohol	0.37	0.60	0.26
Ethylene glycol	0.18	0.14	0.061
Triethylene glycol	0.19	0.35	0.15
Glycerol	0.0083	0.0024	0.0010
Acetone	4.56	8.80	3.72
Furfuraldehyde	0.90	2.24	0.97
Ethyl acetate	3.20	7.48	3.25
Aniline	0.70	0.82	0.36
n-Hexadecane	3.76	1.63	0.71
n-Butyl acetate	2.87	6.12	2.66
n-Butyric acid	0.79	2.53	1.10
Toluene	6.19	6.63	2.88
Methylene chloride	6.53	8.64	3.75
1,1,1-Trichloroethane	4.64	5.65	2.46
Trichloroethylene	8.82	7.87	3.42
Nitrobenzene	2.80	2.29	1.00
Pyridine	2.73	3.83	1.6

†The association factors chosen were 1.9 for CH_3OH, 1.5 for C_2H_5OH, 1.2 for the propyl alcohols, and 1.0 for all other solvents.

parameter be set equal to 0.7.   Amourdam and Laddha [5] measured infinite-dilution diffusion coefficients for 25 diverse systems and found good agreement between experimental values and those estimated with the values of ϕ recommended by Wilke and Chang.   Alkanes (hexane and heptane) diffusing in hexanol or heptanol were poorly predicted by the Wilke-Chang method unless $\phi > 1.5$ [159].   Hayduk and Buckley [95] suggest that the exponent of 0.6 on V_A in Eq. (11-10.1) is too low and a better correlation would result if it were increased to about 0.7.

Example 11-5   Estimate the infinite-dilution diffusion coefficient of adipic acid in methyl alcohol at 30°C.   The viscosity of methyl alcohol at 30°C is 0.514 cP, and the molal volume of adipic acid at the boiling point is 173.8 cm³/g-mol.   The molecular weight of methyl alcohol is 32.04.

solution   From Eq. (11-10.1), with $\phi = 1.5$,

$$\mathcal{D}^{\circ}_{AB} = (7.4 \times 10^{-8}) \frac{[(1.5)(32.04)]^{1/2}(303.2)}{(0.514)(173.8)^{0.6}}$$

$$= 1.37 \times 10^{-5} \text{ cm}^2/\text{s}$$

The experimental value is $1.38 \times 10^{-5} \text{ cm}^2/\text{s}$ [5].

Hayduk et al. [96] reported poor agreement with experimental data when infinite-dilution diffusion coefficients were estimated for propane in various organic solvents.   The agreement became progressively worse as the viscosity of the solvent increased.   Example 11-6 illustrates one case.

Example 11-6   With the Wilke-Chang relation, determine the value of the infinite-dilution diffusion coefficient of propane in chlorobenzene at 0°C. $M_B = 112.56$ (chlorobenzene), $\eta = 1.05$ cP (chlorobenzene), $V_A = 74.5$ cm³/g-mol (propane), $T = 273.2$ K, $\phi = 1.0$.
   solution

$$\mathcal{D}^{\circ}_{AB} = (7.4 \times 10^{-8}) \frac{(112.56)^{1/2}(273)}{(1.05)(74.5)^{0.6}}$$

$$= 1.54 \times 10^{-5} \text{ cm}^2/\text{s}$$

The experimental value is $2.77 \times 10^{-5} \text{ cm}^2/\text{s}$ [96].

The Wilke-Chang equation has been modified for the diffusion of water, alcohols, and other solutes in monohydroxyl alcohols [138].

Scheibel Correlation   Scheibel [192] has proposed that the Wilke-Chang relation be modified to eliminate the association factor ϕ

$$\mathcal{D}^{\circ}_{AB} = \frac{KT}{\eta_B V_A^{1/3}} \tag{11-10.2}$$

where the symbols are defined in Eq. (11-10.1). K is determined by

$$K = (8.2 \times 10^{-8}) \left[1 + \left(\frac{3 V_B}{V_A} \right)^{2/3} \right] \tag{11-10.3}$$

except for *water* as a solvent, if $V_A < V_B$, use $K = 25.2 \times 10^{-8}$; for *benzene* as a solvent, if $V_A < 2V_B$, use $K = 18.9 \times 10^{-8}$; for other solvents, if $V_A < 2.5 V_B$, use $K = 17.5 \times 10^{-8}$.

Reddy-Doraiswamy Correlation   In this proposed relation [183] the association factor is not included, and

$$\mathcal{D}^{\circ}_{AB} = \frac{K' M_B^{1/2} T}{\eta_B (V_A V_B)^{1/3}} \quad \text{where } K' = \begin{cases} 10 \times 10^{-8} & \frac{V_B}{V_A} \leq 1.5 \\ \\ 8.5 \times 10^{-8} & \frac{V_B}{V_A} \geq 1.5 \end{cases} \tag{11-10.4}$$

with the remainder of the terms defined in Eq. (11-10.1).   Reddy and Doraiswamy found average errors of less than 20 percent for 96 binary systems, and even for the difficult cases of water as a solute the diffusion coefficient was usually predicted within 25 percent.

Example 11-7   Using the Scheibel and Reddy-Doraiswamy modifications to the Wilke-Chang equation, estimate the infinite-dilution diffusion coefficient of bromobenzene in ethylbenzene at 7.3°C.   The experimental value is 1.44×10^{-5} cm^2/s [183].

   solution   From Table 3-12, the value of V_A (bromobenzene) is given as 120 cm^3/g-mol.   For ethylbenzene, from Table 3-11, $V_B = (6)(C) + (10)(H) +$ six-member ring $= (6)(14.8) + (10)(3.7) - 15.0 = 110.8$ cm^3/g-mol.   The viscosity of ethylbenzene at 7.3°C is about 0.81 cP, and $M = 106.2$.

   Scheibel.   Since $V_A < 2.5 V_B$, $K = 18.9 \times 10^{-8}$; and from Eq. (11-10.2),

$$\mathscr{D}_{AB}^{\circ} = \frac{(18.9 \times 10^{-8})(280.5)}{(0.81)(110.8)^{1/3}}$$

$$= 1.36 \times 10^{-5} \text{ cm}^2/\text{s}$$

   Reddy-Doraiswamy.   Since $V_B/V_A < 1.5$, $K' = 10 \times 10^{-8}$; and from Eq. (11-10.4),

$$\mathscr{D}_{AB}^{\circ} = \frac{(10 \times 10^{-8})(106.2)^{1/2}(280.5)}{(0.81)[(120)(110.8)]^{1/3}}$$

$$= 1.51 \times 10^{-5} \text{ cm}^2/\text{s}$$

Both estimations are in reasonable agreement with the experimental value.   For this case, the Wilke-Chang relation would have predicted $\mathscr{D}_{AB}^{\circ} = 1.49 \times 10^{-5}$ cm^2/s.

In general, most investigators have found that the Wilke-Chang, Scheibel, and Reddy-Doraiswamy relations predict similar values of the infinite-dilution diffusion coefficient—except, perhaps, for water acting as a solute, in which case the Reddy-Doraiswamy relation is preferable.

   Lusis-Ratcliff Correlation   To estimate the mutual infinite-dilution diffusion coefficient in *organic solvents*, Lusis and Ratcliff [137] have proposed

$$\mathscr{D}_{AB}^{\circ} = \frac{(8.52 \times 10^{-8}) T}{\eta_B V_B^{1/3}} \left[1.40 \left(\frac{V_B}{V_A} \right)^{1/3} + \frac{V_B}{V_A} \right] \qquad (11\text{-}10.5)$$

This equation is similar in form to the Reddy-Doraiswamy relation although no solvent molecular weight is included.   The authors report that, with extensive testing, the errors found were less than those from the Wilke-Chang method and no association factor was necessary.   It is not recommended for systems where there are strong solute-solvent interactions.   For water diffusing as a solute, Olander's hypothesis that water behaves like a tetramer was found to be satisfactory, i.e., V_W should be 4 times the monomer value.   In addition, for acids diffusing into inert solvents, the molal volume should be doubled, as most acids are assumed to diffuse as dimers.   Equation (11-10.5) is *not* recommended for diffusion of solutes in water.   Hayduk and Cheng [97] indicate that large errors may result from using this relation for the diffusion of small molecules, e.g., ethane, in high-molecular-weight solvents, e.g., dodecane; however, good predictions were noted by Lo [130] for alkane-alkane systems.

King, Hsueh, and Mao Correlation [121]

$$\mathscr{D}^{\circ}_{AB} = 4.4 \times 10^{-8} \frac{T}{\eta_B} \left(\frac{V_B}{V_A}\right)^{1/6} \left(\frac{\Delta H_{v_B}}{\Delta H_{v_A}}\right)^{1/2} \tag{11-10.6}$$

with ΔH_v in calories per gram mole. Equation (11-10.6) is not suitable for cases where the solvent is viscous (the recommended limit of $\mathscr{D}^{\circ}_{AB}\eta_B/T$ is 1.5×10^7 cP cm^2/s K); also, it is not particularly accurate for diffusion coefficients in aqueous systems.

Hayduk-Laudie Correlation   Infinite dilution diffusion coefficients of nonelectrolytes in water were correlated by Hayduk and Laudie [98] in an equation very similar to one proposed earlier by Othmer and Thakar [172]

$$\mathscr{D}^{\circ}_{AW} = 13.26 \times 10^{-5} \, \eta_W^{-1.4} V_A^{-0.589} \tag{11-10.7}$$

where η_W = viscosity of water, cP

V_A = solute molal volume at normal boiling point, cm^3/g-mol

$\mathscr{D}^{\circ}$ = binary diffusion coefficient at infinite dilution, cm^2/s

To derive Eq. (11-10.7), Hayduk and Laudie used *only* data published since 1950. Note that no explicit temperature dependence is included. Average errors were reported to be only 5.9 percent. The authors state that equally good results are obtained with the Wilke-Chang relation (11-10.1) when the association factor for water is set equal to 2.26.

Example 11-8   Estimate the diffusion coefficient of ethyl acetate into water at 20°C using the Hayduk-Laudie correlation.

  solution   At 20°C, $\eta_W = 1.002$ cP and the molal volume of ethyl acetate at the normal boiling point is 106 cm^3/g-mol. With Eq. (11-10.7),

$$\mathscr{D}^{\circ}_{AW} = (13.26 \times 10^{-5})(1.002)^{-1.4}(106^{-0.589})$$
$$= 0.85 \times 10^{-5} \text{ cm}^2/\text{s}$$

The experimental value is $1.0 \pm 0.07 \times 10^{-5}$ cm^2/s [127].

A graph of the Hayduk-Laudie correlation is shown in Fig. 11-9. The predicted diffusion coefficient becomes quite sensitive to the value of the solute volume at low values of V_A.

A number of other infinite-dilution correlations have been suggested, but they are less accurate or less convenient than those noted above [1, 21, 25, 55, 56, 65, 70, 72, 108, 114–116, 177, 198, 213, 217].

Effect of Solvent Viscosity   Most of the estimation techniques

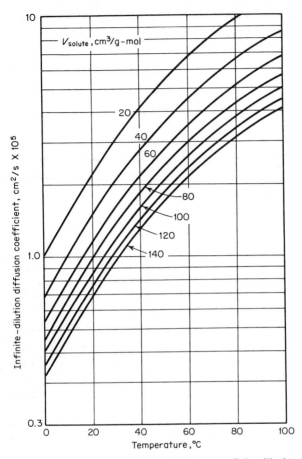

Fig. 11-9 Hayduk-Laudie correlation for infinite-dilution diffusion coefficients in aqueous solution.

introduced earlier in this section have assumed that $\mathscr{D}^{\circ}_{AB}$ varies inversely with the viscosity of the solvent, e.g., Eq. (11-10.1). This inverse dependence originated from the Stokes-Einstein relation for a large (spherical) molecule diffusing through a continuum solvent (small molecules). If, however, the solvent is viscous, one may legitimately question whether this simple relationship is applicable. Davies et al. [49] found, for CO_2, that in various solvents, $\mathscr{D}^{\circ}_{AB}\eta_B{}^{0.45} \approx$ constant for solvents ranging in viscosity from 1 to 27 cP, and these authors noted that in 1930, Arnold [10] had proposed an empirical estimation scheme where $\mathscr{D}^{\circ}_{AB} \propto \eta^{-0.5}$. Hayduk and Cheng [97] investigated the effect of solvent viscosity more extensively and proposed that for nonaqueous systems

$$\mathscr{D}^{\circ}_{AB} = A\eta_B{}^q \qquad (11\text{-}10.8)$$

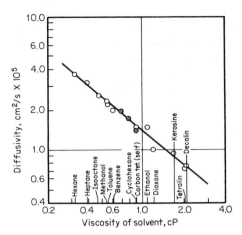
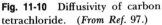

Fig. 11-10 Diffusivity of carbon tetrachloride. (*From Ref. 97.*)

where the constants A and q are particular for a given solute. In Fig. 11-10 the diffusion coefficient of carbon tetrachloride is shown as a function of solvent viscosity on a log-log plot. The applicability of Eq. (11-10.8) is clearly evident. Hayduk and Cheng list values of A and q for a few solutes [97]. A rough correlation exists between A and q, indicating that solutes with lower diffusion coefficients vary in a more pronounced manner with solvent viscosity (or temperature). In Fig. 11-11 data for CO_2 diffusion in various solvents are shown. The solvent-viscosity range is reasonably large, and the correlation for organic solvents is not bad. For contrast, the data for water are also

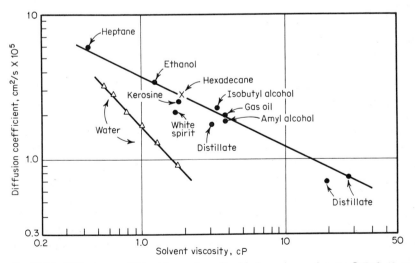

Fig. 11-11 Diffusion coefficient for carbon dioxide in various solvents; ● Ref. 49, × Ref. 97, △ Ref. 100.

shown [100].   These data fall well below the organic-solvent curve and have a slope close to -1.   Hiss and Cussler [102] measured diffusion coefficients of n-hexane and naphthalene in hydrocarbons with viscosities ranging from 0.5 to 5000 cP and report that $\mathcal{D}_{AB}^{\circ} \propto \eta_{B}^{-2/3}$ whereas Hayduk et al. [96] found for methane, ethane, and propane, $\mathcal{D}_{AB}^{\circ} \propto \eta_{B}^{-0.545}$.   Moore and Wellek [159] measured limiting diffusion coefficients of n-heptane and n-decane in various alkanes and alkanols.   When they correlated their data in the manner suggested by Eq. (11-10.8), they found that different values of A and q were required for a given solute, depending upon whether the solvent was an alkane or an alkanol.

These studies and others [72, 136, 226] show clearly that over wide temperature or solvent-viscosity ranges, simple empirical correlations, as presented earlier in this section, are inadequate.   The diffusion coefficient does not decrease in proportion to an increase in solvent viscosity, but $\mathcal{D}_{AB}^{\circ} \propto \eta^{q}$, where q varies, usually from -0.5 to -1.

Discussion and Recommendations: Diffusion Coefficients in Liquids: Aqueous Solutions of Nonelectrolytes   The Wilke-Chang, Scheibel, Hayduk-Laudie, Reddy-Doraiswamy, and King-Hsueh-Mao methods were tested.   The last two were found to be less accurate.   Results for the first three are shown in Table 11-4; they can all be reduced, for aqueous solutions, to the

TABLE 11-4   Diffusion Coefficients in Aqueous Solutions at Infinite Dilution†

				Percent error‡		
Solute	T, °C	$\mathcal{D}_{AW}^{\circ}$ (exp.), cm²/s × 10⁵	Ref.	Wilke and Chang, Eq. (11-10.1)	Scheibel, Eq. (11-10.2)	Hayduk and Laudie, Eq. (11-10.7)
Hydrogen	25	4.8	222	-28	-28	-34
Oxygen	25	2.41	222	0.4	3.7	-7.0
	29.6	3.49§	123	-21	-19	-27
Nitrogen	29.6	3.47§	123	-30	-30	-35
Nitrous oxide	25	2.67	48	-27	-28	-31
Carbon dioxide	25	2.00	222	2.0	1.5	-5.5
Ammonia	12	1.64	110	1.2	4.9	-6.7
Methane	2	0.85	237	10	9.4	2.3
	20	1.49	237	13	11	4.7
	60	3.55	237	15	13	5.0
n-Butane	4	0.50	237	22	12	14
	20	0.89	237	12	4.5	5.6
	60	2.51	237	-2.4	-9.6	-10
Propylene	25	1.44	222	-5.6	-11	-12
Methylcyclo-pentane	2	0.48	237	0	-8.3	-6.3
	10	0.59	237	6.8	-1.7	0

Solute	T, °C	$\mathscr{D}^{\circ}_{AW}$ (exp.), cm²/s × 10⁵	Ref.	Percent error‡ Wilke and Chang, Eq. (11-10.1)	Scheibel, Eq. (11-10.2)	Hayduk and Laudie, Eq. (11-10.7)
	20	0.85	237	0	−7.1	−5.9
	60	1.92	237	7.3	0	−1.0
Benzene	2	0.58	237	−6.9	−14	−14
	10	0.75	237	−5.3	−12	−12
	20	1.02	237	−6.9	−13	−12
	60	2.55	237	−8.6	−16	−16
Ethylbenzene	2	0.44	237	−2.3	−9.1	−9.1
	10	0.61	237	−6.6	−13	−13
	20	0.81	237	−6.2	−12	−11
	60	1.95	237	−4.6	−10	−12
Methyl alcohol	15	1.26	113	7.1	4.0	0
Ethyl alcohol	10	0.84	113	12	6.0	4.5
	15	1.00	110	11	4.0	3.0
	25	1.24	87, 113	18	11	10
n-Propyl alcohol	15	0.87	113	4.6	−2.3	−2.3
Isoamyl alcohol	15	0.69	110	1.4	−4.3	−4.3
Allyl alcohol	15	0.90	110	6.7	0	1.1
Benzyl alcohol	20	0.82	73	−1.2	−7.3	−6.1
Ethylene glycol	20	1.04	73	14	7.7	7.7
	25	1.16	23	31	10	10
	40	1.71	23	14	7.0	6.4
	55	2.26	23	17	10	8.0
	70	2.75	23	26	18	14
Glycerol	15	0.72	110	15	6.9	8.3
Acetic acid	20	1.19	127	2.5	−3.4	−4.2
Oxalic acid	20	1.53	110	−28	−33	−33
Benzoic acid	25	1.21	34	−25	−30	−29
Ethyl acetate	20	1.00	127	−10	−16	−15
Urea	20	1.20	110	5.0	−1.7	−2.5
	25	1.38	132	4.3	−1.4	−2.9
Diethylamine	20	0.97	127	−10	−14	−14
Acetonitrile	15	1.26	110	−10	−15	−16
Aniline	20	0.92	127	−4.3	−11	−10
Furfural	20	1.04	127	−7.7	−14	−13
Pyridine	15	0.58	110	33	24	26
Vinyl chloride	25	1.34	98	3.0	−3.0	−3.7
	50	2.42	98	0.8	−5.4	−7.0
	75	3.67	98	3.5	−3.0	−6.5
Absolute average error				10.7	10.6	11.1

†Experimental molar volumes at the boiling point were used wherever possible. If not available, the Bas volumes from Table 3-11 were estimated. The values calculated by the Wilke-Chang method employed $\phi = 2.6$, which they recommend. Hayduk and Laudie report that the average error by this method is reduced if ϕ for water is taken as 2.26.

‡% Error = [(calc. − exp.)/exp.] × 100.

§These values seem suspect.

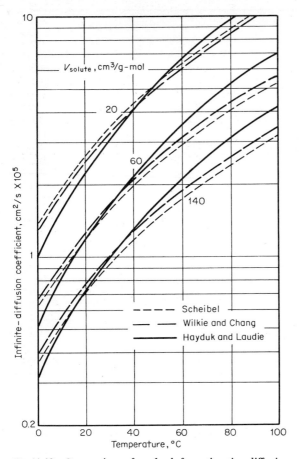

Fig. 11-12 Comparison of methods for estimating diffusion coefficients in aqueous solutions at infinite dilution.

form

$$\mathscr{D}_{AW}^{\circ} = f(T, V_A)$$

and in Fig. 11-12, this correlation is shown for values of $V_A = 20$, 60, and 140 cm^3/g-mol. It is obvious that, except for low values of V_A, all methods yield about the same predicted value of $\mathscr{D}_{AW}^{\circ}$ at a given temperature. Within the scatter of the data shown in Table 11-4, it is evident that any of these three latter methods will normally lead to values of $\mathscr{D}_{AW}^{\circ}$ within 10 to 15 percent.

The Hayduk-Laudie relation (11-10.7) is the simplest to use and is recommended here; the Wilke-Chang equation (11-10.1) gives similar results if ϕ for water is taken as 2.26 instead of 2.6.

TABLE 11-5 Diffusion Coefficients in Organic Liquids at Infinite Dilution

Solute A	Solvent B	T, °C	$\mathcal{D}^{\circ}_{AB}$ (exp.), cm²/s × 10⁵	Ref.	Percent error†				
					Wilke and Chang,‡ Eq. (11-10.1)	Scheibel, Eq. (11-10.2)	Reddy and Doraiswamy, Eq. (11-10.4)	Lusis and Ratcliff, Eq. (11-10.5)	King, Hsueh, and Mao, Eq. (11-10.6)
Acetone	Chloroform	25	2.35	86	39	−3.8	33	18	5.1
		40	2.90	86	39	−3.4	33	3.4	5.5
Benzene		15	2.51	113	0.4	−27	1.2	−15	−19
		25	2.89	190, 191	−1.0	−27	0.5	−16	−19
		40	3.55	190, 191	−0.3	−27	1.4	−15	−19
		55	4.25	190, 191	−1.9	−28	−0.5	−17	−20
n-Butyl acetate		25	1.71	181	27	5.3	46	8.2	14
Ethyl alcohol		15	2.20	113	53	−1.4	16	32	−10
Ethyl ether		15	2.07	113	15	−14	19	−2.9	4.3
		25	2.13	190, 191	27	−5.1	32	7.5	15
		25	2.15	6	26	−5.1	31	6.6	14
Ethyl acetate		25	2.02	181	34	1.0	40	13	9.9
Methyl ethyl ketone		25	2.13	181	34	−1.4	36	14	8.0
Acetic acid	Benzene	25	2.09	33	26	11	13	35	−3.3
Aniline		25	1.96	179	−3.1	−1.5	0.5	−1.5	−9.7
Benzoic acid		25	1.38	33	25	32	35	32	8.0
Bromobenzene		7.5	1.45	203	−10	−7.6	−6.2	−5.5	−6.9
Cyclohexane		25	2.09	190, 191	−12	−9.6	−7.6	−7.2	0.5
		40	2.65	190, 191	−11	−8.3	−6.4	−6.0	1.9
		60	3.45	190, 191	−7.0	−4.2	−2.1	−1.6	6.8
Ethyl alcohol		15	2.25	113	3.1	−11	−23	3.6	−22
Formic acid		25	2.28	33	40	13	−2.2	54	22
n-Heptane		25	2.10	26	−28	−20	−18	−23	−7.6
		45	2.75	26	−24	−15	−13	−19	−2.2
		65	3.65	26	−12	−13	−11	−18	−0.5
		80.1	4.25	26	−19	−9.8	−8.0	−15	3.2
Methyl ethyl ketone		30	2.09	5	7.1	5.3	7.6	14	11

TABLE 11-5 Diffusion Coefficients in Organic Liquids at Infinite Dilution (Continued)

Solute A	Solvent B	T, °C	$\mathcal{D}_{AB}^{\circ}$ (exp.), $cm^2/s \times 10^5$	Ref.	Percent error†				
					Wilke and Chang,‡ Eq. (11-10.1)	Scheibel, Eq. (11-10.2)	Reddy and Doraiswamy, Eq. (11-10.4)	Lusis and Ratcliff, Eq. (11-10.5)	King, Hsueh, and Mao, Eq. (11-10.6)
Oxygen		29.6	2.89	123	72	18	2.4	>100	>100
Naphthalene		7.5	1.19	203	-9.2	1.7	3.4	-3.4	0
Toluene		25	1.85	190, 191	-1.1	2.3	4.5	4.5	8.3
		40	2.39	190, 191	-2.1	1.5	3.6	4.0	7.3
1,2,4-Trichlorobenzene		7.5	1.34	203	-18	-9.0	-6.7	-13	-7.5
Vinyl chloride		7.5	1.77	203	5.6	-6.8	-5.1	13	9.0
Acetic acid	Acetone	15	2.92	15	39	31	34	61	16
		25	3.31	33	37	29	32	59	34
		40	4.04	233	33	25	28	61	11
Benzoic acid		25	2.62	33	13	29	32	32	6.8
Formic acid		25	3.77	33	56	31	13	85	44
Nitrobenzene		20	2.94	183	0.7	12	14	17	1.4
Water		25	4.56	169	§	40	22	>100	1.1
Bromobenzene	n-Hexane	7.3	2.60	233	16	6.1	8.5	30	19
Carbon tetrachloride		25	3.70	88	8.6	-2.2	0	21	21
Dodecane		25	2.73	217	-13	-1.5	0.7	-2.2	17
n-Hexane		25	4.21	139	-16	-20	-18	-6.9	4.3
Methyl ethyl ketone		30	3.74	5	23	5.9	8.3	39	24
Propane		25	4.87	96	6.0	-15	-25	22	24
Toluene		25	4.21	33	-6.9	-15	-13	22	0
Allyl alcohol	Ethyl alcohol	20	0.98	110	15	4.1	4.1	14	4.1
Isoamyl alcohol		20	0.81	110	1.2	4.9	4.9	2.5	12
Benzene		25	1.81	138	-40	-41	-41	-40	-31
Bromoform		20	0.97	207	-2.1	-5.1	-5.1	-2.1	7.2
Camphor		20	0.70	110	-14	-10	-10	-8.5	
Carbon dioxide		17	3.2	110	-48	-62	-68	-49	
Glycerol		20	0.51	110	90	82	82	90	29

Solute	Solvent	Temp.							
Iodine		25	1.32	33	−1.5	−12	−12	−2.3	−17
Iodobenzene		20	1.00	110	−19	−16	−16	−18	−8
Oxygen		29.6	2.64	123	1.0	−32	−42	4.5	38
Pyridine		20	1.10	110	−17	−18	−18	−18	−8.2
Urea		12	0.54	110	>100	70	70	>100	17
Water		25	1.24	124	§	47	24	>100	−16
Carbon tetrachloride		25	1.50	138	−29	−31	−31	−29	23
Adipic acid	n-Butyl alcohol	30	0.399	5	5.5	5.3	10	3.2	−38
Benzene		25	0.988	138	−48	−55	−53	−49	−30
Biphenyl		25	0.627	138	−44	−44	−41	−45	
Butyric acid		30	0.512	5	6.8	−6.3	−2.3	3.7	8.6
p-Dichlorobenzene		25	0.817	138	−49	−52	−50	−51	−39
Methyl alcohol		30	0.59	138	66	13	31	71	28
Oleic acid		30	0.251	5	1.1	7.6	1.7	5.6	51
Propane		0	1.02	96	−73	−78	−77	−73	−63
		25	1.57	15	−61	−69	−68	−62	−48
n-Propyl alcohol		30	0.40	138	65	35	40	63	57
Tartaric acid		30	0.405	5	12	8.6	13	9.3	
Water		25	0.56	138	§	35	21	>100	24
Acetic acid	Ethyl acetate	20	2.18	198	68	28	23	74	27
Acetone		20	3.18	198	2.5	−17	−6.6	7.5	−1.6
Ethyl benzoate		20	1.85	5	7.6	8.1	22	11	19
Methyl ethyl ketone		30	2.93	198	16	−1.0	12	19	17
Nitrobenzene		20	2.25	124	9.3	0	13	12	1.3
Water		25	3.20		§	39	34	>100	11
Benzene	n-Heptane	25	3.40	26	7.4	−14	−24	13	4.4
		45	4.40	26	5.0	−16	−25	11	2.3
		65	6.05	26	−4.1	−23	−31	1.0	−6.6
		98.4	8.40	26	3.1	−17	−26	8.4	0
Absolute average error					23.2	19.5	21.3	28.6	16.6

† % error = [(calc. − exp.)/exp.] × 100.

‡ In the Wilke-Chang method, φ = 1.0 except for ethyl alcohol, where φ = 1.5, and n-butanol, where φ = 1.2.

§ As noted in the text, the Wilke-Chang equation is not to be used when water is the solute.

Organic Solvents.   Experimental infinite-dilution liquid diffusion coefficients in eight different solvents are shown in Table 11-5. The percentage errors found for five estimation techniques are also given.

For solutes diffusing into organic solvents, we recommend the Scheibel method, Eq. (11-10.2).   Errors are, on the average, less than 20 percent.   Note that the temperature range of the data shown extends only from about 0 to 100°C with most values in the range between 15 to 30°C.   For *water* diffusing into organic solvents, $\mathscr{D}^\circ_{AW}$ can be estimated by calculating its value from the Wilke-Chang method and dividing by 2.3.   (See, however, Table 11-3.)

The relative merits of the several estimation methods, as implied by the average percentage errors shown in Tables 11-4 and 11-5, would doubtless be modified if other published data had been employed.   The tabulated list of diffusion coefficients, although lengthy, represents but a fraction of the data available in the literature and may not be sufficiently representative.   No attempt has been made to make a critical study of the experimental techniques used to obtain the values selected for the tables, and some may involve appreciable error.

No estimation methods for self- or tracer-diffusion coefficients were covered in this section.   For the interested reader, some pertinent references are 56, 58, 64, 87, 117, 128, 211, 214, 216, and 219.

11-11   Concentration Dependence of Binary Liquid Diffusion Coefficients

In Sec. 11-2 it was suggested that the diffusion coefficient $\mathscr{D}_{AB}$ may be proportional to $[(\partial \ln a)/(\partial \ln x)]_{T,P}$ in a binary mixture; a is the activity and x the mole fraction.   By virtue of the Gibbs-Duhem equation, $(\partial \ln a)/(\partial \ln x)$ is the same whether written for A or B.   In dilute solutions, as considered in Sec. 11-10, it is unity.   Defining

$$D_{AB} \equiv \frac{\mathscr{D}_{AB}}{[(\partial \ln a)/(\partial \ln x)]_{T,P}} \tag{11-11.1}$$

means that D_{AB} is presumably less sensitive to composition than $\mathscr{D}_{AB}$.

Several liquid models indicate how D_{AB} might be expected to vary with mole fraction; e.g., the Darken equation [47, 76] predicts that

$$\frac{D_{AB}}{D^*_A x_A + D^*_B x_B} = 1.0 \tag{11-11.2}$$

where D^*_A and D^*_B are tracer-diffusion coefficients at x_A, x_B.   McCall and Douglass [140] tested this hypothesis for four binary systems, and their results are shown in Fig. 11-13.   Vignes [220] shows a similar plot.   Although in some cases the Darken equation does provide a reasonable prediction of how $\mathscr{D}_{AB}$ varies with composition, there are

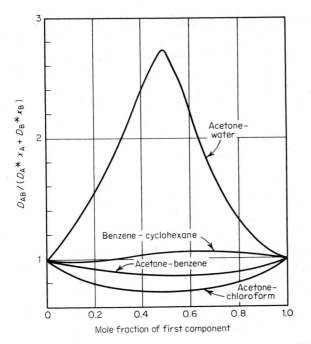

Fig. 11-13   Test of Darken's equation.   (*Data from Ref.* 140.)

many instances where large deviations are noted.   It has been suggested [27, 29, 120, 157] that the Darken form is valid except where association compounds are formed in the binary; correction factors have been derived to account for this assumed association.   Hardt et al. [89] also show several comparisons between the experimental diffusion coefficient and that predicted from Darken's relation; for quite nonideal mixtures, the agreement was poor.   For n-octane–n-dodecane systems, the Darken equation has been shown to correlate the data well [218].   Diffusivity measurements by Ghai and Dullien [75] on the systems toluene-CCl_4, n-hexane–CCl_4, and n-hexane–toluene have resulted in a 4 to 8 percent maximum deviation from the Darken equation.   As there is no association in these systems, dissimilarities in molecular shapes have been assumed to be the main factor responsible for the observed effects.   The Darken equation probably applies only in the limit when the two components are similar, e.g., benzene-diphenyl or n-octane–n-decane.

Another theory predicts that the group $D_{AB}\eta$ should be a linear function of mole fraction [6, 15, 23].   Vignes [220] shows graphs indicating this is not even approximately true for the systems acetone-water and acetone-chloroform.   Rao and Bennett [179] studied several very nonideal mixtures and found that while the group $D_{AB}\eta$ did not

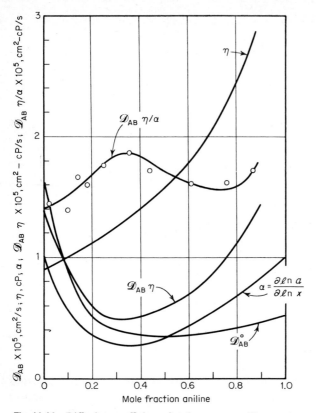

Fig. 11-14 Diffusion coefficients for the system aniline–carbon tetrachloride at 25°C. (*From Ref. 179.*)

vary appreciably with composition, no definite trends could be discerned. One of the systems studied (aniline–carbon tetrachloride) is shown in Fig. 11-14. In this case, $\mathscr{D}_{AB}$, η, α [$= (\partial \ln a)/(\partial \ln x)$], and $\mathscr{D}_{AB}\eta$ varied widely; the group $\mathscr{D}_{AB}\eta/\alpha$ ($= D_{AB}\eta$) also showed an unusual variation with composition. Carman and Stein [30] showed that this type of concentration dependence is satisfactory for the nearly ideal system benzene–carbon tetrachloride and for the nonideal system acetone-chloroform but failed for ethyl alcohol–water. Haluska and Colver [86] and Rathbun and Babb [182] also tested this theory and reported poor results. The latter authors suggest that the $(\partial \ln a)/(\partial \ln x)$ term is overcorrecting the concentration dependence of $\mathscr{D}_{AB}$ and indicate they found better results if this derivative were raised to a fractional power. For associated systems, the exponent found best was 0.6; for systems exhibiting negative deviations from Raoult's law, an exponent of 0.3 was recommended.

Vignes [220] suggested a convenient way of correlating the composi-

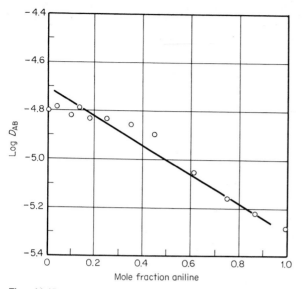

Fig. 11-15 Vignes plot for the system aniline–carbon tetrachloride at 25°C.

tion effect on liquid diffusion coefficient:

$$D_{AB} = (\mathscr{D}^{\circ}_{AB})^{x_B}(\mathscr{D}^{\circ}_{BA})^{x_A} \qquad (11\text{-}11.3)$$

and therefore a plot of $\log D_{AB}$ vs. mole fraction should be linear. He illustrated this relation with many systems, and, with the exception of strongly associated mixtures, excellent results were obtained. Figure 11-15 shows the same aniline–carbon tetrachloride system plotted earlier in Fig. 11-14. While it is not perfect, there is a good agreement with Eq. (11-11.3).

Dullien [57] carried out a statistical test of Vignes' correlation. While it was found to fit experimental data extremely well for ideal or nearly ideal mixtures, there were several instances where it was not particularly accurate for nonideal, nonassociating solutions. Other authors report that the Vignes correlation is satisfactory for benzene and n-heptane [26], toluene and methylcyclohexane [85], but not for benzene and cyclohexane [131].

The Vignes relation can be derived from absolute-rate theory, and a logical modification of this equation is found [125] to be

$$D_{AB}\eta = (\mathscr{D}^{\circ}_{AB}\eta_B)^{x_B}(\mathscr{D}^{\circ}_{BA}\eta_A)^{x_A} \qquad (11\text{-}11.4)$$

A test with 11 systems showed that this latter form was marginally better in fitting experimental data. In Fig. 11-16 we have plotted both $\log D_{AB}\eta$ and $\log D_{AB}$ as a function of composition for the aniline-

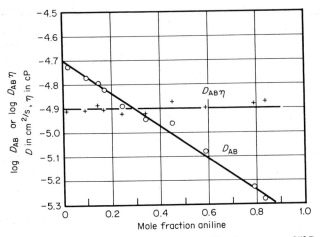

Fig. 11-16 Vignes plot for the system aniline-benzene at 25°C. (*Data from Ref. 179.*)

benzene system.   The original Vignes equation fits the data well, but so does Eq. (11-11.4); in fact, for the latter $D_{AB}\eta$ is essentially constant.

Several other correlation methods have been proposed.   Haluska and Colver [86] recommend an equation which relates D_{AB} to composition but requires $\mathscr{D}^{\circ}_{AB}$, $\mathscr{D}^{\circ}_{BA}$, the solvent, solute, and mixture viscosities, and the solute and solvent molar volumes.   In a test with seven very nonideal binaries, their relation correlated experimental data better than the Vignes equation in five cases.   Ratcliff and Holdcroft [180] suggest that for the diffusion of gases into electrolytes, a Henry's-law form be employed to modify the diffusion coefficient as the solution viscosity changes with electrolyte strength.

Anderson [8] finds that for the effect of concentration on the diffusion coefficients of *macromolecules*, the volume fraction of solute is important. For hemoglobin or serum albumin diffusing in salt solutions

$$\frac{\mathscr{D}_{AB}}{\mathscr{D}^{\circ}_{AB}} = (1 - \Phi)^{6.5} \tag{11-11.5}$$

where $\mathscr{D}^{\circ}_{AB}$ = infinite-dilution coefficient
Φ = solute volume fraction

No activity correction was found necessary.

Other correlation methods have been proposed [7, 36, 42, 71, 74], but they are either less accurate or less general than those discussed above.

Summary and Recommendations   No single correlation is always satisfactory for estimating the concentration effect on liquid diffusion coefficients.   The Vignes method [Eq. (11-11.3)] is recommended here as a well-tested and generally accurate correlation. It is also the easiest to apply, and no mixture viscosities are necessary. Still, the quantity $(\partial \ln a)/(\partial \ln x)$ must be known or estimated from other sources such as vapor-liquid equilibrium data. This correlation should not be used for strongly associating mixtures, but no other method works in such cases.

11-12 The Effect of Temperature on Diffusion in Liquids

Most of the infinite-dilution liquid-diffusion-estimation techniques covered in Sec. 11-10 accounted for temperature variations by assuming that the group $\mathcal{D}_{AB}^{\circ}\eta_B/T$ was constant. This assumption works reasonably well, as evidenced by the comparison of predicted and experimental values of $\mathcal{D}_{AB}^{\circ}$ in Tables 11-4 and 11-5. Most experimental data cover only a small temperature range, however.

Robinson, Edmister, and Dullien [187] investigated the temperature dependence of the system 1-heptanol and cyclohexanone from 10 to 90°C. They found an excellent correlation between $\ln \mathcal{D}_{AB}$ and T^{-1} over the entire temperature range at several compositions.† The group $\mathcal{D}_{AB}\eta/T$ was found to vary with temperature as shown below:

	Temperature, °C			
	10	25	55	90
Mole fraction 1-heptanol	$\mathcal{D}_{AB}\eta/T \times 10^{-8}$, cm² cP/s K			
0	6.82	6.62	5.33	4.45
1	3.88	3.89	3.68	3.54

The exponential dependence of $\mathcal{D}_{AB}$ on T^{-1} has been suggested by several authors [80, 109, 132, 141, 212], and from plots of $\ln \mathcal{D}_{AB}$ vs. T^{-1} activation energies for diffusion are determined.

There are, at present, too few data over a wide temperature range to allow definitive conclusions to be drawn. Though Robinson et al. found $\mathcal{D}_{AB}\eta/T$ to vary, Vadovic and Colver [217] found $\mathcal{D}_{AB}^{\circ}\eta/\rho T$ to be essentially constant (ρ is the density). Over small temperature ranges, all these groups are nearly invariant. Since η is normally a strong function of temperature, however, it follows that $\mathcal{D}_{AB}^{\circ}$ is likewise; the rates of diffusional processes increase significantly with increasing temperature. A few $\mathcal{D}_{AB}^{\circ}$ values from Tables 11-4 and 11-5 are plotted against T^{-1} in Fig. 11-17 to indicate the strong variation. Although

---

†An equally good correlation was obtained when $\ln \mathcal{D}_{AB}$ was plotted vs. $\ln \eta$, but $\ln \eta$ was itself shown to be proportional to T^{-1}.

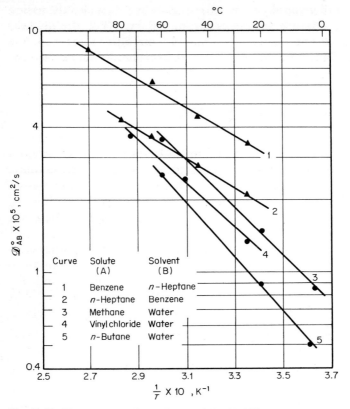

Fig. 11-17   Temperature dependence of liquid diffusion coefficients.

straight lines were drawn for all systems, in most, a curve would have been more suitable.

11-13   Diffusion in Multicomponent Liquid Mixtures

In a binary liquid mixture, as indicated in Secs. 11-2 and 11-9, a single diffusion coefficient was sufficient to express the proportionality between the flux and concentration gradient.   In multicomponent systems, the situation is considerably more complex, and the flux of a given component depends upon the gradient of $n-1$ components in the mixture.   For example, in a ternary system of A, B, and C, the flux of A can be expressed as

$$J_A = \bar{\mathscr{D}}_{AA} \nabla C_A + \bar{\mathscr{D}}_{AB} \nabla C_B \qquad (11\text{-}13.1)$$

Similar relations can be written for J_B and J_C.   The coefficients $\bar{\mathscr{D}}_{AA}$ and

$\bar{\mathscr{D}}_{BB}$ are called the *main* coefficients; they are *not* self-diffusion coefficients. $\bar{\mathscr{D}}_{AB}$, $\bar{\mathscr{D}}_{BA}$, etc., are *cross*-coefficients, as they relate the flux of a component i to a gradient of j. $\bar{\mathscr{D}}_{ij}$ is not normally equal to $\bar{\mathscr{D}}_{ji}$ for multicomponent systems. Thus for a ternary system, there are *four* diffusion coefficients.

Frames of reference in multicomponent systems must be clearly defined. Yon and Toor [240] discuss some limitations in this context.

Diffusion in Mixed Solvents   One important case of multicomponent diffusion results when a solute diffuses through a homogeneous solution of mixed solvents. When the solute is dilute, there are no concentration gradients for the solvent species and one can speak of a single-solute diffusivity with respect to the mixture $\mathscr{D}^{\circ}_{Am}$. This problem has been discussed by several authors [45, 103, 175, 206] and empirical relations proposed for $\mathscr{D}^{\circ}_{Am}$.† Perkins and Geankoplis [175] evaluated most of the methods with available data and suggested a simple correlation of their own:

$$\mathscr{D}^{\circ}_{Am}\eta_m{}^{0.8} = \sum_{\substack{j=1 \\ j \neq A}}^{n} x_j \mathscr{D}^{\circ}_{Aj}\eta_j{}^{0.8} \qquad (11\text{-}13.2)$$

where $\mathscr{D}^{\circ}_{Am}$ = *effective* diffusion coefficient for dilute solute A into mixture, cm²/s

$\mathscr{D}^{\circ}_{Aj}$ = infinite-dilution binary diffusion coefficient of solute A into solvent j, cm²/s

x_j = mole fraction of j

η_m = mixture viscosity, cP

η_j = pure-component viscosity, cP

When this was tested with data for eight ternary systems, errors were normally less than 20 percent, except for the case where CO_2 diffused into an ethyl alcohol–water solvent. These same authors also suggested that the Wilke-Chang equation (11-10.1) might be modified to include the mixed-solvent case, i.e.,

$$\mathscr{D}^{\circ}_{Am} = 7.4 \times 10^{-8} \frac{(\phi M)^{1/2} T}{\eta_m V_A{}^{0.6}} \qquad (11\text{-}13.3)$$

$$\phi M = \sum_{\substack{j=1 \\ j \neq A}}^{n} x_j \phi_j M_j \qquad (11\text{-}13.4)$$

While not extensively tested, Eq. (11-13.3) may provide a rapid, reasonably accurate estimation method.

---

†Equation (11-8.4), which treats a similar situation for gases, is not applicable for liquid mixtures.

Example 11-9 Estimate the diffusion coefficient of acetic acid diffusing into a mixed solvent containing 20.7 mole percent ethyl alcohol in water. The acetic acid concentration is small. Assume the temperature is 25°C.

Data. Let E = ethyl alcohol, W = water, and A = acetic acid. At 25°C, $\eta_E = 1.096$ cP, $\eta_W = 0.8937$ cP, $\mathscr{D}_{AE}^{\circ} = 1.032 \times 10^{-5}$ cm²/s, $\mathscr{D}_{AW}^{\circ} = 1.295 \times 10^{-5}$ cm²/s, and for the solvent mixture under consideration, $\eta_m = 2.350$ cP.

solution From Eq. (11-13.2)

$$\mathscr{D}_{Am}^{\circ} = \left(\frac{1}{2.350}\right)^{0.8} [(0.207)(1.032 \times 10^{-5})(1.096)^{0.8}$$
$$+ (0.793)(1.295 \times 10^{-5})(0.8937)^{0.8}]$$
$$= 0.59 \times 10^{-5} \text{ cm}^2/\text{s}$$

The experimental value reported by Perkins and Geankoplis is 0.571×10^{-5} cm²/s. Note that this value is significantly below the two limiting binary values. The decrease appears to be closely related to the increase in solvent-mixture viscosity relative to the pure components.

Had the modified Wilke-Chang equation been used, with $V_A = 64.1$ cm³/g-mol (Table 3-12), $\phi_E = 1.5$, and $\phi_W = 2.6$, and with $M_E = 46$ and $M_W = 18$, using Eqs. (11-13.3) and (11-13.4) would give

$$\phi M = (0.207)(1.5)(46) + (0.793)(2.6)(18) = 51.39$$
$$\mathscr{D}_{Am}^{\circ} = (7.4 \times 10^{-8}) \frac{(51.39)^{1/2}(298)}{(2.350)(64.1)^{0.6}}$$
$$= 0.55 \times 10^{-5} \text{ cm}^2/\text{s}$$

Multicomponent Diffusion Coefficients There are no convenient and simple methods available for estimating multicomponent-liquid diffusion coefficients in the general case. Cullinan [39, 44]† has discussed the possibility of adapting the Vignes correlation [Eq. (11-11.3)] to ternary systems, but the requirement for extensive thermodynamic-activity data for the mixture has severely limited the applicability of the method. Kett and Anderson [118, 119] apply hydrodynamic theory to estimate ternary diffusion coefficients. Although some success was achieved, the method again requires extensive data on activities, pure-component and mixture volumes, and viscosities as well as tracer and binary diffusion coefficients.

Other authors [22, 28, 43, 126, 161, 162] have also discussed the problem of obtaining multicomponent-liquid diffusion coefficients. The problem is complex, and simple, practical estimation methods have not yet been developed.

11-14 Diffusion in Electrolyte Solutions

When a salt dissociates in solution, it is the ions rather than the molecules that diffuse. In the absence of an electric potential, however, the diffusion of a single salt may be treated as molecular diffusion.

†Interesting correspondence relative to the ideas developed may be found in Refs. 40, 41, 160, and 221.

The theory of diffusion of salts at low concentrations is well developed. At concentrations encountered in most industrial processes, one normally resorts to empirical corrections, with a concomitant loss in generality and accuracy. A comprehensive discussion of this subject is available [167].

For dilute solutions of a *single* salt, the diffusion coefficient is given by the Nernst-Haskell equation

$$\mathscr{D}^{\circ}_{AB} = \frac{RT}{\mathscr{F}^2} \frac{1/n_+ + 1/n_-}{1/\lambda^{\circ}_+ + 1/\lambda^{\circ}_-} \qquad (11\text{-}14.1)$$

where $\mathscr{D}^{\circ}_{AB}$ = diffusion coefficient at infinite dilution, based on molecular concentration, cm^2/s

T = temperature, K

R = gas constant, J/g-mol K = 8.314

$\lambda^{\circ}_+, \lambda^{\circ}_-$ = limiting (zero-concentration) ionic conductances, A/cm^2 (V/cm) (g-equiv/cm^3)

n_+, n_- = valences of cation and anion, respectively

$\mathscr{F}$ = faraday = 96,500 C/g-equiv

Values of λ°_+ and λ°_- can be obtained for many ionic species at 25°C from Table 11-6 or from alternate sources [158, 185]. If values of λ°_+ and λ°_- at other temperatures are needed, an approximate correction factor is $T/334\eta_w$, where η_w is the viscosity of water at T in centipoises.

TABLE 11-6 Limiting Ionic Conductances in Water at 25°C [91]

$A/(cm^2)(V/cm)(g\text{-}equiv/cm^3)$

Cation	λ°_+	Anion	λ°_-
H^+	349.8	OH^-	197.6
Li^+	38.7	Cl^-	76.3
Na^+	50.1	Br^-	78.3
K^+	73.5	I^-	76.8
NH_4^+	73.4	NO_3^-	71.4
Ag^+	61.9	ClO_4^-	68.0
Tl^+	74.7	HCO_3^-	44.5
$\frac{1}{2}Mg^{2+}$	53.1	HCO_2^-	54.6
$\frac{1}{2}Ca^{2+}$	59.5	$CH_3CO_2^-$	40.9
$\frac{1}{2}Sr^{2+}$	50.5	$ClCH_2CO_2^-$	39.8
$\frac{1}{2}Ba^{2+}$	63.6	$CNCH_2CO_2^-$	41.8
$\frac{1}{2}Cu^{2+}$	54	$CH_3CH_2CO_2^-$	35.8
$\frac{1}{2}Zn^{2+}$	53	$CH_3(CH_2)_2CO_2^-$	32.6
$\frac{1}{3}La^{3+}$	69.5	$C_6H_5CO_2^-$	32.3
$\frac{1}{3}Co(NH_3)_6^{3+}$	102	$HC_2O_4^-$	40.2
		$\frac{1}{2}C_2O_4^{2-}$	74.2
		$\frac{1}{2}SO_4^{2-}$	80
		$\frac{1}{3}Fe(CN)_6^{3-}$	101
		$\frac{1}{4}Fe(CN)_6^{4-}$	111

As the salt concentration becomes finite and increases, the diffusion coefficient decreases rapidly and then usually rises, often becoming greater than $\mathscr{D}_{AB}^\circ$ at high normalities. Figure 11-18 illustrates the typical trend for three simple salts. The initial decrease at low concentrations is proportional to the square root of the concentration, but deviations from this trend are usually significant above 0.1 N.

No reliable method has yet been proposed to relate $\mathscr{D}_{AB}^\circ$ to concentration. Gordon [81], however, has proposed an empirical equation which has been applied to systems at concentrations up to $2N$:

$$\mathscr{D}_{AB} = \mathscr{D}_{AB}^\circ \frac{1}{\rho_s \bar{V}_s} \frac{\eta_s}{\eta} \left(1 + m \frac{\partial \ln \gamma_\pm}{\partial m}\right) \qquad (11\text{-}14.2)$$

where $\mathscr{D}_{AB}^\circ =$ diffusion coefficient at infinite dilution [Eq. (11-14.1)], cm^2/s

$\rho_s =$ molal density of solvent, g-mol/cm^3

$\bar{V}_s =$ partial molal volume of solvent, cm^3/g-mol

$\eta_s =$ viscosity of solvent, cP

$\eta =$ viscosity of solution, cP

$m =$ molality of solute, g-mol/1000 g solvent

$\gamma_\pm =$ mean ionic activity coefficient of solute

In many cases, the product $\rho_s \bar{V}_s$ is close to unity, as is the viscosity ratio η_s/η, so that Gordon's relation provides an activity correction to the diffusion coefficient at infinite dilution. Though Harned and Owen [91] tabulate $\gamma_\pm$ as a function of m for many aqueous solutions, there now exist several semiempirical correlation techniques for relating $\gamma_\pm$ to concentration. Bromley [20] presents an analytical relation, and Meissner et al. [147–151] show generalized graphical correlations.

The solid lines in Fig. 11-18 are calculated from Eq. (11-14.2). The agreement is good.

Other equations have been suggested for relating $\mathscr{D}_{AB}$ to concentration [90, 236], but Eq. (11-14.2) is reasonably reliable and not difficult to apply.

Data on the diffusion of CO_2 into electrolyte solutions have been reported by Ratcliff and Holdcroft [180]. The diffusion coefficient was found to decrease linearly with an increase in salt concentration.

Recommendations   For very dilute solutions of electrolytes, employ Eq. (11-14.1). Where values of the limiting ionic conductances in water are not available at the desired temperature, use those given in Table 11-6 for 25°C and multiply $\mathscr{D}_{AB}^\circ$ at 25°C by $T/334\eta_w$, where η_w is the viscosity of water at T in centipoises.

For concentrated solutions, use Eq. (11-14.2). If values of $\gamma_\pm$ and λ° are not available at T, calculate $\mathscr{D}_{AB}$ at 25°C and multiply this by

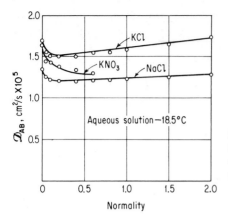

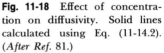

Fig. 11-18 Effect of concentration on diffusivity. Solid lines calculated using Eq. (11-14.2). (*After Ref.* 81.)

$(T/298)[(\eta \text{ at } 25°)/(\eta \text{ at } T)]$. If necessary, the ratio of the solution viscosity at 25°C to that at T may be assumed to be the same as the corresponding ratio for water.

Example 11-10 Estimate the diffusion coefficient of NaOH in a $2N$ aqueous solution at 15°C.

solution From the data on densities of aqueous solutions of NaOH given in Perry [176], it is evident that, up to 12 weight percent NaOH (about $3N$), the density increases almost exactly in inverse proportion to the weight fraction of water; i.e., the ratio of moles of water per 1000 cm³ is essentially constant at 55.5. Thus both V/n_1 and $\bar{V}_1$ are very nearly 55.5 and cancel in Eq. (11-14.2). In this case, the molality m is essentially identical with the normality.

In plotting the values of $\gamma_\pm$ for NaOH at 25°C [91] vs. the molality m, the slope at $2m$ is approximately 0.047. Hence

$$m\frac{\partial \ln \gamma_\pm}{\partial m} = \frac{m}{\gamma_\pm}\frac{\partial \gamma_\pm}{\partial m} = \frac{2}{0.698} \times 0.047 = 0.135$$

The value 0.698 is the mean activity coefficient at $m = 2$.

The viscosities of water and $2N$ NaOH solution at 25°C are 0.894 and 1.42 cP, respectively. Substituting in Eqs. (11-14.1) and (11-14.2) gives

$$\mathscr{D}_{AB}^\circ = \frac{(2)(8.314)(298)}{(\frac{1}{50} + \frac{1}{198})(96{,}500)^2}$$

$$= 2.12 \times 10^{-5} \text{ cm}^2/\text{s}$$

$$\mathscr{D}_{AB} = 2.12 \times 10^{-5}\frac{55.5}{55.5}\frac{0.894}{1.42}[1 + (2)(0.135)]$$

$$= 1.70 \times 10^{-5} \text{ cm}^2/\text{s} \qquad \text{at } 25°\text{C}$$

At 15°C the viscosity of water is 1.144 cP, and so the estimated value of $\mathscr{D}_{AB}$ at 15°C is

$$1.70 \times 10^{-5}\frac{288}{334 \times 1.144} = 1.28 \times 10^{-5} \text{ cm}^2/\text{s}$$

which may be compared with the ICT [110] value of $1.36 \times 10^{-5} \text{ cm}^2/\text{s}$.

Diffusion in Mixed Electrolytes In a system of *mixed electrolytes*, as in the simultaneous diffusion of HCl and NaCl in water, the faster-moving

H^+ ion may move ahead of its Cl^- partner, the electric current being maintained at zero by the lagging behind of the slower-moving Na^+ ions.   In such systems, the unidirectional diffusion of each ion species results from a combination of electric and concentration gradients:

$$N_+ = \frac{\lambda_+}{\mathscr{F}^2}\left(-RT\frac{\partial c_+}{\partial y} - \mathscr{F}c_+\frac{\partial E}{\partial y}\right) \tag{11-14.3}$$

$$N_- = \frac{\lambda_-}{\mathscr{F}^2}\left(-RT\frac{\partial c_-}{\partial y} - \mathscr{F}c_-\frac{\partial E}{\partial y}\right) \tag{11-14.4}$$

where N_+, N_- = diffusion flux densities of cation and anion, respectively, g-equiv/cm^2 s

c_+, c_- = corresponding ion concentrations, g-equiv/cm^3

$\partial E/\partial y$ = gradient of electric potential

This gradient may be imposed externally but is present in the ionic solution even if there is no external electrostatic field, owing to the small separation of charges which result from diffusion itself.   Collision effects, ion complexes, and activity corrections are neglected.

One equation for each cation and one for each anion can be combined with the requirement of zero current at any y: $\Sigma N_+ = \Sigma N_-$.   Solving for the unidirectional flux densities [181] gives

$$n_+N_+ = -\frac{RT\lambda_+}{\mathscr{F}^2 n_+}\left(G_+ - n_+c_+\frac{\Sigma\lambda_+G_+/n_+ - \Sigma\lambda_-G_-/n_-}{\Sigma\lambda_+c_+ + \Sigma\lambda_-c_-}\right) \tag{11-14.5}$$

$$n_-N_- = \frac{RT\lambda_-}{\mathscr{F}^2 n_-}\left(G_- + n_-c_-\frac{\Sigma\lambda_+G_+/n_+ - \Sigma\lambda_-G_-/n_-}{\Sigma\lambda_+c_+ + \Sigma\lambda_-c_-}\right) \tag{11-14.6}$$

where G_+ and G_- are the concentration gradients $\partial c/\partial y$ in the direction of diffusion.

Vinograd and McBain [221a] have used these relations to represent their data on diffusion in multi-ion solutions. $\mathscr{D}_{AB}$ for the hydrogen ion was found to decrease from 12.2 to 4.9×10^{-5} cm^2/s in a solution of HCl and BaCl$_2$ when the ratio of H^+ to Ba^{2+} was increased from zero to 1.3; $\mathscr{D}_{AB}$ at the same temperature is 9.03×10^{-5} for the free H^+ ion, and 3.3×10^{-5} for HCl in water.   The presence of the slow-moving Ba^{2+} accelerates the H^+ ion, the potential existing with zero current causing it to move in dilute solution even faster than it would as a free ion with no other cation present, i.e., electrical neutrality is maintained by the hydrogen ions moving ahead of the chlorine, faster than they would as free ions, while the barium diffuses more slowly than as a free ion.

The interaction of ions in a multi-ion system is important when the several ion conductances differ greatly, as they do when H^+ or OH^- are diffusing.   Where the diffusion of these two ions is not involved, no great error is introduced by the use of "molecular" diffusion coefficients for the salt species present.

The relations proposed by Vinograd and McBain are not adequate to represent a ternary system, where, as indicated in Sec. 11-13, four independent diffusion coefficients must be known to predict fluxes. The subject of ion diffusion in multicomponent systems is covered in detail in the papers by Wendt [229] and Miller [155, 156] in which it is demonstrated how one can obtain multicomponent ion-diffusion coefficients, although the data required are usually not available.

NOTATION

a = activity

c = concentration, usually g-mol/cm³; c_A (c_B), for component A (B); c_+, c_-, ion concentrations, g-equiv/cm³

$\mathscr{D}_{AB}$ = binary diffusion coefficient of A diffusing into B, cm²/s; $\mathscr{D}_{AB}^\circ$, at infinite dilution of A in B; $\widetilde{\mathscr{D}}_{AB}$, cross-coefficient in multicomponent mixtures; $\mathscr{D}_{im}$, of i into a homogeneous mixture; $\mathscr{D}_{AB,}$, reduced coefficient in Eq. (11-4.2)

$\mathscr{D}_A^*$ = tracer-diffusion coefficient of A, cm²/s

$\mathscr{D}_{AA}$ = self-diffusion coefficient of A, cm²/s; $\widetilde{\mathscr{D}}_{AA}$, main coefficient for A in multicomponent diffusion

D_{AB} = activity-corrected diffusion coefficient defined by Eq. (11-2.9), cm²/s

E = electric potential

$\Delta E_\eta, \Delta E_D$ = activation energies for viscosity and diffusion, cal/g-mol

f_D = correction term in Eq. (11-3.1)

$\mathscr{F}$ = faraday = 96,500 C/g-equiv

G_+, G_- = $\partial c_+/\partial y$ or $\partial c_-/\partial y$

ΔH_v = heat of vaporization at normal boiling point, usually cal/g-mol

J_A = flux of A, g-mol/cm² s; J_A^M, flux relative to a plane of no net mole flow; J_A^V, flux relative to a plane of no net volume flow

k = Boltzmann's constant, 1.3805×10^{-23} J/K

K = Scheibel constant in Eq. (11-10.2)

K' = Reddy-Doraiswamy constant in Eq. (11-10.4)

m = molality of solute, g-mol/1000 g solvent

M_A = molecular weight of A

n = number density of molecules

n_+, n_- = valences of cation and anion, respectively

N_A = flux of A relative to fixed coordinate plane, g-mol/cm² s

N_0 = Avogadro's number, 6.023×10^{23}

P = pressure, usually atm; P_c, critical pressure; P_r, reduced pressure, P/P_c

r = distance of separation between molecules, Å

r_A = molecular radius in Stokes-Einstein equation

R = gas constant, 8.314 J/mol K or 1.987 cal/mol K

T = temperature, K; T_b, at normal boiling point; T_c, critical temperature; T_r, reduced temperature T/T_c

T^* = kT/ϵ

V = volume, cm³/g-mol; V_b, at normal boiling point for pure liquid; V_c, critical volume; V_r, reduced volume V/V_c; $\bar{V}_A$, partial molal volume of A

V_A = solute molal volume, use V_{A_b} where known

$\Sigma(v)_A$ = atomic diffusion volumes from Table 11-1, cm³/g-mol

x_A = mole fraction of A

y = distance, cm

Greek

$\alpha = (\partial \ln a)/(\partial \ln x)$

$\beta = M^{1/2} P_c^{1/3} / T_c^{5/6}$

γ = activity coefficient; $\gamma_\pm$, mean ionic activity coefficient

δ = polar parameter defined in Eq. (11-3.9)

ϵ = characteristic energy parameter; ϵ_A, for pure A; ϵ_{AB}, for an A-B interaction

η = viscosity, usually cP; η_A, pure A

$\lambda_+^\circ, \lambda_-^\circ$ = limiting (zero-concentration) ionic conductances, A/cm^2 (V/cm) (g-equiv/cm^3)

μ_A = chemical potential of A, cal/g-mol

ρ = density, g/cm^3 or g-mol/cm^3; ρ_c, critical density; ρ_r, reduced density, ρ/ρ_c

σ = characteristic-length parameter, A; σ_A for pure A; σ_{AB}, for an A-B interaction

ϕ = association parameter for the solvent, Eq. (11-10.1)

Φ_A = volume fraction of A

$\psi(r)$ = intermolecular potential energy of interaction

Ω_D = collision integral for diffusion; Ω_v, for viscosity

Superscripts

$\circ$ = infinite dilution or low pressure

$*$ = tracer value

Subscripts

A, B = components A and B

m = mixture

$_w$ = water

s = solvent

REFERENCES

1. Akgerman, A., and J. L. Gainer: *J. Chem. Eng. Data*, **17**: 372 (1972).
2. Amdur, I., and T. F. Schatzki: *J. Chem. Phys.*, **27**: 1049 (1957).
3. Amdur, I., and L. M. Shuler: *J. Chem. Phys.*, **38**: 188 (1963).
4. Amdur, I., J. Ross, and E. A. Mason: *J. Chem. Phys.*, **20**: 1620 (1952).
5. Amourdam, M. J., and G. S. Laddha: *J. Chem. Eng. Data*, **12**, 389 (1967).
6. Anderson, D. K., and A. L. Babb: *J. Phys. Chem.*, **65**: 1281 (1961).
7. Anderson, D. K., J. R. Hall, and A. L. Babb: *J. Phys. Chem.*, **62**: 404 (1958).
8. Anderson, J. L.: *Ind. Eng. Chem. Fundam.*, **12**: 490 (1973).
9. Arnold, J. H.: *Ind. Eng. Chem.*, **22**: 1091 (1930).
10. Arnold, J. H.: *J. Am. Chem. Soc.*, **52**: 3937 (1930).
11. Bailey, R. G.: *Chem. Eng.*, **82**(6): 86 (1975).
12. Bearman, R. J.: *J. Chem. Phys.*, **32**: 1308 (1960).
13. Bearman, R. J.: *J. Phys. Chem.*, **65**: 1961 (1961).
14. Berry, V. J., Jr., and R. C. Koeller: *AIChE J.*, **6**: 274 (1960).
15. Bidlack, D. L., and D. K. Anderson: *J. Phys. Chem.*, **68**: 3790 (1964).
16. Bird, R. B.: *Adv. Chem. Eng.*, **1**: 156 (1956).
17. Bird, R. B., W. E. Stewart, and E. N. Lightfoot: "Transport Phenomena," chap. 16, Wiley, 1960.
18. Blanc, A.: *J. Phys.*, **7**: 825 (1908).
19. Brokaw, R. S.: *Ind. Eng. Chem. Process Des. Dev.*, **8**: 240 (1969).
20. Bromley, L. A.: *AIChE J.*, **19**: 313 (1973).
21. Brunet, J., and M. H. Doan: *Can. J. Chem. Eng.*, **48**: 441 (1970).
22. Burchard, J. K., and H. L. Toor: *J. Phys. Chem.*, **66**: 2015 (1962).

23. Byers, C. H., and C. J. King: *J. Phys. Chem.*, **70**: 2499 (1966).
24. Byrne, J. J., D. Maguire, and J. K. A. Clarke: *J. Phys. Chem.*, **71**: 3051 (1967).
25. Caldwell, C. S., and A. L. Babb: *J. Phys. Chem.*, **60**: 14, 56 (1956).
26. Calus, W. F., and M. T. Tyn: *J. Chem. Eng. Data*, **18**: 377 (1973).
27. Carman, P. C.: *J. Phys. Chem.*, **71**: 2565 (1967).
28. Carman, P. C.: *Ind. Eng. Chem. Fundam.*, **12**: 484 (1973).
29. Carman, P. C., and L. Miller: *Trans. Faraday Soc.*, **55**: 1838 (1959).
30. Carman, P. C., and L. H. Stein: *Trans. Faraday Soc.*, **52**: 619 (1956).
31. Carmichael, L. T., B. H. Sage, and W. N. Lacey: *AIChE J.*, **1**: 385 (1955).
32. Carswell, A. J., and J. C. Stryland: *Can. J. Phys.*, **41**: 708 (1963).
33. Chang, P., and C. R. Wilke: *J. Phys. Chem.*, **59**: 592 (1955).
34. Chang, S. Y.: S.M. thesis, Massachusetts Institute of Technology, Cambridge, Mass., 1959.
35. Chen, N. H., and D. P. Othmer: *J. Chem. Eng. Data*, **7**: 37 (1962).
36. Cram, R. R., and A. W. Adamson: *J. Phys. Chem.*, **64**: 199 (1960).
37. Crider, W. L.: *J. Am. Chem. Soc.*, **78**: 924 (1956).
38. Cullinan, H. T., Jr.: *Ind. Eng. Chem. Fundam.*, **4**: 133 (1965).
39. Cullinan, H. T., Jr.: *Can. J. Chem. Eng.*, **45**: 377 (1967).
40. Cullinan, H. T., Jr.: *Ind. Eng. Chem. Fundam.*, **6**: 616 (1967).
41. Cullinan, H. T., Jr.: *Ind. Eng. Chem. Fundam.*, **7**: 331 (1968).
42. Cullinan, H. T., Jr.: *Can. J. Chem. Eng.*, **49**: 130 (1971).
43. Cullinan, H. T., Jr.: *Can. J. Chem. Eng.*, **49**: 632 (1971).
44. Cullinan, H. T., Jr., and M. R. Cusick: *Ind. Eng. Chem. Fundam.*, **6**: 72 (1967).
45. Cullinan, H. T., Jr., and M. R. Cusick: *AIChE J.*, **13**: 1171 (1967).
46. Curtiss, C. F., and J. O. Hirschfelder: *J. Chem. Phys.*, **17**: 550 (1949).
47. Darken, L. S.: *Trans. Am. Inst. Mining Metall. Eng.*, **175**: 184 (1948).
48. Davidson, J. F., and E. J. Cullen: *Trans. Inst. Chem. Eng. Lond.*, **35**: 51 (1957).
49. Davies, G. A., A. B. Ponter, and K. Craine: *Can. J. Chem. Eng.*, **45**: 372 (1967).
50. Dawson, R., F. Khoury, and R. Kobayashi: *AIChE J.*, **16**: 725 (1970).
51. Di Pippo, R., J. Kestin, and K. Oguchi: *J. Chem. Phys.*, **46**: 4986 (1967).
52. Dole, M.: *J. Chem. Phys.*, **25**: 1082 (1956).
53. Duda, J. L., and J. S. Vrentas: *J. Phys. Chem.*, **69**: 3305 (1965).
54. Dullien, F. A. L.: *Nature*, **190**: 526 (1961).
55. Dullien, F. A. L.: *Trans. Faraday Soc.*, **59**: 856 (1963).
56. Dullien, F. A. L.: Transport Properties and Molecular Parameters in Liquids, paper presented at *62d Annu. Meet. AIChE, Washington, Nov. 16–20, 1969.*
57. Dullien, F. A. L.: *Ind. Eng. Chem. Fundam.*, **10**: 41 (1971).
58. Dullien, F. A. L.: *AIChE J.*, **18**: 62 (1972).
59. Dullien, F. A. L.: private communication, January 1974.
60. Duncan, J. B., and H. L. Toor: *AIChE J.*, **8**: 38 (1962).
61. Durbin, L., and R. Kobayashi: *J. Chem. Phys.*, **37**: 1643 (1962).
62. Elliott, R. W., and H. Watts: *Can. J. Chem.*, **50**: 31 (1972).
63. Ellis, C. S., and J. N. Holsen: *Ind. Eng. Chem. Fundam.*, **8**: 787 (1969).
64. Ertl, H., and F. A. L. Dullien: *AIChE J.*, **19**: 1215 (1973).
65. Eyring, H., and T. Ree: *Proc. Natl. Acad. Sci.*, **47**: 526 (1961).
66. Fair, J. R., and B. J. Lerner: *AIChE J.*, **2**: 13 (1956).
67. Fairbanks, D. F., and C. R. Wilke: *Ind. Eng. Chem.*, **42**: 471 (1950).
68. Fuller, E. N., and J. C. Giddings: *J. Gas Chromatogr.*, **3**: 222 (1965).
69. Fuller, E. N., P. D. Schettler, and J. C. Giddings: *Ind. Eng. Chem.*, **58**(5): 18 (1966).
70. Gainer, J. L.: *Ind. Eng. Chem. Fundam.*, **5**: 436 (1966).
71. Gainer, J. L.: *Ind. Eng. Chem. Fundam.*, **9**: 381 (1970).
72. Gainer, J. L., and A. B. Metzner: *AIChE-IChemE Symp. Ser.* no. 6, p. 74, 1965.
73. Garner, F. H., and P. J. M. Marchant: *Trans. Inst. Chem. Eng. Lond.*, **39**: 397 (1961).

74. Gautreaux, M. F., W. T. Davis, and J. Coates: Determination of Diffusion Coefficients in Binary Solutions of Non-electrolytes, paper presented at *42d Natl. Meet. AIChE, Atlanta, Ga., Feb. 21–24, 1960.*

75. Ghai, R. K., and F. A. L. Dullien: *J. Phys. Chem.*, in press, 1976.

76. Ghai, R. K., H. Ertl, and F. A. L. Dullien: *AIChE J.*, **19**: 881 (1973).

77. Ghai, R. K., H. Ertl, and F. A. L. Dullien: *AIChE J.*, **20**: 1 (1974).

78. Gilliland, E. R.: *Ind. Eng. Chem.*, **26**: 681 (1934).

79. Glasstone, S., K. J. Laidler, and H. Eyring: "The Theory of Rate Processes," chap. 9, McGraw-Hill, New York, 1941.

80. Goddard, R. R., G. H. F. Gardner, and M. R. J. Wyllis: *Proc. Symp. Interaction Fluids Particles, London, June 20–22, 1962,* p. 326, Institute of Chemical Engineers.

81. Gordon, A. R.: *J. Chem. Phys.*, **5**: 522 (1937).

82. Gotoh, S., M. Manner, J. P. Sørensen, and W. E. Stewart: *Ind. Eng. Chem.*, **12**: 119 (1973).

83. Gotoh, S., M. Manner, J. P. Sørensen, and W. E. Stewart: *J. Chem. Eng. Data*, **19**: 169, 172 (1974).

84. Gupta, G. P., and S. C. Saxena: *AIChE J.*, **14**: 519 (1968).

85. Haluska, J. L., and C. P. Colver: *AIChE J.*, **16**: 691 (1970).

86. Haluska, J. L., and C. P. Colver: *Ind. Eng. Chem. Fundam.*, **10**: 610 (1971).

87. Hammond, B. R., and R. H. Stokes: *Trans. Faraday Soc.*, **49**: 890 (1953).

88. Hammond, B. R., and R. H. Stokes: *Trans. Faraday Soc.*, **51**: 1641 (1955).

89. Hardt, A. P., D. K. Anderson, R. Rathbun, B. W. Mar, and A. L. Babb: *J. Phys. Chem.*, **63**: 2059 (1959).

90. Harned, H. S.: *Chem. Rev.*, **40**: 461 (1947).

91. Harned, H. S., and B. B. Owen: The Physical Chemistry of Electrolytic Solutions, *ACS Monogr.* 95, 1950.

92. Hartland, A., and M. Lipsicas: *Phys. Rev.*, **133A**: 665 (1964).

93. Hartley, G. S., and J. Crank: *Trans. Faraday Soc.*, **45**: 801 (1949).

94. Hattikudur, U. R., and G. Thodos: *J. Chem. Phys.*, **52**: 4313 (1970).

95. Hayduk, W., and W. D. Buckley: *Chem. Eng. Sci.*, **27**: 1997 (1972).

96. Hayduk, W., R. Castañeda, H. Bromfield, and R. R. Perras: *AIChE J.*, **19**: 859 (1973).

97. Hayduk, W., and S. C. Cheng: *Chem. Eng. Sci.*, **26**: 635 (1971).

98. Hayduk, W., and H. Laudie: *AIChE J.*, **20**: 611 (1974).

99. Hildebrand, J. H.: *Science*, **174**: 490 (1971).

100. Himmelblau, D. M.: *Chem. Rev.*, **64**: 527 (1964).

101. Hirschfelder, J. O., C. F. Curtiss, and R. B. Bird: "Molecular Theory of Gases and Liquids," Wiley, New York, 1954.

102. Hiss, T. G., and E. L. Cussler: *AIChE J.*, **19**: 698 (1973).

103. Holmes, J. T., D. R. Olander, and C. R. Wilke: *AIChE J.*, **8**: 646 (1962).

104. Holson, J. N., and M. R. Strunk: *Ind. Eng. Chem. Fundam.*, **3**: 163 (1964).

105. Horrocks, J. K., and E. McLaughlin: *Trans. Faraday Soc.*, **58**: 1367 (1962).

106. Hsu, H.-W., and R. B. Bird: *AIChE J.*, **6**: 516 (1960).

107. Hudson, G. H., J. C. McCoubrey, and A. R. Ubbelohde: *Trans. Faraday Soc.*, **56**: 1144 (1960).

108. Ibrahim, S. H., and N. R. Kuloor: *Br. Chem. Eng.*, **5**: 795 (1960).

109. Innes, K. K., and L. F. Albright: *Ind. Eng. Chem.*, **49**: 1793 (1957).

110. "International Critical Tables," McGraw-Hill, New York, 1926–1930.

111. Islam, M., and J. C. Stryland: *Physica*, **45**: 115 (1969).

112. Johnson, D. W., and C. P. Colver: *Hydrocarbon Process.*, **48**(3): 113 (1969).

113. Johnson, P. A., and A. L. Babb: *Chem. Rev.*, **56**: 387 (1956).

114. Kamal, M. R., and L. N. Canjar: *AIChE J.*, **8**: 329 (1962).

115. Kamal, M. R., and L. N. Canjar: *Chem. Eng.*, **69**(12): 159 (1962).

116. Kamal, M. R., and L. N. Canjar: *Chem. Eng. Prog.*, **62**(1): 82 (1966).

117. Kasper, G. P., and R. E. Wagner: Effective Radius Method for Self-Diffusion in Liquids, paper presented at *67th Natl. Meet. AIChE, Atlanta, Ga.*, Feb. 15–18, 1970.
118. Kett, T. K., and D. K. Anderson: *J. Phys. Chem.*, **73**: 1262 (1969).
119. Kett, T. K., and D. K. Anderson: *J. Phys. Chem.*, **73**: 1268 (1969).
120. Kett, T. K., G. B. Wirth, and D. K. Anderson: Effect of Association on Liquid Diffusion, paper presented at *61st Annu. Meet. AIChE, Los Angeles*, Dec. 1–5, 1968.
121. King, C. J., L. Hsueh, and K.-W. Mao: *J. Chem. Eng. Data*, **10**: 348 (1965).
122. Knuth, E. L.: *Phys. Fluids*, **2**: 340 (1959).
123. Krieger, I. M., G. W. Mulholland, and C. S. Dickey: *J. Phys. Chem.*, **71**: 1123 (1967).
124. Lees, F. P., and P. Sarram: *J. Chem. Eng. Data*, **16**: 41 (1971).
125. Leffler, J., and H. T. Cullinan, Jr.: *Ind. Eng. Chem. Fundam.*, **9**: 84, 88 (1970).
126. Lenczyk, J. P., and H. T. Cullinan, Jr.: *Ind. Eng. Chem. Fundam.*, **10**: 600 (1971).
127. Lewis, J. B.: *J. Appl. Chem. Lond.*, **5**: 228 (1955).
128. Li, J. C. M., and P. Chang: *J. Chem. Phys.*, **23**: 518 (1955).
129. Lispicas, M.: *J. Chem. Phys.*, **36**: 1235 (1962).
130. Lo, H. Y.: *J. Chem. Eng. Data*, **19**: 236 (1974).
131. Loflin, T., and E. McLaughlin: *J. Phys. Chem.*, **73**: 186 (1969).
132. Longsworth, L. G.: *J. Phys. Chem.*, **58**: 770 (1954).
133. Longsworth, L. G.: *J. Phys. Chem.*, **67**: 689 (1963).
134. Lugg, G. A.: *Anal. Chem.*, **40**: 1072 (1968).
135. Lusis, M. A.: *Chem. Proc. Eng.*, **5**: 27 (May 1971).
136. Lusis, M. A.: *Chem. Ind. Devel. Bombay*, January 1972: 48; *AIChE J.*, **20**: 207 (1974).
137. Lusis, M. A., and G. A. Ratcliff: *Can. J. Chem. Eng.*, **46**: 385 (1968).
138. Lusis, M. A., and G. A. Ratcliff: *AIChE J.*, **17**: 1492 (1971).
139. McCall, D. W., and D. C. Douglas: *Phys. Fluids*, **2**: 87 (1959).
140. McCall, D. W., and D. C. Douglas: *J. Phys. Chem.*, **71**: 987 (1967).
141. McCall, D. W., D. C. Douglas, and E. W. Anderson: *Phys. Fluids*, **2**: 87 (1959), **4**: 162 (1961).
142. Marrero, T. R., and E. A. Mason: *J. Phys. Chem. Ref. Data*, **1**: 3 (1972).
143. Marrero, T. R., and E. A. Mason: *AIChE J.*, **19**: 498 (1973).
144. Mason, E. A., and L. Monchick: *J. Chem. Phys.*, **36**: 2746 (1962).
145. Mathur, G. P., and G. Thodos: *AIChE J.*, **11**: 613 (1965).
146. Mathur, G. P., and S. C. Saxena: *Ind. J. Pure Appl. Phys.*, **4**: 266 (1966).
147. Meissner, H. P., and C. L. Kusik: *AIChE J.*, **18**: 294 (1972).
148. Meissner, H. P., and C. L. Kusik: *Ind. Eng. Chem. Process Des. Dev.*, **12**: 205 (1973).
149. Meissner, H. P., C. L. Kusik, and J. W. Tester: *AIChE J.*, **18**: 661 (1972).
150. Meissner, H. P., and N. A. Peppas: *AIChE J.*, **19**: 806 (1973).
151. Meissner, H. P., and J. W. Tester: *Ind. Eng. Chem. Process Des. Dev.*, **11**: 128 (1972).
152. Mian, A. A., J. Coates, and J. B. Cordiner: *Can. J. Chem. Eng.*, **47**: 499 (1969).
153. Mickley, H. S., R. C. Ross, A. L. Squyers, and W. E. Stewart: *NACA Tech. Note* 3208, 1954.
154. Mifflin, T. R., and C. O. Bennett: *J. Chem. Phys.*, **29**: 975 (1958).
155. Miller, D. G.: *J. Phys. Chem.*, **70**: 2639 (1966).
156. Miller, D. G.: *J. Phys. Chem.*, **71**: 616 (1967).
157. Miller, L., and P. C. Carman: *Trans. Faraday Soc.*, **57**: 2143 (1961).
158. Moelwyn-Hughes, E. A.: "Physical Chemistry," Pergamon, London, 1957.
159. Moore, J. W., and R. M. Wellek: *J. Chem. Eng. Data*, **19**: 136 (1974).
160. Mortimer, R. G.: *Ind. Eng. Chem. Fundam.*, **7**: 330 (1968).
161. Mortimer, R. G.: *Ind. Eng. Chem. Fundam.*, **12**: 492 (1973).
162. Mortimer, R. G., and N. H. Clark: *Ind. Eng. Chem. Fundam.*, **10**: 604 (1971).
163. Mrazek, R. V., C. E. Wicks, and K. N. S. Prabhu: *J. Chem. Eng. Data*, **13**: 508 (1968).
164. Naghizadeh, J., and S. A. Rice: *J. Chem. Phys.*, **36**: 2710 (1962).
165. Nain, V. P. S., and J. R. Ferron: *Ind. Eng. Chem. Fundam.*, **11**: 420 (1972).

166. Neufeld, P. D., A. R. Janzen, and R. A. Aziz: *J. Chem. Phys.*, **57**: 1100 (1972).
167. Newman, J. S.: in C. W. Tobias (ed.), "Advances in Electrochemistry and Electrochemical Engineering," vol. 5, Interscience, New York, 1967.
168. O'Hern, H. A., and J. J. Martin: *Ind. Eng. Chem.*, **47**: 2081 (1968).
169. Olander, D. R.: *AIChE J.*, **7**: 175 (1961).
170. Olander, D. R.: *AIChE J.*, **9**: 207 (1963).
171. Othmer, D. F., and H. T. Chen: *Ind. Eng. Chem. Process Des. Dev.*, **1**: 249 (1962).
172. Othmer, D. F., and M. S. Thakar: *Ind. Eng. Chem.*, **45**: 589 (1953).
173. Pasternak, A. D., and D. R. Olander: *AIChE J.*, **13**: 1052 (1967).
174. Paul, R., and I. B. Srivastava: *J. Chem. Phys.*, **35**: 1621 (1961).
175. Perkins, L. R., and C. J. Geankoplis: *Chem. Eng. Sci.*, **24**: 1035 (1969).
176. Perry, J. H.: "Chemical Engineers' Handbook," 3d ed., McGraw-Hill, New York, 1950.
177. Pratap, R., and L. K. Doraiswamy: *Ind. J. Tech.*, **5**(5): 145 (1967).
178. Ramamurthy, A. V., and G. Narsimhan: *Can. J. Chem. Eng.*, **48**: 297 (1970).
179. Rao, S. S., and C. O. Bennett: *AIChE J.*, **17**: 75 (1971).
180. Ratcliff, G. A., and J. G. Holdcroft: *Trans. Inst. Chem. Eng. Lond.*, **41**: 315 (1963).
181. Ratcliff, G. A., and M. A. Lusis: *Ind. Eng. Chem. Fundam.*, **10**: 474 (1971).
182. Rathbun, R. E., and A. L. Babb: *Ind. Eng. Chem. Process Des. Dev.*, **5**: 273 (1966).
183. Reddy, K. A., and L. K. Doraiswamy: *Ind. Eng. Chem. Fundam.*, **6**: 77 (1967).
184. Ree, F. H., T. Ree, and H. Eyring: *Ind. Eng. Chem.*, **50**: 1036 (1958).
185. Robinson, R. A., and R. H. Stokes: "Electrolyte Solutions," 2d ed., Academic, New York, 1959.
186. Robinson, R. C., and W. E. Stewart: *Ind. Eng. Chem. Fundam.*, **7**: 90 (1968).
187. Robinson, R. L., Jr., W. C. Edmister, and F. A. L. Dullien: *Ind. Eng. Chem. Fundam.*, **5**: 75 (1966).
188. Saksena, M. P., and S. C. Saxena: *Int. J. Pure Appl. Phys.*, **4**: 109 (1966).
189. Sandler, S., and E. A. Mason: *J. Chem. Phys.*, **48**: 2873 (1968).
190. Sanni, S. A., and P. Hutchinson: *J. Chem. Eng. Data*, **18**: 317 (1973).
191. Sanni, S. A., C. J. D. Fell, and P. Hutchison: *J. Chem. Eng. Data*, **16**: 424 (1971).
192. Scheibel, E. G.: *Ind. Eng. Chem.*, **46**: 2007 (1954).
193. Schwartz, F. A., and J. E. Brow: *J. Chem. Phys.*, **19**: 640 (1951).
194. Scott, D. S., and K. E. Cox: *Can. J. Chem. Eng.*, **38**: 201 (1960).
195. Seager, S. L., L. R. Geertson, and J. C. Giddings: *J. Chem. Eng. Data*, **8**: 168 (1963).
196. Shain, S. A.: *AIChE J.*, **7**: 17 (1961).
197. Shrier, A. L.: *Chem. Eng. Sci.*, **22**: 1391 (1967).
198. Sitaraman, R., S. H. Ibrahim, and N. R. Kuloor: *J. Chem. Eng. Data*, **8**: 198 (1963).
199. Slattery, J. C., and R. B. Bird: *AIChE J.*, **4**: 137 (1958).
200. Srivastava, B. N., and K. P. Srivastava: *J. Chem. Phys.*, **30**: 984 (1959).
201. Srivastava, B. N., and I. B. Srivastava: *J. Chem. Phys.*, **36**: 2616 (1962).
202. Srivastava, B. N., and I. B. Srivastava: *J. Chem. Phys.*, **38**: 1183 (1963).
203. Stearn, A. E., E. M. Irish, and H. Eyring: *J. Phys. Chem.*, **44**: 981 (1940).
204. Stiel, L. I., and G. Thodos: *Can. J. Chem. Eng.*, **43**: 186 (1965).
205. Sundelöf, L.-O., and I. Södervi: *Arkiv Kem.*, **21**: 143 (1963).
205a. Takahashi, T.: *Chem. Eng. Jpn.*, **7**: 417 (1974).
206. Tang, Y. P., and D. M. Himmelblau, *AIChE J.*, **11**: 54 (1965).
207. Taylor, H. S., *J. Chem. Phys.*, **6**: 311 (1938).
208. Toor, H. L.: *AIChE J.*, **3**: 198 (1957).
209. Trappeniers, N. J., and P. H. Oosting: *Phys. Lett.*, **23**: 445 (1966).
210. Turner, J. C. R.: *Chem. Eng. Sci.*, **30**: 151 (1975).
211. Tyn, M. T.: *AIChE J.*, **20**: 1038 (1974).
212. Tyn, M. T.: *Chem. Eng.*, **82**(12): 106 (1975).

213. Tyn, M. T., and W. F. Calus: *J. Chem. Eng. Data,* **20**: 106 (1975).

214. Tyn, M. T., and W. F. Calus: *Trans. Inst. Chem. Eng.,* **53**: 44 (1975).

215. Tyrell, H. J. V.: "Diffusion and Heat Flow in Liquids," Butterworth, London, 1961.

216. Vadovic, C. J., and C. P. Colver: *AIChE J.,* **18**: 1264 (1972).

217. Vadovic, C. J., and C. P. Colver: *AIChE J.,* **19**: 546 (1973).

218. Van Geet, A. L., and A. W. Adamson: *J. Phys. Chem.,* **68**: 238 (1964).

219. Van Geet, A. L., and A. W. Adamson: *Ind. Eng. Chem.,* **57**(7): 62 (1965).

220. Vignes, A.: *Ind. Eng. Chem. Fundam.,* **5**: 189 (1966).

221. Vignes, A.: *Ind. Eng. Chem. Fundam.,* **6**: 614 (1967).

221a. Vinograd, J. R., and J. W. McBain: *J. Am. Chem. Soc.,* **63**: 2008 (1941).

222. Vivian, J. E., and C. J. King: *AIChE J.,* **10**: 220 (1964).

223. Walker, R. E., N. de Haas, and A. A. Westenberg: *J. Chem. Phys.,* **32**: 1314 (1960).

224. Walker, R. E., and A. A. Westenberg: *J. Chem. Phys.,* **29**: 1139 (1958).

225. Walker, R. E., and A. A. Westenberg: *J. Chem. Phys.,* **32**: 436 (1960).

226. Way, P.: Ph.D. thesis, Massachusetts Institute of Technology, Cambridge, Mass., 1971.

227. Weissman, S.: *J. Chem. Phys.,* **40**: 3397 (1964).

228. Weissman, S., and E. A. Mason: *J. Chem. Phys.,* **37**: 1289 (1962).

229. Wendt, R. P.: *J. Phys. Chem.,* **69**: 1227 (1965).

230. Westenberg, A. A., and G. Frazier: *J. Chem. Phys.,* **36**: 3499 (1962).

231. Westenberg, A. A., and R. E. Walker: in Y. S. Touloukian (ed.), "Thermodynamic and Transport Properties of Gases, Liquids, and Solids," ASME and McGraw-Hill, New York, 1959.

232. Wilke, C. R.: *Chem. Eng. Prog.,* **46**: 95 (1950).

233. Wilke, C. R., and P. Chang: *AIChE J.,* **1**: 264 (1955).

234. Wilke, C. R., and C. Y. Lee: *Ind. Eng. Chem.,* **47**: 1253 (1955).

235. Wise, D. L., and G. Houghton: *Chem. Eng. Sci.,* **21**: 999 (1966).

236. Wishaw, B. F., and R. H. Stokes: *J. Am. Chem. Soc.,* **76**: 2065 (1954).

237. Witherspoon, P. A., and L. Bonoli: *Ind. Eng. Chem. Fundam.,* **8**: 589 (1969).

238. Woessner, D. E., B. S. Snowden, R. A. George, and J. C. Melrose: *Ind. Eng. Chem. Fundam.,* **8**: 779 (1969).

239. Yabsley, M. A., P. J. Carlson, and P. J. Dunlop: *J. Phys. Chem.,* **77**: 703 (1973).

240. Yon, C. M., and H. L. Toor: *Ind. Eng. Chem. Fundam.,* **7**: 319 (1968).

Chapter Twelve

Surface Tension

12-1 Scope

The surface tensions of both pure liquids and liquid mixtures are considered in this chapter.   For the former, methods based on the law of corresponding states and upon the parachor are judged most accurate when estimated values are compared with experimental determinations.   For mixtures, extensions of the pure-component methods are presented, as is a method based upon a thermodynamic analysis of the system.   Interfacial tensions for liquid-liquid or liquid-solid systems are not included.

12-2 Introduction

The boundary layer between a liquid phase and a gas phase may be considered a third phase with properties intermediate between those of a liquid and a gas.   A qualitative picture of the microscopic surface layer shows that there are unequal forces acting upon the molecules; i.e., at low gas densities, the surface molecules are attracted sidewise and toward the bulk liquid but experience less attraction in the direction of the bulk gas.   Thus the surface layer is in tension and tends to contract to the smallest area compatible with the mass of material, container restraints, and external forces, e.g., gravity.

A quantitative index of this tension can be presented in various ways; the most common is the surface tension σ, defined as the force exerted in the plane of the surface per unit length.   We can consider a reversible isothermal process whereby surface area A is increased by pulling the surface apart and allowing the molecules from the bulk liquid to enter at constant temperature and pressure.   The differential reversible work is $\sigma \, dA$; in this case σ is the surface Gibbs energy per unit of area.   As equilibrium systems tend to a state of minimum Gibbs energy, the product σA also tends to a minimum.   For a fixed σ, equilibrium is a state of minimum area consistent with the restraints of the system.

Surface tension is usually expressed in dynes per centimeter; surface Gibbs energy per unit area has units of ergs per square centimeter. These units and numerical values of σ are identical.   In SI units, $1 \text{ erg/cm}^2 = 1 \text{ mJ/m}^2 = 1 \text{ mN/m}$.

The thermodynamics of surface layers furnishes a fascinating subject for study.   Guggenheim [20], Gibbs [18], and Modell and Reid [38] have formulated treatments which differ considerably but reduce to similar equations relating macroscopically measurable quantities.   In addition to the thermodynamic aspect, treatments of the physics and chemistry of surfaces have been published [1, 2, 3, 7, 49].   These subjects are not covered here; instead, the emphasis is placed upon the few reliable methods available to estimate σ either from semitheoretical or empirical equations.

12-3 Estimation of the Surface Tension of a Pure Liquid

As the temperature is raised, the surface tension of a liquid in equilibrium with its own vapor decreases and becomes zero at the critical point [47].   In the reduced-temperature range 0.45 to 0.65, σ for most organic liquids ranges from 20 to 40 dyn/cm, but for some low-molecular-weight, dense liquids such as formamide, $\sigma > 50$ dyn/cm. For water $\sigma = 72.8$ dyn/cm at 20°C; for liquid metals σ is between 300 and 600 dyn/cm; e.g., mercury at 20°C has a value of about 476.

A recent, thorough critical evaluation of experimental surface tensions has been prepared by Jasper [29].

Essentially all useful estimation techniques for the surface tension of a liquid are empirical; only two are discussed in any detail here, although others are briefly noted at the end of this section.

Macleod-Sugden Correlation   Macleod [33] in 1923 suggested a relation between σ and the liquid and vapor densities:

$$\sigma^{1/4} = [P](\rho_L - \rho_v) \qquad (12\text{-}3.1)$$

Sugden [57, 58] has called the temperature-independent parameter $[P]$ the *parachor* and indicated how it might be estimated from the structure of the molecule.   Quale [44] employed experimental surface-tension and density data for many compounds and calculated parachors.   From these, he suggested an additive scheme to correlate $[P]$ with structure, and a modified list of his values is shown in Table 12-1.   When using $[P]$ values determined in this manner, the surface tension is given in dynes per centimeter and the densities are expressed in gram moles per cubic centimeter.   The method is illustrated in Example 12-1, and calculated values of σ are compared with experimental surface tensions in Table 12-2.

TABLE 12-1   Structural Contributions for Calculation of the Parachor†

Carbon-hydrogen:		Special groups (*Continued*):	
C	9.0	Four carbon atoms	20.0
H	15.5	Five carbon atoms	18.5
CH₃—	55.5	Six carbon atoms	17.3
—CH₂—	40.0‡	—CHO	66
CH₃—CH(CH₃)—	133.3	O (not noted above)	20
CH₃—CH₂—CH(CH₃)—	171.9	N (not noted above)	17.5
CH₃—CH₂—CH₂—CH(CH₃)—	211.7	S	49.1
CH₃—CH(CH₃)—CH₂—	173.3	P	40.5
CH₃—CH₂—CH(C₂H₅)—	209.5	F	26.1
CH₃—C(CH₃)₂—	170.4	Cl	55.2
CH₃—CH₂—C(CH₃)₂—	207.5	Br	68.0
CH₃—CH(CH₃)—CH(CH₃)—	207.9	I	90.3
CH₃—CH(CH₃)—C(CH₃)₂—	243.5		
C₆H₅—	189.6	Ethylenic bond:	
		Terminal	19.1
Special groups:		2,3 position	17.7
—COO—	63.8	3,4 position	16.3
—COOH	73.8		
—OH	29.8	Triple bond	40.6
—NH₂	42.5		
—O—	20.0	Ring closure:	
—NO₂ (nitrite)	74	Three-membered	12.5
—NO₃ (nitrate)	93	Four-membered	6.0
—CO(NH₂)	91.7	Five-membered	3.0
=O (ketone)		Six-membered	0.8
Three carbon atoms	22.3		

†As modified from Ref. 44.
‡If $n > 12$ in (—CH₂—)$_n$, increase increment to 40.3.

H 7

C

/, O

H₃C-C-C≺OH

C⁻

(+

Example 12-1 Use the Macleod-Sugden correlation to estimate the surface tension of isobutyric acid at 60°C.   The experimental value quoted by Jasper [29] is 21.36 dyn/cm.

solution At 60°C, the liquid density is 0.912 g/cm³ [28], and, with $M = 88.107$, $\rho_L = 0.912/88.107 = 1.035 \times 10^{-2}$ g-mol/cm³.   At 60°C, isobutyric acid is well below the boiling point, and at this low pressure $\rho_v \ll \rho_L$ and the vapor-density term is neglected.

To determine the parachor from Table 12-1,

$$[P] = CH_3—CH(CH_3)—+—COOH$$
$$= 133.3 + 73.8 = 207.1$$

Then, with Eq. (12-3.1),

$$\sigma = [(207.1)(1.035 \times 10^{-2})]^4 = 21.10 \text{ dyn/cm}$$

$$\text{Error} = \frac{21.10 - 21.36}{21.36} \times 100 = -1.2\%$$

Since σ is proportional to $([P]\rho_L)^4$, Eq. (12-3.1) is *very* sensitive to the values of the parachor and liquid density chosen.   It is remarkable that the estimated values are as accurate as shown in Table 12-2.

Instead of employing experimental densities, correlations relating ρ to T, given in Chap. 3, may be used.   One technique not covered in that chapter is given by Goldhammer [19] and discussed by Gambill [17]:

$$\rho_L - \rho_v = \rho_{L_b} \left(\frac{1 - T_r}{1 - T_{b_r}} \right)^n \tag{12-3.2}$$

ρ_{L_b} is the molal liquid density at the normal boiling point in gram moles per cubic centimeter.   The exponent n ranges from 0.25 to 0.31; Fishtine [15] suggests the following values:

	n
Alcohols	0.25
Hydrocarbons and ethers	0.29
Other organic compounds	0.31

With Eq. (12-3.2), Eq. (12-3.1) becomes

$$\sigma = ([P]\rho_{L_b})^4 \left(\frac{1 - T_r}{1 - T_{b_r}} \right)^{4n} \tag{12-3.3}$$

where $4n$ varies between 1.0 and 1.24.   As shown later, other correlations predict a similar temperature dependence; that is, σ decreases with temperature at a rate somewhat exceeding that predicted by a linear relation.

Corresponding-States Correlation The group $\sigma/P_c^{2/3} T_c^{1/3}$ is dimensionless except for a numerical constant which depends upon the units

TABLE 12-2   Comparison of Calculated and Experimental Values of Surface Tension of Pure Liquids

Compound	T, °C	σ (exp.),† dyn/cm	Percent error in method‡	
			Macleod and Sugden, Eq. (12-3.1)	Brock and Bird, Eq. (12-3.6)
Acetic acid	20	27.59	−4.6	
	60	23.62	−3.2	
Acetone	25	24.02	−5.4	2.4
	35	22.34	−3.9	3.7
	45	21.22	−4.5	2.7
Aniline	20	42.67	−3.2	11
	40	40.50	−6.0	10
	60	38.33	−7.1	9.1
	80	36.15	−8.8	8.0
Benzene	20	28.88	−5.1	−2.0
	40	26.25	−5.6	−1.9
	60	23.67	−5.0	−0.4
	80	21.20	−3.9	−1.9
Benzonitrile	20	39.37	−2.0	1.2
	50	35.89	−3.4	1.1
	90	31.26	−4.2	1.3
Bromobenzene	20	35.82	−0.7	−0.2
	50	32.34	−1.4	0
	100	26.54	−0.7	0.7
n-Butane	−70	23.31	11	1.6
	−40	19.69	5.2	0.9
	20	12.46	1.5	0.9
Carbon disulfide	20	32.32	3.8	3.0
	40	29.35	3.8	3.1
Carbon tetrachloride	15	27.65	−1.1	−4.9
	35	25.21	−1.2	−5.1
	55	22.76	−1.0	−5.0
	75	20.31	0.1	−4.9
	95	17.86	2.5	−4.4
Chlorobenzene	20	33.59	−0.6	−1.7
	50	30.01	0.7	−1.7
	100	24.06	5.8	−1.2
p-Cresol	40	34.88	0.5	
	100	29.32	−0.3	
Cyclohexane	20	25.24	−3.9	−4.8
	40	22.87	−3.5	−4.5
	60	20.49	−2.2	−4.3
Cyclopentane	20	22.61	−5.6	−2.0
	40	19.68	−2.4	0.1
Diethyl ether	15	17.56	0	0
	30	16.20	0.4	−2.5
2,3-Dimethylbutane	20	17.38	−0.9	−0.2
	40	15.38	0.6	−0.5
Ethyl acetate	20	23.97	−4.6	1.1
	40	21.65	−4.8	0
	60	19.32	−4.3	−0.9
	80	17.00	−2.7	−1.4
	100	14.68	0.5	−2.0

Compound	T, °C	σ (exp.),† dyn/cm	Percent error in method‡	
			Macleod and Sugden, Eq. (12-3.1)	Brock and Bird, Eq. (12-3.6)
Ethyl benzoate	20	35.04	−1.9	2.7
	40	32.92	−2.7	2.9
	60	30.81	−3.1	2.9
Ethyl bromide	10	25.36	−5.3	15
	30	23.04	−6.1	13
Ethyl mercaptan	15	23.87	−6.7	3.5
	30	22.68	−9.1	−1.1
Formamide	25	57.02	−8.8	§
	65	53.66	−15	
	100	50.71	−20	
n-Heptane	20	20.14	−0.6	0.3
	40	18.18	0.7	0.1
	60	16.22	3.1	0.3
	80	14.26	6.8	0.8
Isobutyric acid	20	25.04	1.2	
	40	23.20	0.5	
	60	21.36	−1.2	
	90	18.60	−3.5	
Methyl formate	20	24.62	−7.6	4.4
	50	20.05	−7.2	4.6
	100	12.90	−7.4	4.3
	150	6.30	−8.8	5.4
	200	0.87	−14	21
Methyl alcohol	20	22.56	−13	
	40	20.96	−15	
	60	19.41	−17	
Phenol	40	39.27	−6.7	
	60	37.13	−7.3	
	100	32.86	−7.8	
n-Propyl alcohol	20	23.71	−0.6	
	40	22.15	−1.9	
	60	20.60	−3.3	
	90	18.27	−4.0	
n-Propyl benzene	20	29.98	0.2	−1.1
	40	26.83	0.8	−0.7
	60	24.68	2.1	−0.1
	80	22.53	3.9	0.7
	100	20.38	6.6	1.9
Pyridine	20	37.21	−2.8	§
	40	34.60	−3.6	
	60	31.98	−4.1	

†Experimental values from Ref. 29, except for methyl formate, where experimental values were taken from Ref. 33. Surface tensions quoted by Jasper are smoothed values obtained after plotting σ vs. T. Normally a linear relation was assumed over small temperature ranges. See discussion with Eq. (12-4.2).

‡Error = [(calc. − exp.)/exp.] × 100.

§Critical properties not known.

of σ, P_c, and T_c.† Van der Waals suggested in 1894 [63] that this group could be correlated with $1 - T_r$. Brock and Bird [6] developed this idea for nonpolar liquids and proposed that

$$\frac{\sigma}{P_c^{2/3} T_c^{1/3}} = (0.133\alpha_c - 0.281)(1 - T_r)^{11/9} \tag{12-3.4}$$

where P_c is in atmospheres and T_c in kelvins and α_c is the Riedel parameter, introduced in Sec. 6-5. However, using a suggestion by Miller [37] to relate α_c to T_{b_r} and P_c,

$$\alpha_c = 0.9076 \left(1 + \frac{T_{b_r} \ln P_c}{1 - T_{b_r}}\right) \tag{12-3.5}$$

it can then be shown that

$$\sigma = P_c^{2/3} T_c^{1/3} Q (1 - T_r)^{11/9} \tag{12-3.6}$$

$$Q = 0.1207 \left(1 + \frac{T_{b_r} \ln P_c}{1 - T_{b_r}}\right) - 0.281 \tag{12-3.7}$$

Equations (12-3.6) and (12-3.7) were used to compute σ values for nonpolar liquids in Table 12-2. The accuracy is similar to that for the Macleod-Sugden relation discussed earlier. However, the corresponding-states method is not applicable to compounds exhibiting strong hydrogen bonding (alcohols, acids) and quantum liquids (H_2, He, Ne).

A similar correlation was proposed by Riedel [48]. More recently, to broaden the approach to include polar liquids, Hakim et al. [23] introduced the Stiel polar factor (Sec. 2-6) and proposed the following equation:

$$\sigma = P_c^{2/3} T_c^{1/3} Q_p \left(\frac{1 - T_r}{0.4}\right)^m \tag{12-3.8}$$

where σ = surface tension of polar liquid, dyn/cm
P_c = critical pressure, atm
T_c = critical temperature, K
$Q_p = 0.1574 + 0.359\omega - 1.769X - 13.69X^2 - 0.510\omega^2 + 1.298\omega X$
$m = 1.210 + 0.5385\omega - 14.61X - 32.07X^2 - 1.656\omega^2 + 22.03\omega X$
X = Stiel polar factor, Sec. 2-6
ω = acentric factor, Sec. 2-3

Values of X can be obtained for a few materials from Table 2-4, or they can be estimated from the reduced vapor pressure at $T_r = 0.6$, $P_{vp_r}(0.6)$,

$$X = \log P_{vp_r}(0.6) + 1.70\omega + 1.552 \tag{12-3.9}$$

The general reliability of Eq. (12-3.8) is not known. The six constants

---

†The fundamental dimensionless group is $\sigma V_c^{2/3}/RT_c$ [21, 64], but the proportionality $V_c \propto RT_c/P_c$ has been used and the gas constant dropped from the group.

in Q_p and m were obtained from data on only 16 polar compounds, some of which were only slightly polar (diethyl ether, dimethyl ether, ethyl mercaptan, acetone, etc.).   X values are available for only a few substances, and estimated values of σ are sensitive to the value of X chosen.

Example 12-2   Using Eqs. (12-3.6) and (12-3.7), estimate the surface tension of liquid ethyl mercaptan at 30°C.   The experimental value is 22.68 dyn/cm [29].

  solution   From Appendix A, for ethyl mercaptan, $T_c = 499$ K, $T_b = 308.2$ K, $P_c = 54.2$ atm.   Thus $T_{b_r} = 308.2/499 = 0.618$.   With Eq. (12-3.7)

$$Q = 0.1207 \left[1 + \frac{(0.618)(\ln 54.2)}{1 - 0.618}\right] - 0.281 = 0.619$$

$$\sigma = (54.2)^{2/3}(499)^{1/3}(0.619)(1 - \tfrac{308}{499})^{11/9} = 22.4 \text{ dyn/cm}$$

$$\text{Error} = \frac{22.44 - 22.68}{22.68} \times 100 = -1.1\%$$

Other Estimation Methods   Statistical-mechanical theories of liquids yield reasonable results for surface tensions of simple liquids [42].   The surface tension has also been correlated with the molar refraction and refractive index [61], liquid compressibility [34], and viscosity [8, 39, 41, 52, 62].   Schonhorn [52] expanded upon an earlier idea of Pelofsky [41] and showed that $\ln \sigma$ is linearly related to $(\eta_L - \eta_V)^{-1}$, where η is the viscosity.   Ramana et al. [45] proposed a linear relation between $\log \sigma_b$ and T_{b_r} for different homologous series (σ_b is the surface tension at T_b).

For rapid estimations of σ, Hadden [22] has presented a nomograph for relating σ and T for hydrocarbons.   A similar one from Othmer et al. [40] covers a broader class of materials, as shown in Fig. 12-1 and Table 12-3.   Data for light hydrocarbons and their mixtures have been correlated by Porteous [43].

Surface tensions for several cryogenic liquids have been reported by Sprow and Prausnitz [54]; they were correlated by a modification of Eq. (12-3.4)

$$\sigma = \sigma_0(1 - T_r)^p \tag{12-3.10}$$

where σ_0 and p were fitted to the data by a least-squares analysis. Values of p were close to $\tfrac{11}{9}$, but the best value of σ_0 was often slightly larger than $P_c^{2/3}T_c^{1/3}Q$, as predicted by Eq. (12-3.6).

Recommendations   For surface tensions of organic liquids, use the data collection of Jasper [29].   Two estimation methods were presented in this section, and calculated values are compared with experiment in Table 12-2.   For non-hydrogen-bonded liquids, use the corresponding-states method [Eqs. (12-3.6) and (12-3.7)].   The normal boiling point, critical temperature, and critical pressure are required.   Errors are normally less than 5 percent.

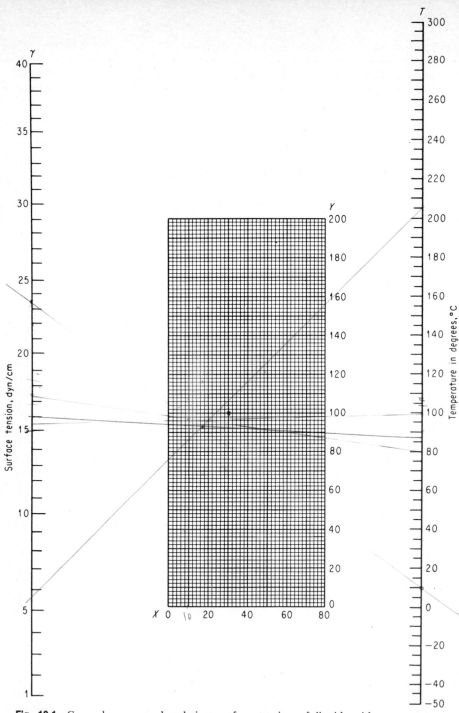

Fig. 12-1 General nomograph relating surface tension of liquids with temperature. Values for X and Y, taken from Table 12-3, are used for any particular compound to give a point on the central grid.   By using this point as a pivot for a straight line, the value of the surface tension can be read from the intersection of this line with the σ scale against the value of the temperature read from the intersection of this line with the T scale.   (*From Ref.* 40.)

TABLE 12-3 *X* and *Y* Values for Surface-Tension Nomograph [40]

Compound	X	Y	Compound	X	Y
Acetal	19	88	Ethyl isobutyrate	20.9	93.7
Acetaldehyde	33	78	Ethyl ether	27.5	64
Acetaldoxime	23.5	127	Ethyl formate	30.5	88.8
Acetamide	17	192.5	Ethyl iodide	28	113.2
Acetic acid	17.1	116.5	Ethyl mercaptan	35	81
Acetic anhydride	25	129	Ethyl propionate	22.6	97
Acetone	28	91	Ethylamine	11.2	83
Acetonitrile	33.5	111	Ethylbenzene	22	118
Acetophenone	18	163	Ethylene chloride	32	120
Allyl alcohol	12	111.5	Ethylene oxide	42	83
Anethole	13	158.1	*n*-Hexane	22.7	72.2
Aniline	22.9	171.8	Hydrogen cyanide	30.6	66
Anisole	24.4	138.9	Mesitylene	17	119.8
Ammonia	56.2	63.5	Methyl acetate	34	90
Isoamyl acetate	16.4	103.1	Methyl alcohol	17	93
Isoamyl alcohol	6	106.8	Methyl *n*-butyrate	25	88
Benzene	30	110	Methyl isobutyrate	24	93.8
Benzonitrile	19.5	159	Methyl chloride	45.8	53.2
Benzylamine	25	156	Methyl ether	44	37
Bromobenzene	23.5	145.5	Methyl ethyl ketone	23.6	97
Isobutyl acetate	16	97.2	Methyl formate	38.5	88
n-Butyl alcohol	9.6	107.5	Methyl propionate	29	95
Isobutyl alcohol	5	103	Methylamine	42	58
n-Butyric acid	14.5	115	Naphthalene	22.5	165
Isobutyric acid	14.8	107.4	Nitroethane	25.4	126.1
n-Butyronitrile	20.3	113	Nitromethane	30	139
Carbon disulfide	35.8	117.2	Nitrosyl chloride	38.5	93
Carbon tetrachloride	26	104.5	Nitrous oxide	62.5	0.5
Chloral	30	113	*n*-Octane	17.7	90
Chlorine	45.5	59.2	Paraldehyde	22.3	103.8
Chlorobenzene	23.5	132.5	Phenetole	20	134.2
p-Chlorobromobenzene	14	162	Phenol	20	168
Chloroform	32	101.3	Phosphorus oxychloride	26	125.2
p-Chlorotoluene	18.7	134	Piperidine	24.7	120
o-Cresol	20	161	Propionic acid	17	112
m-Cresol	13	161.2	Propionitrile	23	108.6
p-Cresol	11.5	160.5	*n*-Propyl acetate	23	97
Cyclohexane	42	86.7	*n*-Propyl alcohol	8.2	105.2
Diethyl ketone	20	101	*n*-Propyl formate	24	97
Diethyl oxalate	20.5	130.8	*n*-Propylamine	25.5	87.2
Diethyl sulfate	19.5	139.5	*p*-Isopropyl toluene	12.8	121.2
Diethylaniline	17	142.6	Pyridine	34	138.2
Dimethyl sulfate	23.5	158	Quinoline	19.5	183
Dimethylamine	16	66	Thiophene	35	121
Dimethylaniline	20	149	Toluene	24	113
Ethyl acetate	27.5	92.4	Triethylamine	20.1	83.9
Ethyl acetoacetate	21	132	Trimethylamine	21	57.6
Ethyl alcohol	10	97	Tripalmitin	2	151
Ethyl benzoate	14.8	151	Triphenyl methane	12.5	182.7
Ethyl bromide	31.6	90.2	*m*-Xylene	20.5	118
Ethyl *n*-butyrate	17.5	102	*p*-Xylene	19	117

For hydrogen-bonded liquids, use the Macleod-Sugden form [Eq. (12-3.1)] with the parachor determined from group contributions in Table 12-1. Either experimental saturated-liquid and saturated-vapor densities may be used, or, with slightly less accuracy, the modified temperature form, Eq. (12-3.3), may be substituted. Errors are normally less than 5 to 10 percent. If reliable values of ω and X are available, however, Eq. (12-3.8) is somewhat more accurate for alcohols.

12-4 Variation of Surface Tension with Temperature

Equation (12-3.3) indicates that

$$\sigma \propto (1 - T_r)^{4n} \tag{12-4.1}$$

where n varies from 0.25 to 0.31. In Fig. 12-2, $\log \sigma$ is plotted against $\log (1 - T_r)$ with experimental data for acetic acid, diethyl ether, and

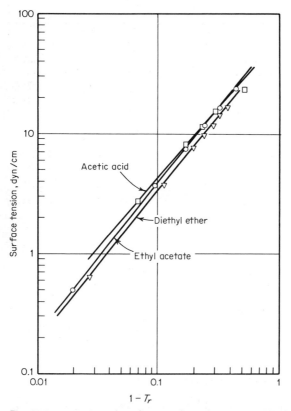

Fig. 12-2 Variation of surface tension with temperature. (*Data from Ref. 33.*)

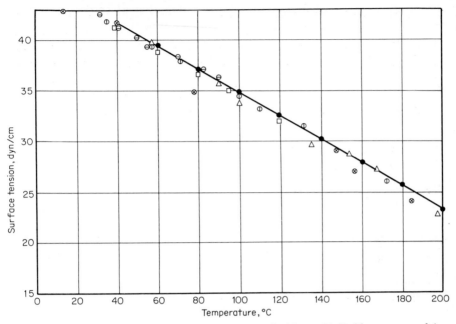

Fig. 12-3 Surface tension of nitrobenzene: ● R. B. Badachhope, M. K. Gharpurey, and A. B. Biswas, *J. Chem. Eng. Data*, **10**: 143 (1965); △ Ref. 58; ⊖ R. Kremann and R. Meirgast, *Monatsh. fuer Chemie*, **35**: 1332 (1914); ① F. M. Jaeger, *Z. Anorg. Allgem. Chem.*, **101**: 1 (1917); □ W. Hückel and W. Jahnentz, *Chem. Ber.*, **75B**: 1438 (1942); ⊗ W. Ramsay and J. Shields, *J. Chem. Soc.*, **63**: 1089 (1893). (*From Ref. 29.*)

ethyl acetate.   For the latter two, the slope is close to 1.25 for a value of $n = 0.31$, while for acetic acid, the slope is 1.16 or $n = 0.29$.   For most organic liquids, this encompasses the range normally found for n although for alcohols, n may be slightly less.   The corresponding-states correlation predicts a slope of 1.22, giving $n = 0.305$.

For values of T_r between 0.4 and 0.7, Eq. (12-4.1) indicates that $d\sigma/dT$ is almost constant, and often the surface-tension–temperature relation is represented by a linear equation

$$\sigma = a + bT \tag{12-4.2}$$

As an example, for nitrobenzene, between 40 and 200°C, data from Jasper [29] are plotted in Fig. 12-3.   The linear approximation is satisfactory.   Jasper lists values of a and b for many materials.

12-5 Surface Tensions of Nonaqueous Mixtures

The surface tension of a liquid mixture is not a simple function of the surface tensions of the pure components because in a mixture the composition of the surface is not the same as that of the bulk. In a typical situation, we know the bulk composition but not the surface composition.

The surface tension of a mixture σ_m is usually less than that calculated from a mole-fraction average of the surface tensions of the pure components. Also, the derivative $d\sigma_m/dx$ usually increases with composition for that component with the largest pure-component surface tension.

The techniques suggested for estimating σ_m can conveniently be divided into two categories: those based on empirical relations suggested earlier for pure liquids and those derived from thermodynamics.

Macleod-Sugden Correlation   Applying Eq. (12-3.1) to mixtures gives

$$\sigma_m^{1/4} = \sum_{i=1}^{n} [P_i](\rho_{L_m}x_i - \rho_{v_m}y_i) \tag{12-5.1}$$

where σ_m = surface tension of mixture, dyn/cm
   $[P_i]$ = parachor of component i
   x_i, y_i = mole fraction of i in liquid and vapor phases
   ρ_{L_m} = liquid-mixture density, g-mol/cm^3
   ρ_{v_m} = vapor-mixture density, g-mol/cm^3

At low pressures, the term involving the vapor density and composition may be neglected; when this simplification is possible, Eq. (12-5.1) has been employed to correlate mixture surface tensions for a wide variety of organic liquids [5, 16, 24, 36, 48] with reasonably good results. Most authors, however, do not obtain $[P_i]$ from general group-contribution tables such as Table 12-1; instead, they regress data to obtain the best value of $[P_i]$ for each component in the mixture. This same procedure has been used with success for gas-liquid systems under high pressure where the vapor term is significant. Weinaug and Katz [65] showed that Eq. (12-5.1) correlates methane-propane surface tensions from -15 to $90°C$ and from 2.7 to 102 atm. Deam and Mattox [11] also employed the same equation for the methane-nonane mixture from -34 to $24°C$ and 1 to 100 atm. Some smoothed data are shown in Fig. 12-4. At any temperature, σ_m decreases with increasing pressure as more methane dissolves in the liquid phase. The effect of temperature is more unusual; instead of decreasing with rising temperature, σ_m increases, except at the lowest pressures. This phenomenon illustrates the fact that at the lower temperatures methane is more soluble in nonane and the effect of liquid composition is more important in determining σ_m than the effect of temperature.

When correlating these data with Eq. (12-5.1), Deam and Mattox

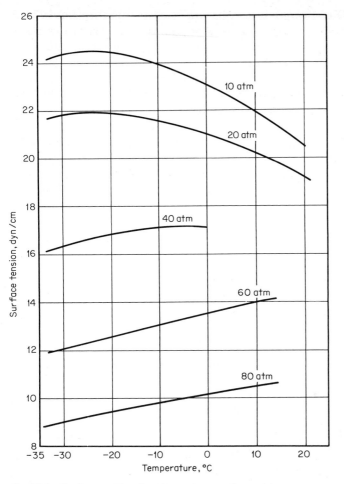

Fig. 12-4 Surface tension for the system methane-nonane.

found that the best results were obtained when $[P_{CH_4}] = 81.0$ and $[P_{C_9H_{20}}] = 387.6$. Note that if Table 12-1 is used, $[P_{CH_4}] = 71.0$ and $[P_{C_9H_{20}}] = 391$.

Other authors have also used Eq. (12-5.1) for correlating high-pressure surface-tension data; e.g., Stegemeier† [56] studied the methane-pentane and methane-decane systems, Reno and Katz [46] studied the nitrogen-butane (and heptane) systems, and Lefrançois and Bourgeois [31] investigated the effect of inert-gas pressure on many organic liquids as well as the pressure of N_2 and H_2 on the surface tension of liquid ammonia.

---

†In this case, the exponent chosen for σ_m was $\frac{3}{11}$ rather than $\frac{1}{4}$.

When the Macleod-Sugden correlation is used, mixture liquid and vapor densities and compositions must be known. Errors at low pressures rarely exceed 5 to 10 percent and can be much less if $[P_i]$ values are obtained from experimental data.

Corresponding-States Correlation Equation (12-3.6) has seen little application for mixtures because the composition of the surface is not the same as that of the bulk. Some mixing rule has to be assumed for P_{c_m}, T_{c_m}, and Q_m. A number of possible rules can be obtained from Chap. 4, or simple mole-fraction averages can be used. A limited test of the latter technique for nonpolar mixtures indicated that the accuracy was of the same order as that obtained with the Macleod-Sugden mixture relation with $[P_i]$ values from Table 12-1.

Other Empirical Methods Often, when only approximate estimates of σ_m are necessary, one chooses the general form

$$\sigma_m{}^r = \sum_{j=1}^{n} x_j \sigma_j{}^r \tag{12-5.2}$$

Hadden [22] recommends $r = 1$ for most hydrocarbon mixtures,† but much closer agreement is found if $r = -1$ to -3. Equation (12-5.1) may also be written in a form similar to Eq. (12-5.2), using Eq. (12-3.1) to express the parachor $[P_i]$. Neglecting the vapor term (low pressures), we have

$$\sigma_m{}^{1/4} = \rho_{L_m} \sum_{i=1}^{n} x_i \sigma_i{}^{1/4} \Big/ \rho_{L_i} \tag{12-5.3}$$

Equation (12-5.3) has an advantage over Eq. (12-5.1) in that at the extremes of composition it yields the correct values ($\sigma_m \to \sigma_i$ as $x_i \to 1.0$), whereas, depending upon the $[P_i]$ chosen, Eq. (12-5.1) may or may not reduce to the appropriate limiting value. The $\frac{1}{4}$ power on σ_i, however, is usually insufficient to provide the necessary curvature.

These empirical rules are illustrated in Example 12-3. The problem is that with no theoretical basis, it is difficult to generalize, and although one rule may correlate data well for one system, it may fail on one quite similar. Figure 12-5 shows mixture surface tensions for several systems. All illustrate the nonlinearity of the σ_m-vs.-x relation but to different degrees. The surface tension of the acetophenone-benzene system is almost linear in composition, whereas the nitromethane-benzene and nitrobenzene–carbon tetrachloride systems are decidedly nonlinear and the diethyl ether–benzene case is intermediate.

---

†In this reference, an empirical, *very* approximate method is suggested for mixtures when the temperature exceeds the critical temperature of at least one component.

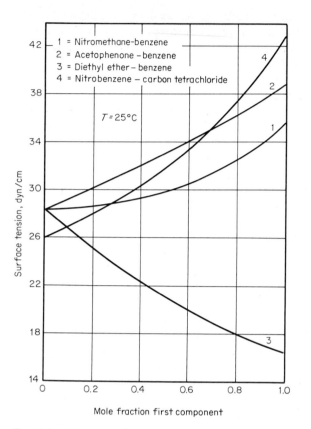

Fig. 12-5 in graph:

1 = Nitromethane-benzene
2 = Acetophenone – benzene
3 = Diethyl ether – benzene
4 = Nitrobenzene – carbon tetrachloride

$T = 25°C$

Surface tension, dyn/cm (vertical axis)

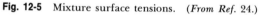

Mole fraction first component

Fig. 12-5   Mixture surface tensions.   (*From Ref. 24.*)

Example 12-3 Estimate the surface tension of a mixture of diethyl ether and benzene containing 42.3 mole percent of the former.   The temperature is 25°C.

Hammick and Andrew [24] list densities and surface tensions for this system at 25°C; other data are from Appendix A.

	Benzene	Diethyl ether	Mixture with 42.3 mol % diethyl ether
Density, g/cm³	0.8722	0.7069	0.7996
Surface tension, dyn/cm	28.23	16.47	21.81
Molecular weight	78.114	74.123	76.426
Critical temperature, K	562.1	466.7	
Critical pressure, atm	48.3	35.9	
Normal boiling point, K	353.3	307.7	

solution *Macleod-Sugden Correlation.* To employ Eq. (12-5.1) the parachors of benzene and diethyl ether are required. From Table 12-1,

$$[P_{benzene}] = C_6H_5— + H = 189.6 + 15.5 = 205.1$$

$$[P_{ether}] = (2)(CH_3—) + (2)(—CH_2—) + —O—$$

$$= (2)(55.5) + (2)(40.0) + 20.0 = 211$$

$$M_m = (0.423)(74.123) + (0.577)(78.114) = 76.426$$

When the vapor term is neglected, Eq. (12-5.1) becomes

$$\sigma_m^{1/4} = \frac{0.7996}{76.426}[(0.423)(211) + (0.577)(205.1)]$$

$$\sigma_m = 22.25 \text{ dyn/cm}$$

$$\text{Error} = \frac{22.25 - 21.81}{21.81} \times 100 = 2.0\%$$

With the parachor values estimated above and with pure-component densities, Eq. (12-3.1) yields surface tensions of 27.5 dyn/cm for pure benzene and 16.40 dyn/cm for pure diethyl ether, compared with experimental values of 28.23 and 16.47 dyn/cm, respectively.

Modified Macleod-Sugden Correlation. With Eq. (12-5.3),

$$\sigma_m^{1/4} = \frac{0.7996}{76.426}\left[\frac{(0.423)(16.47)^{1/4}}{0.7069/74.123} + \frac{(0.577)(28.23)^{1/4}}{0.8722/78.114}\right]$$

$$\sigma_m = 22.63 \text{ dyn/cm}$$

$$\text{Error} = \frac{22.63 - 21.81}{21.81} \times 100 = 3.8\%$$

Although Eq. (12.5-3) yields the correct pure-component values of σ, it invariably yields larger errors for σ_m than Eq. (12-5.1).

Corresponding-States Correlation. With Eqs. (12-3.6) and (12-3.7) applied to mixtures, assume, for simplicity, mole-fraction averages for P_{c_m}, T_{c_m}, and Q_m:

$$P_{c_m} = (0.423)(35.9) + (0.577)(48.3) = 43.1 \text{ atm}$$

$$T_{c_m} = (0.423)(466.7) + (0.577)(562.1) = 522 \text{ K}$$

With Q (diethyl ether) $= 0.680$ and Q (benzene) $= 0.634$, we have

$$Q_m = (0.423)(0.680) + (0.577)(0.634) = 0.653$$

Then, with $T_{r_m} = (25 + 273)/522 = 0.571$, we get

$$\sigma_m = (43.1)^{2/3}(522)^{1/3}(0.653)(1 - 0.571)^{11/9}$$

$$= 22.97 \text{ dyn/cm}$$

$$\text{Error} = \frac{22.97 - 21.81}{21.81} \times 100 = 5.3\%$$

Power Relation. With Eq. (12-5.2), various values of r can be selected. The results shown below indicate the errors.

r	Calculated σ_m, dyn/cm	% error
1	23.25	6.6
0†	22.48	3.1
−1	21.68	−0.6
−2	20.92	−4.0
−3	20.19	−7.4

†When $q = 0$, $\sigma_m = \sigma_1 \exp[x_2 \ln(\sigma_2/\sigma_1)]$.

Thermodynamic Correlations   The estimation procedures introduced earlier in this section are empirical; all employ the bulk-liquid (and sometimes vapor) composition to characterize a mixture.   However, the "surface phase" differs in composition from that of the bulk liquid and vapor, and it is reasonable to suppose that in mixture surface-tension relations, surface compositions play a more important role than bulk compositions.   The fact that σ_m is almost always less than the bulk mole-fraction average is interpreted as indicating that the component or components with the lower pure-component values of σ preferentially concentrate in the surface phase.   Eberhart [12] assumes that σ_m is given by the surface-composition–mole-fraction average.

Both classical and statistical thermodynamics have been employed to derive expressions for σ_m [4, 9, 13, 14, 25, 26, 27, 30, 50, 51, 53,† 55, 59]; the results differ in some aspects, but most arrive at a result similar to

$$\sum_{i=1}^{n} \left(\frac{x_i^B \gamma_i^B}{\gamma_i^\sigma}\right) \exp\frac{\mathscr{A}_i(\sigma_m - \sigma_i)}{RT} = 1 \qquad (12\text{-}5.4)$$

where x_i^B = mole fraction of i in bulk liquid

$\quad\quad \gamma_i^B$ = activity coefficient of i in bulk liquid normalized so that $\gamma_i^B \to 1$ as $x_i \to 1$

$\quad\quad \gamma_i^\sigma$ = activity coefficient of i in surface phase normalized so that $\gamma_i^\sigma \to 1$ as surface phase becomes identical to that for pure i

$\quad\quad \mathscr{A}_i$ = partial molar surface area of i, cm^2/g-mol

$\quad \sigma_m, \sigma_i$ = surface tension of mixture and of component i

The activity coefficients γ_i^B are usually obtained from vapor-liquid equilibrium data or from some liquid model, as discussed in Chap. 8.   For the surface phase, however, no direct measurements appear possible, and a liquid model must be assumed.   Hildebrand and Scott [26] treat the case of a "perfect" surface phase, $\gamma_i^\sigma = 1$, while Eckert and Prausnitz [13] and Sprow and Prausnitz [55] have used regular solution

†Hansen and Sogor [25] have shown that the thermodynamic expression derived by Shereshefsky [53] and used by Cotton [10] is in error.

theory to describe the same phase.   In all cases, the partial molal area $\mathcal{A}_i$ has been approximated as $(V_i)^{2/3}(N_0)^{1/3}$, where V_i is the *pure*-liquid molal volume of i and N_0 is Avogadro's number.   Sprow and Prausnitz have successfully applied Eq. (12-5.4) to a variety on nonpolar binary systems and have estimated σ_m accurately.   Even for polar mixtures, some success was achieved, although in this case a modified Wilson activity-coefficient expression was introduced for the surface phase and one empirical constant was retained to give a good fit between experimental and calculated values of σ_m.

Although the Sprow-Prausnitz approach is more realistic, it is not uncommon to find Eq. (12-5.4) simplified assuming an ideal-liquid mixture, that is, $\gamma_i^{B} = \gamma_i^{\sigma} = 1$.   For a binary system, it can then be shown that this equation simplifies to

$$\sigma_m = x_A\sigma_A + x_B\sigma_B - \frac{\mathcal{A}}{2RT}(\sigma_A - \sigma_B)^2 x_A x_B \qquad (12\text{-}5.5)$$

where the terms are as defined above (with x_A, x_B bulk mole fractions) and $\mathcal{A}$ is an average surface area for the molecules constituting the system.   This simplified form clearly indicates that σ_m is less than a mole-fraction average.

Example 12-4   Repeat Example 12-3 using the simplified (ideal) form of the thermodynamic correlation.

  solution   Let subscript A stand for diethyl ether and subscript B for benzene.   Then $x_A = 0.423$, $\sigma_A = 16.47$, $x_B = 0.577$, and $\sigma_B = 28.23$, giving

$$RT = (8.314 \times 10^7)(298) = 2.478 \times 10^{10} \text{ ergs/g-mol K}$$

$$\mathcal{A}_A = \left(\frac{74.123}{0.7069}\right)^{2/3}(6.023 \times 10^{23})^{1/3}$$

$$= 1.775 \times 10^9 \text{ cm}^2/\text{g-mol}$$

$$\mathcal{A}_B = \left(\frac{78.114}{0.8722}\right)^{2/3}(6.023 \times 10^{23})^{1/3}$$

$$= 1.690 \times 10^9 \text{ cm}^2/\text{g-mol}$$

With $\mathcal{A} \approx (\mathcal{A}_A + \mathcal{A}_B)/2 = 1.743 \times 10^9 \text{ cm}^2/\text{g-mol}$, with Eq. (12-5.5) we have

$$\sigma_m = (0.423)(16.47) + (0.577)(28.23)$$

$$- \frac{1.783 \times 10^9}{(2)(2.478 \times 10^{10})} \times (16.47 - 28.23)^2(0.423)(0.577)$$

$$= 22.03 \text{ dyn/cm}$$

$$\text{Error} = \frac{22.03 - 21.81}{21.81} \times 100 = 1\%$$

Although Example 12-4 shows but a small error when Eq. (12-5.5) is applied to the diethyl ether–benzene system at 25°C, this is more fortuitous than representative.   The more accurate form, Eq. (12-5.4), is preferable, though considerably more effort must be expended to utilize it.

Recommendations   For *nonpolar* systems, use Eq. (12-5.4) as prescribed by Sprow and Prausnitz; even polar systems may be treated if at least one mixture value of σ_m is available.   Errors are typically less than 2 to 3 percent.   For less accurate estimates, there is little choice between the Macleod-Sugden correlation (12-5.1), the corresponding-states correlation (12-3.6) (see Example 12-3), and the ideal thermodynamic relation (12-5.5).   Errors are generally less than 5 to 10 percent.

For mixtures containing one or more polar components, the corresponding-states method cannot be used; also, none of the thermodynamic methods is applicable without introducing a mixture constant.   For polar (nonaqueous) systems, therefore, the Macleod-Sugden relation [Eq. (12-5.1)] is the only method available.   When employed to correlate the mixture surface-tension data of Ling and van Winkle [32], only moderate agreement (5 to 15 percent) was achieved for the polar-polar and polar-nonpolar systems studied.

12-6   Surface Tensions of Aqueous Solutions

Whereas for nonaqueous solutions the mixture surface tension is often approximated by a linear dependence on mole fraction, aqueous solutions show pronounced nonlinear characteristics.   A typical case is shown in Fig. 12-6 for acetone-water at 50°C.   The surface tension of the mixture is represented by an approximately straight line on semilogarithmic coordinates.   This behavior is typical of organic-aqueous systems, where small concentrations of the organic material may significantly affect the mixture surface tension.   The hydrocarbon portion of the organic molecule behaves like a hydrophobic material and tends to be rejected from the water phase by preferentially concentrating on the surface.   In such a case, the bulk concentration is very different from the surface concentration.   Unfortunately, the latter is not easily amenable to direct measurement.   Meissner and Michaels [36] show graphs similar to Fig. 12-6 for a variety of dilute solutions of organic materials in water and suggest that the general behavior is approximated

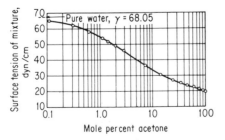

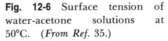

Fig.   12-6 Surface tension of water-acetone solutions at 50°C.   (*From Ref.* 35.)

by the Szyszkowski equation, which they modify to the form

$$\frac{\sigma_m}{\sigma_W} = 1 - 0.411 \log \left(1 + \frac{x}{a}\right) \tag{12-6.1}$$

where σ_W = surface tension of pure water
$\quad\quad x$ = mole fraction of organic material
$\quad\quad a$ = constant characteristic of organic material

Values of a are listed in Table 12-4 for a few compounds. This equation should not be used if the mole fraction of the organic solute exceeds 0.01.

The method of Tamura, Kurata, and Odani [60] may be used to estimate surface tensions of aqueous binary mixtures over wide concentration ranges of the dissolved organic material and for both low- and high-molecular-weight organic-aqueous systems. Equation (12-5.1) is assumed as a starting point, but the significant densities and concentrations are taken to be those characteristic of the surface layer, that is, $(V^\sigma)^{-1}$ replaces ρ_{L_m}, where V^σ is a hypothetical molal volume of the surface layer. V^σ is estimated with

$$V^\sigma = \sum_j x_j^\sigma V_j \tag{12-6.2}$$

where x_j^σ is the mole fraction of j in the surface layer. V_j, however, is chosen as the pure-liquid molal volume of j. Then, with Eq. (12-5.1), assuming $\rho_L \gg \rho_v$,

$$V^\sigma \sigma_m^{1/4} = x_W^\sigma [P_W] + x_O^\sigma [P_O] \tag{12-6.3}$$

where the subscripts W and O represent water and the organic

TABLE 12-4 Constants for the Szyszkowski Equation (12-6.1) [36]

Compound	$a \times 10^4$	Compound	$a \times 10^4$
Propionic acid	26	Ethyl propionate	3.1
n-Propyl alcohol	26	Propyl acetate	3.1
Isopropyl alcohol	26		
Methyl acetate	26	n-Valeric acid	1.7
		Isovaleric acid	1.7
n-Propyl amine	19	n-Amyl alcohol	1.7
Methyl ethyl ketone	19	Isoamyl alcohol	1.7
n-Butyric acid	7.0	Propyl propionate	1.0
Isobutyric acid	7.0	n-Caproic acid	0.75
n-Butyl alcohol	7.0	n-Heptanoic acid	0.17
Isobutyl alcohol	7.0	n-Octanoic acid	0.034
		n-Decanoic acid	0.0025
Propyl formate	8.5		
Ethyl acetate	8.5		
Methyl propionate	8.5		
Diethyl ketone	8.5		

component.   To eliminate the parachor, however, Tamura et al. introduce Eq. (12-3.1); the result is

$$\sigma_m^{1/4} = \psi_w^{\sigma} \sigma_w^{1/4} + \psi_o^{\sigma} \sigma_o^{1/4} \tag{12-6.4}$$

In Eq. (12-6.4), ψ_w^{σ} is the superficial volume fraction water in the surface layer

$$\psi_w^{\sigma} = \frac{x_w^{\sigma} V_w}{V^{\sigma}} \tag{12-6.5}$$

and similarly for ψ_o^{σ}.

Equation (12-6.4) is the final correlation.   To obtain values of the superficial surface volume fractions ψ_w^{σ} and ψ_o^{σ}, equilibrium is assumed between the surface and bulk phases.   Tamura's equation is complex, and after rearrangement it can be written in the following set of equations:

$$\mathcal{B} = \log \frac{\psi_w^q}{\psi_o} \tag{12-6.6}$$

$$\mathcal{C} = \log \frac{(\psi_w^{\sigma})^q}{\psi_o^{\sigma}} \tag{12-6.7}$$

$$\mathcal{C} = \mathcal{B} + \mathcal{W} \tag{12-6.8}$$

$$\mathcal{W} = 0.441 \frac{q}{T} \left(\frac{\sigma_o V_o^{2/3}}{q} - \sigma_w V_o^{2/3} \right) \tag{12-6.9}$$

where ψ_w^{σ} is defined by Eq. (12-6.5) and ψ_w, ψ_o are the superficial bulk volume fractions of water and organic material, i.e.,

$$\psi_w = \frac{x_w V_w}{x_w V_w + x_o V_o} \qquad \psi_o = \frac{x_o V_o}{x_w V_w + x_o V_o} \tag{12-6.10}$$

where $x_w, x_o = $ bulk mole fraction of pure water and pure organic
         component
$V_w, V_o = $ molal volume of pure water and pure organic component
$\sigma_w, \sigma_o = $ surface tension of pure water and pure organic component
$T = $ temperature, K
$q = $ constant depending upon type and size of organic constituent

Materials	q	Example
Fatty acids, alcohols	Number of carbon atoms	Acetic acids, $q = 2$
Ketones	One less than the number of carbon atoms	Acetone, $q = 2$
Halogen derivatives of fatty acids	Number of carbons times ratio of molal volume of halogen derivative to parent fatty acid	Chloroacetic acid, $q = 2 \dfrac{V_b \text{ (chloroacetic acid)}}{V_b \text{ (acetic acid)}}$

The method is illustrated in Example 12-5. Tamura et al. [60] tested the method with some 14 aqueous systems and 2 alcohol-alcohol systems;† the percentage errors are less than 10 percent when q is less than 5 and within 20 percent for q greater than 5. The method cannot be applied to multicomponent mixtures.

Example 12-5 Estimate the surface tension of a mixture of methyl alcohol and water at 30°C when the mole fraction alcohol is 0.122. The experimental value reported is 46.1 dyn/cm [60].

  solution At 30°C (O represents methyl alcohol, W water), $\sigma_W = 71.18$ dyn/cm, $\sigma_O = 21.75$ dyn/cm, $V_W = 18$ cm³/g-mol, $V_O = 41$ cm³/g-mol, and $q =$ number of carbon atoms = 1. From Eqs. (12-6.10),

$$\frac{\psi_W}{\psi_O} = \frac{(0.878)(18)}{(0.122)(41)} = 3.16$$

and from Eq. (12-6.6),

$$\mathscr{B} = \log 3.16 = 0.50$$

$$\mathscr{W}[\text{from Eq. (12-6.9)}] = (0.441)(\tfrac{1}{303})[(21.75)(41)^{2/3} - (71.18)(18)^{2/3}]$$
$$= -0.34$$

Hence

$$\mathscr{C}\ [\text{from Eq. (12-6.8)}] = \mathscr{B} + \mathscr{W} = 0.50 - 0.34 = 0.16$$

$$= \log \frac{\psi_W^\sigma}{\psi_O^\sigma} \quad \text{from Eq. (12-6.7)}$$

Since $\psi_W^\sigma + \psi_O^\sigma = 1$, we have

$$\frac{\psi_W^\sigma}{1 - \psi_W^\sigma} = 10^{0.16} = 1.45$$

$$\psi_W^\sigma = 0.59 \qquad \psi_O^\sigma = 0.41$$

Finally, from Eq. (12-6.4)

$$\sigma_m = [(0.59)(71.18)^{1/4} + (0.41)(21.75)^{1/4}]^4 = 46\ \text{dyn/cm}$$

$$\text{Error} = \frac{46 - 46.1}{46.1} \times 100 = -0.2\%$$

Recommendations For estimating the surface tensions of binary organic-aqueous mixtures, use the method of Tamura, Kurata, and Odani, as given by Eqs. (12-6.2) to (12-6.10) and illustrated in Example 12-5. The method may also be used for other highly polar solutes, such as alcohols. Errors normally do not exceed 10 to 15 percent. If the solubility of the organic compound in water is low, the Szyszkowski equation (12-6.1), as developed by Meissner and Michaels, may be used.

NOTATION

$a =$ parameter in Eq. (12-6.1) and obtained from Table 12-4

$A =$ area

$\mathscr{A} =$ surface area; $\mathscr{A}_i$, partial molal area of component i on the surface

---

†For nonaqueous mixtures comprising polar molecules, the method is unchanged except that $q =$ ratio of molal volumes of the solute to solvent.

$\mathscr{B}$ = parameter in Eq. (12-6.6)
$\mathscr{C}$ = parameter in Eq. (12-6.7)
m = parameter in Eq. (12-3.8)
n = parameter in Eq. (12-3.2)
N_0 = Avogadro's number
p = parameter in Eq. (12-3.10)
$[P_i]$ = parachor of component i (see Table 12-1)
P_{vp} = vapor pressure, atm
P_c = critical pressure, atm; P_{c_m}, pseudocritical pressure
q = parameter in Eq. (12-6.9)
Q = parameter in Eq. (12-3.7)
Q_p = parameter in Eq. (12-3.8)
r = parameter in Eq. (12-5.2)
R = gas constant
T = temperature, K; T_c, critical temperature; T_b, normal boiling point, T_r, reduced
 temperature T/T_c; T_{b_r}, T_b/T_c; T_{c_m}, pseudocritical mixture temperature
V = liquid molal volume, cm³/g-mol; V^σ, for the surface phase
V_c = critical volume, cm³/g-mol
$\mathscr{W}$ = parameter in Eq. (12-6.9)
x_i = liquid mole fraction; x_i^B, in the bulk phase; x_i^σ, in the surface phase
y_i = vapor mole fraction

Greek

α_c = Riedel factor (see Sec. 6-5)
γ_i = activity coefficient of component i; γ_i^B, in the bulk liquid phase; γ_i^σ, in the surface phase
η = liquid or vapor viscosity, cP
ρ = liquid or vapor density, g-mol/cm³
σ = surface tension, dyn/cm; σ_m, for a mixture; σ_O, a parameter in Eq. (12-3.10) or
 representing an organic component
ψ_i = volume fraction of i in the bulk liquid; ψ_i^σ, in the surface phase

Subscripts

b = normal boiling point
L = liquid
m = mixture
O = organic component in aqueous solution
r = reduced value, i.e., the property divided by its value at the critical point
v = vapor
W = water

REFERENCES

1. Adamson, A. W.: "Physical Chemistry of Surfaces," Interscience, New York, 1960.
2. Aveyard, R., and D. A. Haden: "An Introduction to Principles of Surface Chemistry," Cambridge University Press, London, 1973.
3. Barbulescu, N.: *Rev. Roum. Chem.*, **19**: 169 (1974).
4. Belton, J. W., and M. G. Evans: *Trans. Faraday Soc.*, **41**: 1 (1945).
5. Bowden, S. T., and E. T. Butler: *J. Chem. Soc.*, **1939**: 79.
6. Brock, J. R., and R. B. Bird: *AIChE J.*, **1**: 174 (1955).
7. Brown, R. C.: *Contemp. Phys.*, **15**: 301 (1974).
8. Buehler, C. A.: *J. Phys. Chem.*, **42**: 1207 (1938).
9. Clever, H. L., and W. E. Chase, Jr.: *J. Chem. Eng. Data*, **8**: 291 (1963).

10. Cotton, D. J.: *J. Phys. Chem.*, **73**: 270 (1969).
11. Deam, J. R., and R. N. Mattox: *J. Chem. Eng. Data*, **15**: 216 (1970).
12. Eberhart, J. G.: *J. Phys. Chem.*, **70**: 1183 (1966).
13. Eckert, C. A., and J. M. Prausnitz: *AIChE J.*, **10**: 677 (1964).
14. Evans, H. B., Jr., and H. L. Clever: *J. Phys. Chem.*, **68**: 3433 (1964).
15. Fishtine, S. H.: *Ind. Eng. Chem. Fundam.*, **2**: 149 (1963).
16. Gambill, W. R.: *Chem. Eng.*, **64**(5): 143 (1958).
17. Gambill, W. R.: *Chem. Eng.*, **66**(23): 193 (1959).
18. Gibbs, J. W.: "The Collected Works of J. Willard Gibbs," vol. I, "Thermodynamics," Yale University Press, New Haven, Conn., 1957.
19. Goldhammer, D. A.: *Z. Phys. Chem.*, **71**: 577 (1910).
20. Guggenheim, E. A.: "Thermodynamics," 4th ed., North-Holland, Amsterdam, 1959.
21. Guggenheim, E. A.: *Proc. Phys. Soc.*, **85**: 811 (1965).
22. Hadden, S. T.: *Hydrocarbon Process.*, **45**(10): 161 (1966).
23. Hakim, D. I., D. Steinberg, and L. I. Stiel: *Ind. Eng. Chem. Fundam.*, **10**: 174 (1971).
24. Hammick, D. L., and L. W. Andrew: *J. Chem. Soc.*, **1929**: 754.
25. Hansen, R. S., and L. Sogor: *J. Colloid Interface Sci.*, **40**: 424 (1972).
26. Hildebrand, J. H., and R. L. Scott: "The Solubility of Nonelectrolytes," 3d ed., chapter 21, Dover, New York, 1964.
27. Hoar, T. P., and D. A. Melford: *Trans. Faraday Soc.*, **53**: 315 (1957).
28. "International Critical Tables," vol. III, p. 28, McGraw-Hill, New York, 1928.
29. Jasper, J. J.: *J. Phys. Chem. Ref. Data*, **1**: 841 (1972).
30. Kim, S. W., M. S. Jhon, T. Ree, and H. Eyring: *Proc. Natl. Acad. Sci.*, **59**: 336 (1968).
31. Lefrançois, B., and Y. Bourgeois: *Chim. Ind. Genie Chim.*, **105**(15): 989 (1972).
32. Ling, T. D., and M. van Winkle: *J. Chem. Eng. Data*, **3**: 88 (1958).
33. Macleod, D. B.: *Trans. Faraday Soc.*, **19**: 38 (1923).
34. Mayer, S. W.: *J. Phys. Chem.*, **67**: 2160 (1963).
35. McAllister, R. A., and K. S. Howard: *AIChE J.*, **3**: 325 (1957).
36. Meissner, H. P., and A. S. Michaels: *Ind. Eng. Chem.*, **41**: 2782 (1949).
37. Miller, D. G.: *Ind. Eng. Chem. Fundam.*, **2**: 78 (1963).
38. Modell, M., and R. C. Reid: "Thermodynamics and Its Applications," Prentice-Hall, Englewood Cliffs, N.J., 1974.
39. Móritz, P.: *Period. Polytech.*, **3**: 167 (1959).
40. Othmer, D. F., S. Josefowitz, and Q. F. Schmutzler, *Ind. Eng. Chem.*, **40**: 886 (1948).
41. Pelofsky, A. H.: *J. Chem. Eng. Data*, **11**: 394 (1966).
42. Plesner, I. W., and O. Platz: *J. Chem. Phys.*, **48**: 5361, 5364 (1968).
43. Porteous, W.: *J. Chem. Eng. Data*, **20**: 339 (1975).
44. Quale, O. R.: *Chem. Rev.*, **53**: 439 (1953).
45. Rao, M. V., K. A. Reddy, and S. R. S. Sastri: *Hydrocarbon Process.*, **47**(1): 151 (1958).
46. Reno, G. J., and D. L. Katz: *Ind. Eng. Chem.*, **35**: 1091 (1943).
47. Rice, O. K.: *J. Phys. Chem.*, **64**: 976 (1960).
48. Riedel, L.: *Chem. Ing. Tech.*, **27**: 209 (1955).
49. Ross, S. (Chairman): "Chemistry and Physics of Interfaces," American Chemical Society, 1965.
50. Schmidt, R. L.: *J. Phys. Chem.*, **71**: 1152 (1967).
51. Schmidt, R. L., J. C. Randall, and H. L. Clever: *J. Phys. Chem.*, **70**: 3912 (1966).
52. Schonhorn, H.: *J. Chem. Eng. Data*, **12**: 524 (1967).
53. Shereshefsky, J. L.: *J. Colloid Interface Sci.*, **24**: 317 (1967).
54. Sprow, F. B., and J. M. Prausnitz: *Trans. Faraday Soc.*, **62**: 1097 (1966).
55. Sprow, F. B., and J. M. Prausnitz: *Trans. Faraday Soc.*, **62**: 1105 (1966); *Can. J. Chem. Eng.*, **45**: 25 (1967).
56. Stegemeier, G. L.: Ph.D. dissertation, University of Texas, Austin, Tex., 1959.

57. Sugden, S.: *J. Chem. Soc.*, **1924**: 32.
58. Sugden, S.: *J. Chem. Soc.*, **1924**: 1177.
59. Suri, S. K., and V. Ramkrishna: *J. Phys. Chem.*, **72**: 3073 (1968).
60. Tamura, M., M. Kurata, and H. Odani: *Bull. Chem. Soc. Jpn.*, **28**: 83 (1955).
61. Tripathi, R. C.: *J. Indian Chem. Soc.*, **18**: 411 (1941).
62. Tripathi, R. C.: *J. Indian Chem. Soc.*, **19**: 51 (1942).
63. van der Waals, J. D.: *Z. Phys. Chem.*, **13**: 716 (1894).
64. Watkinson, A. P., and J. Lielmezs: *J. Chem. Phys.*, **47**: 1558 (1967).
65. Weinaug, C. F., and D. L. Katz: *Ind. Eng. Chem.*, **35**: 239 (1943).

Appendix A

Property Data Bank

Compounds are grouped by number of carbon atoms. Within each group they are listed by the number of hydrogen atoms and subdivided further by additional atoms in alphabetical order. Names are those most commonly employed. Only 21 characters are allowed. See the Compound Dictionary (Appendix B) for the full name and common synonyms. Most values were taken from compilations of experimental data; a few, however, were estimated by techniques described in this book.

MOLWT = molecular weight
TFP = normal freezing point, K
TB = normal boiling point, K
TC = critical temperature, K
PC = critical pressure, atm
VC = critical volume, cm^3/g-mol
ZC = critical compressibility
OMEGA = Pitzer's acentric factor
LIQDEN = liquid density at TDEN, g/cm^3
TDEN = reference temperature for LIQDEN, K
DIPM = dipole moment, debyes
CPVAP A, CPVAP B, CPVAP C, CPVAP D = constants in ideal-gas heat-capacity equation, with CP in calories per gram-mole kelvin and T in kelvins:

$$CP = CPVAP\ A + (CPVAP\ B)*T + (CPVAP\ C)*T**2 + (CPVAP\ D)*T**3$$

VISB, VISTO = constants in liquid viscosity equation, with viscosity in centipoises and T in kelvins:

$$LOG(VISCOSITY) = (VISB)*((1/T) - (1/VISTO))$$

DELHG = standard enthalpy of formation at 298 K, kcal/g-mol
DELGF = standard Gibbs energy of formation at 298 K for ideal gas at 1 atm, kcal/g-mol
ANTA, ANTB, ANTC = Antoine vapor-pressure-equation coefficients, with vapor pressure in millimeters of mercury and T in kelvins:

$$LN(VAPOR\ PRESSURE) = ANTA - ANTB/(T + ANTC)$$

TMX = maximum temperature for Antoine equation, K
TMN = minimum temperature for Antoine equation, K
HARA, HARB, HARC, HARD = Harlacher vapor-pressure-equation coefficients, with vapor pressure in millimeters of mercury, T in kelvins, and PVP as the vapor pressure:

$$LN(PVP) = HARA + HARB/T + HARC*LN(T) + HARD*PVP/T**2$$

HV = heat of vaporization at normal boiling point, cal/g-mol

NO	FORMULA	NAME	MOLWT	TFP	TB	TC	PC	VC	ZC	OMEGA	LIQDEN	TDEN	DIPM
1	AR	ARGON	39.948	83.8	87.3	150.8	48.1	74.9	0.291	-.004	1.373	90.	0.0
2	BCL3	BORON TRICHLORIDE	117.169	165.9	285.7	452.0	38.2	127.	0.270	0.150	1.35	284.	0.0
3	BF3	BORON TRIFLUORIDE	67.805	146.5	173.3	260.8	49.2			0.042			0.0
4	BR2	BROMINE	159.808	266.0	331.9	584.	102.	127.		0.132	3.119	293.	0.2
5	CLNO	NITROSYL CHLORIDE	65.459	213.5	267.7	440.	90.	139.	0.35	0.318	1.42	261.	1.8
6	CL2	CHLORINE	70.906	172.2	238.7	417.	76.	124.	0.275	0.073	1.563	239.1	0.2
7	CL3P	PHOSPHORUS TRICHLORID	137.333	161.	349.	563.	37.	260.	0.29	0.264	1.574	294.	0.9
8	CL4SI	SILICON TETRACHLORIDE	169.898	204.3	330.4	507.	37.	326.	0.29	0.264	1.48	293.	0.0
9	D2	DEUTERIUM	4.032	18.7	23.7	38.4	16.4	60.3	0.314	-.13	0.165	22.7	0.0
10	D2O	DEUTERIUM OXIDE	20.031	277.0	374.6	644.0	213.8	55.6	0.225		0.105	293.	1.9
11	F2	FLUORINE	37.997	53.5	85.0	144.3	51.5	66.2	0.288	0.048	1.51	85.	
12	F3N	NITROGEN TRIFLUORIDE	71.002	66.4	144.1	234.	44.7	144.		0.132	1.537	144.	0.2
13	F4SI	SILICON TETRAFLUORIDE	104.080	183.0	187.	259.0	36.7	178.			1.66	178.	0.0
14	F6S	SULFUR HEXAFLUORIDE	146.050	222.5	209.3	318.7	37.1	198.	0.281	0.286	1.83	223.	0.0
15	HBR	HYDROGEN BROMIDE	80.912	187.1	206.1	363.2	84.4	100.0	0.283	0.063	2.16	216.	0.8
16	HCL	HYDROGEN CHLORIDE	36.461	159.0	188.1	324.6	82.0	81.0	0.249	0.12	1.193	188.1	1.1
17	HF	HYDROGEN FLUORIDE	20.006	190.	292.7	461.	64.	69.	0.12	0.372	0.967	293.	
18	HI	HYDROGEN IODIDE	127.912	222.4	237.6	424.0	82.0	131.	0.309	0.05	2.803	237.	0.5
19	H2	HYDROGEN	2.016	14.0	20.4	33.2	12.8	65.0	0.305	-.22	0.071	20.	0.0
20	H2O	WATER	18.015	273.2	373.2	647.3	217.6	56.0	0.229	0.344	0.998	293.	1.8
21	H2S	HYDROGEN SULFIDE	34.080	187.6	212.8	373.2	88.2	98.5	0.284	0.100	0.993	213.6	0.9
22	H3N	AMMONIA	17.031	195.4	239.7	405.6	111.3	72.5	0.242	0.250	0.639	273.2	1.5
23	H4N2	HYDRAZINE	32.045	274.7	386.7	653.	145.	96.1	0.260	0.328	1.008	293.	3.0
24	HE(4)	HELIUM-4	4.003		4.21	5.19	2.24	57.3	0.301	-.387	0.123	4.3	0.0
25	I2	IODINE	253.808	386.8	457.5	819.	115.	155.	0.265	-.229	3.74	453.2	1.3
26	KR	KRYPTON	83.800	115.8	119.8	209.4	54.3	91.2	0.288	-.002	2.42	120.	0.0
27	NO	NITRIC OXIDE	30.006	109.5	121.4	180.	64.	58.	0.25	0.607	1.28	121.	0.2
28	NO2	NITROGEN DIOXIDE	46.006	261.9	294.3	431.4	100.	170.	0.480	0.86	1.447	292.9	0.4
29	N2	NITROGEN	28.013	63.3	77.4	126.2	33.5	89.5	0.290	0.040	0.804	78.1	0.0
30	N2O	NITROUS OXIDE	44.013	182.3	184.7	309.6	71.5	97.4	0.274	0.160	1.226	183.6	0.2
31	NE	NEON	20.183	24.5	27.0	44.4	27.2	41.7	0.311	0.00	1.204	27.	0.0
32	O2	OXYGEN	31.999	54.4	90.2	154.6	49.8	73.4	0.288	0.021	1.149	90.	0.0
33	O2S	SULFUR DIOXIDE	64.063	197.7	263.	430.8	77.8	122.	0.268	0.251	1.455	263.	1.6
34	O3	OZONE	47.998	80.5	161.3	261.0	55.0	88.9	0.228	0.215	1.356	161.3	0.6
35	O3S	SULFUR TRIOXIDE	80.058	290.	318.	491.0	81.	130.	0.26	0.041	1.78	318.	0.6
36	XE	XENON	131.300	161.3	165.0	289.7	57.6	118.	0.286	0.002	3.06	165.	0.0
37	CBRF3	TRIFLUOROBROMOMETHANE	148.910	92.0	214.	340.2	39.2	200.	0.280	0.180	1.750	158.	0.7
38	CCLF3	CHLOROTRIFLUOROMETHAN	104.459	92.0	191.7	302.0	38.7	180.	0.282	0.176			0.5
39	CCL2F2	DICHLORODIFLUOROMETHA	120.914	115.4	243.4	385.0	40.7	217.	0.280	0.204	1.381	293.	0.5
40	CCL2O	PHOSGENE	98.916	145.	280.8	455.	56.	190.	0.28				1.1

NO	FORMULA	NAME	CPVAP A	CPVAP B	CPVAP C	CPVAP D	VISB	VISTO	DELHG	DELGF
1	AR	ARGON	4.969	-.767 E-5	1.234 E-8	0.0	107.57	58.76	0.0	0.0
2	BCL3	BORON TRICHLORIDE								
3	BF3	BORON TRIFLUORIDE								
4	BR2	BROMINE	8.087	2.688 E-3	-2.846 E-6	1.083 E-9	387.82	292.79	0.0	0.0
5	CLNO	NITROSYL CHLORIDE	8.144	1.068 E-2	-7.977 E-6	2.424 E-9	191.96	172.35	12.57	16.00
6	CL2	CHLORINE	6.432	8.082 E-3	-9.241 E-6	3.695 E-9			0.0	0.0
7	CL3P	PHOSPHORUS TRICHLORID								
8	CL4SI	SILICON TETRACHLORIDE								
9	D2	DEUTERIUM	7.225	-.158 E-2	2.794 E-6	-0.88 E-9	19.67	8.38	0.0	0.0
10	D2O	DEUTERIUM OXIDE					757.92	304.58	-59.57	-56.08
11	F2	FLUORINE	5.545	8.734 E-3	-8.269 E-6	2.876 E-9	84.20	52.52	0.0	0.0
12	F3N	NITROGEN TRIFLUORIDE								
13	F4SI	SILICON TETRAFLUORIDE							-29.78	-30.38
14	F6S	SULFUR HEXAFLUORIDE							-291.8	-267.0
15	HBR	HYDROGEN BROMIDE	7.320	-.226 E-2	4.114 E-6	-1.490 E-9	251.29	180.75	-8.66	-12.73
16	HCL	HYDROGEN CHLORIDE	7.235	-.172 E-2	2.976 E-6	-0.931 E-9	88.08	166.32	-22.06	-22.77
17	HF	HYDROGEN FLUORIDE	6.941	1.579 E-4	-4.854 E-7	0.598 E-9	372.78	277.74	-64.80	-65.30
18	HI	HYDROGEN IODIDE	7.442	-.341 E-2	7.099 E-6	-3.232 E-9	438.74	199.62	6.30	0.38
19	H2	HYDROGEN	6.483	2.215 E-3	-3.298 E-6	1.826 E-9	155.15	5.39	0.0	0.0
20	H2O	WATER	7.701	4.595 E-4	2.521 E-6	-0.859 E-9	13.82	283.16	-57.80	-54.64
21	H2S	HYDROGEN SULFIDE	7.629	3.431 E-4	5.809 E-6	-2.810 E-9	658.25	165.54	-4.82	-7.90
22	H3N	AMMONIA	6.524	5.692 E-3	4.078 E-6	-2.830 E-9	342.79	169.63	-10.92	-3.86
23	H4N2	HYDRAZINE	2.333	4.525 E-2	-3.958 E-5	1.439 E-8	349.04	290.80	22.75	37.89
24	HE(4)	HELIUM-4					524.98		0.0	0.0
25	I2	IODINE	8.501	1.556 E-3	-1.669 E-6	0.677 E-9	559.62	520.55	0.0	0.0
26	KR	KRYPTON							0.0	0.0
27	NO	NITRIC OXIDE	7.009	-.224 E-3	2.328 E-6	-1.000 E-9			21.60	20.72
28	NO2	NITROGEN DIOXIDE	5.788	1.155 E-2	-4.970 E-6	0.070 E-9	406.20	230.21	8.09	12.42
29	N2	NITROGEN	7.440	-.324 E-2	6.400 E-6	-2.790 E-9	90.30	46.14	0.0	0.0
30	N2O	NITROUS OXIDE	5.164	1.739 E-2	-1.380 E-5	4.371 E-9			19.49	24.77
31	NE	NEON							0.0	0.0
32	O2	OXYGEN	6.713	-.879 E-6	4.170 E-6	-2.544 E-9	85.68	51.50	0.0	0.0
33	O2S	SULFUR DIOXIDE	5.697	1.600 E-2	-1.185 E-5	3.172 E-9	397.85	208.42	-70.95	-71.74
34	O3	OZONE	4.907	1.913 E-2	-1.491 E-5	4.054 E-9	313.79	120.34	34.1	38.91
35	O3S	SULFUR TRIOXIDE					1372.8	315.99	-94.47	-88.52
36	XE	XENON							0.0	0.0
37	CBRF3	TRIFLUOROBROMOMETHANE	5.449	4.565 E-2	-3.765 E-5	1.065 E-8			-155.1	-148.8
38	CCLF3	CHLOROTRIFLUOROMETHAN	7.547	4.257 E-2	-3.603 E-5	1.037 E-8	215.09	165.55	-166.0	-156.3
39	CCL2F2	DICHLORODIFLUOROMETHA							-115.0	-105.7
40	CCL2O	PHOSGENE	6.709	3.250 E-2	-3.281 E-5	1.211 E-8			-52.80	-49.42

NO	FORMULA	NAME	ANTA	ANTB	ANTC	TMX	TMN	HARA	HARB	HARC	HARD	HV
1	AR	ARGON	15.2330	700.51	-5.84	94	81	31.173	-1039.64	-2.832	0.264	1560
2	BCL3	BORON TRICHLORIDE						52.723	-4443.16	-5.404	2.97	
3	BF3	BORON TRIFLUORIDE						67.758	-3481.19	-7.963	0.768	
4	BR2	BROMINE	15.8441	2582.32	-51.56	354	259	36.380	-3748.59	-2.819	1.20	7210.
5	CLNO	NITROSYL CHLORIDE	16.9505	2520.70	-23.46	285	210	42.217	-3412.28	-3.894	1.27	6140.
6	CL2	CHLORINE	15.9610	1978.32	-27.01	264	172					4880.
7	CL3P	PHOSPHORUS TRICHLORID	15.8019	2634.16	-43.15	364	238					6580.
8	CL4SI	SILICON TETRACHLORIDE	13.2954	157.89	0.00	25	19					292.
9	D2	DEUTERIUM										9880.
10	D2O	DEUTERIUM OXIDE						30.772	-1040.27	-2.683	0.210	
11	F2	FLUORINE						39.219	-1970.37	-3.81	0.679	
12	F3N	NITROGEN TRIFLUORIDE	15.6700	714.10	-6.00			28.102	-2394.35	-1.843	0.871	9880.
13	F4SI	SILICON TETRAFLUORIDE	15.6107	1155.69	-15.37	91	59	38.614	-2626.67	-3.443	0.717	
14	F6S	SULFUR HEXAFLUORIDE	19.3785	2524.78	-11.16	155	103	26.160	-3496.52	-1.338	1.84	1560.
15	HBR	HYDROGEN BROMIDE	14.4687	1242.53	-47.86	220	159	33.884	-3013.08	-2.673	1.23	4220.
16	HCL	HYDROGEN CHLORIDE	16.5040	1714.25	-14.45	221	184	12.050	-114.95	0.048	0.048	3860.
17	HF	HYDROGEN FLUORIDE	17.6958	3404.49	15.06	313	207	55.336	-6869.50	-5.115	1.05	1600.
18	HI	HYDROGEN IODIDE	12.9149	957.96	-85.06	256	215	42.687	-3132.31	-3.985	0.871	4724.
19	H2	HYDROGEN	13.6333	164.90	3.19	25	14					216.
20	H2O	WATER	18.3036	3816.44	-46.13	441	284	51.947	-4104.67	-5.146	0.820	9717.
21	H2S	HYDROGEN SULFIDE	16.1040	1768.69	-26.06	230	190					4460.
22	H3N	AMMONIA	16.9481	2132.50	-32.98	261	179	56.096	-6951.84	-5.286	1.63	5580.
23	H4N2	HYDRAZINE	17.9899	3877.65	-45.15	343	288					10700.
24	HE(4)	HELIUM-4	12.2514	33.7329	1.79	4.3	3.7	8.622	-12.23	0.433	0.007	22.
25	I2	IODINE	15.1597	3709.75	-68.16	487	383	30.717	-1408.77	-2.579	0.448	10000.
26	KR	KRYPTON	15.2677	958.75	-8.71	129	113					2309.
27	NO	NITRIC OXIDE	20.1314	1572.52	-4.88	140	95	61.514	-2465.78	-7.211	0.279	3300.
28	NO2	NITROGEN DIOXIDE	20.5033	4141.29	3.65	320	230	61.862	-6073.34	-6.094	1.04	4555.
29	N2	NITROGEN	14.9542	588.72	-6.60	90	54	31.927	-924.86	-3.075	0.264	1333.
30	N2O	NITROUS OXIDE	16.1271	1506.49	-25.99	200	144	46.444	-2867.98	-4.655	0.743	3955.
31	NE	NEON	14.0099	180.47	-2.61	29	24	26.181	-295.44	-2.645	0.041	440.
32	O2	OXYGEN	15.4075	734.55	-5.45	100	63	31.041	-1082.52	-2.761	0.265	1630.
33	O2S	SULFUR DIOXIDE	16.7680	2302.35	-35.97	280	195	55.502	-4552.50	-5.666	1.32	5955.
34	O3	OZONE	15.7427	1272.18	-22.16	174	109					2670.
35	O3S	SULFUR TRIOXIDE	20.8403	3995.70	-36.66	332	290	139.56	-10420.1	-17.38	1.60	9716.
36	XE	XENON	15.2958	1303.92	-14.50	178	158	31.429	-1951.76	-2.544	0.804	3108.
37	CBRF3	TRIFLUOROBROMOMETHANE	15.7565	2167.31	-43.15	341	213					3706.
38	CCLF3	CHLOROTRIFLUOROMETHAN										4772.
39	CCL2F2	DICHLORODIFLUOROMETHA										5830.
40	CCL2O	PHOSGENE						44.255	-2769.96	-4.6415	1.30	

NO	FORMULA	NAME	MOLWT	TFP	TB	TC	PC	VC	ZC	OMEGA	LIQDEN	TDEN	DIPM
41	CCL3F	TRICHLOROFLUOROMETHAN	137.368	162.0	297.0	471.2	43.5	248.	0.279	0.188			0.5
42	CCL4	CARBON TETRACHLORIDE	153.823	250.	349.7	556.4	45.0	276.	0.272	0.194	1.584	298.	0.0
43	CF4	CARBON TETRAFLUORIDE	88.005	86.4	145.2	227.6	36.9	140.	0.277	0.191			0.0
44	CO	CARBON MONOXIDE	28.010	68.1	81.7	132.9	34.5	93.1	0.295	0.049	0.803	81.	0.1
45	COS	CARBONYL SULFIDE	60.070	134.3	222.9	375.	58.	140.	0.26	0.099	1.274	173.7	0.7
46	CO2	CARBON DIOXIDE	44.010	216.6	194.7	304.2	72.8	94.0	0.274	0.225	0.777	293.	0.0
47	CS2	CARBON DISULFIDE	76.131	161.3	319.4	552.	78.0	170.	0.293	0.115	1.293	273.	0.0
48	CHCLF2	CHLORODIFLUOROMETHANE	86.469	113.	232.4	369.2	49.1	165.	0.267	0.215	1.23	289.	1.3
49	CHCL2F	DICHLOROMONOFLUOROMET	102.923	138.	282.	451.6	51.0	197.	0.272	0.202	1.38	282.	1.1
50	CHCL3	CHLOROFORM	119.378	209.6	334.3	536.4	54.0	239.	0.293	0.216	1.489	293.	1.0
51	CHN	HYDROGEN CYANIDE	27.025	259.9	298.9	456.8	53.2	139.	0.197	0.407	0.688	293.	3.0
52	CH2BR2	DIBROMOMETHANE	173.835	220.6	370.	583.	71.				2.50	293.	1.4
53	CH2CL2	DICHLOROMETHANE	84.933	178.1	313.0	510.	60.0	193.	0.277	0.193	1.317	298.	1.6
54	CH2O	FORMALDEHYDE	30.026	156.	254.	408.	65.			0.253	0.815	253.	2.3
55	CH2O2	FORMIC ACID	46.025	281.5	373.8	580.					1.226	288.	1.5
56	CH3BR	METHYL BROMIDE	94.939	179.5	276.7	464.	85.	139.	0.268	0.273	1.737	268.	1.8
57	CH3CL	METHYL CHLORIDE	50.488	175.4	248.9	416.3	65.9	124.	0.275	0.156	0.915	293.	1.9
58	CH3F	METHYL FLUORIDE	34.033	131.4	194.8	317.8	58.0	190.	0.275	0.190	0.843	213.	1.8
59	CH3I	METHYL IODIDE	141.939	206.7	315.6	528.	65.	173.	0.224	0.172	2.279	293.	1.6
60	CH3NO2	NITROMETHANE	61.041	244.6	374.4	588.	62.3		0.224	0.346	1.138	293.	3.1
61	CH4	METHANE	16.043	90.7	111.7	190.6	45.4	99.0	0.288	0.008	0.425	111.7	0.0
62	CH4O	METHANOL	32.042	175.5	337.8	512.6	79.9	118.	0.224	0.559	0.791	293.	1.7
63	CH4S	METHYL MERCAPTAN	48.107	150.	279.1	470.0	71.4	145.	0.268	0.155	0.866	293.	1.5
64	CH5N	METHYL AMINE	31.058	179.7	266.8	430.	73.6	140.	0.292	0.275	0.703	259.6	1.3
65	CH6N2	METHYL HYDRAZINE	46.072	179.7	364.	567.	79.3	271.	0.462	0.275			1.7
66	C2CLF5	CHLOROPENTAFLUOROETHA	154.467	167.	234.	353.2	31.2	252.	0.271	0.253	1.455	298.	0.3
67	C2CL2F4	1,1-DICHLORO-1,2,2,2-	170.922	179.	277.0	418.6	32.6	294.	0.279	0.255	1.48	277.	
68	C2CL2F4	1,2-DICHLORO-1,1,2,2-	170.922	179.3	276.9	418.9	32.2	293.	0.275	0.252	1.580	289.	0.5
69	C2CL3F3	1,2,2-TRICHLORO-1,1,2	187.380	238.2	320.7	487.2	33.7	304.	0.256		1.62	293.	
70	C2CL4	TETRACHLOROETHYLENE-	165.834	251.	394.3	620.	44.	290.	0.25		1.645	298.	0.0
71	C2CL4F2	1,1,2,2-TETRACHLORO-1	203.831	298.	367.7	551.					1.590	195.	
72	C2F4	PERFLUOROETHENE	100.016	130.7	197.5	306.4	38.9	175.	0.271	0.226			0.0
73	C2F6	PERFLUOROETHANE	138.012	172.4	194.9	292.8	29.4	224.	0.25				0.2
74	C2N2	CYANOGEN	52.035	245.3	252.0	400.	59.	256.	0.265	0.24			
75	C2HCL3	TRICHLOROETHYLENE	131.389	186.8	360.4	571.	48.5			0.213	1.462	293.	2.3
76	C2HF3O2	TRIFLUOROACETIC ACID	114.024	257.9	345.6	491.3	32.2		0.271		1.535	273.	0.0
77	C2H2	ACETYLENE	26.038	192.4	189.2	308.3	60.6	113.	0.273	0.184	0.615	189.	0.0
78	C2H2F2	1,1-DIFLUOROETHYLENE	64.035	138.	232.	302.8	44.0	154.	0.30	0.207			1.4
79	C2H2O	KETENE	42.038	138.	232.	380.	64.	145.	0.265	0.122			1.5
80	C2H3CL	VINYL CHLORIDE	62.499	119.4	259.8	429.7	55.3	169.	0.265		0.969	259.	

NO	FORMULA	NAME	CPVAP A	CPVAP B	CPVAP C	CPVAP D	VISB	VISTO	DELHG	DELGF
41	CCL3F	TRICHLOROFLUOROMETHAN	9.789	3.893 E-2	-3.383 E-5	9.903 E-9			-68.0	-58.64
42	CCL4	CARBON TETRACHLORIDE	9.725	4.893 E-2	-5.421 E-5	2.112 E-8	540.15	290.84	-24.00	-13.92
43	CF4	CARBON TETRAFLUORIDE	3.339	4.838 E-2	-3.882 E-5	1.078 E-8			-223.0	-212.34
44	CO	CARBON MONOXIDE	7.373	-.307 E-2	6.662 E-6	-3.037 E-9	94.06	48.90	-26.42	-32.81
45	COS	CARBONYL SULFIDE	5.529	1.907 E-2	-1.676 E-5	5.860 E-9			-33.08	-39.59
46	CO2	CARBON DIOXIDE	4.728	1.754 E-2	-1.338 E-5	4.097 E-9	578.08	185.24	-94.05	-94.26
47	CS2	CARBON DISULFIDE	6.555	1.941 E-2	-1.831 E-5	6.384 E-9	274.08	200.22	27.98	15.99
48	CHCLF2	CHLORDIFLUOROMETHANE	4.132	3.865 E-2	-2.794 E-5	7.305 E-9			-119.9	-112.47
49	CHCL2F	DICHLOROMONOFLUOROMET	5.652	3.777 E-2	-2.866 E-5	7.795 E-9			-71.4	-64.10
50	CHCL3	CHLOROFORM	5.733	4.522 E-2	-4.397 E-5	1.590 E-8	394.81	246.50	-24.2	-16.38
51	CHN	HYDROGEN CYANIDE	5.222	1.448 E-2	-1.185 E-5	4.336 E-9	194.70	145.31	31.20	28.71
52	CH2BR2	DIBROMOMETHANE					428.91	294.57	-1.0	-1.34
53	CH2CL2	DICHLOROMETHANE	3.094	3.877 E-2	-3.110 E-5	1.005 E-8	359.55	225.13	-22.80	-16.46
54	CH2O	FORMALDEHYDE	5.607	7.540 E-3	7.130 E-6	-5.494 E-9	319.83	171.35	-27.70	-26.27
55	CH2O2	FORMIC ACID	2.798	3.243 E-2	-2.009 E-5	4.817 E-9	729.35	325.72	-90.49	-83.89
56	CH3BR	METHYL BROMIDE	3.446	2.606 E-2	-1.290 E-5	2.389 E-9	298.15	211.15	-9.00	-6.73
57	CH3CL	METHYL CHLORIDE	3.914	2.422 E-2	-9.288 E-6	0.613 E-9	426.45	193.56	-20.63	-15.03
58	CH3F	METHYL FLUORIDE	3.302	2.058 E-2	-4.946 E-6	-0.474 E-6			-55.90	-50.19
59	CH3I	METHYL IODIDE	2.581	2.318 E-2	-2.487 E-5	8.325 E-9	336.19	229.95	3.34	3.74
60	CH3NO2	NITROMETHANE	1.773	4.782 E-2	-2.583 E-5	4.980 E-9	452.50	261.21	-17.86	-1.666
61	CH4	METHANE	4.598	1.245 E-2	2.860 E-6	-2.703 E-9	114.14	57.60	-17.89	-12.15
62	CH4O	METHANOL	5.052	1.594 E-2	6.179 E-6	-6.811 E-9	555.30	260.64	-48.08	-38.84
63	CH4S	METHYL MERCAPTAN	3.169	3.479 E-2	-2.041 E-5	4.956 E-9			-5.49	-2.37
64	CH5N	METHYL AMINE	2.741	3.409 E-2	-1.274 E-5	1.135 E-9			2.4	7.71
65	CH6N2	METHYL HYDRAZINE					311.80	176.30		42.51
66	C2CLF5	CHLOROPENTAFLUOROETHA	6.648	8.340 E-2	-6.904 E-5	1.944 E-8			-214.6	
67	C2CL2F4	1,1-DICHLORO-1,2,2,2-	9.662	7.830 E-2	-6.572 E-5	1.868 E-8				
68	C2CL2F4	1,2-DICHLORO-1,1,2,2-	9.262	8.216 E-2	-7.047 E-5	2.032 E-8				
69	C2CL3F3	1,2,2-TRICHLORO-1,1,2	14.603	6.865 E-2	-5.780 E-5	1.649 E-8			-178.10	
70	C2CL4	TETRACHLOROETHYLENE	10.980	5.387 E-2	-5.478 E-5	2.002 E-8	392.58	281.82	-2.9	5.04
71	C2CL4F2	1,1,2,2-TETRACHLORO-1								
72	C2F4	PERFLUOROETHENE	6.929	5.439 E-2	-4.863 E-5	1.619 E-8			-157.40	-149.07
73	C2F6	PERFLUOROETHANE	6.405	8.259 E-2	-6.883 E-5	1.943 E-8			-321.00	-300.52
74	C2N2	CYANOGEN	8.583	2.210 E-2	-1.946 E-5	7.045 E-9			73.84	71.03
75	C2HCL3	TRICHLOROETHYLENE	7.207	5.462 E-2	-5.324 E-5	1.969 E-8	145.67	196.60	-16.40	4.75
76	C2HF3O2	TRIFLUOROACETIC ACID								
77	C2H2	ACETYLENE	6.406	1.810 E-2	-1.196 E-5	3.373 E-9			54.19	50.00
78	C2H2F2	1,1-DIFLUOROETHYLENE	0.734	5.839 E-2	-5.014 E-5	1.677 E-8			-82.50	-76.84
79	C2H2O	KETENE	1.525	3.913 E-2	-2.950 E-5	6.445 E-9			-14.60	-14.41
80	C2H3CL	VINYL CHLORIDE	1.421	4.823 E-2	-3.669 E-5	1.140 E-8	276.90	167.04	8.40	12.31

NO	FORMULA	NAME	ANTA	ANTB	ANTC	TMX	TMN	HARA	HARB	HARC	HARD	HV
41	CCL3F	TRICHLOROFLUOROMETHAN	15.8516	2401.61	-36.3	300	240	48.709	-4464.14	-4.753	2.85	5920.
42	CCL4	CARBON TETRACHLORIDE	15.8742	2808.19	-45.99	374	253	51.009	-5386.51	-4.953	3.82	7170.
43	CF4	CARBON TETRAFLUORIDE	16.0543	1244.55	-13.06	148	93	32.981	-997.18	-3.216	0.284	2860.
44	CO	CARBON MONOXIDE	14.3686	530.22	-13.15	108	63	41.853	-3137.78	-3.914	1.30	1444.
45	COS	CARBONYL SULFIDE										
46	CO2	CARBON DIOXIDE	22.5898	3103.39	-0.16	204	154	52.703	-3146.64	-5.572	0.705	4100.
47	CS2	CARBON DISULFIDE	15.9844	2690.85	-31.62	342	228	37.401	-4255.99	-3.027	2.21	6390.
48	CHCLF2	CHLORODIFLUOROMETHANE	15.5602	1704.80	-41.3	240	225	52.663	-3768.03	-5.474	1.55	4826.
49	CHCL2F	DICHLOROMONOFLUOROMET						54.563	-4629.02	-5.59	2.22	5960.
50	CHCL3	CHLOROFORM	15.9732	2696.79	-46.16	370	260	52.872	-5359.56	-5.2	2.96	7100.
51	CHN	HYDROGEN CYANIDE	16.5138	2585.80	-37.15	330	234	37.742	-4183.37	-3.004	2.18	6027.
52	CH2BR2	DIBROMOMETHANE										
53	CH2CL2	DICHLOROMETHANE	16.3029	2622.44	-41.70	332	229	53.767	-5110.2	-5.364	2.41	6690.
54	CH2O	FORMALDEHYDE	16.4775	2204.13	-30.15	271	185	45.118	-3873.26	-4.42	3.41	5500.
55	CH2O2	FORMIC ACID	16.9882	3599.58	-26.09	409	271	55.295	-4467.46	-5.788	2.35	5240.
56	CH3BR	METHYL BROMIDE	16.0252	2271.71	-34.83	326	215	43.660	-3642.21	-4.064	1.46	5715.
57	CH3CL	METHYL CHLORIDE	16.1052	2077.97	-29.55	266	180	43.063	-2890.54	-4.102	0.906	5120.
58	CH3F	METHYL FLUORIDE	16.3428	1704.41	-19.27	209	141	47.781	-4686.90	-4.577	2.84	6500.
59	CH3I	METHYL IODIDE	16.0905	2639.55	-3.5	325	260					
60	CH3NO2	NITROMETHANE	16.2193	2972.64	-64.15	409	278	50.133	-5996.30	-4.641	3.08	8225.
61	CH4	METHANE	16.2243	597.84	-7.16	120	93	30.715	-1300.61	-2.641	0.442	1955.
62	CH4O	METHANOL	18.5875	3626.55	-34.29	364	257	72.258	-7064.20	-7.68	1.86	8426.
63	CH4S	METHYL MERCAPTAN	16.1909	2338.38	-34.44	300	200	46.610	-4233.88	-4.408	1.71	5870.
64	CH5N	METHYL AMINE	17.2622	2484.83	-32.92	311	212	62.306	-4954.32	-6.642	1.40	6210.
65	CH6N2	METHYL HYDRAZINE	15.1424	2319.84	-9.47	400	270					
66	C2CLF5	CHLOROPENTAFLUOROETHA	15.7343	1848.90	-30.88	230	175	51.878	-3659.53	-5.433	2.25	4650.
67	C2CL2F4	1,1-DICHLORO-1,2,2,2-										
68	C2CL2F4	1,2-DICHLORO-1,1,2,2-	15.8424	2532.61	-45.67	360	250	52.316	-4327.01	-5.35	3.02	5560.
69	C2CL3F3	1,1,2-TRICHLORO-1,1,2	16.1642	3229.29	-52.15	460	307	57.097	-5249.75	-5.913	3.91	6570.
70	C2CL4	TETRACHLOROETHYLENE						75.315	-7113.72	-8.344	4.95	8300.
71	C2CL4F2	1,1,2,2-TETRACHLORO-1										
72	C2F4	PERFLUOROETHENE	15.8800	1574.60	-27.22	210	140	51.903	-3165.74	-5.537	1.34	
73	C2F6	PERFLUOROETHANE	15.6422	1512.94	-26.94	200	170	48.373	-2969.90	-5.032	1.53	3860.
74	C2N2	CYANOGEN						58.323	-4390.8	-6.185	1.51	
75	C2HCL3	TRICHLOROETHYLENE	16.1827	3028.13	-43.15	400	260	53.842	-5776.65	-5.295	3.70	
76	C2HF3O2	TRIFLUOROACETIC ACID						46.122	-2891.04	-4.612	0.863	7500.
77	C2H2	ACETYLENE	16.3481	1637.14	-19.77	202	194					4050.
78	C2H2F2	1,1-DIFLUOROETHYLENE										
79	C2H2O	KETENE	16.0197	1849.21	-35.15	255	170					4930.
80	C2H3CL	VINYL CHLORIDE	14.9601	1803.84	-43.15	290	185	48.672	-3955.89	-4.823	1.85	4930.

NO	FORMULA	NAME	MOLWT	TFP	TB	TC	PC	VC	ZC	OMEGA	LIQDEN	TDEN	DIPM
81	C2H3CLF2	1-CHLORO-1,1-DIFLUORO	100.496	142.	263.4	410.2	40.7	231.	0.279		1.10	303.	2.1
82	C2H3CLO	ACETYL CHLORIDE	78.498	160.2	323.9	508.	58.	204.	0.28	0.344	1.104	293.	2.4
83	C2H3CL3	1,1,2-TRICHLOROETHANE	133.405	236.5	386.9	602.	41.	294.	0.24	0.22	1.441	293.	1.7
84	C2H3F	VINYL FLUORIDE	46.044	130.0	235.5	327.8	51.7	144.	0.277				1.4
85	C2H3F3	1,1,1-TRIFLUOROETHANE	84.041	161.9	225.5	346.2	37.1	221.	0.289	0.257	0.782	293.	2.3
86	C2H3N	ACETONITRILE	41.053	229.3	354.8	548.	47.7	173.	0.184	0.321	0.958	293.	3.5
87	C2H3NO	METHYL ISOCYANATE	57.052		312.	491.	55.	129.	0.276	0.278	0.577	163.	0.0
88	C2H4	ETHYLENE	28.054	104.0	169.4	282.4	49.7	129.	0.276	0.085	1.168	298.	2.0
89	C2H4CL2	1,1-DICHLOROETHANE	98.960	176.2	330.4	523.	50.	240.	0.28	0.248	1.250	289.	1.8
90	C2H4CL2	1,2-DICHLOROETHANE	98.960	237.5	356.6	561.	53.	220.	0.25	0.286			2.3
91	C2H4F2	1,1-DIFLUOROETHANE	66.051	150.2	248.4	386.6	44.4	181.	0.253	0.266			2.5
92	C2H4O	ACETALDEHYDE	44.054	150.2	293.6	461.	55.	154.	0.22	0.303	0.778	293.	1.9
93	C2H4O	ETHYLENE OXIDE	44.054	161.	283.5	469.	71.0	140.	0.258	0.200	0.899	273.	1.3
94	C2H4O2	ACETIC ACID	60.052	289.8	391.1	594.4	57.1	171.	0.200	0.454	1.049	293.	1.8
95	C2H4O2	METHYL FORMATE	60.052	174.2	304.9	487.2	59.2	172.	0.255	0.252	0.974	293.	2.0
96	C2H5BR	ETHYL BROMIDE	108.966	154.6	311.5	503.8	61.5	215.	0.320	0.254	1.451	298.	2.0
97	C2H5CL	ETHYL CHLORIDE	64.515	136.9	285.4	460.4	52.0	199.	0.274	0.190	0.896	293.	2.0
98	C2H5F	ETHYL FLUORIDE	48.060	129.9	235.4	375.3	49.6	169.	0.272	0.238			0.0
99	C2H5N	ETHYLENE IMINE	43.069	195.	329.8	469.					0.833	298.	
100	C2H6	ETHANE	30.070	89.9	184.5	305.4	48.2	148.	0.285	0.098	0.548	183.	0.0
101	C2H6O	DIMETHYL ETHER	46.069	131.7	248.3	400.0	53.0	178.	0.287	0.192	0.667	293.	1.3
102	C2H6O	ETHANOL	46.069	159.1	351.5	516.2	63.0	167.	0.248	0.635	0.789	293.	1.7
103	C2H6O2	ETHYLENE GLYCOL	62.069	260.2	470.4	645.	76.	186.	0.27		1.114	293.	2.2
104	C2H6S	ETHYL MERCAPTAN	62.134	125.3	308.2	499.	54.2	207.	0.274	0.190	0.839	293.	1.5
105	C2H6S	DIMETHYL SULFIDE	62.130	174.9	310.5	503.0	54.6	201.	0.266	0.190	0.848	293.	1.5
106	C2H7N	ETHYL AMINE	45.085	192.	289.7	456.	55.5	178.	0.264	0.284	0.683	293.	1.3
107	C2H7N	DIMETHYL AMINE	45.085	181.0	280.0	437.6	52.4	187	0.272	0.288	0.656	293.	
108	C2H7NO	MONOETHANOLAMINE	61.084	283.5	443.5	614.	44.	196.	0.17		1.016	293.	2.6
109	C2H8N2	ETHYLENEDIAMINE	60.099	284.	390.4	593.	62.	206.	0.26	0.51	0.896	293.	1.9
110	C3H3N	ACRYLONITRILE	53.064	189.5	350.5	536.	45.	210.	0.21	0.35	0.806	293.	3.5
111	C3H4	PROPADIENE	40.065	136.9	238.7	393.	54.0	162.	0.271	0.313	0.658	238.	0.2
112	C3H4	METHYL ACETYLENE	40.065	170.5	250.0	402.4	55.5	164.	0.276	0.218	0.706	223.	0.7
113	C3H4O	ACROLEIN	56.064	186.	326.	506.	51.	210.	0.23	0.33	0.839	293.	2.9
114	C3H4O2	ACRYLIC ACID	72.064	285.	414.	615.	56.	210.	0.31	0.56	1.051	293.	
115	C3H4O2	VINYL FORMATE	72.064	235.5	319.6	475.	57.	234.	0.26	0.55	0.963	293.	
116	C3H5CL	ALLYL CHLORIDE	76.526	138.7	318.3	514.	47.		0.25	0.13	0.937	293.	2.0
117	C3H5CL3	1,2,3-TRICHLOROPROPAN	147.432	258.5	429.	651.	39.	348.	0.205	0.31	1.389	293.	
118	C3H5N	PROPIONITRILE	55.080	180.5	370.5	564.4	41.3	230.	0.205	0.318	0.782	293.	3.7
119	C3H6	CYCLOPROPANE	42.081	145.7	240.4	397.8	54.2	170.	0.282	0.264	0.563	288.	0.0
120	C3H6	PROPYLENE	42.081	87.9	225.4	365.0	45.6	181.	0.275	0.148	0.612	223.	0.4

NO	FORMULA	NAME	CPVAP A	CPVAP B	CPVAP C	CPVAP D	VISB	VISTO	DELHG	DELGF
81	C2H3CLF2	1-CHLORO-1,1-DIFLUORO	4.017	6.584 E-2	-4.758 E-5	1.267 E-8			-58.30	-49.29
82	C2H3CLO	ACETYL CHLORIDE	5.976	4.086 E-2	-2.354 E-5	5.300 E-9			-33.10	-18.52
83	C2H3CL3	1,1,2-TRICHLOROETHANE	1.510	8.194 E-2	-7.064 E-5	2.300 E-8	346.72	304.43		
84	C2H3F	VINYL FLUORIDE								
85	C2H3F3	1,1,1-TRIFLUOROETHANE	1.372	7.502 E-2	-6.203 E-5	2.010 E-8	334.91	210.05	-178.20	-162.23
86	C2H3N	ACETONITRILE	4.892	2.857 E-2	-1.073 E-5	0.765 E-9	616.78	227.47	21.00	25.24
87	C2H3NO	METHYL ISOCYANATE	8.542	2.483 E-2	-1.390 E-5	-4.030 E-9	168.98	93.94		
88	C2H4	ETHYLENE	0.909	3.740 E-2	-1.994 E-5	4.192 E-9			12.50	16.28
89	C2H4CL2	1,1-DICHLOROETHANE	2.979	5.518 E-2	-4.896 E-5	1.505 E-8	412.27	239.10	-31.05	-17.47
90	C2H4CL2	1,2-DICHLOROETHANE	4.893	5.722 E-2	-3.435 E-5	8.094 E-9	473.95	277.98	-31.00	-17.65
91	C2H4F2	1,1-DIFLUOROETHANE	2.072	4.353 E-2	-3.480 E-5	8.107 E-9	319.27	186.56	-118.00	-104.26
92	C2H4O	ACETALDEHYDE	1.843	5.308 E-2	-2.404 E-5	5.685 E-9	368.70	192.82	-39.76	-31.86
93	C2H4O	ETHYLENE OXIDE	-1.796	6.087 E-2	-3.001 E-5	6.190 E-9			-12.58	-3.13
94	C2H4O2	ACETIC ACID	1.156	6.449 E-2	-4.187 E-5	1.182 E-8	341.88	194.22	-103.93	-90.03
95	C2H4O2	METHYL FORMATE	0.342	5.608 E-2	-4.656 E-5	1.362 E-8	600.94	306.21	-83.60	-71.03
96	C2H5BR	ETHYL BROMIDE	1.590	6.285 E-2	-3.517 E-5	9.086 E-9	363.19	212.70	-15.30	-6.29
97	C2H5CL	ETHYL CHLORIDE	-0.132	5.207 E-2	-4.393 E-5	1.325 E-8	369.80	220.68	-26.70	-14.34
98	C2H5F	ETHYL FLUORIDE	1.038	7.219 E-2	-2.784 E-5	5.757 E-9	320.94	190.83	-62.50	-50.08
99	C2H5N	ETHYLENE IMINE	-4.961	4.254 E-2	-1.927 E-5	1.349 E-8			29.50	42.54
100	C2H6	ETHANE	1.292	4.277 E-2	-1.657 E-5	2.081 E-8	156.60	95.57	-20.24	-7.87
101	C2H6O	DIMETHYL ETHER	4.064	5.113 E-2	-1.250 E-5	-0.458 E-9			-43.99	-26.99
102	C2H6O	ETHANOL	2.153	5.931 E-2	-2.004 E-5	0.328 E-9	686.64	300.88	-56.12	-40.22
103	C2H6O2	ETHYLENE GLYCOL	8.526	5.615 E-2	-3.576 E-5	7.190 E-9	136.50	402.41	-93.05	-72.77
104	C2H6S	ETHYL MERCAPTAN	3.564	4.478 E-2	-3.239 E-5	7.552 E-9	419.60	206.21	-11.02	-1.12
105	C2H6S	DIMETHYL SULFIDE	5.805	6.572 E-2	-1.642 E-5	0.979 E-9	267.34	184.24	-8.97	1.66
106	C2H7N	ETHYL AMINE	0.882	6.438 E-2	-3.781 E-5	9.096 E-9	340.54	192.44	-11.00	8.91
107	C2H7N	DIMETHYL AMINE	-0.041	7.188 E-2	-3.175 E-5	5.587 E-9			-4.50	16.25
108	C2H7NO	MONOETHANOLAMINE	2.224	5.749 E-2	-4.342 E-5	1.112 E-8	198.41	367.03	-48.18	
109	C2H8N2	ETHYLENEDIAMINE	9.147		-1.036 E-5	-9.430 E-9	839.76	316.41		
110	C3H3N	ACRYLONITRILE	2.554	5.273 E-2	-3.739 E-5	1.099 E-8	343.31	210.42	44.20	46.68
111	C3H4	PROPADIENE	2.366	4.723 E-2	-2.822 E-5	6.645 E-9			45.92	48.37
112	C3H4	METHYL ACETYLENE	3.513	4.453 E-2	-2.803 E-5	7.701 E-9			44.32	46.47
113	C3H4O	ACROLEIN	2.889	5.029 E-2	-2.557 E-5	4.552 E-9			-16.94	-15.57
114	C3H4O2	ACRYLIC ACID	0.416	7.621 E-2	-5.618 E-5	1.666 E-8	388.17	217.14	-80.36	-68.37
115	C3H4O2	VINYL FORMATE	6.643	4.392 E-2	-8.502 E-6	-5.577 E-8	733.02	307.15		
116	C3H5CL	ALLYL CHLORIDE	0.604	7.277 E-2	-5.442 E-5	1.742 E-8	428.40	224.83	-0.15	10.42
117	C3H5CL3	1,2,3-TRICHLOROPROPAN	6.421	8.651 E-2	-6.656 E-5	2.099 E-8	368.62	210.60	-44.40	-23.37
118	C3H5N	PROPIONITRILE	3.679	5.363 E-2	-6.628 E-5	4.667 E-9	818.63	342.88	12.10	22.98
119	C3H6	CYCLOPROPANE	-8.417	9.108 E-2	-6.882 E-5	2.158 E-8	366.77	225.86	12.74	24.95
120	C3H6	PROPYLENE	0.886	5.602 E-2	-2.771 E-5	5.266 E-9	273.84	131.63	4.88	14.99

NO	FORMULA	NAME	ANTA	ANTB	ANTC	TMX	TMN	HARA	HARB	HARC	HARD	HV
81	C2H3CLF2	1-CHLORO-1,1-DIFLUORO	15.7514	2447.33	-55.53	355	237					6850.
82	C2H3CLO	ACETYL CHLORIDE	16.0381	3110.79	-56.16	428	302					7960.
83	C2H3CL3	1,1,2-TRICHLOROETHANE										
84	C2H3F	VINYL FLUORIDE										
85	C2H3F3	1,1,1-TRIFLUOROETHANE	15.8965	1814.91	-29.92	300	270	50.589	-3540.17	-5.223	1.79	4580.
86	C2H3N	ACETONITRILE	16.2874	2945.47	-49.15	390	260	47.394	-5392.43	-4.357	3.49	7500.
87	C2H3NO	METHYL ISOCYANATE	16.3258	2480.37	-56.31	340	230					7070.
88	C2H4	ETHYLENE	15.5368	1347.01	-18.15	182	120	38.961	-2282.37	-3.678	0.881	3237.
89	C2H4CL2	1,1-DICHLOROETHANE	16.0842	2697.29	-45.03	352	242	56.233	-5422.68	-5.726	3.17	6860.
90	C2H4CL2	1,2-DICHLOROETHANE	16.1764	2927.17	-50.22	373	240	51.956	-5712.66	-4.991	3.30	7650.
91	C2H4F2	1,1-DIFLUOROETHANE	16.1871	2095.35	-29.16	273	238	48.591	-3837.61	-4.811	1.87	5100.
92	C2H4O	ACETALDEHYDE	16.2481	2465.15	-37.15	320	210					6150.
93	C2H4O	ETHYLENE OXIDE	16.7400	2567.61	-29.01	310	200					6120.
94	C2H4O2	ACETIC ACID	16.8080	3405.57	-56.34	430	290	57.834	-6841.98	-5.647	3.44	5660.
95	C2H4O2	METHYL FORMATE	16.5104	2590.87	-42.60	324	225	57.840	-5228.90	-5.939	2.23	6740.
96	C2H5BR	ETHYL BROMIDE	15.9338	2251.68	-41.44	333	226	37.985	-4246.27	-3.09	2.27	6330.
97	C2H5CL	ETHYL CHLORIDE	15.9800	2332.01	-36.48	310	200	48.665	-4364.03	-4.733	2.26	5900.
98	C2H5F	ETHYL FLUORIDE	16.0686	1966.89	-27.00	252	170					7660.
99	C2H5N	ETHYLENE IMINE	16.4227	2610.44	-63.15	359	248					
100	C2H6	ETHANE	15.6637	1511.42	-17.16	199	130	38.759	-2464.42	-3.601	1.073	3515.
101	C2H6O	DIMETHYL ETHER	15.8467	2351.44	-17.10	265	179	48.857	-3840.19	-4.856	1.71	5140.
102	C2H6O	ETHANOL	18.9119	3803.98	-41.168	369	270	83.319	-7994.90	-9.201	2.35	9260.
103	C2H6O2	ETHYLENE GLYCOL	20.2501	6022.18	-28.25	494	364					12550.
104	C2H6S	ETHYL MERCAPTAN	16.0077	2497.23	-41.77	330	226	51.954	-4900.34	-5.139	2.55	6440.
105	C2H6S	DIMETHYL SULFIDE	16.0001	2511.56	-42.35	331	215					6440.
106	C2H7N	ETHYL AMINE	17.0073	2618.73	-37.30	316	218	64.056	-5352.01	-6.875	2.08	6700.
107	C2H7N	DIMETHYL AMINE	16.2653	2358.77	-35.15	310		67.611	-5350.44	-7.435	2.03	6330.
108	C2H7NO	MONOETHANOLAMINE	17.8174	3988.33	-86.93	477	344					12000.
109	C2H8N2	ETHYLENEDIAMINE	16.4082	3108.49	-72.15	425	292					10000.
110	C3H3N	ACRYLONITRILE	15.9253	2782.21	-51.15	385	255					7800.
111	C3H4	PROPADIENE	13.1563	1054.72	-77.08	257	174					4450.
112	C3H4	METHYL ACETYLENE	15.6227	1850.66	-44.07	267	183					5290.
113	C3H4O	ACROLEIN	15.9057	2606.53	-45.15	360	235					6770.
114	C3H4O2	ACRYLIC ACID	16.5617	3319.18	-80.15	450	315					11000.
115	C3H4O2	VINYL FORMATE	16.6531		-63.15	350	240					7680.
116	C3H5CL	ALLYL CHLORIDE	15.9772	2531.92	-47.15	350	230					6475.
117	C3H5CL3	1,2,3-TRICHLOROPROPAN	16.1246	3417.27	-69.15	470	315	53.398	-5937.37	-5.2	4.28	9180.
118	C3H5N	PROPIONITRILE	15.9571	2940.86	-55.15	405	270					7710.
119	C3H6	CYCLOPROPANE	15.8599	1971.04	-26.65	245	180					4790.
120	C3H6	PROPYLENE	15.7027	1807.53	-26.15	240	160	44.794	-3260.31	-4.379	1.63	4400.

NO	FORMULA	NAME	MOLWT	TFP	TB	TC	PC	VC	ZC	OMEGA	LIQDEN	TDEN	DIPM
121	C3H6CL2	1,2-DICHLOROPROPANE	112.987	172.7	369.5	577.	44.4	226.	0.21	0.24	1.15	293.	1.9
122	C3H6O	ACETONE	58.080	178.2	329.4	508.1	46.4	209.	0.232	0.309	0.790	293.	2.9
123	C3H6O	ALLYL ALCOHOL	58.080	144.	370.	545.	56.4	203.	0.256	0.63	0.855	288.	
124	C3H6O	PROPIONALDEHYDE	58.080	193.	321.	496.	47.	223.	0.26	0.313	0.797	293.	2.7
125	C3H6O	PROPYLENE OXIDE	58.080	161.	307.5	482.2	48.6	186.	0.228	0.269	0.829	293.	2.0
126	C3H6O	VINYL METHYL ETHER	58.080	151.5	278.	436.	47.	205.	0.27	0.34	0.750	293.	
127	C3H6O2	PROPIONIC ACID	74.080	252.5	414.0	612.	53.0	230.	0.242	0.536	0.993	293.	1.5
128	C3H6O2	ETHYL FORMATE	74.080	193.8	327.4	508.4	46.8	229.	0.257	0.283	0.927	289.	2.0
129	C3H6O2	METHYL ACETATE	74.080	175.	330.1	506.8	46.3	228.	0.254	0.324	0.934	293.	1.7
130	C3H7CL	PROPYL CHLORIDE	78.542	150.4	319.6	503.	45.2	254.	0.278	0.230	0.891	293.	2.0
131	C3H7CL	ISOPROPYL CHLORIDE	78.542	156.0	308.9	485.	46.6	230.	0.269	0.232	0.862	293.	2.1
132	C3H8	PROPANE	44.097	85.5	231.1	369.8	41.9	203.	0.281	0.152	0.582	231.	0.0
133	C3H8O	1-PROPANOL	60.096	146.9	370.4	536.7	51.0	218.5	0.253	0.624	0.804	293.	1.7
134	C3H8O	ISOPROPYL ALCOHOL	60.096	184.7	355.4	508.3	47.0	220.	0.248		0.786	293.	1.7
135	C3H8O	METHYL ETHYL ETHER	60.096	134.	280.5	437.8	43.4	221.	0.267	0.236	0.700	291.	1.2
136	C3H8O2	METHYLAL	76.096	168.	315.	497.					0.888	291.	1.0
137	C3H8O2	1,2-PROPANEDIOL	76.096	213.	460.5	625.	60.	237.	0.28		1.036	293.	3.6
138	C3H8O2	1,3-PROPANEDIOL	76.096	246.4	487.6	658.	59.	241.	0.26		1.053	293.	3.7
139	C3H8O3	GLYCEROL	92.095	291.	563.	726.	66.	255.	0.28		1.261	293.	3.0
140	C3H8S	METHYL ETHYL SULFIDE	76.157	167.2	339.8	533.	42.				0.837	293.	
141	C3H9N	N-PROPYL AMINE	59.112	190.	321.8	497.0	46.8	233.	0.267	0.229	0.717	293.	1.3
142	C3H9N	ISOPROPYL AMINE	59.112	177.9	305.6	476.	50.	229.	0.29	0.297	0.688	293.	
143	C3H9N	TRIMETHYL AMINE	59.112	156.	276.1	433.2	40.2	254.	0.287	0.195	0.633	293.	0.6
144	C4H2O3	MALEIC ANHYDRIDE	98.058	326.	472.8	455.	49.	202.	0.26	0.092	1.31	333.	4.0
145	C4H4	VINYLACETYLENE	52.076	227.6	278.1	455.	54.0	218.	0.294	0.204	0.710	273.	
146	C4H4O	FURAN	68.075	187.5	304.5	490.2	56.2	219.	0.259	0.200	0.938	293.	0.7
147	C4H4S	THIOPHENE	84.136	234.9	357.3	579.4	56.2	219.	0.259	0.200	1.071	289.	0.5
148	C4H5N	ALLYL CYANIDE	67.091	186.7	392.	585.	39.	265.	0.22	0.39	0.835	293.	3.4
149	C4H5N	PYRROLE	67.091		403.	640.					0.967	294.	1.8
150	C4H6	1-BUTYNE	54.092	147.4	281.2	463.7	46.5	220.	0.27	0.050	0.650	289.	0.8
151	C4H6	2-BUTYNE	54.092	240.9	300.2	488.6	50.2	221.	0.277	0.124	0.691	293.	0.8
152	C4H6	1,2-BUTADIENE	54.092	137.0	284.0	443.7	44.4	219.	0.267	0.255	0.652	293.	0.4
153	C4H6	1,3-BUTADIENE	54.092	164.3	268.7	425.	42.7	221.	0.270	0.195	0.621	293.	0.0
154	C4H6O2	VINYL ACETATE	86.091	173.	346.	525.	43.	265.	0.26	0.34	0.932	293.	1.7
155	C4H6O3	ACETIC ANHYDRIDE	102.089	199.	412.	569.	46.2	290.	0.287		1.087	293.	3.0
156	C4H6O4	DIMETHYL OXALATE	118.090	327.	436.6	628.					1.15	288.	
157	C4H6O4	SUCCINIC ACID	118.090	456.	508.								
158	C4H7N	BUTYRONITRILE	69.107	161.	391.	582.2	37.4	285.	0.223	0.371	0.792	293.	2.2
159	C4H7O2	METHYL ACRYLATE	86.091	196.7	353.5	536.	42.	265.	0.25	0.35	0.956	293.	3.8
160	C4H8	1-BUTENE	56.108	87.8	266.9	419.6	39.7	240.	0.277	0.187	0.595	293.	0.3

NO	FORMULA	NAME	CPVAP A	CPVAP B	CPVAP C	CPVAP D	VISB	VISTO	DELHG	DELGF
121	C3H6CL2	1,2-DICHLOROPROPANE	2.496	8.729E-2	-6.219E-5	1.849E-8	514.36	281.03	-39.6	-19.86
122	C3H6O	ACETONE	1.505	6.224E-2	-2.992E-5	4.867E-9	367.25	209.68	-52.00	-36.58
123	C3H6O	ALLYL ALCOHOL	-0.264	7.515E-2	-4.853E-5	1.271E-8	793.52	307.26	-31.55	-17.03
124	C3H6O	PROPIONALDEHYDE	-2.800	6.244E-2	-3.105E-5	5.078E-9	343.44	219.33	-45.90	-31.18
125	C3H6O	PROPYLENE OXIDE	-2.020	7.779E-2	-4.750E-5	1.152E-8	377.43	213.36	-22.17	-6.16
126	C3H6O	VINYL METHYL ETHER	3.733	5.592E-2	-2.316E-5	2.537E-9	318.41	180.98		
127	C3H6O2	PROPIONIC ACID	1.354	8.811E-2	-6.842E-5	2.359E-8	535.04	299.32	-108.78	-88.27
128	C3H6O2	ETHYL FORMATE	5.893	5.532E-2	-5.063E-6	-1.280E-8	400.91	226.23	-88.74	
129	C3H6O2	METHYL ACETATE	3.953	5.363E-2	-1.037E-5	6.961E-8	408.62	224.03	-97.86	
130	C3H7CL	PROPYL CHLORIDE	-0.799	8.660E-2	-5.991E-5	1.779E-8	374.77	215.00	-31.10	-12.11
131	C3H7CL	ISOPROPYL CHLORIDE	0.440	8.330E-2	-5.359E-5	1.400E-8	306.25	212.24	-35.00	-14.94
132	C3H8	PROPANE	-1.009	7.315E-2	-3.789E-5	7.678E-8	222.67	133.41	-24.82	-5.61
133	C3H8O	1-PROPANOL	0.590	7.926E-2	-4.431E-5	1.026E-8	951.04	327.83	-61.28	-38.67
134	C3H8O	ISOPROPYL ALCOHOL	7.745	4.502E-2	1.530E-5	-2.212E-8	1139.7	323.44	-65.11	-41.44
135	C3H8O	METHYL ETHYL ETHER	4.459	6.414E-2	-2.447E-5	2.138E-8	303.82	171.66	-51.73	-28.12
136	C3H8O2	METHYLAL	0.151	1.006E-1	-7.121E-5	2.138E-8				
137	C3H8O2	1,2-PROPANEDIOL	1.975	8.779E-2	-5.163E-5	1.207E-8			-101.33	
138	C3H8O2	1,3-PROPANEDIOL	2.012	1.061E-1	-7.545E-5	2.240E-8			-97.71	
139	C3H8O3	GLYCEROL	4.654	6.904E-2	-2.888E-5	3.073E-9			-139.8	
140	C3H8S	METHYL ETHYL SULFIDE	1.598	8.356E-2	-4.352E-5	8.565E-9	433.64	228.46	-14.25	2.73
141	C3H9N	N-PROPYL AMINE	-1.788	9.973E-2	-6.749E-5	1.994E-8			-17.30	9.51
142	C3H9N	ISOPROPYL AMINE							-20.02	
143	C3H9N	TRIMETHYL AMINE	-1.960	9.486E-2	-5.299E-5	1.104E-8			-5.70	23.64
144	C4H2O3	MALEIC ANHYDRIDE	-3.123	8.323E-2	-5.217E-5	1.156E-8	952.48	365.81		
145	C4H4	VINYLACETYLENE	1.614	6.785E-2	-5.410E-5	1.782E-8			72.80	73.13
146	C4H4O	FURAN	-8.486	1.032E-1	-8.251E-5	2.566E-8	389.40	222.70	-8.29	0.221
147	C4H4S	THIOPHENE	-7.310	1.070E-1	-9.009E-5	2.992E-8	498.60	264.90	27.66	30.30
148	C4H5N	ALLYL CYANIDE	5.183	6.142E-2	-2.847E-5	2.936E-9	521.30	252.03		
149	C4H5N	PYRROLE	2.997	6.553E-2	-3.690E-5	8.240E-9			25.88	
150	C4H6	1-BUTYNE	3.804	5.688E-2	-2.555E-5	4.188E-9			39.48	48.30
151	C4H6	2-BUTYNE	2.675	6.505E-2	-3.507E-5	7.378E-9			34.97	44.32
152	C4H6	1,2-BUTADIENE	-0.403	8.165E-2	-5.589E-5	1.513E-8	300.59	163.12	38.77	47.43
153	C4H6	1,3-BUTADIENE	3.621	6.676E-2	-2.103E-5	-3.965E-8	457.89	235.35	26.33	36.01
154	C4H6O2	VINYL ACETATE	-5.524	1.215E-1	-8.551E-5	2.349E-8	502.33	286.04	-75.5	
155	C4H6O3	ACETIC ANHYDRIDE	3.600	1.120E-1	-7.508E-5	1.896E-8			-137.60	-113.93
156	C4H6O4	DIMETHYL OXALATE								
157	C4H6O4	SUCCINIC ACID								
158	C4H7N	BUTYRONITRILE	3.633	7.657E-2	-3.912E-5	7.123E-8	438.04	256.84	8.14	25.97
159	C4H6O2	METHYL ACRYLATE	3.622	6.678E-2	-2.103E-5	-3.966E-8	451.02	245.30		
160	C4H8	1-BUTENE	-0.715	8.436E-2	-4.754E-5	1.066E-8	255.30	151.86	-0.03	17.04

NO	FORMULA	NAME	ANTA	ANTB	ANTC	TMX	TMN	HARA	HARB	HARC	HARD	HV
121	C3H6CL2	1,2-DICHLOROPROPANE	16.0385	2985.07	-52.16	408	288					7500.
122	C3H6O	ACETONE	16.6513	2940.46	-35.93	350	241					5960.
123	C3H6O	ALLYL ALCOHOL	16.9066	2928.20	-85.15	400	286					9550.
124	C3H6O	PROPIONALDEHYDE	16.2315	2659.02	-44.15	350	235					6760.
125	C3H6O	PROPYLENE OXIDE	15.3227	2107.58	-64.87	340	225					6450.
126	C3H6O	VINYL METHYL ETHER	14.4602	1980.22	-25.15	315	190					4550.
127	C3H6O2	PROPIONIC ACID	17.3789	3723.42	-67.48	450	315	76.490	-8619.48	-8.139	3.93	7700.
128	C3H6O2	ETHYL FORMATE	16.1611	2603.30	-54.15	360	240	60.604	-5724.26	-6.305	3.07	7200.
129	C3H6O2	METHYL ACETATE	16.1295	2601.92	-56.15	360	245	61.268	-5840.56	-6.374	3.08	7200.
130	C3H7CL	PROPYL CHLORIDE	15.9594	2581.48	-42.95	350	230					6510.
131	C3H7CL	ISOPROPYL CHLORIDE	16.0384	2490.48	-43.15	340	225					6280.
132	C3H8	PROPANE	15.7260	1872.46	-25.16	249	164	43.492	-3266.92	-4.179	1.81	4487.
133	C3H8O	1-PROPANOL	17.5439	3166.38	-80.15	400	285	101.82	-9416.25	-11.79	3.13	9980.
134	C3H8O	ISOPROPYL ALCOHOL	18.6929	3640.20	-53.54	374	273	74.838	-5631.77	-8.549	2.045	9520.
135	C3H8O	METHYL ETHYL ETHER	13.5435	1161.63	-112.44	310	205					5900.
136	C3H8O2	METHYLAL	15.8237	2415.92	-52.58	315	270					
137	C3H8O2	1,2-PROPANEDIOL	20.5324	6091.95	-22.46	483	357					12940.
138	C3H8O2	1,3-PROPANEDIOL	17.2392	3888.84	-123.2	525	380					13500.
139	C3H8O3	GLYCEROL		4487.04	-140.2	600	440					14600.
140	C3H8S	METHYL ETHYL SULFIDE	15.9765	2722.95	-48.37	360	250					7050.
141	C3H9N	N-PROPYL AMINE	15.9957	2551.72	-49.15	350	235	50.869	-4261.51	-5.127	2.59	7100.
142	C3H9N	ISOPROPYL AMINE	16.3637	2582.35	-40.15	337	239					6500.
143	C3H9N	TRIMETHYL AMINE	16.0499	2230.51	-39.15	305	215					5760.
144	C4H2O3	MALEIC ANHYDRIDE	16.2747	3765.65	-82.15	516	352					5850.
145	C4H4	VINYLACETYLENE	16.0100	2203.57	-43.15	305	200					6474.
146	C4H4O	FURAN	16.0612	2442.70	-45.41	363	238					7520.
147	C4H4S	THIOPHENE	16.0243	2869.07	-51.80	380	260					8200.
148	C4H5N	ALLYL CYANIDE	16.0019	3128.75	-58.15	430	400					
149	C4H5N	PYRROLE	16.7966	3457.47	-62.73	440	330					5970.
150	C4H6	1-BUTYNE	16.0605	2271.42	-40.30	300	200					6370.
151	C4H6	2-BUTYNE	16.2871	2536.78	-37.34	320	240					5800.
152	C4H6	1,2-BUTADIENE	16.1039	2397.26	-30.88	305	245					5370.
153	C4H6	1,3-BUTADIENE	15.7727	2142.66	-34.30	290	215					
154	C4H6O2	VINYL ACETATE	16.1003	2744.68	-56.15	379	255					
155	C4H6O3	ACETIC ANHYDRIDE	16.3982	3287.56	-75.11	437	308					9850.
156	C4H6O4	DIMETHYL OXALATE										
157	C4H6O4	SUCCINIC ACID										
158	C4H7N	BUTYRONITRILE	16.2092	3202.21	-56.16	433	307	56.605	-6476.68	-5.599	5.03	8220.
159	C4H7O2	METHYL ACRYLATE	16.1088	2788.43	-59.15	390	260	48.333	-3996.80	-4.788	2.46	7650.
160	C4H8	1-BUTENE	15.7564	2132.42	-33.15	295	190					5238.

NO	FORMULA	NAME	MOLWT	TFP	TB	TC	PC	VC	ZC	OMEGA	LIQDEN	TDEN	DIPM
161	C4H8	CIS-2-BUTENE	56.108	134.3	276.9	435.6	41.5	234.	0.272	0.202	0.621	293.	0.3
162	C4H8	TRANS-2-BUTENE	56.108	167.6	274.0	428.6	40.5	238.	0.274	0.214	0.604	293.	0.0
163	C4H8	CYCLOBUTANE	56.108	182.4	285.7	459.9	49.2	210.	0.274	0.209	0.694	293.	0.5
164	C4H8	ISOBUTYLENE	56.108	132.8	266.3	417.9	39.5	239.	0.275	0.190	0.594	293.	0.5
165	C4H8O	N-BUTYRALDEHYDE	72.107	176.8	348.0	524.	40.	278.	0.26	0.352	0.802	293.	2.6
166	C4H8O	ISOBUTYRALDEHYDE	72.107	208.2	337.	513.	41.	274.	0.27	0.35	0.789	293.	2.6
167	C4H8O	METHYL ETHYL KETONE	72.107		352.8	535.6	41.0	267.	0.249	0.329	0.805	293.	3.0
168	C4H8O	TETRAHYDROFURAN	72.107	164.7	339.1	540.2	51.2	224.	0.259		0.889	293.	1.7
169	C4H8O	VINYL ETHYL ETHER	72.107	157.9	308.8	475.	40.2	260.	0.27		0.793	293.	1.3
170	C4H8O2	N-BUTYRIC ACID	88.107	267.9	436.4	628.	52.0	292.	0.295	0.67	0.958	293.	1.5
171	C4H8O2	1,4-DIOXANE	88.107	285.	374.5	587.	51.4	238.	0.254	0.288	1.033	293.	0.4
172	C4H8O2	ETHYL ACETATE	88.107	189.6	350.3	523.2	37.8	286.	0.252	0.363	0.901	293.	1.9
173	C4H8O2	ISOBUTYRIC ACID	88.107	227.2	427.9	609.	40.	292.	0.23	0.61	0.968	293.	1.3
174	C4H8O2	METHYL PROPIONATE	88.107	185.7	353.0	530.6	39.5	282.	0.256	0.352	0.915	293.	1.7
175	C4H8O2	N-PROPYL FORMATE	88.107	180.3	353.7	538.0	40.1	285.	0.259	0.315	0.911	289.	1.9
176	C4H9CL	1-CHLOROBUTANE	92.569	150.1	352.6	542.	36.4	312.	0.255	0.218	0.886	293.	2.0
177	C4H9CL	2-CHLOROBUTANE	92.569	141.8	341.4	520.6	39.	305.	0.28	0.30	0.873	293.	2.1
178	C4H9CL	TERT-BUTYL CHLORIDE	92.569	247.8	324.	507.	39.	295.	0.28	0.19	0.842	293.	2.1
179	C4H9N	PYRROLIDINE	71.123		359.7	568.6	54.	249.	0.296		0.852	295.	1.5
180	C4H9NO	MORPHOLINE	87.122	268.4	401.4	618.	55.4	253.	0.27	0.37	1.000	293.	1.5
181	C4H10	N-BUTANE	58.124	134.8	272.7	425.2	37.5	255.	0.274	0.193	0.579	293.	0.1
182	C4H10	ISOBUTANE	58.124	113.6	261.3	408.1	36.0	263.	0.283	0.176	0.557	293.	0.1
183	C4H10O	N-BUTANOL	74.123	183.9	390.9	562.9	43.6	274.	0.259	0.590	0.810	293.	1.8
184	C4H10O	2-BUTANOL	74.123	158.5	372.7	536.0	41.4	268.	0.252	0.576	0.807	293.	1.7
185	C4H10O	ISOBUTANOL	74.123	165.2	381.0	547.7	42.4	273.	0.257	0.588	0.802	293.	1.7
186	C4H10O	TERT-BUTANOL	74.123	298.8	355.6	506.2	39.2	275.	0.259	0.618	0.787	293.	1.7
187	C4H10O	ETHYL ETHER	74.123	156.9	307.7	466.7	35.9	280.	0.262	0.281	0.713	293.	1.3
188	C4H10O2	1,2-DIMETHOXYETHANE	90.123	202.	358.6	536.	38.2	271.	0.235	0.371	0.867	293.	
189	C4H10O3	DIETHYLENE GLYCOL	106.122	265.	519.	681.	46.	316.	0.26		1.116	293.	
190	C4H10S	DIETHYL SULFIDE	90.184	169.2	365.3	557.	39.1	318.	0.272	0.300	0.837	293.	1.6
191	C4H10S2	DIETHYL DISULFIDE	122.244	171.7	427.2	642.					0.998	293.	2.0
192	C4H11N	N-BUTYL AMINE	73.139	224.1	350.6	524.	41.	288.	0.27	0.396	0.739	293.	1.3
193	C4H11N	ISOBUTYL AMINE	73.139	188.	340.6	516.	42.	284.	0.28			293.	1.2
194	C4H11N	DIETHYL AMINE	73.139	223.4	328.6	496.6	36.6	301.	0.270	0.299	0.707	293.	1.1
195	C5H5N	PYRIDINE	79.102	231.5	388.5	620.0	55.6	254.	0.277	0.24	0.983	293.	2.3
196	C5H8	CYCLOPENTENE	68.119	138.1	317.4	506.0					0.772	293.	0.9
197	C5H8	1,2-PENTADIENE	68.119	135.9	318.0	503.	40.2	276.	0.269	0.173	0.693	293.	
198	C5H8	1-TRANS-3-PENTADIENE	68.119	185.7	315.2	496.	39.4	275.	0.266	0.175	0.676	293.	0.7
199	C5H8	1,4-PENTADIENE	68.119	124.9	299.1	478.	37.4	276.	0.263	0.104	0.661	293.	0.4
200	C5H8	1-PENTYNE	68.119	167.5	313.3	493.4	40.	278.	0.275	0.164	0.690	293.	0.9

NO	FORMULA	NAME	CPVAP A	CPVAP B	CPVAP C	CPVAP D	VISB	VISTO	DELHG	DELGF
161	C4H8	CIS-2-BUTENE	0.105	7.054 E-2	-2.431 E-5	-0.147 E-9	268.94	155.34	-1.67	15.74
162	C4H8	TRANS-2-BUTENE	4.375	6.123 E-2	-1.675 E-5	-2.147 E-9	259.01	153.30	-2.67	15.05
163	C4H8	CYCLOBUTANE	-12.003	1.200 E-1	-8.498 E-5	2.501 E-8			6.37	26.30
164	C4H8	ISOBUTYLENE	3.834	6.698 E-2	-2.607 E-5	2.173 E-9	472.31	233.42	-4.04	13.88
165	C4H8O	N-BUTYRALDEHYDE	3.363	8.257 E-2	-4.115 E-5	6.896 E-9	464.06	253.64	-49.00	-27.43
166	C4H8O	ISOBUTYRALDEHYDE	5.843	8.015 E-2	-4.913 E-5	1.521 E-8	423.84	231.67	-51.56	-29.00
167	C4H8O	METHYL ETHYL KETONE	2.614	8.501 E-2	-4.538 E-5	9.362 E-9	419.79	244.46	-56.97	-34.91
168	C4H8O	TETRAHYDROFURAN	-4.563	1.633 E-1	-9.868 E-5	3.473 E-8	349.95	189.02	-44.03	
169	C4H8O	VINYL ETHYL ETHER	2.804	9.881 E-2	-5.804 E-5	1.321 E-8			-33.50	
170	C4H8O2	N-BUTYRIC ACID	4.127	7.729 E-2	-3.514 E-5	5.134 E-9	660.36	321.13	-113.73	
171	C4H8O2	1,4-DIOXANE	-12.796	1.630 E-1	-9.757 E-5	2.537 E-8	640.42	308.77	-75.30	-43.21
172	C4H8O2	ETHYL ACETATE	1.728	9.725 E-2	-4.996 E-5	6.818 E-9	427.38	235.94	-105.86	-78.25
173	C4H8O2	ISOBUTYRIC ACID	2.344	1.115 E-1	-8.884 E-5	3.225 E-8	588.65	311.24	-115.66	
174	C4H8O2	METHYL PROPIONATE	4.348	7.699 E-2	-2.234 E-5	-4.365 E-9	442.68	238.39		
175	C4H8O2	N-PROPYL FORMATE	-0.624	1.074 E-1	-7.014 E-5	1.930 E-8	452.97	246.09		
176	C4H9CL	1-CHLOROBUTANE	-0.820	1.089 E-1	-7.119 E-5	1.972 E-8	783.72	260.03	-35.20	-9.27
177	C4H9CL	2-CHLOROBUTANE	-0.939	1.111 E-1	-6.893 E-5	1.880 E-8	480.77	237.30	-38.60	-12.78
178	C4H9CL	TERT-BUTYL CHLORIDE	-12.308	1.275 E-1	-7.738 E-5	1.798 E-8	543.41	253.35	-43.80	-15.32
179	C4H9N	PYRROLIDINE	-10.223	1.287 E-1	-6.368 E-5	1.003 E-8	914.14	332.75	-0.86	27.41
180	C4H9NO	MORPHOLINE	-0.332	7.913 E-2	-2.647 E-5	-0.674 E-9				
181	C4H10	N-BUTANE	0.780	9.189 E-2	-4.409 E-5	6.915 E-9	265.84	160.20	-30.15	-4.10
182	C4H10	ISOBUTANE	2.266	9.984 E-2	-5.354 E-5	1.119 E-8	302.51	170.20	-32.15	-4.99
183	C4H10O	N-BUTANOL	-1.841	1.014 E-1	-5.561 E-5	6.140 E-8	984.54	341.12	-65.65	-36.04
184	C4H10O	ISOBUTANOL	-11.611	1.120 E-1	-6.888 E-5	1.727 E-8	144.17	331.50	-69.94	-40.06
185	C4H10O	2-BUTANOL	7.699	1.713 E-1	-1.692 E-4	6.974 E-4	1199.1	343.85	-67.69	-39.99
186	C4H10O	TERT-BUTANOL	17.450	8.022 E-2	-2.473 E-5	2.235 E-8	972.10	363.38	-74.67	-42.46
187	C4H10O	ETHYL ETHER	3.247	9.457 E-2	-3.190 E-5	2.006 E-9	353.14	190.58	-60.28	-29.24
188	C4H10O2	1,2-DIMETHOXYETHANE	6.424	8.266 E-2	-3.506 E-5	4.410 E-9				
189	C4H10O2	DIETHYLENE GLYCOL	1.213	1.099 E-1	-4.251 E-5	6.327 E-9	1943.0	385.24	-136.5	
190	C4H10O3	DIETHYL SULFIDE	2.267	1.058 E-1	-6.472 E-5	1.426 E-8	407.59	233.32	-19.95	4.25
191	C4H10S	DIETHYL DISULFIDE	0.487	1.069 E-1	-5.749 E-5	1.815 E-8	472.06	246.98	-17.84	5.32
192	C4H10S2	N-BUTYL AMINE		1.058 E-1	-5.039 E-5	5.572 E-8	473.89	229.29	-22.00	11.76
193	C4H11N	ISOBUTYL AMINE		1.177 E-1	-5.214 E-5	8.725 E-9				
194	C4H11N	DIETHYL AMINE		1.106 E-1	-8.498 E-5	2.399 E-8			-17.30	17.23
195	C5H5N	PYRIDINE	9.504	9.267 E-2	-6.160 E-5	1.298 E-8	618.50	291.58	33.50	45.46
196	C5H8	CYCLOPENTENE	-9.915	6.714 E-2	-5.446 E-5	1.253 E-8	396.83	218.66	7.87	26.48
197	C5H8	1,2-PENTADIENE	2.108	9.438 E-2	-1.603 E-5	-5.617 E-9			34.80	50.29
198	C5H8	1-TRANS-3-PENTADIENE	7.330	8.386 E-2	-5.670 E-5	1.337 E-8			18.60	35.07
199	C5H8	1,4-PENTADIENE	1.671		-4.570 E-5	9.787 E-9			25.20	40.69
200	C5H8	1-PENTYNE	4.315						34.50	50.25

NO	FORMULA	NAME	ANTA	ANTB	ANTC	TMX	TMN	HARA	HARB	HARC	HARD	HV
161	C4H8	CIS-2-BUTENE	15.8171	2210.71	-36.15	305	200	49.609	-4217.05	-4.938	2.58	5580.
162	C4H8	TRANS-2-BUTENE	15.8177	2212.32	-33.15	300	200	50.137	-4174.56	-5.041	2.66	5439.
163	C4H8	CYCLOBUTANE	15.9254	2359.09	-31.78	290	200					5780.
164	C4H8	ISOBUTYLENE	15.7528	2125.75	-33.15	290	190	50.832	-4104.56	-5.157	2.46	5286.
165	C4H8O	N-BUTYRALDEHYDE	15.1668	2839.09	-50.15	380	255					7530.
166	C4H8O	ISOBUTYRALDEHYDE	15.9888	2676.98	-51.15	370	247					7500.
167	C4H8O	METHYL ETHYL KETONE	15.5986	3150.42	-36.65	376	257					7460.
168	C4H8O	TETRAHYDROFURAN	16.1069	2768.38	-46.90	370	270					7070.
169	C4H8O	VINYL ETHYL ETHER	15.8911	2449.26	-46.15	340	225					6330.
170	C4H8O2	N-BUTYRIC ACID	17.9240	4130.93	-70.55	470	335	47.683	-5328.22	-4.426	3.88	10040.
171	C4H8O2	1,4-DIOXANE	16.1327	2966.88	-62.15	410	275					8690.
172	C4H8O2	ETHYL ACETATE	16.1516	2790.50	-57.15	385	260	73.806	-9015.33	-7.651	4.22	6330.
173	C4H8O2	ISOBUTYRIC ACID	16.7792	3385.49	-94.15	465	330	65.669	-6394.77	-6.965	4.01	9830.
174	C4H8O2	METHYL PROPIONATE	16.1693	3385.06	-58.92	385	260	82.657	-9222.72	-8.986	5.15	7780.
175	C4H8O2	N-PROPYL FORMATE	15.7671	2893.95	-69.69	360	280	65.367	-6419.79	-6.915	3.98	7760.
176	C4H9CL	1-CHLOROBUTANE	15.9750	2826.26	-49.05	385	255	63.318	-6292.56	-6.635	4.01	7170.
177	C4H9CL	2-CHLOROBUTANE	15.9907	2753.43	-47.15	375	250					6980.
178	C4H9CL	TERT-BUTYL CHLORIDE	15.8121	2567.15	-44.15	360	235					6550.
179	C4H9N	PYRROLIDINE	15.9444	2717.03	-67.90	400	300					9000.
180	C4H9NO	MORPHOLINE	16.2364	3171.35	-71.15	440	300					
181	C4H10	N-BUTANE	15.6782	2154.90	-34.42	290	195	48.334	-4065.57	-4.781	2.68	5352.
182	C4H10	ISOBUTANE	15.5381	2032.73	-33.15	280	187	46.141	-3771.21	-4.509	2.57	5090.
183	C4H10O	N-BUTANOL	17.2160	3137.02	-94.43	404	288					10300.
184	C4H10O	2-BUTANOL	17.2102	3026.03	-86.65	393	298					9750.
185	C4H10O	ISOBUTANOL	16.8712	2874.73	-100.3	388	293					10050.
186	C4H10O	TERT-BUTANOL	16.8548	2658.29	-95.50	376	293					9330.
187	C4H10O	ETHYL ETHER	16.0823	2511.29	-41.95	340	225	57.260	-5105.90	-5.945	3.40	6380.
188	C4H10O2	1,2-DIMETHOXYETHANE	16.0261	2869.79	-53.15	393	262					7510.
189	C4H10O3	DIETHYLENE GLYCOL	17.0326	4122.52	-122.5	560	402					13670.
190	C4H10S	DIETHYL SULFIDE	15.9531	2896.27	-54.49	390	260					7590.
191	C4H10S2	DIETHYL DISULFIDE	16.0607	3421.57	-64.19	455	312					9010.
192	C4H11N	N-BUTYL AMINE	16.6085	3012.70	-48.96	373	259					7670.
193	C4H11N	ISOBUTYL AMINE	16.1419	2704.16	-56.15	373	251	64.890	-5912.65	-6.955	3.73	7400.
194	C4H11N	DIETHYL AMINE	16.0545	2595.01	-53.15	350	242					6650.
195	C5H5N	PYRIDINE	16.0910	3095.13	-61.15	425	285					8400.
196	C5H8	CYCLOPENTENE	15.9356	2583.07	-39.70	378	244					6450.
197	C5H8	1,2-PENTADIENE	15.9297	2544.34	-44.30	340	250					6590.
198	C5H8	1-TRANS-3-PENTADIENE	15.9182	2541.69	-41.43	340	250					6460.
199	C5H8	1,4-PENTADIENE	15.7392	2344.02	-41.69	320	240					6010.
200	C5H8	1-PENTYNE	16.0429	2515.62	-45.97	335	230					

NO	FORMULA	NAME	MOLWT	TFP	TB	TC	PC	VC	ZC	OMEGA	LIQDEN	TDEN	DIPM
201	C5H8	2-METHYL-1,3-BUTADIEN	68.119	127.2	307.2	484.	38.0	276.	0.264	0.164	0.681	293.	0.3
202	C5H8	3-METHYL-1,2-BUTADIEN	68.119	159.5	314.0	496.	40.6	267.	0.266	0.160	0.686	293.	3.0
203	C5H8O	CYCLOPENTANONE	84.118	222.5	403.9	626.	53.	268.	0.28	0.35	0.950	293.	
204	C5H8O2	ETHYL ACRYLATE	100.118	201.	373.	552.	37.0	320.	0.261	0.40	0.921	293.	0.0
205	C5H10	CYCLOPENTANE	70.135	179.3	322.4	511.6	44.5	260.	0.276	0.192	0.745	293.	0.4
206	C5H10	1-PENTENE	70.135	107.9	303.1	464.7	40.0	300.	0.31	0.245	0.640	293.	
207	C5H10	CIS-2-PENTENE	70.135	121.8	310.1	476.	36.0	300.	0.28	0.240	0.656	293.	
208	C5H10	TRANS-2-PENTENE	70.135	132.9	309.5	475.	36.1	300.	0.28	0.237	0.649	293.	
209	C5H10	2-METHYL-1-BUTENE	70.135	135.6	304.3	465.	34.0	294.	0.262	0.232	0.650	293.	0.5
210	C5H10	2-METHYL-2-BUTENE	70.135	139.4	311.7	470.	34.0	318.	0.280	0.285	0.662	293.	
211	C5H10	3-METHYL-1-BUTENE	70.135	104.7	293.3	450.	34.7	300.	0.282	0.209	0.627	293.	
212	C5H10O	VALERALDEHYDE	86.134	182.	376.	554.	35.	333.	0.26	0.40	0.810	293.	2.6
213	C5H10O	METHYL N-PROPYL KETON	86.134	196.	375.5	564.0	38.4	301.	0.250	0.348	0.806	293.	2.5
214	C5H10O	METHYL ISOPROPYL KETO	86.134	181.	367.4	553.4	38.0	310.	0.259	0.349	0.803	293.	2.8
215	C5H10O	DIETHYL KETONE	86.134	234.2	375.1	561.0	36.9	336.	0.269	0.347	0.814	293.	2.7
216	C5H10O2	N-VALERIC ACID	102.134	239.	458.7	651.	38.	340.	0.24	0.616	0.939	293.	
217	C5H10O2	ISOBUTYL FORMATE	102.134	178.	371.6	551.	38.3	350.	0.296	0.390	0.885	293.	1.9
218	C5H10O2	N-PROPYL ACETATE	102.134	178.	374.8	549.4	32.9	345.	0.252	0.392	0.887	293.	1.8
219	C5H10O2	ETHYL PROPIONATE	102.134	199.3	372.	546.0	33.2	345.	0.257	0.395	0.895	289.	1.8
220	C5H10O2	METHYL BUTYRATE	102.134	188.4	375.8	554.4	34.3	340.	0.257	0.382	0.898	293.	1.7
221	C5H10O2	METHYL ISOBUTYRATE	102.134	185.4	365.4	540.8	33.9	339.	0.259	0.367	0.891	293.	2.0
222	C5H11N	PIPERIDINE	85.150	262.7	379.7	594.0	47.	289.	0.28	0.25	0.862	293.	1.2
223	C5H12	N-PENTANE	72.151	143.4	309.2	469.6	33.3	304.	0.261	0.251	0.626	293.	0.0
224	C5H12	2-METHYL BUTANE	72.151	113.3	301.0	460.4	33.4	306.	0.271	0.227	0.620	293.	0.0
225	C5H12	2,2-DIMETHYL PROPANE	72.151	256.6	282.6	433.8	31.6	303.	0.269	0.197	0.591	293.	0.0
226	C5H12O	1-PENTANOL	88.150	195.0	411.0	586.	38.	326.	0.26	0.58	0.815	293.	1.7
227	C5H12O	2-METHYL-1-BUTANOL	88.150	203.	401.9	571.5	38.	322.	0.26	0.70	0.819	293.	
228	C5H12O	3-METHYL-1-BUTANOL	88.150	156.	404.4	579.	38.	329.	0.26	0.58	0.810	293.	1.8
229	C5H12O	2-METHYL-2-BUTANOL	88.150	264.4	375.2	545.	39.	319.	0.28	0.50	0.809	293.	1.9
230	C5H12O	2,2-DIMETHYL-1-PROPAN	88.150	327.	386.3	549.	39.	319.	0.28		0.783	327.	
231	C5H12O	ETHYL PROPYL ETHER	88.150	146.4	336.8	500.6	32.1			0.331	0.733	293.	1.2
232	C6F6	PERFLUOROBENZENE	186.056	186.0	353.4	516.7	32.6	442.	0.224	0.40			0.0
233	C6F12	PERFLUOROCYCLOHEXANE	300.047	256.1	325.7	457.2	24.	360.	0.255	0.073			
234	C6F14	PERFLUORO-N-HEXANE	338.044	186.0	330.3	451.7	18.8	359.	0.24	0.272			
235	C6H4CL2	O-DICHLOROBENZENE	147.004	256.1	453.6	697.3	40.5	372.	0.255	0.26	1.306	293.	2.3
236	C6H4CL2	M-DICHLOROBENZENE	147.004	248.4	446.	684.	38.	324.	0.24	0.27	1.288	293.	1.4
237	C6H4CL2	P-DICHLOROBENZENE	147.004	326.3	447.3	685.	39.		0.26	0.27	1.248	328.	
238	C6H5BR	BROMOBENZENE	157.010	242.3	429.2	670.	44.6	324.	0.249	0.249	1.495	293.	1.5
239	C6H5CL	CHLOROBENZENE	112.559	227.6	404.9	632.4	44.6	308.	0.265	0.249	1.106	293.	1.6
240	C6H5F	FLUOROBENZENE	96.104	234.0	358.5	560.1	44.9	271.	0.265	0.245	1.024	293.	1.4

NO	FORMULA	NAME	CPVAP A	CPVAP B	CPVAP C	CPVAP D	VISB	VISTO	DELHG	DELGF
201	C5H8	2-METHYL-1,3-BUTADIEN	-0.815	1.095 E-1	-7.971 E-5	2.389 E-8	328.49	182.48	18.10	34.86
202	C5H8	3-METHYL-1,2-BUTADIEN	3.508	8.593 E-2	-4.719 E-5	1.018 E-8			31.00	47.47
203	C5H8O	CYCLOPENTANONE	-9.707	1.248 E-1	-7.462 E-5	1.703 E-8	574.71	303.44	-46.04	
204	C5H8O2	ETHYL ACRYLATE	-4.015	8.813 E-1	-3.300 E-5	-1.369 E-8	438.04	256.84		
205	C5H10	CYCLOPENTANE	-12.808	1.296 E-1	-7.239 E-5	1.549 E-8	406.69	231.67	-18.46	9.23
206	C5H10	1-PENTENE	-0.032	1.034 E-1	-5.534 E-5	1.118 E-8	305.25	174.70	-5.00	18.91
207	C5H10	CIS-2-PENTENE	-3.414	1.099 E-1	-6.068 E-5	1.303 E-8	305.31	175.72	-6.71	17.17
208	C5H10	TRANS-2-PENTENE	0.455	9.988 E-2	-5.201 E-5	1.052 E-8	349.33	176.62	-7.59	16.71
209	C5H10	2-METHYL-1-BUTENE	2.525	9.547 E-2	-4.648 E-5	7.915 E-9	369.27	193.39	-8.68	15.68
210	C5H10	2-METHYL-2-BUTENE	2.819	8.381 E-2	-2.667 E-5	-1.387 E-9	322.47	180.43	-10.17	14.26
211	C5H10	3-METHYL-1-BUTENE	5.193	9.290 E-2	-4.794 E-5	9.579 E-9			-6.92	17.87
212	C5H10O	VALERALDEHYDE	3.401	9.409 E-2	-5.033 E-5	7.553 E-9	521.30	252.03	-54.45	-25.88
213	C5H10O	METHYL N-PROPYL KETON	0.274	1.147 E-1	-5.731 E-5	1.591 E-8	437.94	243.03	-61.82	-32.76
214	C5H10O	METHYL ISOPROPYL KETO	-0.696	1.192 E-1	-7.009 E-5	1.592 E-8				
215	C5H10O	DIETHYL KETONE	7.168	9.409 E-2	-4.554 E-5	8.115 E-9	409.17	235.65	-61.82	-32.33
216	C5H10O2	N-VALERIC ACID	3.198	1.202 E-1	-7.001 E-5	1.581 E-8	729.09	341.13	-117.20	-85.37
217	C5H10O2	ISOBUTYL FORMATE	4.741	9.634 E-2	-3.431 E-5	-1.768 E-9	489.53	255.83	-111.31	
218	C5H10O2	N-PROPYL ACETATE	3.683	1.075 E-1	-4.027 E-5	-3.437 E-9	463.31	248.72	-112.3	-77.32
219	C5H10O2	ETHYL PROPIONATE	4.742	9.636 E-2	-3.432 E-5	-1.768 E-9	479.35	249.66		
220	C5H10O2	METHYL BUTYRATE					451.21	246.09		
221	C5H10O2	METHYL ISOBUTYRATE								
222	C5H11N	PIPERIDINE	-12.675	1.502 E-1	-8.020 E-5	1.535 E-8	772.79	313.49	-11.71	
223	C5H12	N-PENTANE	-0.866	1.164 E-1	-6.163 E-5	1.267 E-8	313.66	182.48	-35.00	-2.00
224	C5H12	2-METHYL BUTANE	-2.275	1.210 E-1	-6.519 E-5	1.337 E-8	367.32	191.58	-36.92	-3.54
225	C5H12	2,2-DIMETHYL PROPANE	-3.963	1.326 E-1	-7.897 E-5	1.823 E-8	355.54	196.35	-39.67	-3.64
226	C5H12O	1-PENTANOL	0.924	1.205 E-1	-6.304 E-5	1.223 E-8	1151.61	349.62	-71.4	-34.49
227	C5H12O	2-METHYL-1-BUTANOL	-2.265	1.356 E-1	-8.315 E-5	2.063 E-8	1259.4	349.85	-72.3	-39.58
228	C5H12O	3-METHYL-1-BUTANOL	-2.279	1.357 E-1	-8.323 E-5	2.066 E-8	1148.8	349.51	-72.2	-39.5
229	C5H12O	2-METHYL-2-BUTANOL	-2.887	1.456 E-1	-1.004 E-4	2.934 E-8	1502.0	336.75	-78.8	-29.98
230	C5H12O	2,2-DIMETHYL-1-PROPAN	2.903	1.289 E-1	-7.547 E-5	1.701 E-8				
231	C5H12O	ETHYL PROPYL ETHER							-70.00	
232	C6F6	PERFLUOROBENZENE	8.656	1.258 E-1	-1.086 E-4	3.477 E-8	399.87	213.39	-228.64	-210.18
233	C6F12	PERFLUOROCYCLOHEXANE								
234	C6F14	PERFLUORO-N-HEXANE								
235	C6H4CL2	O-DICHLOROBENZENE	-3.416	1.315 E-1	-1.078 E-4	3.414 E-8	554.35	319.07	7.16	19.76
236	C6H4CL2	M-DICHLOROBENZENE	-3.246	1.312 E-1	-1.076 E-4	3.408 E-8	402.20	300.89	6.32	18.78
237	C6H4CL2	P-DICHLOROBENZENE	-3.426	1.322 E-1	-1.089 E-4	3.458 E-8	483.82	312.03	5.50	18.44
238	C6H5BR	BROMOBENZENE	-6.880	1.278 E-1	-9.746 E-5	2.894 E-8	508.18	302.42	25.10	33.11
239	C6H5CL	CHLOROBENZENE	6.094	1.345 E-1	-1.080 E-4	3.407 E-8	477.76	276.22	12.39	23.70
240	C6H5F	FLUOROBENZENE	-9.250	1.354 E-1	-1.059 E-4	3.237 E-8	452.06	252.89	-27.86	-16.50

NO	FORMULA	NAME	ANTA	ANTB	ANTC	TMX	TMN	HARA	HARB	HARC	HARD	HV
201	C5H8	2-METHYL-1,3-BUTADIEN	15.8548	2467.40	-39.64	330	250					6230.
202	C5H8	3-METHYL-1,2-BUTADIEN	15.9880	2541.83	-42.26	335	250					6510.
203	C5H8O	CYCLOPENTANONE	16.0897	3193.92	-66.15	440	300					8740.
204	C5H8O2	ETHYL ACRYLATE	16.0890	2974.94	-58.15	409	274					7950.
205	C5H10	CYCLOPENTANE	15.8574	2588.48	-41.79	345	230					6524.
206	C5H10	1-PENTENE	15.7646	2405.96	-39.63	325	220	51.816	-4694.26	-5.202	3.42	6022.
207	C5H10	CIS-2-PENTENE	15.8251	2459.05	-42.56	330	220	55.199	-4985.30	-5.668	3.51	6240.
208	C5H10	TRANS-2-PENTENE	15.9011	2495.97	-40.18	330	220	56.420	-5028.79	-5.853	3.62	6230.
209	C5H10	2-METHYL-1-BUTENE	15.8260	2426.42	-40.36	325	220	60.581	-5160.84	-6.474	3.47	6094.
210	C5H10	2-METHYL-2-BUTENE	15.9238	2521.53	-40.31	335	226	55.255	-5010.98	-5.671	3.71	6287.
211	C5H10	3-METHYL-1-BUTENE	15.7179	2333.61	-36.33	315	210					5760.
212	C5H10O	VALERALDEHYDE	16.1623	3030.20	-58.15	412	277					8040.
213	C5H10O	METHYL N-PROPYL KETON	16.0031	2994.87	-62.25	410	275					8000.
214	C5H10O	METHYL ISOPROPYL KETO	14.1779	1993.12	-103.2	406	271	111.20	-9773.63	-13.26	4.73	7320.
215	C5H10O	DIETHYL KETONE	16.8138	3410.51	-40.15	400	275					8060.
216	C5H10O2	N-VALERIC ACID	17.6306	4092.15	-86.65	495	350					11900.
217	C5H10O2	ISOBUTYL FORMATE	16.2292	2980.47	-64.15	409	278	58.420	-6314.51	-5.879	4.41	8170.
218	C5H10O2	N-PROPYL ACETATE	16.2291	2980.47	-64.15	410	280	69.656	-7028.88	-7.475	5.10	8170.
219	C5H10O2	ETHYL PROPIONATE	16.1620	2935.11	-64.16	396	276	67.631	-6889.83	-7.193	4.98	8180.
220	C5H10O2	METHYL BUTYRATE						65.898	-5819.11	-6.941	4.98	8145.
221	C5H10O2	METHYL ISOBUTYRATE						66.160	-6637.51	-7.016	4.79	7974.
222	C5H11N	PIPERIDINE	16.1004	3015.46	-61.15	416	280					8180.
223	C5H12	N-PENTANE	15.8333	2477.07	-39.94	330	220	52.682	-4827.08	-5.313	3.68	6160.
224	C5H12	2-METHYL BUTANE	15.6338	2348.67	-40.05	322	216	50.428	-4565.64	-5.021	3.55	5900.
225	C5H12	2,2-DIMETHYL PROPANE	15.2069	2034.15	-45.37	305	260	49.600	-4213.21	-4.977	3.31	5438.
226	C5H12O	1-PENTANOL	16.5270	3026.89	-105.0	411	310					10600.
227	C5H12O	2-METHYL-1-BUTANOL	16.2708	2752.19	-116.3	402	307					10800.
228	C5H12O	3-METHYL-1-BUTANOL	16.7127	3026.43	-104.1	426	298					10540.
229	C5H12O	2-METHYL-2-BUTANOL	15.0113	1988.08	-137.8	375	298					9700.
230	C5H12O	2,2-DIMETHYL-1-PROPAN	18.1336	3694.96	-65.00	406	328					10300.
231	C5H12O	ETHYL PROPYL ETHER	15.4539	2423.41	-62.28	360	246	58.911	-5663.85	-6.1	4.33	7290.
232	C6F6	PERFLUOROBENZENE	16.1940	2827.53	-57.66	390	270	74.686	-6815.04	-8.318	5.39	9480.
233	C6F12	PERFLUOROCYCLOHEXANE	13.9087	1374.07	-136.8	400	280	119.20	-8611.09	-14.89	6.04	9230.
234	C6F14	PERFLUORO-N-HEXANE	15.8307	2488.59	-59.73	330	270	90.505	-7074.74	-10.78	7.33	9270.
235	C6H4CL2	O-DICHLOROBENZENE	16.2799	3798.23	-59.84	483	331					
236	C6H4CL2	M-DICHLOROBENZENE	16.8173	4104.13	-43.15	475	326					
237	C6H4CL2	P-DICHLOROBENZENE	16.1135	3622.83	-64.64	477	327					
238	C6H5BR	BROMOBENZENE	15.7972	3313.00	-67.71	450	320	56.566	-7005.23	-5.548	5.59	
239	C6H5CL	CHLOROBENZENE	16.0676	3295.12	-55.60	420	320	57.251	-6684.47	-5.686	4.98	8735.
240	C6H5F	FLUOROBENZENE	16.5487	3181.78	-37.59	370	250	55.141	-5819.21	-5.489	3.88	

NO	FORMULA	NAME	MOLWT	TFP	TB	TC	PC	VC	ZC	OMEGA	LIQDEN	TDEN	DIPM
241	C6H5I	IODOBENZENE	204.011	241.8	461.4	721.	44.6	351.	0.265	0.246	1.855	277.	1.4
242	C6H6	BENZENE	78.114	278.7	353.3	562.1	48.3	259.	0.271	0.212	0.885	289.	0.0
243	C6H6O	PHENOL	94.113	314.0	455.0	694.2	60.5	229.	0.24	0.440	1.059	313.	1.6
244	C6H7N	ANILINE	93.129	267.	457.5	699.	52.4	270.	0.247	0.382	1.022	293.	1.6
245	C6H7N	4-METHYL PYRIDINE	93.129	276.9	418.5	646.	44.	311.	0.26	0.27	0.955	293.	
246	C6H10	1,5-HEXADIENE	82.146	132.	332.6	507.	34.	328.	0.26	0.16	0.692	293.	0.6
247	C6H10	CYCLOHEXENE	82.146	169.7	356.1	560.4	42.9	292.	0.27	0.210	0.816	289.	3.1
248	C6H10O	CYCLOHEXANONE	98.145	242.0	428.8	629.	38.	312.	0.23	0.443	0.951	288.	0.3
249	C6H12	CYCLOHEXANE	84.162	279.7	353.9	553.4	40.2	308.	0.273	0.213	0.779	293.	0.0
250	C6H12	METHYLCYCLOPENTANE	84.162	130.7	345.0	532.7	37.4	319.	0.273	0.239	0.754	289.	0.4
251	C6H12	1-HEXENE	84.162	133.3	336.6	504.0	31.3	350.	0.26	0.285	0.673	293.	
252	C6H12	CIS-2-HEXENE	84.162	132.0	342.0	518.	32.4	351.	0.27	0.256	0.687	293.	
253	C6H12	TRANS-2-HEXENE	84.162	140.	341.0	516.	32.3	351.	0.27	0.242	0.678	293.	
254	C6H12	CIS-3-HEXENE	84.162	135.3	339.6	517.	32.4	350.	0.27	0.225	0.680	293.	0.3
255	C6H12	TRANS-3-HEXENE	84.162	159.7	340.3	519.9	32.1	350.	0.26	0.227	0.677	293.	0.0
256	C6H12	2-METHYL-2-PENTENE	84.162	138.1	340.5	518.	32.4	351.	0.27	0.229	0.691	289.	
257	C6H12	3-METHYL-CIS-2-PENTEN	84.162	138.3	340.9	518.	32.4	351.	0.27	0.269	0.694	293.	
258	C6H12	3-METHYL-TRANS-2-PENT	84.152	134.7	343.6	521.	32.5	350.	0.27	0.207	0.698	293.	
259	C6H12	4-METHYL-CIS-2-PENTEN	84.162	139.	329.6	490.	30.	360.	0.27	0.29	0.669	293.	
260	C6H12	4-METHYL-TRANS-2-PENT	84.162	132.	331.7	493.	30.	360.	0.27	0.29	0.669	293.	
261	C6H12	2,3-DIMETHYL-1-BUTENE	84.162	115.9	328.8	501.	32.0	343.	0.27	0.221	0.678	294.	
262	C6H12	2,3-DIMETHYL-2-BUTENE	84.162	198.9	346.4	524.	33.2	351.	0.27	0.239	0.708	293.	
263	C6H12	3,3-DIMETHYL-1-BUTENE	84.162	158.	314.4	490.	32.1	340.	0.27	0.121	0.653	293.	
264	C6H12O	CYCLOHEXANOL	100.161	298.	434.3	625.	37.	327.	0.24	0.55	0.942	303.	1.7
265	C6H12O	METHYL ISOBUTYL KETON	100.161	189.	389.6	571.	32.3	371.	0.26	0.400	0.801	293.	2.8
266	C6H12O2	N-BUTYL ACETATE	116.160	199.7	399.2	579.	31.	400.	0.26	0.417	0.898	273.	1.8
267	C6H12O2	ISOBUTYL ACETATE	116.160	174.3	390.	551.	30.	414.	0.27	0.479	0.875	293.	1.9
268	C6H12O2	ETHYL BUTYRATE	116.160	180.	394.	566.	31.	395.	0.26	0.47	0.879	293.	1.8
269	C6H12O2	ETHYL ISOBUTYRATE	116.160	185.	384.2	553.	30.	410.	0.27	0.427	0.869	293.	2.1
270	C6H12O2	N-PROPYL PROPIONATE	116.161	197.3	395.7	578.					0.881	293.	1.8
271	C6H14	N-HEXANE	86.178	177.8	341.9	507.4	29.3	370.	0.260	0.296	0.659	293.	0.0
272	C6H14	2-METHYL PENTANE	86.178	119.5	333.4	497.5	29.7	367.	0.267	0.279	0.653	293.	
273	C6H14	3-METHYL PENTANE	86.178	155.	336.4	504.4	30.8	367.	0.273	0.275	0.664	293.	
274	C6H14	2,2-DIMETHYL BUTANE	86.178	173.3	322.9	488.7	30.4	359.	0.272	0.231	0.649	293.	
275	C6H14	2,3-DIMETHYL BUTANE	86.178	144.6	331.2	499.9	30.9	358.	0.270	0.247	0.662	293.	
276	C6H14O	1-HEXANOL	102.177	229.2	430.2	610.	40.	381.	0.30	0.56	0.819	293.	1.8
277	C6H14O	ETHYL BUTYL ETHER	102.177	170.	365.4	531.	30.	390.	0.27	0.40	0.749	293.	1.2
278	C6H14O	DIISOPROPYL ETHER	102.177	187.7	341.5	500.0	28.4	386.	0.267	0.34	0.724	293.	1.2
279	C6H15N	DIPROPYLAMINE	101.193	210.	382.4	550.	31.	407.	0.28	0.455	0.738	293.	1.0
280	C6H15N	TRIETHYLAMINE	101.193	158.4	362.7	535.	30.	390.	0.27	0.329	0.728	293.	0.9

NO	FORMULA	NAME	CPVAP A	CPVAP B	CPVAP C	CPVAP D	VISB	VISTO	DELHG	DELGF
241	C6H5I	IODOBENZENE	-6.992	1.329 E-1	-1.077 E-4	3.447 E-8	565.72	331.21	38.85	44.88
242	C6H6	BENZENE	-8.101	1.133 E-1	-7.206 E-5	1.703 E-8	545.64	265.34	19.82	30.99
243	C6H6O	PHENOL	-8.561	1.429 E-1	-1.153 E-4	3.647 E-8	1405.5	370.07	-23.03	-7.86
244	C6H7N	ANILINE	-9.677	1.525 E-1	-1.226 E-4	3.901 E-8	1074.6	357.07	20.76	39.84
245	C6H7N	4-METHYL PYRIDINE	-4.163	1.166 E-1	-6.663 E-5	1.302 E-8	500.97	285.50	24.43	
246	C6H10	1,5-HEXADIENE							20.0	20.0
247	C6H10	CYCLOHEXENE	-16.397	1.732 E-1	-1.293 E-4	3.927 E-8	506.92	264.54	-1.28	25.54
248	C6H10O	CYCLOHEXANONE	-9.030	1.323 E-1	-4.665 E-5	-3.664 E-9	787.38	336.47	-55.00	-21.69
249	C6H12	CYCLOHEXANE	-13.027	1.460 E-1	-6.027 E-5	3.156 E-9	653.62	290.84	-29.43	7.59
250	C6H12	METHYLCYCLOPENTANE	-11.968	1.524 E-1	-8.699 E-5	1.914 E-8	440.52	243.24	-25.50	8.55
251	C6H12	1-HEXENE	-0.417	1.268 E-1	-6.933 E-5	1.446 E-8	357.43	197.74	-9.96	20.90
252	C6H12	CIS-2-HEXENE	-2.343	1.268 E-1	-6.490 E-5	1.153 E-8	344.33	197.95	-12.51	18.22
253	C6H12	TRANS-2-HEXENE	-7.864	1.655 E-1	-1.342 E-4	4.788 E-8	344.33	197.95	-12.88	18.27
254	C6H12	CIS-3-HEXENE	-5.190	1.388 E-1	-8.029 E-5	1.761 E-8	344.33	197.95	-11.38	19.84
255	C6H12	TRANS-3-HEXENE	-1.036	1.316 E-1	-7.840 E-5	1.922 E-8	344.33	197.95	-13.01	18.55
256	C6H12	2-METHYL-2-PENTENE	-3.523	1.354 E-1	-7.979 E-5	1.902 E-8			-14.28	17.02
257	C6H12	3-METHYL-CIS-2-PENTEN	-3.523	1.354 E-1	-7.979 E-5	1.902 E-8			-13.80	17.50
258	C6H12	3-METHYL-TRANS-2-PENT	-3.523	1.354 E-1	-7.979 E-5	1.902 E-8			-14.02	17.04
259	C6H12	4-METHYL-CIS-2-PENTEN	-0.400	1.284 E-1	-7.271 E-5	1.613 E-8			-12.03	19.63
260	C6H12	4-METHYL-TRANS-2-PENT	3.016	1.231 E-1	-7.182 E-5	1.750 E-8			-12.99	19.03
261	C6H12	2,3-DIMETHYL-1-BUTENE	1.678	1.334 E-1	-8.828 E-5	2.539 E-8			-13.32	18.89
262	C6H12	2,3-DIMETHYL-2-BUTENE	0.548	1.153 E-1	-5.252 E-5	7.265 E-9			-14.15	18.13
263	C6H12	3,3-DIMETHYL-1-BUTENE	-2.999	1.310 E-1	-6.963 E-5	1.244 E-8			-10.31	23.46
264	C6H12O	CYCLOHEXANOL	-13.264	1.723 E-1	-9.760 E-5	1.967 E-8			-70.40	-28.18
265	C6H12O	METHYL ISOBUTYL KETON	0.930	1.351 E-1	-7.925 E-5	1.966 E-8	473.65	259.03	-67.84	
266	C6H12O2	N-BUTYL ACETATE	3.253	1.311 E-1	-5.442 E-5	10.189 E-9	537.58	272.30	-116.26	
267	C6H12O2	ISOBUTYL ACETATE	1.746	1.371 E-1	-6.152 E-5	2.630 E-9	533.99	270.49	-118.34	
268	C6H12O2	ETHYL BUTRATE	5.137	1.177 E-1	-4.629 E-5	0.850 E-9	489.95	264.22		
269	C6H12O2	ETHYL ISOBUTYRATE								
270	C6H12O2	N-PROPYL PROPIONATE								
271	C6H14	N-HEXANE	-1.054	1.390 E-1	-7.449 E-5	1.551 E-8	362.79	207.09	-39.96	-0.06
272	C6H14	2-METHYL PENTANE	-2.524	1.477 E-1	-8.533 E-5	1.931 E-8	384.13	208.27	-41.66	-1.20
273	C6H14	3-METHYL PENTANE	-0.570	1.359 E-1	-6.854 E-5	1.202 E-8	372.11	207.55	-41.02	-0.51
274	C6H14	2,2-DIMETHYL BUTANE	-3.973	1.503 E-1	-8.314 E-5	1.636 E-8	438.44	226.67	-44.35	-2.30
275	C6H14	2,3-DIMETHYL BUTANE	-3.489	1.469 E-1	-8.063 E-5	1.629 E-8	434.19	228.86	-42.49	-0.98
276	C6H14O	1-HEXANOL	1.149	1.407 E-1	-7.189 E-5	1.296 E-8	1179.44	354.94	-75.9	-32.4
277	C6H14O	ETHYL BUTYL ETHER	5.643	1.282 E-1	-6.039 E-5	9.928 E-9	443.32	234.68		
278	C6H14O	DIISOPROPYL ETHER	1.792	1.397 E-1	-7.229 E-5	1.396 E-8	410.58	219.67	-76.20	-29.13
279	C6H15N	DIPROPYLAMINE	1.543	1.503 E-1	-8.098 E-5	1.689 E-8	561.11	257.39		
280	C6H15N	TRIETHYLAMINE	-4.402	1.709 E-1	-1.049 E-4	2.609 E-8	355.52	214.48	-23.80	26.36

NO	FORMULA	NAME	ANTA	ANTB	ANTC	TMX	TMN	HARA	HARB	HARC	HARD	HV
241	C6H5I	IODOBENZENE	16.1454	3776.53	-64.38	470	290	57.691	-7589.50	-5.646	6.046	9440.
242	C6H6	BENZENE	15.9008	2788.51	-52.36	377	280	52.010	-5557.61	-5.072	3.61	7352.
243	C6H6O	PHENOL	16.4279	3490.89	-98.59	481	345	72.558	-9072.60	-7.516	4.442	10900.
244	C6H7N	ANILINE	16.6748	3857.52	-73.15	500	340	65.861	-8442.37	-6.662	5.18	10000.
245	C6H7N	4-METHYL PYRIDINE	16.2143	3409.40	-62.65	460	300					8950.
246	C6H10	1,5-HEXADIENE	16.1351	2728.54	-45.45	350	300					6561.
247	C6H10	CYCLOHEXENE	15.8243	2813.53	-49.98	360	282					7280.
248	C6H10O	CYCLOHEXANONE					300					9500.
249	C6H12	CYCLOHEXANE	15.7527	2766.63	-50.50	380	280	53.451	-5562.12	-5.303	4.22	7160.
250	C6H12	METHYLCYCLOPENTANE	15.8023	2731.00	-47.11	375	250					6950.
251	C6H12	1-HEXENE	15.8089	2654.81	-47.30	360	240	55.909	-5423.07	-5.705	4.54	6760.
252	C6H12	CIS-2-HEXENE	15.2057	2897.97	-39.30	370	245					6960.
253	C6H12	TRANS-2-HEXENE	15.8727	2701.72	-48.62	365	245	60.438	-5734.51	-6.348	4.73	6910.
254	C6H12	CIS-3-HEXENE	15.8384	2680.52	-48.40	365	245					6860.
255	C6H12	TRANS-3-HEXENE	15.9288	2718.68	-47.77	365	245					6920.
256	C6H12	2-METHYL-2-PENTENE	15.9423	2725.89	-47.64	370	245					6930.
257	C6H12	3-METHYL-CIS-2-PENTEN	15.9124	2731.79	-46.47	364	245					6890.
258	C6H12	3-METHYL-TRANS-2-PENT	15.9484	2750.50	-48.33	366	248					7000.
259	C6H12	4-METHYL-CIS-2-PENTEN	15.7527	2580.52	-46.56	352	250					6590.
260	C6H12	4-METHYL-TRANS-2-PENT	15.8425	2631.57	-46.00	354	238					6680.
261	C6H12	2,3-DIMETHYL-1-BUTENE	15.8012	2612.69	-43.78	360	240					6550.
262	C6H12	2,3-DIMETHYL-2-BUTENE	16.0043	2798.63	-47.71	375	235					7083.
263	C6H12	3,3-DIMETHYL-1-BUTENE	15.3755	2326.80	-48.24	340	250					6130.
264	C6H12O	CYCLOHEXANOL					225	86.548	-9573.09	-9.539	5.86	10870.
265	C6H12O	METHYL ISOBUTYL KETON	15.7165	2893.66	-70.75	425	285					8500.
266	C6H12O2	N-BUTYL ACETATE	16.1836	3151.09	-59.15	435	295					8600.
267	C6H12O2	ISOBUTYL ACETATE	16.1714	3092.83	-56.15	427	289					8568.
268	C6H12O2	ETHYL BUTYRATE	15.9987	3127.60	-60.15	432	288					8200.
269	C6H12O2	ETHYL ISOBUTYRATE	16.8641	3558.18	-47.86	420	292	74.336	-7477.19	-8.108	5.66	8365.
270	C6H12O2	N-PROPYL PROPIONATE				370	245					8690.
271	C6H14	N-HEXANE	15.8366	2697.55	-48.78	370	240	57.279	-5587.42	-5.885	4.778	6896.
272	C6H14	2-METHYL PENTANE	15.7476	2614.38	-46.58	370	240	55.352	-5301.22	-5.65	4.911	6640.
273	C6H14	3-METHYL PENTANE	15.7701	2653.43	-46.02	365	240	54.479	-5323.33	-5.509	4.579	6710.
274	C6H14	2,2-DIMETHYL BUTANE	15.5536	2489.50	-43.81	350	230	51.470	-4910.28	-5.134	4.323	6287.
275	C6H14	2,3-DIMETHYL BUTANE	15.6802	2595.44	-44.25	354	235	51.700	-5061.44	-5.138	4.472	6520.
276	C6H14O	1-HEXANOL	18.0994	4055.45	-76.49	430	308					11600.
277	C6H14O	ETHYL BUTYL ETHER	16.0477	2921.52	-55.15	400	265					7600.
278	C6H14O	DIISOPROPYL ETHER	16.3417	2895.73	-43.15	364	249					7010.
279	C6H15N	DIPROPYLAMINE	15.5939	2895.08	-55.15	422	302					8840.
280	C6H15N	TRIETHYLAMINE	15.8853	2882.38	-51.15	400	260					7500.

NO	FORMULA	NAME	MOLWT	TFP	TB	TC	PC	VC	ZC	OMEGA	LIQDEN	TDEN	DIPM
281	C7F14	PERFLUOROMETHYLCYCLOH	350.055	195.	349.5	486.8	23.	664.	0.273	0.482	1.733	293.	
282	C7F16	PERFLUORO-N-HEPTANE	388.051	260.	355.7	474.8	16.0			0.56			
283	C7H5N	BENZONITRILE	103.124	216.	464.	699.4	41.6	341.	0.25	0.36	1.010	288.	3.5
284	C7H6O	BENZALDEHYDE	106.124	216.	452.	695.	46.			0.32	1.045	293.	2.8
285	C7H6O2	BENZOIC ACID	122.124	395.6	523.	752.	45.				1.075	403.	1.7
286	C7H8	TOLUENE	92.141	178.	383.8	591.7	40.6	316.	0.264	0.257	0.867	293.	0.4
287	C7H8O	METHYL PHENYL ETHER	108.140	235.7	426.8	641.	41.2	334.	0.28		0.996	293.	1.2
288	C7H8O	BENZYL ALCOHOL	108.140	257.8	478.6	677.	46.	282.	0.24	0.71	1.041	298.	1.7
289	C7H8O	O-CRESOL	108.140	304.1	464.2	697.6	49.4	310.	0.241	0.443	1.028	313.	1.6
290	C7H8O	M-CRESOL	108.140	285.4	475.4	705.8	45.0			0.464	1.019	293.	1.8
291	C7H8O	P-CRESOL	108.140	307.9	475.1	704.6	50.8			0.515			1.6
292	C7H9N	2,3-DIMETHYLPYRIDINE	107.156		434.	655.4					0.942	298.	2.2
293	C7H9N	2,5-DIMETHYLPYRIDINE	107.156		430.2	644.2					0.938	273.	2.2
294	C7H9N	3,4-DIMETHYLPYRIDINE	107.156		452.3	683.8					0.954	298.	1.9
295	C7H9N	3,5-DIMETHYLPYRIDINE	107.156		445.1	667.2					0.939	298.	2.6
296	C7H9N	METHYLPHENYLAMINE	107.156	216.	469.1	701.	51.3	343.	0.26	0.435	0.989	293.	1.7
297	C7H9N	O-TOLUIDINE	107.156	258.4	473.3	694.	37.	343.	0.24	0.406	0.998	293.	1.6
298	C7H9N	M-TOLUIDINE	107.156	242.8	476.5	709.	41.				0.989	323.	1.5
299	C7H9N	P-TOLUIDINE	107.156	316.9	473.8	667.					0.964	293.	1.6
300	C7H14	CYCLOHEPTANE	98.189	265.	391.9	589.	36.7	390.	0.30	0.336	0.810	293.	
301	C7H14	1,1-DIMETHYLCYCLOPENT	98.189	203.4	361.0	547.	34.0	360.	0.27	0.273	0.759	289.	
302	C7H14	CIS-1,2-DIMETHYLCYCLO	98.189	219.3	372.7	564.8	34.0	368.	0.27	0.269	0.777	289.	
303	C7H14	TRANS-1,2-DIMETHYLCYC	98.189	155.6	365.0	553.2	34.0	362.	0.27	0.269	0.756	289.	
304	C7H14	ETHYLCYCLOPENTANE	98.189	134.7	376.6	569.5	33.5	375.	0.269	0.283	0.771	289.	
305	C7H14	METHYLCYCLOHEXANE	98.189	146.6	374.1	572.1	34.3	368.	0.28	0.233	0.774	289.	
306	C7H14	1-HEPTENE	98.189	154.3	366.8	537.2	28.	440.	0.26	0.358	0.697	293.	0.0
307	C7H14	2,3,3-TRIMETHYL-1-BUT	98.189	163.3	351.0	533.	28.6	400.	0.263	0.192	0.705	293.	0.3
308	C7H16	N-HEPTANE	100.205	182.6	371.6	540.2	27.0	432.	0.261	0.351	0.684	293.	
309	C7H16	2-METHYLHEXANE	100.205	154.9	363.2	530.3	27.0	421.	0.257	0.330	0.679	293.	0.0
310	C7H16	3-METHYLHEXANE	100.205	100.	365.0	535.2	27.0	404.	0.267	0.324	0.687	293.	0.0
311	C7H16	2,2-DIMETHYLPENTANE	100.205	149.4	352.4	520.4	27.4	416.	0.256	0.289	0.674	293.	
312	C7H16	2,3-DIMETHYLPENTANE	100.205	154.	362.9	537.3	28.7	393.	0.265	0.299	0.695	293.	0.0
313	C7H16	2,4-DIMETHYLPENTANE	100.205		353.7	519.7	27.0	418.	0.274	0.306	0.673	293.	0.0
314	C7H16	3,3-DIMETHYLPENTANE	100.205	138.7	359.2	536.3	29.1	414.	0.267	0.270	0.693	293.	0.0
315	C7H16	3-ETHYLPENTANE	100.205	154.6	366.6	540.6	28.5	416.	0.267	0.310	0.698	293.	0.0
316	C7H16	2,2,3-TRIMETHYLBUTANE	100.205	248.3	354.0	531.1	29.2	398.	0.25	0.251	0.690	293.	0.0
317	C7H16O	1-HEPTANOL	116.204	239.2	449.5	633.	30.	435.	0.26	0.56	0.822	293.	1.7
318	C8H4O3	PHTHALIC ANHYDRIDE	148.118	404.	560.	810.	47.	368.					5.3
319	C8H8	STYRENE	104.152	242.5	418.3	647.	39.4			0.257	0.906	293.	0.1
320	C8H8O	METHYL PHENYL KETONE	120.151	292.8	474.9	701.	38.	376.	0.25	0.42	1.032	288.	3.0

NO	FORMULA	NAME	CPVAP A	CPVAP B	CPVAP C	CPVAP D	VISB	VISTO	DELHG	DELGF
281	C7F14	PERFLUOROMETHYLCYCLOH							-692.2	-737.87
282	C7F16	PERFLUORO-N-HEPTANE							-808.9	
283	C7H5N	BENZONITRILE	-6.221	1.369 E-1	-1.058 E-4	3.222 E-8	686.84	314.66	52.30	62.35
284	C7H6O	BENZALDEHYDE	-2.900	1.185 E-1	-6.794 E-5	1.234 E-8	2617.6	407.88	-8.79	5.35
285	C7H6O2	BENZOIC ACID	-12.251	1.503 E-1	-1.012 E-4	2.537 E-8	467.33	255.24	-69.36	-50.29
286	C7H8	TOLUENE	-5.817	1.224 E-1	-6.605 E-5	1.173 E-8	388.84	325.85	11.95	29.16
287	C7H8O	METHYL PHENYL ETHER								
288	C7H8O	BENZYL ALCOHOL	-1.767	1.309 E-1	-8.019 E-5	1.856 E-8	1088.0	367.21	-22.47	
289	C7H8O	O-CRESOL	-7.709	1.673 E-1	-1.415 E-4	5.073 E-8	1533.4	365.61	-30.74	-8.86
290	C7H8O	M-CRESOL	-10.750	1.735 E-1	-1.440 E-4	4.962 E-8	1785.6	370.75	-31.63	-9.69
291	C7H8O	P-CRESOL	-9.705	1.685 E-1	-1.375 E-4	4.699 E-8	1826.9	372.68	-29.97	-7.38
292	C7H9N	2,3-DIMETHYLPYRIDINE							16.31	
293	C7H9N	2,5-DIMETHYLPYRIDINE							15.87	
294	C7H9N	3,4-DIMETHYLPYRIDINE							16.73	
295	C7H9N	3,5-DIMETHYLPYRIDINE							17.39	
296	C7H9N	METHYLPHENYLAMINE	-3.819	1.357 E-1	-7.244 E-5	1.109 E-8	915.12	332.74	20.4	47.61
297	C7H9N	O-TOLUIDINE					1085.1	356.46		
298	C7H9N	M-TOLUIDINE					928.12	354.07		
299	C7H9N	P-TOLUIDINE					738.90	356.02		
300	C7H14	CYCLOHEPTANE	-18.197	1.879 E-1	-1.004 E-4	1.806 E-8			-28.52	15.06
301	C7H14	1,1-DIMETHYLCYCLOPENT	-13.827	1.832 E-1	-1.075 E-4	2.413 E-8			-33.05	9.33
302	C7H14	CIS-1,2-DIMETHYLCYCLO	-13.290	1.819 E-1	-1.071 E-4	2.422 E-8			-30.96	10.93
303	C7H14	TRANS-1,2-DIMETHYLCYC	-13.022	1.813 E-1	-1.070 E-4	2.429 E-8			-32.67	9.17
304	C7H14	ETHYLCYCLOPENTANE	-13.211	1.794 E-1	-1.050 E-4	2.398 E-8	433.81	249.72	-30.37	10.65
305	C7H14	METHYLCYCLOHEXANE	-14.789	1.673 E-1	-1.060 E-4	2.237 E-8	528.41	271.58	-36.99	6.52
306	C7H14	1-HEPTENE	-0.789	1.504 E-1	-8.388 E-5	1.817 E-8	368.69	214.32	-14.89	22.90
307	C7H14	2,3,3-TRIMETHYL-1-BUT							-20.67	
308	C7H16	N-HEPTANE	-1.229	1.615 E-1	-8.720 E-5	1.829 E-8	436.73	232.53	-44.88	1.91
309	C7H16	2-METHYLHEXANE	-9.408	2.064 E-1	-1.502 E-4	4.386 E-8	417.46	225.13	-46.59	0.77
310	C7H16	3-METHYLHEXANE	-1.683	1.633 E-1	-8.919 E-5	1.871 E-8			-45.96	1.10
311	C7H16	2,2-DIMETHYLPENTANE	-11.966	2.139 E-1	-1.519 E-4	4.146 E-8	417.37	226.19	-49.27	0.02
312	C7H16	2,3-DIMETHYLPENTANE	-1.683	1.633 E-1	-8.919 E-5	1.871 E-8			-47.62	0.16
313	C7H16	2,4-DIMETHYLPENTANE	-1.683	1.633 E-1	-8.919 E-5	1.871 E-8			-48.28	0.74
314	C7H16	3,3-DIMETHYLPENTANE	-1.683	1.633 E-1	-8.919 E-5	1.871 E-8			-48.17	0.63
315	C7H16	3-ETHYLPENTANE	-1.683	1.633 E-1	-8.919 E-5	1.871 E-8			-45.33	2.63
316	C7H16	2,2,3-TRIMETHYLBUTANE	-5.480	1.796 E-1	-1.056 E-4	2.400 E-8			-48.95	1.02
317	C7H16O	1-HEPTANOL	1.172	1.619 E-1	-8.232 E-5	1.444 E-8	1287.0	361.83	-79.3	-28.9
318	C8H4O3	PHTHALIC ANHYDRIDE	-1.064	1.562 E-1	-1.023 E-4	2.411 E-8			-88.8	
319	C8H8	STYRENE	-6.747	1.471 E-1	-9.609 E-5	2.373 E-8	528.64	276.71	35.22	51.10
320	C8H8O	METHYL PHENYL KETONE	-7.065	1.531 E-1	-9.724 E-5	2.322 E-8	1316.4	310.82	-20.76	0.44

NO	FORMULA	NAME	ANTA	ANTB	ANTC	TMX	TMN	HARA	HARB	HARC	HARD	HV
281	C7F14	PERFLUOROMETHYLCYCLOH	15.7130	2510.57	-61.93	385	290	51.689	-5514.04	-5.004	5.47	
282	C7F16	PERFLUORO-N-HEPTANE	15.9747	2719.68	-64.50	390	270	83.896	-7348.95	-9.644	7.82	
283	C7H5N	BENZONITRILE						59.774	-7912.31	-5.881	6.53	
284	C7H6O	BENZALDEHYDE	16.3501	3748.62	-66.12	460	300					10200.
285	C7H6O2	BENZOIC ACID	17.1634	4190.70	-125.2	560	405					12100.
286	C7H8	TOLUENE	16.0137	3096.52	-53.67	410	280					7930.
287	C7H8O	METHYL PHENYL ETHER	16.2394	3430.82	-69.58	440	370	56.785	-6283.50	-5.681	4.84	
288	C7H8O	BENZYL ALCOHOL	17.4582	4384.81	-73.15	603	385					12070.
289	C7H8O	O-CRESOL	15.9148	3305.37	-108.0	480	370	75.616	-9341.59	-7.959	5.47	10800.
290	C7H8O	M-CRESOL	17.2878	4274.42	-74.09	480	370	79.796	-9885.80	-8.509	6.14	11330.
291	C7H8O	P-CRESOL	15.1989	3479.39	-111.3	480	370	64.083	-8825.19	-6.316	5.42	11340.
292	C7H9N	2,3-DIMETHYLPYRIDINE	17.1492	4219.74	-33.04	440	420					
293	C7H9N	2,5-DIMETHYLPYRIDINE	16.3046	3545.14	-63.59	435	350					
294	C7H9N	3,4-DIMETHYLPYRIDINE	16.9517	4237.04	-41.65	460	400					
295	C7H9N	3,5-DIMETHYLPYRIDINE	16.8850	4106.95	-44.45	460	400					
296	C7H9N	METHYLPHENYLAMINE	16.3066	3756.28	-80.71	480	320					10835.
297	C7H9N	O-TOLUIDINE	16.7834	4072.58	-72.15	500	375					10900.
298	C7H9N	M-TOLUIDINE	16.7498	4080.32	-73.15	500	355					10700.
299	C7H9N	P-TOLUIDINE	16.6968	4041.04	-72.15	500	350					7900.
300	C7H14	CYCLOHEPTANE	15.7818	3066.05	-56.80	435	330					7240.
301	C7H14	1,1-DIMETHYLCYCLOPENT	15.6973	2807.94	-51.20	390	260					7576.
302	C7H14	CIS-1,2-DIMETHYLCYCLO	15.7729	2922.30	-52.94	400	270					7375.
303	C7H14	TRANS-1,2-DIMETHYLCYC	15.7594	2861.53	-51.46	390	260					7715.
304	C7H14	ETHYLCYCLOPENTANE	15.8581	2990.13	-52.47	402	270					7440.
305	C7H14	METHYLCYCLOHEXANE	15.7105	2926.04	-51.75	400	265					7430.
306	C7H14	1-HEPTENE	15.8894	2895.51	-53.97	400	253	52.902	-5797.19	-5.199	5.23	6900.
307	C7H14	2,3,3-TRIMETHYL-1-BUT	15.6536	2719.47	-49.56	375	270	60.035	-6147.41	-6.211	5.70	7576.
308	C7H16	N-HEPTANE	15.8737	2911.32	-56.51	390	264	61.276	-6303.87	-6.373	6.00	7330.
309	C7H16	2-METHYLHEXANE	15.8261	2845.06	-53.60	390	265	60.131	-6074.01	-6.244	5.79	7360.
310	C7H16	3-METHYLHEXANE	15.8133	2855.66	-53.93	390	265	59.325	-6059.25	-6.123	5.72	6970.
311	C7H16	2,2-DIMETHYLPENTANE	15.6917	2740.15	-49.85	378	262	55.514	-5590.61	-5.636	5.49	7263.
312	C7H16	2,3-DIMETHYLPENTANE	15.7815	2850.64	-51.33	388	256	57.249	-5882.73	-5.843	5.58	7050.
313	C7H16	2,4-DIMETHYLPENTANE	15.7190	2744.78	-51.52	378	260					7086.
314	C7H16	3,3-DIMETHYLPENTANE	15.8317	2889.10	-47.83	385	266					7399.
315	C7H16	3-ETHYLPENTANE	15.6398	2882.44	-53.26	392	254	54.572	-5634.72	-5.487	5.49	6919.
316	C7H16	2,2,3-TRIMETHYLBUTANE		2764.42	-47.10	379						
317	C7H16O	1-HEPTANOL	15.3068	2626.42	-146.6	449	333	52.761	-5431.67	-5.251	5.37	11500.
318	C8H4O3	PHTHALIC ANHYDRIDE	15.9984	4467.01	-83.15	615	409					11850.
319	C8H8	STYRENE	16.0193	3328.57	-63.72	460	305					8800.
320	C8H8O	METHYL PHENYL KETONE	16.2384	3781.07	-81.15	520	350					

NO	FORMULA	NAME	MOLWT	TFP	TB	TC	PC	VC	ZC	OMEGA	LIQDEN	TDEN	DIPM
321	C8H8O2	METHYL BENZOATE	136.151	260.8	472.2	692.	36.8	396.	0.25	0.43	1.086	293.	1.9
322	C8H10	O-XYLENE	106.168	248.0	417.6	630.2	36.8	369.	0.263	0.314	0.880	293.	0.5
323	C8H10	M-XYLENE	106.168	225.3	412.3	617.0	35.0	376.	0.260	0.331	0.864	293.	0.3
324	C8H10	P-XYLENE	106.168	286.4	411.5	616.2	34.7	379.	0.260	0.324	0.861	293.	0.1
325	C8H10	ETHYLBENZENE	106.168	178.2	409.3	617.1	35.6	374.	0.263	0.301	0.867	293.	0.4
326	C8H10O	O-ETHYLPHENOL	122.167	269.8	477.7	703.0					1.037	273.	
327	C8H10O	M-ETHYLPHENOL	122.167	269.	491.6	716.4					1.025	273.	
328	C8H10O	P-ETHYLPHENOL	122.167	318.	491.	716.4							
329	C8H10O	PHENETOLE	122.167	243.	443.0	647.					0.979	277.	1.2
330	C8H10O	2,3-XYLENOL	122.167	348.	490.1	722.8							
331	C8H10O	2,4-XYLENOL	122.167	298.	484.0	707.6							2.0
332	C8H10O	2,5-XYLENOL	122.157	348.	484.3	723.0							1.5
333	C8H10O	2,6-XYLENOL	122.167	322.	474.1	701.							
334	C8H10O	3,4-XYLENOL	122.167	338.	500.	729.8							1.7
335	C8H10O	3,5-XYLENOL	122.167	337.	494.8	715.6							1.8
336	C8H11N	N,N-DIMETHYLANILINE	121.183	275.6	466.7	687.	35.8				0.956	293.	1.6
337	C8H16	1,1-DIMETHYLCYCLOHEXA	112.216	239.7	392.7	591.	29.3	416.	0.25	0.238	0.785	289.	
338	C8H16	CIS-1,2-DIMETHYLCYCLO	112.216	223.1	402.9	606.	29.3			0.236	0.796	293.	
339	C8H16	TRANS-1,2-DIMETHYLCYC	112.216	185.0	396.6	596.	29.3			0.242	0.776	293.	
340	C8H16	CIS-1,3-DIMETHYLCYCLO	112.216	197.6	393.3	591.	29.3			0.224	0.766	293.	
341	C8H16	TRANS-1,3-DIMETHYLCYC	112.216	183.0	397.8	598.	29.3			0.189	0.785	293.	
342	C8H16	CIS-1,4-DIMETHYLCYCLO	112.216	185.7	397.5	598.	29.3			0.234	0.783	293.	
343	C8H16	TRANS-1,4-DIMETHYLCYC	112.216	236.2	392.5	590.	29.3			0.242	0.763	293.	
344	C8H16	ETHYLCYCLOHEXANE	112.216	161.8	404.9	609.	29.9	450.	0.27	0.243	0.788	293.	0.0
345	C8H16	1,1,2-TRIMETHYLCYCLOP	112.216		386.9	579.5	29.0			0.252			
346	C8H16	1,1,3-TRIMETHYLCYCLOP	112.216		378.0	569.5	27.9			0.211			
347	C8H16	CIS,CIS,TRANS-1,2,4-T	112.216		391.	579.	28.4			0.277			
348	C8H16	CIS,TRANS,CIS-1,2,4-T	112.216		382.4	571.	27.7			0.246			
349	C8H16	1-METHYL-1-ETHYLCYCLO	112.216		394.7	592.	29.5			0.250			
350	C8H16	N-PROPYLCYCLOPENTANE	112.216	155.8	404.1	603.	29.6	425.	0.25	0.335	0.781	289.	
351	C8H16	ISOPROPYLCYCLOPENTANE	112.216	160.5	399.6	601.	29.6			0.240	0.776	293.	
352	C8H16	1-OCTENE	112.216	171.4	394.4	566.6	25.9	464.	0.26	0.386	0.715	293.	0.3
353	C8H16	TRANS-2-OCTENE	112.216	185.4	398.1	580.	27.3			0.350	0.720	293.	
354	C8H18	N-OCTANE	114.232	216.4	398.8	568.8	24.5	492.	0.259	0.394	0.703	289.	0.0
355	C8H18	2-METHYLHEPTANE	114.232	164.	390.8	559.6	24.5	488.	0.260	0.378	0.702	289.	
356	C8H18	3-METHYLHEPTANE	114.232	152.7	392.1	563.6	25.1	464.	0.252	0.369	0.706	293.	
357	C8H18	4-METHYLHEPTANE	114.232	152.2	390.9	561.7	25.1	476.	0.259	0.369	0.705	293.	
358	C8H18	2,2-DIMETHYLHEXANE	114.232	152.	382.0	549.8	25.1	478.	0.264	0.338	0.695	293.	
359	C8H18	2,3-DIMETHYLHEXANE	114.232		388.0	563.4	25.9	468.	0.262	0.346	0.712	293.	
360	C8H18	2,4-DIMETHYLHEXANE	114.232		382.6	553.5	25.2	472.	0.262	0.343	0.700	293.	

NO	FORMULA	NAME	CPVAP A	CPVAP B	CPVAP C	CPVAP D	VISB	VISTO	DELHG	DELGF
321	C8H8O2	METHYL BENZOATE	-5.066	1.314 E-1	-4.298 E-5	1.057 E-8	768.94	332.33	-60.68	29.18
322	C8H10	O-XYLENE	-3.786	1.424 E-1	-8.224 E-5	1.798 E-8	513.54	277.98	4.54	28.41
323	C8H10	M-XYLENE	-6.966	1.504 E-1	-8.950 E-5	2.025 E-8	453.42	257.18	4.12	28.41
324	C8H10	P-XYLENE	-5.993	1.443 E-1	-8.058 E-5	1.629 E-8	475.16	261.40	4.29	28.95
325	C8H10	ETHYLBENZENE	-10.294	1.689 E-1	-1.149 E-4	3.107 E-8	472.82	264.22	7.12	31.21
326	C8H10O	O-ETHYLPHENOL							-34.82	
327	C8H10O	M-ETHYLPHENOL							-35.01	
328	C8H10O	P-ETHYLPHENOL							-34.55	
329	C8H10O	PHENETOLE					646.88	305.91		
330	C8H10O	2,3-XYLENOL							-37.58	
331	C8H10O	2,4-XYLENOL							-38.88	
332	C8H10O	2,5-XYLENOL							-38.58	
333	C8H10O	2,6-XYLENOL							-38.68	
334	C8H10O	3,4-XYLENOL							-37.38	
335	C8H10O	3,5-XYLENOL							-38.57	
336	C8H11N	N,N-DIMETHYLANILINE	-17.222	2.149 E-1	-1.199 E-4	2.461 E-8	553.02	320.03	20.10	55.26
337	C8H16	1,1-DIMETHYLCYCLOHEXA	-16.330	2.143 E-1	-1.227 E-4	2.624 E-8			-43.26	8.42
338	C8H16	CIS-1,2-DIMETHYLCYCLO	-16.356	2.179 E-1	-1.279 E-4	2.821 E-8			-41.15	9.85
339	C8H16	TRANS-1,2-DIMETHYLCYC	-15.554	2.111 E-1	-1.178 E-4	2.436 E-8			-43.02	8.24
340	C8H16	CIS-1,3-DIMETHYLCYCLO	-15.323	2.108 E-1	-1.198 E-4	2.552 E-8			-44.16	7.13
341	C8H16	TRANS-1,3-DIMETHYLCYCLO	-15.323	2.108 E-1	-1.198 E-4	2.552 E-8			-42.20	8.68
342	C8H16	CIS-1,4-DIMETHYLCYCLO	-16.806	2.181 E-1	-1.268 E-4	2.758 E-8			-42.22	9.07
343	C8H16	TRANS-1,4-DIMETHYLCYC	-15.260	2.124 E-1	-1.220 E-4	2.634 E-8			-44.12	7.58
344	C8H16	ETHYLCYCLOHEXANE					506.43	280.76	-41.05	9.38
345	C8H16	1,1,2-TRIMETHYLCYCLOP								
346	C8H16	1,1,3-TRIMETHYLCYCLOP								
347	C8H16	CIS,CIS,TRANS-1,2,4-T								
348	C8H16	CIS,TRANS,CIS-1,2,4-T								
349	C8H16	1-METHYL-1-ETHYLCYCLO								
350	C8H16	N-PROPYLCYCLOPENTANE	-13.359	2.018 E-1	-1.176 E-4	2.669 E-8	454.23	264.22	-35.39	12.57
351	C8H16	ISOPROPYLCYCLOPENTANE								
352	C8H16	1-OCTENE	-0.979	1.729 E-1	-9.641 E-5	2.072 E-8	418.82	237.63	-19.82	24.91
353	C8H16	TRANS-2-OCTENE	-3.062	1.799 E-1	-1.061 E-4	2.509 E-8	427.64	240.32	-22.59	22.15
354	C8H18	N-OCTANE	-1.456	1.842 E-1	-1.002 E-4	2.115 E-8	473.70	251.71	-49.82	3.92
355	C8H18	2-METHYLHEPTANE	-21.435	2.967 E-1	-2.808 E-4	1.103 E-7	643.61	259.51	-51.50	3.05
356	C8H18	3-METHYLHEPTANE	-2.201	1.877 E-1	-1.051 E-4	2.316 E-8			-50.82	3.28
357	C8H18	4-METHYLHEPTANE	-2.201	1.877 E-1	-1.051 E-4	2.316 E-8			-50.69	4.00
358	C8H18	2,2-DIMETHYLHEXANE	-2.201	1.877 E-1	-1.051 E-4	2.316 E-8			-53.71	2.56
359	C8H18	2,3-DIMETHYLHEXANE	-2.201	1.877 E-1	-1.051 E-4	2.316 E-8			-51.13	4.23
360	C8H18	2,4-DIMETHYLHEXANE	-2.201	1.877 E-1	-1.051 E-4	2.316 E-8			-52.44	2.80

NO	FORMULA	NAME	ANTA	ANTB	ANTC	TMX	TMN	HARA	HARB	HARC	HARD	HV
321	C8H8O2	METHYL BENZOATE	16.2272	3751.83	-81.15	516	350	61.763	-7149.21	-6.302	6.11	10300.
322	C8H10	O-XYLENE	16.1156	3395.57	-59.46	445	305	55.493	-6666.23	-5.436	6.08	8800.
323	C8H10	M-XYLENE	16.1390	3366.99	-58.04	440	300	56.175	-6673.70	-5.543	6.19	8690.
324	C8H10	P-XYLENE	16.0963	3346.65	-57.84	440	300	58.100	-6792.54	-5.802	5.75	8600.
325	C8H10	ETHYLBENZENE	16.0195	3279.47	-59.95	450	300					8500.
326	C8H10O	O-ETHYLPHENOL	17.9610	4928.36	-45.75	500	350					11490.
327	C8H10O	M-ETHYLPHENOL	17.1955	4272.77	-86.08	500	370					12140.
328	C8H10O	P-ETHYLPHENOL	19.0905	5579.62	-44.15	500	370					12100.
329	C8H10O	PHENETOLE	16.1673	3473.20	-78.66	460	385					
330	C8H10O	2,3-XYLENOL	16.2424	3724.58	-102.4	500	420					11300.
331	C8H10O	2,4-XYLENOL	16.2456	3655.26	-103.8	500	410					11260.
332	C8H10O	2,5-XYLENOL	16.2368	3667.32	-102.4	490	410					11200.
333	C8H10O	2,6-XYLENOL	16.2809	3749.35	-85.55	480	400					10600.
334	C8H10O	3,4-XYLENOL	16.3004	3733.53	-113.9	520	430					11900.
335	C8H10O	3,5-XYLENOL	16.4192	3775.91	-109.0	500	410					11800.
336	C8H11N	N,N-DIMETHYLANILINE	16.9647	4276.08	-52.80	480	345	52.143	-6026.09	-5.055	6.20	7790.
337	C8H16	1,1-DIMETHYLCYCLOHEXA	15.6535	3043.34	-55.30	420	283	56.130	-6641.82	-5.595	6.56	8040.
338	C8H16	CIS-1,2-DIMETHYLCYCLO	15.7438	3117.43	-57.31	430	290	53.523	-6162.66	-5.245	6.38	7860.
339	C8H16	TRANS-1,2-DIMETHYLCYC	15.7337	3081.95	-54.02	424	286					7840.
340	C8H16	CIS-1,3-DIMETHYLCYCLO	15.7470	3093.95	-55.08	420	284	56.097	-6271.67	-5.615	6.29	8090.
341	C8H16	TRANS-1,3-DIMETHYLCYC	15.7371	3098.39	-57.76	425	288	53.571	-6219.26	-5.233	6.29	8070.
342	C8H16	CIS-1,4-DIMETHYLCYCLO	15.7333	3063.44	-57.00	420	287	52.909	-6071.72	-5.163	6.20	7790.
343	C8H16	TRANS-1,4-DIMETHYLCYC	15.6984		-54.57	433	283					8200.
344	C8H16	ETHYLCYCLOHEXANE	15.8125	3183.25	-56.15	414	293					7790.
345	C8H16	1,1,2-TRIMETHYLCYCLOP	15.7084	3015.51	-54.59	404	279					7570.
346	C8H16	1,1,3-TRIMETHYLCYCLOP	15.6794	2938.09	-53.25	418	273					7900.
347	C8H16	CIS,CIS,TRANS-1,2,4-T	15.7543	3073.95	-54.42	417	283					7900.
348	C8H16	CIS,TRANS,CIS-1,2,4-T	15.7756	3009.70	-53.23	422	282					8040.
349	C8H16	1-METHYL-1-ETHYLCYCLO	15.8222	3120.66	-55.06	431	286					8152.
350	C8H16	N-PROPYLCYCLOPENTANE	15.8969	3187.67	-59.99	427	294					8150.
351	C8H16	ISOPROPYLCYCLOPENTANE	15.8561	3176.22	-55.18	420	289					8070.
352	C8H16	1-OCTENE	15.9630	3116.52	-60.39	425	289	64.487	-6883.34	-6.765	6.98	8200.
353	C8H16	TRANS-2-OCTENE	15.8554	3134.97	-58.00	425	289					
354	C8H18	N-OCTANE	15.9426	3120.29	-63.63	425	292	66.639	-7100.69	-7.053	7.31	8225.
355	C8H18	2-METHYLHEPTANE	15.9278	3079.63	-59.46	417	285	65.685	-6865.40	-6.957	7.12	8080.
356	C8H18	3-METHYLHEPTANE	15.8865	3065.96	-60.74	418	286	64.371	-6817.44	-6.763	7.02	8100.
357	C8H18	4-METHYLHEPTANE	15.8893	3057.05	-60.59	417	285	64.394	-6799.54	-6.759	6.98	8100.
358	C8H18	2,2-DIMETHYLHEXANE	15.7431	2932.56	-58.08	405	276	61.971	-6425.90	-6.475	6.72	7710.
359	C8H18	2,3-DIMETHYLHEXANE	15.8189	3029.06	-58.99	415	283	61.855	-6587.23	-6.425	6.79	7936.
360	C8H18	2,4-DIMETHYLHEXANE	15.7797	2965.44	-58.36	408	278	62.103	-6487.48	-6.482	6.74	7790.

NO	FORMULA	NAME	MOLWT	TFP	TB	TC	PC	VC	ZC	OMEGA	LIQDEN	TDEN	DIPM
361	C8H18	2,5-DIMETHYLHEXANE	114.232	181.9	382.3	550.0	24.5	482.	0.262	0.352	0.693	293.	
362	C8H18	3,3-DIMETHYLHEXANE	114.232	147.	385.1	562.0	26.2	443.	0.252	0.321	0.710	293.	
363	C8H18	3,4-DIMETHYLHEXANE	114.232		390.9	568.8	26.6	466.	0.265	0.338	0.719	293.	
364	C8H18	3-ETHYLHEXANE	114.232		391.7	565.4	25.7	455.	0.252	0.361	0.716	289.	0.0
365	C8H18	2,2,3-TRIMETHYLPENTAN	114.232	160.9	383.0	563.4	26.9	436.	0.254	0.297	0.692	293.	
366	C8H18	2,2,4-TRIMETHYLPENTAN	114.232	165.8	372.4	543.9	25.3	468.	0.266	0.303	0.692	293.	
367	C8H18	2,3,3-TRIMETHYLPENTAN	114.232	172.5	387.9	573.5	27.8	455.	0.269	0.290	0.726	293.	
368	C8H18	2,3,4-TRIMETHYLPENTAN	114.232	163.9	386.6	566.3	26.7	461.	0.267	0.317	0.719	293.	
369	C8H18	2-METHYL-3-ETHYLPENTA	114.232	158.2	388.8	567.0	26.7	443.	0.254	0.330	0.719	293.	
370	C8H18	3-METHYL-3-ETHYLPENTA	114.232	182.3	391.4	576.5	27.7	455.	0.267	0.304	0.727	293.	
371	C8H18O	1-OCTANOL	130.231	257.7	468.4	658.	34.	490.	0.31	0.53	0.826	293.	2.0
372	C8H18O	2-OCTANOL	130.231	241.2	452.9	637.	27.	494.	0.26	0.52	0.821	293.	1.6
373	C8H18O	2-ETHYLHEXANOL	130.231	203.2	457.8	613.	27.2	494.	0.267		0.833	293.	1.8
374	C8H18O	BUTYL ETHER	130.231	175.3	415.6	580.	25.	500.	0.25	0.50	0.768	293.	1.2
375	C8H19N	DIBUTYLAMINE	129.247	211.	432.8	596.	25.	517.	0.26	0.59	0.767	293.	1.1
376	C9H10	ALPHA-METHYL STYRENE	118.179	238.3	438.5	654.	33.6	397.	0.25	0.48	0.911	293.	1.9
377	C9H10O2	ETHYL BENZOATE	150.178	173.7	485.9	697.	32.	451.	0.25	0.344	1.046	293.	0.4
378	C9H12	N-PROPYLBENZENE	120.195	177.1	432.4	638.3	31.6	440.	0.265	0.335	0.862	293.	0.4
379	C9H12	ISOPROPYLBENZENE	120.195	192.3	425.6	631.0	31.7	428.	0.26	0.294	0.881	293.	
380	C9H12	1-METHYL-2-ETHYLBENZE	120.195	177.6	438.3	651.	30.0	460.	0.26	0.360	0.881	293.	
381	C9H12	1-METHYL-3-ETHYLBENZE	120.195	177.6	434.5	637.	28.0	490.	0.26	0.322	0.865	293.	
382	C9H12	1-METHYL-4-ETHYLBENZE	120.195	210.8	435.2	640.	29.0	470.	0.26	0.322	0.861	293.	
383	C9H12	1,2,3-TRIMETHYLBENZEN	120.195	247.7	449.2	664.5	34.1	430.	0.27	0.39	0.894	293.	0.0
384	C9H12	1,2,4-TRIMETHYLBENZEN	120.195	227.	442.5	649.1	31.9	430.	0.258	0.398	0.880	289.	0.6
385	C9H12	1,3,5-TRIMETHYLBENZEN	120.195	228.4	437.9	637.3	30.9	433.	0.26	0.398	0.865	293.	0.3
386	C9H18	N-PROPYLCYCLOHEXANE	126.243	178.7	429.9	639.	27.7			0.258	0.793	293.	0.1
387	C9H18	ISOPROPYLCYCLOHEXANE	126.243	183.4	427.7	640.	28.			0.237	0.802	293.	
388	C9H18	1-NONENE	126.243	191.8	420.0	592.	23.1	580.	0.28	0.430	0.745	273.	0.0
389	C9H20	N-NONANE	128.259	219.7	424.0	594.6	22.8	548.	0.26	0.444	0.718	293.	
390	C9H20	2,2,3-TRIMETHYLHEXANE	128.259		406.8	588.	24.6			0.332	0.720	289.	
391	C9H20	2,2,4-TRIMETHYLHEXANE	128.259	153.	399.7	573.7	23.4			0.321	0.717	289.	
392	C9H20	2,2,5-TRIMETHYLHEXANE	128.259	167.4	397.3	568.	23.0	519.	0.26	0.357	0.752	293.	0.0
393	C9H20	3,3-DIETHYLPENTANE	128.259		419.3	610.	26.4			0.338			
394	C9H20	2,2,3,3-TETRAMETHYLPE	128.259		413.4	607.6	27.0			0.279			
395	C9H20	2,2,3,4-TETRAMETHYLPE	128.259		406.2	592.7	25.7			0.311			
396	C9H20	2,2,4,4-TETRAMETHYLPE	128.259	206.	395.4	577.7	24.6			0.315	0.719	293.	0.0
397	C9H20	2,3,3,4-TETRAMETHYLPE	128.259		414.7	607.4	26.8			0.299			
398	C10H8	NAPHTHALENE	128.174	353.5	491.1	748.4	40.0	410.	0.267	0.302	0.971	363.	0.0
399	C10H12	1,2,3,4-TETRAHYDRONAP	132.206	242.	480.7	719.	34.7			0.303	0.973	293.	
400	C10H14	N-BUTYLBENZENE	134.222	185.2	456.4	660.5	28.5	497.	0.261	0.392	0.860	293.	0.4

NO	FORMULA	NAME	CPVAP A	CPVAP B	CPVAP C	CPVAP D	VISB	VISTO	DELHG	DELGF
361	C8H18	2,5-DIMETHYLHEXANE	-2.201	1.877 E-1	-1.051 E-4	2.316 E-8			-53.21	2.50
362	C8H18	3,3-DIMETHYLHEXANE	-2.201	1.877 E-1	-1.051 E-4	2.316 E-8	446.20	244.67	-52.61	3.17
363	C8H18	3,4-DIMETHYLHEXANE	-2.201	1.877 E-1	-1.051 E-4	2.316 E-8	437.60	238.33	-50.91	4.14
364	C8H18	3-ETHYLHEXANE	-2.201	1.877 E-1	-1.051 E-4	2.316 E-8	474.57	257.61	-50.40	3.95
365	C8H18	2,2,3-TRIMETHYLPENTAN	-1.782	1.858 E-1	-1.024 E-4	2.191 E-8	467.04	246.43	-52.61	4.09
366	C8H18	2,2,4-TRIMETHYLPENTAN	-2.201	1.877 E-1	-1.051 E-4	2.316 E-8			-53.57	3.27
367	C8H18	2,3,3-TRIMETHYLPENTAN	-2.201	1.877 E-1	-1.051 E-4	2.316 E-8			-51.73	4.52
368	C8H18	2,3,4-TRIMETHYLPENTAN	-2.201	1.877 E-1	-1.051 E-4	2.316 E-8			-51.97	4.52
369	C8H18	2-METHYL-3-ETHYLPENTA	-2.201	1.877 E-1	-1.051 E-4	2.316 E-8			-50.48	5.08
370	C8H18	3-METHYL-3-ETHYLPENTA	-2.201	1.877 E-1	-1.051 E-4	2.316 E-8			-51.38	4.76
371	C8H18O	1-OCTANOL	1.474	1.817 E-1	-9.069 E-5	1.496 E-8	1312.1	369.97	-86.0	-28.7
372	C8H18O	2-OCTANOL	6.181	1.825 E-1	-1.009 E-4	2.165 E-8				
373	C8H18O	2-ETHYLHEXANOL	-3.581	2.067 E-1	-1.261 E-4	3.068 E-8	1798.0	351.17	-87.31	-21.16
374	C8H18O	BUTYL ETHER	1.446	1.846 E-1	-9.758 E-5	1.931 E-8	473.50	266.56	-79.80	
375	C8H19N	DIBUTYLAMINE	2.332	1.930 E-1	-1.049 E-4	2.209 E-8	581.42	286.54		
376	C9H10	ALPHA-METHYL STYRENE	-5.811	1.656 E-1	-1.082 E-4	2.820 E-8	354.34	270.80		32.80
377	C9H10O2	ETHYL BENZOATE	4.937	1.645 E-1	-8.618 E-5	1.209 E-8	746.50	338.47		32.74
378	C9H12	ISOPROPYLBENZENE	-7.473	1.788 E-1	-1.099 E-4	2.582 E-8	527.45	282.65	1.87	31.33
379	C9H12	N-PROPYLBENZENE	-9.402	1.873 E-1	-1.215 E-4	3.084 E-8	517.17	276.22	0.94	30.22
380	C9H12	1-METHYL-2-ETHYLBENZE	-3.928	1.671 E-1	-9.841 E-5	2.228 E-8			0.29	30.28
381	C9H12	1-METHYL-3-ETHYLBENZE	-6.926	1.742 E-1	-1.042 E-4	2.388 E-8			-0.46	29.77
382	C9H12	1-METHYL-4-ETHYLBENZE	-6.523	1.714 E-1	-1.009 E-4	2.679 E-8	463.17	266.08	-0.49	27.95
383	C9H12	1,2,3-TRIMETHYLBENZEN	-1.658	1.513 E-1	-7.945 E-5	1.579 E-8	872.74	297.75	-2.29	28.19
384	C9H12	1,2,4-TRIMETHYLBENZEN	-1.115	1.490 E-1	-7.793 E-5	1.523 E-8	437.52	268.27	-3.33	11.31
385	C9H12	1,3,5-TRIMETHYLBENZEN	-4.679	1.606 E-1	-8.819 E-5	1.839 E-8	549.08	293.93	-3.84	
386	C9H18	N-PROPYLCYCLOHEXANE	-14.932	2.362 E-1	-1.384 E-4	3.084 E-8			-46.20	
387	C9H18	ISOPROPYLCYCLOHEXANE	-0.888	1.940 E-1	-1.077 E-4	2.318 E-8				
388	C9H18	1-NONENE	0.751	1.618 E-1	-4.606 E-5	-7.121 E-9	471.00	258.92	-24.74	26.93
389	C9H20	N-NONANE	-10.899	2.521 E-1	-1.713 E-4	4.745 E-8	525.56	272.12	-54.74	5.93
390	C9H20	2,2,3-TRIMETHYLHEXANE	-14.405	2.638 E-1	-1.842 E-4	5.225 E-8			-57.65	5.86
391	C9H20	2,2,4-TRIMETHYLHEXANE	-12.923	2.615 E-1	-1.850 E-4	5.385 E-8			-58.13	5.38
392	C9H20	2,2,5-TRIMETHYLHEXANE	-16.067	2.690 E-1	-1.908 E-4	5.508 E-8			-60.71	3.21
393	C9H20	3,3-DIETHYLPENTANE	-13.037	2.601 E-1	-1.808 E-4	5.116 E-8				8.38
394	C9H20	2,2,3,3-TETRAMETHYLPE	-13.037	2.601 E-1	-1.808 E-4	5.116 E-8			-56.70	8.20
395	C9H20	2,2,3,4-TETRAMETHYLPE	-16.099	2.790 E-1	-2.057 E-4	6.147 E-8			-56.64	7.80
396	C9H20	2,2,4,4-TETRAMETHYLPE	-13.117	2.606 E-1	-1.816 E-4	5.154 E-8			-57.83	8.13
397	C9H20	2,3,3,4-TETRAMETHYLPE							-56.46	8.15
398	C10H8	NAPHTHALENE	-16.433	2.030 E-1	-1.554 E-4	4.731 E-8			36.08	53.44
399	C10H12	1,2,3,4-TETRAHYDRONAP	-5.491	1.895 E-1	-1.050 E-4	2.047 E-8	873.32	352.57	6.6	39.90
400	C10H14	N-BUTYLBENZENE					563.84	296.01	-3.30	34.58

NO	FORMULA	NAME	ANTA	ANTB	ANTC	TMX	TMN	HARA	HARB	HARC	HARD	HV
361	C8H18	2,5-DIMETHYLHEXANE	15.7954	2964.06	-58.74	408	278	62.872	-6532.90	-6.590	6.84	7800.
362	C8H18	3,3-DIMETHYLHEXANE	15.7755	3011.51	-55.71	411	279	59.518	-6352.78	-6.118	6.69	7760.
363	C8H18	3,4-DIMETHYLHEXANE	15.8415	3062.52	-58.29	417	284	61.319	-6588.72	-6.344	6.76	7953.
364	C8H18	3-ETHYLHEXANE	15.8671	3057.57	-60.55	418	286					8033.
365	C8H18	2,2,3-TRIMETHYLPENTAN	15.7162	2981.56	-54.73	409	277	58.179	-6218.74	-5.942	6.54	7650.
366	C8H18	2,2,4-TRIMETHYLPENTAN	15.6850	2899.28	-52.41	398	269	58.265	-6039.34	-5.988	6.48	7411.
367	C8H18	2,3,3-TRIMETHYLPENTAN	15.7578	3057.94	-52.77	415	280	56.436	-6186.92	-5.685	6.56	7730.
368	C8H18	2,3,4-TRIMETHYLPENTAN	15.7818	3028.09	-55.62	413	280	58.957	-6346.90	-6.003	6.61	7823.
369	C8H18	2-METHYL-3-ETHYLPENTA	15.8040	3035.08	-57.84	415	282					7879.
370	C8H18	3-METHYL-3-ETHYLPENTA	15.8126	3102.06	-53.47	418	283					7838.
371	C8H18O	1-OCTANOL	15.7428	3017.81	-137.1	468	343					12100.
372	C8H18O	2-OCTANOL	14.7108	2441.66	-150.7	453	345					10600.
373	C8H18O	2-ETHYLHEXANOL	15.3614	2773.46	-140.0	458	348					11130.
374	C8H18O	BUTYL ETHER	16.0778	3296.15	-66.15	455	305					8900.
375	C8H19N	DIBUTYLAMINE	16.7307	3721.90	-64.15	459	322					9500.
376	C9H10	ALPHA-METHYL STYRENE	16.3308	3644.30	-67.15	493	348					9150.
377	C9H10O2	ETHYL BENZOATE	16.2065	3845.09	-84.15	531	361					10700.
378	C9H12	N-PROPYLBENZENE	16.0062	3433.84	-66.01	463	316					9140.
379	C9H12	ISOPROPYLBENZENE	15.9722	3363.60	-63.37	454	311	46.941	-6285.25	-4.227	6.86	8970.
380	C9H12	1-METHYL-2-ETHYLBENZE	16.1253	3535.33	-65.85	467	321	64.337	-7662.94	-6.617	7.18	9290.
381	C9H12	1-METHYL-3-ETHYLBENZE	16.1545	3521.08	-64.64	463	318	65.670	-7678.11	-6.815	7.20	9210.
382	C9H12	1-METHYL-4-ETHYLBENZE	16.1135	3516.31	-64.23	463	318	61.404	-7422.59	-6.212	7.23	9180.
383	C9H12	1,2,3-TRIMETHYLBENZEN	16.2121	3670.22	-66.07	479	329					9570.
384	C9H12	1,2,4-TRIMETHYLBENZEN	16.2190	3622.58	-64.59	471	324	56.241	-7256.56	-5.459	7.27	9330.
385	C9H12	1,3,5-TRIMETHYLBENZEN	16.2893	3614.19	-63.57	466	321	58.041	-7326.78	-5.706	7.22	9330.
386	C9H18	N-PROPYLCYCLOHEXANE	15.8567	3363.62	-65.21	459	330					8820.
387	C9H18	ISOPROPYLCYCLOHEXANE	15.8260	3346.12	-63.71	440	330					
388	C9H18	1-NONENE	16.0118	3305.03	-67.61	448	308	69.085	-7626.91	-7.339	8.38	8680.
389	C9H18	N-NONANE	15.9671	3291.45	-71.33	452	312	73.133	-7969.42	-7.89	8.69	8823.
390	C9H20	2,2,3-TRIMETHYLHEXANE	15.8017	3164.17	-61.66	436	297					8310.
391	C9H20	2,2,4-TRIMETHYLHEXANE	15.7639	3084.08	-61.94	428	291					8130.
392	C9H20	2,2,5-TRIMETHYLHEXANE	15.7445	3052.17	-62.24	420	315					8070.
393	C9H20	3,3-DIETHYLPENTANE	15.8709	3341.62	-57.57	440	350					8600.
394	C9H20	2,2,3,3-TETRAMETHYLPE	15.7280	3220.55	-59.31	440	328	64.104	-7011.38	-6.731	8.46	8430.
395	C9H20	2,2,3,4-TETRAMETHYLPE	15.7363	3167.42	-58.21	430	328					8190.
396	C9H20	2,2,4,4-TETRAMETHYLPE	15.6648	3049.98	-57.13	413	313					7850.
397	C9H20	2,3,3,4-TETRAMETHYLPE	15.8029	3269.07	-58.19	425	325					8350.
398	C10H8	NAPHTHALENE	16.1426	3992.01	-71.29	525	360					10340.
399	C10H12	1,2,3,4-TETRAHYDRONAP	16.2805	4009.49	-64.89	500	365					9490.
400	C10H14	N-BUTYLBENZENE	16.0793	3633.40	-71.77	486	335					9380.

NO	FORMULA	NAME	MOLWT	TFP	TB	TC	PC	VC	ZC	OMEGA	LIQDEN	TDEN	DIPM
401	C10H14	ISOBUTYLBENZENE	134.222	221.7	445.9	650.	31.	480.		0.378	0.853	293.	0.3
402	C10H14	SEC-BUTYLBENZENE	134.222	197.7	446.5	664.	29.1		0.28	0.274	0.862	293.	0.4
403	C10H14	TERT-BUTYLBENZENE	134.222	215.3	442.3	660.	29.3			0.265	0.867	293.	0.5
404	C10H14	1-METHYL-2-ISOPROPYLB	134.222		451.5	670.	28.6			0.277	0.876	293.	
405	C10H14	1-METHYL-3-ISOPROPYLB	134.222		448.3	666.	29.0			0.279	0.861	293.	
406	C10H14	1-METHYL-4-ISOPROPYLB	134.222		450.3	653.	27.9			0.371	0.857	293.	
407	C10H14	1,4-DIETHYLBENZENE	134.222	200.	456.9	657.9	27.7	480.	0.25	0.403	0.862	293.	0.0
408	C10H14	1,2,4,5-TETRAMETHYLBE	134.222	231.	470.0	675.	29.	480.	0.25	0.426	0.838	354.	0.1
409	C10H15N	N-BUTYLANILINE	149.236	352.	513.9	721.	28.	518.	0.25		0.932	293.	
410	C10H18	CIS-DECALIN	138.254	259.	468.9	702.2	31.		0.23	0.23	0.897	293.	0.0
411	C10H18	TRANS-DECALIN	138.254	230.	460.4	690.0	31.		0.27	0.27	0.870	293.	0.0
412	C10H19N	CAPRYLONITRILE	153.269	242.8	516.	622.0	32.1			0.362	0.820	293.	
413	C10H20	N-BUTYLCYCLOHEXANE	140.270	255.3	454.1	667.	31.			0.319	0.799	293.	
414	C10H20	ISOBUTYLCYCLOHEXANE	140.270	198.4	452.5	659.	30.8			0.264	0.795	293.	
415	C10H20	SEC-BUTYLCYCLOHEXANE	140.270		444.7	669.	26.4			0.252	0.813	293.	
416	C10H20	TERT-BUTYLCYCLOHEXANE	140.270	232.0	444.7	659.	26.3				0.813	293.	0.0
417	C10H20	1-DECENE	140.270	206.9	443.7	615.	21.8	650.	0.28	0.491	0.741	293.	
418	C10H22	N-DECANE	142.286	243.5	447.3	617.6	20.8	603.	0.247	0.490	0.730	293.	0.0
419	C10H22	3,3,5-TRIMETHYLHEPTAN	142.286		428.8	609.6	22.9			0.388			
420	C10H22	2,2,3,3-TETRAMETHYLHE	142.286		433.5	623.1	24.8			0.360			
421	C10H22	2,2,5,5-TETRAMETHYLHE	142.286		410.6	581.5	21.6			0.374			
422	C10H22O	1-DECANOL	158.285	280.1	503.4	700.	22.	600.	0.23		0.830	293.	1.8
423	C11H10	1-METHYLNAPHTHALENE	142.201	242.7	517.8	772.	35.2	445.	0.25	0.334	1.020	293.	0.5
424	C11H10	2-METHYLNAPHTHALENE	142.201	307.7	514.2	761.	34.6	462.	0.26	0.382	0.990	313.	0.4
425	C11H14O2	BUTYL BENZOATE	178.232	251.	523.	723.	26.	561.	0.25	0.58	1.006	293.	
426	C11H22	N-HEXYLCYCLOPENTANE	154.297	224.	476.3	660.1	21.1			0.476	0.751	293.	
427	C11H22	1-UNDECENE	154.297	247.6	465.8	637.	19.7	660.	0.24	0.518	0.740	293.	0.0
428	C11H24	N-UNDECANE	156.313		469.1	638.8	19.4			0.535		293.	
429	C12H10	DIPHENYL	154.212	342.4	528.4	789.	38.	502.	0.295	0.364	0.990	347.	0.0
430	C12H10O	DIPHENYL ETHER	170.211	300.	531.2	766.	31.			0.444	1.066	303.	1.1
431	C12H24	N-HEPTYLCYCLOPENTANE	168.324		497.3	679.	19.2			0.515	0.758	293.	
432	C12H24	1-DODECENE	168.324	238.0	486.5	657.	18.3			0.558	0.748	293.	
433	C12H26	N-DODECANE	170.340	263.6	489.5	658.3	18.0			0.562	0.794	293.	0.0
434	C12H26O	DIHEXYL ETHER	186.339	230.	499.6	657.	18.	713.	0.24	0.70	0.835	293.	
435	C12H26O	DODECANOL	186.339	297.1	533.1	679.	19.	720.	0.24		0.779	293.	1.6
436	C12H27N	TRIBUTYLAMINE	185.355		486.6	643.	18.	718.	0.24			293.	0.8
437	C13H12	DIPHENYLMETHANE	168.239	300.	537.5	767.	29.4			0.471	1.006	293.	0.4
438	C13H26	N-OCTYLCYCLOPENTANE	182.351		516.9	694.	17.7			0.564		293.	
439	C13H26	1-TRIDECENE	182.351	250.1	505.9	674.	16.8			0.598	0.766	293.	
440	C13H28	N-TRIDECANE	184.367	267.8	508.6	675.8	17.0	780.	0.24	0.623	0.756	293.	

NO	FORMULA	NAME	CPVAP A	CPVAP B	CPVAP C	CPVAP D	VISB	VISTO	DELHG	DELGF
401	C10H14	ISOBUTYLBENZENE	-15.560	2.363 E-1	-1.723 E-4	5.140 E-8	582.66	295.82	-5.15	
402	C10H14	SEC-BUTYLBENZENE	-20.541	2.632 E-1	-2.089 E-4	6.751 E-8			-4.17	
403	C10H14	TERT-BUTYLBENZENE							-5.42	
404	C10H14	1-METHYL-2-ISOPROPYLB								
405	C10H14	1-METHYL-3-ISOPROPYLB	-11.646	2.165 E-1	-1.446 E-4	3.887 E-8			-7.00	
406	C10H14	1-METHYL-4-ISOPROPYLB								
407	C10H14	1,4-DIETHYLBENZENE	-8.937	2.071 E-1	-1.328 E-4	3.370 E-8			-5.32	32.95
408	C10H14	1,2,4,5-TETRAMETHYLBE	3.946	1.557 E-1	-6.876 E-5	7.779 E-9			-10.82	28.55
409	C10H15N	N-BUTYLANILINE	-8.137	2.184 E-1	-1.328 E-4	3.075 E-8	1111.1	341.28		
410	C10H18	CIS-DECALIN	-26.860	2.671 E-1	-1.578 E-4	3.432 E-8	702.27	339.66	-40.38	20.51
411	C10H18	TRANS-DECALIN	-23.328	2.495 E-1	-1.308 E-4	2.145 E-8			-43.57	17.55
412	C10H19N	CAPRYLONITRILE								
413	C10H20	N-BUTYLCYCLOHEXANE	-15.037	2.582 E-1	-1.506 E-4	3.344 E-8	598.30	311.39	-50.95	13.49
414	C10H20	ISOBUTYLCYCLOHEXANE								
415	C10H20	SEC-BUTYLCYCLOHEXANE								
416	C10H20	TERT-BUTYLCYCLOHEXANE								
417	C10H20	1-DECENE	-1.114	2.168 E-1	-1.208 E-4	2.616 E-8	518.37	277.80	-29.67	28.93
418	C10H20	N-DECANE	-1.890	2.295 E-1	-1.263 E-4	2.701 E-8	558.61	288.37	-59.67	7.94
419	C10H22	3,3,5-TRIMETHYLHEPTAN	-16.808	2.943 E-1	-2.065 E-4	5.864 E-8			-61.80	8.02
420	C10H22	2,2,3,3-TETRAMETHYLHE	-14.052	2.941 E-1	-2.101 E-4	6.174 E-8				
421	C10H22	2,2,5,5-TETRAMETHYLHE	-14.890	2.973 E-1	-2.139 E-4	6.253 E-8				
422	C10H22O	1-DECANOL	3.480	2.137 E-1	-9.365 E-5	8.242 E-9	1481.8	380.00	-96.0	-24.9
423	C11H10	1-METHYLNAPHTHALENE	-15.482	2.242 E-1	-1.658 E-4	4.814 E-8	862.89	361.76	27.93	52.03
424	C11H10	2-METHYLNAPHTHALENE	-13.499	2.149 E-1	-1.545 E-4	4.395 E-8	699.42	351.79	27.75	51.66
425	C11H14O2	BUTYL BENZOATE	-4.148	2.072 E-1	-1.101 E-4	1.728 E-8	882.36	350.34		
426	C11H22	N-HEXYLCYCLOPENTANE	-13.930	2.694 E-1	-1.561 E-4	3.518 E-8	617.57	318.65	-50.07	18.69
427	C11H22	1-UNDECENE	-1.334	2.395 E-1	-1.338 E-4	2.905 E-8	566.26	294.89	-34.60	30.94
428	C11H24	N-UNDECANE	-2.005	2.517 E-1	-1.385 E-4	2.954 E-8	605.50	305.01	-64.60	9.94
429	C12H10	DIPHENYL	-23.184	2.641 E-1	-2.115 E-4	6.664 E-8	733.87	369.58	43.52	66.94
430	C12H10O	DIPHENYL ETHER	-14.505	2.217 E-1	-1.402 E-4	3.245 E-8	1146.0	379.29	11.94	20.70
431	C12H24	N-HEPTYLCYCLOPENTANE	-14.155	2.922 E-1	-1.692 E-4	3.813 E-8	654.77	333.12	-55.00	32.96
432	C12H24	1-DODECENE	-1.563	2.622 E-1	-1.470 E-4	3.203 E-8	615.67	310.07	-39.52	11.96
433	C12H26	N-DODECANE	-2.228	2.564 E-1	-1.516 E-4	3.246 E-8	631.63	318.78	-69.52	
434	C12H26O	DIHEXYL ETHER	8.010	2.564 E-1	-1.322 E-4	4.007 E-8	723.43	323.35		
435	C12H26O	DODECANOL	2.203	2.635 E-1	-1.275 E-4	1.858 E-8	1417.8	398.89	-105.84	-20.81
436	C12H27N	TRIBUTYLAMINE	1.909	2.861 E-1	-1.601 E-4	3.460 E-8	889.06	312.48		
437	C13H12	DIPHENYLMETHANE								
438	C13H26	N-OCTYLCYCLOPENTANE	-14.319	3.145 E-1	-1.818 E-4	4.080 E-8	695.83	346.19	-59.92	22.72
439	C13H26	1-TRIDECENE	-1.700	2.845 E-1	-1.594 E-4	3.466 E-8	658.16	323.71	-44.45	34.96
440	C13H28	N-TRIDECANE	-2.499	2.974 E-1	-1.651 E-4	3.558 E-8	664.10	332.10	-74.45	13.97

NO	FORMULA	NAME	ANTA	ANTB	ANTC	TMX	TMN	HARA	HARB	HARC	HARD	HV
401	C10H14	ISOBUTYLBENZENE	15.9524	3512.47	-69.03	476	326					9040.
402	C10H14	SEC-BUTYLBENZENE	15.9999	3544.19	-68.10	476	325					9070.
403	C10H14	TERT-BUTYLBENZENE	15.9300	3462.28	-69.87	472	323					8990.
404	C10H14	1-METHYL-2-ISOPROPYLB	15.9809	3564.52	-70.00	481	330					9110.
405	C10H14	1-METHYL-3-ISOPROPYLB	15.9811	3543.79	-69.22	478	328	67.726	-8033.58	-7.076	8.39	
406	C10H14	1-METHYL-4-ISOPROPYLB	15.9424	3539.21	-70.10	480	329	63.225	-7800.97	-6.432	8.41	
407	C10H14	1,4-DIETHYLBENZENE	16.1140	3657.22	-71.18	487	335					9410.
408	C10H14	1,2,4,5-TETRAMETHYLBE	16.3023	3850.91	-71.72	500	361	64.139	-8300.92	-6.478	8.80	10080.
409	C10H15N	N-BUTYLANILINE	16.3994	4079.72	-96.15	560	385					11690.
410	C10H18	CIS-DECALIN	15.8312	3671.61	-69.74	495	368					9400.
411	C10H18	TRANS-DECALIN	15.7989	3610.66	-66.49	470	363					9200.
412	C10H19N	CAPRYLONITRILE										
413	C10H20	N-BUTYLCYCLOHEXANE	15.9116	3542.57	-72.32	485	332					9200.
414	C10H20	ISOBUTYLCYCLOHEXANE	15.8141	3437.99	-69.99	455	355					
415	C10H20	SEC-BUTYLCYCLOHEXANE	15.8670	3524.57	-70.78	470	360					
416	C10H20	TERT-BUTYLCYCLOHEXANE	15.7884	3457.85	-67.04	450	357					
417	C10H20	1-DECENE	16.0129	3448.18	-76.09	460	356	73.938	-8380.48	-7.95	9.90	9240.
418	C10H22	N-DECANE	16.0114	3456.80	-78.67	476	330	75.475	-8563.64	-8.149	10.20	9388.
419	C10H22	3,3,5-TRIMETHYLHEPTAN	15.7848	3305.20	-67.66	458	313					8760.
420	C10H22	2,2,3,3-TETRAMETHYLHE	15.7598	3371.05	-64.09	463	314					8690.
421	C10H22	2,2,5,5-TETRAMETHYLHE	15.8446	3172.92	-66.15	438	300					8430.
422	C10H22O	1-DECANOL	15.9395	3389.43	-139.	503	376					12000.
423	C11H10	1-METHYLNAPHTHALENE	16.2008	4206.70	-78.15	551	380					11100.
424	C11H10	2-METHYLNAPHTHALENE	16.2758	4237.37	-74.75	548	377					11000.
425	C11H14O2	BUTYL BENZOATE	16.3363	4158.47	-94.15	570	390					11700.
426	C11H22	N-HEXYLCYCLOPENTANE	16.0140	3702.56	-81.55	507	351					9840.
427	C11H22	1-UNDECENE	16.0412	3597.72	-83.41	496	345	78.295	-9105.75	-8.489	11.46	9770.
428	C11H24	N-UNDECANE	16.0541	3614.07	-85.45	498	348	80.121	-9305.80	-8.729	11.75	9920.
429	C12H10	DIPHENYL	16.6832	4602.23	-70.42	545	343					10900.
430	C12H10O	DIPHENYL ETHER	16.3459	4310.25	-87.31	598	418					11260.
431	C12H24	N-HEPTYLCYCLOPENTANE	16.0589	3850.38	-88.15	529	368					10360.
432	C12H24	1-DODECENE	16.0610	3729.87	-90.88	517	361	82.968	-9846.99	-9.073	13.10	10270.
433	C12H26	N-DODECANE	16.1134	3774.56	-91.31	520	364	84.248	-10012.5	-9.236	13.37	10430.
434	C12H26O	DIHEXYL ETHER	16.3372	3982.78	-89.15	545	373					10900.
435	C12H26O	DODECANOL	15.2638	3242.04	-157.1	580	407					
436	C12H27N	TRIBUTYLAMINE	16.2878	3865.58	-86.15	531	362					10600.
437	C13H12	DIPHENYLMETHANE	14.4856	2902.44	-167.9	563	473	88.010	-10609.4	-9.709	15.00	10850.
438	C13H26	N-OCTYLCYCLOPENTANE	16.0941	3983.01	-95.85	549	385					10750.
439	C13H26	1-TRIDECENE	16.0850	3856.23	-97.84	537	377					10910.
440	C13H28	N-TRIDECANE	16.1355	3892.91	-98.93	540	380					

NO	FORMULA	NAME	MOLWT	TFP	TB	TC	PC	VC	ZC	OMEGA	LIQDEN	TDEN	DIPM
441	C14H10	ANTHRACENE	178.234	489.7	614.4	883.							0.0
442	C14H10	PHENANTHRENE	178.234	373.7	612.6	878.							0.0
443	C14H28	N-NONYLCYCLOPENTANE	196.378		535.3	710.5	16.3			0.610			
444	C14H28	1-TETRADECENE	196.378	260.3	526.3	689.	15.4			0.644	0.786	273.	
445	C14H30	N-TETRADECANE	198.394	279.0	526.7	694.	16.	830.	0.23	0.679	0.763	293.	
446	C15H30	N-DECYLCYCLOPENTANE	210.405		552.5	723.8	15.0			0.654			
447	C15H30	1-PENTADECENE	210.405	269.4	541.5	704.	14.4			0.682	0.791	273.	
448	C15H32	N-PENTADECANE	212.421	283.	543.8	707.	15.	880.	0.23	0.706	0.769	293.	
449	C16H22O4	DIBUTYL-O-PHTHALATE	278.350	238.	608.						1.047	293.	
450	C16H32	N-DECYLCYCLOHEXANE	224.432		570.8	750.	13.4			0.583	0.788	283.	
451	C16H32	1-HEXADECENE	224.432	277.3	558.0	717.	13.2			0.721	0.773	293.	
452	C16H34	N-HEXADECANE	226.448	291.	560.	717.	14.			0.742			
453	C17H34	N-DODECYLCYCLOPENTANE	238.459		584.1	750.	12.8			0.719			
454	C17H36O	HEPTADECANOL	256.474	327.	597.	736.	14.				0.848	327.	
455	C17H36	N-HEPTADECANE	240.475	295.	575.2	733.	13.	1000.	0.22	0.770	0.778	293.	
456	C18H14	O-TERPHENYL	230.310	330.	605.	891.0	38.5	769.	0.405				
457	C18H14	M-TERPHENYL	230.310	360.	638.	924.8	34.6	784.	0.358				
458	C18H14	P-TERPHENYL	230.310	485.	649.	926.0	32.8	779.	0.336				
459	C18H36	1-OCTADECENE	252.486	290.8	588.0	739.	11.2			0.807	0.789	293.	0.7
460	C18H36	N-TRIDECYLCYCLOPENTAN	252.486		598.6	761.	11.9						
461	C18H38	N-OCTADECANE	254.502	301.3	589.5	745.	11.9			0.755	0.777	301.	
462	C18H38O	1-OCTADECANOL	270.501	331.	608.	747.	14.			0.790	0.812	332.	1.7
463	C19H38	N-TETRADECYLCYCLOPENT	266.513		599.	772.	11.1			0.789	0.789	305.	
464	C19H40	N-NONADECANE	268.529	305.	603.1	756.	11.0			0.827			
465	C20H40	N-PENTADECYLCYCLOPENT	280.540		625.	780.	10.1			0.833			
466	C20H42	N-EICOSANE	282.556	310.	617.0	767.	11.0			0.907	0.775	313.	
467	C20H42O	1-EICOSANOL	298.555	339.	629.	770.	12.						
468	C21H42	N-HEXADECYLCYCLOPENTA	294.567		637.	791.	9.6			0.861			

NO	FORMULA	NAME	CPVAP A	CPVAP B	CPVAP C	CPVAP D	VISB	VISTO	DELHG	DELGF
441	C14H10	ANTHRACENE	-14.087	2.402 E-1	-1.575 E-4	3.835 E-8	513.28	405.81	53.7	
442	C14H10	PHENANTHRENE	-14.087	2.402 E-1	-1.575 E-4	3.835 E-8			48.4	
443	C14H28	N-NONYLCYCLOPENTANE	-14.524	3.372 E-1	-1.948 E-4	4.372 E-8	735.19	357.74	-64.85	24.72
444	C14H28	1-TETRADECENE	-1.903	3.071 E-1	-1.722 E-4	3.748 E-8	697.49	336.13	-49.36	36.99
445	C14H30	N-TETRADECANE	-2.623	3.195 E-1	-1.773 E-4	3.817 E-8	689.85	344.21	-79.38	15.97
446	C15H30	N-DECYLCYCLOPENTANE	-14.790	3.601 E-1	-2.082 E-4	4.679 E-8	771.74	368.30	-69.78	26.73
447	C15H30	1-PENTADECENE	-2.198	3.302 E-1	-1.859 E-4	4.067 E-8	739.13	347.46	-54.31	38.97
448	C15H32	N-PENTADECANE	-2.846	3.422 E-1	-1.904 E-4	4.108 E-8	718.51	355.92	-84.31	17.98
449	C16H22O4	DIBUTYL-O-PHTHALATE	0.449	2.995 E-1	-1.462 E-4	1.665 E-8	2588.1	336.24		
450	C16H32	N-DECYLCYCLOHEXANE	-16.484	3.951 E-1	-2.296 E-4	5.118 E-8	925.84	378.69		
451	C16H32	1-HEXADECENE	-2.318	3.523 E-1	-1.982 E-4	4.324 E-8	767.48	357.85	-59.23	40.99
452	C16H34	N-HEXADECANE	-3.109	3.652 E-1	-2.039 E-4	4.418 E-8	738.30	366.11	-89.23	20.00
453	C17H34	N-DODECYLCYCLOPENTANE	-15.110	4.049 E-1	-2.333 E-4	5.220 E-8	853.90	385.53	-80.28	30.10
454	C17H36O	HEPTADECANOL	-1.861	3.948 E-1	-2.242 E-4	4.881 E-8			-130.47	-10.67
455	C17H36	N-HEPTADECANE	-3.336	3.879 E-1	-2.169 E-4	4.710 E-8	757.88	375.90	-94.15	22.01
456	C18H14	O-TERPHENYL					1094.1	461.27		
457	C18H14	M-TERPHENYL					940.58	460.94		
458	C18H14	P-TERPHENYL					911.01	461.10		
459	C18H36	1-OCTADECENE	-2.706	3.975 E-1	-2.239 E-4	4.893 E-8	816.19	376.93	-69.08	45.01
460	C18H36	N-TRIDECYLCYCLOPENTAN	-15.336	4.276 E-1	-2.465 E-4	5.516 E-8	891.80	392.78	-84.55	32.74
461	C18H38	N-OCTADECANE	-3.456	4.101 E-1	-2.291 E-4	4.964 E-8	777.40	385.00	-99.08	24.02
462	C18H38O	1-OCTADECANOL	-2.079	4.174 E-1	-2.360 E-4	5.153 E-8			-135.39	-8.65
463	C19H38	N-TETRADECYLCYCLOPENT	-15.508	4.501 E-1	-2.592 E-4	5.794 E-8	924.60	399.62	-89.48	34.77
464	C19H40	N-NONADECANE	-3.700	4.329 E-1	-2.424 E-4	5.267 E-8	793.62	393.54	-104.00	26.03
465	C20H40	N-PENTADECYLCYCLOPENT	-15.786	4.730 E-1	-2.724 E-4	6.090 E-8	950.57	406.33	-94.41	36.78
466	C20H42	N-EICOSANE	-5.246	4.632 E-1	-2.667 E-4	6.039 E-8	811.29	401.67	-108.93	28.04
467	C20H42O	1-EICOSANOL	-3.005	4.657 E-1	-2.671 E-4	6.009 E-8			-145.25	-4.64
468	C21H42	N-HEXADECYLCYCLOPENTA	-15.927	4.954 E-1	-2.851 E-4	6.373 E-8	977.42	412.29	-99.33	38.79

NO	FORMULA	NAME	ANTA	ANTB	ANTC	TMX	TMN	HARA	HARB	HARC	HARD	HV
441	C14H10	ANTHRACENE	17.6701	6492.44	-26.13	655	490					13500.
442	C14H10	PHENANTHRENE	16.7187	5477.94	-69.39	655	450					13300.
443	C14H28	N-NONYLCYCLOPENTANE	16.1089	4096.30	-103.0	569	400					11290.
444	C14H28	1-TETRADECENE	16.1643	4018.01	-102.7	557	392	92.474	-11329.2	-10.27	17.07	11210.
445	C14H30	N-TETRADECANE	16.1480	4008.52	-105.4	560	394	91.172	-11322.9	-10.07	16.66	11380.
446	C15H30	N-DECYLCYCLOPENTANE	16.1261	4203.94	-109.7	586	413					11710.
447	C15H30	1-PENTADECENE	16.1539	4103.15	-110.6	574	406	98.920	-12205.3	-11.09	19.16	11630.
448	C15H32	N-PENTADECANE	16.1724	4121.51	-111.8	577	408	95.000	-11995.6	-10.54	18.45	11820.
449	C16H22O4	DIBUTYL-O-PHTHALATE	16.9539	4852.47	-138.1	657	469					18900.
450	C16H32	N-DECYLCYCLOHEXANE	16.1627	4373.37	-111.8	573	463					12040.
451	C16H32	1-HEXADECENE	16.2203	4245.00	-115.2	592	420	105.90	-13117.0	-11.99	21.68	12050.
452	C16H34	N-HEXADECANE	16.1841	4214.91	-118.7	594	423	95.680	-12411.3	-10.58	20.27	12240.
453	C17H34	N-DODECYLCYCLOPENTANE	16.1915	4395.87	-124.2	619	441					12570.
454	C17H36O	HEPTADECANOL	15.6161	3672.62	-188.1	656	464					14590.
455	C17H36	N-HEPTADECANE	16.1510	4294.55	-124.0	610	434					12640.
456	C18H14	O-TERPHENYL										
457	C18H14	M-TERPHENYL										
458	C18H14	P-TERPHENYL										
459	C18H36	1-OCTADECENE	16.2221	4416.13	-127.2	623	444					12970.
460	C18H36	N-TRIDECYLCYCLOPENTAN	16.2270	4483.13	-131.3	634	453					12980.
461	C18H38	N-OCTADECANE	16.1232	4361.79	-129.9	625	445					13020.
462	C18H38O	1-OCTADECANOL	15.6898	3757.82	-193.1	658	474					
463	C19H38	N-TETRADECYLCYCLOPENT	16.2632	4439.38	-138.1	648	465					13380.
464	C19H40	N-NONADECANE	16.1533	4450.44	-135.6	639	456					13390.
465	C20H40	N-PENTADECYLCYCLOPENT	16.3092	4642.01	-145.1	661	476					13780.
466	C20H42	N-EICOSANE	16.4685	4680.46	-141.1	652	471					13740.
467	C20H42O	1-EICOSANOL	15.8233	3912.10	-203.1	679	492					15600.
468	C21H42	N-HEXADECYLCYCLOPENTA	16.3553	4715.69	-152.1	674	488					14180.

Appendix B

Compound Dictionary

1	AR	ARGON
2	BCL3	BORON TRICHLORIDE
3	BF3	BORON TRIFLUORIDE
4	BR2	BROMINE
5	CLNO	NITROSYL CHLORIDE
6	CL2	CHLORINE
7	CL3P	PHOSPHORUS TRICHLORIDE
8	CL4SI	SILICON TETRACHLORIDE
9	D2	DEUTERIUM(NORMAL)
10	D2O	DEUTERIUM OXIDE
		HEAVY WATER
11	F2	FLUORINE
12	F3N	NITROGEN TRIFLUORIDE
13	F4SI	SILICON TETRAFLUORIDE
14	F6S	SULFUR HEXAFLUORIDE
15	HBR	HYDROGEN BROMIDE
16	HCL	HYDROGEN CHLORIDE
17	HF	HYDROGEN FLUORIDE
18	HI	HYDROGEN IODIDE
19	H2	HYDROGEN(NORMAL)
20	H2O	WATER
21	H2S	HYDROGEN SULFIDE
22	H3N	AMMONIA
23	H4N2	HYDRAZINE
24	HE-4	HELIUM-4
25	I2	IODINE
26	KR	KRYPTON
27	NO	NITRIC OXIDE
		NITROGEN OXIDE
28	NO2	NITROGEN DIOXIDE
		NITROGEN PEROXIDE
29	N2	NITROGEN
30	N2O	NITROUS OXIDE
31	NE	NEON
32	O2	OXYGEN
33	O2S	SULFUR DIOXIDE
34	O3	OZONE
35	O3S	SULFUR TRIOXIDE
36	XE	XENON
37	CBRF3	TRIFLUOROBROMOMETHANE
		BROMOTRIFLUOROMETHANE
38	CCLF3	CHLOROTRIFLUOROMETHANE
		FREON 13
39	CCL2F2	DICHLORODIFLUOROMETHANE
		FREON 12
40	CCL2O	PHOSGENE
		CARBON OXYCHLORIDE
		CARBONYL CHLORIDE
41	CCL3F	TRICHLOROFLUOROMETHANE
42	CCL4	CARBON TETRACHLORIDE
		PERCHLOROMETHANE
		TETRACHLOROMETHANE
43	CF4	CARBON TETRAFLUORIDE
		PERFLUOROMETHANE
		TETRAFLUOROMETHANE
44	CO	CARBON MONOXIDE
45	COS	CARBONYL SULFIDE
		CARBON OXYSULFIDE
46	CO2	CARBON DIOXIDE
47	CS2	CARBON DISULFIDE
48	CHCLF2	CHLORODIFLUOROMETHANE
		FREON 22
49	CHCL2F	DICHLOROMONOFLUOROMETHANE
		FREON 21
50	CHCL3	CHLOROFORM
		TRICHLOROMETHANE
51	CHN	HYDROGEN CYANIDE
52	CH2BR2	DIBROMOMETHANE
		METHYLENE BROMIDE
53	CH2CL2	DICHLOROMETHANE
		METHYLENE CHLORIDE
54	CH2O	FORMALDEHYDE
		METHANAL

COMPOUND DICTIONARY (CONTINUED)

```
55   CH2O2      FORMIC ACID
                    METHANOIC ACID
56   CH3BR      METHYL BROMIDE
                    BROMOMETHANE
57   CH3CL      METHYL CHLORIDE
                    CHLOROMETHANE
58   CH3F       METHYL FLUORIDE
                    FLUOROMETHANE
59   CH3I       METHYL IODIDE
                    IODOMETHANE
60   CH3NO2     NITROMETHANE
61   CH4        METHANE
62   CH4O       METHANOL
63   CH4S       METHYL MERCAPTAN
                    METHANETHIOL
64   CH5N       METHYL AMINE
                    AMINOMETHANE
                    MONOMETHYLAMINE
65   CH6N2      METHYL HYDRAZINE


66   C2CLF5     CHLOROPENTAFLUOROETHANE
67   C2CL2F4    1,1-DICHLORO-1,2,2,2-TETRAFLUOROETHANE
68   C2CL2F4    1,2-DICHLORO-1,1,2,2-TETRAFLUOROETHANE
                    FREON 114
                    SYM-DICHLORODIFLUOROETHANE
69   C2CL3F3    1,2,2-TRICHLORO-1,1,2-TRIFLUOROETHANE
                    FREON 113
70   C2CL4      TETRACHLOROETHYLENE
                    PERCHLOROETHYLENE
71   C2CL4F2    1,1,2,2-TETRACHLORO-1,2-DIFLUOROETHANE
72   C2F4       PERFLUOROETHENE
                    TETRAFLUOROETHYLENE
73   C2F6       PERFLUOROETHANE
                    HEXAFLUOROETHANE
74   C2N2       CYANOGEN
75   C2HCL3     TRICHLOROETHYLENE
                    ETHINYL TRICHLORIDE
                    TRICHLOROETHENE
76   C2HF3O2    TRIFLUOROACETIC ACID
77   C2H2       ACETYLENE
                    ETHYNE
78   C2H2F2     1,1-DIFLUOROETHYLENE
                    1,1-DIFLUOROETHENE
                    VINYLIDINE
79   C2H2O      KETENE
                    ETHONONE
80   C2H3CL     VINYL CHLORIDE
                    CHLOROETHYLENE
81   C2H3CLF2   1-CHLORO-1,1-DIFLUOROETHANE
82   C2H3CLO    ACETYL CHLORIDE
83   C2H3CL3    1,1,2-TRICHLOROETHANE
84   C2H3F      VINYL FLUORIDE
                    FLUOROETHENE
                    FLUOROETHYLENE
85   C2H3F3     1,1,1-TRIFLUOROETHANE
86   C2H3N      ACETONITRILE
                    ETHANENITRILE
87   C2H3NO     METHYL ISOCYANATE
                    METHYLCARBYLAMINE
88   C2H4       ETHYLENE
                    ETHENE
89   C2H4CL2    1,1-DICHLOROETHANE
                    ETHYLIDENE CHLORIDE
90   C2H4CL2    1,2-DICHLOROETHANE
91   C2H4F2     1,1-DIFLUOROETHANE
                    ETHYLIDENE FLUORIDE
92   C2H4O      ACETALDEHYDE
                    ETHANAL
93   C2H4O      ETHYLENE OXIDE
                    EPOXYETHANE
                    OXIRANE
94   C2H4O2     ACETIC ACID
                    ETHANOIC ACID
95   C2H4O2     METHYL FORMATE
```

COMPOUND DICTIONARY (CONTINUED)

```
 96   C2H5BR     ETHYL BROMIDE
                    BROMOETHANE
 97   C2H5CL     ETHYL CHLORIDE
                    CHLOROETHANE
 98   C2H5F      ETHYL FLUORIDE
                    FLUOROETHANE
 99   C2H5N      ETHYLENE IMINE
                    AZIRANE
                    DIHYDROAZIRINE
100   C2H6       ETHANE
101   C2H6O      DIMETHYL ETHER
                    METHYL ETHER
102   C2H6O      ETHANOL
                    ETHYL ALCOHOL
103   C2H6O2     ETHYLENE GLYCOL
                    DIHYDROXYETHANE
                    1,2-ETHANEDIOL
104   C2H6S      ETHYL MERCAPTAN
                    ETHANETHIOL
105   C2H6S      DIMETHYL SULFIDE
                    METHYL SULFIDE
                    2-THIAPROPANE
106   C2H7N      ETHYL AMINE
                    ETHANE,AMINO
107   C2H7N      DIMETHYLAMINE
108   C2H7NO     MONOETHANOLAMINE
                    2-AMINOETHANOL
                    ETHANOLAMINE
109   C2H8N2     ETHYLENEDIAMINE
                    1,2-DIAMINOETHANE


110   C3H3N      ACRYLONITRILE
111   C3H4       PROPADIENE
                    ALLENE
                    DIMETHYLENEMETHANE
112   C3H4       METHYL ACETYLENE
                    PROPYNE
113   C3H4O      ACROLEIN
114   C3H4O2     ACRYLIC ACID
115   C3H4O2     VINYL FORMATE
116   C3H5CL     ALLYL CHLORIDE
                    3-CHLOROPROPENE
117   C3H5CL3    1,2,3-TRICHLOROPROPANE
118   C3H5N      PROPIONITRILE
                    PROPANENITRILE
119   C3H6       CYCLOPROPANE
                    TRIMETHYLENE
120   C3H6       PROPYLENE
                    PROPENE
121   C3H6CL2    1,2-DICHLOROPROPANE
122   C3H6O      ACETONE
                    2-PROPANONE
123   C3H6O      ALLYL ALCOHOL
                    2-PROPENOL
                    PROPEN-1-OL-2
124   C3H6O      N-PROPIONALDEHYDE
                    1-PROPANAL
125   C3H6O      PROPYLENE OXIDE
126   C3H6O      VINYL METHYL ETHER
127   C3H6O2     PROPIONIC ACID
                    PROPANOIC ACID
128   C3H6O2     ETHYL FORMATE
129   C3H6O2     METHYL ACETATE
                    METHYL METHANOATE
130   C3H7CL     PROPYL CHLORIDE
                    1-CHLOROPROPANE
131   C3H7CL     ISOPROPYL CHLORIDE
                    2-CHLOROPROPANE
132   C3H8       PROPANE
133   C3H8O      1-PROPANOL
                    N-PROPYL ALCOHOL
134   C3H8O      ISOPROPYL ALCOHOL
                    ISOPROPANOL
                    2-PROPANOL
```

COMPOUND DICTIONARY (CONTINUED)

135	C3H8O	METHYL ETHYL ETHER
136	C3H8O2	METHYLAL
		DIMETHOXYMETHANE
		FORMAL
		FORMALDEHYDE DIMETHYLACETAL
		METHYLENE DIMETHYL ETHER
137	C3H8O2	PROPANEDIOL-1,2
		PROPYLENE GLYCOL (ALPHA)
138	C3H8O2	1,3-PROPANEDIOL
139	C3H8O3	GLYCEROL
		1,2,3-PROPANETRIOL
140	C3H8S	METHYL ETHYL SULFIDE
		2-THIABUTANE
141	C3H9N	N-PROPYL AMINE
142	C3H9N	ISOPROPYL AMINE
143	C3H9N	TRIMETHYL AMINE
144	C4H2O3	MALEIC ANHYDRIDE
145	C4H4	VINYLACETYLENE
		1-BUTEN-3-YNE
146	C4H4O	FURAN
147	C4H4S	THIOPHENE
		THIOFURAN
148	C4H5N	ALLYL CYANIDE
		BETA-BUTENONITRILE
		3-BUTENENITRILE
		VINYLACETONITRILE
149	C4H5N	PYRROLE
		AZOLE
150	C4H6	1-BUTYNE
		ETHYLACETYLENE
151	C4H6	2-BUTYNE
		DIMETHYLACETYLENE
152	C4H6	1,2-BUTADIENE
		METHYLALLENE
153	C4H6	1,3-BUTADIENE
		BIVINYL
154	C4H6O2	VINYL ACETATE
155	C4H6O3	ACETIC ANHYDRIDE
156	C4H6O4	DIMETHYL OXALATE
157	C4H6O4	SUCCINIC ACID
		BUTANEDIOIC ACID
158	C4H7N	BUTYRONITRILE
		BUTANENITRILE
		PROPYL CYANIDE
159	C4H7O2	METHYL ACRYLATE
160	C4H8	1-BUTENE
161	C4H8	CIS-2-BUTENE
162	C4H8	TRANS-2-BUTENE
163	C4H8	CYCLOBUTANE
		TETRAMETHYLENE
164	C4H8	ISOBUTYLENE
		ISOBUTENE
		2-METHYLPROPENE
165	C4H8O	N-BUTYRALDEHYDE
		BUTANAL
166	C4H8O	ISOBUTYRALDEHYDE
		2-METHYL-PROPANAL
		2-METHYLPROPIONALDEHYDE
167	C4H8O	METHYL ETHYL KETONE
		2-BUTANONE
168	C4H8O	TETRAHYDROFURAN
		1,4-EPOXYBUTANE
		TETRAMETHYLENE OXIDE
169	C4H8O	VINYL ETHYL ETHER
		ETHOXYACETYLENE
		ETHYL ETHENYL ETHER
		ETHYL VINYL ETHER
170	C4H8O2	N-BUTYRIC ACID
		BUTANOIC ACID
171	C4H8O2	1,4-DIOXANE
172	C4H8O2	ETHYL ACETATE
173	C4H8O2	ISOBUTYRIC ACID
		2-METHYLPROPANOIC ACID

COMPOUND DICTIONARY (CONTINUED)

```
174  C4H8O2    METHYL PROPIONATE
                 METHYL PROPANOATE
175  C4H8O2    N-PROPYL FORMATE
                 PROPYL METHANOATE
176  C4H9CL    1-CHLOROBUTANE
                 BUTYL CHLORIDE
177  C4H9CL    2-CHLOROBUTANE
                 2-BUTYL CHLORIDE
                 SEC-BUTYL CHLORIDE
178  C4H9CL    TERT-BUTYL CHLORIDE
                 2-CHLORO-2-METHYL PROPANE
179  C4H9N     PYRROLIDINE
                 1-AZACYCLOPENTANE
                 TETRAHYDROPYRROLE
                 TETRAMETHYLENEIMINE
180  C4H9NO    MORPHOLINE
                 DIETHYLENE IMIDE OXIDE
                 TETRAHYDRO-1,4-OXAZINE
181  C4H10     N-BUTANE
182  C4H10     ISOBUTANE
                 2-METHYLPROPANE
183  C4H10O    N-BUTANOL
                 1-BUTANOL
                 N-BUTYL ALCOHOL
184  C4H10O    2-BUTANOL
                 SEC-BUTYL ALCOHOL
185  C4H10O    ISOBUTANOL
                 ISOBUTYL ALCOHOL
                 2-METHYL-1-PROPANOL
186  C4H10O    TERT-BUTYL ALCOHOL
                 2-METHYL-2-PROPANOL
187  C4H10O    ETHYL ETHER
                 DIETHYL ETHER
                 ETHOXYETHANE
188  C4H10O2   1,2-DIMETHOXYETHANE
                 ACETALDEHYDE DIMETHYL ACETAL
                 DIMETHYL ACETAL
                 ETHYLENE GLYCOL DIMETHYL ETHER
                 ETHYLIDENE DIMETHYL ETHER
189  C4H10O3   DIETHYLENE GLYCOL
                 2,2'-DIHYDROXYDIETHYL ETHER
                 2,2'-ORYDIETHANOL
190  C4H10S    DIETHYL SULFIDE
                 ETHYL SULFIDE
                 3-THIAPENTANE
191  C4H10S2   DIETHYL DISULFIDE
                 3,4-DITHIAHEXANE
                 ETHYL DISULFIDE
192  C4H11N    N-BUTYL AMINE
193  C4H11N    ISOBUTYL AMINE
194  C4H11N    DIETHYL AMINE


195  C5H5N     PYRIDINE
196  C5H8      CYCLOPENTENE
197  C5H8      1,2-PENTADIENE
                 ETHYLALLENE
198  C5H8      1-TRANS-3-PENTADIENE
                 1,3-PENTADIENE,TRANS
199  C5H8      1,4-PENTADIENE
                 ALLYLETHYLENE
                 DIVINYLMETHANE
200  C5H8      1-PENTYNE
                 N-PROPYLACETYLENE
201  C5H8      2-METHYL-1,3-BUTADIENE
                 ISOPRENE
202  C5H8      3-METHYL-1,2-BUTADIENE
                 ASYM-DIMETHYLALLENE
203  C5H8O     CYCLOPENTANONE
                 ADIPIC KETONE
                 KETO-PENTAMETHYLENE
204  C5H8O2    ETHYL ACRYLATE
                 ETHYL ESTER OF PROPENOIC ACID
205  C5H10     CYCLOPENTANE
                 PENTAMETHYLENE
```

COMPOUND DICTIONARY (CONTINUED)

```
206   C5H10      1-PENTENE
207   C5H10      CIS-2-PENTENE
                   CIS-BETA-N-AMYLENE
208   C5H10      TRANS-2-PENTENE
209   C5H10      2-METHYL-1-BUTENE
210   C5H10      2-METHYL-2-BUTENE
211   C5H10      3-METHYL-1-BUTENE
212   C5H10O     VALERALDEHYDE
                   N-VALERALDEHYDE
                   1-PENTANAL
213   C5H10O     METHYL N-PROPYL KETONE
                   2-PENTANONE
214   C5H10O     METHYL ISOPROPYL KETONE
                   3-METHYL-2-BUTANONE
215   C5H10O     DIETHYL KETONE
                   3-PENTANONE
216   C5H10O2    N-VALERIC ACID
                   PENTANOIC ACID
217   C5H10O2    ISOBUTYL FORMATE
                   ISOBUTYL METHANOATE
218   C5H10O2    N-PROPYL ACETATE
                   N-PROPYL ETHANOATE
219   C5H10O2    ETHYL PROPIONATE
                   ETHYL PROPANOATE
220   C5H10O2    METHYL BUTYRATE
                   METHYL BUTANOATE
221   C5H10O2    METHYL ISOBUTYRATE
                   METHYL-2-METHYLPROPANOATE
222   C5H11N     PIPERIDINE
223   C5H12      N-PENTANE
224   C5H12      2-METHYL BUTANE
                   ISOPENTANE
225   C5H12      2,2-DIMETHYL PROPANE
                   NEOPENTANE
226   C5H12O     1-PENTANOL
227   C5H12O     2-METHYL-1-BUTANOL
                   ACTIVE AMYL ALCOHOL
                   AMYL ALCOHOL
                   2-METHYLBUTANOL
228   C5H12O     3-METHYL-1-BUTANOL
                   ISOAMYL ALCOHOL
                   3-METHYLBUTANOL
229   C5H12O     2-METHYL-2-BUTANOL
                   2-METHYLBUTANOL-2
                   TERT-AMYL ALCOHOL
                   TERT-PENTYL ALCOHOL
230   C5H12O     2,2-DIMETHYL-1-PROPANOL
                   TERT-BUTYLCARBINOL
                   TERT-BUTYL METHYL ALCOHOL
                   NEOPENTYL ALCOHOL
231   C5H12O     ETHYL PROPYL ETHER


232   C6F6       PERFLUOROBENZENE
                   HEXAFLUOROBENZENE
233   C6F12      PERFLUOROCYCLOHEXANE
234   C6F14      PERFLUORO-N-HEXANE
235   C6H4CL2    O-DICHLOROBENZENE
                   1,2-DICHLOROBENZENE
236   C6H4CL2    M-DICHLOROBENZENE
                   1,3-DICHLOROBENZENE
237   C6H4CL2    P-DICHLOROBENZENE
                   1,4-DICHLOROBENZENE
238   C6H5BR     BROMOBENZENE
                   PHENYL BROMIDE
239   C6H5CL     CHLOROBENZENE
                   PHENYL CHLORIDE
240   C6H5F      FLUOROBENZENE
                   PHENYL FLUORIDE
241   C6H5I      IODOBENZENE
                   PHENYL IODIDE
242   C6H6       BENZENE
243   C6H6O      PHENOL
244   C6H7N      ANILINE
                   PHENYLAMINE
```

COMPOUND DICTIONARY (CONTINUED)

245	C6H7N	4-METHYLPYRIDINE
		GAMMA-PICOLINE
246	C6H10	1,5-HEXADIENE
		BIALLYL
247	C6H10	CYCLOHEXENE
		1,2,3,4-TETRAHYDROBENZENE
248	C6H10O	CYCLOHEXANONE
249	C6H12	CYCLOHEXANE
		HEXAHYDROBENZENE
		HEXAMETHYLENE
250	C6H12	METHYLCYCLOPENTANE
		METHYLPENTAMETHYLENE
251	C6H12	1-HEXENE
252	C6H12	CIS-2-HEXENE
253	C6H12	TRANS-2-HEXENE
254	C6H12	CIS-3-HEXENE
255	C6H12	TRANS-3-HEXENE
256	C6H12	2-METHYL-2-PENTENE
257	C6H12	3-METHYL-CIS-2-PENTENE
258	C6H12	3-METHYL-TRANS-2-PENTENE
259	C6H12	4-METHYL-CIS-2-PENTENE
260	C6H12	4-METHYL-TRANS-2-PENTENE
261	C6H12	2,3-DIMETHYL-1-BUTENE
262	C6H12	2,3-DIMETHYL-2-BUTENE
263	C6H12	3,3-DIMETHYL-1-BUTENE
264	C6H12O	CYCLOHEXANOL
		HEXAHYDROPHENOL
		HEXALIN
265	C6H14O	METHYL ISOBUTYL KETONE
		4-METHYL-2-PENTANONE
266	C6H12O2	N-BUTYL ACETATE
		N-BUTYL ETHANOATE
267	C6H12O2	ISOBUTYL ACETATE
		ISOBUTYL ETHANOATE
		2-METHYL PROPYL ETHANOATE
268	C6H12O2	ETHYL BUTYRATE
		ETHYL BUTANOATE
269	C6H12O2	ETHYL ISOBUTYRATE
		ETHYL 2-METHYL PROPANOATE
270	C6H12O2	N-PROPYL PROPIONATE
271	C6H14	N-HEXANE
272	C6H14	2-METHYL PENTANE
273	C6H14	3-METHYL PENTANE
274	C6H14	2,2-DIMETHYL BUTANE
		NEOHEXANE
275	C6H14	2,3-DIMETHYL BUTANE
276	C6H14O	1-HEXANOL
		N-HEXYL ALCOHOL
277	C6H14O	ETHYL BUTYL ETHER
		1-ETHOXYBUTANE
278	C6H14O	DIISOPROPYL ETHER
		ISOPROPYL ETHER
279	C6H15N	DIPROPYLAMINE
		DI-N-PROPYLAMINE
280	C6H15N	TRIETHYLAMINE
281	C7F14	PERFLUOROMETHYLCYCLOHEXANE
		UNDECAFLUORO(TRIFLUOROMETHYL)CYCLOHEXANE
282	C7F16	PERFLUORO-N-HEPTANE
		HEXADECAFLUOROHEPTANE
283	C7H5N	BENZONITRILE
		PHENYL CYANIDE
284	C7H6O	BENZALDEHYDE
285	C7H6O2	BENZOIC ACID
286	C7H8	TOLUENE
		METHYLBENZENE
287	C7H8O	METHYL PHENYL ETHER
		ANISOLE
		METHOXYBENZENE
288	C7H8O	BENZYL ALCOHOL
289	C7H8O	O-CRESOL
		2-HYDROXYTOLUENE
		2-METHYLPHENOL

COMPOUND DICTIONARY (CONTINUED)

```
290  C7H8O    M-CRESOL
                3-HYDROXYTOLUENE
                3-METHYLPHENOL
291  C7H8O    P-CRESOL
                4-HYDROXYTOLUENE
                4-METHYLPHENOL
292  C7H9N    2,3-DIMETHYLPYRIDINE
                2,3-LUTIDINE
293  C7H9N    2,5-DIMETHYLPYRIDINE
                2,5-LUTIDINE
294  C7H9N    3,4-DIMETHYLPYRIDINE
                3,4-LUTIDINE
295  C7H9N    3,5-DIMETHYLPYRIDINE
                3,5-LUTIDINE
296  C7H9N    METHYLPHENYLAMINE
                N-METHYLANILINE
297  C7H9N    O-TOLUIDINE
                2-AMINOTOLUENE
                2-METHYLANILINE
298  C7H9N    M-TOLUIDINE
                3-AMINOTOLUENE
                3-METHYLANILINE
299  C7H9N    P-TOLUIDINE
                4-AMINOTOLUENE
                4-METHYLANILINE
300  C7H14    CYCLOHEPTANE
                HEPTAMETHYLENE
                SUBERANE
301  C7H14    1,1-DIMETHYLCYCLOPENTANE
302  C7H14    CIS-1,2-DIMETHYLCYCLOPENTANE
                1,CIS-2-DIMETHYLCYCLOPENTANE
303  C7H14    TRANS-1,2-DIMETHYLCYCLOPENTANE
                1,TRANS-2-DIMETHYLCYCLOPENTANE
304  C7H14    ETHYLCYCLOPENTANE
305  C7H14    METHYLCYCLOHEXANE
                HEXAHYDROTOLUENE
306  C7H14    1-HEPTENE
                N-HEPTENE
307  C7H14    2,3,3-TRIMETHYL-1-BUTENE
                TRIPTENE
308  C7H16    N-HEPTANE
309  C7H16    2-METHYLHEXANE
                ISOHEPTANE
310  C7H16    3-METHYLHEXANE
311  C7H16    2,2-DIMETHYLPENTANE
312  C7H16    2,3-DIMETHYLPENTANE
313  C7H16    2,4-DIMETHYLPENTANE
314  C7H16    3,3-DIMETHYLPENTANE
315  C7H16    3-ETHYLPENTANE
                TRIETHYLMETHANE
316  C7H16    2,2,3-TRIMETHYLBUTANE
                TRIPTAN
317  C7H16O   1-HEPTANOL
                N-HEPTYL ALCOHOL


318  C8H4O3   PHTHALIC ANHYDRIDE
                PHTHALANDIONE
319  C8H8     STYRENE
                VINYLBENZENE
320  C8H8O    METHYL PHENYL KETONE
                ACETOPHENONE
321  C8H8O2   METHYL BENZOATE
322  C8H10    O-XYLENE
                1,2-DIMETHYLBENZENE
323  C8H10    M-XYLENE
                1,3-DIMETHYLBENZENE
324  C8H10    P-XYLENE
                1,4-DIMETHYLBENZENE
325  C8H10    ETHYLBENZENE
326  C8H10O   O-ETHYLPHENOL
                1-ETHYL-2-HYDROXYBENZENE
                2-ETHYLPHENOL
                PHLOROL
```

COMPOUND DICTIONARY (CONTINUED)

```
327   C8H10O    M-ETHYLPHENOL
                    1-ETHYL-3-HYDROXYBENZENE
                    3-ETHYLPHENOL
328   C8H10O    P-ETHYLPHENOL
                    1-ETHYL-4-HYDROXYBENZENE
                    4-ETHYLPHENOL
329   C8H10O    PHENETOLE
                    ETHOXYBENZENE
                    ETHYL PHENYL ETHER
330   C8H10O    2,3-XYLENOL
                    1,2-DIMETHYL-O-HYDROXYBENZENE
                    2,3-DIMETHYLPHENOL
331   C8H10O    2,4-XYLENOL
                    1,3-DIMETHYL-5-HYDROXYBENZENE
                    2,4-DIMETHYLPHENOL
332   C8H10O    2,5-XYLENOL
                    1,4-DIMETHYL-2-HYDROXYBENZENE
                    2,5-DIMETHYLPHENOL
333   C8H10O    2,6-XYLENOL
                    1,3-DIMETHYL-2-HYDROXYBENZENE
                    2,6-DIMETHYLPHENOL
334   C8H10O    3,4-XYLENOL
                    3,4-DIMETHYL-4-HYDROXYBENZENE
                    3,4-DIMETHYLPHENOL
335   C8H10O    3,5-XYLENOL
                    3,5-DIMETHYL-6-HYDROXYBENZENE
                    3,5-DIMETHYLPHENOL
336   C8H11N    N,N-DIMETHYLANILINE
337   C8H16     1,1-DIMETHYLCYCLOHEXANE
338   C8H16     CIS-1,2-DIMETHYLCYCLOHEXANE
339   C8H16     TRANS-1,2-DIMETHYLCYCLOHEXANE
340   C8H16     CIS-1,3-DIMETHYLCYCLOHEXANE
341   C8H16     TRANS-1,3-DIMETHYLCYCLOHEXANE
342   C8H16     CIS-1,4-DIMETHYLCYCLOHEXANE
343   C8H16     TRANS-1,4-DIMETHYLCYCLOHEXANE
344   C8H16     ETHYLCYCLOHEXANE
345   C8H16     1,1,2-TRIMETHYLCYCLOPENTANE
346   C8H16     1,1,3-TRIMETHYLCYCLOPENTANE
347   C8H16     CIC,CIS,TRANS-1,2,4-TRIMETHYLCYCLOPENTANE
348   C8H16     CIS,TRANS,CIS-1,2,4-TRIMETHYLCYCLOPENTANE
349   C8H16     1-METHYL-1-ETHYLCYCLOPENTANE
350   C8H16     N-PROPYLCYCLOPENTANE
351   C8H16     ISOPROPYLCYCLOPENTANE
352   C8H16     1-OCTENE
353   C8H16     TRANS-2-OCTENE
354   C8H18     N-OCTANE
355   C8H18     2-METHYLHEPTANE
356   C8H18     3-METHYLHEPTANE
357   C8H18     4-METHYLHEPTANE
358   C8H18     2,2-DIMETHYLHEXANE
359   C8H18     2,3-DIMETHYLHEXANE
360   C8H18     2,4-DIMETHYLHEXANE
361   C8H18     2,5-DIMETHYLHEXANE
362   C8H18     3,3-DIMETHYLHEXANE
363   C8H18     3,4-DIMETHYLHEXANE
364   C8H18     3-ETHYLHEXANE
365   C8H18     2,2,3-TRIMETHYLPENTANE
366   C8H18     2,2,4-TRIMETHYLPENTANE
367   C8H18     2,3,3-TRIMETHYLPENTANE
368   C8H18     2,3,4-TRIMETHYLPENTANE
369   C8H18     2-METHYL-3-ETHYLPENTANE
                    3-ETHYL-2-METHYLPENTANE
370   C8H18     3-METHYL-3-ETHYLPENTANE
371   C8H18O    1-OCTANOL
                    N-OCTYL ALCOHOL
372   C8H18O    2-OCTANOL
                    CAPRYL ALCOHOL
373   C8H18O    2-ETHYLHEXANOL
                    2-ETHYL-1-HEXANOL
374   C8H18O    BUTYL ETHER
                    DIBUTYL ETHER
375   C8H19N    DIBUTYLAMINE
```

COMPOUND DICTIONARY (CONTINUED)

```
376   C9H10     ALPHA-METHYL STYRENE
377   C9H10O2   ETHYL BENZOATE
378   C9H12     N-PROPYLBENZENE
379   C9H12     ISOPROPYLBENZENE
                  CUMENE
                  2-PROPYLBENZENE
380   C9H12     1-METHYL-2-ETHYLBENZENE
                  1-ETHYL-2-METHYLBENZENE
                  O-ETHYLTOLUENE
381   C9H12     1-METHYL-3-ETHYLBENZENE
                  1-ETHYL-3-METHYLBENZENE
                  M-ETHYLTOLUENE
382   C9H12     1-METHYL-4-ETHYLBENZENE
                  1-ETHYL-4-METHYLBENZENE
                  P-ETHYLTOLUENE
383   C9H12     1,2,3-TRIMETHYLBENZENE
                  HEMELLITENE
384   C9H12     1,2,4-TRIMETHYLBENZENE
                  PSEUDOCUMENE
385   C9H12     1,3,5-TRIMETHYLBENZENE
                  MESITYLENE
386   C9H18     N-PROPYLCYCLOHEXANE
387   C9H18     ISOPROPYLCYCLOHEXANE
                  HEXAHYDROCUMENE
                  NORMENTHANE
388   C9H18     1-NONENE
389   C9H20     N-NONANE
390   C9H20     2,2,3-TRIMETHYLHEXANE
391   C9H20     2,2,4-TRIMETHYLHEXANE
392   C9H20     2,2,5-TRIMETHYLHEXANE
393   C9H20     3,3-DIETHYLPENTANE
                  TETRAETHYLPENTANE
394   C9H20     2,2,3,3-TETRAMETHYLPENTANE
395   C9H20     2,2,3,4-TETRAMETHYLPENTANE
396   C9H20     2,2,4,4-TETRAMETHYLPENTANE
397   C9H20     2,3,3,4-TETRAMETHYLPENTANE


398   C10H8     NAPHTHALENE
399   C10H12    1,2,3,4-TETRAHYDRONAPHTHALENE
                  TETRALIN
400   C10H14    N-BUTYLBENZENE
401   C10H14    ISOBUTYLBENZENE
402   C10H14    SEC-BUTYLBENZENE
403   C10H14    TERT-BUTYLBENZENE
404   C10H14    1-METHYL-2-ISOPROPYLBENZENE
                  O-CYMENE
405   C10H14    1-METHYL-3-ISOPROPYLBENZENE
                  M-CYMENE
406   C10H14    1-METHYL-4-ISOPROPYLBENZENE
                  P-CYMENE
407   C10H14    1,4-DIETHYLBENZENE
                  P-DIETHYLBENZENE
408   C10H14    1,2,4,5-TETRAMETHYLBENZENE
                  DURENE
409   C10H15N   N-BUTYLANILINE
410   C10H18    CIS-DECALIN
                  CIS-BICYCLO(4,4,0)DECANE
                  CIS-DECAHYDRONAPHTHALENE
                  NAPHTHANE
411   C10H18    TRANS-DECALIN
                  TRANS-BICYCLO(4,4,0)DECANE
                  TRANS-DECAHYDRONAPHTHALENE
412   C10H19N   CAPRYLONITRILE
                  CAPRINITRILE
                  DECANENITRILE
                  DECANOIC ACID,NITRILE
                  NONYL CYANIDE
413   C10H20    N-BUTYLCYCLOHEXANE
                  1-CYCLOHEXYLBUTANE
414   C10H20    ISOBUTYLCYCLOHEXANE
415   C10H20    SEC-BUTYLCYCLOHEXANE
                  2-CYCLOHEXYLBUTANE
416   C10H20    TERT-BUTYLCYCLOHEXANE
417   C10H20    1-DECENE
                  N-DECYLENE
418   C10H22    N-DECANE
```

COMPOUND DICTIONARY (CONTINUED)

```
419   C10H22    3,3,5-TRIMETHYLHEPTANE
420   C10H22    2,2,3,3-TETRAMETHYLHEXANE
421   C10H22    2,2,5,5-TETRAMETHYLHEXANE
422   C10H22O   1-DECANOL
                N-DECYL ALCOHOL


423   C11H10    1-METHYLNAPHTHALENE
                ALPHA-METHYLNAPHTHALENE
424   C11H10    2-METHYLNAPHTHALENE
425   C11H14O2  BUTYL BENZOATE
426   C11H22    N-HEXYLCYCLOPENTANE
                1-CYCLOPENTYLHEXANE
427   C11H22    1-UNDECENE
                1-HENDECENE
428   C11H24    N-UNDECANE
                HENDECANE


429   C12H10    DIPHENYL
                BIPHENYL
                PHENYLBENZENE
430   C12H10O   DIPHENYL ETHER
                PHENYL ETHER
431   C12H24    N-HEPTYLCYCLOPENTANE
                1-CYCLOPENTYLHEPTANE
432   C12H24    1-DODECENE
                ALPHA-DODECYLENE
433   C12H26    N-DODECANE
434   C12H26O   DIHEXYLETHER
435   C12H26O   DODECANOL
                DODECYL ALCOHOL
436   C12H27N   TRIBUTYLAMINE


437   C13H12    DIPHENYLMETHANE
                DITAN
438   C13H26    N-OCTYLCYCLOPENTANE
                1-CYCLOPENTYLOCTANE
439   C13H26    1-TRIDECENE
440   C13H28    N-TRIDECANE


441   C14H10    ANTHRACENE
442   C14H10    PHENANTHRENE
443   C14H28    N-NONYLCYCLOPENTANE
                1-CYCLOPENTYLNONANE
444   C14H28    1-TETRADECENE
                1-TETRADECYLENE
445   C14H30    N-TETRADECANE


446   C15H30    N-DECYLCYCLOPENTANE
                1-CYCLOPENTYLDECANE
447   C15H30    1-PENTADECENE
448   C15H32    N-PENTADECANE


449   C16H22O4  DIBUTYL-O-PHTHALATE
450   C16H32    N-DECYLCYCLOHEXANE
                CYCLOHEXADECANE
451   C16H32    1-HEXADECENE
                CETENE
                CETYLENE
452   C16H34    N-HEXADECANE
                CETANE


453   C17H34    N-DODECYLCYCLOPENTANE
                1-CYCLOPENTYLDODECANE
454   C17H36O   HEPTADECANOL
                1-HEPTADECANOL
455   C17H36    N-HEPTADECANE
```

COMPOUND DICTIONARY (CONTINUED)

```
456   C18H14     O-TERPHENYL
                    1,2-DIPHENYL BENZENE
457   C18H14     M-TERPHENYL
                    1,3-DIPHENYL BENZENE
458   C18H14     P-TERPHENYL
459   C18H36     1-OCTADECENE
460   C18H36     N-TRIDECYLCYCLOPENTANE
                    1-CYCLOPENTYLTRIDECANE
461   C18H38     N-OCTADECANE
462   C18H38O    1-OCTADECANOL
                    STEARYL ALCOHOL


463   C19H38     N-TETRADECYLCYCLOPENTANE
                    1-CYCLOPENTYLTETRADECANE
464   C19H40     N-NONADECANE


465   C20H40     N-PENTADECYLCYCLOPENTANE
                    1-CYCLOPENTYLPENTADECANE
466   C20H42     N-EICOSANE
467   C20H42O    1-EICOSANOL


468   C21H42     N-HEXADECYLCYCLOPENTANE
                    1-CYCLOPENTYLHEXADECANE
```

Appendix C

Lennard-Jones Potentials as Determined from Viscosity Data[†]

Substance		b_0,[‡] cm³/g-mol	σ, Å	ϵ/k, K
Ar	Argon	46.08	3.542	93.3
He	Helium	20.95	2.551§	10.22
Kr	Krypton	61.62	3.655	178.9
Ne	Neon	28.30	2.820	32.8
Xe	Xenon	83.66	4.047	231.0
Air	Air	64.50	3.711	78.6
AsH_3	Arsine	89.88	4.145	259.8
BCl_3	Boron chloride	170.1	5.127	337.7
BF_3	Boron fluoride	93.35	4.198	186.3
$B(OCH_3)_3$	Methyl borate	210.3	5.503	396.7
Br_2	Bromine	100.1	4.296	507.9
CCl_4	Carbon tetrachloride	265.5	5.947	322.7
CF_4	Carbon tetrafluoride	127.9	4.662	134.0
$CHCl_3$	Chloroform	197.5	5.389	340.2
CH_2Cl_2	Methylene chloride	148.3	4.898	356.3
CH_3Br	Methyl bromide	88.14	4.118	449.2
CH_3Cl	Methyl chloride	92.31	4.182	350
CH_3OH	Methanol	60.17	3.626	481.8
CH_4	Methane	66.98	3.758	148.6
CO	Carbon monoxide	63.41	3.690	91.7
COS	Carbonyl sulfide	88.91	4.130	336.0
CO_2	Carbon dioxide	77.25	3.941	195.2
CS_2	Carbon disulfide	113.7	4.483	467
C_2H_2	Acetylene	82.79	4.033	231.8
C_2H_4	Ethylene	91.06	4.163	224.7
C_2H_6	Ethane	110.7	4.443	215.7
C_2H_5Cl	Ethyl chloride	148.3	4.898	300
C_2H_5OH	Ethanol	117.3	4.530	362.6
C_2N_2	Cyanogen	104.7	4.361	348.6
CH_3OCH_3	Methyl ether	100.9	4.307	395.0

Substance		$b_0,\ddagger$ cm^3/g-mol	σ, Å	ϵ/k, K
CH_2CHCH_3	Propylene	129.2	4.678	298.9
CH_3CCH	Methylacetylene	136.2	4.761	251.8
C_3H_6	Cyclopropane	140.2	4.807	248.9
C_3H_8	Propane	169.2	5.118	237.1
n-C_3H_7OH	n-Propyl alcohol	118.8	4.549	576.7
CH_3COCH_3	Acetone	122.8	4.600	560.2
CH_3COOCH_3	Methyl acetate	151.8	4.936	469.8
n-C_4H_{10}	n-Butane	130.0	4.687	531.4
iso-C_4H_{10}	Isobutane	185.6	5.278	330.1
$C_2H_5OC_2H_5$	Ethyl ether	231.0	5.678	313.8
$CH_3COOC_2H_5$	Ethyl acetate	178.0	5.205	521.3
n-C_5H_{12}	n-Pentane	244.2	5.784	341.1
$C(CH_3)_4$	2,2-Dimethylpropane	340.9	6.464	193.4
C_6H_6	Benzene	193.2	5.349	412.3
C_6H_{12}	Cyclohexane	298.2	6.182	297.1
n-C_6H_{14}	n-Hexane	265.7	5.949	399.3
Cl_2	Chlorine	94.65	4.217	316.0
F_2	Fluorine	47.75	3.357	112.6
HBr	Hydrogen bromide	47.58	3.353	449
HCN	Hydrogen cyanide	60.37	3.630	569.1
HCl	Hydrogen chloride	46.98	3.339	344.7
HF	Hydrogen fluoride	39.37	3.148	330
HI	Hydrogen iodide	94.24	4.211	288.7
H_2	Hydrogen	28.51	2.827	59.7
H_2O	Water	23.25	2.641	809.1
H_2O_2	Hydrogen peroxide	93.24	4.196	289.3
H_2S	Hydrogen sulfide	60.02	3.623	301.1
Hg	Mercury	33.03	2.969	750
$HgBr_2$	Mercuric bromide	165.5	5.080	686.2
$HgCl_2$	Mercuric chloride	118.9	4.550	750
HgI_2	Mercuric iodide	224.6	5.625	695.6
I_2	Iodine	173.4	5.160	474.2
NH_3	Ammonia	30.78	2.900	558.3
NO	Nitric oxide	53.74	3.492	116.7
$NOCl$	Nitrosyl chloride	87.75	4.112	395.3
N_2	Nitrogen	69.14	3.798	71.4
N_2O	Nitrous oxide	70.80	3.828	232.4
O_2	Oxygen	52.60	3.467	106.7
PH_3	Phosphine	79.63	3.981	251.5
SF_6	Sulfur hexafluoride	170.2	5.128	222.1
SO_2	Sulfur dioxide	87.75	4.112	335.4
SiF_4	Silicon tetrafluoride	146.7	4.880	171.9
SiH_4	Silicon hydride	85.97	4.084	207.6
$SnBr_4$	Stannic bromide	329.0	6.388	563.7
UF_6	Uranium hexafluoride	268.1	5.967	236.8

†R. A. Svehla, *NASA Tech. Rep.* R-132, Lewis Research Center, Cleveland, Ohio, 1962.
‡$b_0 = \frac{2}{3}\pi N_0 \sigma^3$, where N_0 is Avogadro's number.
§The parameter σ was determined by quantum-mechanical formulas.

Index of Recommendations

Subject Index